First Aid

Report all accidents to your instructor immediately! These include cuts, chemicals in eyes, burns, fires, explosions, and all other types of accidents. Following treatment according to protocols in place at your institution, explain to your instructor exactly what happened so that a report can be filed. **All students should be enrolled in a health insurance plan prior to registering as a laboratory student.**

- **Cuts:** Go to the sink and wash the cut with copious amounts of cold water and inform your instructor of the accident. Apply a bandage as appropriate, after the bleeding has subsided. If bleeding is serious, apply a bandage and pressure and seek emergency medical attention. Do not attempt to apply a tourniquet.

- **Chemicals in eyes:** Go immediately to the eye-wash station. Have someone guide you there if you can't see. Bathe your eyes in cool water from the eye-wash fountain for at least 15 minutes. If you are using an eye-wash bottle, lie down and have someone pour cool water into your eyes for at least 15 minutes. Seek emergency medical attention in all cases.

- **Burns (from fire or chemicals):** For burns over large areas of your skin, go to the safety shower and drench the burned area immediately. Remove clothing as necessary. For minor burns, go to the sink and wash your burned skin with cold water for at least 15 minutes. Wash chemical burns with water and detergent if desired, but not with other chemicals. Seek emergency medical attention.

- **Bench fire:** If a fire erupts on your lab bench, step back, away from the fire. Be sure that others around you step away from the fire. Assess the size of the fire and its potential for spreading. Move all nearby solvents and chemicals away from the fire. Turn off any nearby burners. Small fires frequently extinguish themselves and do not require a fire extinguisher. Try covering the fire with a beaker or a large watch glass. If the fire poses a hazard, locate the closest powder or carbon dioxide extinguisher. Pull the pin on the extinguisher and discharge at the base of the fire. Discharge the extinguisher again as needed until the fire is out. For large fires, clear the area and follow emergency procedures.

- **Person on fire:** Roll the person on the floor to smother the flames. Get the victim to the emergency shower as soon as possible. Only if a shower is unavailable or inoperable, put out flames with a fire blanket. Do not use an extinguisher on a person. Seek medical attention.

- **Explosion:** Get the victim to lie down. If there are burns and if the person is conscious, treat as for burns. If there are no burns, cover the victim with a fire blanket. Summon emergency personnel. Comfort and aid the victim until help arrives.

- **Swallowed chemical:** Report the incident immediately. Seek medical attention.

Refer to Material Safety Data Sheets (MSDS) available for each chemical used in the laboratory for First Aid information pertaining to a specific chemical.

Microscale and Miniscale Organic Chemistry Laboratory Experiments

Second Edition

Allen M. Schoffstall

*The University of Colorado
at Colorado Springs*

and

Barbara A. Gaddis

*The University of Colorado
at Colorado Springs*

with

Melvin L. Druelinger

Colorado State University-Pueblo

 Higher Education

Boston   Burr Ridge, IL   Dubuque, IA   Madison, WI   New York   San Francisco   St. Louis
Bangkok   Bogotá   Caracas   Kuala Lumpur   Lisbon   London   Madrid   Mexico City
Milan   Montreal   New Delhi   Santiago   Seoul   Singapore   Sydney   Taipei   Toronto

Higher Education

MICROSCALE AND MINISCALE ORGANIC CHEMISTRY LAB EXPERIMENTS
SECOND EDITION

Published by McGraw-Hill, a business unit of The McGraw-Hill Companies, Inc., 1221 Avenue of the Americas, New York, NY 10020. Copyright © 2004, 2000 by The McGraw-Hill Companies, Inc. All rights reserved. No part of this publication may be reproduced or distributed in any form or by any means, or stored in a database or retrieval system, without the prior written consent of The McGraw-Hill Companies, Inc., including, but not limited to, in any network or other electronic storage or transmission, or broadcast for distance learning.

Some ancillaries, including electronic and print components, may not be available to customers outside the United States.

This book is printed on acid-free paper.

6 7 8 9 0 VNH/VNH 0 9 8

ISBN-13: 978-0-07-242456-0
ISBN-10: 0-07-242456-7

Publisher: *Kent A. Peterson*
Sponsoring editor: *Thomas D. Timp*
Senior developmental editor: *Shirley R. Oberbroeckling*
Senior marketing manager: *Tamara L. Good-Hodge*
Project manager: *Joyce Watters*
Lead production supervisor: *Sandy Ludovissy*
Senior media project manager: *Stacy A. Patch*
Senior media technology producer: *Jeffry Schmitt*
Senior coordinator of freelance design: *Michelle D. Whitaker*
Cover/interior designer: *Rokusek Design*
Cover image: *Rokusek Design*
Senior photo research coordinator: *Lori Hancock*
Compositor: *Precision Graphics*
Typeface: *10/12 Times Roman*
Printer: *Von Hoffmann Corporation*

"Permission for the publication herein of Sadtler Standard Spectrar has been granted, and all rights are reserved, by BIO-RAD Laboratories, Sadtler Division."

"Permission for the publication of Aldrich/ACD Library of FT NMR Spectra has been granted and all rights are reserved by Aldrich Chemical."

All experiments contained in this laboratory manual have been performed safely by students in college laboratories under the supervision of the authors. However, unanticipated and potentially dangerous reactions are possible due to failure to follow proper procedures, incorrect measurement of chemicals, inappropriate use of laboratory equipment, and other reasons. The authors and the publisher hereby disclaim any liability for personal injury or property damage claimed to have resulted from the use of this laboratory manual.

Library of Congress Cataloging-in-Publication Data

Schoffstall, Allen M.
    Microscale and miniscale organic chemistry laboratory experiments / Allen M.
  Schoffstall, Barbara A. Gaddis, Melvin L. Druelinger.—2nd ed.
        p.   cm.
    Includes bibliographical references and index.
        ISBN 0–07–242456–7 (acid-free paper)
        1. Chemistry, Organic—Laboratory manuals.   I. Gaddis, Barbara A.   II. Druelinger,
  Melvin L.   III. Title.

QD261 .S34   2004
547.0078—dc21   2003008663
                                                          CIP

www.mhhe.com

Dedication
To Carole, Larry, and Judy for their patience, help, and encouragement.

To organic students who develop a passion for doing and learning from organic laboratory experiments and to the instructors who make laboratory learning meaningful.

Brief Contents

Contents

Preface to Second Edition

This book is a comprehensive introductory treatment of the organic chemistry laboratory. The student will be guided in doing numerous exercises to learn basic laboratory techniques. The student will then use many proven traditional experiments normally performed in the two-semester organic laboratory course.

Several trends in organic laboratory education have emerged since publication of the first edition. These trends are recognition of the pedagogical value of discovery experiments, the increased emphasis on molecular modeling and computer simulations, and the development of green experiments. All of these trends are incorporated into this book along with the use of traditional experiments.

DISCOVERY EXPERIMENTS

Discovery experiments are given a special label in the Table of Contents and in each chapter where they appear. Discovery experiments incorporate the pedagogical advantages of inductive inquiry experiments with the ease of design found in expository experiments. Discovery experiments (or guided inquiry experiments) have a specific procedure designed to give a predetermined but unspecified result. Students use a deductive thought process to arrive at a desired conclusion. Students are "guided" by inferring a general scientific principle. Discovery experiments have been employed successfully in large laboratory sections, as well as in small classroom environments. Student interest is increased during discovery experiments because the result of the experiment is unknown to the student. The desired goal of discovery experiments is increased student learning. Discovery experiments can also provide the opportunity for individual reflection and class discussion and may involve students in developing and interpreting laboratory procedures. These features and advantages of discovery experiments have caused your text authors to emphasize discovery experiments in this edition of the text.

MOLECULAR MODELING

Molecular modeling by computer saw a revolution in the late 1990s with the advent of affordable, sufficiently fast personal computers with adequate memory. Computer modeling enhances the benefits of assembling molecular models using model kits. Use of these kits is still encouraged. However, gone are the days where students had to depend only on molecular model kits to represent molecules in three dimensions. While these models still have their uses, computer modeling programs now provide exciting visualization of molecules and calculation of physical properties and thermodynamic parameters. Where possible, it is desirable to incorporate computer modeling into organic laboratory programs. The exercises in this book can be done using relatively inexpensive commercial software from one or more providers.

COMPUTER SIMULATION OF EXPERIMENTS

Another use of computers is for simulation of laboratory procedures and experiments. Demonstrations of laboratory techniques are available as clips on the CD accompanying this text. Simulations of experiments are useful as prelab exercises to familiarize students with the experiment and to enhance learning in the laboratory. Simulations are also useful as illustrations of experiments that are difficult to carry out in the undergraduate laboratory environment. Experiments that require special equipment, inert gaseous environments, or especially noxious and toxic reagents can be experienced by students through virtual experiments on the computer. Examples of such experiments are available on the CD accompanying this text.

GREEN CHEMISTRY

Academic and industrial organic chemists have led an initiative to replace the use of organic solvents with aqueous solvents. They have encouraged the recycling of chemicals in order to reduce production requirements of chemicals. They have encouraged use of environmentally benign reagents in place of hazardous and toxic reagents where possible. In this text, there have been efforts to reduce quantities of toxic reagents and solvents wherever possible and to develop "green" experiments. For example, new Experiment 14.2 is on the use of indium reagents in aqueous solvents to accomplish coupling reactions similar to Grignard reactions. Another objective of green chemistry is to prevent waste. In this book, microscale and miniscale experiments are used in order to help minimize waste.

MICROSCALE AND MINISCALE TECHNIQUES

Microscale and miniscale organic techniques were first introduced two decades ago. However, changing over to new, smaller glassware and equipment has been slow in some laboratories for a number of reasons. One reason is the initial cost, but most institutions benefit by reduced costs of chemicals and hazardous waste disposal. The decision of whether to use a microscale procedure or a miniscale procedure often depends on the methods of characterization chosen by the instructor. This governs how much product is required for analysis. If a distillation is desired, a miniscale procedure is often chosen because of difficulties associated with distilling very small quantities of liquid. If an analysis of liquid products is to be done only by gas chromatographic analysis, a microscale procedure will cut down on costs of waste disposal.

NEW FEATURES IN THE SECOND EDITION

Accompanying a new section on molecular modeling, significant additions to this edition include expanded coverage of Diels-Alder chemistry, inclusion of enone chemistry with a chapter on enols, a new chapter on dicarbonyl compounds, and expanded coverage of heterocycles in the chapter on amines. New experiments and new options within experiments are included in many chapters. Many are discovery experiments. Among these are

Experiment 3.3, Relationships Between Structure and Physical Properties;
Experiment 3.8, Purification and Analysis of a Liquid Mixture;
Experiment 5.1B, Miniscale Synthesis of Alkenes Via Acid-catalyzed Dehydration of 3,3-Dimethyl-2-butanol;

Experiment 9.1C, Microscale Reaction of Cyclopentadiene with Maleic Anhydride;
Experiment 9.1E, Reaction of Anthracene with Maleic Anhydride;
Experiment 14.2, Using Indium Intermediates: Reaction of Allyl Bromide with an Aldehyde;
Experiment 15.3, Photochemical Oxidation of Benzyl Alcohol;
Experiment 16.2, Nucleophilic Aliphatic Substitution Puzzle: Substitution Versus Elimination;
Experiment 17.1C, Microscale Horner-Emmons Reaction of Diethylbenzyl Phosphonate and Benzaldehyde;
Experiment 18.2A, Microscale Reduction of 2-Cyclohexenone;
Experiment 18.2B, Microscale Reduction of *trans*-4-Phenyl-3-buten-2-one;
Experiment 18.3, Catalytic Transfer Hydrogenation Miniscale Reaction of Cyclohexenone;
Experiment 21.1, Base-Catalyzed Condensations of Dicarbonyl Compounds;
Experiment 22.2, Synthesis of Pyrazole and Pyrimidine Derivatives;
Experiment 24.1, Exploring Structure-function Relationships of Phenols;
Experiment 26.1, Soap from a Spice: Isolation, Identification and Hydrolysis of a Triglyceride;
Experiment 26.2, Preparation of Esters of Cholesterol and Determination of Liquid Crystal Behavior;
Experiment 29.2, Multistep Synthesis of Sulfanilamide Derivatives as Growth Inhibitors;
Experiment 29.3, Structural Determination of Isomers Using Decoupling and Special NMR Techniques.

INSTRUCTOR'S MANUAL

An instructor's manual is available on the website accompanying this text. This manual includes directions for laboratory preparators, instructor's notes for each experiment, solutions to problems, and prelab and postlab assignments. Test questions about many experiments are available on the web CT.

COURSE WEBSITE

The website http://www.mhhe.com/schoffstall2 offers supportive backup for the organic laboratory course. It presents updated helpful hints for lab preparators and instructors, typical schedules, sample electronic report forms, sample quiz and exam questions, examples of lab lecture or material for self-paced prelab student preparation, and relevant links to other websites. Some additional experiments are available on the website.

ACKNOWLEDGMENTS

We wish to acknowledge several individuals who have contributed to the second edition. Connie Pitman, laboratory technician at the University of Colorado Springs, has made numerous valuable comments about the experiments. She has also coauthored the Instructor's Manual and Solutions Guide. Shirley Oberbroeckling has served as the Developmental editor and Joyce Watters as the Project Manager for this edition of the text. The following faculty and students have contributed to the second edition by testing experiments and suggesting improvements:

Robert A. Banaszak, Anna J. Espe, Sam T. Seal, Shannon J. Coleman, Shannon R. Gilkes, Molly M. Simbric, Daniela Dumitru, Patricia D. Gromko, Amy M. Scott,

Tomasz Dziedzic, Paul J. Lunghofer, Rafael A. Vega, Justin A. Russok, Darush Fathi, and Michael Slogic.

We are grateful to the following individuals who served as reviewers for this edition. They are:

Monica Ali, Oxford College
Steven W. Anderson, University of Wisconsin - Whitewater
Satinder Bains, Arkansas State University - Beebe
David Baker, Delta College
John Barbaro, University of Alabama - Birmingham
George Bennett, Milikin University
Cliff Berkman, San Francisco State University
Lea Blau, Stern College for Women
Lynn M. Bradley, The College of New Jersey
Bruce S. Burnham, Rider College
Patrick E. Canary, West Virginia Northern Community College
G. Lynn Carlson, University of Wisconsin - Kenosha
Jeff Charonnat, California State University - Northridge
Wheeler Conover, Southeast Community College
Wayne Counts, Georgia Southwestern State University
Tammy A. Davidson, East Tennessee State University
David Forbes, University of South Alabama - Mobile
Eric Fossum, Wright State University - Dayton
Nell Freeman, St. Johns River Community College
Edwin Geels, Dordt College
Jack Goldsmith, University of South Carolina - Aiken
Ernest E. Grisdale, Lord Fairfax Community College
Tracy Halmi, Pennsylvania State Behrend - Erie
C. E. Heltzel, Transylvania University
Gary D. Holmes, Butler County Community College
Harvey Hopps, Amarillo College
William C. Hoyt, St. Joseph's College
Chui Kwong Hwang, Evergreen Valley College
George F. Jackson, University of Tampa
Tony Kiessling, Wilkes University
Maria Kuhn, Madonna University
Andrew Langrehr, Jefferson College
Elizabeth M. Larson, Grand Canyon University
John Lowbridge, Madisonville Community College
William L. Mancini, Paradise Valley Community College
John Masnovi, Cleveland State University
Anthony Masulaitis, New Jersey City University
Ray Miller, York College
Tracy Moore, Louisiana State University - Eunice
Michael D. Mosher, University of Nebraska - Kearney
Michael J. Panigot, Arkansas State University
Neil H. Potter, Susquehanna University
Walda J. Powell, Meredith College
John C. Powers, Pace University
Steve P. Samuel, SUNY - Old Westbury
Greg Spyridis, Seattle University

Paris Svoronos, Queensboro Community College
Eric L. Trump, Emporia State University
Patibha Varma Nelson, St. Xavier University
Chad Wallace, Asbury College
David Wiendenfeld, University of North Texas
Linfeng Xie, University of Wisconsin - Oshkosh

We hope you find your laboratory experience profitable and stimulating.

Walkthrough

Microscale and Miniscale Organic Chemistry Laboratory Experiments offers a comprehensive introduction to organic laboratory techniques that is flexible, engaging, and user-friendly. It provides techniques for handling glassware and equipment, safety in the laboratory, micro- and miniscale experimental procedures, theory of reactions and techniques, relevant background information, applications, and spectroscopy.
This text features:

Flexible Content

Along with the traditional experiments, it offers the flexibility of choosing starting materials, characterization methods, and either microscale or miniscale procedures. The organization of the material is clearly defined, allowing the instructor the flexibility to coordinate the lecture with the laboratory experiments

Discovery Experiments

Discovery experiments incorporate pedagogical advantages of inductive inquiry experiments with the ease of design found in expository experiments. Discovery experiments (or guided inquiry experiments) have a specific procedure designed to give a pre-determined, but unspecified result. Discovery-oriented experiments require students to do "detective work" because the results are not always a foregone conclusion.

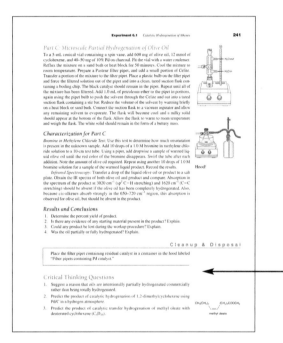

Molecular Modeling

While model kits still have their use, computer-modeling programs now provide exciting visualization of molecules and calculation of physical properties and thermodynamic parameters. Where possible, it is desirable to incorporate computer modeling into organic laboratory programs. The exercises in this book can be done using relatively inexpensive commercial software from one or more providers.

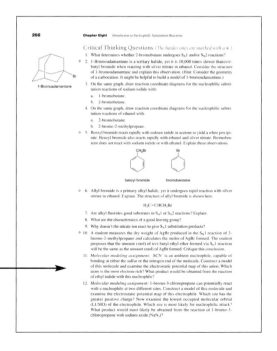

Green Chemistry

Green chemistry encourages recycling of chemicals in order to reduce production requirements of chemicals. In this text, there have been efforts to reduce quantities of toxic reagents and solvents wherever possible and to develop "green" experiments. Green experiments are given a special label in the Table of Contents and in each chapter where they appear.
The media includes:

CD-ROM

Demonstrations of laboratory techniques are available as clips on the CD that accompanies this text. Simulations of experiments are useful as pre-lab exercises to familiarize the students with the experiment and to enhance learning in the laboratory. Simulations are also useful as illustrations of experiments that are difficult to carry out in the undergraduate laboratory environment.

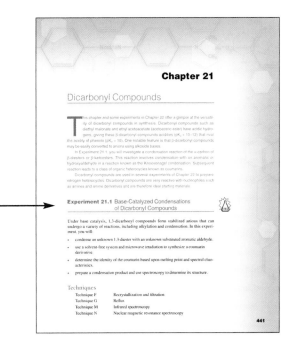

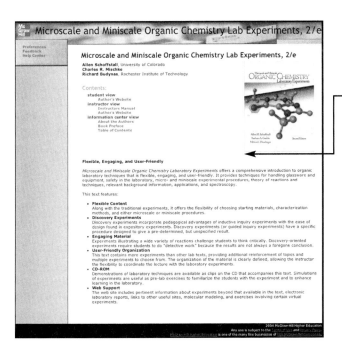

Web Support

The website includes pertinent information about experiments beyond what is available in the text, electronic laboratory reports, and links to other useful sites, molecular modeling, and sample quiz and exam questions. The website also includes the Instructor Manual with directions for laboratory preparators, instructor's notes for each experiment, solutions to problems and prelab and postlab assignments.

Introduction

Welcome to the Organic Chemistry Laboratory! In this introduction, you will learn about:

◆ important features of the organic lab.

◆ goals for the organic laboratory.

◆ working in the laboratory.

◆ laboratory safety.

◆ material safety data sheets.

◆ your laboratory notebook.

◆ laboratory reports.

◆ how to be a successful organic laboratory student.

Important Features of the Organic Lab

It is educational and enlightening. Understanding the principles behind an experiment can determine whether you have a good experience or an unsatisfactory one. Many students learn as much or more from the lab as from the classroom.

It includes some discovery experiments and identification of compounds. Student interest and learning are enhanced when the outcome of an experiment is not a foregone conclusion.

It teaches techniques and the practicalities of organic synthesis. Experimental work is dealing with the realities of performing techniques and chemical reactions in the laboratory. Details of reactions may be overlooked in organic textbooks, but it is often these details that reveal the true nature, beauty, and challenge of the subject. In the organic lab, you will find out quickly about yields of products and side products as you attempt to maximize the amount and purity of your product.

It teaches safe practices in the laboratory. In the organic lab there are some risks. There are some experiments that use hazardous chemicals. To minimize risks, good planning and preparedness are required. It is always necessary to think about safety in the laboratory.

It encourages active participation. Because you are actively involved you will most likely have questions about procedure and about theory. You will have opportunities to discuss these questions with your instructor.

It teaches efficiency in the laboratory. Accomplishing tasks in a timely manner is important. It will often be necessary to work on two or more tasks in the same time period.

It encourages cooperation and teamwork. Cooperation is vital when working with several people in a laboratory setting. Consideration of others, taking turns, and being courteous are part of working in the lab.

Goals for the Organic Laboratory

There are several specific goals and objectives for the organic laboratory. By the end of the course, you should be able to

- understand theory and principles of organic laboratory techniques and be able to interpret results and answer questions about experiments;
- follow experimental procedures carefully and use good laboratory technique, rigorously adhering to the rules regarding safety;
- keep a neat and up-to-date notebook, written using correct grammar, that represents an accurate accounting of work done;
- design experiments to synthesize, isolate, and characterize organic products and complete flow schemes for reaction workup procedures.

Students who master these objectives in a timely fashion will have a successful laboratory experience. Overall, the goals of the laboratory course enhance and support goals of the organic chemistry lecture course.

Working in the Laboratory

Microscale organic experiments became common in the introductory organic laboratory during the 1980s. Prior to that time, macroscale equipment was used in most organic laboratories. In macroscale laboratory procedures, reactions are performed using 5–20 grams or more of reagents. Macroscale experiments are still important today, particularly when it is necessary to prepare a compound for multistep reaction sequences and when it is necessary to purify each product along the way before beginning the next step.

Most laboratory texts today use miniscale or microscale experiments or both, as in this text. Miniscale experiments employ scaled-down macroscale glassware and use approximately 0.3–5 grams of starting materials. Microscale reactions are performed using special microscale glassware and generally use less than 300 mg of starting material. In this text, you will be introduced to both microscale and miniscale experiments.

The importance and popularity of microscale and miniscale experiments as compared to macroscale experiments are due to a number of factors.

Small-scale experiments produce little waste and are environmentally friendly. There is little waste produced in microscale organic experiments and relatively little waste produced in miniscale experiments. Minimal amounts of solvent are used. What little waste there is can be easily managed. Since the costs of disposing of chemicals have increased very dramatically in recent years, the less waste that is produced, the better. It is frequently more expensive to dispose of a chemical than it is to purchase it initially.

Small-scale experiments are less expensive. Less starting material and less reagents are used. Fewer chemicals are needed during the workup procedures and less solvent is used for the reaction and workup.

Small-scale experiments are based on a wide array of starting materials. Chemicals can be used that are relatively expensive, which would be prohibitive for experiments performed on a larger scale. This widens the range of experiments that can be done.

Small-scale experiments require less time than larger scale experiments. It takes less time to bring a small reaction mixture to the proper temperature. Workup procedures can be accomplished in less time. Purification procedures also require less time.

Small-scale experiments require careful laboratory technique. Students are required to carefully measure out chemicals and isolate and purify small amounts of products. This encourages development of good laboratory practice.

Small-scale experiments are safer. The smaller quantities of chemicals used in small-scale experiments reduce the risk of contact, if safety precautions are followed.

However, if toxic and corrosive chemicals are to be employed for an experiment, there is still danger even when using smaller amounts and it is still necessary to adhere strictly to safety precautions. Also, it is generally possible to avoid having offensive odors in the lab if small quantities of corrosive and odiferous chemicals are used. When these are used in the hood, this problem is further reduced.

Laboratory Safety

Small-scale experiments may be safer than macroscale experiments, but accidents can occur. It is important to plan ahead, to recognize potential hazards, and to rigorously follow safety rules. Your lab instructor will explain rules for safely working in the organic laboratory. Important safety rules that must be followed are listed here.

1. **Wear approved eye protection at all times in the laboratory to avoid eye injury.** The goggles and safety glasses will protect your eyes from flying glass particles or caustic chemicals. It is important that you wear eye protection at all times while working in the laboratory. Wear eye protection around your eyes, not propped on top of your head. It is inadvisable to wear contact lenses in the lab, since solvent vapors or splashed chemicals may get underneath the lens and cause damage before the lens can be removed.

   In the event of chemicals splashing in your eye, use the eye wash fountain to rinse out your eye. Know the location of the eye wash fountain and know how to use it—the few minutes required to do so could save a lifetime of vision.

2. **Dress properly while in the laboratory to avoid chemical burns.** Wear clothing that is approved for your laboratory. Do not wear loose-fitting clothing that can catch on glassware or reagent bottles and cause breakage. Wearing a lab coat or vinyl apron can keep spills and splashes off your skin and clothing.

   In the event that acidic, corrosive, irritating or toxic chemicals are splashed on you, quickly rinse with water. If the chemical is spilled on your hands or arms, it is often easiest to rinse off the chemical using the faucet and sink. If the chemical is spilled on your legs, or if the chemical spill occupies a wide area of your body, use the safety shower. Strip off your outer clothing—forget about modesty for the moment—and wash off the chemical. The safety shower releases many gallons of water in a short period of time, so it is effective at rinsing off a chemical quickly. Report **all** injuries or accidents to your instructor immediately and seek appropriate medical attention.

3. **Work under the hood when using toxic or irritating chemicals to avoid breathing their vapors.** If hoods are not available, work in well-ventilated areas to avoid local buildup of hazardous vapors. The hoods draw away the vapors and vent them away from students. Do not smell any of the chemicals. It should be obvious that tasting chemicals is strictly forbidden. However, touching the lab bench and then touching your mouth can cause you to ingest chemicals. Wear latex gloves whenever handling corrosive, toxic, or irritating chemicals. Wipe the outside of reagent bottles before picking them up to use. Always wash your hands after being in the organic chemistry lab, even if you wore plastic gloves. Never eat or drink and do not open food containers in the laboratory.

   In the event that you breathe in a chemical, immediately seek fresh air to replace the chemical vapors in your lungs with air. Should you ingest any chemical, tell the instructor immediately what was ingested and seek medical help.

4. **Do not have any open flames in the organic lab and exercise extreme care when heating volatile organic liquids.** Many organic solvents and chemicals are flammable, with very low flash points. Flammable volatile solvents such as diethyl ether and petroleum ether are particularly dangerous, since their vapors disperse

around the lab. Volatile solvents should not be heated directly on a hot plate, since they can ignite easily if spilled on the hot surface or if a spark from the thermostat ignites the vapors. Before using electrical equipment, such as Variacs, heating mantles, or hot plates, make certain that there are no frayed electrical cords.

Most organic chemistry laboratories have chemical fire extinguishers, such as liquid carbon dioxide fire extinguishers, or dry chemical fire extinguishers, such as sodium bicarbonate or ammonium phosphate. These fire extinguishers work by laying CO_2 or the inorganic powder over the fire, thereby removing the oxygen source and smothering the fire. Although you will probably never have to use them, you should know where the fire extinguishers are located and how to activate them in the event of a fire.

5. **Handle chemicals properly.** Always read the label on the reagent bottle before using to make certain that you are using the correct chemical. Before mixing any chemicals, check again to make certain that you have obtained the correct reagents. Never use a chemical from an unlabeled bottle or beaker. Transfer out only the amount needed; if an excess is inadvertently measured out, ask the instructor for disposal instructions. Do not return reagents to the stock bottle. Always make certain that reagent bottles and dispensing containers are wiped clean before picking them up. This is especially important when working with strong acids or bases that can cause severe burns. When finished with a reagent bottle, wipe off the outside and replace caps and lids. Spills on the floor or bench top must be cleaned immediately; notify the instructor.

When finished with an experiment, the chemicals must be disposed of properly. Acidic or basic aqueous solutions should be neutralized and washed down the drain with water. Halogenated and nonhalogenated organic solvents must be placed in separate containers for recovery or disposal. Some specialized chemicals have specific requirements for disposal. Always check the Cleanup and Disposal section in each experiment and follow directions carefully.

Most important, follow the directions in the experimental procedure. Doing unauthorized experiments is strictly forbidden.

6. **Know the properties of the chemicals to be used in an experiment.** You will be working with a variety of organic compounds in the lab. Some of the chemicals have little or no danger associated with their use, while others are more hazardous and require the use of gloves and hoods. Some chemicals are irritants, which means that they may cause a rash. Others may be toxic or corrosive. Still others may be lachrymatory (causing eyes to tear) or carcinogenic (cancer causing). Many of the organic solvents you use will be flammable. To have a safe laboratory experience you must know the properties of the compounds, how to handle them, and how to dispose of the chemicals once you are finished with the experiment.

Knowing the properties of the chemicals will help you understand how to work with these chemicals safely. Before coming to lab, you must look up the physical properties and hazards of all of the compounds you are using in lab. Information about physical properties of the chemicals, such as boiling point, melting point, and density, can be found in handbooks such as the *Merck Index* and the *Handbook of Chemistry and Physics*. Safety information about specific chemicals can be found in a catalog from a chemical supply company, such as the Aldrich Catalog, which lists brief safety descriptors of the chemicals sold. Even more information can be obtained from the NFPA label on a chemical, which evaluates the hazard of the chemical toward fire (top red quadrant), reactivity (yellow quadrant on the right side), and health effects due to exposure to vapor or to skin contact (blue quadrant on the left). Symbols may be written in the bottom white quadrant to indicate specific hazards, such as OX for oxidizer and COR for corrosive. A number in each quadrant indicates the degree of hazard, with a 0 representing no hazard and

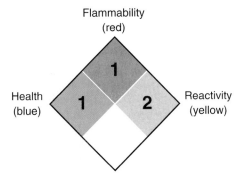

a 4 representing extreme hazard. An example of how to obtain safety information from a label is illustrated. For this fictitious chemical, a 1 in the fire quadrant means that the compound is not very flammable, but may ignite if heated very strongly. A 2 in the reactivity quadrant means that the chemical is stable at normal temperature and pressure, but will become unstable at high temperature and high pressures. In the health category a 1 means that the chemical may cause irritation if not treated.

The labels on the chemical bottles provide a succinct overview of the hazards associated with use of the chemical. A more exhaustive source of safety information on each chemical used in the laboratory is the information provided by chemical manufacturers in a Material Safety Data Sheet (MSDS) for every marketed compound.

Material Safety Data Sheet (MSDS)

The federal government requires chemical manufacturers to provide an MSDS for every chemical sold. An MSDS provides an abundance of information about physical properties, toxicity, permissible levels of exposure, health consequences of exposure, first aid, and protocols for safe handling, storage, and disposal of the chemical. The format for the MSDS varies by chemical supplier, although the information provided is similar. As an example of the type of information available, the MSDS of diethyl ether reveals that diethyl ether may be harmful if inhaled, ingested, or absorbed by your skin; that it can cause skin irritation, chest pains, nausea, headache, and vomiting; and that the lethal dose (LD_{50}) for a human is 260 mg/kg of body weight. From the MSDS you will also learn that diethyl ether is extremely flammable and must be kept away from any sparks. Being aware of the hazards of the chemicals you are working with will help you become a more conscientious laboratory worker.

The MSDS also provides valuable information on how to dispose of a chemical after use. Doing microscale experiments reduces the amount of waste produced, but does not entirely eliminate waste. Part of being a good lab student is knowing how to minimize waste and effectively dispose of the waste you create. In each experiment, you will be given explicit directions about how to dispose of the waste in the Cleanup and Disposal section of the experimental procedure. These procedures are based upon information in the MSDS. Be sure to read and follow the directions carefully. Refer to Appendix D for further information concerning the MSDS supplied for each chemical and for a sample MSDS.

These safety instructions are not comprehensive; your facility may have its own set of laboratory safety rules. Your laboratory instructor will explain these rules to you and show you where safety equipment (fire extinguishers, eye wash fountain, safety shower) can be found. With the proper precautions, your organic lab course will be safe, challenging, and enlightening.

Your Laboratory Notebook

In addition to knowing how to work in the lab safely, you also need to know how to keep a laboratory notebook. Laboratory courses foster making careful observations and keeping good records. Use a black pen and a bound notebook. Begin each experiment prior to coming to the laboratory by writing the statement of purpose of the lab. Then do all required calculations and answer assigned prelab questions. Prepare a table of all reagents, listing recommended amounts (weight and mol), relevant physical properties, and any hazards associated with use.

During the lab period, record in your notebook the amounts of reagents you actually use and calculate the theoretical yield of product calculated based upon these amounts. Describe the steps of the procedure and how you characterized the product. In a section for results, tabulate data or calculate your percent yield. If you make an error in recording, cross out the error and then make the correct entry. Draw conclusions based upon your results. The key features of any good experimental account are neatness, brevity, clarity, completeness, accuracy, and timeliness.

Laboratory Reports

There are two main types of experiments in this lab text: (1) preparative experiments, in which an organic starting material (substrate) is converted to an organic product; and (2) investigative experiments, in which a given property or technique is studied.

Preparative Experiments

Reports for writing up preparative experiments generally include the following:

1. Name and date of the experiment
2. Introduction: title and purpose of the experiment
3. Reference to procedure used
4. Balanced reaction and important side reactions
5. Mechanism, if applicable
6. Physical properties of reagents, products, and side products in tabular form to include chemical name, molecular weight, boiling point, melting point, color, density, solubility, quantities used, number of moles used, safety information, and hazardous properties of compounds
7. Flow scheme of all operations in an experimental procedure (for selected experiments). Illustrate each step of the experiment using an arrow, starting with all reactants. Write all substances produced during each step to the right of the arrow. Write all substances removed in each step beneath the arrow. Write the desired materials to the right of the arrow and the unwanted materials below the arrow. Designate top and bottom layers of extraction steps, as well as distillates, pot residues, filtered solids, and filtrates. A sample flow scheme is shown in Appendix C.
8. Experimental procedure listing the steps to be followed in the experiment
9. Observations and any changes in the procedure
10. Results and conclusions, giving percent yield and characterization methods
11. Spectra, if applicable
12. Prelab and postlab assignments
13. Critical thinking questions

The notebook should be completed through Step 8 and assigned prelab questions should be answered before starting work in the lab. Your instructor will provide the specific format preferred for your lab. A sample laboratory report can be found in Appendix F.

Investigative Experiments

Investigative experiments generally follow a similar format as a preparative experiment, but usually include a statement of purpose, a description of the property being studied in the experiment, tabulation of data, analysis of the results, thorough discussion of the implication of the experimental results, and answers to prelab and postlab assignments. Your instructor will provide a specific format for investigative experiments.

How to Be a Successful Organic Laboratory Student

Rigorously following safety rules, preparing for laboratory work in advance, and writing a good lab notebook are important for achieving a successful laboratory experience. Here are some other hints to help you achieve success.

- Clean all glassware **by the end** of each experiment so it will be clean and dry for the next experiment.
- Double-check your calculations to make sure you are using correct amounts of reagents.
- Keep your bench top clean to avoid spillage and breakage.
- **Save** everything from an experiment until the end, when you are sure you don't need these materials any longer.
- Outline the steps in the experimental procedure. As you perform each step, think about why you are doing the step. Be thinking ahead to the next step of the reaction and how to allocate your time effectively.
- Be prepared to begin work as soon as you come to the lab. Prepare your laboratory notebook ahead of time, including prelab assignments. When you come into the lab, know what chemicals, glassware, and equipment you are going to use.

In summary, plan ahead, work hard, but most of all, have a safe and enjoyable experience.

Chapter 1

Techniques in the Organic Chemistry Laboratory

Basic laboratory operations used in the organic laboratory are introduced in this chapter. Mastery of these techniques will allow you to perform numerous organic experiments later in the course. Brief laboratory exercises are included as practical illustrations of each technique. Each exercise is designed to focus on a single technique. Until you have mastered the techniques, you may wish to use this chapter as a reference as you encounter each technique in different experiments throughout the course.

Technique A: Glassware and Equipment *Heating and Cooling*

Introduction

Checking out an equipment locker in the organic laboratory is like receiving presents during the holidays—everyone has lots of new toys to try out! Most of the items in the locker are made of glass and all are used in some way to perform reactions, work up reaction mixtures, or purify and analyze products.

Two types of glassware are used in the organic laboratory. **Microscale** glassware is stored in a case of about the same size as a laptop computer. Each item of glassware has its own individually shaped space, allowing for easy storage between use. Plastic screwcaps and liners are used as fast and efficient connectors or lids for the glassware.

A second type of glassware is **miniscale** glassware. The joints are ground-glass standard-taper joints that are made to fit snugly together. Miniscale glassware is often stored in a case, but if this glassware is stored loose in a drawer, it is important to arrange the items so that they don't bump into one another when the locker is opened.

Other useful items of glassware commonly found in the locker are thermometers, Erlenmeyer flasks, graduated cylinders, and beakers. Thermometers should always be stored in their plastic or cardboard containers. Beakers are best stored as nests. Graduated cylinders should be laid on their sides. If items are kept in the same place in the drawer after each lab period, it is easy to find the equipment you need.

Microscale Glassware and Related Equipment

Items commonly found in a threaded microscale glassware kit are shown in Figure 1A-1. A **Claisen adapter** (a) is a special piece of glassware that allows for placement of more than one item on top of a reaction vessel. A **distillation head** (b) is for distillation

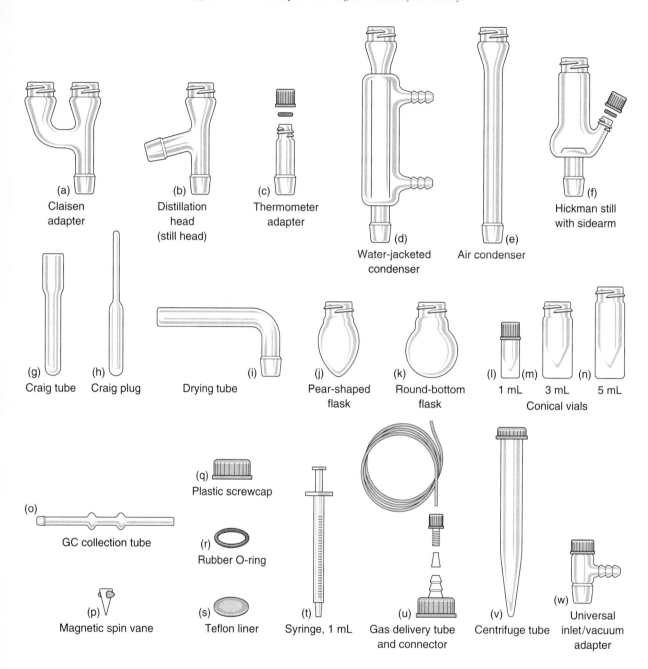

(a) Claisen adapter

(b) Distillation head (still head)

(c) Thermometer adapter

(d) Water-jacketed condenser

(e) Air condenser

(f) Hickman still with sidearm

(g) Craig tube

(h) Craig plug

(i) Drying tube

(j) Pear-shaped flask

(k) Round-bottom flask

(l) 1 mL (m) 3 mL (n) 5 mL Conical vials

(o) GC collection tube

(p) Magnetic spin vane

(q) Plastic screwcap

(r) Rubber O-ring

(s) Teflon liner

(t) Syringe, 1 mL

(u) Gas delivery tube and connector

(v) Centrifuge tube

(w) Universal inlet/vacuum adapter

Figure 1A-1 Threaded microscale glassware kit

of 5 mL or more of a liquid. (Not all kits are equipped with this piece.) A **thermometer adapter** (c) is included so that a thermometer may be mounted on top of the distillation head. A **water-jacketed condenser** (d) allows cold water to flow around the outer compartment of the condenser. This condenser is used when volatile reagents and solvents are used during a reaction. It can also be used as an air condenser.

An **air condenser** (e) has no outer compartment. It may be packed and used as a fractional distilling column or it may be placed on top of a vial to heat high-boiling liquids. The **Hickman still** (f) consists of a tube, similar to the air condenser, but with a circular lip to trap liquid during a micro-distillation. The Hickman still may have a side arm (as shown here) for easy removal of condensed liquid. The **Craig tube and plug** (g, h) are used for isolating small quantities of crystals. In some kits, both pieces are made of glass. Great care must be taken so as not to break either of the parts during use. The bent tube is a **drying tube** (i), which is designed to fit on top of a condenser. The tube is packed with fresh drying agent prior to use.

Some kits contain a 10-mL **pear-shaped flask** (j), used for distillations. A 10-mL **round-bottom flask** (k) is used for distillation or for carrying out reactions that require no more than 5–8 mL total volume of solution. A magnetic stir bar (not shown) is sometimes used with the round-bottom flask. The **conical vials** (l,m,n) are the most used items of equipment in the kit. The vials have capacities of 1, 3, and 5 mL. The **gas-liquid chromatography (GC) collection tube** (o) is used for collecting condensed liquid samples at the exit ports of the gas chromatography instrument. A **magnetic spin vane** (p) is designed for use with the conical vials. Each conical vial is equipped with a **plastic screwcap** (q). The screwcaps have openings that allow insertion of other items, such as condensers. Each conical vial is equipped with a **rubber O-ring** (r). Round **Teflon liners** (s) may be placed inside the plastic screwcap to close off vials from the outside atmosphere. The Teflon liner may be penetrated with a syringe needle when adding a solution to a vial via syringe while the system is closed.

The kit should also contain a 1-mL **syringe** (t). The syringe may be made of glass with a Teflon plunger. This glass syringe is reusable and should be cleaned between usage. Some kits contain disposable syringes made of plastic. Needles are generally not included in the kits. They will be distributed when needed. Most kits contain a **gas delivery tube** assembly (u) that includes four parts. The tube is used to transport any gas formed during a reaction to a separate container. A **centrifuge tube** (v) can be used with both the Craig tube and the glass tubing. The kit also includes a **universal inlet/vacuum adapter** (w).

Some kits may vary in content; it is not necessary to have all of the items in each kit. Other glassware used in microscale experiments is shown in Figure 1A-2, including a **Hirsch funnel** (a) containing a porous plug, to be used with a **suction flask** (b) to collect solid products by filtration, a separate **filter adapter** (c), a microscale **chromatography column** (d), a **pipet** (e), and a **spatula** (f).

Microscale experiments can also be done using test tubes, small Erlenmeyer flasks, Pasteur pipets, small beakers, and other glassware.

Miniscale Glassware

Using stardard-taper glassware may be new to you. If this glassware is furnished in your locker in the form of a kit, it will look similar to the kit shown in Figure 1A-3. The kit consists of some items of glassware that have standard-taper joints. The joint sizes may be 14/20 or 19/22. The designation 14/20 means that the inside width is 14 mm and the joint is 20 mm long. Glassware with 14/20 and 19/22 joints is known as miniscale. A similar but larger form of this glassware has 24/40 joints and is called macroscale glassware. A dozen or more items make up the typical miniscale kit (see Figure 1A-3).

A **bleed tube** (a) is used for vacuum distillations. A **Claisen adapter** (b) allows for placement of more than one item on top of a reaction vessel. The **distilling head** (c) holds a thermometer and passes distillate into a condenser. A **thermometer adapter** (d) fits in the distilling head to hold a thermometer. A **bent vacuum adapter** (e) connects the condenser to the receiver in distillation. A **condenser** (f) is required to condense vapors during distillation. A **distillation column** (g) is used for reflux and also as a second condenser. Several **round-bottom flasks** (h–l) of various sizes are included. Extractions are accomplished by using a **separatory funnel** (m). A **stopper** (n) is necessary for the top of the separatory funnel.

Figure 1A-2

Additional glassware
for microscale
experiments

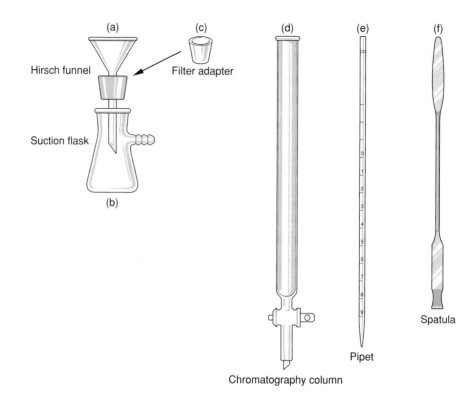

Figure 1A-3   Miniscale glassware kit

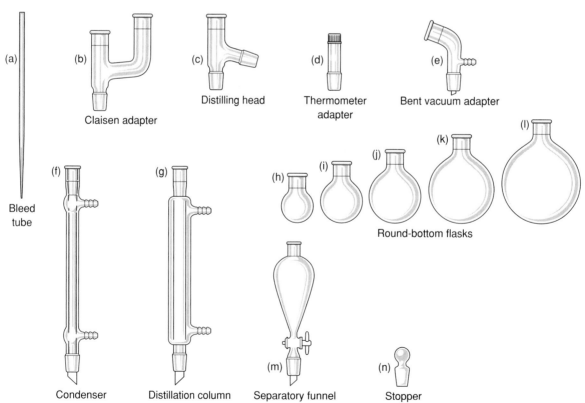

Additional Glassware and Equipment

Common glassware items, such as beakers, funnels, and Erlenmeyer flasks, are also generally stocked in student lockers. These and other items commonly found in student lockers are illustrated in Figures 1A-4 and 1A-5.

How to Clean Glassware

Dispose of all chemical residues properly and then wash glassware with soap and water. Use small amounts of acetone to clean organic residues that are difficult to remove. Use petroleum ether or a similar nonpolar solvent to remove grease from joints. Use a brush, if necessary, to scrub away charred matter. Treat flasks that are very resistant to cleaning with 3 M or 6 M nitric acid. **Clean all glassware immediately after use. Never put dirty glassware away in the locker.**

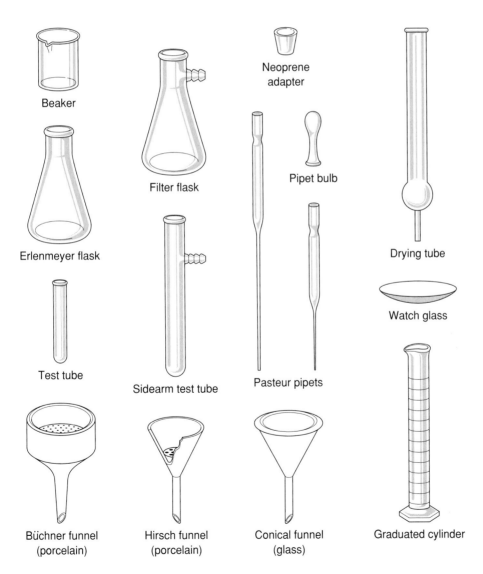

Figure 1A-4   Common glassware and equipment in the organic laboratory

Figure 1A-5

Common glassware
and equipment in the
organic laboratory

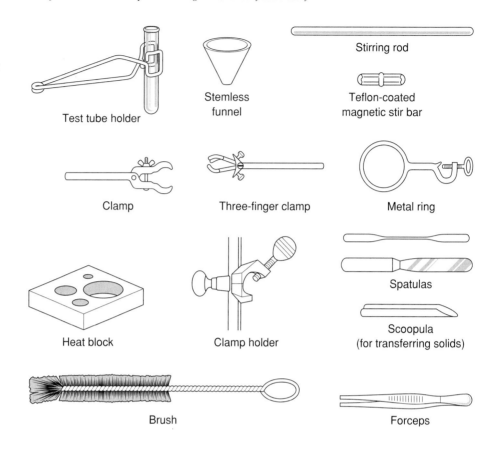

How to Heat and Cool Glass Reaction Vessels

There are three methods for heating glassware in the laboratory.

Refer to Figures 1A-6
and 1A-7.

1. Use a **water bath** or **steam bath** to heat an Erlenmeyer flask or beaker by placing the item on or just above the bath. **Add a boiling chip, spin vane, or spin bar** and heat contents of the containers to any temperature as necessary up to just under the boiling point of water. Use a **hot plate** to heat a water bath and a steam line to heat a steam bath.

The temperature of sand
in a sand bath is hard to
control. Use a minimal
amount of sand.

2. Use a **heat block** or a **sand bath** to heat vials in microscale experiments. **Add a boiling chip, spin vane, or spin bar** and place vials containing solutions in a metal heat block or in a sand bath (a glass crystallizing dish or metal container containing sand, for example) with a hot plate underneath as the heat source. Heat the sand bath 20–30°C above the boiling point of the solvent being evaporated or distilled because heat transfer through sand is not very efficient. **Be very careful when touching items near or on a hot plate.** It is hard to know if a heat block or sand bath is hot or cold.

Do not plug a heating
mantle directly into an
electrical outlet.

3. Use a variable **transformer (Variac)** and a **heating mantle** of the proper size to heat miniscale round-bottom flasks. Variacs and heating mantles may be new to you. To assemble, set the on/off switch to "off" and set the dial to a reading of zero. Insert the plug of the power cord of the Variac into a standard electrical outlet. Connect the second cord from the Variac to the heating mantle. Connect the heating mantle to the cord and use a clockwise rotation to lock the connection. Unlock only by a counter-clockwise rotation. Remember to "twist on" and "twist off" to connect or disconnect the cord to a heating mantle. **Add a boiling chip or spin bar** to the

flask being heated. Then add the chemicals to the flask. Turn on and adjust the Variac from 0 to 110 volts. The desired setting depends upon the volatility of the solvent being heated. For example, a setting of 20 volts will cause diethyl ether to boil, but a much higher setting of 70–80 volts must be used to boil water.

Common heating devices are shown in Figure 1A-6. They include a steam bath/hot water bath, a hot plate/stirrer with heat block, and a Variac transformer and heating mantle. An alternate to a fiberglass heating mantle is a Thermowell (Laboratory Craftsmen) heater. The Thermowell looks much like a fiberglass heating mantle, but is made of hard ceramic material. The Thermowell offers the advantage of not requiring an exact fit between the size of flask and well. A larger Thermowell will accommodate smaller-sized flasks. Use boiling chips or magnetic stirrer when heating liquids to boiling using a Thermowell heater.

Examples of setups for heating are shown in Figure 1A-7. These include a heat block, a sand bath, and a heating mantle. A steam bath can also be used as a hot water bath.

Glassware apparatus may also require cooling to moderate an exothermic reaction or when trying to obtain crystalline products. Ice baths are generally very effective for cooling. To make an ice bath, simply add ice to a beaker or water bath and partially submerge the vessel to be cooled. Add a stir bar or spin vane to the vessel to perform a reaction at cold temperatures with stirring.

Add sodium chloride or calcium chloride to an ice bath to make an ice-salt bath for those situations where temperatures from 0 to $-10°C$ are required. To obtain even lower temperatures, place small chunks of dry ice in a dry metal or glass container to prepare special baths at $-78°C$. Carefully pour isopropyl alcohol or acetone over the dry ice. White carbon dioxide vapors will form during this process and some frothing may occur. Partially submerge the reaction flask to be cooled in the dry ice bath.

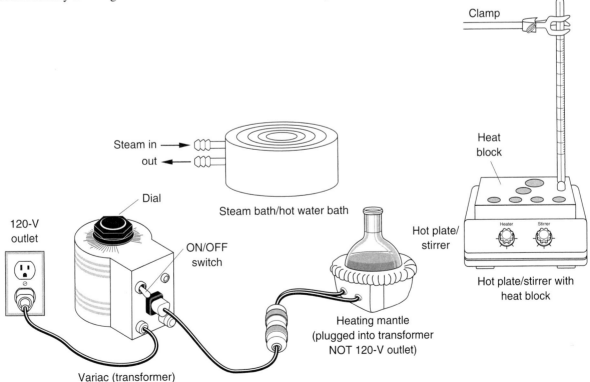

Figure 1A-6 Heating devices and accessories

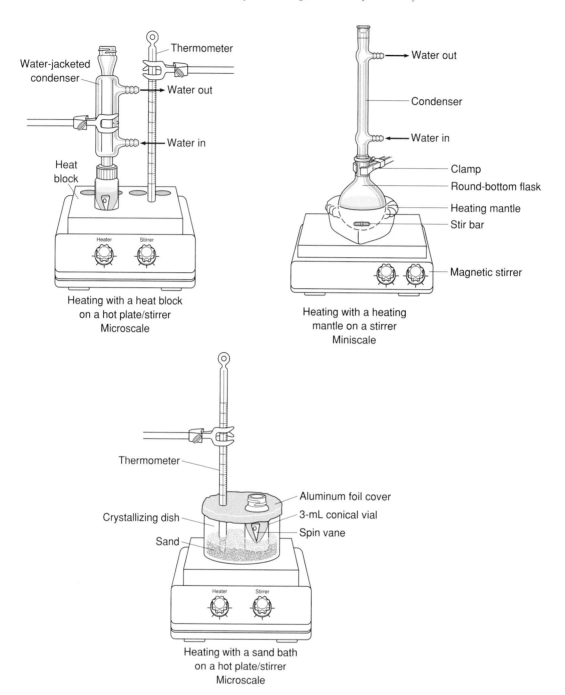

Figure 1A-7   Heating setups using a heat block, a sand bath, and a heating mantle

Questions

1. What types of reaction vessels are used in microscale experiments? Are these vessels different in any way from miniscale equipment?

2. An organic solvent has a boiling point of 120°C. What heating device should be used to remove the solvent?

3. Give a reason for not having a graduated cylinder included with microscale glassware.

Technique B: Weighing and Measuring

Introduction to Weighing

Because of the small quantities used in microscale experiments, solid and liquid reagents are usually weighed to three-place accuracy using a three-place or four-place, top-loading balance. For miniscale experiments, a two-place balance is often sufficient.

How to Weigh Solids and Liquids

Weigh solids into glass containers or onto weighing paper using a spatula. Weigh liquids in containers that can be capped. Transfer liquids with a Pasteur pipet. Place the container or paper on the balance pan of a top-loading balance and push the tare button. The scale of a three-place balance should read 0.000 g. Add the desired amount of substance in portions up to the desired weight. Record the actual amount used in your notebook. Use an amount that is within 1–2% of the weight given in each experiment. If too much sample is weighed out, transfer any excess material to an appropriate bottle provided by the instructor. **Clean the balance and the area around the balance when you have finished weighing.** A three-place, top-loading balance is shown in Figure 1B-1.

Do not weigh hot or warm objects.

Introduction to Measuring Volumes of Liquids

Volatile liquids should be measured with a calibrated glass pipet or syringe (by volume instead of weight) to avoid loss of liquid through evaporation. Three different kinds of pipets are available: a calibrated Pasteur pipet, a calibrated glass pipet (see Figure 1B-2), and an automatic delivery pipet (see Figure 1B-4). Use the calibrated Pasteur pipet for transfer of liquids where the exact amount of liquid isn't important. The pipet consists of two parts: a glass tube drawn to a blunt tip and a latex bulb. The glass tube is disposable, although it can generally be used a number of times. The bulb is used to draw liquid into the glass tube. Calibrated glass pipets are designed to measure liquids in increments of 0.01 mL to 50 mL. The 0.1-mL, 0.5-mL, and 1.0-mL pipets will be most useful in this course.

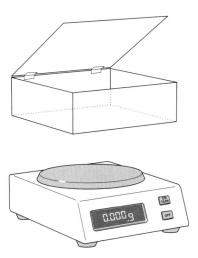

Figure 1B-1

A three-place, top-loading balance with draft shield

Figure 1B-2

Calibrated Pasteur
pipet and calibrated
glass pipet

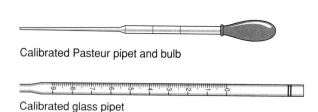

Calibrated Pasteur pipet and bulb

Calibrated glass pipet

Figure 1B-3

Steps for filling and dispensing 0.50 mL of a
liquid using a
calibrated pipet

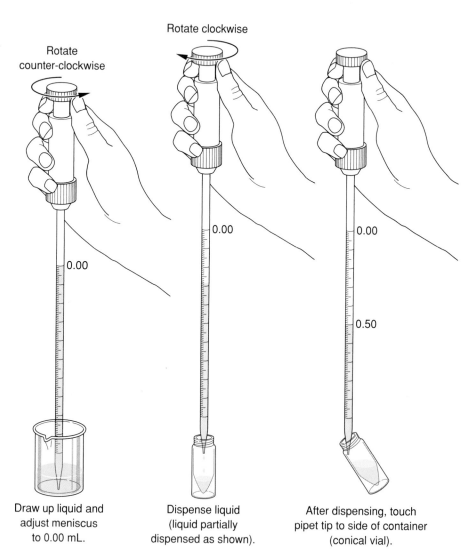

Rotate clockwise

Rotate
counter-clockwise

0.00

0.00

0.00

0.50

Draw up liquid and
adjust meniscus
to 0.00 mL.

Dispense liquid
(liquid partially
dispensed as shown).

After dispensing, touch
pipet tip to side of container
(conical vial).

How to Use a Calibrated Glass Pipet

Do not pipet by mouth!   Use a pipet with either a bulb or a pipet pump. If you are using a pipet bulb, attach the
bulb to the wide end of the pipet. Squeeze the pipet bulb. Place the tip of the pipet
below the surface of liquid. Gently release pressure on the bulb allowing liquid to be
drawn up into the pipet slightly above the 0-mL mark. Adjust the volume of liquid so
that the meniscus of the liquid reads 0.00 mL. Place the tip of the pipet into the top of

the receiving vessel and apply pressure to the bulb to dispense the liquid. Deliver the last drop of the desired volume by touching the tip against the wall of the vessel.

If dispensing using a pipet pump, see the steps shown in Figure 1B-3. Attach the pump to the wide end of the pipet. Place the tip of the pipet below the surface of liquid. Turn the knob at the top of the pump counter-clockwise to draw liquid into the pipet slightly above the 0-mL mark. Adjust the volume of liquid so that the meniscus of the liquid reads 0.00 mL. Place the tip of the pipet into the top of the receiving vessel and turn the pump knob clockwise to dispense 0.50 mL of the liquid. Deliver the last drop of the desired volume by touching the tip against the wall of the vessel.

If this procedure is difficult, which is often the case if the solvent is diethyl ether or a similar highly volatile solvent, it will be necessary to draw up the solvent a few times before dispensing. Transfer a slight excess of the amount of the solvent you want to pipet to a clean Erlenmeyer flask. Using a clean, dry pipet and filter bulb, draw up a portion of solvent and release to allow it to return to the flask. Repeat this process several times to equilibrate the temperature of the pipet and solvent. Then draw the solvent up to the mark on the pipet for delivery.

How to Use an Automatic Delivery Pipet

An automatic delivery pipet, illustrated in Figure 1B-4, is a device that can be set to deliver an exact volume of liquid (usually aqueous) by dialing in the desired volume at the top of the pipet. Pipets are available for different volumes and most pipets are adjustable to allow delivery of different volumes. The pipets are very expensive, so be careful when using them. Delivering volatile liquids may lead to inaccuracies.

The automatic delivery pipet consists of two parts, the main body and a disposable plastic tip. The tip is the only part of the pipet that touches the surface of the liquid. Liquid should never touch the main body of the pipet and even the tip should reach only just below the surface of the liquid being transferred. Differences in surface tension of organic liquids may affect the amount of a liquid drawn up at a particular pipet setting, so automatic delivery pipets are used most often to measure out aqueous solutions.

The steps for dispensing a liquid using an automatic delivery pipet are shown in Figure 1B-5. Attach a plastic tip to the bottom of the pipet and dial in the desired volume at the top of the pipet and lock in the values. Depress the piston to the first stop (first point of resistance). Do this before dipping the tip of the pipet below the surface of the liquid being dispensed. Next, insert the plastic tip 1–2 cm below the surface of the liquid and slowly release the plunger to load the pipet. This procedure prevents formation of bubbles. To dispense the liquid to a new container, touch the tip of the pipet against the container and depress the piston slowly to the first stop. Then push the piston to the second stop (all the way) to dispense the last drop of liquid. The same tip can be used again to dispense another sample of the same liquid. Replace the plastic tip to dispense a different liquid.

How to Use a Syringe

Use a syringe when a liquid must be injected with a needle. Clean and dry all parts of the syringe. Transfer a portion of liquid from the stock bottle to a vial. With the plunger inserted all the way into the barrel of the syringe, introduce the tip of the needle below the surface of the liquid in the vial. Carefully pull back on the plunger so that liquid is drawn up into the barrel to the desired volume. Drain the syringe into another container or inject the liquid through a septum into another vial. Remember that a dirty needle will contaminate the liquid.

To add a liquid to a reaction mixture for a reaction that must be run under a dry or inert atmosphere, push the needle of a loaded syringe through the cap liner or septum.

Figure 1B-4

The automatic delivery pipet (used mainly for dispensing aqueous solutions)

Figure 1B-5

Steps for dispensing a liquid using an automatic delivery pipet

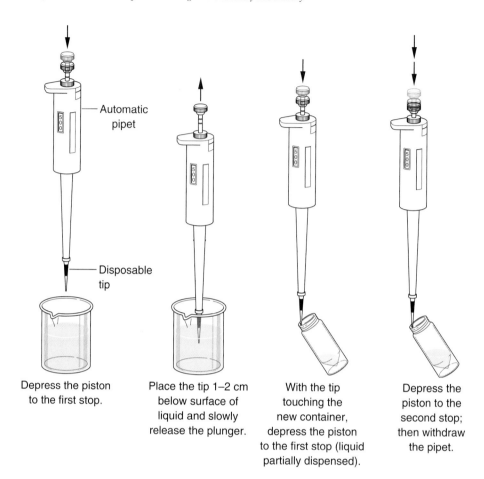

Depress the piston to the first stop.

Place the tip 1–2 cm below surface of liquid and slowly release the plunger.

With the tip touching the new container, depress the piston to the first stop (liquid partially dispensed).

Depress the piston to the second stop; then withdraw the pipet.

Add the liquid in the syringe at a convenient rate by applying pressure to the plunger. Be careful when pushing a needle through a septum. The septum will offer some resistance, but most needles are sharp enough to penetrate a septum if a slow, steady pressure is applied. Obtain a new needle if excessive pressure is required.

Clean the syringe by drawing up a wash solvent (such as alcohol or acetone) and expelling the solvent into a suitable container. Do this several times. Set the syringe and plunger aside to dry. The needle should be detached and dried in an oven (placed in a beaker) if it is to be reused in the same lab period.

Be very careful when working with syringes. It is not uncommon to "get poked" by a needle. Be careful when placing a syringe on the lab bench. Syringes can roll off the bench easily and be damaged. They are expensive to replace. Disposable syringes are used in many laboratories. These syringes need not be cleaned, but must be properly discarded.

Safety First!

Always wear eye protection in the laboratory.

Exercise B.1: Determining Density of an Aqueous Solution

The balance pan should be clean and dry. Turn on and tare the balance. Place an empty capped 3-mL or 5-mL conical vial on the pan and record the weight. Attach a plastic tip to an automatic delivery pipet. Be sure that the setting at the top of the pipet corresponds to the desired volume of 0.500 mL. With the pipet held upright, push down on the piston to the first stop position. Place the tip of the pipet just below the surface of the stock bottle of an aqueous solution assigned by your instructor. Release the piston gradually to draw the liquid into the tip. Transfer the liquid from the pipet to the vial. Record the weight. Transfer the sample in the vial to a bottle labeled "recovered aqueous solutions" using a Pasteur pipet. Clean up any spilled liquid before leaving the balance area. Calculate the density of the solution.

Exercise B.2: Determining Density of an Organic Liquid

The balance pan should be clean and dry. Turn on and tare the balance. Place an empty capped 3-mL or 5-mL vial on the pan and record the weight. Use a calibrated or volumetric pipet and bulb or pipet pump to draw up 0.500 mL of an organic liquid from a sample of organic liquid issued by your instructor. Do this by initially drawing up more than 0.500 mL and then releasing a small amount back into the reservoir so that the meniscus reads 0.500 mL. Transfer the liquid from the pipet to the vial. Record the weight. Transfer the sample in the vial to a bottle labeled "recovered organic liquids" using a Pasteur pipet. Clean up any spilled liquid before leaving the balance area. Calculate the density of the liquid. Use Table 1B-1 to determine the identity of the liquid.

Safety First!

Always wear eye protection in the laboratory.

Exercise B.3: Calibrating a Pasteur Pipet

Calibrate a Pasteur pipet using a file or a marking pencil, at heights of liquid equal to 0.5, 1.0, 1.5, and 2.0 mL or other volumes if desired. Use a 1-mL syringe or a calibrated pipet to furnish the amounts of liquid to be measured. Mark three Pasteur pipets in case one gets broken. Insert a plug of cotton into one the pipets to make a "filter pipet." Filter pipets are used to trap solids during transfers from the pipet and to keep volatile, low-boiling solvents from being involuntarily released from the pipet. The calibrated Pasteur pipets will be used in future experiments.

Safety First!

Always wear eye protection in the laboratory.

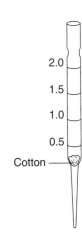

Questions

1. A microscale procedure requires 2.5 mL of diethyl ether. Describe the best method to measure this amount.
2. Describe the procedure for measuring out:
   a. 50 mg of an organic solid for a reaction to be run using a 3-mL vial.
   b. 1.00 mL of an organic liquid.
   c. approximately 0.5 mL of water.
3. Contrast and compare the capabilities and use of an automatic delivery pipet versus a syringe.
4. Why are liquids not poured from one container into another in microscale experiments?
5. Why might it be preferable to use a syringe rather than a pipet?

Table 1B-1   Densities of Selected
Organic Liquids

Organic Compound	Density$^{20/4}$(g/mL)
Hexane	0.6603
3-Methylpentane	0.6645
Heptane	0.6837
Cyclohexane	0.7785
1-Propanol	0.8035
3-Pentanone	0.8138
p-Xylene	0.8611
Toluene	0.8669
Tetrahydrofuran	0.8892
Ethyl acetate	0.9003

Technique C: Melting Points

Introduction

Most organic compounds are molecular substances that have melting points below 300°C. The melting point of a solid compound is a physical property that can be measured as a method of identification. Melting points of pure compounds are recorded in handbooks of physical data, such as the *Handbook of Chemistry and Physics* (CRC). The reported melting point is that temperature at which solid and liquid phases exist in equilibrium. A melting point of a solid is actually a melting range, which starts from the temperature at which the first drop of liquid appears and ends at the temperature at which the entire sample is liquid. Students should report melting points as melting ranges. Sometimes only a single temperature is reported in the tables of physical properties; in this case, the value represents the upper temperature of the melting range.

The measured melting range gives a rough indication of the purity of the compound: the purer the compound, the higher its melting point and the narrower its melting range. Melting a solid requires energy to overcome the crystal lattice energy. Impurities disrupt the crystal lattice, so less energy is required to break intermolecular attractions; impurities thus generally lower and broaden the melting point. Since the decrease in melting point of a solid is generally proportional to the amount of impurity present, the difference between the expected melting point of the pure compound and the experimentally measured melting point gives a rough approximation of purity.

Figure 1C-1 is a phase diagram of a mixture of phenol and diphenylamine. The phase diagram is a graph of temperature versus composition. The top convex line represents the temperature at which the entire solid is melted, and the bottom concave line represents the temperature at which the solid just begins to melt. The distance between these two lines represents the melting point range. Pure phenol (at 0 mol percent diphenylamine) exhibits a sharp melting point at 41°C; that is, both the liquid and the solid line converge at 41°C. A sample of phenol containing 5 mol percent of diphenylamine begins to melt at 35°C and the solid is all melted by 40°C, giving a melting point range of 35–40°C. A sample of phenol containing 10 mol percent of diphenylamine has a melting point range of 30–37°C. As the amount of impurities increases, the melting point range becomes proportionally lower and broader.

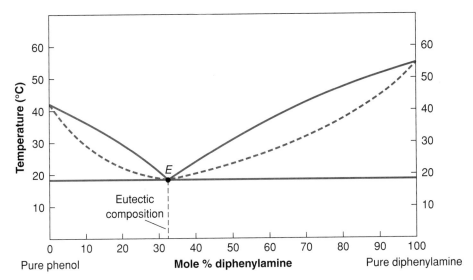

Figure 1C-1

Temperature vs. compo-
sition diagram of a
phenol/diphenylamine
binary mixture

It would be tempting to say that addition of impurities always lowers the melting
point and broadens the melting point range. However, this is not the case. The lowest melting point on the phase diagram occurs at 32 mol percent diphenylamine. A sample of phenol containing this amount of diphenylamine melts sharply at 19°C. For a binary
(two-phase system), the minimum melting point temperature is called the eutectic point,
and the composition at this temperature is called the eutectic mixture. For the
phenol/diphenylamine system, the eutectic point occurs at 32 mol percent diphenylamine.
A mixture of 68 mol percent phenol and 32 mol percent diphenylamine (the eutectic composition) behaves like a pure compound, exhibiting a sharp melting point. Addition of
more diphenylamine beyond the eutectic composition actually increases the melting point:
a sample of phenol containing 40 mol percent diphenylamine begins to melt around 21°C.

Not all two-phase mixtures exhibit eutectic behavior; a sharp melting point is usually indicative of purity. However, a sharp melting point may also be obtained if the
compound and impurity form a eutectic mixture. For nearly pure compounds, the presence of small amounts of impurities generally lowers the melting point. Purification
generally affords crystals having a higher and sharper melting point than the impure
solid. In rare cases, the melting point may rise when a certain additive is present.

Mixed Melting Behavior

Because impurities change (usually lower) the melting point, a mixed melting point can
be used to determine whether two compounds are identical or different. Suppose that
there are two beakers of a white solid sitting on the bench top. The melting point of each
solid is found to be 133°C. Are the two solids the same or are they different? In a mixed
melting point, samples of the two compounds are thoroughly mixed and a new melting
point is measured. If the two compounds are not identical, the melting point of the mixture generally will be depressed and broadened. If the melting point is unchanged, the
two compounds are most probably identical.

Melting Behavior of Solids

Several physical changes occur as a solid melts. In many cases, the crystals will soften
and shrink immediately before melting. The crystals may appear to "sweat" as traces
of solvent or air bubbles are released. These are normal occurrences, but shouldn't be

considered as "melting." The melting point is measured beginning at the time the first free liquid is seen. Sometimes compounds decompose rather than melt. Decomposition usually involves changes such as darkening or gas evolution. Decomposition may occur at or even below the melting point for compounds that are thermally labile. In general, decomposition occurs over a fairly broad temperature range. For this reason, a sample should not be used in two consecutive melting point determinations. The sample may decompose, causing the second measured melting point to be lower than the first.

Some compounds having fairly high vapor pressures can change directly from a solid to a gas without passing through the liquid phase. This is called sublimation. In order to measure a melting point (more accurately, a sublimation point) for such compounds, a sealed, evacuated melting point capillary tube must be used.

Calibration of the Thermometer

Frequently, thermometers are not equally accurate at all temperatures and must be calibrated to obtain accurate melting points. To calibrate a thermometer, the melting points of pure solid samples are measured and a graph is made of the measured (observed) melting point temperatures versus the difference between the observed and the expected melting points. When reporting experimental melting points, corrections should be made by adding the appropriate value from the graph to the observed melting point. A sample thermometer calibration curve is shown in Figure 1C-2. An observed melting point of 190°C would be reported as 188°C (observed temperature + correction factor = 190°C + [−2°C] = 188°C).

Apparatus for Measuring Melting Points

Several commercial devices that measure melting points are available. These include the Mel-Temp apparatus and the Fisher-Johns Block, which are heated metal block devices. The Mel-Temp uses a closed-end capillary tube. The Fisher-Johns uses two glass plates horizontally placed to sandwich the substance between them. The Thomas-Hoover device is a mechanically stirred oil bath that also uses capillary tubes. These devices are shown in Figure 1C-3.

How to Determine a Melting Point

Preparing the sample. Use 1–2 mg of dry solid. For a mixed melting point determination, thoroughly mix approximately equal amounts of the two samples using a mortar and pestle.

Figure 1C-2

Temperature calibration curve

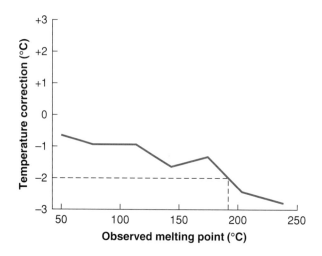

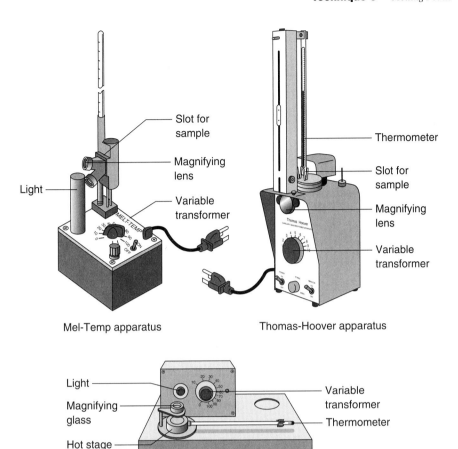

Mel-Temp apparatus

Thomas-Hoover apparatus

Fisher-Johns block

Loading the capillary. Do not load the capillary tube with too much sample: this causes the melting point to be wide and slightly high because the temperature will continue to rise while the compound continues to melt. Place 1–2 mg of the sample on a watch glass or a piece of weighing paper. Push the open end of the capillary down onto the sample and tap on the solid sample. Then invert the tube. Gently tap the bottom of the tube on the bench or drop the tube through a short 2-ft piece of glass tubing; this causes the sample to pack more tightly and give a more accurate melting point. This process is illustrated in Figure 1C-4.

Setting the heating rate. If the melting point is unknown, heat the sample rapidly to establish an approximate melting point. Turn off the apparatus as soon as the compound melts and note the temperature. Let the temperature drop until it is approximately 10°C below the observed melting point and repeat the melting point determination with a new sample. Heat the sample rapidly to within 10°C of the known melting point. Then slow to 1–2°C per minute. Heating too rapidly results in inaccurate, usually wider, melting point measurements. An appropriate heating rate can also be determined by referring to the heating-rate curve that often accompanies the melting point apparatus.

Observe the sample through the magnifying eyepiece as the sample melts. Record the melting point as the temperature from the start of melting until all solid is converted to liquid. Remember that shrinking, sagging, color change, texture changes, and sweating are not melting. When the sample has melted, turn off the melting point apparatus and remove the capillary tube. Discard the capillary after use into the glass disposal container.

Figure 1C-4

Loading a sample into a melting point capillary tube

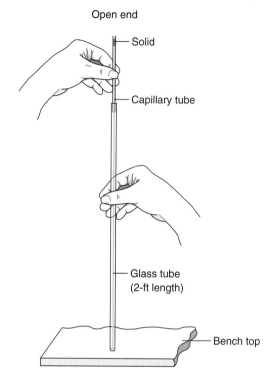

Exercise C.1: Calibration of a Thermometer

Safety First!

Always wear eye protection in the laboratory.

Determine the melting points of a series of pure solids. Suggested compounds are diphenylamine (53°C), *m*-dinitrobenzene (90°C), benzoic acid (122°C), salicyclic acid (159°C), succinic acid (189°C), and 3,5-dinitrobenzoic acid (205°C). Calculate the difference between the known melting point and the measured melting point for each compound. The differences will be positive or negative. Plot the measured melting point on the *x* axis and the correction factor on the *y* axis as in Figure 1C-2.

Exercise C.2: Melting Point of an Unknown Solid

Safety First!

Always wear eye protection in the laboratory.

Melting points should be recorded as corrected or uncorrected.

Calibrate the thermometer, if directed to do so by the instructor (Exercise C.1). Obtain a sample of an unknown solid and record the number of the unknown in your lab notebook. Load a capillary tube with a small amount of the solid. Prepare another sample in the same manner. Place the tube in a melting point apparatus. Heat rapidly to get an estimate of the actual melting point. Turn off the apparatus immediately when the compound melts and note the temperature. Let the temperature drop until it is approximately 10°C below the observed melting point. Then place the second tube in the melting point apparatus and start heating at a rate of 1–2°C per minute. Record the melting point range of the sample (corrected, if necessary). The unknown sample may be identified by comparing the melting point with a list of unknowns in Table 1C-1. Report the melting point range and the identity of the unknown solid. Dispose of the capillary tube in a glass disposal container and return any unused solid to the instructor.

Table 1C-1 Melting Points of Selected
Organic Solids

Organic compound	Melting point (°C)
Benzophenone	48
2-Naphthaldehyde	60
Benzhydrol	68
Vanillin	80
Benzil	95
o-Toluic acid	104
4-Hydroxyacetophenone	109
4-Hydroxybenzaldehyde	115
Benzoic acid	122
trans-Cinnamic acid	133
3-Nitrobenzoic acid	140
2-Nitrobenzoic acid	146
Adipic acid	153
Camphor	178
p-Anisic acid	184

Exercise C.3: Mixed Melting Point

Obtain two vials from the instructor: one will be a sample of cinnamic acid and the other will be an unknown, which will be either cinnamic acid or urea. Record the number of the unknown in your lab notebook. Measure the melting points of cinnamic acid and the other sample and record the melting points in your lab notebook. Use a spatula to transfer 1–2 mg each of cinnamic acid and the unknown solid to a pestle. With a mortar, grind the solids together to mix thoroughly. Measure and record the melting point of the mixture in your lab notebook. Determine whether the unknown is cinnamic acid or urea. Justify your conclusions.

Safety First!

Always wear eye protection in the laboratory.

Questions

1. A student measures the melting point of benzoic acid and reports the melting point in the lab notebook as 122°C. Explain what the student did wrong.

2. Two substances obtained from different sources each melt at 148–150°C. Are they the same? Explain.

3. A substance melts sharply at 135°C. Is it a pure compound? Explain.

4. Benzoic acid and 2-naphthol each melt at 122°C. A sample of unknown solid melts around 122°C. The solid is either benzoic acid or 2-naphthol. Describe a method to determine the identity of the unknown compound.

5. Explain why atmospheric pressure affects boiling points of liquids, but does not affect melting points of solids.

6. Refer to the temperature vs. composition diagram of the phenol/diphenylamine binary system (Figure 1C-1). Estimate the melting point range for a mixture of 85 mol percent phenol/15 mol percent diphenylamine.

7. Do impurities always lower the melting point of an organic compound? Explain.

Technique D: Boiling Points

Introduction

Boiling point is a useful physical property for identifying pure liquids as well as for evaluating the purity of a substance. The boiling point of a liquid is defined as the temperature at which its vapor pressure equals atmospheric pressure. Vapor pressure is a measure of the tendency of molecules to escape from the surface of a liquid. Liquids with higher vapor pressures have lower boiling points than liquids with lower vapor pressures.

The boiling point of a liquid can be measured using a capillary tube as described in Exercises D.1 and D.2. Alternately, the boiling point can be measured during miniscale distillation (Technique G). The temperature at which most of the material distills is recorded as the boiling point. The barometric pressure at the time of the distillation should be recorded, so that a correction can be applied to determine the normal boiling point. The normal boiling point is the boiling point when atmospheric pressure is 1 atmosphere (760 torr) at sea level. Reference books list normal boiling points unless indicated otherwise.

Boiling point varies with atmospheric pressure and with elevation. For example, in Denver, at 5,280 feet, boiling points of common solvents are 4–5°C lower than at sea level. The atmospheric pressure may vary by a few degrees from day to day (or hour to hour) because of weather changes and passage of weather fronts. These differences must be taken into account when comparing boiling points measured in different labs and measured at different times. In general, a correction to the boiling point can be made by allowing approximately 0.35°C per 10 torr deviation from 760 torr (the standard pressure) in the vicinity of 760 torr.

It is also possible to estimate the boiling point of a substance at various pressures using a pressure-temperature nomograph. For example, the reported normal boiling point of an organic liquid is 78.3°C. The nomograph in Figure 1D-1 can be used to determine the boiling point of this liquid in a lab where the atmospheric pressure is 650 torr. A straight line is drawn between the reported normal boiling point (78.3°C) and the pressure (650 torr). The extension of this line crosses the ΔT correction axis at 4°C. At 650 torr this liquid is predicted to boil at 74.3°C (78.3° – 4°C). If the normal boiling point is known, the expected boiling point can be estimated at any desired reduced pressure as low as 600 torr.

The nomograph may also be used to correct an observed (or measured) boiling point to the boiling point at 760 torr (the reported or normal boiling point). To do this, draw a straight line between the pressure and the temperature corresponding to the observed boiling point. The extension of this line intersects the correction factor that must be added to the observed boiling point. For example, the boiling point of a liquid that boils at 122°C in a lab where the ambient pressure is 685 torr would be reported as 125°C (observed boiling point + ΔT = 122°C + 3°C = 125° C).

Intermolecular Attractions

The boiling point of an organic liquid depends upon molecular properties. Boiling points increase with increasing molecular weight in a homologous series of molecules, e.g., pentane < hexane < heptane << decane. This phenomenon is explained in part by the fact that larger molecules have greater intermolecular attractions and require more heat to vaporize. Another trend is the increase in boiling point with decreasing branching of isomeric molecules, if the functional groups are the same. This is due to decreased surface area and intermolecular attractions for more highly branched molecules, e.g., 2,2-dimethylpropane < 2-methylbutane < pentane. For molecules of similar molecular weights, boiling points increase with increasing polarity of the molecules,

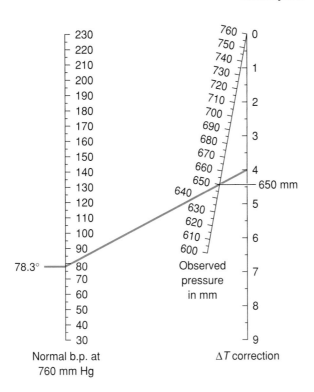

e.g., pentane < 2-butanone < 1-butanol < propanoic acid. More polar molecules have stronger intermolecular forces of attraction. Molecules of similar molecular weight that can hydrogen-bond have even stronger intermolecular interactions.

How to Do a Microscale Boiling Point Determination

Use a syringe to inject 3–5 μL of liquid sample into a capillary tube. Centrifuge, if necessary, to get the entire sample to the bottom of the capillary. Prepare a bell by drawing a melting point capillary tube over a micro burner to make a very fine capillary. Cut a small length of this and carefully seal at one end to make a bell top. (See Figure 1D-2.) Place the bell capillary, sealed end up, into the melting point capillary containing the liquid. Centrifuge, if necessary, to force the bell down to the bottom of the capillary and into the liquid. The open end of the bell should be straight, resulting from a careful break using a file. Ideally, the bell should be about 1½ times the height of the sample liquid. Place the entire system in a heating device, such as that used for melting point measurements (e.g., Mel-Temp, Thomas-Hoover, or similar system). Apply heat until a rapid and steady stream of bubbles flows from the inverted capillary. Stop heating and watch the flow of bubbles. Bubbling will stop and the liquid will flow back up into the inverted capillary. The temperature at which the liquid reenters the bell is recorded as the boiling point.

How to Do a Miniscale Boiling Point Determination

Add 0.3–0.5 mL of the liquid to a small diameter test tube or Craig tube that contains a boiling chip. Clamp the tube in a heat block or sand bath. Suspend the thermometer so that the bottom of the thermometer bulb is 0.5 cm above the level of the liquid and clamp in place. Position the thermometer so that it does not contact the sides of the tube. (See Figure 1D-3.) Gently heat the liquid to boiling. Continue to heat slowly until the refluxing vapor forms a ring of condensate about 1–2 cm above the tip of the bulb.

Figure 1D-2

Microscale boiling
point determination
using a glass bell

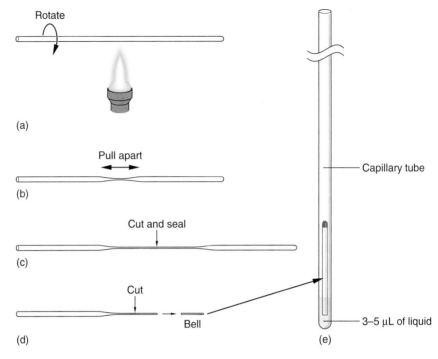

Micro boiling point apparatus

When the temperature stabilizes for at least one minute, record the temperature. This
procedure works well for liquids that boil higher than 50°C.

Microscale

Exercise D.1: Determining the Micro Boiling Point of an Unknown Liquid

Safety First!

**Always wear eye
protection in the
laboratory.**

Obtain a sample of an unknown liquid from the instructor and record the number of the
unknown in your laboratory notebook. The instructor will demonstrate how to make a
glass bell or one will be supplied. Heat the capillary tube fairly rapidly until a steady
stream of bubbles emerges from the glass bell. Turn down the heat. Watch carefully
until the last bubble collapses and liquid enters the glass bell. Note the temperature. The
temperature at which the liquid reenters the bell is the observed boiling point. Record
this temperature. Because the boiling point of a liquid depends upon pressure, a correction must be applied. To do this, use a nomograph (Figure 1D-1). Record the observed
boiling point and the corrected boiling point. Use Table 1D-1 to determine the identity
of the unknown liquid. Calculate the percent error between the corrected boiling point and
the literature (table) value. (See Appendix B). Report the corrected boiling point, the identity of the unknown, and the percent error. Return any unused sample to the instructor.

Miniscale

Exercise D.2: Determining the Boiling Point of an Unknown Liquid

Obtain a sample of an unknown liquid from the instructor. Record the unknown number
in your laboratory notebook. Add 0.3–0.5 mL of the liquid and a boiling chip to a small

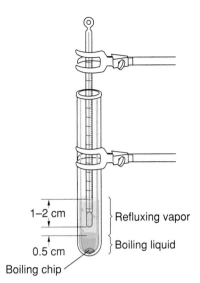

Figure 1D-3

Apparatus for miniscale boiling point determination

1–2 cm
} Refluxing vapor

0.5 cm
} Boiling liquid

Boiling chip

diameter test tube or Craig tube. Clamp the tube in a heat block or sand bath. Suspend the thermometer so that the bottom of the thermometer bulb is 0.5 cm above the level of the liquid and clamp in place. Position the thermometer so that it does not contact the sides of the tube. (See Figure 1D-3.) Gently heat the liquid to boiling. Continue to heat slowly until the refluxing vapor forms a ring of condensate about 1–2 cm above the tip of the bulb. When the temperature stabilizes for at least one minute, record the temperature. Correct the measured boiling point for atmospheric pressure using a nomograph (Figure 1D-1). Use Table 1D-1 to determine the identity of the unknown liquid. Calculate the percent error between the corrected boiling point and the literature (table) value. Report the corrected boiling point, the identity of the unknown, and the percent error. (See Appendix B). Return any unused sample to the instructor.

Safety First!

Always wear eye protection in the laboratory.

Questions

1. What is the relationship between the volatility of a liquid and its vapor pressure?
2. What is the relationship between the vapor pressure of a liquid and its temperature?
3. Define the terms "boiling point" and "normal boiling point."

Table 1D-1   Boiling Points of Selected Organic Liquids

Organic compound	Boiling point (°C)
Hexane	69.0
Ethyl acetate	77.1
2-Propanol	82.4
2,3-Dimethylpentane	89.8
1-Propanol	97.4
2-Pentanone	102.0
Toluene	110.6
1-Butanol	117.2
p-Xylene	138.3

4. How is boiling point affected by elevation?

5. A distillate was collected between 78–82°C at sea level. The reported boiling point was 80°C. Comment on the correctness of this report.

6. Explain the effects of polarity of molecules on boiling point.

7. The normal boiling point of toluene (at 760 torr) is 110°C. Use the boiling point nomograph (Figure 1D-1) to determine the boiling point of toluene at 600 torr.

8. A liquid boils at 102°C in a lab in Colorado Springs, where the ambient pressure is 600 torr. What is the corrected boiling point of the liquid?

Technique E: Index of Refraction

Introduction

Light travels at different speeds through different media. It travels slower in air than in a vacuum. It travels slower yet in liquids and solids. As light passes from one medium to another, it also changes directions: this is called refraction as shown in Figure 1E-1.

The index of refraction n is defined as the ratio of sin i to sin r, where i is the angle of incidence and r is the angle of refraction.

$$n = \frac{\sin i}{\sin r}$$

The resulting number, which is unitless, is always greater than 1 since the angle of incident light is always greater than the angle of refracted light as light passes from a less dense to a more dense medium.

The refractive index is an important physical property of organic liquids. The refractive index for many organic liquids can be found in chemical reference books such as the *Handbook of Chemistry and Physics* (CRC) and the *Merck Index*. The literature values are referenced to the **D** line of a sodium lamp at 20°C. A correction factor must be applied if the refractive index (n_{obs}) is measured at temperatures, t, other than 20°C.

$$n_D{}^{20} = n_{obs} + 0.00045 \, (t - 20)$$

The refractive index of a liquid can be measured very accurately. Since impurities affect the refractive index of a liquid, refractometry is a quick and easy method of deter-

Figure 1E-1

Index of refraction

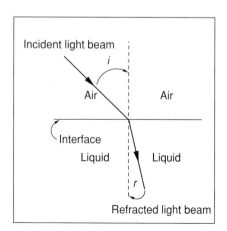

mining the purity of the liquid. Relatively small amounts of sample are required for a measurement (100–250 µL).

How to Use the Abbe Refractometer

The most common instrument for measuring refractive index is the Abbe refractometer, which is shown in Figure 1E-2.

> It is important to avoid scratching the prism surface.

To use the refractometer, open the hinged sample prisms and be sure that the stage area (lower prism surface) is clean and dry. Clean, if necessary, using a tissue and alcohol. Add 4 drops of liquid to the stage and move the upper prism over it. Close the prisms. The light source illuminates the upper prism. Adjust the swivel arm of the light source if necessary, then tighten the swivel arm. Turn the large-scale adjustment knob back and forth and look for a split optical field of light and dark that is centered exactly in the crosshairs. This is shown in Figure 1E-3.

Adjust the drum for maximum contrast between the lighter and the darker half circles. Press and hold the scale/sample field switch. The refractive index scale will appear. The upper scale may be read to four decimal places. The scale shown in Figure 1E-4 reads 1.4652. Correct the measured refractive index for temperature, if it differs from 20°C.

Exercise E.1: Measuring the Refractive Index of an Unknown Liquid

Safety First!

Always wear eye protection in the laboratory.

Obtain a sample of a pure liquid unknown. Record the identification number of the unknown. Place 4 drops of the liquid between the plates of an Abbe refractometer. Measure and record the refractive index of the liquid. Record the ambient temperature. If the

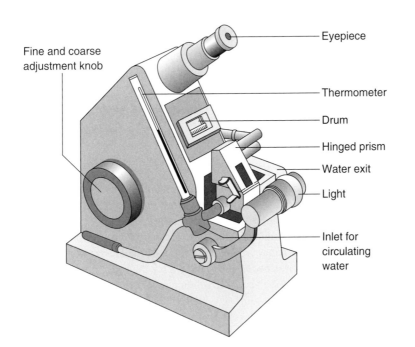

Figure 1E-2   The Abbe refractometer

Fine and coarse adjustment knob

Eyepiece

Thermometer

Drum

Hinged prism

Water exit

Light

Inlet for circulating water

Figure 1E-3

Optical field

Figure 1E-4

Refractive index scale

Table 1E-1   Refractive Indices of Selected
Organic Liquids

Organic compound	Refractive index (n_D^{20})
Acetonitrile	1.3442
Hexane	1.3751
1-Propanol	1.3850
1-Butanol	1.3993
Cyclohexane	1.4266
Toluene	1.4961

temperature differs from 20°C, apply a correction factor to the refractive index. Record the corrected refractive index. Identify the unknown liquid by comparing the refractive index with a list of unknowns in Table 1E-1. Calculate the percent error between the corrected refractive index and the literature (table) value. Report the corrected refractive index, the identity of the unknown liquid, and the percent error. (Refer to Appendix B). Return any unused sample to the instructor.

Questions

1. A student measures the refractive index of an organic liquid. The value is 1.2821 at 18°C. What is the corrected refractive index?

2. Is refractometry useful in the microscale organic chemistry lab? Explain.

3. Do impurities raise or lower the refractive index of a liquid? Explain.

4. The refractive indices of cyclohexane and toluene are 1.4264 and 1.4967, respectively, at 20°C. A mixture of these two liquids has a refractive index of 1.4563 at that same temperature. Assuming that the relationship between refractive index and concentration is linear, calculate the composition of the mixture.

5. If two samples of organic liquids give the same refractive index reading, are the two liquids the same? Explain.

Technique F: Recrystallization, Filtration, and Sublimation

Introduction to Recrystallization

The physical property that is most useful for purification of solids is differential solubility in an appropriate solvent. When crystals form during a reaction or following an extraction, impurities may become trapped within the crystal lattice or upon the surface of the solid. Washing the crystals with cold solvent can remove adsorbed impurities from the surface, but this process cannot remove the trapped (occluded) impurities. To remove these by recrystallization, it is necessary to redissolve the solid in hot solvent, filter off any insoluble impurities, and then cool the solution to let the material crystallize again.

Theory

Organic solids are usually more soluble in hot solvent than in a comparable volume of cold solvent. In recrystallization, a saturated solution is formed by carefully adding an amount of hot solvent just necessary to dissolve a given amount of solid. A slight excess of solvent may be required for hot gravity filtration used in miniscale experiments. As the solution cools, the solubility of the solid decreases and the solid crystallizes. Unavoidably, some of the solid remains dissolved in the cold solvent, so that not all of the crystals dissolved originally are recovered.

In order to be successfully separated from the impurities in a certain solvent, the solid and the impurities should have differing solubilities. Impurities that remain undissolved in the solvent can be removed by hot gravity filtration of the solution prior to cooling. Impurities that are more soluble will remain in solution after the solid crystallizes. If the impurities and the compound have very similar solubility characteristics, repeated recrystallizations may be required or a different solvent may be required for purification. A crystal of the desired compound (called a "seed" crystal) can be added to the cooling solution to encourage crystallization if crystals fail to form initially.

Solvents for Recrystallization

Selecting a solvent is crucial for the successful recrystallization of an organic solid. An ideal recrystallization solvent should:

- dissolve all of the compound at the boiling point of the solvent.
- dissolve very little or none of the compound when the solvent is at room temperature.
- have different solubilities for the compound and the impurities.
- have a boiling point below the melting point of the compound, so that the compound actually dissolves, not melts, in the hot solvent.
- have a relatively low boiling point (60–100°C) so that the solvent is easily removed from the crystals and the crystals are easy to air dry.
- be nonreactive with the compound, nontoxic, and not have an offensive odor.
- be relatively inexpensive.

A good recrystallization solvent should possess many, if not all, of these properties. The importance of selecting an appropriate recrystallization solvent cannot be overstated.

The compound being recrystallized should dissolve in a reasonable amount of hot solvent but be insoluble in the cold solvent. Recall the adage, **"like dissolves like."** Polar compounds will dissolve in polar solvents, but not in nonpolar solvents. The opposite is true for nonpolar compounds. If the polarities of the compound and the solvent are too similar, the compound will be readily soluble in that solvent and will not crystallize. The compound and the solvent cannot have radically different polarities, or the compound will not dissolve at all. This implies that the compound and the solvent should have somewhat different—but not totally different—polarities. It is obvious that an understanding of polarity is crucial to the selection of a recrystallization solvent.

Polarity is determined by dipole moments of substances. Compounds that contain only carbon and hydrogen are nonpolar. This includes compounds such as alkanes, alkenes, alkynes, and aromatic compounds. Somewhat more polar are ethers and halogenated compounds, followed by ketones and esters. More polar yet are alcohols. This trend is summarized in Table 1F-1.

Factors that influence polarity and solubility of an organic compound are molecular weight and proportion of hydrocarbon in the molecule. Compounds with higher molecular weight are less soluble in general than those with lower molecular weight with the same functional group. The higher the proportion of hydrocarbon to functional group within the molecule, the less polar the compound.

Common recrystallization solvents listed in Table 1F-2 are arranged in order of decreasing solvent polarity, as measured by the dielectric constant. Dielectric constant

Table 1F-1   Classes of Organic Solvents and Examples Used
for Recrystallization

Polarity	Class	Example
Most Polar	Alcohols	Methanol, ethanol, 2-propanol
↑	Ketones	Acetone, 2-butanone
	Esters	Ethyl acetate
	Halogenated alkanes	1, 2-dichloroethane
	Ethers	Diethyl ether, tetrahydrofuran
	Aromatic compounds	Toluene
Least Polar	Alkanes	Hexane, petroleum ether

is a measure of a solvent's ability to moderate the force of attraction between oppositely charged particles (standard = 1.0 for a vacuum).

Chemical reference books, such as the *Handbook of Chemistry and Physics* or the *Merck Index*, are sources of valuable information about solubility. The abbreviations that are used are "i" (indicating that the compound is insoluble), "s" (soluble), "δ" (slightly soluble), "h" (soluble in hot solvent), and "v" (very soluble). A solvent that is listed as slightly soluble or soluble in hot solvent should be a good recrystallization solvent for that compound. Sometimes the handbook will give the exact solubility of a compound in a given solvent. The units of solubility are typically given as grams of compound dissolved per 100 mL of solvent. Unless otherwise noted, the solubilities are measured at 25°C. If solubility information is available, it is possible to calculate the relative amount of solvent needed to dissolve a given amount of compound.

The best way to be certain that a solvent will be a good recrystallization solvent is to try it and see! Test the solvent on a small amount of the compound. If that solvent doesn't dissolve the crystals when hot, or if it dissolves the crystals at room temperature, try another solvent. Keep trying until you find a solvent that dissolves the compound when the solvent is hot, but not when the solvent is cold.

Since it is usually easier to use a single solvent than a solvent pair, it is well worth the time and effort to try to find a single solvent for recrystallization. However, sometimes it happens that no solvent is found that will dissolve the compound when hot, but not when cold. In this case, a solvent pair must be used. A solvent pair consists of two miscible solvents, which have rather different polarities. A solvent pair then consists of two solvents, one in which the compound is soluble, the other in which the compound is insoluble. The most common solvent pair is ethanol-water.

Recrystallization using a single solvent is done by dissolving the compound in the **minimum amount of hot solvent,** filtering any insoluble impurities, then letting the solution cool to room temperature. The rate at which the crystals form affects the purity of the crystals. Cooling too rapidly (such as immersing the warm solution in an ice bath) results in the formation of very small crystals that adsorb impurities from the solution. If crystals are too large, solution and impurities can be trapped (occluded) within the crystal lattice. The optimal crystals are obtained by leaving the flask undisturbed until crystallization occurs. When crystals do appear, the mixture should be cooled further in an ice bath to ensure complete crystallization. The crystals are then suction-filtered, rinsed with several small portions of ice cold solvent to remove traces of surface impurities, and allowed to dry.

Frequently, a second crop of crystals can be obtained from the filtrate (called the mother liquor). This entails heating the filtrate solution to reduce the volume, then letting

Table 1F-2   Properties of Common Recrystallization Solvents

Solvent	Boiling point (°C)	Comments	Dielectric constant
Very polar solvent			
Water	100	Good for moderately polar and polar compounds; difficult to remove solvent from crystals	78.5
Polar solvents			
Methanol	65	Can sometimes be used in place of ethanol	33.0
Ethanol	78	Usually used as 95% ethanol; good for relatively nonpolar compounds	24.3
Acetone	60	Low boiling, but inexpensive; 2-butanone has similar properties and a higher boiling point (80°C)	21.2
2-Propanol	82	Inexpensive; good for relatively nonpolar compounds	18.3
Moderately polar solvent			
Ethyl acetate	77	Good for relatively nonpolar compounds	6.0
Nonpolar solvents			
Toluene	110	High boiling point makes it difficult to remove from crystals; good for moderately polar compounds	2.4
Petroleum ether/ Ligroin (high boiling)	90–110	Very flammable; easy to remove from crystals; cheaper than heptane	2.0
Petroleum ether/ Ligroin (medium boiling)	60–90	Also available as "hexanes"; very flammable; easy to remove from crystals; cheaper than heptane; useful in solvent pair recrystallization	2.0
Petroleum ether/ Ligroin (low boiling)	30–60	Also available as "pentanes"; very flammable	2.0
Cyclohexane	81	Good solvent for less polar compounds	1.9
Hexane	69	Good solvent for less polar compounds; useful in solvent pair recrystallization	1.9

the solution cool slowly as crystallization occurs. The crystals obtained from this second crop are usually not as pure as the initial crop. Melting points should be obtained and the purity evaluated before the two crops of crystals are combined. Since impurities usually lower the melting point and increase the melting point range, the measured melting points are good indications of the purity of the crystals.

The purity of the crystals is one measure of the success of a recrystallization procedure. A second measure is the percent recovery, that is, how many grams of pure compound are obtained relative to the amount of the impure crystals. The percent recovery for a recrystallization process can be determined by measuring the mass of the crystals before recrystallization and after. The percent recovery is given by the equation below.

$$\frac{\text{mass of pure crystals}}{\text{mass of impure crystals}} \times 100\% = \text{Percent Recovery}$$

Low percent recoveries may be due to using the wrong solvent, using too much solvent, incomplete crystallization (not enough time or low enough temperature), or inefficient filtration technique. Since the compound usually has some solubility in the cold solvent, there is always some loss of product.

Choosing a Solvent

The first step (and most important step) is choosing a recrystallization solvent or solvent pair. This is necessary whether doing a microscale or a miniscale recrystallization.

General Procedure

You should narrow the choice of solvents by considering the polarity of the compound being recrystallized. For nonpolar compounds, try solvents of moderate polarity. For moderately polar compounds, try solvents of higher polarity, such as water. For polar compounds, try water or solvents of low to moderate polarity. (Refer to Table 1F-2 for solvent polarities.)

Place about 50 mg of the solid (about the tip of a spatula) into a small test tube. Add 1 mL of the solvent to be tested. Mix the contents thoroughly and observe. If most of the solid dissolves at room temperature, the compound is soluble in this solvent and the solvent will not be a good recrystallization solvent. Start over with a new portion of the solid and a new solvent.

If, on the other hand, most of the solid does not dissolve at room temperature, gently heat the test tube to boiling in a sand bath or water bath. Observe.

- If the solid dissolves in the hot solvent, the solvent will be a good recrystallization solvent.
- If only some of the solid dissolves, try adding another 1 mL portion of the solvent and heating in a sand bath or water bath. If more of the solid dissolves, the compound is slightly soluble and this solvent may work as a recrystallization solvent. Add enough **hot** solvent to dissolve all of the crystals. Then place the test tube in an ice bath. If many crystals form, this solvent will be good for recrystallization.
- If the solid does not at least partially dissolve in the hot solvent, the compound is insoluble in the hot solvent and the solvent will not be a good recrystallization solvent. Start over with a new portion of the solid and a new solvent.

This process is illustrated in Figure 1F-1.

Choosing a Solvent Pair

If no solvent is found that will dissolve the compound when hot but not cold, you must perform a solvent-pair recrystallization. This is easier to do if accurate and detailed notes have been recorded. Select a solvent that dissolved the compound and a different solvent that failed to dissolve the compound. **Remember that the two solvents selected must be miscible.** This means they mix in all proportions, such as ethanol and water. To ensure that the two solvents selected are miscible, put 1-mL portions of each solvent in a test tube and shake. If the solution is homogenous (one phase, clear), the solvents are miscible. If two liquid phases appear, the solvents are immiscible: they cannot be used for a solvent-pair recrystallization.

To test the solvent pair, place about 100 mg of the solid in a test tube. Add dropwise the hot solvent in which the compound is more soluble. Add just enough hot solvent to dissolve the solid, no more. Then add dropwise the hot solvent in which the compound is less soluble. If the solution turns cloudy due to the crystallization of the solid, this

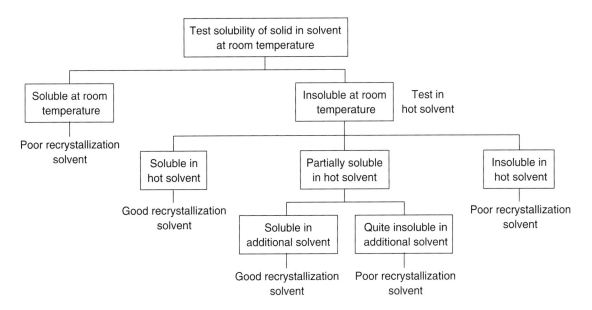

Figure 1F-1 Flow scheme for determining the suitability of a solvent for recrystallization of an organic solid

will be an acceptable solvent pair to use for recrystallization. Once the solvent or solvent system has been determined, proceed with the recrystallization process.

How to Do a Microscale Recrystallization
Microscale Single-Solvent Recrystallization

Microscale single-solvent recrystallization should be used to recrystallize 50 mg to 300 mg of material. Put the crude crystals into a 10-mL Erlenmeyer flask that contains a boiling chip. With a pipet, add approximately 0.2–0.3 mL aliquots of the chosen hot solvent. After each addition of solvent, swirl the crystals and heat the solution on a warm sand bath or heat block. Continue to add the hot solvent until the entire solid has dissolved. Be patient. Don't heat the solution too strongly or the solvent will boil away. Adjust the temperature of the heat source to keep the solution just under the boiling point of the solvent. If there is still undissolved solid, skip ahead to the next paragraph. Otherwise, continue. When the solid has dissolved, remove the Erlenmeyer flask from the sand bath and set it aside to cool to room temperature undisturbed. A large beaker packed with a paper towel or packing material works well to insulate the flask. Place the Erlenmeyer flask down into the beaker. After 10–15 minutes or when the flask is at room temperature, place the flask in a beaker filled with ice. Let the flask stand 5–10 minutes or until the flask feels cold to the touch. Place a small beaker of solvent in the ice bath to cool. If crystallization does not occur even after the solution is cold, scratch the inside of the flask with a glass rod or add a crystal of product (a seed crystal) to the solution. If crystallization still does not occur, reduce the volume of solution and let it cool to room temperature. After crystallization is complete, suction filter the crystals using a Hirsch funnel and small filter flask. Rinse the crystals sparingly with several drops of the ice-cold solvent and air dry the crystals.

Hood!

If further addition of the hot solvent does not dissolve more of the solid, the solid particles are probably insoluble impurities. If this is the case, it is necessary to filter the mixture. Prepare a Pasteur filter pipet (a Pasteur pipet fitted with a cotton plug). Hold

Microscale filtration apparatus
(See Figure 1F-3).

the pipet over the flask for a minute or two to warm (to prevent crystallization within the pipet). Obtain a clean 10-mL Erlenmeyer flask or a 5-mL conical vial. Add 5–10 drops of the hot solvent to the flask or vial and set it on the warm sand bath or heat block. Squeeze on the bulb to remove the air and place the pipet in the mixture. Slowly release the bulb to draw up the solution into the Pasteur pipet. Try not to draw up any of the solid impurities. Transfer the solution to a clean flask or vial. Any solid that gets drawn up in the pipet will stick to the cotton. (See Figure 1F-4.) The pipet may be rinsed with one or two small portions of solvent. Sometimes crystals form immediately upon contact with the container. If this happens, add more hot solvent until all of the solid dissolves. Then remove the container from the heat source, allow it to cool to room temperature, and proceed as above.

Hot filtration should also be done if the solution is highly colored (when the pure compound is white!). In this case, it will be necessary to add a small amount of decolorizing carbon (such as Norit or charcoal) to the warm (not boiling!) solution. Decolorizing carbon has a very large surface area that enables it to adsorb colored impurities. To use decolorizing carbon, add 5–6 more drops of solvent and let the solution cool a little below the boiling point so as to avoid splattering. Add a spatula-tip full or less of the decolorizing carbon and swirl. Heat the black mixture for several minutes, then filter using a filter pipet. Transfer the clear solution to a clean container and proceed as described above. If small particles of carbon can be seen, it may be necessary to refilter using a new Pasteur pipet.

Microscale Solvent-Pair Recrystallization

This procedure should be used if no single solvent is found that will work for the recrystallization procedure.

Hood!

Heat two small flasks, each containing one of the solvents to be used for the recrystallization. Place the crystals in a 10-mL Erlenmeyer flask or 5-mL conical vial. With a pipet, add dropwise the hot solvent that was found to best dissolve the compound. Swirl after each addition of hot solvent. When the crystals are dissolved, add dropwise the hot solvent that does not dissolve the compound. Observe carefully after each addition. Continue adding the solvent until the solution appears cloudy or crystal formation is observed. This is called the "cloud point." Then add 1–2 drops more of the first solvent to just redissolve the crystals and cause the cloudiness to disappear. Remove the container from the heat source and allow to cool to room temperature, undisturbed. Chill further in an ice bath, then suction filter the crystals. Wash with several small portions of ice-cold solvent (the solvent in which the compound is insoluble!). Air dry the crystals or place in a drying oven.

If some of the solid does not dissolve during addition of the first hot solvent, remove the solid impurities by drawing up the solution with a Pasteur filter pipet. Transfer the solution to a clean vial or Erlenmeyer flask. Add a little more of the first hot solvent if necessary. Then add the second hot solvent dropwise until the solution appears cloudy.

If no crystals are obtained upon cooling, reheat the solution. Again add dropwise the hot solvent that does not dissolve the compound. At the cloud point, cool the solution and wait for crystallization.

Ultra-Microscale Recrystallization and Use of a Craig Tube

Hood!

For very small amounts of product, it is necessary to use a Craig tube for recrystallization. This procedure should be used for recrystallization of less than 50 mg of material.

Place the crude crystals in the bottom of a 1-mL or 2-mL Craig tube. Dissolve the crystals in a minimum amount of hot solvent. Fit the Craig plug into the Craig tube.

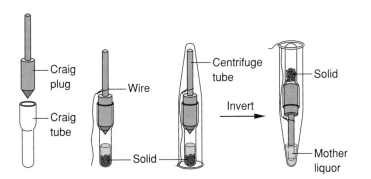

Figure 1F-2

Craig tube
recrystallization

Cool the solution to room temperature. Place the tube in a beaker filled with ice to ensure complete precipitation. To prepare the tube for crystallization, wrap a piece of wire around the neck of the Craig tube. (See Figure 1F-2.) The wire must be longer than the centrifuge tube. Fit a centrifuge tube over the Craig tube. Holding the wire firmly, invert the centrifuge tube.

Centrifuge the mixture, making certain that it is balanced. The liquid will be forced down into the tip of the centrifuge tube, leaving the crystals on the plug. Carefully remove the Craig tube from the centrifuge tube using the wire "handle." Over a clean watch glass, disassemble the Craig tube. Scrape the crystals onto the watch glass and let them dry thoroughly before weighing or taking a melting point. This process is illustrated in Figure 1F-2.

How to Do a Miniscale Recrystallization
Miniscale Single-Solvent Recrystallization

This procedure should be used for recrystallizing 300 mg or more of a solid. Transfer the impure solid into a 10-mL or larger Erlenmeyer flask. Heat the solvent on a hot sand bath or steam bath in the hood. Most organic solvents are flammable, so use extreme caution when heating. Add the minimum amount of the hot solvent required to dissolve the crude crystals. Swirl to dissolve, while heating. Continue to add small portions of the hot solvent until all of the solid dissolves. If it becomes apparent that insoluble impurities are present, any solids that do not dissolve may be removed by hot gravity filtration. Obtain a second Erlenmeyer flask and add 1–2 mL of the hot solvent. Set a stemless funnel fitted with a piece of fluted filter paper (see Figure 1F-6) on top of the Erlenmeyer. Set the funnel in an iron ring or place a paper clip between the funnel and neck of the Erlenmeyer. Heat the flask and funnel so hot vapors fill the system. To keep the solid in solution, add 1–2 mL of the hot solvent to the first Erlenmeyer flask. Add a spatula-tip full of decolorizing carbon if the solution is deeply colored. Keeping everything hot, pour the solution through the filter paper in small batches. Continue pouring until all of the solution is transferred. The solution should be clear and colorless. If crystallization occurs in the flask or filter paper, add more of the hot solvent to dissolve the crystals. When all the solid is dissolved, remove the solution from the heat and let it stand undisturbed. Cool to room temperature (about 15–20 minutes). After crystallization occurs, place the flask in an ice bath to ensure complete crystallization. If no crystals form, scratch the inside of the flask with a glass rod or add a seed crystal. If crystallization still doesn't occur, add a boiling chip and heat the solution in the hood to reduce the volume. After the solution has been cooled in the ice bath, filter by suction filtration, washing with several small portions of ice-cold solvent. Dry the crystals on the filter by drawing air through using continued suction or by placing in a drying oven.

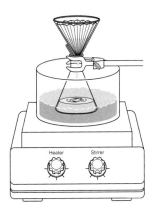

Miniscale Solvent-Pair Recrystallization

This procedure should be used when a single recrystallization solvent cannot be found. Place 300 mg or more of the crude crystals in a 25-mL or larger Erlenmeyer flask. On a hot sand bath or steam bath, heat two flasks containing the solvents. One flask will contain solvent in which the compound is soluble. The other flask will contain solvent in which the compound is insoluble. Add dropwise the solvent in which the compound is soluble. Add just enough to dissolve the compound, no more. Then, add dropwise the solvent in which the compound is less soluble. Observe carefully. Add just enough until the solution turns cloudy (the cloud point) and remains so even after swirling. This is the point at which the crystals are beginning to come out of solution. Now add 1 or 2 drops (no more!) of the solvent in which the compound is soluble. Add just enough to dissipate the cloudiness. Remove the flask from the heat source and let the solution stand undisturbed to cool to room temperature. When crystallization occurs, chill the flask in an ice bath and suction filter the crystals. If crystallization does not occur, scratch the inside of the flask with a glass rod or add a seed crystal. If crystallization still does not occur, heat the solution in the hood. Again add dropwise the hot solvent that does not dissolve the compound. At the cloud point, cool the solution and wait for crystallization. After crystallization occurs, cool the flask in an ice bath and suction filter. Wash the crystals with several small portions of ice-cold solvent (the one in which the compound is insoluble). Allow the crystals to dry before taking a melting point.

If some of the solid does not dissolve during addition of the first hot solvent, filter the insoluble material by gravity filtration and continue.

Important Tips Concerning Recrystallization

1. Use an Erlenmeyer flask in the hood for recrystallization, not a beaker. The solvent can too easily boil away from a beaker or be splashed out.
2. If no precipitate forms even after the solution has been standing in an ice bath, it may be necessary to induce crystallization. Try one or more of the following techniques:
   a. Scratch the inside surface of the Erlenmeyer flask with a glass rod. The scratched glass acts to induce crystallization.
   b. Add a seed crystal of the compound (if available) to the cooled solution. This can act as a template to initiate crystallization.
   c. If neither of the previous suggestions works, it is probable that too much solvent was added initially. Reduce the volume by heating until the solution appears cloudy. This is the point at which crystallization is occurring. Add a few more drops of the hot solvent until the solution is clear, then allow to cool slowly.
   d. If crystals have still not formed, it is possible that the wrong solvent was selected, one in which the compound was too soluble. Evaporate the solvent by boiling, leaving the impure crystals. Then try again with a different solvent.
3. Experimental procedures generally do not specify the exact amount of solvent to use for recrystallization of a crude product. If a volume of solvent is specified, that amount is based on an average yield. If the mass of crystals actually obtained is significantly different from the average yield, the amount of solvent used should be scaled accordingly. **Always use the minimum amount of hot solvent to dissolve the crystals.**
4. Whenever an experimental procedure specifies two solvents (separated by a hyphen) to be used for recrystallization, it implies a solvent-pair recrystallization. The solvent listed first is the solvent in which the compound is soluble. The second solvent is the one in which the compound is insoluble. For example, "recrystallize the solid from ethanol-water" means that the solid is soluble in ethanol and not soluble in water.

Hood!

5. Oils are sometimes obtained instead of crystals. This is a common problem, especially if the solid has a low melting point. An oil can form if the boiling point of the solvent is greater than the melting point of the compound. When the oil solidifies, impurities are trapped within the crystal lattice of the crystals. If this happens, try one or more of the suggestions listed below:

   a. Redissolve the oil by heating gently and let the solution cool back down to room temperature. If an oil starts to re-form, shake the mixture vigorously and cool in an ice bath. Repeat until solidification occurs.

   b. Add a little more solvent to the oil. If using a solvent pair, add a little more of the solvent in which the compound is insoluble. If this doesn't work, try adding a little more of the solvent in which the compound is soluble. Let stand. If a seed crystal is available, add one now.

   c. Triturate the oil. To triturate, add a small amount of a solvent in which the product is expected to be insoluble. With a glass rod, gently mash the oil in the solvent. If impurities in the oil are soluble in the solvent, then the oil may crystallize using this procedure.

   d. If the above methods don't work, start over, using a new solvent for recrystallization. Use a solvent with a lower boiling point.

Introduction to Filtration

Once crystals have formed, they must be separated by the process of filtration. In filtration, a porous barrier, usually filter paper or sintered glass, allows the liquid but not the solid to pass through. The two most important methods of filtration are gravity filtration and suction (vacuum) filtration. Gravity filtration is done when the desired substance is in the solution; the solid (such as decolorizing carbon or a drying agent) is collected on the paper and discarded and the filtrate is collected for further use. Suction filtration is chosen when the desired substance is the solid and it is necessary to isolate it as a dry solid.

How to Do a Microscale Suction Filtration

This technique is used to isolate a dry solid product as in recrystallization. The equipment required is a heavy-walled, sidearm suction flask, filter ring or filter adapter (neoprene), a Hirsch funnel, a piece of appropriately sized filter paper, and a safety trap connected with thick-walled vacuum tubing. The flask should be clamped to avoid breakage (see Figure 1F-3).

The filter paper should be large enough that the holes in the funnel are just covered. If the filter paper selected is too large (larger than the diameter of the funnel), a tight seal cannot be obtained. Additionally, some of the mixture may flow over the unsealed

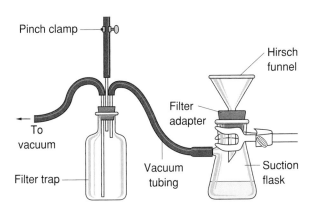

Figure 1F-3
Microscale filtration apparatus

Pinch clamp — To vacuum — Filter trap — Vacuum tubing — Filter adapter — Hirsch funnel — Suction flask

edge and result in loss of product as well as contamination of the filtrate. A small quantity of the solvent to be filtered is used to wet the paper, and suction is applied, usually via a water aspirator. The maximum suction should always be applied, so the aspirator is fully turned on. This procedure prevents solids from getting under the filter paper, which could lead to clogging of the funnel and/or loss of the solid into the filtrate. The mixture to be filtered is swirled and then rapidly poured into the funnel. After the liquid has been drawn through, the resulting crystals are usually washed with a minimum amount of cold solvent. If the solvent is high boiling, traces may be washed out by using a more volatile solvent, as long as it does not dissolve the filtered crystals. Drawing air through the filter for a few minutes further dries the crystals. It is important to remove all of the solvent or the yield and melting point will be inaccurate. It may be necessary to press out the last remnants of the solvent with a spatula by pressing down on the filter cake. The resulting dry, solid product is carefully removed from the filter paper, taking care not to scrape off pieces of the paper fibers and contaminate the product.

A safety trap must be used during the suction filtration to prevent backup of water into the filtrate. This can easily occur if there is a sudden drop in pressure, for example, if nearby aspirators are turned on. **It is also important to break the vacuum connection at the side arm of the suction flask before the aspirator vacuum is turned off or a water backup may occur.**

How to Use a Microscale Filter Pipet

Use of a filter pipet is a quick and easy method for separating a liquid from an unwanted solid, such as a drying agent. Prepare the filter pipet by placing a small piece of cotton or glass wool in the neck of a Pasteur pipet (see Figure 1F-4). Use a piece of wire or an applicator stick to push the cotton down into the tip of the pipet. Draw up the solution to be filtered into the filter pipet. Apply pressure to the pipet bulb and completely force all of the liquid out of the filter pipet into a clean flask. The drying agent or solid impurities will adhere to the cotton. Rinse the filter pipet with fresh solvent and add this rinse to the flask. Alternatively, the solution may be drawn up in a clean Pasteur pipet and drained through the filter pipet.

How to Do a Miniscale Suction Filtration

The procedure for suction filtration is very similar to that of microscale suction filtration. A Büchner funnel is used in place of a Hirsch funnel. The size of the suction flask will depend upon the quantity of solution to be filtered. Because the apparatus tends to be very top-heavy, it should be clamped to a ring stand. The setup is shown in Figure 1F-5.

Place the filter paper in the funnel and pour a little of the solvent through the funnel to wet the filter paper. Turn on the aspirator or vacuum source full force to seal the filter paper to the funnel. Then carefully pour the solution in the center of the funnel. Rinse the reaction flask with a very small amount of the cold solvent for complete transfer. Rinse the crystals with a minimum amount of the cold solvent. Continue to apply vacuum to help dry the crystals. Break the vacuum, then turn off the aspirator. When the crystals are completely dry, carefully scrape the crystals from the filter paper.

How to Do a Miniscale Gravity Filtration

This simple technique requires only a filter paper, a short-stem glass funnel, and a receiving vessel. A stemless funnel is used for hot filtration. The filter paper should be folded into quarters or preferably fluted to maximize the surface area and the rate of flow through it. Fluted filter paper can easily be prepared by folding the filter paper in

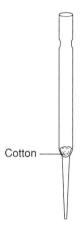
Cotton

Figure 1F-4
Filter pipet

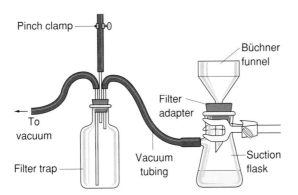

half and then in half again, resulting in quarters. Each of these quarters is folded again to give the fluted paper in eighths. This procedure is illustrated in Figure 1F-6.

Proper choice of the size of filter paper will result in a fluted funnel that fits into the glass funnel and is just slightly below the rim of the funnel. The funnel is held above the collection flask by an iron ring, wire, clay triangle, or paper clip. The funnel should not be flush against the receiving flask; an air space should be provided to avoid slow or incomplete filtration. (See Figure 1F-7.) In some cases it is desirable to preheat both the funnel and filter when filtration of a hot solution or mixture is required. This is most commonly accomplished by placing the glass funnel and filter paper over a hot (boiling) solvent until they are bathed in the hot vapors. Filtration is then quickly carried out. This is to prevent premature crystallization of the solid in the filter paper or stem of the funnel. As noted above, this technique is used primarily to remove undissolved solids during a hot filtration in a recrystallization procedure or to remove a drying agent or decolorizing carbon.

Important Tips Concerning Filtration

1. Always clamp the suction filtration apparatus to a ring stand, so that the apparatus does not tip over.
2. Make sure that the filter paper is just big enough to cover the holes of the funnel. If the paper is too big, solids can get underneath and spill into the flask along with the filtrate.
3. Placing a rubber sheet over the top of the funnel and securing it with a rubber band can help dry the crystals. The suction forces the rubber sheet down on top of the crystals, helping to dry the crystals.
4. Crystals should be thoroughly dried before being carefully scraped off the filter paper. Otherwise, tiny paper fibers will contaminate the crystals.
5. With high-boiling solvents, it is best to let the crystals air dry overnight. To do this, carefully lift the filter paper and crystals out of the funnel and place on a large watch glass.

Introduction to Sublimation

Some solids, which have high vapor pressures, do not melt into a liquid, but are converted directly to a gas. The process of converting from a solid phase into a gas phase without going through the liquid phase is called **sublimation.** A familiar example of sublimation is dry ice (solid CO_2). A piece of dry ice sitting on the lab bench will disappear into a fog of carbon dioxide vapor, leaving no trace of liquid. Why does this occur? As a solid is heated, its vapor pressure increases. At the melting-point temperature, if the vapor pressure of the solid is less than ambient pressure, the solid will melt into a liquid.

Figure 1F-6

Steps for fluting filter paper

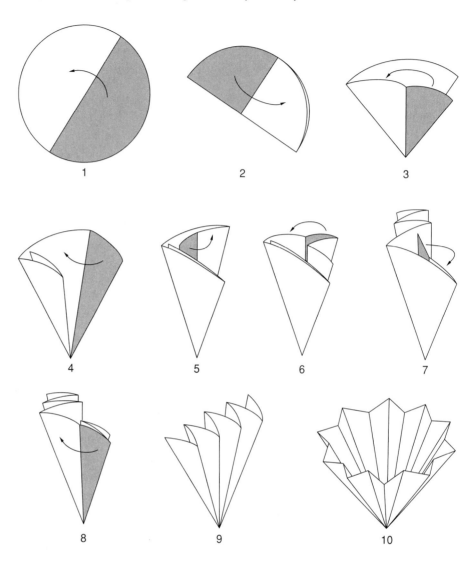

Figure 1F-7

Miniscale hot gravity filtration

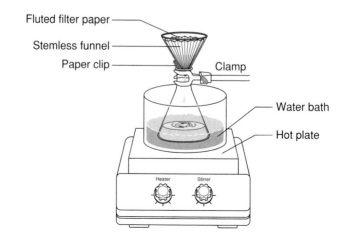

This is the case for most organic solids. If, however, the vapor pressure of the solid is greater than ambient pressure, the solid will sublime. This is the case with dry ice.

Organic compounds that sublime tend to be relatively nonpolar and fairly symmetrical. They have small dipole moments and relatively low molecular weight. Examples of compounds that sublime under reduced pressure are shown in Table 1F-3. Structures of these compounds are shown in Figure 1F-8. Compounds with higher vapor pressures at their melting points are easier to sublime.

Because few organic compounds have high enough vapor pressures to sublime at atmospheric pressure, this technique is generally performed under reduced pressure using a water aspirator, house vacuum line, or vacuum pump. For example, caffeine melts at 235°C, but sublimes at only 51°C if the pressure is reduced to 15 torr. Sublimation is relatively simple to perform. The sample to be sublimed is placed in the bottom of a tube or flask that is fitted with an inner tube (called a cold finger). The cold finger is cooled with running water or filled with ice. The tube or flask is evacuated and the sample is heated in a sand bath to a temperature just below its melting point. As it is heated, the sample will condense onto the cold finger, where it will resolidify, leaving the nonvolatile impurities in the bottom of the sublimation chamber. When sublimation is complete, the vacuum is released slowly and the cold finger carefully withdrawn to avoid displacing the crystals. Examples of sublimation setups are shown in Figure 1F-9.

Recrystallization and sublimation are both methods of purifying organic solids. Which technique you choose will depend upon the organic compound you are trying to

Table 1F-3 Sublimation Data for Selected Organic Compounds

Compound	Melting point (°C)	Vapor pressure (torr) at melting point
Hexachloroethane	186	780
Camphor	179	370
Anthracene	218	41
1,4-Dichlorobenzene	53	8.5
1,4-Dibromobenzene	87	9

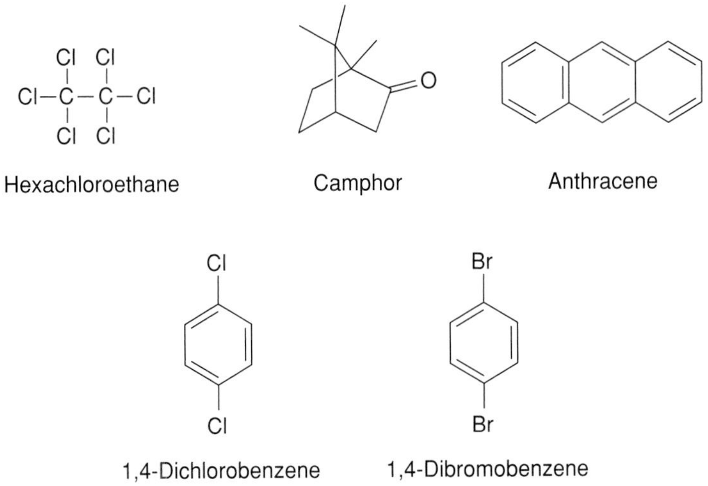

Hexachloroethane

Camphor

Anthracene

1,4-Dichlorobenzene

1,4-Dibromobenzene

Figure 1F-8

Structures of compounds in Table 1F-3

Figure 1F-9

Microscale and mini-scale sublimation setups

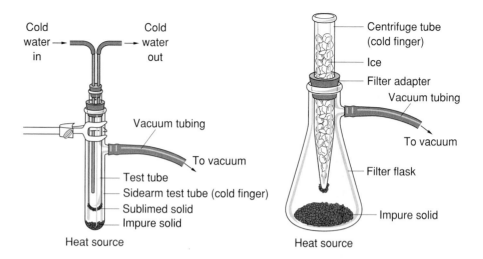

purify and the nature of the impurities present. In general, recrystallization is more applicable for a wider range of compounds, but sublimation is an alternative for those specific compounds that have high vapor pressures at their melting points.

Microscale

Safety First!

Always wear eye protection in the laboratory.

Exercise F.1: Recrystallizing an Impure Solid

Obtain a sample of an impure solid from your instructor. The solid will be either acetanilide or benzoic acid. Record the sample number in your lab notebook.

Add 5–6 mL of water to a 10-mL Erlenmeyer flask containing a boiling chip and heat to boiling on a sand bath. Weigh out 100 mg of the crude solid and place in a 10-mL Erlenmeyer flask that contains a boiling chip. Add approximately 1 mL of hot water to the solid and swirl. Keeping the solution hot, use a pipet to add small aliquots (about 0.1 mL) of hot solvent and swirl after each addition. Continue to add solvent until all the solid dissolves. If necessary, filter off insoluble impurities using a filter pipet.

When the solid has dissolved, remove the flask from the heat, place a small watch glass on top of the flask, and allow it to cool. When the flask is at room temperature, place the flask in an ice bath for 5–10 minutes to ensure complete crystallization. If an oil forms, add a little more water, heat to boiling, and cool. If crystallization does not occur, scratch the inside of the flask with a glass rod. If necessary, place the flask on a hot sand bath or water bath and reduce the volume by boiling off excess solvent.

When crystallization is complete, vacuum filter the crystals using a Hirsch funnel. Wash the crystals with 0.1–0.2 mL of ice-cold water. When the crystals are completely dry, record the weight and the melting point. Calculate the percent recovery. Transfer the pure crystals into a labeled vial and turn in to the instructor.

Miniscale

Safety First!

Always wear eye protection in the laboratory.

Exercise F.2: Recrystallizing an Impure Solid

Obtain a sample of an impure solid from your instructor. The solid will be either acetanilide or benzoic acid. Record the sample number in your lab notebook.

Add approximately 20–30 mL of water to a 50-mL Erlenmeyer flask and heat to the boiling point on a hot plate. Weigh 500 mg of crude solid and place in a 25-mL Erlenmeyer flask that contains a boiling chip. Add approximately 5 mL of hot water to the solid and swirl. Place the flask on a hot plate and slowly heat to boiling. Add small aliquots (about 0.5–1 mL) of hot water and swirl well after each addition. Continue to add solvent until all the solid dissolves. If the solution is deeply colored, or if the solid is not fully dissolved after the addition of 10–15 mL of hot water, insoluble impurities may be present. Perform a hot gravity filtration (see Figure 1F-7), only if instructed to do so.

When the solid has dissolved, remove the flask from the heat, place a small watch glass on top of the flask, and allow it to cool. When the flask is at room temperature, place the flask in an ice bath for 5–10 minutes to ensure complete crystallization. If an oil forms, add a little more water, heat to boiling, and recool. If crystallization does not occur, scratch the inside of the flask with a glass rod. If necessary, place the flask on a hot plate or steam bath in the hood and reduce the volume by boiling off excess solvent.

When the flask is cold and crystallization is complete, suction filter the crystals. Wash the crystals with small portions of ice-cold water. Air dry the crystals or place in a drying oven. When the crystals are completely dry, record the weight and the melting point. Identify the unknown solid. Calculate the percent recovery. Transfer the pure crystals into a labeled vial and turn in to the instructor.

Exercise F.3: Recrystallizing an Impure Solid with Hot Gravity Filtration

Miniscale

Safety First!

Always wear eye protection in the laboratory.

Add approximately 20–25 mL of water to a 125-mL Erlenmeyer flask and heat to the boiling point on a hot plate. Weigh 500 mg of crude acetanilide or benzoic acid and place in a 25-mL Erlenmeyer flask that contains a boiling chip. Add approximately 5 mL of hot water to the solid and swirl. Place the flask on a hot plate and slowly heat to boiling. Add small aliquots (about 1 mL) of hot water and swirl well after each addition. Continue to add solvent until all the acetanilide dissolves. If the solution is colored, the solid contains impurities and it is necessary to do a hot gravity filtration.

Remove the flask from the heat; cool to slightly below the boiling point (or the solution will splatter when the charcoal is added). Add a spatula-tip full of activated charcoal to the solution and swirl. Add additional charcoal to remove the color from the solution, but don't add more charcoal than necessary, since its high surface area adsorbs the solid, reducing the percent recovery. Add 1–2 mL additional hot water to the flask and reheat the solution to the boiling point.

As the solution is reheating, set up another 25-mL Erlenmeyer flask containing a boiling chip with a stemless funnel and fluted filter paper. Pour 1–2 mL of hot water through the funnel to fill the flask with solvent vapors and put the flask on a hot plate. Pour the solution in small batches through the filtration system. The solution should be colorless. If it is not, add additional charcoal and repeat the filtration process. When the filtration is complete, boil off 2–3 mL of the excess solvent added. Then remove the flask from the heat, place a small watch glass on top of the flask, and allow it to cool. When the flask is at room temperature, place the flask in an ice bath for 5–10 minutes to ensure complete crystallization. If an oil forms, add a little more water, heat to boiling, and cool. If crystallization does not occur, scratch the inside of the flask with a glass rod or add a seed crystal of the solid. If necessary, place the flask on a hot plate and reduce the volume by boiling off excess solvent.

When the flask is cold and crystallization is complete, suction filter the crystals. Wash the crystals with small portions of ice-cold water. Air dry the crystals or place them in a drying oven. When the crystals are completely dry, record the weight and the melting point. Identify the unknown solid. Calculate the percent recovery. Transfer the pure solid into a labeled vial and turn in to the instructor.

Microscale

Exercise F.4: Purifying an Unknown Solid by Solvent-Pair Recrystallization

Obtain an unknown solid from the instructor. The solid will be phenanthrene (melting point = 100°C), anthracene (melting point = 218°C), or *trans*-cinnamic acid (melting point = 133°C). Weigh approximately 100 mg of the solid into a 25-mL Erlenmeyer flask containing a boiling chip. Record the exact mass used. Add 0.5 mL of hot 95% ethanol to the solid, and swirl to dissolve. Heat the solution on a hot plate, but do not heat so strongly that the solvent boils away. While keeping the solution warm, dissolve the solid in a minimum amount of hot 95% ethanol by adding it from a pipet dropwise until the solid just dissolves to produce a clear solution. Add hot water dropwise until the solution becomes cloudy. Then add a few more drops of hot 95% ethanol until the solution is clear. Remove the flask from the heat. Let cool to room temperature, undisturbed, to induce crystallization. Chill the flask in an ice bath. Isolate the crystals by suction filtration, rinsing the crystals with a minimum amount of ice-cold water. Air dry. Weigh the crystals and calculate the percent recovery. Measure the melting point and identify the unknown solid.

Microscale

Exercise F.5: Sublimation of Caffeine

Safety First!

Always wear eye protection in the laboratory.

Weigh about 200 mg of impure caffeine and record the mass. Assemble the sublimation chamber from a 20 × 150-mm sidearm test tube and a 15 × 125-cm test tube, as shown in Figure 1F-9. The inner test tube (the cold finger) may be filled with ice chips or it may be cooled with running water. If using water cooling, fit the cold finger with a two-holed neoprene adapter in which glass tubing has been inserted. Attach rubber tubing to the glass tubing; connect one piece of rubber tubing to the cold water faucet and insert the other piece of rubber tubing firmly into the drain. Put the impure caffeine sample in the bottom of the sidearm test tube and clamp in place. Attach thick-walled rubber tubing to the sidearm. Connect the tubing to a safety trap and connect the trap to an aspirator or house vacuum. Fit the cold finger with neoprene adapter into the sidearm test tube. Fill the cold finger with ice chips, or connect the rubber tubing and gently turn on the cold water. Turn on the aspirator or house vacuum. Heat gently using a warm sand bath or heat block. As the sample is heated, crystals will form on the cold finger. Continue heating until sublimation is complete and no more crystals form on the cold finger. Remove the heat source and allow the apparatus to cool to room temperature. **Cautiously break the vacuum at the trap.** Turn off the cold water, if necessary. **Very carefully, remove the cold finger** from the sublimation chamber, so as not to dislodge any of the crystals. Scrape the crystals onto a tared piece of weighing paper and reweigh. Record the mass of pure caffeine.

An alternative sublimation chamber can be made from a 125-mL filter flask rather than a sidearm test tube.

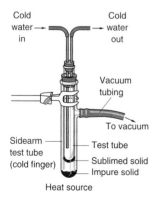

Cold water in → ← Cold water out

Vacuum tubing

To vacuum

Sidearm test tube (cold finger)

Test tube

Sublimed solid

Impure solid

Heat source

Exercise F.6: Sublimation of Caffeine

Weigh about 500 mg of impure caffeine and record the exact mass used. Assemble the sublimation chamber from a flask and cold finger, as shown in Figure 1F-9. Put the impure caffeine sample in the bottom of the flask and clamp in place. Attach thick-walled rubber tubing to the vacuum outlet. Connect the tubing to a trap and connect the trap to an aspirator or house vacuum. Attach rubber tubing to the water inlet and outlet of the cold finger and gently turn on the cold water. Turn on the aspirator or house vacuum. Heat the sublimation chamber gently using a warm sand bath or oil bath. As the sample is heated, crystals will form on the cold finger. Continue heating until sublimation is complete and no more crystals form on the cold finger. Remove the heat source and allow the apparatus to cool to room temperature. **Cautiously break the vacuum at the trap** and, if necessary, turn off the cold water. **Very carefully, remove the cold finger** from the sublimation chamber, so as not to dislodge any of the crystals. Scrape the crystals onto a tared piece of weighing paper and reweigh. Record the mass of pure caffeine.

Safety First!

Always wear eye protection in the laboratory.

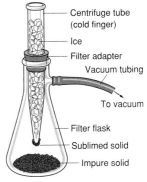

Centrifuge tube (cold finger)
Ice
Filter adapter
Vacuum tubing
To vacuum
Filter flask
Sublimed solid
Impure solid
Heat source

Questions

1. List the most important criteria for selecting a recrystallization solvent.

2. When is it necessary to use a solvent-pair recrystallization?

3. Why should the recrystallization solvent have a fairly low boiling point?

4. What problems might occur if crystallization occurs too rapidly?

5. Why should the recrystallization mixture be cooled in an ice bath prior to filtering?

6. Why should the boiling point of the solvent be lower than the melting point of the compound being recrystallized?

7. What is a "cloud point"? What does it signify when doing a solvent-pair recrystallization?

8. Sometimes crystallization does not occur, even after cooling the solution in the ice bath. Describe the probable cause of this problem and explain what steps might be taken to induce crystallization.

9. After recrystallization, a student obtained a very low percent recovery of a solid. Describe a probable cause of the low yield and explain what steps might be taken to increase the recovery.

10. A compound has a solubility in ethanol of 4.24 g/100 mL at 78°C and of 0.86 g/100 mL at 0°C.

    a. What volume of hot ethanol will be necessary to dissolve 50 mg of the compound? (Assume no impurities.)

    b. How much of the compound will remain dissolved in the solvent after recrystallization is complete? (Assume no loss of crystals due to faulty technique!)

    c. What is the maximum percent recovery that could be attained?

11. Is it better to use a long-stem or short-stem funnel when doing a hot gravity filtration? Explain.

12. During a suction filtration, what are the consequences of using a filter paper that is larger than the funnel diameter and using one that is too small? What are the consequences for the use of each?

13. What problem occurs if a student turns off the aspirator before detaching the tubing at a connecting point?

14. What criteria should you use to determine whether recrystallization or sublimation will be a better method of purification?

15. Why is sublimation usually done under reduced pressure conditions?

16. An organic solid has a vapor pressure of 900 torr at its melting point (100°C). Explain how you would purify this compound.

17. Another organic solid has a vapor pressure of 25 torr at its melting point (100°C). Explain how you would purify this compound.

18. Hexachloroethane has a vapor pressure of 780 torr at its melting point, 186°C. What difficulties might be encountered when attempting to measure the melting point of this compound in a melting point determination apparatus?

Technique G: Distillation and Reflux

Introduction to Simple Distillation

Distillation is a process used to purify liquids. **Simple distillation** is the condensation of vapors from a boiling liquid and collection of the condensed vapors in a receiving vessel as shown by the diagram in Figure 1G-1.

Following the path of a typical liquid molecule in the distilling flask, the molecule circulates in the solution until it passes into the vapor state due to heating. Once in the vapor state, the molecule returns to the solution in the flask or travels up through the still head and over through the condenser to the receiving vessel. Distillation of ocean water to furnish pure water is an example of a commercial application of simple distillation.

Theory of Simple Distillation

A pure liquid, such as acetic acid, exhibits a vapor pressure that is temperature dependent. At ambient temperature and pressure, the aroma of acetic acid can be detected due to molecules in the vapor state above the surface of the liquid. Upon heating, the vapor pressure increases slowly and then more rapidly near the boiling point, as shown in Figure 1G-2. Vapor-pressure curves of other common liquids are also shown in the diagram. More volatile liquids, such as diethyl ether, exhibit higher vapor pressures at all temperatures than less volatile liquids.

Distillation of a pure liquid occurs at the boiling point of the liquid. The boiling point is defined as the temperature at which the vapor pressure above the solution is equal to the atmospheric pressure (or internal pressure in vacuum distillation). Normal boiling point is defined as the boiling point of a pure liquid when the atmospheric pressure is one atmosphere (760 torr).

The temperature reading of the thermometer in the still head remains at room temperature until the first vapors of liquid contact the bulb. The temperature rises rapidly until it reaches the boiling point of the liquid, then remains constant as long as distillate condenses in the condenser. When little liquid remains in the distilling flask, the head temperature drops and no more liquid is obtained as distillate. A plot of temperature (in the distillation head) versus volume of distillate for the distillation of cyclohexane (bp 81°C) is shown in Figure 1G-3.

It is common to distill a mixture of liquids, such as cyclohexane and toluene. Just as for pure cyclohexane, the solution of cyclohexane and toluene boils when the vapor

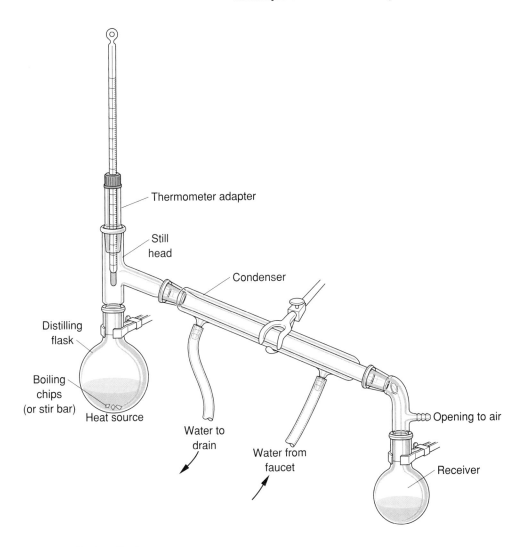

Figure 1G-1   Simple distillation apparatus

pressure above the solution (P_{tot}) is equal to the atmospheric pressure (P_{atm}). The contributions of each component to the total pressure are called partial pressures, $P_{cyclohexane}$ and $P_{toluene}$. Dalton's Law of Partial Pressures states that the partial pressures are additive. A solution will boil when $P_{atm} = P_{tot}$; that is, at the boiling point:

$$P_{atm} = P_{tot} = P_{cyclohexane} + P_{toluene} \qquad (1.1)$$

Raoult's Law states that there is a simple proportionality between solvent mol fraction (X) and its partial pressure (P) at a given temperature. For mixtures of liquids that obey Raoult's Law, the partial pressure of each of the components of a solution is equal to the product of the vapor pressure of the pure component ($P°$) at a particular temperature and the mol fraction (X) of the component. Mol fraction is equal to the number of moles of a component divided by the total number of moles of all components.

Raoult's Law:

$$\text{partial pressure of cyclohexane} = P_{cyclohexane} = P°_{cyclohexane} X_{cyclohexane} \qquad (1.2)$$

$$\text{partial pressure of toluene} = P_{toluene} = P°_{toluene} X_{toluene} \qquad (1.3)$$

Figure 1G-2

Vapor pressure-
temperature diagram

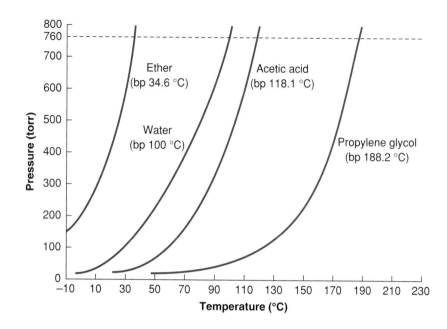

Figure 1G-3

Simple distillation of
30 mL of cyclohexane

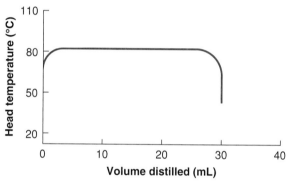

Equation (1.1) may be rewritten as the commonly used expression of Raoult's Law:

$$(1.4) \qquad P_{tot} = P°_{cyclohexane} X_{cyclohexane} + P°_{toluene} X_{toluene}$$

In Equation (1.4), the total vapor pressure of the solution is equal to the product of the vapor pressure of pure cyclohexane at a particular temperature and the mol fraction of cyclohexane plus the product of the vapor pressure of pure toluene at the same temperature and its mol fraction.

Example 1: Calculate the vapor pressure of a solution of 0.600 mol fraction of toluene and 0.400 mol fraction of cyclohexane at 32°C and determine whether the solution will boil. The vapor pressure of pure cyclohexane is 125 torr and the vapor pressure of pure toluene is 40.2 torr at 32°C.

Solution: Calculate the total vapor pressure using Equation (1.4).

$$P_{tot} = P°_{cyclohexane} X_{cyclohexane} + P°_{toluene} X_{toluene}$$

$$= (125 \text{ torr})(0.400) + (40.2 \text{ torr})(0.600)$$

$$= 50.0 \text{ torr} + 24.1 \text{ torr}$$

$$= 74.1 \text{ torr}$$

Since the total vapor pressure is below 760 torr, the solution will not boil at 32°C.

Example 2: Calculate the vapor pressure of pure toluene in a solution of 0.500 mol fraction of toluene and cyclohexane where the solution begins to boil (94°C) at sea level (760 torr). The vapor pressure (see CRC) of cyclohexane is 1100 torr at 94°C.

Solution: Use Equation (1.4), to solve for the vapor pressure of pure toluene.

$$P_{atm} = P_{tot} = P°_{cyclohexane}X_{cyclohexane} + P°_{toluene}X_{toluene}$$

$$760 \text{ torr} = (1100 \text{ torr})(0.500) + P°_{toluene}(0.500)$$

$$P°_{toluene} = (760 - 550)/0.500 \text{ torr} = 420 \text{ torr}$$

The solution in Example 2 starts to boil at a temperature that lies between the boiling points of the two components. This is represented graphically for binary mixtures of cyclohexane and toluene in Figure 1G-4. The composition of the initial distillate formed in Example 2 can be determined from a plot of temperature versus composition. The vertical axis on the left is pure toluene and the axis on the right is pure cyclohexane. To find the composition of the initial distillate for the solution in Example 2, draw a vertical line from the mol fraction of the solution (0.50 mol fraction) until it intersects the liquid curve at point A. From A, draw a horizontal line (a tie line) to intersect the vapor curve at point B. Then, draw a perpendicular from B to the baseline. This gives the approximate composition of the first drops of distillate (about 0.75 mol fraction of cyclohexane) for the distillation of a solution that is 0.50 mol fraction in each component. The tie line A-B intersects the liquid and vapor curves and shows the boiling point of the mixture.

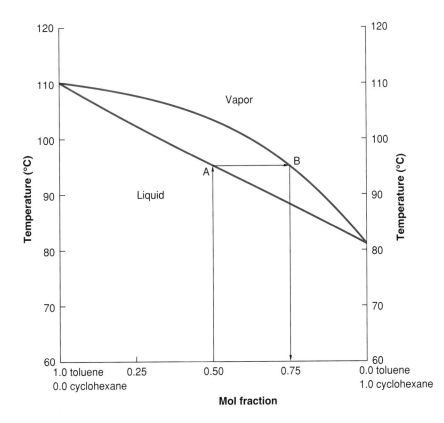

Figure 1G-4

Temperature-composition diagram for the toluene-cyclohexane system

The experimentally determined composition of the vapor above a solution is not the same as the composition of the solution. The vapor is always richer in the more volatile component than the solution. A toluene/cyclohexane solution having a mol fraction of 0.50 of cyclohexane has a vapor composition that is about 0.75 mol fraction cyclohexane.

The boiling point of the distillate changes during the distillation as diagrammed in Figure 1G-5. Early in the distillation, the boiling point of the distillate is closer to that of cyclohexane and the composition of the distillate is closer to that of cyclohexane. Examination of the temperature-composition diagram indicates that the initial distillate is about 75% cyclohexane. As the distillation progresses, the composition of the distillate changes. The distillate becomes richer in toluene and the boiling point of the distillate increases. Pure cyclohexane and toluene cannot be obtained by this method, but simple distillation will work to give pure compounds if the difference in boiling points is at least 100°C.

Microscale Apparatus for Simple Distillation and Assembly

Microscale

Examples of typical microscale distillation equipment are shown in Figure 1G-6. Assemble as shown in the inset. In Apparatus A, a Hickman still is employed. This is a glass tube that has a collar to catch condensed distillate as it flows from the top of the still downward toward the distilling flask. Some Hickman stills have a small exit port near the collar that is opened to allow convenient removal of distillate at appropriate intervals. Apparatus B is similar to Apparatus A except that a condenser is placed above the Hickman still. This apparatus is used for distilling solutions of volatile liquids (bp < 100°C), which require a cold condenser to trap and condense the vapors. The condensed liquid falls from the condenser back into the distilling vial. A drying tube may be attached to the top of the condenser, if necessary, to keep moisture away from the distillate. Unless directed otherwise, do not grease joints.

Several methods of heating are used for microscale distillation. A metal heating block on a hot plate is the most popular heating method. A hot sand bath, using heat from a hot plate, is another option. Hot water or steam baths can be used for those distillations where the boiling point of the distillate is below 100°C.

How to Do a Simple Microscale Distillation

Microscale

Place a hot plate stirrer on the bench if you are using a spin vane or a spin bar or a hot plate if using boiling chips. Place a sand bath or heat block on the hot plate. Add a spin

Figure 1G-5

Distillate temperature vs. volume distilled

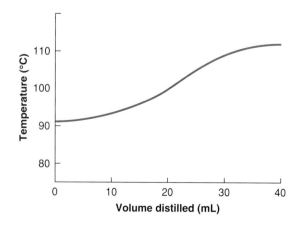

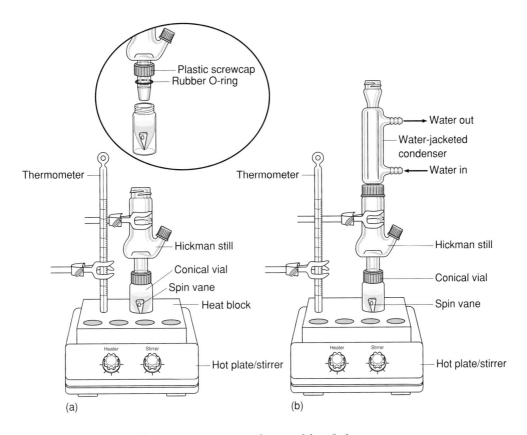

Figure 1G-6 Microscale distillation apparatus and assembly of glassware

vane or a boiling chip to a 3-mL or 5-mL conical vial. Use a 3-mL conical vial for volumes of 0.5–2.0 mL and a 5-mL conical vial for volumes of 2.0–3.5 mL. Clamp the distilling vial to a ring stand. Add the liquid or solution to be distilled to the vial. The vial should not be more than two-thirds full. Attach a Hickman still to the top of the vial and tighten the screwcap. Use an O-ring to seal the joint. If desired, a condenser may be used on top of the Hickman still. Use a water condenser for volatile liquids or an air condenser for high-boiling liquids (bp > 150°C). Fit a thermometer into the top of the apparatus using a clamp. Leave the system open to the outside atmosphere. **Never heat a totally closed system.** Position the thermometer inside the Hickman still so that the thermometer bulb reaches down into the neck of the vial but does not touch the sides of the vial. A typical apparatus is shown in Figure 1G-7.

Turn on the heat source. If using a sand bath, the temperature of the sand bath should be about 20°C higher than the boiling point of the liquid or the boiling point of the lowest boiling component of a solution before distillation will occur. Turn up the heat until the liquid begins to boil. Drops will appear on the thermometer bulb and the temperature of the thermometer inside the apparatus will increase. Liquid will collect in the collar of the Hickman still. Discard the first few drops of distillate. Using a Pasteur pipet, transfer condensed liquid at intervals from the collar to a clean vial. Record the temperature and volume of each aliquot. The observed boiling temperature will be lower than the reported boiling point of the liquid. This is because the small volumes of liquid being distilled do not thoroughly bathe the thermometer bulb. An accurate boiling point can be obtained by performing a boiling point or micro boiling point determination.

Figure 1G-7

Simple microscale distillation apparatus

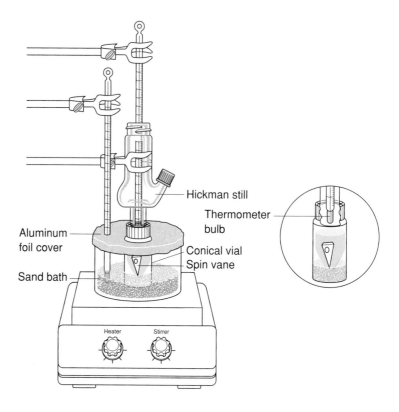

Never heat the vial or flask to dryness since explosions can occur and because flasks containing decomposed organic matter are difficult to clean. When only a few drops of liquid remain in the vial, turn off the hot plate and allow the apparatus to cool. Loosen the caps and raise the apparatus by loosening the clamp and moving it up an inch or two and retightening it. Cap and label all sample vials; store in your locker or in a refrigerator until the distillate fractions can be analyzed. Cover with parafilm also, if directed.

Microscale

Important Tips Concerning Microscale Distillation

1. **Never heat a system that is totally closed from the outside atmosphere because an explosion is very likely to occur as pressure builds up.**
2. **Never heat a system to dryness because of possible explosion.**
3. Do not fill the distillation vial more than two-thirds full. Overfilling may cause the liquid in the vial to bump and contaminate the distillate.
4. Add boiling chips or a spin vane to reduce bumping. Never add a boiling chip to a solution that is already boiling. Doing so will cause the liquid to bump violently.
5. Place the thermometer properly in the neck of the microscale vial when using a Hickman still. For other distilling heads, place the bulb of the thermometer below the exit for the condenser. Remember that the observed thermometer reading in a microscale distillation might be lower than the reported boiling point because very small amounts are being distilled.
6. Control the rate of heating carefully so that the distillation vial does not become overheated. An overheated distilling vial may lead to bumping.
7. Be very careful when connecting tubing to condensers. Do not use excessive force. Use a lubricant, such as water or glycerine, or try another section of tubing

if there is a problem. Be sure that the **correct diameter tubing** is being used. Attach the tubing **before** connecting the condenser to the vial. Tubing can be fastened using wire.
8. Handle a hot plate near the base, rather than at the top. The top may be hot!

Miniscale Apparatus for Simple Distillation

Miniscale

Refer to the typical miniscale distillation setup in Figure 1G-8. Important pieces are the Variac, heating mantle, distilling flask, stir bar or boiling chips, still head, thermometer, condenser, bent adapter, and receiving flask.

How to Do a Simple Miniscale Distillation

Assemble the apparatus from the bottom up. Choose the correct size heating mantle. Connect the electrical cords from the mantle to the Variac and from the Variac to the electrical outlet. Do not turn on the Variac until setup is complete. Clamp the distilling flask to a ring stand, insert the stir bar or boiling chips, and add the distilling head and thermometer. Position the bulb of the thermometer below the opening in the still head leading to the condenser. Do not grease standard-taper joints except when alkali is used and then grease joints only lightly.

Attach tubing to the condenser (one for water in and the other for water out) and clamp the condenser to a second ring stand. Tighten the clamp enough to hold the condenser, but loosely enough that the condenser can slide onto the standard-taper joint of the

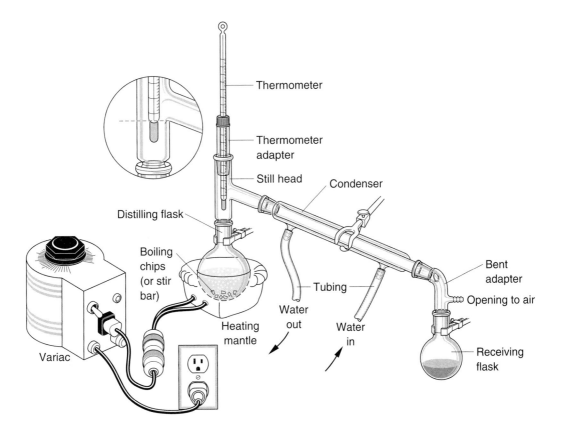

Figure 1G-8   Apparatus for simple miniscale distillation

still head. Attach a bent adapter to the condenser and place a receiving vessel beneath the adapter to collect the distillate. Use a third ring stand, if necessary, to position and clamp the receiving vessel. When all parts are connected, remove the distilling head and add the solution to be distilled. Replace the still head and reconnect the condenser. (See Figure 1G-8.) Alternatively, place the liquid to be distilled in the distilling flask when attaching the flask to the ring stand initially. Check to make sure that all joints are tightly fitted and that there are no leaks. If the liquid is high boiling (bp > 125°C), insulate the distilling flask and still head using aluminum foil or glass wool. Do not wrap the condenser.

Turn on the stirrer if a spin bar is being used. Turn the Variac to a low setting (~30) and then increase the setting as necessary to bring the liquid to boiling. Turn on the water in the condenser before the distillate begins to come over through the condenser. Discard the first few drops of distillate as they often contain low-boiling impurities or water. Adjust the Variac so that the rate of distillation is 1 to 2 drops per second.

If a mixture of liquids is being distilled, it will be necessary to turn up the setting of the Variac during the distillation. This may have to be done more than once depending on the boiling points of the components of the mixture. Divide the distillate, if desired, into several fractions by periodically changing receiving vessels. Collect the distillate at a given temperature range and change receiving vessels when the temperature changes. (Correct boiling points for atmospheric pressure.) The purest liquid is that which is distilled at constant temperature at the boiling point of the liquid. Label all receiving vessels and record the boiling range of each. Record the volumes or weights of each fraction. Cover each vessel until such time as the fractions can be analyzed.

As the liquid in the distilling flask reaches a small volume, turn off the Variac and stop the distillation. Never heat a distilling flask to dryness! There may be danger of explosion in certain cases. Also, distilling flasks that have been heated to dryness are frequently difficult to clean. Disassemble the cooled equipment as soon as possible in the order opposite to which it was assembled and then wash it and set aside to dry.

Miniscale

Important Tips Concerning Miniscale Distillation

1. **Never heat a system that is totally closed from the outside atmosphere because an explosion may occur due to pressure buildup.**
2. **Never heat a system to dryness because of possible explosion.**
3. Use a calibrated thermometer so that accurate temperature readings can be obtained.
4. Do not fill the distilling flask more than two-thirds full. Overfilling causes bumping, which will contaminate the distillate.
5. Control the heating rate carefully to avoid overheating the distilling flask. An overheated flask may cause bumping and may give an observed boiling point that is too high due to superheating of the solution.
6. Add boiling chips or a stir bar to the flask prior to distillation to minimize bumping. Never add boiling chips to a solution that is already boiling. Doing so causes violent bumping.
7. Turn up the heat gradually during distillation. A certain amount of heat is necessary to start the distillation. After some of the liquid has distilled, turn up the heat to ensure a steady flow of condensate, because the boiling point of distillate may increase, particularly in the case of mixtures.
8. When distilling a high-boiling component, wrap the distilling flask and still head with aluminum foil or glass wool for insulation.
9. Compounds having high boiling points (> 150°C) do not require a flow of cold water through the condenser.

Introduction to Reflux

Reflux is continual boiling of a solution in a vial or flask where the solvent is continually returned to the reaction vessel from a condenser atop the vial or flask. The condenser is usually water cooled and often a drying tube is connected to the top of the condenser to keep moisture away from the solution.

Reflux is one of the most common techniques for carrying out organic reactions. Using this technique it is possible to heat a reaction mixture at the boiling point of the solvent without losing the solvent through evaporation. Molecules of liquid are vaporized due to heating the vial or flask. As they escape from the flask, they encounter the cooling effect of the condenser and they condense to be returned to the flask. Reactions can be heated for several hours, days, or even weeks as long as water pressure is maintained to keep cool water flowing through the condenser. In this text, reflux times are 75 minutes or less.

Reaction rates of most organic reactions are faster at higher temperature. A rough rule of thumb is that the reaction rate doubles for every 10°C rise in the reaction temperature. Raising the reaction temperature has a dramatic effect on reaction rates of reactions that have high activation energies. The effect is less dramatic for reactions that have low activation barriers. If it takes an hour to accomplish a reaction in boiling diethyl ether, the reaction should take only about 5–10 minutes in boiling tetrahydrofuran, a cyclic ether with a boiling point 30°C higher than diethyl ether. Reflux setups are shown in Figure 1G-9.

How to Do a Microscale Reaction with Reflux

Microscale

Obtain a hot plate/stirrer if using a spin vane or a hot plate if using boiling chips. Place a sand bath or heat block on the hot plate. Attach a vial containing a boiling chip or spin vane to a clamp connected to a ring stand. Place the solution to be refluxed in the

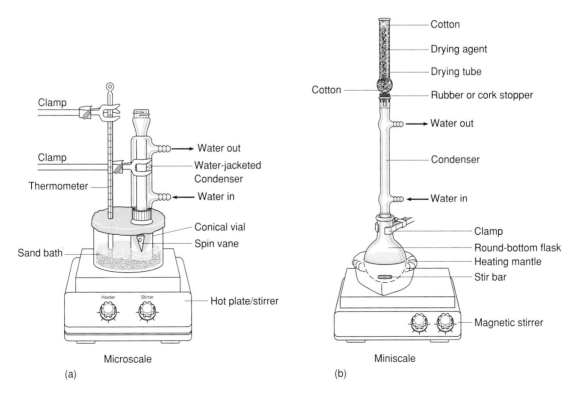

Figure 1G-9 Microscale and miniscale reflux equipment

vial. The vial should not be more than two-thirds full. If a sand bath is used, push the vial down into the sand so that it is partially immersed in the sand, guaranteeing a good transfer of heat from the sand to the vial. Connect tubing to the condenser. Attach the condenser with rubber o-ring and plastic cap to the vial as shown in Figure 1G-9. If moisture must be excluded from the reaction, attach a calcium chloride drying tube to the top of the condenser. Add a wad of cotton to the tube, then the drying agent and a second wad of cotton to contain the drying agent. Turn on the water to the condenser. Turn on the hot plate (and stirrer, if applicable) and adjust to a setting that assures vapors from the solution going up into the condenser and condensing to return to the solution. Vapors should not escape from the top of the condenser. Reduce the heat setting if vapors condense more than one-third up into the condenser. Reflux the solution for the designated time. Check the apparatus periodically. When finished, turn off the hot plate, but allow stirring to continue to avoid bumping. When at room temperature, turn off the water and disassemble the apparatus. Wash all glassware.

Miniscale

How to Do a Miniscale Reaction with Reflux

Attach a round-bottom flask containing a boiling chip or stir bar to a clamp connected to a ring stand. Place the solution to be refluxed in the flask. Fill the flask no more than two-thirds full. Fit a heating mantle of the proper size to the flask and connect to a Variac. Connect the Variac to the outlet. Do not turn on the Variac until setup is complete. Attach a water condenser, with hoses already connected, to the flask. Wire the hoses, if directed by your instructor. If necessary, attach a calcium chloride drying tube to the top of the condenser. Add a wad of cotton to the tube, then the drying agent and a second wad of cotton to contain the drying agent. Do not grease standard-taper joints unless the solution is basic. Even then, apply only a thin film of grease. Start water flowing through the condenser and turn on the Variac. Adjust to a setting that sends solvent vapors up into the condenser and condensing liquid back into the solution. Vapors should not escape from the top of the condenser. Reduce the heat setting if vapors condense more than one-third up into the condenser. Reflux the solution for the designated time. Check the apparatus periodically. When finished, turn off the Variac, but allow stirring to continue to avoid bumping. When at room temperature, turn off the water and disassemble the apparatus. Wash all glassware.

Important Tips on Reflux

1. Always clamp the vial or flask to a ring stand.
2. Connect tubing to the condenser prior to assembling the reflux apparatus.
3. Add a boiling chip, spin vane, or spin bar.
4. Do not fill the distillation vial or flask over two-thirds full.
5. Apply enough heat to achieve reflux, but not excessive heat. Often the vapors can be seen condensing in the condenser. Condensation should occur near the bottom of the condenser.
6. Be sure to turn on the water to the condenser before starting to heat.
7. Check the apparatus periodically during reflux to be sure that solvent vapors are not escaping and that the proper amount of heat is being applied.
8. Recover the spin bar or spin vane as soon as possible after reflux as these are easily lost down the drain.

Evaporation of Solvents

Isolating a solid from solution often requires evaporation of the solvent. The microscale evaporation setups in Figure 1G-10(a), (b) or (c) can be used for evaporation of a volatile solvent. Solvent can be driven off under the hood with a stream of air or nitrogen or otherwise drawn off with a source of vacuum, such as a water aspirator. Miniscale evaporation can be done in a similar manner as shown in Figure 1G-10(b) or (c). For the apparatus in Figure 1G-10(c), solvent is evaporated from a suction flask under vacuum. This setup is sometimes also used in microscale procedures. In each setup, boiling chips or a spin vane or stir bar should be used to prevent bumping.

Exercise G.1: Distilling a Cyclohexane/Toluene Mixture

Pipet 1.25 mL each of cyclohexane and toluene into a 5-mL conical vial that contains a spin vane or boiling chip. Connect a Hickman still to the vial and assemble the apparatus as shown in Figure 1G-7. Position a thermometer in the thermometer adapter so that the thermometer is centered in the middle of the joint of the Hickman still. Turn on the heat source and slowly raise the temperature until vapors can be seen in the still. These vapors will recondense in the collar of the Hickman still. Use a pipet, calibrated in 0.1-mL divisions, to withdraw the liquid collected in the collar and to estimate the volume of distillate recovered. Record the temperature at the time that the aliquot was taken. Collect distillate at intervals from the lip of the Hickman still. Continue the distillation until there is only about 0.5 mL of liquid remaining in the conical vial. Remove the heat source and let the apparatus cool to room temperature. Place all organic liquids in an appropriately labeled container. Correct the boiling points to standard pressure. Estimate and record the temperature ranges at which fractions were collected and estimate the volumes of distillate collected at each temperature. Graph the corrected boiling temperature versus volume (mL) of distillate.

Microscale

Safety First!

Always wear eye protection in the laboratory.

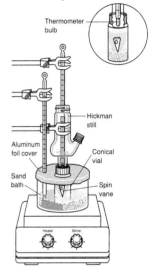

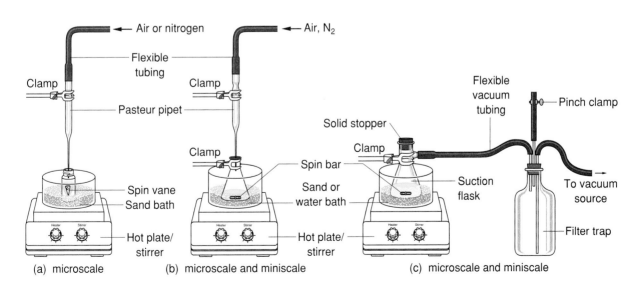

(a) microscale   (b) microscale and miniscale   (c) microscale and miniscale

Figure 1G-10 Setups for evaporation of volatile solvents (in hood)

Miniscale

Exercise G.2: Distilling a Mixture of Cyclohexane and Toluene

Lay out all of the glassware needed to perform a simple distillation. Make certain that each piece of glassware is clean, dry, and free of star cracks. Refer to Figure 1G-8 for an illustration of the apparatus for simple miniscale distillation.

Pipet 10.0 mL each of cyclohexane and toluene into a 50-mL round-bottom flask that contains a spin bar or boiling chip. Assemble the apparatus for simple distillation as shown in Figure 1G-8. Position the bulb of the thermometer below the neck of the distilling head. Be certain that the bent adapter fits well into the neck of a graduated cylinder to minimize evaporation of the distillate. Gently turn on the water to the condenser. Turn on the Variac and heat the solution until vapors reach the thermometer bulb. Increase the rate of heating. Discard the first few drops of distillate, which may contain water or volatile impurities. Collect distillate in the receiver at the rate of 1–2 drops per second. Record the temperature versus volume (mL) of distillate during the entire distillation. Take readings after every 1–2 mL, or more frequently if the temperature is changing rapidly. As the distillation proceeds, cyclohexane will be depleted and the pot will become enriched in toluene. It may be necessary to increase the heat to continue the distillation at a suitable rate. Stop the distillation when 1–2 mL remains in the pot. Remove the heat source and let the apparatus cool to room temperature. Disassemble the distillation apparatus. Place all organic liquids in an appropriately labeled container. Wash and dry all glassware. Correct the boiling points to standard pressure. Estimate and record the temperature ranges at which fractions were collected and estimate the volumes of distillate collected at each temperature. Graph the corrected boiling temperature versus volume (mL) of distillate.

Questions

1. What is the relationship between volatility and the vapor pressure of a solvent?

2. Explain what happens to the vapor pressure of a liquid as temperature increases. Are temperature and vapor pressure directly proportional? If not, explain the relationship by drawing a graph.

3. Explain Dalton's Law of Partial Pressures. Is Dalton's law dependent upon ideal behavior? Explain.

4. Why is it preferable to allow cold water to enter at the bottom of a condenser and exit at the top rather than vice versa?

5. Explain the difference between the vapor pressure of a pure solvent ($P^°_{solvent}$) and its partial pressure ($P_{solvent}$).

6. Can the vapor pressure of a liquid be greater than 1 atm? Explain.

7. What possible hazard might occur if a distillation vessel is heated all the way to dryness?

8. Refer to Figure 1G-4. Predict the boiling point of the initial drops of distillate from the distillation of a mixture that contains 0.80 mol fraction of toluene and 0.20 mol fraction of cyclohexane.

9. Is it proper procedure to add the boiling chips down through the condenser once a solution has started to reflux? What is expected to happen if this procedure is followed?

10. A certain reaction requires 80 minutes to reach completion in refluxing diethyl ether (bp 35°C). Estimate the reflux time required if the reaction is run in tetrahydrofuran (bp 65°C).

Technique H: Fractional Distillation and Steam Distillation

Introduction to Fractional Distillation

Fractional distillation improves separation of components of a mixture beyond that which is possible in a simple distillation. Simple distillation is generally unsatisfactory unless the components have vastly different boiling points (about a 100°C difference). A mixture of cyclohexane and toluene cannot be separated into pure cyclohexane and pure toluene by simple distillation because the boiling points of these compounds are too close together. Better separation can be achieved by fractional distillation. The key to an efficient fractional distillation is the number of vaporization-condensation cycles provided by the apparatus. Each cycle is equivalent to a separate simple distillation. A simple distillation gives a single vaporization-condensation cycle. (Refer to Figure 1G-4.) To understand the greater efficiency of fractional distillation, refer to the temperature-composition diagram (Figure 1H-1) for fractional distillation of a solution containing 0.50 mol fraction each of toluene and cyclohexane.

The diagram shows the expected initial boiling point and vapor composition for a fractional distillation apparatus that gives two vaporization-condensation cycles. The two-step sequence can be traced by drawing a vertical line shown from the 0.50-mol fraction mark on the *x* axis up to point A on the liquid curve. Connecting the tie line A-B gives the vapor composition for the first cycle at point B. In a simple distillation, the distillate will contain 0.75 mol fraction of cyclohexane and boil at 94°C. However, in the fractional distillation, the vapor recondenses (B to point C). The second tie line C-D gives the vapor composition at point D (from the intersection of a vertical line drawn from D to the *x* axis) after the second vaporization-condensation cycle. For this distillation, the initial distillate should contain 0.87 mol fraction of cyclohexane and boil at 86°C. The boiling point and composition of the initial distillate from a three-cycle distillation of a mixture of cyclohexane and toluene will boil at 84°C and contain 0.97 mol fraction of cyclohexane. Most fractional distillations carried out in the first-year organic laboratory achieve two- or three-cycle efficiency. Special columns, called spinning band columns, are required to obtain greater efficiency. Fractional distillation requires the same equipment as simple distillation, plus one extra piece of glassware, called a fractionating column, as shown in Figure 1H-2.

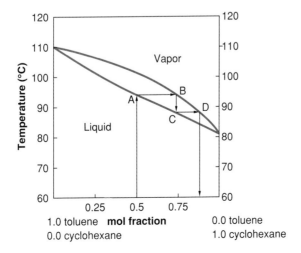

Figure 1H-1

Temperature-composition diagram for the toluene-cyclohexane system

Figure 1H-2

Miniscale fractional
distillation setup

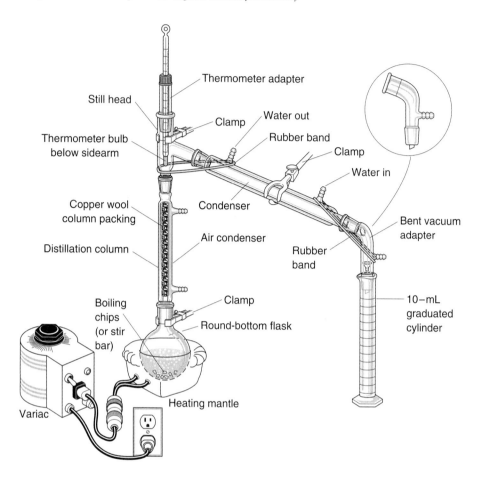

The fractionating column fits directly into the distilling flask. The most common fractionating columns are glass columns, such as a Vigreux column, that are specially designed to contain a high surface area, affording a greater equilibration of the vapor above the distilling flask. Alternatives are packed glass tubes containing metal sponge, glass helices, or glass beads. Equilibration differentiates between more-volatile and less-volatile molecules. The effect is much the same as doing multiple simple distillations back to back, but more efficiently. Examples of fractionating columns and packing materials are shown in Figure 1H-3.

The metal or glass packing in the fractionating columns serves to provide more surface area for vapor to condense. A drawback to the increased area is that there is more "hold-up" of the vapor. This can be a serious problem in columns that are packed with glass beads or glass helices. Liquid that is expected to distill can become "lost" in the fractionating column. Columns that have significant hold-up should be used only for distilling large volumes of liquid. A comparison of hold-up and efficiencies of different types of columns is shown in Table 1H-1.

Vigreux columns work well in miniscale distillation because of small hold-up. However, unless a mixture of liquids is used that has at least a 60°C difference in boiling points, only partial separation can be achieved. A Vigreux column or a column using one of the packings listed in Table 1H-1 can be used for miniscale fractional distillation as described later in Exercise H.2.

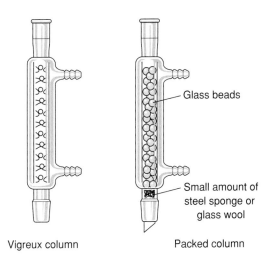

Figure 1H-3

Fractional distillation columns

Glass beads

Small amount of steel sponge or glass wool

Vigreux column   Packed column

Table 1H-1   Properties of Fractionating Columns Used for Miniscale Distillation

Column packing	Distillation cycles	Boiling point difference required	Hold-up
Metal sponge	2	70	moderate
Vigreux (no packing)	2	60	small
Glass beads	3	55	moderate
Glass helices	5	35	moderate

How to Do a Miniscale Fractional Distillation

Miniscale

Glassware used in miniscale fractional distillation is shown in Figure 1H-2 on page 66. Clamp the distilling flask to a ring stand and add a stir bar or boiling chips and the liquid to be distilled. Attach the fractionating column. Place the still head and thermometer on top of the column. Position the bulb of the thermometer adjacent to the arm of the still head leading to the condenser. Do not grease standard-taper joints except when alkali is used and then grease joints very lightly upon assembling the apparatus. Attach tubing to the condenser (bottom for water coming in and top for water going out) and clamp the condenser to a second ring stand. Tighten the clamp enough to hold the condenser, but loosely enough that the condenser can slide onto the standard-taper joint of the still head. Attach a bent adapter to the condenser and place a receiving flask beneath the adapter to collect the distillate. Turn on the stirrer if a spin bar is being used. Turn on the Variac to a low setting (~30) and increase the setting as necessary to bring the liquid to boiling. High-boiling liquids may require insulation of the glass parts that are being heated. Wrap the distilling flask, fractionating column, and still head with aluminum foil.

Turn on water to the condenser unless the liquid being distilled is high boiling (>150°C), in which case it is not necessary to use a water-cooled condenser. The temperature of the thermometer in the still head will remain at room temperature until the first vapors of liquid meet the thermometer bulb. It is usually possible to observe a ring of condensate advancing up the fractionating column. The temperature rises rapidly as the distillate reaches the boiling point of the most volatile component. The temperature increases gradually as the lower boiling component distills away and some of the higher boiling component begins to distill.

When a mixture of liquids is being distilled, it will be necessary to turn up the setting of the Variac during the distillation. This may have to be done more than once depending on the boiling points of the components of the mixture. A good rate of distillation is at least 1 to 2 drops per second. An acceptable rate depends on the amount of liquid to be distilled.

When distilling a mixture, the distillate may be divided into several fractions by periodically changing receiving vessels. Collect the distillate at a given temperature and change receiving vessels when the temperature changes.

When the liquid in the distilling flask reaches 1–2 mL, turn off the Variac to stop the distillation, but continue stirring. Do not heat to dryness. When the apparatus is cool, turn off the water and disassemble the equipment in the order opposite to which it was assembled. Record the boiling range and volumes or weights of each fraction. Cap each vessel to preserve the samples for analysis. After the apparatus has cooled, wash the glassware and allow to dry.

Miniscale

Practical Tips about Miniscale Fractional Distillation

The tips given in Technique G for simple distillation also apply to fractional distillation. Additional points of emphasis are given here.

1. Be very careful when assembling the fractional distillation apparatus. The apparatus is tall and high clearance is necessary on the lab bench. Use clamps and ring stands to support the apparatus, not books or notebooks. The apparatus should be stable upon assembly and should not teeter.
2. Don't lay the thermometer down near the edge of the bench. It is very common for thermometers to break during assembly and disassembly of fractional distillation equipment.
3. Be prepared to apply more heat for fractional distillation than for simple distillation of a comparable solution. The vapors of distillate must travel a greater vertical distance before reaching the condenser or still head. It is often useful to wrap the fractionating column with aluminum foil. It is easy to observe the effect of heating the distilling flask as the hot vapors begin to pass up through the fractionating column. Gently touch the outside of the column at various heights of the column. Alternatively, observe a ring of vapor that rises up through the fractionating column as the first vapors are pushed upward. Peel back a section of aluminum foil wrapping if necessary. Ultimately, the vapors come into contact with the thermometer bulb and the thermometer shows a rise in temperature.
4. Sometimes the temperature observed in the still head will decrease during distillation. This means that additional heat must be supplied and it might be a good time to change receiving vessels if multiple fractions are being collected.

Introduction to Steam Distillation

Many organic substances begin to decompose at temperatures above 230°C. In steam distillation, all volatile matter distills together with water at temperatures just below 100°C, thus avoiding high-temperature decomposition. Steam distillation is a useful method for isolating high-boiling liquids, known as oils, from other nonvolatile organic compounds, such as waxes, complex fats, proteins, and sugars.

The technique of steam distillation is based upon the principle that each component of immiscible liquid mixtures contributes to the total vapor pressure as if the other components were not there. Natural oils can be isolated readily by steam distillation. For example, if lemon grass oil and water are placed in a flask with water, and heat or steam is added, the distillate will contain water and volatile oils, mostly a water-insoluble substance called citral.

$$CH_3$$
$$|$$
$$(CH_3)_2CH=CHCH_2CH_2C=CHCHO$$

citral (mixture of cis and trans isomers)

A key feature concerning the theory of steam distillation is that the boiling point of the oily, aqueous distillate will never exceed the boiling point of water. This is because both water and the oily component each contribute to the total vapor pressure as if the other component were not present. The mixture boils when the combined vapor pressures of water and the oil equal the atmospheric pressure. The oil has a small, but significant vapor pressure at 100°C, so that the boiling point of the mixture will be just below the boiling point of water. This is true even though the boiling point of the oil is very high (citral, bp 229°C). Steam distillation allows separation and isolation of the oil at a low temperature (below 100°C) where citral will not decompose. Another advantage of steam distillation is that relatively small amounts of an oil can be successfully isolated. The technique is ideally suited for the isolation of oils from natural products.

Theory of Steam Distillation

Distilling a mixture of water and one or more immiscible organic liquids is called steam distillation. Raoult's Law does not apply to immiscible liquids because such liquids do not exhibit ideal behavior. Instead, each component of the mixture contributes its vapor pressure to the total pressure at a particular temperature as expressed in Equation (1.5):

$$P_{total} = P^{\circ}_{component\ 1} + P^{\circ}_{component\ 2} + P^{\circ}_{component\ n} + P^{\circ}_{water} \qquad (1.5)$$

When the total vapor pressure reaches atmospheric pressure, distillation occurs. P° values of each component are used in Equation (1.5) because each component exhibits a vapor pressure as if the other components were not there. Since the vapor pressure of water is 760 torr at 100°C, the presence of any organic material that contributes a vapor pressure will cause the boiling point to drop below 100°C. Nitrobenzene is a high-boiling liquid (bp 210–211°C). It has a vapor pressure of about 10 torr at 98°C where water has a vapor pressure of 750 torr. The mixture of nitrobenzene and water distills at 98°C, because the sum of their vapor pressures equals 760 torr at that temperature. Any organic liquid that is immiscible with water and exhibits even a small vapor pressure will steam distill at temperatures below 100°C. The mixture that passes through the condenser looks cloudy because it contains two immiscible phases. Generally, steam or boiling water is added to the distilling flask until the organic material has finished distilling. When this occurs, the distillate in the condenser will no longer appear cloudy because the distillate will contain only water.

The amount of water necessary to distill a given amount of organic material can be calculated using Equation (1.6). The mol fraction of organic substance is equal to the vapor pressure of the organic substance at the temperature of the distillate divided by the total pressure (atmospheric pressure).

$$\text{mol fraction of organic substance} = P^{\circ}_{organic\ substance}/P_{total} \qquad (1.6)$$

As an example, for the steam distillation of nitrobenzene, the mol fraction in the distillate is calculated using Equation (1.6).

$$\text{mol fraction of nitrobenzene} = P^{\circ}_{nitrobenzene}/P_{total} = 10\ torr/760\ torr = 1.32 \times 10^{-2}$$

$$\text{mol fraction of water} = 1 - 1.32 \times 10^{-2} = 0.987$$

This means that 1.32×10^{-2} mol of nitrobenzene will be distilled along with 0.987 mol of water. This translates to 17.8 g (= 18.0 g/mol × 0.987 mol) of water for every 1.63 g (= 1.32×10^{-2} mol × 123.11 g/mol) of nitrobenzene that will be distilled.

Miniscale

Exercise H.1: Miniscale Fractional Distillation of a Mixture of Cyclohexane and Toluene

Lay out all of the glassware needed to perform fractional distillation. Make certain it is clean, dry, and free of star cracks. Pipet 10.0 mL each of cyclohexane and toluene into a 50-mL round-bottom flask that contains a spin bar or boiling chip. Alternatively, the instructor may assign different volumes. Assemble the apparatus as shown in Figure 1H-2. Position the bulb of the thermometer below the neck of the still head. Be certain that the bent adapter fits well into the neck of the receiving vessel (25-mL graduated cylinder) to minimize evaporation of the distillate. Gently turn on the water to the condenser. Turn on the heat source and heat the solution until vapors reach the thermometer bulb. This is a slow process. Insulating the column with aluminum foil will speed up the process. Increase the rate of heating so that distillate starts to appear in the receiver. Discard the first few drops of distillate, which may contain water or volatile impurities. Collect distillate in the receiver at the rate of 1 to 2 drops per second. Record the temperature versus volume (mL) of distillate during the entire distillation. Take readings after every mL or more frequently if the temperature is changing rapidly. As the distillation proceeds, cyclohexane will be depleted and the pot will become enriched in toluene. Increase the heat to maintain the distillation at a suitable rate. Stop distilling when about 0.5–1.0 mL remains in the pot. Remove the heat source and let the apparatus cool to room temperature. Turn off the flow of water and disassemble the apparatus. Place organic liquids in an appropriately labeled container. Wash all glassware and let dry. Correct the boiling points to standard pressure. Graph the corrected boiling temperature versus volume (mL) of distillate. Mark the temperature ranges at which pure cyclohexane and pure toluene were collected and estimate the volumes of pure cyclohexane and toluene. Use Raoult's Law to explain why the temperature changes as the distillation progresses. Calculate the initial mol fractions of cyclohexane and toluene in the mixture. Use Figure 1H-1 to determine the number of theoretical plates in the distillation.

Microscale

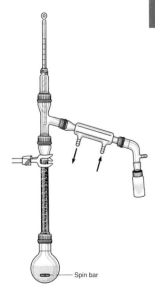

Spin bar

Exercise H.2: Microscale Fractional Distillation of Cyclohexane and Toluene

Place 3.0 mL cyclohexane and 3.0 mL toluene in a 10-mL round-bottom flask containing a spin bar. Prepare a fractional column by packing an air condenser with copper wool. Fit the fractional column to the flask. Place a connecting adapter on top of the column. Fit a thermometer adapter on top of the connecting adapter. Connect a water-jacketed condenser to the sidearm of the connecting adapter. Add a vacuum adapter to the water-jacketed condensor. A 5-mL conical vial will be used as a receiving flask. Place a thermometer into the thermometer adapter. The thermometer bulb must be just below the level of the sidearm of the connecting adapter.

Start the solution stirring. Turn on the heat to the aluminum heat block or sand bath and slowly heat. Discard the first few drops of distillate. Collect the next 3 drops of distillate in a clean vial labeled "fraction 1". Cap the vial tightly and set aside for further analysis. Connect the 5-mL conical vial to the vacuum adapter and collect the remaining distillate at a rate of 1-2 drops per second. Record the temperature and the estimated volume throughout the distillation. Distill until approximately 0.5 mL of liquid remains in the round-bottom flask. Then collect the last 3 drops of distillate in a clean vial labeled "fraction 2". Cap the vial tightly and set aside until ready to inject on the GC. Turn off the heat, let the system cool to room temperature, and disassemble the apparatus.

Correct boiling points for standard pressure. Graph the correcting boiling point temperature versus volume of distillate. Mark the temperature ranges at which pure cyclohexane and pure toluene were collected and estimate the volumes of pure cyclohexane and pure toluene. The composition of the initial and the final distillate can be analyzed, if desired, by refractometry (Technique E) or gas chromatography (Technique J).

Exercise H.3: Steam Distillation of Lemon Grass Oil

Miniscale

Measure 1.0 mL of lemon grass oil (density = 0.87 g/mL) into a 50-mL round-bottom flask. Add 30 mL of water to the flask. Assemble the distillation apparatus for simple distillation with magnetic stirring (Figure 1G-8) or use a setup with a larger round-bottom flask that permits introduction of steam below the surface of the water. Ask your instructor to inspect your apparatus before proceeding. Heat (with stirring if using a simple distillation apparatus) and collect distillate in a 25-mL graduated cylinder until the distillate is no longer oily (after about 15–20 mL of distillate has been collected). Pour the distillate into a 125-mL separatory funnel. (See Technique I, Extraction and Drying.) Extract with two 10-mL portions of methylene chloride. Combine the methylene chloride solutions.

Dry the methylene chloride solution using about 1 g of anhydrous sodium sulfate. Swirl to be sure that no cloudiness remains in the methylene chloride solution. Gravity filter to remove the drying agent. (See Technique F, Recrystallization, Filtration, and Sublimation.) Distill the methylene chloride solution using simple distillation. (See Figure 1G-8.) Stop distilling when the volume in the distilling flask is about 2 mL. Transfer the distilled solvent to a bottle for recovered methylene chloride. Transfer the solution remaining in the distilling flask to a tared Erlenmeyer flask and evaporate the remaining solvent in the hood using a warm water bath or sand bath. There should be an oily residue remaining in the flask. Tilt the flask on its side for a few minutes to allow any remaining solvent vapors to escape. Weigh the flask and determine the amount of oil distilled. Measure the refractive index of the oil and compare the corrected value with the reported refractive index of citral (1.4876). What conclusion can you draw?

Questions

1. Explain how the following factors will affect the separation of cyclohexane from toluene and the overall efficiency of the separation:
   a. Heat is applied too rapidly.
   b. The distilling flask is more than two-thirds full.
   c. The distilling flask is less than one-tenth full.
   d. A very long fractionating column is used.
   e. No fractionating column is used.
2. How would the graph of temperature versus volume distilled appear for a perfect separation of methanol (bp 65.4°C) and toluene (bp 110°C) using an efficient fractionating column?
3. The efficiency of a fractional distillation is given as the number of cycles or theoretical plates. Traversing through steps A → D in Figure 1H-1 is equivalent to having completed two vaporization-condensation cycles or two theoretical plates.

Predict the temperature of the initial distillate for the solution used in Figure 1H-1 for a fractional distillation that is rated at four theoretical plates.

4. Define the following terms and explain how each affects the efficiency of a fractional distillation:

    a. theoretical plate.

    b. column hold-up.

5. A fractionating column filled with glass helices is rated at about 2 theoretical plates. How would the number of plates be affected if the column were used without the helices?

6. Suggest another possible way to obtain citral from lemon grass oil.

7. Does the boiling temperature ever exceed 100°C during steam distillation? Explain.

8. Can steam distillation of ethanol be done using a mixture of ethanol and water? Why or why not?

9. Give one or more reasons why microscale steam distillations may not give good results.

10. The vapor pressure of a water-insoluble organic substance is negligible at 100°C. Can this substance be purified by steam distillation? Explain.

11. Should steam distillation be considered as a method of isolation and purification of an organic liquid that has a vapor pressure of 760 torr at 46°C?

Technique I: Extraction and Drying

Introduction

Extraction is one of the most common methods of separating an organic product from a reaction mixture or a natural product from a plant. In liquid-liquid extraction, equilibrium concentrations of a solute are established between two immiscible solvents. Usually one of the solvents is water and the other is a moderately polar or nonpolar organic solvent, such as methylene chloride, petroleum ether, or diethyl ether. The concentration of solute in each solvent will depend upon the relative solubility of the solute in each of the immiscible solvents. The solute will be more concentrated in the solvent in which it is more soluble.

Extraction is commonly used to separate an organic product from a reaction mixture containing water-soluble impurities. The organic compound is preferentially drawn into the organic solvent in which it is more soluble. Other components, particularly if ionic or very polar, are preferentially drawn into the water. When the two layers are separated, the organic compound in the organic solvent has been freed from most of the water-soluble impurities.

In solid-liquid extraction, the theory is the same. Natural products are composed of a complex mixture of organic compounds. Isolating specific compounds from the solid involves repeated exposure of the solid to a solvent in which one or more components are soluble. These components dissolve in the solvent and are removed from the solid residue. Steeping a cup of tea is an example of a solid-liquid extraction. The hot water dissolves the flavoring and caffeine from the tea bag or tea leaves.

The terms "extraction" and "washing" imply slightly different processes, although the mechanics are the same. Extraction is the process of removing a compound of interest from a solution or a solid mixture. Brewing coffee is another example of a solid-liquid extraction: caffeine and other flavors are extracted from the bean by the hot water.

Washing, on the other hand, usually means removing impurities from the substance of interest. For example, an organic solution containing a neutral product may be washed with aqueous base to remove acidic impurities from the organic solution.

Theory

The organic solvent plays a pivotal role in liquid-liquid extraction. To serve as a good solvent for extraction of an organic compound, the solvent should:

- have high solubility for the organic compound.
- be immiscible with the other solvent (usually water).
- have a relatively low boiling point so as to be easily removed from the compound after extraction.
- be nontoxic, nonreactive, readily available, and inexpensive.

Typical solvents used for extraction are hexane, methylene chloride, diethyl ether, and ethyl acetate. The properties of these solvents are listed in Table 1I-1.

One of the most important properties of the solvent is its solubility for a particular organic compound. In an extraction, the organic compound is distributed between two different liquid phases: the organic solvent and water. The efficiency of extraction will depend upon the solubility of the compound in the two solvents. The ratio of solubilities is called the distribution coefficient (K_D). By usual convention, the distribution coefficient is equal to the solubility of a compound in an organic solvent divided by the solubility in water.

$$K_D = \frac{\text{solubility in organic solvent (g/100 mL)}}{\text{solubility in water (g/100 mL)}}$$

The units of solubility are typically given in grams of compound dissolved per 100 mL of solvent (written as g/100 mL). The magnitude of K_D gives an indication of the efficiency of extraction: the larger the value of K_D, the more efficient the extraction. If K_D is 1, equal amounts of the compound will dissolve in each of the phases. If K_D is much smaller than 1, the compound is more soluble in water than in the organic solvent.

If a compound has a low K_D for a given extraction, it is better to search for a different organic solvent in which the compound is more soluble in order to do a liquid-liquid extraction. If this is not feasible, doing multiple extractions can increase the amount of compound extracted.

Table 1I-1   Physical Properties of Organic Solvents Used in Extraction

Solvent	Solubility in H_2O	Boiling point (°C)	Density (g/mL)	Safety information
Methylene chloride CH_2Cl_2	Very slightly soluble	40	1.3255	Narcotic in high concentrations; suspected carcinogen
Diethyl ether $(CH_3CH_2)_2O$	Slightly soluble	35	0.7134	Very flammable; forms peroxides
Ethyl acetate $CH_3CO_2CH_2CH_3$	Fairly soluble	77	0.902	Flammable
Hexane $CH_3(CH_2)_4CH_3$	Insoluble	69	0.660	Narcotic in high concentrations

How to Do a Microscale Extraction

Place 1 to 2 mL of an aqueous solution containing the compound to be extracted in a 5-mL conical vial. Add 1 to 2 mL of an immiscible organic solvent. Cap and shake the vial to mix the layers thoroughly. Carefully vent by frequently loosening the cap. Repeat until the two phases are thoroughly mixed. Let the vial stand until the two layers have separated. Prepare a Pasteur filter pipet with a small cotton plug in the tip. The cotton plug helps to prevent volatile solvents from leaking out of the pipet. (See Figure 1F-4 in Technique F.)

In microscale extraction, the bottom layer is always removed from the extraction vessel first. (To see why, try to remove the top layer without getting any of the bottom layer.) This means that a different procedure is used, depending upon whether the organic phase is on the top or on the bottom.

If the organic solvent is more dense than water (such as methylene chloride, d = 1.3255), the organic solvent will be the bottom layer in the conical vial. Squeeze the bulb of the filter pipet to remove the air. Then place the tip of the filter pipet at the bottom of the conical vial. Very slowly release the bulb and draw up the organic layer into the pipet.

Figure 1I-1

Steps for the extraction of an aqueous solution using a solvent (CH_2Cl_2) that is more dense than water

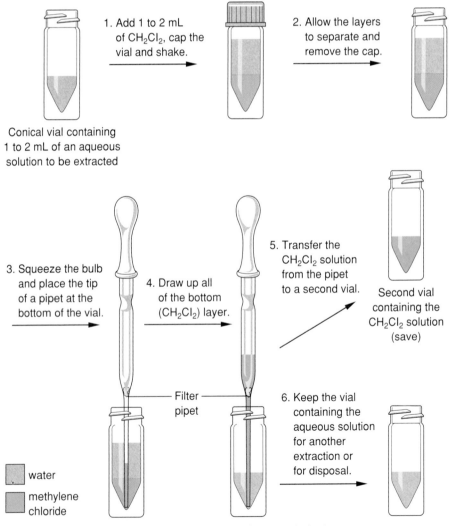

Steps 1 to 5 can be repeated if two or more extractions are desired.

Be careful not to draw up any of the aqueous layer. Transfer the solution in the pipet to a clean conical vial or flask (see Figure 1I-1). Extract the aqueous layer remaining in the vial with additional portions of the organic solvent and combine the organic layers.

If the organic solvent is less dense than water (such as diethyl ether, d = 0.7134; hexane, d = 0.660; or ethyl acetate, d = 0.902), the organic solvent will be the top layer in the conical vial (see Figure 1I-2). Squeeze the bulb of the filter pipet to remove the air. Then place the tip of the filter pipet at the bottom of the conical vial. Very slowly release the bulb and draw up the aqueous layer into the pipet. The organic layer remains in the conical vial. Put the aqueous solution into a separate container. Transfer the organic layer to a clean vial or flask. Transfer the aqueous solution back to the original vial and add a second portion of organic solvent. Shake and vent. Separate the layers as before, but combine the organic layer with the organic layer from the first extraction. Repeat as necessary. Combine all organic layers.

An alternate procedure (not shown) can be used when the organic layer is less dense than water and the total volume of solution is 2 mL or less. In this procedure both of the solvents are drawn up into the pipet. Then the bottom layer is returned to the vial.

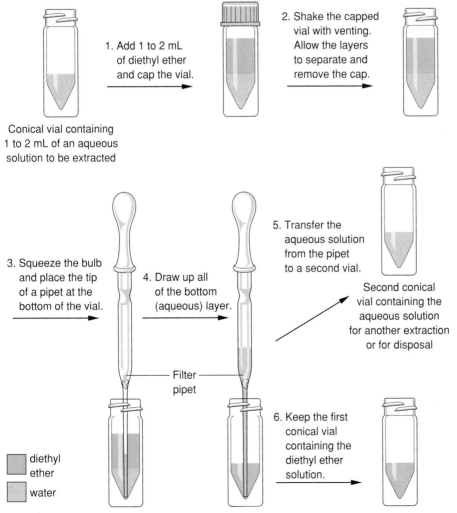

Figure 1I-2

Steps for the extraction of a compound from an aqueous solution using an organic solvent (diethyl ether) that is less dense than water

Squeeze the bulb of the filter pipet to remove the air. Place the tip of the pipet at the bottom of the conical vial. Very slowly release the bulb and draw up all of the solution (both layers) into the filter pipet. Now gently squeeze the bulb to return the bottom aqueous layer to the conical vial. Put the organic layer into a clean conical vial or test tube. Repeat until the entire top layer is in a separate vial. Then add fresh solvent to the aqueous solution and repeat the extraction. Combine the organic layers.

How to Do a Miniscale Extraction

Attach an iron ring to a ring stand. Place a separatory funnel in the ring and an Erlenmeyer flask below the separatory funnel. Check to make sure that the stopcock is closed. Add approximately 1 mL of water to the separatory funnel to check for leaks. Using a stemmed funnel, pour the solution to be extracted into the separatory funnel. Add an approximately equal volume of the other solvent. Replace the stopper and give it a half-turn to seal in place. Cradle the separatory funnel in one hand while keeping the index finger on the top of the stopper. Gently invert the separatory funnel. Make sure that the stem of the separatory funnel is not directed at anyone. Slowly open the stopcock. A "whoosing" sound will be heard as the pressure is released. Close the stopcock. Shake the separatory funnel gently, then again open the stopcock. Repeat this process, but more vigorously, about five to six times until the two layers have thoroughly mixed. Vent one more time. Then invert the separatory funnel and place it in the ring to let the layers separate. Remove the stopper and set it on the lab bench. Allow the layers to clearly separate.

If the organic solvent is less dense than water, the organic solvent will be the top layer. Drain the lower aqueous layer into a clean flask. Pour the organic layer out of the top of the separatory funnel into a different clean flask. To do further extractions, return the aqueous layer to the separatory funnel. Add additional solvent and repeat the procedure. Combine the organic extracts.

If the organic solvent is more dense than water, the organic solvent will be the bottom layer. When the layers have separated, remove the stopcock and drain the lower organic layer into a clean flask. Add fresh solvent to the aqueous layer remaining in the separatory funnel and repeat the extraction procedure. Combine the organic extracts. This process is illustrated in Figure 1I-3.

Important Tips Concerning Extraction

1. **Never discard any layer until the experiment has been completed!** It is impossible to recover product from a layer that was discarded.
2. **Make certain that the phases are thoroughly mixed by vigorously shaking the container.** In order to be extracted from one solvent into another, the solvents must share as much surface area as possible to allow efficient transfer of the compound from one solvent to another.
3. Using one of these methods to determine which layer is the organic layer and which layer is the aqueous layer:
   a. Look up the densities of the solvents in the *Handbook of Chemistry and Physics* or the *Merck Index*; usually the more dense solvent will be on the bottom.
   b. Add 2–3 mL more water to the separatory funnel or 0.1–0.2 mL more water to the vial and see which layer grows; the layer that increases must be the aqueous layer.
   c. Put 1–2 drops from one of the layers onto a watch glass or in a small vial and add several drops of water. If two layers appear or if the mixture appears

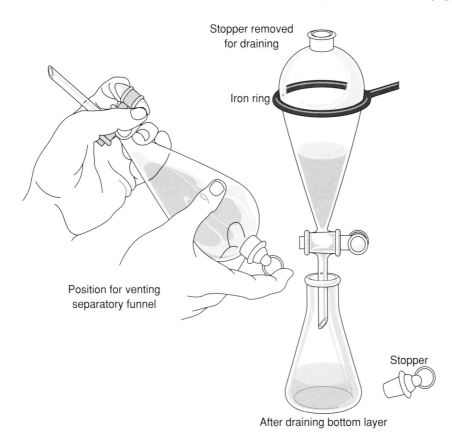

Stopper removed
for draining

Iron ring

Position for venting
separatory funnel

Stopper

After draining bottom layer

   cloudy, the layer was organic. If the solution is clear and only one layer exists,
   the layer was aqueous.
4. Vent frequently, especially with very volatile solvents such as diethyl ether or
   methylene chloride. Failure to do so could result in the top being blown off and the
   contents strewn all over the bench top. One way or the other, the pressure buildup
   will be alleviated! Do not vent the separatory funnel towards another person.
5. If there is only one layer in the conical vial or the separatory funnel and there
   should be two, perhaps the wrong layer was used. Try adding a little water and see
   what happens. If there is still only one layer, try adding some of the organic solvent. **Since nothing has been discarded** (Tip 1) there is no problem. Find the correct layer and repeat the extraction.
6. An emulsion is a suspension that prevents a sharp interface between the layers.
   Should an emulsion form, there are several things to try to break it up:
   a. Stir the mixture gently with a glass rod in the hopes of separating out the lay-
      ers, then let it stand undisturbed.
   b. Add a small portion of saturated sodium chloride solution (brine) or solid
      sodium chloride or other electrolyte to the solution. This increases the ionic
      strength of the aqueous layer, which may aid in the separation.
   c. For miniscale extractions using a separatory funnel, gravity filter the solution
      through filter paper. This should only be done after all else fails since filtering
      is slow and messy.
7. When considering the density of a solution, remember that the density of a dilute
   aqueous solution is approximately the same as the density of water.

Microscale Extraction

8. If it is difficult to keep the organic layer in the filter pipet, try drawing up and expelling several portions of the solvent before filtering the solution.
9. Never draw up liquid into the latex bulb! This contaminates both the liquid and the bulb.

Miniscale Extraction

10. If the liquid is not draining evenly from the separatory funnel, check to make certain that the stopper has been removed.
11. The proper way to hold a Teflon stopcock in a separatory funnel is to attach the Teflon washer first, then the rubber ring, then the nut.
12. To test for leaks, add a little water to the empty separatory funnel and examine the area around the stopcock for signs of leakage.
13. A Teflon stopcock should not be greased. A glass stopcock must be lightly greased to prevent leaking. Be sure that no grease gets in the opening of the stopcock. Glass stoppers should not be greased.

Drying and Drying Agents

During extraction, the organic layer becomes saturated with water. A drying agent must be used to remove water from the organic solvent. Drying agents are anhydrous salts that are used to remove water from "wet" organic liquids. The salts bond to water molecules to form hydrated salts that can easily be removed with a filter pipet (microscale) or by gravity filtration (miniscale).

A good drying agent should be fast acting. It should have a high capacity for reacting with water so that a small quantity of drying agent adsorbs a lot of water. Several common drying agents and their properties are detailed in Table 1I-2.

Because of its large capacity and ease of separation, sodium sulfate is the drying agent most frequently used in microscale experiments. Sodium sulfate and magnesium sulfate are commonly used for miniscale experiments.

How to Do Microscale Drying

Place the solution to be dried in a small Erlenmeyer flask or conical vial. Add a small amount of the drying agent (about a quarter of a tip of a spatula) to the solution and swirl. Observe carefully. If the drying agent clumps, add another small portion of the drying agent until the solution is clear and the additional drying agent stays free-flowing (the small particles stay suspended). The amount of drying agent required will depend upon the solvent. Diethyl ether dissolves more water than does an equal volume of methylene chloride or hexane. Let the solution stand for a few minutes (5–10 minutes if using sodium sulfate). Prepare a Pasteur filter pipet that has been fitted with a cotton plug. Squeeze the bulb to remove the air. Insert the pipet into the bottom of the conical vial or Erlenmeyer flask. Slowly release the bulb and carefully draw the liquid up into the pipet. Try to avoid drawing up the drying agent, if possible. However, any drying agent that is drawn up into the pipet will stick to the cotton plug. Drain the liquid into a clean vial or Erlenmeyer flask. Rinse the drying agent with additional small portions of solvent and combine the washings. Discard the drying agent according to the instructor's directions.

Table 1I-2   Properties of Common Drying Agents*

Drying Agent	Capacity	Speed	Comments
Calcium chloride ($CaCl_2$)	low	rapid	Cannot be used with compounds containing oxygen and nitrogen; used in drying tubes.
Magnesium sulfate ($MgSO_4$)	high	rapid	Suitable for all organic liquids; hydrated salt is fine powder so is sometimes difficult to separate from dried solvent.
Potassium carbonate (K_2CO_3)	moderate	moderate	Forms larger hydrated crystals, so is easily separated from the dried solvent; cannot be used with acidic compounds.
Sodium carbonate (Na_2CO_3)	moderate	moderate	Forms larger hydrated crystals, so is easily separated from the dried solvent; cannot be used for acidic compounds.
Sodium sulfate (Na_2SO_4)	high	relatively slow (5–10 minutes)	Easily filtered, used in both microscale and miniscale experiments; inexpensive.

Do Not Use Hydrated Salts for Drying.

Other drying agents used for drying tubes and solvents:

Drying Agent	Comments
Drierite ($CaSO_4$)	Frequently used in drying tubes to keep moisture from getting into a reaction; an indicator in the reagent changes color from blue (anhydrous form) to pink (hydrated form) when saturated with water.
Molecular sieves (Type 4A)	Composed of silicates that trap water molecules within the lattice; frequently used in bottles of solvents when anhydrous solvents are needed.

*All salts are anhydrous.

How to Do Miniscale Drying

Place the solution to be dried in an Erlenmeyer flask or beaker. Add a spatula-tip full of the drying agent to the solution. Swirl to suspend the particles. In the presence of water, the small particles of the drying agent clump together. Continue adding small portions of the drying agent until the solution is completely clear (no water droplets remain) or when additional drying agent is free flowing. Loosely stopper the flask and let the solution stand for 10–15 minutes. Meanwhile, prepare a gravity filtration apparatus. (See Technique F, Recrystallization, Filtration, and Sublimation). Pour the dried solution carefully through the filter paper. Rinse the flask and filter paper with additional portions of solvent and then combine the solvent washings.

Important Tips Concerning Drying Agents

1. A solvent that has been in contact with water is not considered to be dry unless it has been dried with a drying agent.
2. Exact amounts of drying agent are usually not specified in a procedure. The only way to know for sure how much drying agent to add is by careful addition and observation.

3. Do not add more drying agent than is necessary. The drying agent can adsorb the organic product as well as water.
4. Swirling the drying agent and the solution enhances the speed of the drying.
5. If a liquid dissolves the drying agent, that liquid is probably water. Check to see if the correct layer has been chosen following an extraction or steam distillation.
6. Be aware that different organic solvents absorb differing amounts of water, so they will require different amounts of drying agent. Diethyl ether partially dissolves some water and requires a lot of drying agent.

Microscale

Exercise I.1: Determining the Distribution Coefficient of Caffeine

Safety First!

Always wear eye protection in the laboratory.

Do not breathe methylene chloride vapors. Work in the hood.

Weigh out about 100 mg of caffeine; record the exact mass. Transfer the caffeine to a 5-mL conical vial. Using a calibrated pipet, add 1.0 mL of methylene chloride and 1.0 mL of water to the vial. Cap the vial and shake vigorously to mix the layers thoroughly and dissolve the caffeine. Cautiously unscrew the cap to vent and release the pressure. If all of the caffeine does not dissolve, add another 0.5 mL of both methylene chloride and water and shake again. Allow the layers to separate. Use a filter pipet (Technique F) to transfer the methylene chloride (lower layer) to a clean Erlenmeyer flask or vial. Be sure to transfer all of the methylene chloride, but no water. Add anhydrous sodium sulfate to the methylene chloride solution until no further clumping is observed. (This will take about the tip of a spatula full of sodium sulfate.) Let the solution stand for 5 minutes. With a clean, dry filter pipet, transfer the dried solution into a tared clean, dry 10-mL Erlenmeyer flask that contains a boiling chip. Rinse the drying agent remaining with a fresh portion of methylene chloride (about 0.5 mL); combine this rinse with the methylene chloride solution. Evaporate the solvent by placing the Erlenmeyer flask in a warm sand bath under the hood. Reweigh after cooling to room temperature.

Calculate the mass of caffeine extracted by methylene chloride. Calculate the mass of caffeine remaining in the water. (The mass of caffeine in the water will be the difference between the starting mass of caffeine and the mass in caffeine in methylene chloride.) Calculate the distribution coefficient of caffeine in the methylene chloride/water system.

Miniscale

Exercise I.2: Determining the Distribution Coefficient of Caffeine

Safety First!

Always wear eye protection in the laboratory.

Do not breathe methylene chloride vapors. Work in the hood.

Weigh out about 300 mg of caffeine; record the exact mass. Transfer the caffeine to a 25-mL Erlenmeyer flask. Using a calibrated pipet, add 10.0 mL of water to the flask. Swirl the flask to dissolve the caffeine. Pour the solution into a 50-mL or 125-mL separatory funnel that is supported in a ring stand. Make certain that the stopcock is closed before adding the solution. Now pipet 10.0 mL methylene chloride into the Erlenmeyer flask. Swirl the flask and pour the contents into the separatory funnel. Stopper the funnel. Partially invert the funnel and open the stopcock to release the pressure. Close the stopcock, shake or swirl the funnel gently several times, and vent again by inverting the funnel and opening the stopcock. Repeat this process several times. Replace the funnel in the ring and allow the layers to separate. Remove the stopper and open the stopcock to drain the lower methylene chloride layer into a clean, dry Erlenmeyer flask. Don't dispose of the aqueous layer until the experiment is finished. Use a spatula to add anhydrous sodium sulfate to the methylene chloride solution and swirl the flask. Continue adding small amounts of the drying agent until no further clumping is observed and some of the drying agent is free flowing. Stopper the flask with a cork and let it stand for 5 minutes. Gravity filter the dried solution into a tared clean, dry 25-mL Erlenmeyer flask that contains a boiling chip. Rinse the drying agent with a fresh portion of methylene chloride (about 1–2 mL) and add the rinse to the tared Erlenmeyer flask. Place the flask in a warm sand bath. Evaporate the solvent under the hood. Alternately, reduce the volume of the methylene chloride using a simple distillation apparatus. When the volume is reduced to 3–4 mL, transfer the solution to a tared Erlenmeyer flask that contains a boiling chip and evaporate the rest of the methylene chloride under the hood using a warm sand bath. Reweigh the flask.

Calculate the mass of caffeine extracted into methylene chloride. Calculate the mass of caffeine remaining in the water. (The mass of caffeine in the water will be the difference between the starting mass of caffeine and the mass of caffeine in methylene chloride.) Calculate the distribution coefficient of caffeine in the methylene chloride/water system.

Exercise I.3: Using Distribution Coefficients to Identify an Unknown Solid

Microscale

Safety First!

Always wear eye protection in the laboratory.

Do not breathe methylene chloride vapors. Work in the hood.

This exercise involves the extraction of an unknown solid in two different solvent systems: methylene chloride/water and ethyl acetate/water. The solid, which is either caffeine or resorcinol, will be identified by measuring the amount of the solid that dissolves in the organic solvent, calculating the amount of solid that remains in the aqueous (water) layer, calculating the distribution coefficient for each solvent system, and then comparing the calculated distribution coefficients with known values (shown in Table 1I-3). Caffeine is found in many soft drinks and beverages. Resorcinol is used in cosmetics and as a topical anesthetic.

Extraction Using Methylene Chloride/Water: Weigh out about 100 mg of the assigned solid; record the exact mass and record the unknown number. Transfer the solid to a 5-mL conical vial. Using a calibrated pipet, add 2.0 mL of methylene chloride and 2.0 mL of water to the vial. Cap the vial and shake gently to mix the layers. Cautiously unscrew the cap to vent and release the pressure. Tighten the cap and shake vigorously until all of the solid dissolves. Allow the layers to separate. Use a filter pipet (Technique F) to transfer the methylene chloride (lower layer) to a clean 25-mL Erlenmeyer flask. Be sure to transfer all of the methylene chloride, but no water. Add anhydrous sodium sulfate to the methylene chloride solution until no further clumping is observed. Let the solution stand for 5 minutes. With a clean, dry filter pipet, transfer the dried solution into a tared clean, dry 10-mL Erlenmeyer flask that contains a boiling chip. (If possible, dry this flask in the oven for 10 minutes; let cool to room temperature before weighing the flask.) Rinse the drying agent remaining with a fresh 1-mL portion of methylene chloride; combine this

Table 1I-3 Distribution Coefficient Values for Caffeine and Resorcinol in Two Different Solvent Systems.

Compound	K_D methylene chloride/water	K_D ethyl acetate/water
Caffeine	1.80	9.0
Resorcinol	0.07	6.6

rinse with the methylene chloride solution. Evaporate the solvent under the hood by placing the Erlenmeyer flask on a hot plate turned on low. Remove the flask from the hot plate and let cool to room temperature before weighing.

Extraction in Ethyl Acetate/Water: This extraction procedure varies slightly, because ethyl acetate is less dense than water. Weigh out about 100 mg of the assigned solid; record the exact mass and record the unknown number. Transfer the solid to a 5-mL conical vial. Using a calibrated pipet, add 2.0 mL of ethyl acetate and 2.0 mL of water to the vial. Cap the vial and shake it gently to mix the layers. Cautiously unscrew the cap to vent and release the pressure. Tighten the cap and shake vigorously until all of the solid dissolves. Allow the layers to separate. Ethyl acetate is less dense than water and will be the upper layer. However, in a microscale extraction, it is always the lower layer that is removed from the vial. Use a pipet (Technique F) to transfer the aqueous lower layer to a vial or flask. Then use a clean filter pipet to transfer the ethyl acetate to a clean 25-mL Erlenmeyer flask. Add anhydrous sodium sulfate to the ethyl acetate solution until no further clumping is observed. Let the solution stand for 5 minutes. With a clean, dry filter pipet, transfer the dried solution into a tared clean, dry 10-mL Erlenmeyer flask that contains a boiling chip. (If possible, dry this flask in the oven for 10 minutes; let cool to room temperature before weighing the flask.) Rinse the drying agent remaining with a fresh 1-mL portion of ethyl acetate; combine this rinse with the ethyl acetate solution. Evaporate the solvent under the hood by placing the Erlenmeyer flask on a hot plate turned on low. Remove the flask from the hot plate and let cool to room temperature before weighing.

Analysis: Determine the mass of the solid extracted into methylene chloride. Calculate the mass of solid remaining in the water. (The mass of solid in the water will be the difference between the starting mass of the solid and the mass in the solid in methylene chloride.) Calculate the distribution coefficient of the solid in the methylene chloride/water system. Repeat the analysis for the ethyl acetate system. Compare the calculated values with the actual values shown in Table 1I-3 and identify the unknown.

Questions

1. Approximately 1.0 g of caffeine will dissolve in 28 mL of methylene chloride and in 46 mL of water. Calculate the distribution coefficient of caffeine in this solvent system.

2. The K_D methylene chloride/water of an organic compound is 2.50. A solution is made by dissolving 48 mg of the compound in 10 mL of H_2O. How many milligrams of the compound will be extracted using:
   a. one portion of 10 mL of methylene chloride?
   b. two portions of 5 mL of methylene chloride?

3. What are four characteristics of a good extraction solvent?

4. *Library project*: Search for the method now commonly used to extract caffeine from coffee to manufacture decaffeinated coffee.

5. If too much drying agent is added to a solution, what can be done to correct the situation?

6. It has been suggested that drying agents can be collected after an experiment and the hydrated salt heated in an oven to drive off the water. The recycled drying agent can then be used again for another experiment. Is this a good idea? Give advantages and disadvantages of this proposal.

7. Cite advantages and disadvantages of using the following drying agents: sodium sulfate, magnesium sulfate, sodium carbonate, and potassium carbonate.

Note: Do not breathe ethyl acetate vapors. Work in the hood.

8. Drierite ($CaSO_4$) and calcium chloride are used in drying tubes to keep moisture out of reaction vessels. What properties are desirable for these agents?

9. The density of an aqueous solution increases as solutes such as salts are added. What bearing might this have on predicting whether the aqueous solution will be on the top or on the bottom for an extraction?

10. Look up the structures of resorcinol and caffeine. Which compound is expected to have the larger dipole? Explain.

11. Would it be possible to separate a mixture of caffeine and resorcinol using extraction? If so, suggest a method for doing so.

Technique J: Gas-Liquid Chromatography

Introduction

Chromatography is the separation of components of a mixture by differential adsorption between a stationary phase and a mobile phase. The word "chromatography," coined by Tswett around 1900, means to "write" in "color." Early chromatographic methods utilized paper as the stationary phase. The use of chromatography by analytical chemists began in the early 1950s and organic chemists followed soon afterward. Today, chromatography consists of a diverse collection of separation methods. All chromatographic techniques are based on a similar principle—components of a mixture can often be differentiated by exposure to two competing phases, a stationary phase and a mobile phase. Some components have little attraction for the stationary phase and are carried rapidly through the system with the mobile phase; other components adhere strongly to the stationary phase and have little attraction for the mobile phase.

The chromatographic methods used most frequently by organic chemists are gas-liquid chromatography, column chromatography, high-performance liquid chromatography, and thin-layer chromatography. Gas-liquid chromatography is used for separations of volatile or reasonably volatile organic liquids and solids. In gas-liquid chromatography, the components are partitioned between a liquid coating on the column (the stationary phase) and an inert gas (the mobile phase). In column and thin-layer chromatography, the components are partitioned between a solid stationary phase (like silica gel or alumina) and a liquid mobile phase (the solvent).

Gas-Liquid Chromatography Basics

Gas-liquid chromatography (GC) was invented in the early 1940s, but was not widely used until the late 1950s. It is now used universally in organic and analytical laboratories because most separations are rated at hundreds or thousands of theoretical plates–much superior to fractional distillation. In addition, the equipment is relatively inexpensive ($3,000 to $30,000), analysis times are short, and very small amounts of sample are required. GC can be used for both qualitative separations and quantitative analyses. These characteristics make GC an ideal tool for the microscale and miniscale organic laboratories. The stationary phase for GC is usually an organic polymer coated on the inside of a tube, such as a long capillary tube, and the mobile phase is an inert gas, such as helium.

A small volume (1–10 μL) of a mixture of volatile substances (usually dissolved in a solvent) is injected by syringe onto a heated column through which an inert carrier gas is flowing. The heat applied, as well as the gas flow, help the molecules from the sample travel through the column. Smaller, more volatile molecules generally emerge first from the opposite end of the column through an exit port and are detected. The detector is connected to a recording device, which shows a deflection when a sample passes the detector in proportion to the amount of sample detected. Compounds are eluted through an exit port either in an intact form or as combustion products, depending upon the type of detector used. A schematic diagram of a gas chromatograph is shown in Figure 1J-1.

Columns

Columns are of two main types: metal columns of $1/4$-inch, $1/8$-inch, or $1/16$-inch diameter that contain a polymer-coated, powdered solid support and glass columns that have very small diameters (capillary columns) that contain a fine polymer coating on the inside of the column. Column lengths range from 3 to 30 feet for metal columns and much longer (100 feet) for capillary columns.

The polymers used in GC columns are called stationary phase materials. The polymer coating aids in separating components by partially absorbing the compounds, which helps to selectively slow them down. The polymeric coatings may be polar, semipolar, or nonpolar. Examples are carbowax [$HO(CH_2-CH_2-O)_nH$], which is a polar stationary phase, and silicone polymers, such as SE 30 and Dow-Corning 200 (shown here), which are nonpolar stationary phases.

$$\left(\!\!\begin{array}{c}\overset{\displaystyle CH_3}{\underset{\displaystyle CH_3}{\overset{|}{\underset{|}{Si}}}}-O-\overset{\displaystyle CH_3}{\underset{\displaystyle CH_3}{\overset{|}{\underset{|}{Si}}}}-O\end{array}\!\!\right)_{\!\!n}$$

Detectors

As molecules emerge from the column, they are detected and recorded. A computer may be used to compile data furnished by the detector. Thermal conductivity detectors (TCD), flame ionization detectors (FID), and mass selective detectors (MSD) are common types of detectors.

The thermal conductivity detector, the most common type of detector, senses a difference in thermal conductivity of gases eluting from a GC column. Helium has a very high thermal conductivity and organic materials appearing in the effluent cause a decrease in the thermal conductivity. The lowered thermal conductivity slightly raises the temperature of the detector filament. This causes the electrical resistance of the filament to increase.

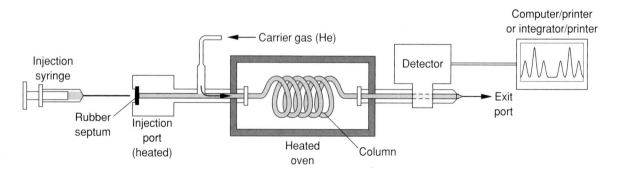

Figure 1J-1   Schematic drawing of a gas chromatograph

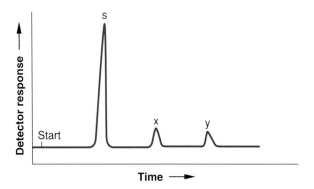

This change in electrical resistance is recorded as a peak on a chromatogram. The change in thermal conductivity is linear according to the number of molecules in the effluent.

Flame ionization detectors consist of a flame fueled by hydrogen gas. As organic molecules are eluted from the column, some are ionized by an ionization source. The intensity of ions is measured as current by an electrometer. The electrometer passes a signal to the recorder causing a peak in the chromaogram. FID detectors are very sensitive and can detect as little as 1 nanomol of an organic substance.

Mass selective detectors used in gas chromatography–mass spectrometry (GC/MS) are also extremely sensitive.

Chromatogram

A chromatogram is a plot of detector response versus time. A chromatogram for separation of two components, x and y, dissolved in solvent s, is shown in Figure 1J-2. The solvent is chosen so that it is usually eluted first. Common solvents are diethyl ether and acetone. Substance x is eluted before substance y.

Retention Time

The retention time (t_R) of a compound is the time elapsed from injection ($t = 0$) to the appearance of the peak. Retention times may be measured in seconds or minutes or cm, where the recorder chart speed is constant. Retention times are useful in identifying components of a mixture if samples of the pure compounds are available for comparison. The standards must be run under the same conditions as the mixture for direct comparison.

Retention times can be changed by switching columns, raising or lowering the column temperature, and increasing or decreasing the flow of carrier gas. It is common practice in the organic lab to note carefully all conditions used in operating the GC instrument.

Efficiency of Separation

Measuring the peak width (w) at the baseline of the peak allows evaluation of the efficiency of the process. The number of theoretical plates (N) and the height equivalent of a theoretical plate (HETP) may be calculated by measuring the height of the column (in cm) and using the equations given here. The smaller the value of y (narrow peak), the higher the efficiency. For high efficiency, N should be large and HETP should be small. Capillary columns are often rated at 1000 plates or more. Metal columns of small diameter ($1/8$ inch, 300 plates) are rated higher than those of larger diameter ($1/4$ inch, 100 plates).

$$N = 16 \, [t_R/w]^2$$

$$HETP = \text{length of column}/N$$

Calculation of the Number of Theoretical Plates

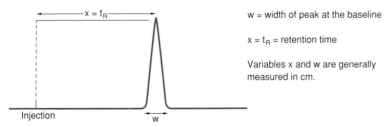

w = width of peak at the baseline

$x = t_R$ = retention time

Variables x and w are generally measured in cm.

GC Separations

There are several factors that affect separation. These include carrier gas flow rate, column temperature, injector and exit port temperatures, detector temperature, amount of sample injected, length of column, diameter of column, and type of compound (polymer) used to coat the inside of the column. Of these factors, those associated with the column are critical.

The inner areas of all columns are coated with a polymeric stationary phase, which can be polar, semipolar, or nonpolar. A polar stationary phase has a high affinity for polar components and retains them on the column more than nonpolar components. Conversely, a nonpolar stationary phase retains nonpolar components more effectively than polar components. Changing the stationary phase will influence rates of elution of components of different polarities, resulting in differences in peak positions on chromatograms.

An example of an idealized separation of cyclohexane (bp 81°C) and ethanol (bp 78°C) illustrates the effect of the nature of the stationary phase. With a polar stationary phase, ethanol is retained and cyclohexane, which is not retained, is eluted first. A nonpolar stationary phase retains the cyclohexane so that ethanol is eluted first.

Longer columns, such as capillary columns, retain components longer and offer better separation. Slower flow rates of the carrier gas (the mobile phase) decrease elution rates of components, which can improve separations. Low temperatures also slow elution rates and can improve separation. Smaller amounts of sample give better separations, particularly if components have very similar retention times.

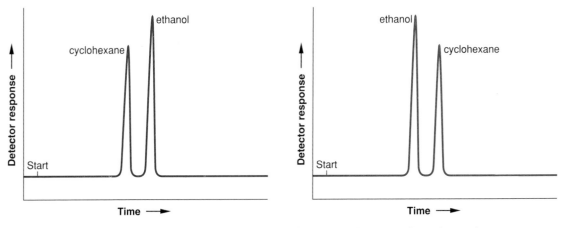

Separation using a polar stationary phase   Separation using a nonpolar stationary phase

When separating a mixture of two substances, it is desirable to achieve baseline separation. Baseline separation means having two totally separate peaks that show no overlap. Many times it is not possible to achieve baseline separation, as in the case of isomers or other compounds that have very similar boiling points. In such a case it is customary to express the degree of separation as resolution (R). Resolution can be calculated as shown, using the retention times of two components and the estimated peak widths at the baseline. It is desirable to have R values of 1.5 or greater for quantitative work. When R is less than 1, the calculated amounts of each component are approximate.

Calculation of Resolution, R

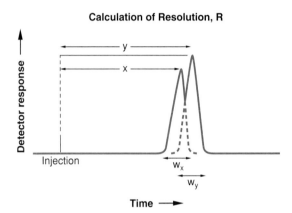

w = width of peak at the baseline

y = retention time of peak y

x = retention time of peak x

$\Delta t_R = y - x$

$R = 2\Delta t_R/(w_x + w_y)$

If R = 1, there is 98% separation.

Quantitative Analysis

The relative amount of each component may be calculated from areas using several different methods. Electronic integration is now common in most laboratories. With electronic integration, area percent values for each component of a mixture are output directly.

Mechanical methods can also be used. One of these methods is triangulation. The area of each peak is determined by measuring the peak height, and the peak width at half height, and then multiplying these two measurements to determine the area of each peak.

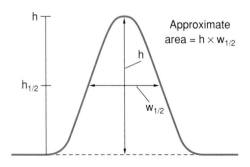

Approximate area = $h \times w_{1/2}$

An earlier mechanical method was to employ a planimeter to trace each peak, noting the reading on the planimeter after measuring the perimeter of each peak. Still another method was to photocopy the chromatogram using high-quality paper. The peaks are each cut out and weighed. The area percent of each component can be found by dividing the weight of each component by the total of all weights and multiplying by 100.

Measurement of areas of overlapping peaks is best done by approximation. (If possible, it is best to avoid overlapping peaks by changing the temperature, the flow rate of carrier gas, or the column in order to increase resolution.) The example on page 87 shows how to use approximation. The inner line of the first peak is extended to the projection of the baseline and the same is done for the inner line of the second peak. Areas are then estimated using triangulation. Electronic integrators perform the approximation automatically.

Calculations Based on Weight Percent

Determining the areas beneath all of the peaks of a chromatogram enables the analyst to assign percentages to each of the components. This is called area normalization. The weight percent of component A is shown here.

$$\%A = \frac{\text{area of A}}{\text{total area}} \times 100$$

If a mixture of three components, A, B, and C, is analyzed, the weight percent of each component may be determined easily from the relative peak areas if the peaks are not overlapping and if the compounds have similar structures. Components that have different functionalities give different detector responses, making it necessary to apply correction factors prior to determining the weight percent of each component.

Relative detector response factors for different compounds can be determined by comparing peak areas of a mixture prepared by measuring out equal weights of each compound in the mixture. Dividing the area percent by the weight percent of each component gives a measure of relative sensitivity to the detector. A component giving a smaller response value means that the detector does not respond as much to it as a component with a higher value.

Sample Problem

A mixture containing acetone and pentane was analyzed by GC using an FID detector. The response factors for acetone and pentane, according to McNair and Bonelli (1960), are 0.49 and 1.04, respectively. The following data were obtained:

compound	area of peak
acetone	4.30 cm^2
pentane	2.20 cm^2

The equation used to solve for weight percent of each component in a mixture is given below:

$$\% \text{ component} = \frac{\text{corrected area of component}}{\sum \text{corrected areas of all components}} \times 100,$$

where the corrected areas are calculated by dividing the measured areas by the detector response

$$\text{corrected area of component} = \frac{\text{measured area of component}}{\text{detector response for that component}}$$

For the acetone-pentane mixture, the corrected area of acetone is 4.30 cm^2/ 0.49 = 8.8 cm^2 and the corrected area of pentane is 2.20 cm^2/1.04 = 2.1 cm^2. The sum of the corrected areas is 8.8 cm^2 + 2.1 cm^2 = 10.9 cm^2.

Therefore, the weight percent of acetone in the mixture is 8.8 cm^2/10.9 cm^2 × 100 = 80.7% and the weight percent of pentane is 19.3%. Without the correction factors the predicted weight percent values would be 66% acetone and 34% pentane, which is obviously very different from the values corrected for differential response of the detector. On the other hand, analysis of a mixture of heptane and pentane would give similar weight percent values with or without the detector response values factored in. Thermal conductivity detectors also have different responses for different substances.

How to Use the Gas Chromatograph

Column, carrier gas flow rate, column temperature, chart speed, and other parameters will usually be already chosen. Be sure to note these conditions. The detector and injection port temperatures are kept at a higher temperature than the column.

Liquids can be run neat or in a solution, while volatile solids are dissolved in a low-boiling solvent. Rinse the syringe several times with the sample or sample solution. Then fill the syringe. Make certain there are no air bubbles in the syringe. Then hold the syringe straight up and expel excess liquid, bringing the volume to the appropriate level (usually 1–5 μL). Insert the needle of the syringe into the center of the rubber septum by holding close to the end of the needle. Push needle all the way into the septum until resistance is felt. Then gently but quickly depress the plunger of the syringe to inject the sample. (Be careful not to bend the needle since syringes are expensive. The instructor will demonstrate this technique.) Start the chart paper as soon as the sample is injected. Remove the syringe by pulling straight out. Rinse the syringe several times with acetone and let dry. When the components have eluted from the column, stop the chart recorder.

If the peaks are too small or too large, adjust the attenuation or change the amount of sample injected. An integrator attached to a recorder will furnish percentages of components for each analysis. If no integrator is available, individual peaks may be cut out from the chart pages and weighed. Alternatively, peak width at half height times the height may be calculated for each peak and relative percentage of each component determined (p. 87). Percentages for closely overlapping peaks are difficult to determine by these methods.

To identify components of a mixture, standards should be injected under the same set of conditions. Alternately, it is possible to "spike" a mixture by adding some of the pure component to the mixture and rerunning the chromatogram. The peak that grows in size relative to the others corresponds to the standard injected.

Exercise J.1: Determining Relative Detector Response Factors in GC

Safety First!

Always wear eye protection in the laboratory.

Obtain a sample of a mixture containing equal masses of 1-butanol (bp 117°C), 2-butanone (bp 82°C), and methyl acetate (bp 58°C). These compounds were chosen because they have different functional groups, but similar molecular weights. Carefully inject 1–2 μL of the mixture and record the chromatogram. Note and record the column temperature, flow rate of the carrier gas, column polarity, and other column properties. If necessary, adjust the attenuation so that all three peaks fit on the paper. Measure the area of each peak. In order to identify the compound corresponding to each peak, inject 1–2 μL aliquots of pure 1-butanol, 2-butanone, and methyl acetate and compare retention times. Determine the relative mass percent of each component. Determine the relative detector response factor of each compound by dividing the experimental mass percent (from normalized peak areas) by the actual mass percent of each component in the mixture (33.3%).

Exercise J.2: Determining Mass Percent of a Mixture Using GC

Obtain a sample of a mixture containing 1-butanol (bp 117°C), 2-butanone (bp 82°C), and methyl acetate (58°C). Record the unknown number of the sample. Carefully inject 1–2 μL of the mixture and record the chromatogram. If necessary, adjust the attenuation so that all three peaks fit on the paper. Record the column temperature, flow rate of the carrier gas, column polarity, and other conditions as noted by the instructor. Measure the area of each peak. If necessary, inject 1–2 μL aliquots of the pure compounds in order to identify the peak associated with each compound. Using the detector response factors for each component determined in Exercise J.1, calculate the corrected area for each component by dividing the measured area of the component by the detector response factor for that component. (If Exercise J.1 was not done, the instructor will provide the detector response factors.) Determine the mass percent of the mixture by dividing the corrected area for each component by the sum of the corrected areas for all components. If more than one GC instrument is available for use, repeat the analysis of the mixture using (1) a higher column temperature; (2) a faster flow rate for the carrier gas; or (3) a different column.

Exercise J.3: Determining Mass Percent of a Mixture of Alcohols

Safety First!

Always wear eye protection in the laboratory.

Obtain a mixture of three alcohols: 2-propanol (bp 82°C), 1-propanol (bp 97°C), and 1-butanol (bp 117°C). Carefully inject 1–2 μL of the mixture and record the chromatogram. If necessary, adjust the attenuation so that all three peaks fit on the paper. Record the column temperature, flow rate of the carrier gas, column polarity, and other conditions as noted by the instructor. Measure the area of each peak. In order to identify the alcohol corresponding to each peak, inject 1–2 μL aliquots of the pure alcohols and compare retention times. Calculate the mass percent of each component in the mixture of alcohols. (Assume the detector response is equal for all of the components.) Explain the trend in order of elution for the column used.

If more than one GC instrument is available for use, repeat the analysis of the mixture using (1) a higher column temperature; (2) a faster flow rate for the carrier gas; or (3) a different column.

Questions

1. Explain the principle of operation of a thermal conductivity detector (TCD).
2. Explain the relationship between HETP and column length.
3. When analyzing the chromatogram obtained from a binary (two-component) mixture, it is necessary to know which peak corresponds to component A and which to component B. Explain how this can be done using the GC instrument.
4. What type of detector should be used if components are to be collected while they are being analyzed by GC? Explain.

5. What is the mobile phase used in gas chromatography? What is the stationary phase? Give an example of a stationary phase.

6. How is resolution affected by lowering the column temperature? By lengthening the column?

7. Explain how the amount of sample injected onto a column affects the separation of a two-component mixture.

8. Calculate resolution (R) and the relative amounts of each component for the chromatogram shown below.

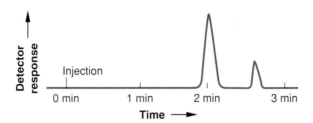

9. A chromatogram of a mixture of components A, B, and C is given below. Suggest methods for improving resolution.

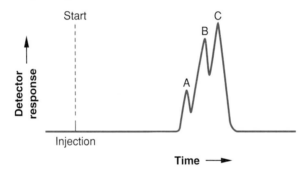

10. For each of the pairs of compounds below, indicate whether the relative detector response factors would be expected to be similar or different. Explain your reasoning.

    a. decane (MW 142) and 1-nonanol (MW 144)

    b. 2-propanol (bp 82°C) and 1-butanol (bp 117°C)

    c. 1-butanol (MW 74, bp 117°C) and diethyl ether (MW 74, bp 35°C)

11. What effect would each of the following have on the retention times?

    a. increase the column temperature

    b. increase the flow rate of the carrier gas

    c. use a more polar column

12. Explain why the alcohols in Exercise J3 are assumed to have equivalent response to the detector.

Reference

McNair, H. M., and Bonelli, E. J. *Basic Gas Chromatography. 5th ed.* Walnut Creek, CA: Varian Aerograph, 1960.

Technique K: Thin-Layer, Column, and High-Performance Liquid Chromatography

Introduction

Chromatographic techniques are used extensively in the organic laboratory for both qualitative separations and quantitative analysis. A thorough understanding of the various types of chromatography is essential. Thin-layer chromatography (TLC) is used to determine the purity of a compound, to evaluate how far a reaction has proceeded, and to analyze the composition of a mixture, while column chromatography is used to physically separate components of a mixture. High-pressure liquid chromatography (HPLC) can be used for both qualitative and quantitative analysis.

All chromatographic techniques are based on a similar principle: components of a mixture can often be differentiated by exposure to two competing phases. In TLC, the stationary phase is a polar adsorbent, such as silica gel or alumina, which has been coated onto a glass or plastic plate. The mobile phase is an organic solvent or mixture of solvents. The liquid moves up the plate by capillary action. Column chromatography is similar to TLC, with a stationary phase of silica gel or alumina and a mobile liquid phase, but differs in that the liquid travels down the column. A polymer-coated adsorbent is the stationary phase in HPLC. The mobile liquid phase in HPLC is forced through a small diameter column by a pump at high pressure.

Introduction to Thin-Layer Chromatography (TLC)

TLC is the separation of moderately volatile or nonvolatile substances based upon differential adsorption on an inert solid (the stationary phase) immersed in an organic solvent or solvent mixture (the mobile phase). The components are distributed between the stationary phase (usually silica gel or alumina) and the solvent depending upon the polarities of the compound and solvent. The compounds are carried up the plate (ascending chromatography) at a rate dependent upon the nature of the compounds and the solvent.

Compounds are separated by adsorption chromatography based upon differential attachment of molecules to the adsorbent and the polarity of the solvent used for the separation as shown in Figure 1K-1. Polar compounds are strongly attracted to and held by a polar adsorbent. Nonpolar compounds are held weakly. When a nonpolar solvent is passed

Figure 1K-1

Adsorption equilibrium for molecules between adsorbent and solvent

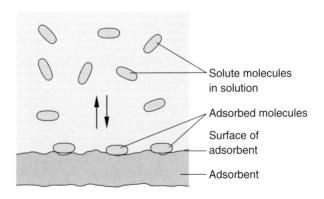

through the adsorbent, nonpolar compounds are released easily, but polar compounds are retained. When a moderately polar solvent is passed through the adsorbent, both nonpolar and polar compounds are released, but the nonpolar compounds move faster because there is still an attraction between the polar compounds and the polar adsorbent. In general, nonpolar compounds move faster than polar compounds for TLC on silica gel or alumina.

Polarity of Solvents Used for Adsorption Chromatography

Solvents are rated according to their polarities. Common solvents are listed below from most polar to least polar.

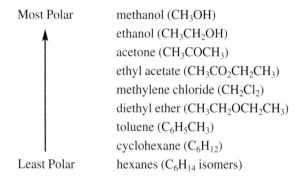

Most Polar → Least Polar:

methanol (CH_3OH)
ethanol (CH_3CH_2OH)
acetone (CH_3COCH_3)
ethyl acetate ($CH_3CO_2CH_2CH_3$)
methylene chloride (CH_2Cl_2)
diethyl ether ($CH_3CH_2OCH_2CH_3$)
toluene ($C_6H_5CH_3$)
cyclohexane (C_6H_{12})
hexanes (C_6H_{14} isomers)

If a polar compound moves too slowly in a nonpolar solvent, switching to a more polar solvent will cause the compound to move faster. If a nonpolar compound moves too fast on TLC, switching to a less polar solvent will cause the compound to move slower. The polarity of the solvent system can also be varied by mixing miscible solvents to give the desired separation.

Calculating R_f Values

The distance traveled by each component is expressed as a rate or retardation factor (R_f). R_f values are calculated by dividing the distance traveled by a component by the distance between the origin and the solvent front (distance traveled by the solvent). Hence all R_f values will fall between 0 and 1. R_f values are measured from the origin where the initial spot was applied to the center of the spot.

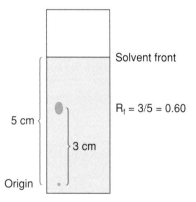

Solvent front

$R_f = 3/5 = 0.60$

5 cm

3 cm

Origin

In general, R_f values on silica gel or alumina are decreased by decreasing the polarity of the solvent and increased by increasing the solvent polarity. The main objective in TLC analysis is to obtain separation of components of a mixture. A solvent is selected to give a range of R_f values. If a single solvent fails to give adequate separation, a solvent mixture should be used. Examples of TLC that show the effect of solvent polarity are given below:

Example 1 The compounds 9-fluorenone, *m*-dinitrobenzene, and azobenzene are moderately polar. The structures are shown below:

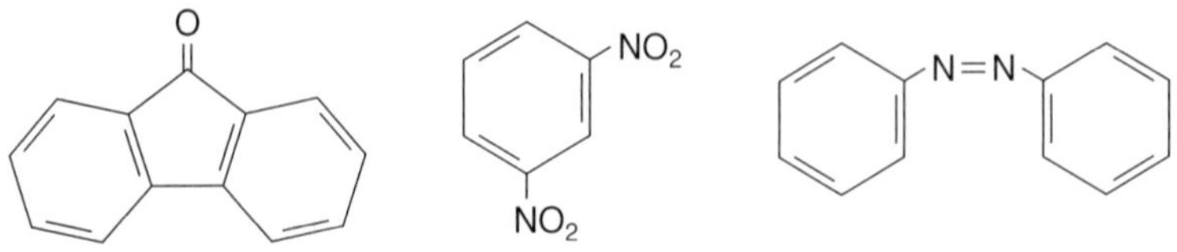

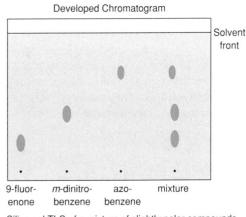

9-fluorenone   *m*-dinitrobenzene   azobenzene

In order to separate a mixture of these compounds effectively, toluene, which is a relatively nonpolar solvent, was chosen. Solutions of each of the three pure compounds were spotted on the silica gel plate as well as a solution containing a mixture of the three compounds.

Successful separation was achieved using toluene. The separation obtained is very satisfactory since all R_f values are considerably different from one another, as shown in Table 1K-1. The silica gel TLC of 9-fluorenone, *m*-dinitrobenzene, azobenzene, and a mixture of the three components eluted with toluene is shown here.

Example 2 For the TLC separation of benzaldoxime (B) and *p*-tolualdoxime (T), hexane failed to move the compounds significantly, but ethyl acetate moved both compounds too fast, giving little separation. In this case, a mixed solvent system containing both hexane and ethyl acetate gave better results, moving both compounds about midway up the plate, while also achieving different R_f values for B and T. The exact per-

Developed Chromatogram

Solvent front

9-fluor-   *m*-dinitro-   azo-   mixture
enone   benzene   benzene

Silica gel TLC of a mixture of slightly polar compounds
and each of the pure components using toluene as solvent

Table 1K-1 R_f Values for Example 1

Compound	TLC in toluene	
	Distance from origin (cm)	R_f value
9-fluorenone	1.6	0.25
m-dinitrobenzene	2.9	0.47
azobenzene	4.3	0.70
solvent front	6.2	1.00

centages of a mixed solvent system are best determined by trial and error. In this case, the optimal mixture was 70% hexane/30% ethyl acetate.

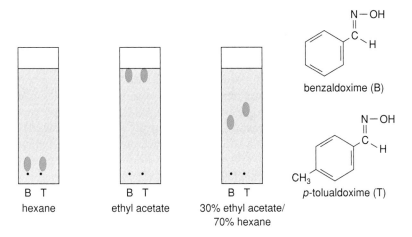

benzaldoxime (B)

p-tolualdoxime (T)

B   T
hexane

B   T
ethyl acetate

B   T
30% ethyl acetate/
70% hexane

Resolution

In TLC, it is highly desirable to avoid overlapping of spots in the developed chromatogram, just as it was desirable to avoid overlapping peaks in GC. In GC, resolution can be improved by lowering the column temperature, decreasing the carrier gas flow rate, and using a different column or longer column. In TLC, resolution can be increased by varying the nature of the eluting solvent or by changing the adsorbent.

   Resolution can be determined from the developed chromatogram by measuring the distance from the origin to the center of each spot (t_R) and measuring the height of each spot (h):

$$\text{Resolution (R)} = 2\left[\frac{\Delta t_R}{h_A + h_B}\right]$$

This is illustrated for the two-component mixture shown below.

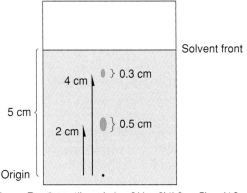

Solvent front

4 cm } 0.3 cm

5 cm

2 cm } 0.5 cm

Origin

Resolution = R = 2 Δt_R/(h_A + h_B) = 2(4 − 2)/(.3 + .5) = 4/.8 = 5

Good separation of components by TLC can be determined by looking at the spots on the developed plate. Calculation of R is not necessary unless results of several different eluting solvents are being compared. The best solvents have the highest R values.

Materials and Methods in TLC

Adsorbent

The most frequently used adsorbents are silica gel (SiO_2) and alumina (Al_2O_3), which are coated onto a plastic or glass support. TLC plates may be purchased from commercial vendors or they may be homemade. Commercial plates have adsorbents bound to a plastic, glass, or aluminum sheet as a backbone. The binder is calcium sulfate. The adsorbent is very uniform, about 0.1–0.2 mm thick. Homemade plates are usually made by dipping glass microscope slides into mixtures of adsorbent and binder in aqueous solution. Homemade plates are not usually as uniform in thickness as commercial plates. However, since TLC is generally used qualitatively rather than quantitatively, it is not so critical that the thickness of the layer be that uniform. Coated microscope slides are wiped clean on one side and the coated side is left to dry. These plates may be used directly for TLC. Commercial plates may be sized appropriately using scissors or a glass cutter.

Sample Application

Dots (in pencil) are marked uniformly along one of the narrow edges of the plate near the bottom at intervals of about 1–2 cm. Solutions of samples are spotted individually at each of the marks and the solvent is allowed to dry. The diameter of the residue from a spotted solution should be kept as small as possible to minimize diffusion effects. Capillary tubes or Pasteur pipets that have been drawn out in a flame are used as applicators. The very small openings in these drawn-out tubes permit application of very small amounts of solution.

Development

The development chamber can be a beaker covered by a watch glass or a jar with a screw-top cap. The chamber should be equilibrated before use by placing the developing solvent and a piece of filter paper in the chamber and capping the chamber. The chamber should stand undisturbed for 5–10 minutes to saturate the chamber with solvent vapors. The plate is put into the developing chamber after making certain that the level of the solvent is below that of the applied compounds on the plate. As the chromatogram is developed, the solvent advances up the plate through capillary action. Unless colored compounds are being separated (such as dyes in inks), the components cannot be visualized. The plate is left in the developing chamber until the level of the solvent is about 2 cm from the top of the plate. Then the plate is removed from the chamber, the solvent line is marked immediately with pencil, and the plate is allowed to dry.

Don't move the solvent chamber during chromatography.

Visualization

Because most organic compounds are not colored, the spots must be visualized using a UV light, an iodine chamber, or an indicator spray. Certain substances, such as aromatic compounds, absorb ultraviolet light and appear as purple spots. Commercial plates often contain a fluorescent dye that gives a light green background when exposed to UV radiation. Compounds that do not absorb UV light must be viewed in another way, such as placing the plate in an enclosed chamber containing solid iodine crystals. Most organic substances form colored complexes with iodine. After a few minutes, organic compounds on the developed plate begin to appear as brown spots. The effect is reversible over time after removal from the iodine chamber, so spots should be circled. Another option for visualizing spots is to spray the plate with a reagent that will cause development of a color. A commonly used spray is anisaldehyde-H_2SO_4, which gives colored spots for alcohols and certain other compounds.

The overall TLC procedure is summarized in Figure 1K-2. (a) A solution is spotted at the origin on a TLC plate, using a drawn capillary tube. (b) The plate is placed in a development chamber containing solvent. The level of the solvent is below the origin. (c) The solvent front is marked immediately upon removing the developed TLC plate from the development chamber. The dried TLC plate is visualized and spots are circled.

Introduction to Column Chromatography

Column chromatography using silica gel or alumina works like TLC except that solvents are passed down through the adsorbent holding the mixture to be separated (descending chromatography). Column chromatography may be thought of as a preparative version of TLC that permits separation and isolation of products by collection of eluted fractions. One major difference is that the sample is loaded at the top of an adsorbent, such as silica gel, and solvent is drained through a "column" of adsorbent, with components of the sample mixture being eluted from the bottom of the column. The relative speeds that components travel are similar to the relative R_f values in TLC. Thus, nonpolar components travel faster down the column than do polar components and are eluted first. There are several different types of column chromatography.

Gravity Column Chromatography

Gravity column chromatography is frequently used in the organic lab. A sample is applied to the top of a column filled with an adsorbent such as silica gel or alumina. Solvent is added to the top of the column. Components of the sample travel down the column at different rates, depending upon the polarity of the compounds, the polarity of the solvent and the nature of the adsorbent. Fractions are collected as the solvent continues to be added at the top and eluted at the bottom. After the first component is completely eluted, the solvent polarity can be increased so that a second component may be eluted faster.

Dry-Column Chromatography

Another type of column chromatography is dry-column chromatography. In dry-column chromatography, a solvent is chosen so as to spread out the components in a series of

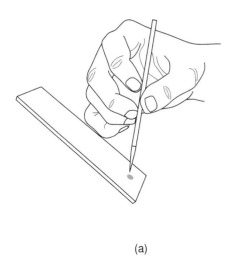

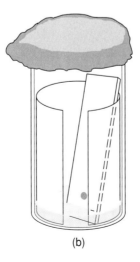

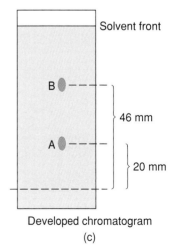

(a)   (b)   (c)

Figure 1K-2   Procedure for TLC

bands. The column is packed and the sample is added to the top of the column. The chosen solvent is added only until the first drops of solvent reach the bottom of the column. No more solvent is then added. The column is then slit from stem to stern. A disposable plastic or nylon column works best here. Nylon is nice because it doesn't absorb UV light. The individual regions of adsorbent that absorb UV light are collected and the components are eluted with an appropriate solvent and filtered. The solvent is evaporated and the individual components are then isolated and purified.

Other Types of Column Chromatography

Ion exchange chromatography and size-exclusion gel permeation chromatography are commonly used in analytical and biochemistry laboratories. They are less frequently used in organic laboratories.

How to Do a Miniscale Gravity Column Chromatography

Preparing the Column

The chromatography column is usually a long glass tube, with a stopcock at the bottom that can be opened or closed.

Make sure that the column is clean and dry. Place a wad of cotton or glass wool in the bottom of the column, then add a layer of sand. The cotton plug and sand prevent loss of adsorbent through the stopcock and also keep the stopcock clear of particles. Next measure out the adsorbent into a beaker. The amount of adsorbent you need depends upon the difficulty of separation. In general, you will need about 10 g of adsorbent per 100 mg of sample to be separated. However, if the separation is an easy one (one polar component mixed with one nonpolar component), you will need less adsorbent. Pour the adsorbent into the column through a funnel. This should be done in the hood to avoid breathing in the fine particles of adsorbent. Tamp the side of the column gently to produce even packing of the adsorbent in the column. Now carefully pour the solvent into the column. Silica gel and alumina swell and give off heat as they take up the solvent. To eliminate any air pockets that may form as the solvent travels through the column, gently tamp on the column. Drain the solvent until the level of the solvent is to the top of the adsorbent. An example of a column prepared in this way is shown in Figure 1K-3.

Applying the Sample to the Column

If the sample is a liquid, add it directly to the top of the adsorbent. If the sample is a solid, dissolve it in a minimum amount of the eluting solvent and apply it carefully and evenly to the adsorbent using a pipet and bulb. Add a second layer of sand to prevent the adsorbent and sample from being disturbed.

Elution of the Sample

Choose an eluting solvent that is relatively nonpolar, but polar enough to move the least polar component of a mixture. Carefully add a pool of solvent and open the stopcock. Drain the solvent into collecting flasks or vials, called fractions. The number of fractions required will vary depending upon the separation. Add fresh solvent as necessary so that the level of solvent is always above the level of the sand. After the first component has been eluted, increase the polarity of the solvent to elute subsequent components.

If the components are colored, you can follow the progress of the separation visually. With colorless compounds, you must use TLC to determine which fractions con-

Miniscale

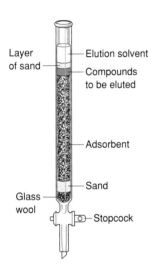

Layer of sand — Elution solvent

Compounds to be eluted

Adsorbent

Sand

Glass wool

Stopcock

Figure 1K-3

Chromatography column

tain the desired components. Pool the appropriate fractions and then remove the solvent by evaporation or distillation, leaving the component behind as a residue. The process of elution and collection is illustrated in Figure 1K-4.

How to Do Microscale Column Chromatography

Microscale

The procedure for microscale column chromatography is similar to that of miniscale. The procedure utilizes a Pasteur pipet, although a straw or plastic tube may be substituted.

Place a small wad of cotton or glass wool inside the tube and push it near the bottom of the tube, using a copper wire. Add a small layer of sand, followed by a layer of the adsorbent. The adsorbent should weigh about 100 times the sample weight (1 g adsorbent per 10 mg of sample). Add solvent until drops come out the bottom and the solvent level is flush with the level of the silica gel.

Immediately apply the sample to the top of the adsorbent. If the sample is a liquid, add it directly to the top of the column. If the sample is a solid or a thick oil, dissolve it in a minimum amount of solvent and add it to the adsorbent, using a pipet.

Add solvent to the top of the tube. The solvent should be polar enough to cause one of the components of the sample mixture to move. Collect fractions in flasks or test tubes. After the first component has eluted, increase solvent polarity to elute each subsequent component. Pool the appropriate fractions and then remove the solvent by evaporation under a hood. A microscale gravity column is illustrated in Figure 1K-5.

Introduction to High-Performance Liquid Chromatography

High-performance liquid chromatography (HPLC) can also be used to separate mixtures. HPLC is most useful as a quantitative analytical method, much the same as GC. The method resembles GC, but instead of a carrier gas, a solvent is used as the mobile phase. The most common detector uses UV detection.

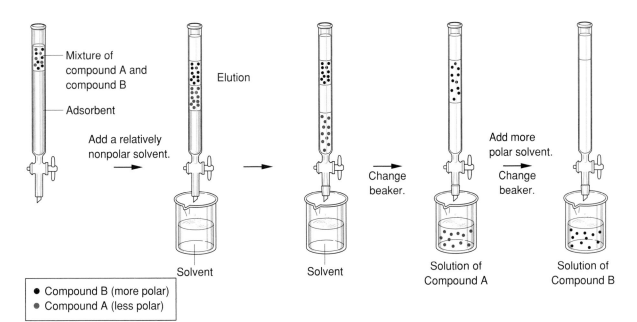

Figure 1K-4   Procedure for chromatographic separation of a two-component mixture

Figure 1K-5

Microscale gravity
column

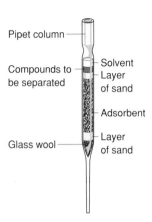

In HPLC, a small volume (1–2 mL) of a sample solution is injected by syringe onto the column, while a solvent mixture (mobile phase) passes through the column that contains a finely divided packing material. Solvent is forced through the column using a high-pressure pumping system. Elution may be accomplished using a simple solvent or solvent mixture (isocratic separation) or by gradually changing the solvent composition (gradient elution). Solvents used for HPLC must be ultra-pure HPLC-grade solvents. Water should be doubly deionized.

A major difference between HPLC separations and adsorption chromatographic separations is that most HPLC separations are accomplished using a reversed-phase stationary phase. (Reversed-phase TLC plates are also commercially available.) This means that polar components are eluted first and nonpolar components last. Reversed-phase columns contain finely divided silica gel particles chemically bonded to C-18 (or C-8) molecules. Nonpolar components dissolve well in the nonpolar C-18 coating. Starting with a very polar solvent such as water encourages elution of polar compounds. Gradually an organic solvent is mixed in, which eventually draws off the less-polar components. Analysis of components of chromatograms is done the same as in GC. A schematic of an HPLC system is shown in Figure 1K-6.

Figure 1K-6

Schematic of HPLC
instrument

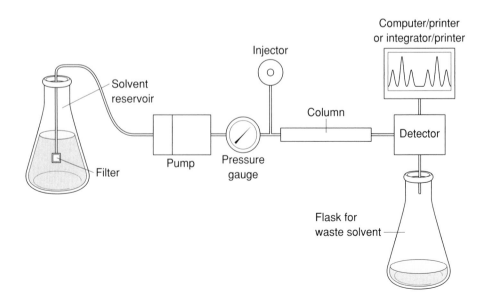

Exercise K.1: Analysis of Analgesic Tablets by TLC

Over-the-counter pain medications, such as Excedrin, Tylenol, or Advil, commonly contain aspirin, ibuprofen, acetaminophen, and/or caffeine (in addition to inert materials). TLC can be used to determine whether an analgesic tablet contains one or more of these active ingredients.

Obtain an analgesic tablet from the instructor. With a mortar and pestle, pulverize the tablet. Dissolve 1–2 mg of the powder in 0.5 mL methylene chloride. If the solid does not dissolve, add ethanol dropwise. Similarly prepare solutions of aspirin, ibuprofen, acetominophen, and caffeine. Obtain a silica gel TLC plate (5–10 cm or similar), development chamber (400-mL beaker, filter paper, and watch glass), and a ruler. Pour 8–10 mL of ethyl acetate into the development chamber and place the watch glass on top.

With a pencil, draw a faint line 1 cm from the bottom of the TLC plate, being careful not to break the silica gel. Make five dots on this line (called the "origin") to indicate where the mixture and standards will be applied. With a capillary tube, apply a tiny drop of each solution at the origin and air dry. Place the TLC plate in the developing chamber, making certain that the level of the solvent is below the origin. Replace the watchglass. When the level of the solvent is near the top of the plate, remove the plate, rapidly mark the solvent front and let the plate air dry. Visualize under UV light or in an iodine chamber and circle each spot. Calculate R_f values of the analgesic and of caffeine, ibuprofen, aspirin, and acetaminophen. List the active ingredient(s) found in the assigned analgesic tablet.

Safety First!

Always wear eye protection in the laboratory.

Caution: Do not look at the UV light.

Exercise K.2: Separating Ferrocene and Acetylferrocene Using Column Chromatography

Obtain a solid mixture of ferrocene and acetylferrocene from the instructor. Weigh out 30–40 mg of the mixture into a tared vial and record the exact mass. Dissolve the mixture in 4–5 drops of methylene chloride. Cap the vial until ready to use.

To prepare the column, obtain a Pasteur pipet, a small piece of glass wool, sand, and silica gel. Place a glass-wool plug in the bottom of the pipet. Pour in a small layer of sand. Fill the pipet about two-thirds full with silica gel. With a clean pipet, apply the solution of unknown evenly to the top of the silica gel. Rinse the vial with 1–2 drops of methylene chloride and add to the column. (Alternatively, add the solid sample directly to the top of the silica gel.) Pour a small layer of sand on top of the sample. This will keep the sample from being disturbed as solvent is added to the column.

Carefully add hexane to the top of the column. The hexane will drain through the column and should be collected in small vials or test tubes. Continue to add solvent so that the column never dries out. As the hexane moves down the column, the sample will begin to separate and will appear as separate bands in the silica gel. When one of the colored bands is at the bottom of the column, change collection vials and collect the colored effluent in a tared vial. When this component has been eluted from the column, the polarity of the solvent should be increased in order to hasten the elution of any other component. Pour a mixture of 1:1 hexane/ethyl acetate in the column and watch the rate at which the band moves through the column. Less time is needed if the solvent is changed to pure ethyl acetate. Collect the effluent in a different vial. Assess the results of the chromatographic separation by running two TLC plates using 10% ethyl

Microscale

Safety First!

Always wear eye protection in the laboratory.

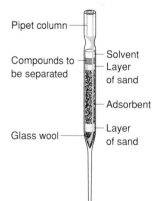

Pipet column

Compounds to be separated

Solvent

Layer of sand

Adsorbent

Layer of sand

Glass wool

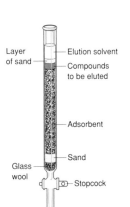

Ferrocene   Acetylferrocene

acetate/hexanes and 50% ethyl acetate/hexanes as eluents. Spot each product fraction and determine the effectiveness of the separation process.

Put the tared vials in a small beaker and set them under the hood while the solvent evaporates. Alternately, the vials may be heated on a sand bath or heat block under the hood. Using a stream of dry air or nitrogen will hasten the evaporation. Reweigh the vials and calculate the mass of each component eluted from the column. Return any unused solid to the instructor. Place solvents in the appropriately labeled bottle for recycling.

Miniscale

Exercise K.3: Separating Ferrocene and Acetylferrocene Using Column Chromatography

Safety First!

Always wear eye protection in the laboratory.

Obtain a solid mixture of ferrocene and acetylferrocene from the instructor. Weigh out 300 mg of the mixture into a small vial and record the exact mass. Dissolve the solid in 0.5 mL of methylene chloride. Cap the vial until ready to use.

To prepare the column, obtain a column or buret, a small piece of glass wool, sand, and silica gel. Make a glass-wool plug in the bottom of the column. Pour in a small layer of sand. Fill the column about two-thirds full with silica gel. Pour hexane into the top of the column, being careful not to disturb the layer of silica gel. Drain the hexane until the eluent level is at the top of the silica gel. With a clean pipet, apply the solution of ferrocene and acetylferrocene evenly to the top of the silica gel. Rinse the vial with 1–2 drops of methylene chloride and add it to the column. Pour a small layer of sand on top of the sample. The sand will keep the sample from being disturbed as solvent is added to the column.

Carefully add hexane to the top of the column. The hexane will drain through the column and should be collected in a flask or beaker. Continue to add solvent so that the column never dries out. As the hexane moves down the column, the sample will begin to separate and will appear as separate bands in the silica gel. When one of the colored bands is at the bottom of the column, change collection flasks and collect the colored effluent in a tared flask. When this component has eluted from the column, the polarity of the solvent should be increased in order to hasten the elution of any other component. Pour a mixture of 1:1 hexane/ethyl acetate into the column and watch the rate at which the band moves through the column. Less time is needed if the solvent is changed to pure ethyl acetate. Collect the effluent in a different flask. Evaporate the solvent under the hood using a warm sand bath. Reweigh the flasks and calculate the mass of each component. Return any unused solid to the instructor. Place solvents in the appropriately labeled bottle for recycling.

Layer of sand — Elution solvent
— Compounds to be eluted
— Adsorbent
— Sand
Glass wool
— Stopcock

Exercise K.4: HPLC Analysis of Benzaldehyde and Benzyl Alcohol

Safety First!

Always wear eye protection in the laboratory.

Dissolve 2 μL of a 1:1 mixture of benzaldehyde and benzyl alcohol in 2 mL of acetonitrile. Similarly prepare a standard solution of benzyl alcohol or benzaldehyde. The instructor will demonstrate the use of the HPLC. Pump the solvent system (acetonitrile/water) through a C-18 reversed-phase HPLC column for 5 minutes at a flow rate of 1 mL/minute. Use a syringe to draw up 1–2 μL of the sample solution according to directions of the instructor and inject it onto the column. Record the chromatogram until both components have eluted from the column. The chromatogram is monitored by UV detection at 254 nm. When the sample has eluted from the HPLC, inject 1–2 μL of the standard solution and record the chromatogram. Based on the retention times of the components of the mixture and the retention times of the standard, identify each component of the mixture. Measure the areas of each peak and determine the relative mass percent of each component. (Molar absorptivities of the two components are about the same at 254 nm.)

Benzaldehyde

Benzyl alcohol

Questions

1. List three factors that affect R_f values in adsorption TLC. Explain.

2. Explain how to prepare sample solutions and capillary applicators.

3. Describe the basic principle underlying all chromatographic processes.

4. Predict the order of elution for the silica gel adsorption TLC separation of the mixture containing benzyl alcohol, benzaldehyde, and benzoic acid using methylene chloride solvent. Explain.

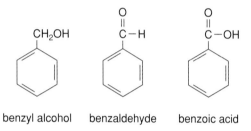

benzyl alcohol   benzaldehyde   benzoic acid

5. The resolution calculated for the TLC separation of a two-component mixture is determined to be equal to 2.0. Does this represent good separation or not? Explain.

6. When should a mixed-solvent system be used for TLC? Explain.

7. Is the order of elution of the components the same for TLC as for column chromatography?

8. Explain how the various fractions can be analyzed from a gravity column.

9. What chromatographic technique would be useful:

   a. to determine whether a nonvolatile compound is pure?

   b. to separate a mixture of two nonvolatile solids?

   c. to determine whether a reaction is complete?

10. Caffeine absorbs UV light. Describe how to determine if a cup of coffee is caffeinated or decaffeinated using HPLC.

Technique L: Polarimetry

Optical activity is a physical property of chiral molecules. Chiral molecules lack planes of symmetry and may exist as two or more stereoisomers. Stereoisomers that rotate the plane of plane-polarized light in equal but opposite directions are called enantiomers. Enantiomers have nonsuperimposable mirror-image structures. Enantiomers have identical physical properties, including the same boiling and melting points, except for rotation of the plane of plane-polarized light. Specific rotations [α] of enantiomers are equal, but opposite. One enantiomer has a positive rotation, indicating rotation of plane-polarized light to the right. The other enantiomer has a negative rotation, indicating rotation of the plane of plane-polarized light to the left.

The technique known as polarimetry is used to measure the optical rotation of liquids and solutions. Only optically active solutions can give readings of optical rotation other than $0°$. Common solvents, such as water, methanol, ethanol, and ethyl acetate are not chiral and are not optically active. They may be used to prepare solutions of optically active substances for measurement of optical rotation using a polarimeter. The yellow light of a sodium lamp (sodium D line) is the accepted type of light to measure optical rotations.

The optical rotation of a chiral liquid or of a chiral substance in solution is called the observed rotation α. The observed rotation is read directly from a polarimeter and is related to the specific rotation $[\alpha]_D$ (a physical constant for the substance) according to Equation (1.7):

$$(1.7) \qquad\qquad [\alpha]_D = \frac{\alpha}{c \times l}$$

where c is concentration in g/mL and l is the length of the polarimeter tube in decimeters. The rotation of an optically active liquid may be measured for the neat liquid or by dissolving the liquid in a solvent. In the case of a neat liquid, c is equal to the density of the liquid in g/mL.

A commercial polarimeter is illustrated in Figure 1L-1. This polarimeter is accurate to $\pm 0.1°$.

Sample tubes of differing lengths may be used, but the most common is the 1-dm (10-cm) tube. The sodium-vapor light source is placed at the end of the instrument away from the eyepiece. The operator peers through the eyepiece and sees the light from the lamp as it traverses through the sample tube. The field is adjusted by turning the wheel to the right or left until it looks homogeneous as shown in Figure 1L-2. Sample fields in the figure on the left and right are in need of adjustment whereas the field in the center is correct and a rotation reading may be taken without further adjustment.

Optical Purity and Enantiomeric Excess

Impure samples of optically active compounds give calculated specific rotations that are less than the expected specific rotation. Optical purity is the ratio of the specific

Figure 1L-1

Placing a sample tube in a (commercial) polarimeter

© Copyright 1996 by Cole-Palmer Instrument Company; used with permission.

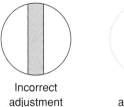 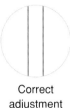 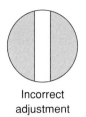

Incorrect   Correct   Incorrect
adjustment   adjustment   adjustment

rotation of the sample over the specific rotation of the pure enantiomer, multiplied by 100%. Optical purity is also equal to the difference between the percentage of the enantiomer in excess and the percentage of the other enantiomer (enantiomeric excess). It often happens that a sample of one enantiomer is contaminated by the other enantiomer. For example, pure L-tartaric acid has a specific rotation of +12.7°. Racemic tartaric acid contains 50% each of the D and L forms and solutions of the racemic form give no optical rotation. If a solution contains 75% of the L form and 25% of the D form, then the optical purity is equal to 75% − 25% = 50%. Rotation of part of the L form (25%) is cancelled by an equal amount of the D form. Therefore, only half of the original tartaric acid sample is causing a rotation. There are two methods of doing the calculation.

Measurement of Optical Rotation and Calculation of Optical Purity

Calculation of optical purity (o.p.) is done by dividing the specific rotation of a mixture containing an enantiomer by the specific rotation of that enantiomer as in Equation (1.8).

$$\text{o.p.} = \frac{[\alpha]_{\text{impure enantiomer}}}{[\alpha]_{\text{pure enantiomer}}} \times 100\% \qquad (1.8)$$

For example, the observed rotation is + 0.95° for a solution of L-tartaric acid that has a concentration of 0.15 g/mL using a 1-dm tube. In order to do the calculation from experimental data, it is necessary to read the experimentally observed rotation and to calculate the observed specific rotation using Equation (1.7).

$$[\alpha] = \frac{+0.95°}{0.15 \text{ g/mL} \times 1 \text{ dm}} = +6.3°$$

The specific rotation of pure L-tartaric acid is 12.7°. Using Equation (1.8), the optical purity of the solution is:

$$\text{o. p.} = 6.3°/12.7° \times 100\% = 50\%$$

Calculation of Enantiomeric Excess

In the second method, enantiomeric excess is calculated from the percentages of each enantiomer in a mixture, if known. The difference between the percentages of the major (En1) and minor (En2) enantiomers gives the enantiomeric excess (e. e.):

$$\text{e.e.} = \%\text{En1} - \%\text{En2}$$

The solution contained 75% of L-tartaric acid and 25% of D-tartaric acid. Therefore, the enantiomeric excess was 50%:

$$\text{e.e.} = 75\% - 25\% = 50\%$$

It can be seen from these two examples that optical purity is equivalent to enantiomeric excess:

$$\text{optical purity} = \text{enantiomeric excess}$$

How to Measure Optical Rotation

If the sample is a liquid, introduce it directly into the sample tube. If the sample is a solid or a liquid in short supply, weigh the sample and dissolve in a specified volume of solvent, such as water, ethanol, or ethyl acetate. Place the liquid or solution in the sample tube, usually a 1-dm tube, which holds about 10 mL of liquid. Fill the tube completely, as any air bubbles may result in incorrect readings. Place the sample tube in the sample compartment and close the compartment door. If the sample tube area is being adjusted to a temperature other than room temperature, allow the system to achieve temperature equilibration.

Focus the polarimeter. Adjust the field by rotating the inner cylinder holding the eyepiece. **Do not adjust the position of the lamp or the instrument. If either has been moved, notify the instructor.** Rotate the wheel holding the eyepiece slowly to the right. The field appears as shown in Figure 1L-2. When the field is adjusted correctly to take a reading, both the outer fields and the inner field should have matching shades. If no match is found initially, try rotating the inner wheel to the left and look for matching fields. When a match is found, read the scale. Refer to Figure 1L-3 for a view of the scale.

Miniscale

Repeat the process on the same sample two or three times. Approach the desired field from opposite directions and take an average of the readings. The averaged reading should be accurate to 0.05°. Repeat using 20 mL of the same solution in the 2-dm

Figure 1L-3

Polarimetry scale

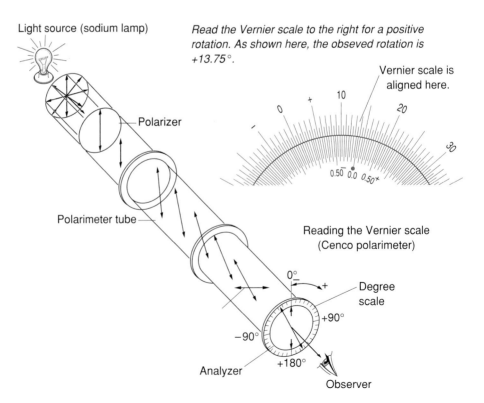

Light source (sodium lamp)

Read the Vernier scale to the right for a positive rotation. As shown here, the obseved rotation is +13.75°.

Vernier scale is aligned here.

Polarizer

Polarimeter tube

Reading the Vernier scale (Cenco polarimeter)

Degree scale

0°

+90°

−90°

+180°

Analyzer

Observer

cell. If the first reading was correct, the second reading should be double the first reading. Rinse the polarimeter tubes with pure solvent and place the solution in the appropriate container marked for this purpose. Add pure solvent to one of the tubes and take several readings. If the readings differ from 0.00°, the observed reading for the sample solution should be corrected to account for the difference.

Remove the solvent and place it in a container marked for used solvent. Drain the tube, rinse with solvent and place back in the box. Be careful when leaving a tube unattended on the lab bench. It could accidentally roll off the bench.

Exercise L.1: Determining Optical Purity of an Unknown Liquid

Be certain that the polarimeter shows a reading of zero prior to making measurements. Pipet 10.0 mL of an unknown optically active liquid or solution into a small Erlenmeyer flask. Weigh the liquid and calculate its density. Fill a 10-mL polarimeter tube with the unknown liquid. Measure the optical rotation of the liquid. This is the observed rotation. Rinse out the polarimeter tube using a small amount of acetone. Calculate and record the specific rotation of the unknown liquid and temperature.

Safety First!

Always wear eye protection in the laboratory.

Exercise L.2: Determining the Melting Points of Enantiomers and Racemates

Obtain three samples of unknown solids: two enantiomers and their racemic form. Place a sample of each into separate melting point capillaries and take the approximate melting points. Repeat this process using a slower heating rate. Record the melting points of each solid. Explain the results obtained in the melting point determinations.

Safety First!

Always wear eye protection in the laboratory.

Questions

1. A solution of L-glyceraldehyde (0.25 g/mL) in water has an observed rotation of −2.17° in a 1-dm tube at 20°C using a sodium lamp. Calculate the specific rotation of L-glyceraldehyde.

2. Predict the observed rotation of a 0.100 M aqueous solution of L-tartaric acid measured in a 1-dm tube at 20°C. The specific rotation of L-tartaric acid is +12.7°. The molecular mass of L-tartaric acid is 150.09 g/mole.

3. D-Tartaric acid is available commercially and has a specific rotation of −12.0°. Give a likely reason that the value is not −12.7°.

4. A student's calculation of specific rotation is found to be a factor of 10 too small. Suggest a likely reason for this error.

5. A student obtains an observed optical rotation of = +45.5°. A second student challenges the reading and says that it should really be −314.5°. Which student is correct? How can the dispute be settled by using another measurement?

6. A solution is found to have an enantiomeric excess of 10% of the D enantiomer of a substance. How much of the D form is present and how much of the L form is present?

7. A student is asked to prepare 10 mL of a 1.0 M solution of a compound (MW = 100) that is 60% optically pure, starting with the pure enantiomers and solvent. Describe what should be done.

8. D-Tartaric acid has a specific rotation recorded in common handbooks, but (+,−)-tartaric acid has no reported specific rotation. Why?

Chapter 2

Spectroscopic Methods and Molecular Modeling

Spectroscopy has revolutionized the world of organic chemistry. Infrared (IR) spectroscopy is now practiced in most laboratories and students can use IR spectrophotometers to record and interpret an IR spectrum. The use of IR spectroscopy and an introduction to interpretation of spectra are described in Technique M.

Nuclear magnetic resonance (NMR) spectroscopy is covered extensively in organic chemistry texts. This is a very powerful tool for elucidating structures of organic compounds. The identities of many simple organic compounds can be deduced from their NMR spectra alone. For time spent, NMR spectroscopy gives the most useful information about structure. Two types of NMR spectroscopy are discussed in Technique N. One is proton NMR spectroscopy and the other is carbon NMR spectroscopy. Recording of spectra is included in the exercises if equipment is available. If not, it is still possible to do the spectral exercises that are included.

Ultraviolet (UV) spectroscopy has limited use in the organic chemistry laboratory. However, there are some advanced, specific applications where UV spectroscopy is the only technique that gives the structural information required. Compounds containing alternating single and double bonds (conjugated) give measurable UV spectra. If the conjugation is very extensive and the compound is colored, visible spectroscopy can be used as an analytical method. UV spectroscopy and visible spectroscopy are described in Technique O. They are used to study reaction rates in this text.

Mass spectrometry (MS), which is covered in Technique P, has seen increased use in many undergraduate laboratories because of greater availability of instrumentation. The combination of gas chromatography and MS in the same instrument (GC/MS) is the most available and most easily operated MS instrument for organic chemistry students. MS gives important information about molecular weight and the complexities of molecules. MS can be used in combination with other spectroscopic methods to elucidate structure.

Molecular modeling using molecular modeling kits or molecular modeling computer software is introduced in Technique Q. The advent of modeling software and personal computers capable of running the software programs has created a new technique for organic chemistry. Molecular modeling is introduced as a technique in this chapter and is applied in several experiments in this book.

Technique M: Infrared Spectroscopy

Theory

When an organic molecule absorbs radiant energy, vibrations of certain bonds within the molecule become excited. If the frequency of incident radiation matches the specific vibrational frequency of the bond, the bond is excited from a lower vibrational state to a higher vibrational state and the amplitude of the vibration increases. Specific vibrations result from absorption of specific frequencies of infrared radiation. Structurally different molecules do not absorb exactly the same energies of infrared radiation so they give different patterns of absorption. Specific bonds and functional groups in the molecule have specific vibrational frequencies, so they absorb characteristic frequency ranges of radiation.

When infrared radiation is passed through a thin-film or wafer-thin sample, some of the light passes directly through the sample. But certain frequencies corresponding to specific vibrational frequencies within the molecule are absorbed. This makes infrared (IR) spectroscopy a valuable tool for identifying different functional groups.

The two most important vibrational modes in IR spectroscopy are "stretching and bending." These are illustrated for vibrational modes of the methylene ($-CH_2-$) group (see Figure 2M-1).

Stretching involves a change in the interatomic distance. Stretching is symmetric if atoms move in opposite directions along the internuclear axis and is asymmetric if the atoms move in the same direction.

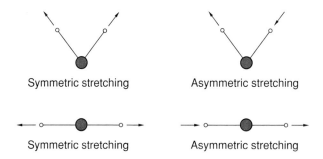

For bending, IR absorption causes a change in bond angles. Types of bending modes include scissoring, rocking, wagging, and twisting. Stretching requires more energy than bending, so it occurs at higher frequencies (shorter wavelengths). Asymmetric stretching requires more energy than symmetric stretching.

Stretching and bending must cause a change in the dipole moment of the bond in order to be IR active. Bonds that show large changes in bond dipole moment give more intense IR absorption than bonds that show small changes in bond moment. For example, IR absorption due to carbon-oxygen double bond stretching is intense, but absorption due to carbon-carbon double bond stretching is weak. IR absorption by the polar carbon-oxygen bonds leads to a change in dipole moment and therefore results in intense peaks.

An IR spectrum is a graph of absorption intensity (given as percent transmittance) as a function of radiation frequency (given as wavenumber in units of reciprocal centimeters—cm^{-1}). Frequency and wavenumber are directly proportional; therefore, wavenumber is also directly proportional to energy: the higher the wavenumber, the greater the energy. The range of the wavenumber scale is 4000 to 600 cm^{-1}. IR spectra may also display a wavelength scale, in microns (μ). The range of wavelengths is 2.5 to 16 μ.

$$E = h\nu = \frac{hc}{\lambda}$$

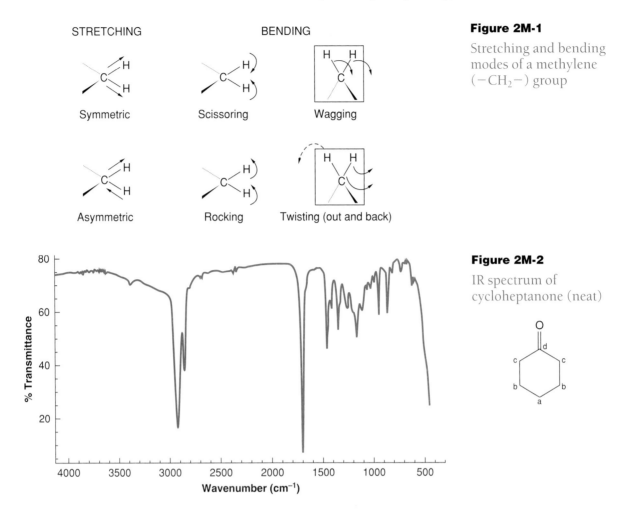

Figure 2M-1

Stretching and bending modes of a methylene ($-CH_2-$) group

Figure 2M-2

IR spectrum of cycloheptanone (neat)

The IR spectrum of cycloheptanone is shown in Figure 2M-2. Absorptions are shown as "valleys," some of which are broader than others. IR absorptions are broadened due to the number of rotational energy states available to each vibrational state.

Bonds give rise to characteristic IR absorption frequencies and intensities. The energy required to excite a given vibrational mode is dependent upon the size of the atoms and strengths of the bonds. The frequency of absorption is proportional to the square root of the force constant for the bond divided by the reduced mass:

$$\bar{v} = \frac{1}{2\pi c}\sqrt{\frac{k}{\mu}}$$

where $\bar{v}$ is the vibrational frequency, k the force constant and μ the reduced mass $[m_1 m_2/(m_1 + m_2)]$. In general, the smaller the atoms, the better the overlap and the higher the frequency of absorption. As atomic size increases, the frequency decreases. Bond strength also influences the frequency of absorption. This means that a carbon-carbon triple bond vibrates (higher force constant, k) at higher frequencies than a carbon-carbon double bond, which vibrates at higher frequencies than a carbon-carbon single bond:

alkyne	carbon-carbon triple bond stretching	$(2250–2100 \text{ cm}^{-1})$
alkene	carbon-carbon double bond stretching	$(1680–1600 \text{ cm}^{-1})$
alkane	carbon-carbon single bond stretching	$(1200–800 \text{ cm}^{-1})$

Similarly, a carbonyl group (C=O) vibrates at higher energy than does a carbon-oxygen single bond:

C=O   (1810–1630 cm^{-1})

C–O   (1300–1000 cm^{-1})

Ranges of absorption do not necessarily imply that there is broad absorption over the entire range. Rather, the maximum absorption may occur anywhere within the range listed. These bond frequencies are used to predict IR absorption frequencies. A compound containing a carbonyl group absorbs IR radiation in the region of 1810–1630 cm^{-1} because the frequency of light matches the vibrational frequency of the carbonyl group.

C≡C–H   3300 cm^{-1}
C=C–H   3100–3000 cm^{-1}
C–C–H   3000–2800 cm^{-1}

Hybridization affects the characteristic absorption frequency of the carbon-hydrogen bond. The greater the s character, the shorter and stronger the bond. Carbon-hydrogen bonds of sp hybridized carbons are stronger than carbon-hydrogen bonds of sp^2 hybridized carbons, which are stronger than carbon-hydrogen bonds of sp^3 hybridized carbons. Therefore sp C–H stretching occurs at higher frequency than sp^2 C–H stretching and sp^3 C–H stretching.

Conjugation also affects the frequency of absorption. Conjugated alkenes and carbonyl compounds absorb at lower frequencies than their nonconjugated counterparts. The shift is about 30 cm^{-1} per unit of conjugation. Conjugation of a double bond lengthens the bond, reducing the force constant.

Three important regions of the IR spectrum indicating functional groups are the region from 3600–3100 cm^{-1} where OH and NH stretching occur, the region around 1700 cm^{-1} where C=O stretching occurs, and the region around 1650 cm^{-1} where C=C stretching occurs. Many important functional classes, such as alcohols, amines, amides, carboxylic acids, ketones, aldehydes, esters, alkenes, and aromatic compounds are identified by the presence (or absence) of absorption in these regions.

Region from 3650 to 3100 cm^{-1} (C–H, N–H, OH)

Although both OH and NH stretching occur in this region, they can be distinguished by the intensity of the peaks. Neat liquid samples of alcohols and phenols give one very broad, very intense absorption around 3650–3200 cm^{-1}, while neat liquid samples of amines or amides may show absorption bands around 3550–3100 cm^{-1} that are somewhat sharper and less intense. Primary amines and unsubstituted amides show two bands in this region while secondary amines show only one band. Tertiary amines do not give absorption in this region. Carboxylic acids show strong and extremely broad OH stretching (3300–2400 cm^{-1}), which sometimes obscures the carbon-hydrogen stretching band. It is generally easy to differentiate between an alcohol, an amine, and a carboxylic acid, as illustrated by the spectra shown in Figure 2M-3. Terminal alkenes, aromatic compounds, and alkynes also show absorption in this region. sp^2 C–H stretching for terminal alkenes and aromatic compounds occurs at higher frequencies than sp^3 C–H stretching, around (3100–3000 cm^{-1}), while sp C–H stretching for terminal alkynes occurs at 3300 cm^{-1}.

Region around 1700 cm^{-1} (C=O)

This is the region that corresponds to carbonyl absorption. The actual absorption frequency depends both upon the type of carbonyl (ketone, aldehyde, ester, amide, anhydride, or acid chloride) and the degree of conjugation. An attached electron-withdrawing substituent raises the energy of absorption and the frequency. Thus, acid chlorides absorb at higher frequencies than do ketones or aldehydes. Amides and carboxylic acids have

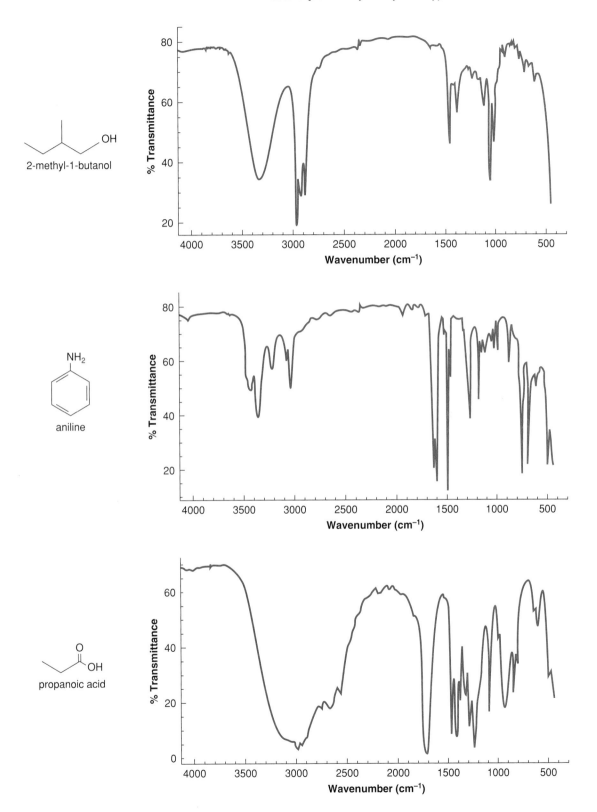

Figure 2M-3 IR spectra of an aliphatic alcohol, a primary aromatic amine, and an aliphatic carboxylic acid (neat liquids)

lower carbonyl absorption frequencies than ketones. In the case of amides, lower absorption frequencies are due to conjugation (resonance). In the case of carboxylic acids, the lowering of energy is due to strong intermolecular hydrogen bonding. This weakens the carbonyl, resulting in a lower energy absorption. This trend in the approximate carbonyl stretching frequency is summarized below for aliphatic carbonyl compounds:

Anhydrides	1810 cm^{-1} and 1760 cm^{-1}	(both very strong)
Acid chlorides	1800 cm^{-1}	(very strong)
Esters	1735 cm^{-1}	(very strong)
Aldehyde	1725 cm^{-1}	(very strong)
Ketone	1715 cm^{-1}	(very strong)
Carboxylic acid	1710 cm^{-1}	(very strong)
Amide	1660 cm^{-1}	(strong)

Conjugation with carbon-carbon double bonds and aromatic rings lowers the absorption frequency. For example, the carbonyl stretching frequency of 3-methyl-2-butanone, an aliphatic ketone, occurs at 1715 cm^{-1}, while the carbonyl stretching frequency of 4-methyl-3-penten-2-one, a conjugated ketone, occurs at 1700 cm^{-1}.

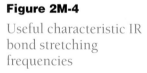

3-methyl-2-butanone   4-methyl-3-penten-2-one

Region around 1600 cm^{-1} (C=C, C=N)

In this region are carbon-carbon (C=C) stretching of unsymmetrical alkenes and aromatics (1680–1430 cm^{-1}) and carbon-nitrogen (C=N) stretching (1690–1500 cm^{-1}). Conjugation lowers the energy of absorption due to a resonance effect.

A summary of important types of bonds and their approximate characteristic absorptions is shown in Figure 2M-4:

Figure 2M-4

Useful characteristic IR bond stretching frequencies

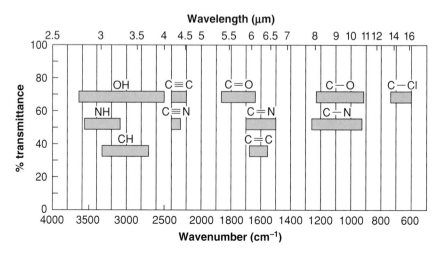

Conjugation and resonance generally decrease absorption frequencies, whereas ring strain increases absorption frequencies and ring strain affects the actual frequency of absorption. The values in the chart in Figure 2M-4 represent ranges. Other functional groups also have characteristic absorption bands. A summary of characteristic absorption bands for various functional groups is given in the correlation chart in Table 2M-1.

Table 2M-1 Absorption Frequencies of Specific Functional Groups

Type of compound	Major IR bands (cm^{-1}) (approximate)	Notes
Alkane	3000 (C–H stretching) 1465 (CH$_2$ scissoring) 1450 (CH$_3$ asymmetric bending) 1375 (CH$_3$ symmetric bending)	Saturated hydrocarbons give simple spectra, with few absorptions. An isopropyl group shows two equally intense peaks around 1385 and 1370 cm^{-1}, while for a *tert*-butyl group, the peaks are of unequal intensity.
Alkene	3100 (**H**–**C**=C stretching) 1630 (weak) (C=C stretching) 1000–650 (**H**–**C**=C out-of-plane bending)	Monosubstituted alkenes show two strong bands at 990 and 910 cm^{-1}. *cis*-Alkenes show one strong band around 700 cm^{-1}, while *trans*-alkenes have one strong band near 890 cm^{-1}. Symmetrical alkenes do not show a C=C stretching absorption.
Alkyne	3300 sharp (**H**–**C**≡**C**) stretching 2150–2100 (C≡C stretching)	Absorption at 2150–2100 cm^{-1} is weak or absent for internal alkynes.
Aromatic	3100 (sp^2 C–H stretching) 2000–1667 (overtone bands) 1600–1475 (C=C ring stretching) 900–690 (out-of-plane bending)	Bands at 900–690 cm^{-1} show the alkyl substitution pattern: mono-substituted—two bands near 750 and 700 cm^{-1}; *ortho*-disubstituted—one band near 750 cm^{-1}; *meta*-disubstituted—three bands near 890, 800, and 690 cm^{-1}; and *para*-disubstituted—one band near 830 cm^{-1}.
Alkyl halide	800–600 (C–Cl stretching)	C–Br and C–I absorb in the far infrared region (<690 cm^{-1}). Aryl chlorides have C–Cl stretching at 1096–1089 cm^{-1}.
Alcohol and phenol	3650–3200 broad (OH stretching) 1250–1000 (C–O stretching)	Neat alcohols absorb near 3300 cm^{-1}. Broad width of the band is due to intermolecular hydrogen bonding. The position of the C–O stretching band varies for a primary, secondary, and tertiary alcohol: 1050 cm^{-1} (primary); 1100 cm^{-1} (secondary); 1150 cm^{-1} (tertiary). In phenols, this band is around 1220–1200 cm^{-1}.
Ether	1300–1000 (C–O stretching)	For aliphatic ethers, this is generally the main band of interest. Aromatic ethers have two bands at 1245 and 1030 cm^{-1}.
Amine	3550–3250 (N–H stretching) 1640–1560 broad (N–H bending—primary amine) 1250–1000 (C–N stretching)	Primary amines show two overlapping absorption bands at 3550–3250 cm^{-1}, while secondary amines have only one. Tertiary amines do not absorb in this region. Aliphatic amines also show weak to moderate C–N stretching at 1100–1000 cm^{-1}. This band is shifted to higher frequencies in aromatic amines (1200–1100 cm^{-1}).
Aldehyde	1725 (C=O stretching) 2750 and 2850 (aldehydic C–H stretching)	Carbonyl stretching for aromatic aldehydes is at lower frequency. Bands at 2750 and 2850 cm^{-1} are sharp, but weak. The higher frequency band may be obscured by aliphatic C–H stretching band.
Ketone	1715 (C=O stretching)	Conjugation decreases the frequency of the carbonyl stretching. Strained cyclic ketones absorb at higher frequencies than nonstrained cyclic ketones (1775–1740 cm^{-1}).
Ester	1735 (C=O stretching) 1300–1000 (C–O stretching)	C–O stretching at 1300–1000 cm^{-1} is actually two bands, one stronger than the other. Small ring lactones have higher carbonyl stretching frequencies.
Carboxylic acid	3300–2400 (O–H stretching) 1710 (C=O stretching) 1320–1210 (C–O stretching)	Carboxylic acids have very broad O–H stretching that may obscure the C–H stretching band. The carbonyl stretching is also broad. The carbonyl group in dimerized saturated aliphatic acids absorbs in the region 1720–1706 cm^{-1} while monomeric aliphatic acids show carbonyl absorption near 1710 cm^{-1}.

continued

Table 2M-1 *continued*

Type of compound	Major IR bands (cm^{-1}) (approximate)	Notes
Amide	3550–3060 (N—H stretching) 1650 (C=O stretching) 1640–1550 (N—H bending)	Unsubstituted amides show two N—H stretching bands, while mono-substituted amides show one. Disubstituted amides do not absorb in the N—H stretching region. The carbonyl stretching frequency is decreased due to resonance.
Anhydride	1810 and 1760 (two C=O stretches) 1300–900 (C—O stretching)	Anhydrides give two strong carbonyl bands, one for symmetric stretching and the other for asymmetric stretching.
Nitrile	2260–2240 (C≡N stretching)	Conjugation decreases the frequency to 2240–2222 cm^{-1}.
Nitro	1550 and 1350 (NO$_2$ stretching)	Symmetric and asymmetric NO$_2$ stretching.

General Approach to Solving an IR Spectrum

IR spectroscopy can be used to identify bond types and functional groups present and bond types and functional groups that are absent. In analyzing an IR spectrum, you should not attempt to identify all of the bands, but look for significant peaks that help you identify functional groups.

A general approach to use is to first look in the region around 1700 cm^{-1} for the presence of carbonyl absorption. If carbonyl absorption is present, note the wavenumber and look for evidence of an aldehyde, ester, carboxylic acid, amide, or anhydride. If there is none, the compound is a ketone. Examine the specified regions for aromaticity and determine if the compound is aromatic or aliphatic. Aromatic compounds and vinylic compounds show sp^2 C—H stretching at 3100 cm^{-1}; unsymmetrical alkenes also show C=C stretching at 1660–1600 cm^{-1}, and aromatic compounds have two absorptions at 1600 and 1475 cm^{-1}. Aromatics also have strong out-of-plane bending at 900–690 cm^{-1}, characteristic of the substitution pattern. If carbonyl absorption is not present, look at the region 3650–3100 cm^{-1} to see if the molecule contains an OH or NH group or if the compound is an alkene or terminal alkyne. This approach is illustrated in the flow scheme in Figure 2M-5.

Once a functional group has been tentatively identified, you should look for confirmatory absorptions. For example, alcohols show broad OH absorption around 3600 cm^{-1}, but will also show strong C—O absorption in the region of 1260–1000 cm^{-1}. The presence of this latter band confirms the identification of the compound as an alcohol. Terminal alkynes show sp C—H stretching at 3300 cm^{-1}, but also show a sharp peak at 2150 cm^{-1}. This latter peak is often the most distinctive peak in the IR spectrum of a terminal alkyne because there are few other absorptions in this region. Other functional groups have similar characteristic absorptions. Consult Table 2M-1 for a detailed description.

You should be careful when trying to distinguish between alkanes, alkyl halides, and internal alkynes and alkenes using IR spectroscopy alone. Other spectroscopic methods, simple chemical tests, or analysis of physical properties may be useful in these cases.

Recording an IR Spectrum

In order to record an IR spectrum of a neat liquid, you will need (1) the infrared spectrophotometer, (2) salt plates and holder or an IR card, and (3) the sample. For a solid sample, you will need a mortar and pestle and Nujol oil (to prepare a Nujol mull) or high-quality, dry KBr and a pellet press. Solids may also be dissolved in a volatile solvent and the solution applied to an IR card.

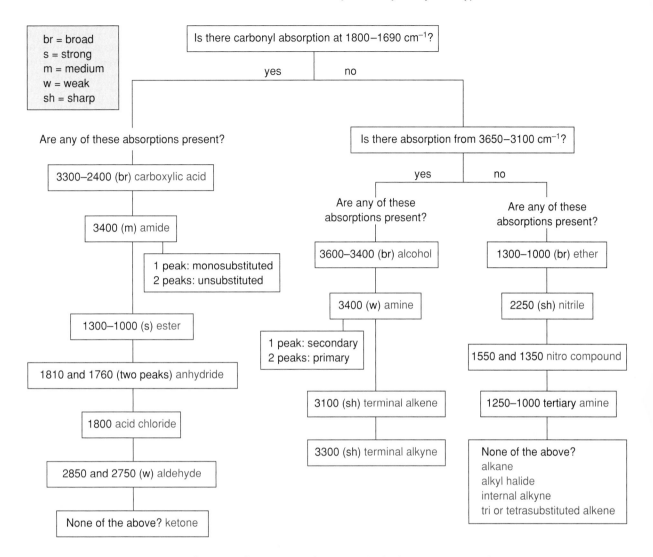

Figure 2M-5   Flow scheme for classifying monofunctional aliphatic compounds using IR spectroscopy

The IR Spectrophotometer

There are two types of IR spectrophotometers, which differ in the mode of radiation. A dispersive IR spectrophotometer contains a prism or grating, which breaks incoming radiation into separate wavelengths. A Fourier transform spectrophotometer (FTIR) provides pulses of radiation containing all wavelengths in the infrared. The two spectrophotometers also differ in schematics.

In a dispersive IR spectrophotometer, a beam of infrared light is split into two beams of equal intensity. The sample cell is in the path of one of the beams; the other goes through a reference cell. The beams pass through a monochromator that systematically varies the wavelength. A detector measures the difference in intensities of absorption between the reference and the sample. The chart recorder measures the difference in intensity as a function of wavenumber (or frequency) as the pen tracks the frequencies from 4000 to 600 cm^{-1} over a period of a few minutes.

In Fourier transform infrared spectroscopy, the entire spectrum is recorded at once and stored in computer memory. The data undergoes a mathematical Fourier transform to convert it to useful data in the normal format. The spectrum can be compared with other spectra stored in memory. For example, a spectrum of an organic sample in methylene chloride solvent can be placed in memory and a prerecorded spectrum of methylene chloride only can be used to subtract the solvent spectrum from that of the sample. The subtraction is generally not perfect, however. For further information about FTIR, see the references at the end of Technique M. All IR spectra in this book are FTIR spectra.

Salt Plates

Salt plates are plates made of large crystals of sodium chloride. Because salt dissolves in water, it is absolutely critical that no water come in contact with the plates, ever! Even the slightest bit of moisture in the sample or on your fingertips is enough to cause pits on the surface of the plate. Handle the salt plates only on the edges. Store them in a desiccator whenever they are not in use.

Salt plates should be cleaned with a few drops of an anhydrous solvent such as methylene chloride or 1,1,2-trichloroethane. Acetone is not a good solvent for cleaning, because it absorbs too much water, which then dissolves the salt plate. To clean the plates, place a few drops of methylene chloride in the middle of each plate. Rub gently with a tissue. Replace in the desiccator when finished.

Hood!

Disposable Cards for IR Spectroscopy

Disposable polyethylene film cards may also be used to obtain spectra of organic compounds. A drop of solution containing an organic sample is applied to the film. The solvent evaporates, producing a very thin film of the organic compound embedded on the polyethylene. These cards are relatively expensive, but they are particularly useful for obtaining IR spectra of solids.

How to Prepare a Sample for IR Analysis

Preparing a Liquid Sample

Take the salt plates from the desiccator and place on a Kim-wipe tissue to avoid scratching the surfaces. With a clean pipet, deposit a drop of the neat liquid (or more, if the liquid is extremely volatile) onto the center of one of the plates. Be certain that the sample is dry and contains no water! Position the other plate directly on top. The liquid will be compressed to a fine film. Make certain there are no air bubbles between the plates. Insert the plates into the cell plate holder. On opposite corners, carefully tighten the bolts. Do not overtighten. The plates are fragile and break easily! Place the cell in the sample side of the spectrometer. Air is the reference.

Preparing a Solid Sample

You can prepare solid samples either using a Nujol mull or by making a KBr pellet. To make a Nujol mull, grind about 10–20 mg of the solid with a mortar and pestle. When the sample is finely ground, add 1–2 drops of Nujol (mineral oil). Grind the sample and Nujol together to make a fine homogenous paste. Spread a dab of the paste on the salt plate, and cover with the second salt plate. The paste should spread evenly between the plates. Make certain there are no air bubbles. Insert the plates into the cell plate holder. On opposite corners, carefully tighten the bolts (Figure 2M-6). Do not overtighten. The plates are fragile and break easily! If the paste is too thick, light cannot get through the

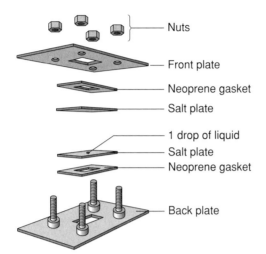

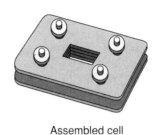

Assembled cell

Salt plates and holder

sample. If this happens, add a little more oil to the mull and mix again. If the paste is too thin, the mull will consist primarily of the Nujol and the IR spectrum of the compound will be weak. If this happens, redo the mull with a higher proportion of compound. Don't forget to subtract out the absorptions due to the Nujol. It is helpful to run a spectrum of pure Nujol to see exactly where the Nujol absorption bands are. Characteristic peaks occur at 2924 cm^{-1}, 1462 cm^{-1}, and 1377 cm^{-1}. The IR spectra of Nujol and acetanilide run as a Nujol mull are shown in Figure 2M-7.

Solid samples can also be run using a KBr pellet. See Figure 2M-8. The KBr must be of extremely high quality and must be extremely dry. It is usually stored in an oven. The handheld or nut/bolt pellet press must also be clean and dry. To make a KBr pellet with the nut/bolt press, insert one bolt into the press so that it goes halfway up into the press. Into the opening put 1–2 mg of the solid and about 100 mg of the finely powdered KBr, which have been previously ground together with a mortar and pestle. Shake the press so that the powder forms a thin, even layer across the bottom. Insert the other bolt and gently tighten. With more force, tighten the bolt, first using hands, and then a wrench. An automotive torque wrench can be used to tighten properly. Do not strip the bolts, but apply enough force so that the sample forms a thin pellet. Then remove both of the bolts. The pellet remains in the press. Carefully examine the pellet. The pellet should be transparent. If it is too thick, light will not be able to get through the sample. If this is the case, remove the pellet and remake it, using slightly less KBr and more force. When an acceptable pellet is formed, place the entire pellet in the holder.

Preparing a Film Card for IR

Although the cost of film cards is prohibitive for large classes, these disposable cards are particularly useful for solid samples. Dissolve 10–20 mg of the solid in a few drops of a volatile solvent in which the solid is soluble. Apply a drop of the solution to the polyethylene film card and allow the solvent to evaporate. Place the card in the holder and record the spectrum. Peaks at 2928, 2850, 1478, and 743 cm^{-1} are due to the polyethylene film. Figure 2M-9 shows the film card and the FTIR spectrum of polyethylene.

The film cards may also be used for nonvolatile liquid samples. Dissolve a drop of liquid sample in a few drops of a volatile solvent, such as diethyl ether or methylene chloride. Apply the solution to the polyethylene film card and allow the solvent to evaporate.

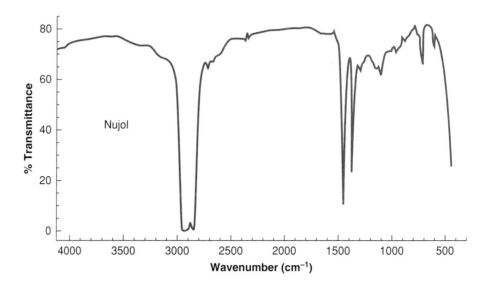

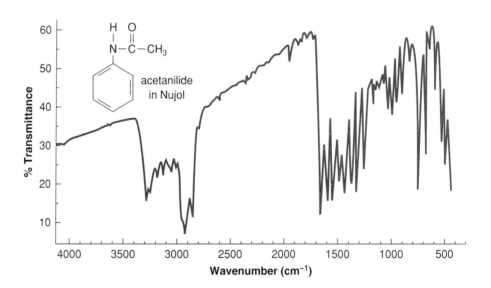

Figure 2M-7   FTIR spectra of Nujol and acetanilide prepared as a Nujol mull

Important Tips Concerning IR

1. Store the recorder pen capped or in a humidifier to keep the ink flowing properly.
2. Store the salt plates in a desiccator to keep water away.
3. Never put a sample that contains water on salt plates.
4. If the peaks are too large and bottom out on the page using dispersive IR spectroscopy, make sure that the Baseline/Offset is set to 90%. Try adjusting the gain. Lower numbers should decrease the amplitude. Try running the spectrum using a smaller amount of sample.

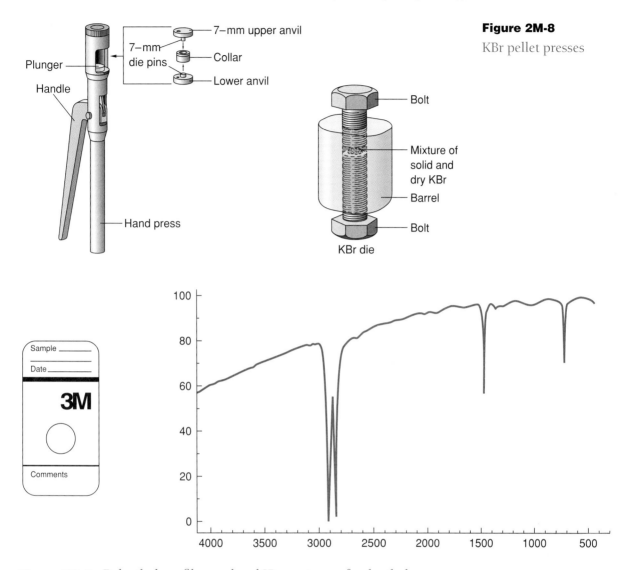

Figure 2M-9 Polyethylene film card and IR spectrum of polyethylene

5. If the peaks are too small, there is not enough sample. Increase the gain. If this is not successful, add more drops of the sample if running a neat liquid. If making a KBr pellet, add more of the sample to the press.

6. If a Nujol mull of a functional compound shows only C—H stretching around 3000 cm^{-1} and two other bands around 1500 and 1400 cm^{-1}, there is too much Nujol and not enough compound. Remake the mull, adding more of the compound.

7. Dispersive-IR spectrophotometers can give inaccurate wavenumbers because of mechanical slippage or paper misalignment and must be calibrated by comparison with reference peaks. To calibrate the instrument, record the IR spectrum of a sample such as polystyrene, which is readily available as a thin transparent film. Polystyrene has a strong absorption at 1601 cm^{-1} that can be used as a reference peak. If the position of this band differs from 1601 cm^{-1}, a correction factor must be applied to the reported wavenumbers of the sample. FTIR instruments do not need to be calibrated.

Exercise M.1: Recording the IR Spectrum of an Organic Liquid

Obtain an organic liquid. Draw the structure of the assigned compound. Obtain an IR spectrum and interpret the major absorption frequencies, using Table 2M-1 of characteristic frequencies.

Exercise M.2: Recording the IR Spectrum of an Organic Solid

Obtain an organic solid. Draw the structure of the assigned compound. Prepare a Nujol mull, a KBr pellet, or an IR card of the solid and obtain an IR spectrum. Interpret the major absorption frequencies, using Table 2M-1 of characteristic frequencies. If a Nujol mull is used, identify the frequencies associated with the mull.

Exercise M.3: Spectroscopic Identification of Unknowns

Look up the structures of the compounds given below. The instructor will assign one or more of the IR spectra, numbered A–F. Identify as many peaks as possible for each spectrum. Then determine the structure of the unknown, which will be one of the following compounds:

acetic anhydride	chlorobenzene	1-octene
allyl alcohol	cyclohexane	2-pentanone
aniline	diethylamine	propanoic acid
anisole	N, N-dimethylformamide	pyridine
benzaldehyde	ethyl benzoate	salicylaldehyde
benzonitrile	ethyl propanoate	styrene
benzyl alcohol	heptane	toluene
benzyl chloride	2-methyl-1-butanol	triethylamine

Questions *(continued on p. 125)*

1. Which absorbs at higher frequencies: a $C-H$ or $C-D$ bond? Explain.
2. Why does H_2 not give an IR spectrum?
3. Is 2,3-dimethyl-2-butene expected to show a $C=C$ stretching absorption? Explain.
4. How can cyclohexane and 2,2-dimethylpropane be differentiated using IR spectroscopy?
5. Explain why the $C=C$ stretch for a *trans*-disubstituted alkene is weaker than for a *cis*-disubstituted alkene.
6. Explain why primary amines and unsubstituted amides have two NH stretching absorptions.
7. Why do anhydrides show two carbonyl peaks?
8. Why is the carbonyl band in carboxylate salts shifted to frequencies below 1710 cm^{-1}?

Exercise M.3

Spectrum A
Major peaks (cm⁻¹):
3350, 3101,
1660, 1431,
1138, 1042

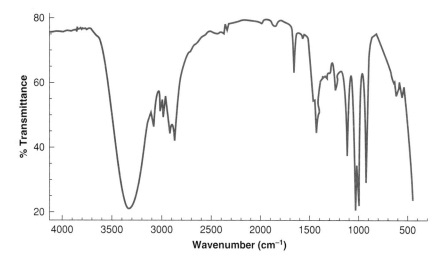

Spectrum B
Major peaks (cm⁻¹):
2980, 1725,
1384, 1175

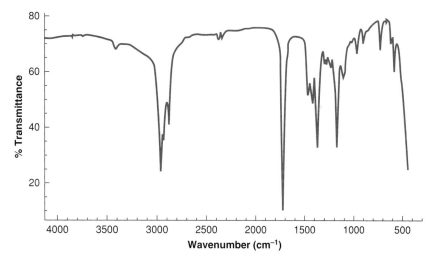

Spectrum C
Major peaks (cm⁻¹):
3044, 2940,
1630, 1516,
758, 700

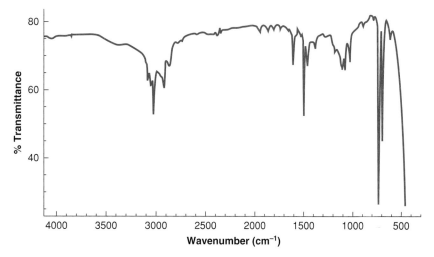

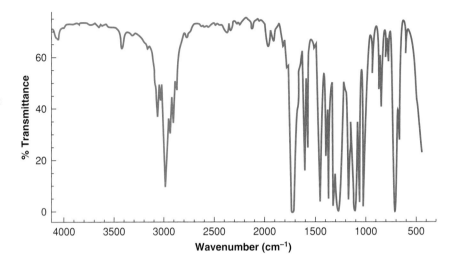

Spectrum D
Major peaks (cm⁻¹):
3091, 2987, 1735,
1621, 1470, 1289,
1127, 720

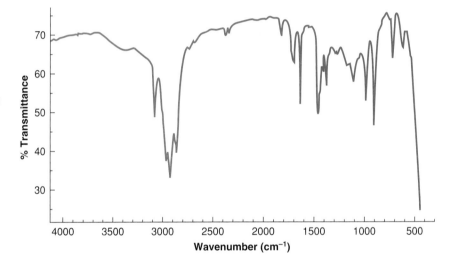

Spectrum E
Major peaks (cm⁻¹):
3100, 1660,
1479, 930

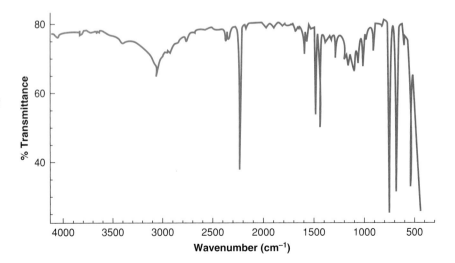

Spectrum F
Major peaks (cm⁻¹):
3091, 2247, 1507,
1470, 767, 700

9. Explain how to distinguish an aliphatic ether from an aromatic ether using IR spectroscopy.
10. Organic liquids must be dried before an IR spectrum is run. Why?
11. Explain how IR spectroscopy can be used to differentiate between an aldehyde, a ketone, and an ester.
12. What identifying features in the IR spectrum help to characterize a tertiary amine?
13. How can *m*-xylene be distinguished from *p*-xylene using IR spectroscopy?
14. How can a nitrile be distinguished from a terminal alkyne in the IR spectrum?
15. Draw the principal resonance forms of 3-buten-2-one. Why does the carbonyl stretching absorption occur at lower frequency than for the carbonyl for acetone?
16. What is Nujol? Explain the IR absorptions given by Nujol.
17. What important factor leads to the strong intensity of carbonyl absorption?

References

Cooper, J. W. *Spectroscopic Techniques for Organic Chemists.* New York: Wiley, 1980.

Kemp, W. *Organic Spectroscopy.* 3d ed. New York: W. H. Freeman and Company, 1991.

Lambert, J. B., Shurvell, H. F., Lightner, D., and Cooks, R. G. *Organic Structural Spectroscopy.* Upper Saddle River, NJ: Prentice-Hall, Inc., 1998.

Pavia, D., Lampman, G., and Kriz, G. *Introduction to Spectroscopy.* 3rd ed. Pacific Grove: Brooks/Cole, 2001.

Silverstein, R. M., Webster, F. X. *Spectrometric Identification of Organic Compounds.* 6th ed. New York: Wiley, 1998.

Technique N: Nuclear Magnetic Resonance Spectroscopy

Introduction

Nuclear magnetic resonance (NMR) spectroscopy is the most important technique for characterization of structure because it reveals specific placement and connectivity of atoms. While infrared (IR) spectroscopy can be used to help determine which functional groups are present in a molecule, NMR spectroscopy can be used to determine the exact location of functional groups. NMR spectroscopy gives valuable information about the carbon skeleton and about the molecule as a whole.

NMR spectroscopy can be applied for any nucleus that has a nuclear spin—that is, any nucleus that has an odd mass number and/or odd atomic number. This includes such nuclei as ^1H, ^{13}C, ^{19}F, ^{15}N, and ^{31}P. ^1H NMR (proton) and ^{13}C NMR are discussed here. For a particular type of atom, such as ^1H, an NMR spectrum shows absorptions, called signals, for each nonequivalent hydrogen: one signal for methane or ethane, two signals for propane or butane, three signals for pentane or hexane, etc. Four valuable types of information may be obtained from a ^1H NMR spectrum:

1. The number of signals in the spectrum indicates the number of nonequivalent protons in the molecule.
2. The position of the signals indicates the magnetic environment of the protons.
3. The intensity of the signals indicates the relative number of protons of each type in the molecule.
4. The splitting of the signal indicates the number of nonequivalent protons on the same or adjacent atoms to the proton being observed.

Theory

To understand the principles behind ^{1}H NMR spectroscopy, it is helpful to think of the nucleus as a tiny bar magnet. When placed in a magnetic field, only two spin states are possible for the hydrogen nucleus. The magnet can be aligned with the magnetic field (spin state = +1/2) or against the field (spin state = –1/2). Since slightly more energy (about 0.0001 calorie more) is required to oppose the magnetic field, there is a slight excess of nuclei aligned with the field (spin state = +1/2). When an electric field is applied at the appropriate radio frequency, these nuclei can absorb energy and "flip" their spins so that their spin now opposes the magnetic field.

How much energy does it take to accomplish this spin flip? The frequency of radiation depends upon the magnetic field strength. For a typical proton, radio-frequency radiation of 300 MHz is needed to reorient spins in a 70,460 Gauss magnetic field. Picture the hydrogen nucleus as being a child's top. The top precesses as it spins in a magnetic field.

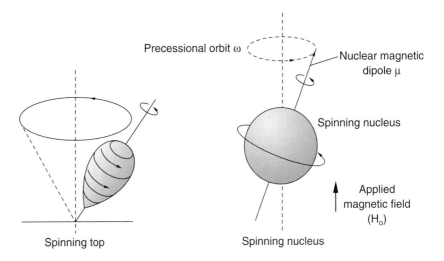

As the magnetic field strength is increased, the angular frequency (ω) increases also. In order to induce a spin flip, the radiation frequency must exactly match the precessional frequency. This condition, called resonance, differs for various protons within the molecule. The exact radio-frequency energy required for resonance of each type of proton depends upon the electronic environment of that proton. When the resonance condition is reached for each type of proton, the detector coil senses absorption of energy and a signal is observed. The number of signals in a ^{1}H NMR spectrum reflects the number of different types of protons in the molecule. The NMR spectrum is a plot of applied magnetic field (H_o) versus signal intensity at a constant radio frequency.

A typical NMR spectrum is shown in Figure 2N-1. The scale on the *x* axis is called the delta (δ) scale. This scale is independent of the type of NMR spectrometer used to record the spectrum:

Chemical shift: The difference in position of the center of a signal and a reference signal in ppm.

$$\text{chemical shift (ppm)} = \frac{\text{shift downfield from TMS (in Hz)}}{\text{spectrometer frequency (in MHz)}}$$

The δ scale on the NMR spectrum in Figure 2N-1 extends from 0 to 10 ppm. This is the range of values in which most protons come into resonance. The strength of the magnetic field increases from left to right. Therefore, going from left to right is considered moving upfield; going from right to left is moving downfield. A higher field strength is required to bring a proton that is upfield into resonance than for a proton that is downfield.

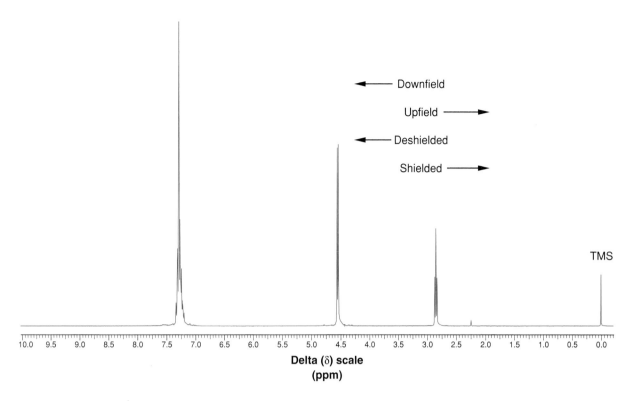

Figure 2N-1   The ^{1}H NMR scale

SOURCE: Reprinted with permission of Aldrich Chemical.

Chemical Shift in ^{1}H NMR Spectroscopy

The positions of the signals give information about the electronic environment of the protons in the molecule. The actual differences between field strengths required to bring individual protons in a molecule into resonance are extremely small—on the order of parts per million (ppm). For this reason, rather than measuring exact field strengths, a process of measuring relative field strengths has been developed. In ^{1}H NMR spectroscopy, the resonance frequency of a known type of proton is used as a reference; the field strengths of all other protons are compared to it. In general, the reference standard used is tetramethylsilane, $(CH_3)_4Si$, more commonly called TMS. This compound is ideal because all of its protons are equivalent and it comes into resonance at higher field strength than most other compounds. A very small amount of TMS is added to a solution of the compound dissolved in a deuterated solvent or other solvent that doesn't contain hydrogens (such as CCl_4). The signal for the TMS is, by definition, at $\delta 0$. The field strengths of all the protons in the molecule can be read directly from the spectrum. A proton giving a signal at $\delta 1.0$ has a chemical shift of $\delta 1.0$. The chemical shift of a proton in a molecule depends upon its electronic environment. Two factors account for the observed chemical shifts, local diamagnetic shielding and the anisotropic effect.

Local Diamagnetic Shielding and Anisotropic Effect

When placed in an external magnetic field, protons are shielded to a certain extent from the applied magnetic field by valence electrons surrounding the nuclei. The electrons establish a weak field that opposes the applied magnetic field. This is called shielding. Shielding is crucial to the utility of NMR as an analytical method. The extent of shielding is dependent upon the electron density around the proton. A higher applied field is needed to bring strongly shielded protons into resonance, so signals from these protons are observed far upfield. Tetramethylsilane is more highly shielded than most other organic molecules. Alkanes and cycloalkanes have small chemical shifts (signals appear at relatively high field strength) because the protons in these molecules are also strongly shielded.

Electronegative substituents bonded to carbon deshield protons bonded to that carbon by withdrawing electron density from around the protons. Because of the decreased electron density, the induced magnetic field is less and the strength of the applied magnetic field required to bring the proton into resonance is lower at a fixed frequency. These deshielded protons come into resonance at lower field strengths (downfield) in ^{1}H NMR spectroscopy (see Figure 2N-2).

In general, protons bonded to carbons bearing electronegative atoms are shifted downfield. The greater the electronegativity of the attached atom, the further downfield the proton. Therefore, the methyl protons of CH_3F appear further downfield (δ4.3) than the methyl protons of CH_3Cl (δ3.1) or CH_3Br (δ2.7). Similarly, the greater the number of electronegative substituents, the further downfield the proton. Thus, the chemical shifts of the methyl protons increase in this order: CH_3Cl (δ3.1), CH_2Cl_2 (δ5.35), $CHCl_3$ (δ7.25).

The chemical shift of a proton can be used to identify its functionality. Figure 2N-2 shows the correlation of chemical shifts of a variety of protons bonded to different functional groups. Notice that protons bonded to oxygen atoms (alcohols) and protons bonded to nitrogen atoms (amines and amides) come into resonance over a very wide range. The exact chemical shift of these protons will depend upon various factors such as the pH of the solution, the concentration of the solution, the temperature, and the degree of hydrogen bonding within the molecule. A detailed correlation of chemical shift values is given in Table 2N-1 and inside the back cover.

Proximity to an electronegative atom does not explain all of the observed chemical shifts. For example, vinyl protons (those attached to a carbon-carbon double bond) have chemical shifts in the range of δ5–6. This is significantly further downfield than both aliphatic protons and acetylenic protons (protons bonded to a carbon–carbon triple bond). Aromatic protons come into resonance even farther downfield—around δ7. Within the molecules, circulating electrons induce a magnetic field that either reinforces the applied field or opposes the applied field, depending upon the geometry of the molecule. This is called an **anisotropic effect.** For alkenyl and aryl protons, the induced field

Figure 2N-2

Relationship between functional group and chemical shift in ^{1}H NMR spectroscopy

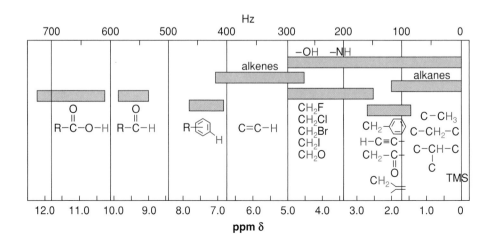

Table 2N-1 ^{1}H NMR Chemical Shift Values for Specific Types of Protons

Type of Proton	Approximate δ Value
Alkane	
methyl ($-CH_3$)	0.9
methylene ($-CH_2^-$), $-CH_2$	1.3
methine ($-CH-$)	1.4
Allyllic: $C=C-CH$	1.6–2.6
Benzylic: $ArC-H$	2–3
α to Carbonyl: $-CO-CH$	2.1–2.5
Alkyne: $RC\equiv CH$	2.5
α to Amines: RCH_2NH_2	2.5–3.0
α to Halide: RCH_2X	3–4
α to Oxygen: RCH_2O-	3–4
Vinyl: $C=C-H$	4.5–6.5
Aromatic: $Ar-H$	6.5–8.5
Aldehyde: $R-CHO$	9–10
Carboxylic acid: RCO_2H	10–12
Alcohols: $RO-H$	Variable: 2–5
Phenols: $ArO-H$	Variable: 4–7

reinforces the applied field and the chemical shifts of these protons are farther downfield. For acetylenic protons, the induced field opposes the applied field; more energy is needed to bring these protons into resonance and their chemical shifts are upfield.

Generally, signals in the chemical shift region of δ7-8 indicate aryl protons and a signal in the region of δ5–6 indicates alkenyl protons. Chemical shift ranges for various types of protons are given in Table 2N-1. Chemical shifts for aromatic protons, alkenyl protons and alkynyl protons result from the positioning of these groups with respect to the applied field. The field detected by the bonded protons may vary according to the region of relative shielding or deshielding as shown.

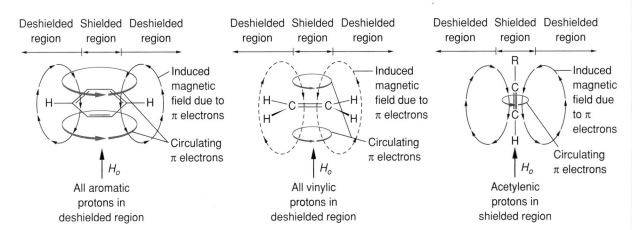

Equivalence of Protons in ^{1}H NMR Spectroscopy

The number of signals in a spectrum is equal to the number of nonequivalent protons in the molecule. Each set of equivalent protons in a molecule gives rise to a signal, the chemical shift of which is dependent upon the electronic environment of the protons.

There are two sets of equivalent protons in butane ($CH_3CH_2CH_2CH_3$): the six equivalent hydrogens on the methyl groups and the four equivalent hydrogens on the two methylene groups. A plane of symmetry divides the molecule into two equal halves. Butane has two ^1H NMR signals.

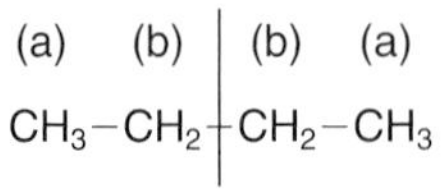

$$
\begin{array}{cc|cc}
\text{(a)} & \text{(b)} & \text{(b)} & \text{(a)} \\
\end{array}
$$
$$CH_3-CH_2\!\mid\!CH_2-CH_3$$

There are three sets of equivalent protons in pentane ($CH_3CH_2CH_2CH_2CH_3$): the six protons in the two methyl groups are equivalent; the four hydrogens on carbon-2 and carbon-4 are equivalent; and the two hydrogens in the middle, on carbon-3, are equivalent. Each of these sets of protons differs from the others. Pentane shows three ^1H NMR signals.

There are two ways to determine easily whether protons are equivalent. The first is to look for a plane of symmetry in the molecule. The second is to mentally substitute another element such as bromine for one of the hydrogen atoms and then compare the two structures by writing the IUPAC names. If the names are the same, the protons are equivalent; if the names are different, the protons are not equivalent. The structures shown here have, respectively, one, two, three, and four types of equivalent protons. The individual sets of equivalent protons are marked.

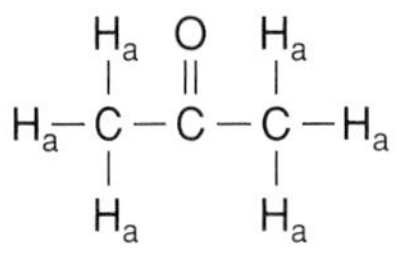

| all protons equivalent | 2 types of protons | 3 types of protons | 4 types of protons |

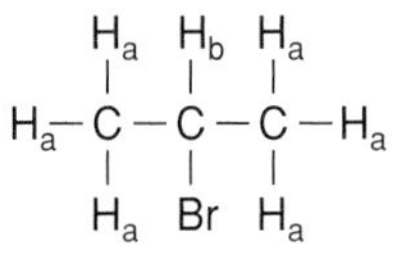

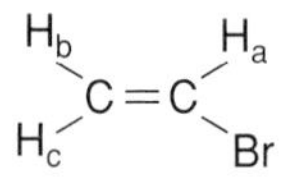

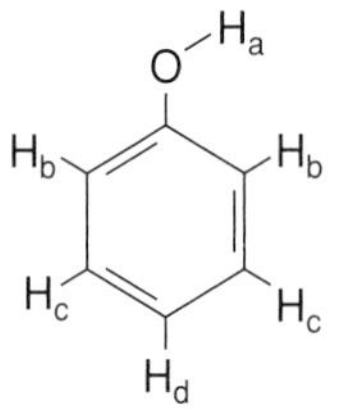

Integration in ^1H NMR Spectroscopy

The intensity of an NMR signal is proportional to the number of protons of each type in the molecule. An FT NMR instrument digitally integrates each signal area and provides ratios of hydrogens for each signal. In continuous-wave NMR (CW NMR), the ratios must be determined experimentally by electronically integrating the signals and measuring the heights of the integrals. There are two ways to determine the number of hydrogens under each signal from the measured heights. If the molecular formula is known, add up the heights of all of the integrals, then divide the heights of each of the integrals by the total heights of all the integrals and multiply by the total number of hydrogens in the formula.

$$\frac{\text{height of integral}}{\text{total heights of all integrals}} \times \frac{\text{total number}}{\text{of hydrogens}} = \frac{\text{number of hydrogens}}{\text{in signal}}$$

The number of hydrogens in a signal can be determined from the integrals. Figure 2N-3 shows the spectrum of a compound having the formula $C_4H_{10}O$, which has been both electronically integrated and measured with a ruler. The electronic integration gives ratios of 0.54 and 0.36; measuring the signals with a ruler gives 74 mm and 48 mm. The signals integrate to 6H and 4H, respectively:

Signal a: [74 mm / (74 mm + 48 mm)] × 10H = 6H
Signal b: [48 mm / (74 mm + 48 mm)] × 10H = 4H

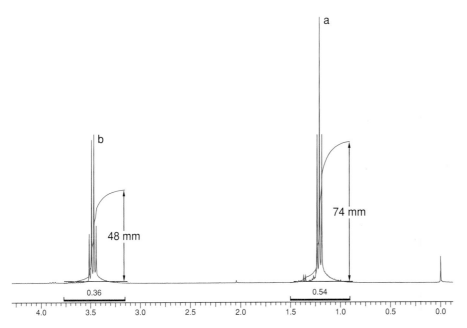

Figure 2N-3

Integrating an NMR spectrum

SOURCE: Reprinted with permission of Aldrich Chemical.

If the formula is not known, integrate by determining the ratio between the signals, rounding off to the nearest whole number or multiplying by a factor to produce a whole number. The electronic integration illustrates this process: signal a = 0.54/0.36 = 1.5; signal b = 0.36/0.36 = 1. The ratio of signal a:b is 1.5H:1H. To get a whole number, multiply by 2. This gives a ratio of signals of 3H:2H. Note that this is just a ratio—without knowing the formula, you can't know the exact number of protons in each signal. For this compound, the actual number of protons under each signal is 6H:4H.

In this book, integrals are generally not plotted so that the nature of each signal can be seen clearly.

Splitting (coupling) in ^{1}H NMR Spectroscopy

Another extremely valuable piece of information can be obtained from an NMR spectrum of a simple organic compound: By evaluating the appearance and number of peaks in each signal, the number of protons on adjacent carbons can be determined. Coupling with nonequivalent protons on adjacent (vicinal) carbon atoms can be observed in ^{1}H NMR spectroscopy. This type of coupling is called "vicinal" coupling. In essence, the protons on a carbon "see" the number of protons on directly adjacent carbon atoms. The number of peaks in a signal equals the number of nonequivalent protons on adjacent carbons (n) plus one more. This is called the "n + 1" rule of spin multiplicity. It applies to cases where vicinal coupling is expected. Magnetically (chemically) equivalent protons do not show spin-spin splitting. For example, in ethane, cyclobutane, and benzene, all of the protons in the molecule are equivalent. In each of these molecules, all protons are equivalent due to symmetry within the molecule.

```
            1
          1   1
        1   2   1
      1   3   3   1
    1   4   6   4   1
  1   5  10  10   5   1
```

Pascal's Triangle

If hydrogens on adjacent carbons are not equivalent, characteristic splitting patterns will be observed. The "n + 1" rule is based on counting the number of nonequivalent protons on directly adjacent carbons to which the proton is coupled and adding one to this number. If there are no nonequivalent protons on directly adjacent carbons, the signal will consist of only one peak (a singlet). With only one nonequivalent proton on adjacent carbons, the signal will be split into two peaks (a doublet); with two nonequivalent protons on adjacent carbons, the signal will be split into three peaks (a triplet); with three, the signal will be split into four peaks (a quartet); with four, the signal will be split into five peaks (a quintet); and with five, the signal will be split into six peaks (a sextet). The number of peaks in a signal and the relative areas of the peaks can be predicted using Pascal's Triangle. Usually anything greater than six peaks is called a multiplet. Since the outermost peaks may shrink into the baseline, it is sometimes difficult to determine the exact number of peaks in a multiplet. Abbreviations are frequently used to describe the peaks, such as "s" for singlet, "d" for doublet, "t" for triplet, "q" for quartet, and "m" for multiplet.

Succinic acid has the structure shown below. The spectrum of succinic acid shows two singlets. The CH_2 groups do not show splitting because the hydrogens are equivalent.

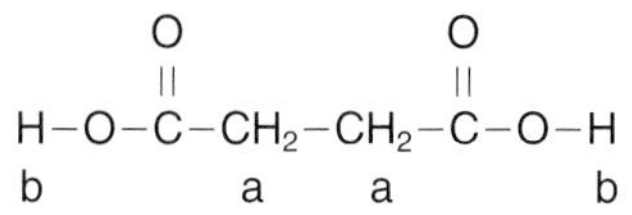

$$
\begin{array}{c}
\quad\quad\quad\quad O \quad\quad\quad\quad\quad\quad O \\
\quad\quad\quad\quad || \quad\quad\quad\quad\quad\quad || \\
H-O-C-CH_2-CH_2-C-O-H \\
b \quad\quad\quad\; a \quad\quad\; a \quad\quad\quad\quad\; b
\end{array}
$$

succinic acid

Butane has two types of hydrogens, labeled "a" and "b." Hydrogens "a" are coupled with hydrogens "b." The signal for protons "a" shows a triplet (n + 1 = 2 + 1 = 3). Hydrogens "b" will also be coupled with hydrogens "a," but will not be coupled with each other since they are equivalent. Therefore the signal for "b" will be a quartet (n + 1 = 3 + 1 = 4).

$$
\begin{array}{c}
CH_3-CH_2-CH_2-CH_3 \\
a \quad\;\; b \quad\;\; b \quad\;\; a
\end{array}
$$

butane

When nonequivalent adjacent carbons each have attached protons, spin-spin splitting is usually observed. This arises because the magnetic field experienced by the proton is increased or decreased by the spin state of a neighboring proton. The field experienced by the proton is slightly increased if the neighboring proton is aligned with the field and is decreased if the neighboring proton is aligned against the field. For a single adjacent proton, there are only two possible spin states. This would result in the signal H_a being split into two equal-sized peaks (a doublet) (see Figure 2N-4).

For a proton with two neighboring nonequivalent protons, there are three possibilities: (1) the neighboring protons may have both spins aligned with the field; (2) the neighboring protons may have one spin aligned with the field and one against the field (two possibilities); or (3) both neighboring protons may have their spins aligned against the field. The signal H_a will be split into a triplet, with the middle peak twice as large as the others. Since there is twice the probability that the two neighboring protons will have opposite spins, the middle peak will be twice as large as the outer peaks (Figure 2N-5).

For a proton with three neighboring nonequivalent protons, there are four possibilities: (1) all of the spins may be aligned with the field; (2) two of the spins may be aligned with the field with one spin aligned against the field (three different possibilities); (3) one of the spins may be aligned with the field and two of the spins may be aligned against the field (three different possibilities); and (4) all of the spins may be aligned against the field. This peak will be split into a quartet, with the middle two peaks approximately three times larger than the outer peaks (Figure 2N-6).

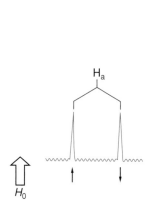

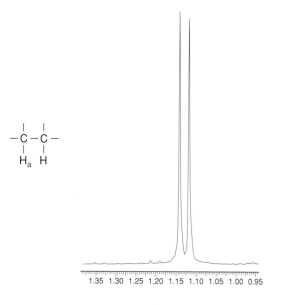

Figure 2N-4

Signal split into a doublet by one adjacent proton

SOURCE: Reprinted with permission of Aldrich Chemical.

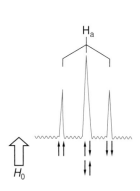

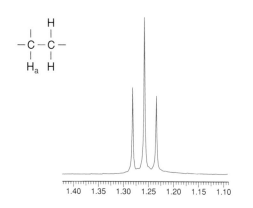

Figure 2N-5

Signal split into a triplet by two adjacent protons

SOURCE: Reprinted with permission of Aldrich Chemical.

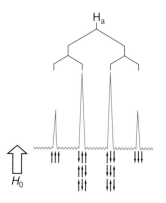

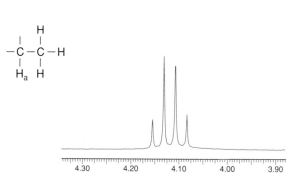

Figure 2N-6

Signal split into quartet by three adjacent protons

SOURCE: Reprinted with permission of Aldrich Chemical.

Some typical splitting patterns are illustrated in the NMR spectrum of ethyl acetate in Figure 2N-7.

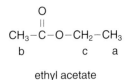

ethyl acetate

Figure 2N-7

^{1}H NMR spectrum of ethyl acetate

SOURCE: Reprinted with permission of Aldrich Chemical.

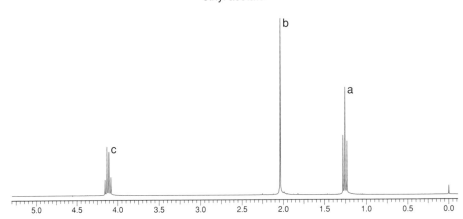

Coupling Constants

The strength of the spin-spin interaction between protons can be determined by measuring the distance between the peaks in a signal. This value, given the symbol J, is called the coupling constant. The J value is independent of field strength. Protons that are more strongly coupled have large J values (expressed in Hertz). For alkanes, J values for vicinal coupling are usually on the order of 7–8 Hz.

The situation is quite different for terminal alkenes. Here coupling can occur between protons on the same carbon (called geminal coupling) in addition to vicinal coupling. For vinyl protons, the coupling constant depends upon the relative position of the protons on the carbon–carbon double bond. Trans coupling constants (J_{trans}) are 12–23 Hz; cis coupling constants (J_{cis}) are around 5–12 Hz; and geminal coupling constants (J_{gem}) are 0–2 Hz. It is usually possible to differentiate between cis and trans protons by measuring the coupling constants. For example, the alkene regions of *cis*-3-chloropropenoic acid and *trans*-3-chloropropenoic acid are shown in Figure 2N-8. Notice that the coupling between protons H_a and H_b is smaller in *cis*-3-chloropropenoic acid than in *trans*-3-chloropropenoic acid. The structures of *cis*-3-chloropropenoic acid and *trans*-3-chloropropenoic acid are shown below:

$$\underset{\text{cis-3-chloropropenoic acid}}{\underset{H_a}{\overset{Cl}{>}}C=C\underset{H_b}{\overset{CO_2H_c}{<}}} \qquad \underset{\text{trans-3-chloropropenoic acid}}{\underset{Cl}{\overset{H_a}{>}}C=C\underset{H_b}{\overset{CO_2H_c}{<}}}$$

cis-3-chloropropenoic acid   *trans*-3-chloropropenoic acid

The acidic proton H_c comes into resonance around $\delta12$, and is not shown in either spectrum.

The magnitude of the coupling constant falls off dramatically with increasing distance. Splitting is observed for protons on the same carbon (if the protons are not equivalent) or for protons on adjacent carbons, unless carbon-carbon double bonds are present. For example, anisole has three sets of nonequivalent aromatic protons. Coupling of hydrogens on adjacent carbons, such as H_a–H_b (ortho coupling), is 7–8 Hz.

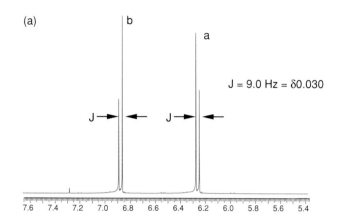

(a)

b

a

J = 9.0 Hz = δ0.030

J →| |← J →| |←

7.6   7.4   7.2   7.0   6.8   6.6   6.4   6.2   6.0   5.8   5.6   5.4

(b)

b

a

J = 13.5 Hz = δ0.057

J →| |← J →| |←

8.1   7.9   7.7   7.5   7.3   7.1   6.9   6.7   6.5   6.3   6.1   5.9   5.7

Figure 2N-8

Expanded ^{1}H NMR spectra (alkene protons) of a) *cis*-3-chloropropenoic acid and b) *trans*-3-chloropropenoic acid

Source: Reprinted with permission of Aldrich Chemical

0.22 0.33

7.5   7.0   6.5   6.0   5.5   5.0   4.5   4.0   3.5   3.0   2.5   2.0   1.5   1.0   0.5   0.0

$O-CH_3$

H_a $H_{a'}$

H_b $H_{b'}$

H_c

anisole

Figure 2N-9

^{1}H NMR spectrum of anisole

Source: Reprinted with permission of Aldrich Chemical.

Coupling for H_a–H_c (meta coupling) is about 3 Hz. Coupling is observed even though H_a and H_c are not on adjacent carbons. This effect is ascribed to conjugation of electron density through π orbitals. Coupling between H_a and $H_{b'}$ (para coupling) is small, but finite (J = 0–1 Hz). The NMR spectrum of anisole is shown in Figure 2N-9.

Protons in a Chiral Environment

There are two signals in the ^{1}H NMR spectrum of chloroethane. The methyl protons are each equivalent and are called homotopic. The methylene protons are equivalent in CDCl$_3$ solution and are called enantiotopic. The distinction between homotopic

CH_3CH_2Cl
chloroethane

Figure 2N-10

¹H NMR spectrum of (R)-(-)-2-methyl-2,4-pentanediol

SOURCE: Reprinted with permission of Aldrich Chemical.

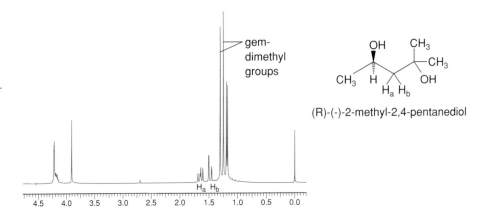

and enantiotopic is that enantiotopic protons are nonequivalent in a chiral environment. Methylene protons adjacent to a stereogenic center are called diastereotopic. They are nonequivalent and they can show different NMR signals. In (R)-(-)-2-methyl-2,4-pentanediol, protons H$_a$ and H$_b$ have different chemical shifts because they are in an achiral environment, being adjacent to a stereogenic center. The NMR spectrum of (R)-(-)-2-methyl-2,4-pentanediol, shown in Figure 2N-10, indicates the complexity of diastereotopic protons. Protons H$_a$ and H$_b$ have different chemical shift values at δ1.46 and δ1.62 and also show different splitting patterns, much different than predicted by the n + 1 rule. The gem-dimethyl groups are also not equivalent but are diastereotopic.

Diastereotopic Protons in Alkenes

Frequently NMR spectra are more complicated than predicted based on first-order approximations. For the examples in this section, the "n + 1" rule generally does not apply. Where it does apply, the coupling constants vary for the peaks within a signal. For example, in the spectrum of vinyl bromide, three signals (for the three types of nonequivalent protons) are expected.

vinyl bromide

Using the "n + 1" rule, H$_c$ should be split into a triplet (since there are two protons on the adjacent carbon); and H$_a$ and H$_b$ should be split into doublets. The actual spectrum of vinyl bromide shows about twelve overlapping peaks. (See Figure 2N-12).

The spectrum of vinyl bromide is complex, because the three vinyl protons are neither chemically equivalent nor magnetically equivalent. To be chemically equivalent all protons must behave identically. To be magnetically equivalent all protons must have the very same chemical shift and be coupled equivalently to all other protons.

In vinyl bromide, protons H$_a$, H$_b$, and H$_c$ are all different. The best way to predict the spectrum of an alkene such as vinyl bromide is with a tree diagram that illustrates the coupling for each proton one at a time. The tree diagram is shown in Figure 2N-11 and the ¹H NMR spectrum of vinyl bromide is shown in Figure 2N-12.

H$_c$ will be split into a doublet by H$_a$. The splitting between the peaks will be around 15 Hz. This coupling is labeled J$_{ac}$ to indicate that the coupling is between protons a and c. The next strongest coupling of H$_c$ is with H$_b$; each of the peaks will be further split into a doublet. The distance between these peaks will be around 6 Hz (J$_{bc}$). The signal for H$_c$ will be a doublet of doublets.

For H$_b$, the strongest coupling will be with H$_c$ (cis coupling). H$_b$ will be split into a doublet by H$_c$ with a coupling constant of 6 Hz. Each line will also be split into a doublet by H$_a$, with a coupling constant of around 2 Hz. The signal for H$_b$ will appear as a doublet of doublets. H$_a$ will be split into a doublet by H$_c$ (trans coupling) and into a further doublet by

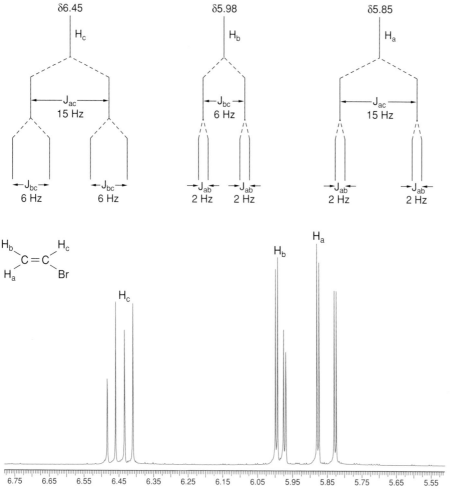

Figure 2N-11

Tree diagram of vinyl bromide

Figure 2N-12

^{1}H NMR spectrum of vinyl bromide

SOURCE: Reprinted with permission of Aldrich Chemical.

H_b (geminal coupling). The signal for H_a will resemble a doublet of doublets. Because the chemical shifts of the protons are so similar, the spectrum appears to be even more complex.

Aromatic compounds, too, show complex splitting due to the magnetic nonequivalence of the protons. In the spectra of some simple alkyl aromatic compounds, such as toluene and ethylbenzene, the signal for all aromatic protons appears to be a singlet. This occurs when the chemical shifts and coupling constants of the protons are very similar. If a splitting pattern is observed, it can yield valuable structural information about the substitution pattern of the aromatic compound. The spectrum for a *para*-disubstituted benzene, in which at least one substituent is not alkyl, commonly appears as an apparent pair of doublets in the region δ7–8. Other substitution patterns give much more complex splitting in this region. The ^{1}H NMR spectra of *p*-bromonitrobenzene and *o*-bromonitrobenzene are shown in Figure 2N-13.

Protons on Heteroatoms
Hydroxyl Protons

Hydroxyl protons (OH) show ^{1}H NMR signals anywhere from δ2 to δ5. The exact chemical shift is concentration dependent. In more concentrated samples, the OH peak is shifted downfield. In more dilute solutions, such as those generally used for 200–300 MHz

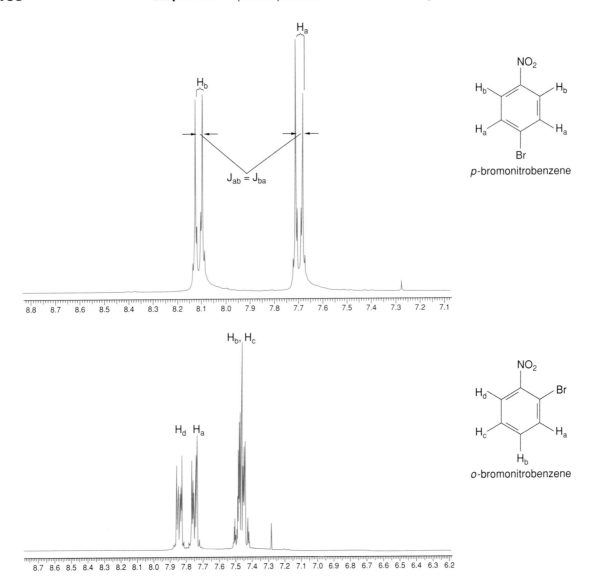

Figure 2N-13   Expanded aromatic regions of *p*-bromonitrobenzene (top) and *o*-bromonitrobenzene (bottom)

SOURCE: Reprinted with permission of Aldrich Chemical.

spectra, the OH signal is observed around $\delta 1–2$. The hydroxyl peak for phenols is also concentration dependent, appearing between $\delta 4$ and $\delta 7.5$. Clearly the chemical shift by itself cannot be used to conclude that the sample is an alcohol or a phenol.

One way to easily identify which peak is due to an OH is to add a drop of D_2O to the sample, shake the tube, and rerun the NMR spectrum. This causes the OH peak to decrease in size or to disappear due to the exchange between the hydrogen and deuterium. (Deuterium has a nuclear spin, but it comes into resonance at different field strengths than protons.)

Coupling may or may not be observed between the hydroxyl proton and other hydrogens on adjacent carbons. If there is rapid exchange of the hydroxylic proton between alcohol molecules, the hydroxyl proton will be a singlet. It is as though the hydroxyl proton weren't on the molecule long enough to couple with hydrogens on adjacent carbons.

There are three conditions under which coupling will be observed: at low temperatures, with extremely pure alcohols, and with dilute solutions. Cooling slows down the rate of exchange, so that under very cold conditions, coupling can be observed. Extremely pure alcohols, which have not even a trace of acid present, will show coupling between hydroxylic protons and hydrogens on adjacent carbons, following the "n + 1" rule. The OH peak for highly purified ethanol is a triplet. However, under room-temperature conditions for alcohols of ordinary purity, the CW NMR spectrum of an alcohol shows a singlet for the hydroxyl proton. For high-field spectra, the OH protons often show coupling because the concentration of sample is very small, limiting hydrogen exchange between OH groups.

Protons on Nitrogen

There are many similarities between the NMR spectra of alcohols and amines. Like protons on oxygen, protons on nitrogen come into resonance anywhere between $\delta 1$ and $\delta 5$. Like alcohols, amines undergo rapid intermolecular exchange, so no splitting is observed between the NH and hydrogens on adjacent carbon atoms in CW NMR spectra. Splitting is usually observed in FT NMR spectra. The amine protons show up typically as a sharp singlet. Protonation of the amine with strong acid (to generate the quaternary ammonium salt) slows the rate of exchange, allowing coupling to be observed. Because nitrogen has a nuclear spin itself, the coupling tends to be complicated.

The NH bond of amides undergoes exchange much more slowly, so coupling between hydrogen on adjacent carbons occurs. However, the NH signal is so broad (due to a phenomenon called "nuclear quadrupole broadening") that splitting may be obscured. Sometimes it is easy to overlook the NH signal of an amide, since it may look like a small "hump" on the baseline. However, integration should still indicate the presence of a hydrogen.

Solvents for ^{1}H NMR Spectroscopy

The most common solvent for NMR spectroscopy is deuterochloroform ($CDCl_3$). Other aprotic solvents can also be used such as deuteroacetone (CD_3COCD_3) and deutero-DMSO (CD_3SOCD_3). Commercially available $CDCl_3$ contains a small amount of $CHCl_3$, which gives a singlet at $\delta 7.2$. Small amounts of water will also appear in the spectrum unless special effort is made to keep the solvent away from water. The chemical shift of the water peak is about $\delta 1.4$ for FT NMR spectra. Similarly, peaks will be seen at $\delta 2.04$ for CHD_2COCD_3 (an impurity in deuteroacetone) and at $\delta 2.49$ for CHD_2SOCD_3 (an impurity in deutero-DMSO). Both signals show closely spaced quintets. Table 2N-2 lists properties of NMR solvents.

Table 2N-2 Properties of NMR Solvents

Deuterated NMR Solvents	Structure	Solvent impurity Chemical shift (δ) and multiplicity	
		^{1}H NMR	^{13}C NMR
Acetone	CD_3COCD_3	2.04(5)	29.8(7), 206.5(7)
Acetonitrile	CD_3CN	1.93(5)	1.3(7), 118.2(7)
Benzene	C_6D_6	7.24(1)	128.0(3)
Chloroform	$CDCl_3$	7.24(1)	77.5(3)
Deuterium oxide	D_2O	4.63–4.67(1)	—
Dimethyl sulphoxide	CD_3SOCD_3	2.49(5)	39.7(7)
Methyl alcohol	CD_3OD	4.78(1), 3.30(5)	49.0(7)

Solvents used for spectra in this book are $CDCl_3$ and CD_3SOCD_3.

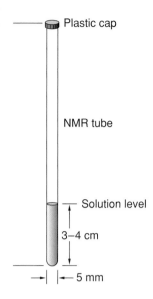

Plastic cap

NMR tube

Solution level

3–4 cm

5 mm

Figure 2N-14

CW NMR tube

How to Prepare a Sample for 60–90 MHz CW ^{1}H NMR Spectroscopy

Wear gloves and work in the hood when preparing solutions for NMR analysis. Prepare solutions that contain about 10 to 15% of the organic solute in the NMR solvent. Measure 40–50 mg of the sample into a small vial. Add 0.4–0.5 mL of the NMR solvent containing TMS with a clean, dry syringe or new Pasteur pipet. Cap and shake the vial to mix thoroughly before transferring the solution to the NMR tube with a clean, dry pipet. Undissolved solids can severely reduce the quality of spectra, so solid samples should be filtered using a Pasteur filter pipet into a clean, dry NMR tube. The volume of solution in the tube should be 0.4–0.5 mL. Add additional pure solvent as necessary.

It is very easy to contaminate the solvent if a dirty Pasteur pipet is used to remove solvent from the supply. **Be sure to use a clean, dry pipet every time that solvent is removed from the solvent bottle.** It is very frustrating to try to determine the composition of an unknown if added impurity peaks must be accounted for as well.

How to Prepare a Sample for FT NMR

Sample preparation for and operation of high-field FT NMR spectrometers is significantly different than for CW instruments. Solutions to be analyzed by FT NMR should be approximately ten times less concentrated than for low-field work. Weigh out 5–10 mg of sample into a clean, tared vial. Add about 0.75 mL of a deuterated solvent, usually $CDCl_3$ containing TMS, and thoroughly mix to dissolve. Using a Pasteur pipet, transfer the dissolved sample to a Pasteur filter pipet that is placed in a clean, dry 5-mm NMR tube. Adjust the volume to the correct column length specified by the instructor by adding pure solvent as needed. Carefully cap the NMR tube and shake to thoroughly mix.

Structural Identification from the NMR Spectrum

The following six steps should be taken to determine the structure of an organic compound from its NMR spectrum.

1. **From the formula, if known, calculate the elements of unsaturation (unsaturation number or degrees of unsaturation).** A good first step is to find out if the organic molecule contains rings or multiple bonds. To do this, calculate the unsaturation number (U) from the molecular formula

$$U = \frac{(2C+2)-H}{2}$$

where C is the number of carbon atoms and H is the number of hydrogen atoms in the molecule. If the molecule contains atoms other than carbon and hydrogen, the following rules apply in calculating the unsaturation number:

Halogens count the same as hydrogens.
Nitrogen atoms count as 1/2 carbon.
All other atoms, such as oxygen or sulfur, are not counted.

Three examples are shown below:

Molecular formula	Unsaturation number
$C_4H_6O_2$	2
$C_6H_4Cl_2$	4
CH_5N	0

The unsaturation number is equal to the total number of rings and/or multiple bonds in a compound. Once the unsaturation number is known, the number of double bonds, triple bonds, or rings can be determined. A ring or a double bond counts as one element of unsaturation, a triple bond counts as two elements of unsaturation, and a benzene ring counts as four elements of unsaturation.

Examples

$U = 1$

CH_3CCH_3 (with O double-bonded)

Unsaturation number	Implication
0	No double bonds, triple bonds, or rings
1	One ring or one double bond
2	One triple bond or two rings or double bonds or some combination thereof
3	One triple bond and one double bond or ring or some combination thereof
4	Often a benzene ring (1 degree for the ring and 1 for each of the double bonds)

$U = 2$

$U = 3$

$U = 4$

2. **Examine the spectrum.** Determine the number of equivalent types of protons from the number of signals for the compound.
3. **Integrate the spectrum.** Determine the ratios of protons.
4. **Determine the types of protons present.** Examination of the number of signals and chemical shifts often reveals the general nature of the compound.

Chemical shift	Implication
$\delta 0.9–1.5$	Methyl, methylene, and methine protons
$\delta 2–2.5$	Protons adjacent to a carbonyl group, benzylic hydrogens, or allylic hydrogens.
$\delta 2.5$	Sharp singlet, (1H) - alkynyl proton
$\delta 3–4$	Protons on a carbon adjacent to an oxygen or halogen
$\delta 5–6$	Alkene protons: coupling constant of 0–2 Hz—geminal alkene; coupling constant of 5–12 Hz—*cis* alkene; coupling constant of 12–23 Hz—*trans* alkene; or proton on carbon attached to two oxygens or halogens
$\delta 7–8$	Aromatic ring protons
$\delta 9–10$	Aldehyde proton
$\delta 10–13$	Carboxylic acid proton

5. **Look for characteristic splitting patterns.** Splitting patterns and coupling constants further enhance the information obtained from the chemical shifts. Characteristic patterns are listed below:
   a. An ethyl group occurs as a triplet (which integrates to 3H) and a quartet (which integrates to 2H). The triplet is always further upfield.

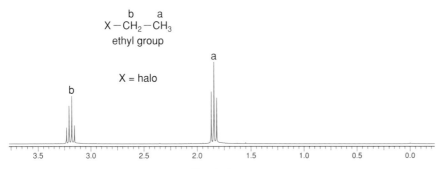

SOURCE: Reprinted with permission of Aldrich Chemical.

b.   An isopropyl group occurs as a doublet (which integrates to 6H) and a multi-plet (a heptet, actually, which integrates to 1H). Often, the outer peaks of multi-plets can't be seen.

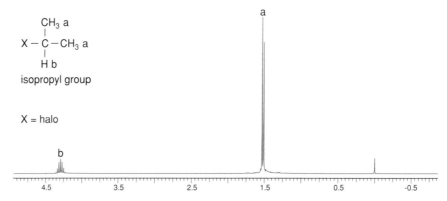

SOURCE: Reprinted with permission of Aldrich Chemical.

c.   A *tert*-butyl group appears as a singlet (which integrates to 9H).

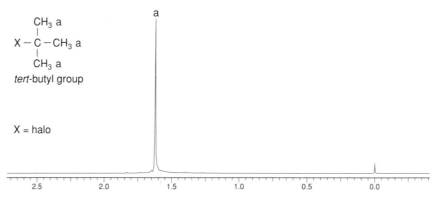

SOURCE: Reprinted with permission of Aldrich Chemical.

d.   Barring overlapping signals, a singlet that integrates to 3H is usually one methyl group; a signal that integrates to 6H is usually two equivalent methyl groups; and a signal that integrates to 9H is usually three equivalent methyl groups.

6. **Put it all together.** Draw a structure that fits the elements of unsaturation, the chemical shifts, and splitting patterns. To practice these skills, try the following problem. Additional problems are given at the end of this exercise.

Example: A compound with the formula $C_5H_{10}O$ gives the ^1H NMR spectrum shown in Figure 2N-15. Determine the structure of the compound.

Solution

Unsaturation Number:	$U = \dfrac{(2C+2)-H}{2} = \dfrac{(2(5)+2)-10}{2} = 1$
Integration:	Signal a is 0.52; signal b is 0.26; signal c is 0.09. Dividing through by smallest number gives a ratio of approximately 6H for signal a, 3H for signal b, and 1H for signal c.
Spectral Analysis:	The presence of three signals indicates three types of hydrogens. One element of unsaturation means there is a ring or a

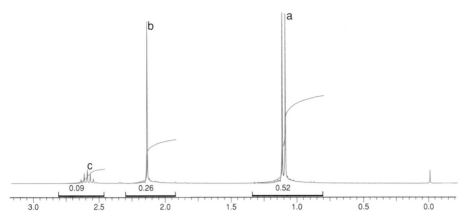

Figure 2N-15

Integrated ^{1}H NMR spectrum of an unknown compound

SOURCE: Reprinted with permission of Aldrich Chemical.

double bond. No signal at δ7–8 indicates the absence of aromatic protons. No signal at δ5–6 indicates the absence of alkenyl protons. A singlet around δ2.1 implies a methyl group attached to a carbonyl since there are no aromatic protons. The absence of signals for protons in the δ3.5–4 region rules out the possibility of a cyclic ether.

Splitting patterns: **Signal a:** a doublet of 6H implies two methyl groups attached to a carbon with one H; **Signal b:** a singlet of 3H implies one methyl group attached to a carbon with no H; **Signal c:** a heptet of 1H implies one hydrogen on a carbon that is adjacent to carbons with 6H (outer peaks of multiplets are often not seen). This splitting pattern is characteristic of an isopropyl group.

Structure: The compound must have a carbonyl group, since it is obviously not an alkene. Since there is no aldehydic proton signal around δ9–10, the compound could be a ketone. The methyl group at δ2.1 must be directly bonded to the carbonyl. The isopropyl group must be bonded on the other side of the carbonyl.

Identity: The compound is 3-methyl-2-butanone.

Practice problems for ^{1}H NMR spectroscopy are given in Exercise N.3.

3-methyl-2-butanone

Introduction to ^{13}C NMR Spectroscopy

In addition to hydrogen (^{1}H), other nuclei that possess an odd number of protons or neutrons (or both) have nuclear spins and are capable of showing resonance. In organic chemistry, the most important of these nuclei is ^{13}C. Like ^{1}H NMR spectroscopy, ^{13}C nuclear magnetic resonance spectroscopy (^{13}C NMR) can give detailed information about the different types of carbon in the compound.

Unfortunately, ^{13}C NMR spectroscopy is roughly 6000 times less sensitive than ^{1}H NMR spectroscopy because ^{13}C is much less abundant than ^{1}H (1.108% abundant versus 99.8%). This means that only about one of every 100 carbon atoms will have a nuclear spin. The remaining 99 atoms are ^{12}C, which do not have a nuclear spin.

Due partly to the low natural abundance of ^{13}C nuclei, ^{13}C NMR spectra must be obtained by Fourier Transform (FT) analysis of the free induction decay. In high-field NMR, the entire ^{13}C spectral width is irradiated briefly leading to excitation of each type of ^{13}C nucleus present. This process is repeated many times over a period of several minutes or hours. The resulting free induction decay is processed using computer-programmed FT methods to give ^{13}C NMR spectra. For further information, consult the references at the end of this Technique M.

The spectral width for ^{13}C nuclei is much larger than for ^1H nuclei (about $\delta 200$). Therefore, it is unusual to have different types of ^{13}C nuclei that have exactly the same chemical shift. Just like ^1H NMR, ^{13}C nuclei have characteristic chemical shift values. For example, carbon atoms of methyl groups have chemical shifts of $\delta 0$–35. Carbons bonded to electronegative atoms such as bromine, oxygen, and nitrogen appear downfield. Aromatic carbons have chemical shifts of approximately $\delta 110$–165 while alkenyl carbons have chemical shifts of approximately $\delta 100$–150. Typical values of ^{13}C chemical shifts are listed in Table 2N-3.

^{13}C Nuclei have a spin of $^1/_2$, meaning that there are two spin states, just as in ^1H NMR. ^{13}C NMR spectra are usually obtained using noise decoupling, a technique of simultaneous irradiation of all carbon and hydrogen nuclei. This technique simplifies ^{13}C spectra by removal of all C−H coupling. This, together with the unlikely positioning of ^{13}C nuclei next to one another, results in singlets for all carbon nuclei. Noise-decoupled ^{13}C spectra are not integrated, as peak areas do not correspond to the number of carbons in the signal.

The ^{13}C NMR spectrum for 2-vinylpyridine in $CDCl_3$ solvent is shown in Figure 2N-16. If you look closely, you can observe seven signals, corresponding to the seven nonequivalent carbons in 2-vinylpyridine. In general, there is one signal for every type

Table 2N-3 ^{13}C NMR Chemical Shifts

Type of carbon	Approximate δ value
Alkyl	$\delta 0$–50
C−Cl, C−Br	$\delta 20$–50
C−N; amines	$\delta 30$–60
C−O; alcohols, ethers	$\delta 50$–275
Alkynyl	$\delta 70$–90
Alkenyl	$\delta 100$–150
Aromatic	$\delta 110$–165
Nitrile	$\delta 115$–125
C=O; carboxylic acids, esters	$\delta 165$–180
C=O; ketones, aldehydes	$\delta 180$–220

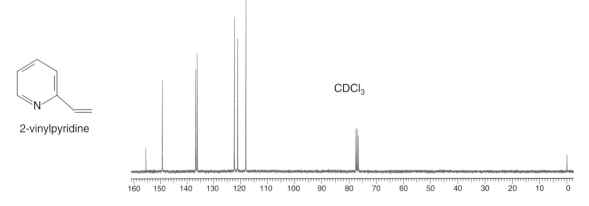

Figure 2N-16 ^{13}C NMR spectrum of 2-vinylpyridine

Source: Reprinted with permission of Aldrich Chemical.

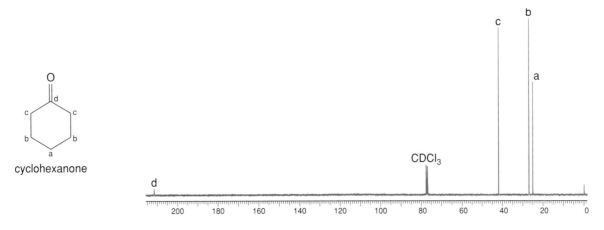

Figure 2N-17   ¹³C NMR spectrum of cyclohexanone

SOURCE: Reprinted with permission of Aldrich Chemical.

of carbon present in the molecule. The three peaks at δ77.5 are due to the solvent, deuterochloroform.

Cyclohexanone has four different types of carbons and four signals in the ¹³C NMR. The ¹³C NMR spectra of cyclohexanone is shown in Figure 2N-17. The carbonyl carbon is far downfield, around δ215, and is very small. This signal may be easily missed. The triplet at δ77.5 is due to deuterochloroform.

Off-Resonance Decoupling

A second decoupling technique is sometimes used in ¹³C NMR spectroscopy. This technique, called off-resonance decoupling, eliminates all coupling between carbons and hydrogens that are farther than one bond apart. This means that each carbon is coupled only to hydrogens that are directly attached to that carbon. Each signal will be a singlet, doublet, triplet, or quartet depending upon whether there are 0, 1, 2, or 3 hydrogens attached to the carbon atom. In the case of very simple organic compounds, this technique can be used to determine the number of hydrogens bonded to each carbon. Splitting patterns in off-resonance decoupled ¹³C NMR spectra are listed here.

Structure	Number of hydrogens bonded	¹³C NMR signal
$-\overset{\textstyle\mid}{\underset{\textstyle\mid}{C}}-$	0	singlet (s)
$-\overset{\textstyle\mid}{\underset{\textstyle\mid}{C}}-H$	1	doublet (d)
$H-\overset{\textstyle\mid}{\underset{\textstyle\mid}{C}}-H$	2	triplet (t)
$H-\overset{\textstyle H}{\underset{\textstyle\mid}{C}}-H$	3	quartet (q)

Solvents for ^{13}C NMR Spectroscopy

The most common solvent in NMR is $CDCl_3$. In the spectrum of 2-vinylpyridine, notice that the signal for $CDCl_3$ consists of three peaks of equal height, due to C−D coupling, at $\delta 77.5$. Deuterium has a spin of 1, resulting in three spin states. High-field instruments lock on the deuterium NMR signal. Other common deuterated solvents for ^{13}C NMR include C_6D_6 and CD_3COCD_3, which give signals at $\delta 128.0$ for deuterobenzene and at $\delta 29.8$ and $\delta 206.5$ for deuteroacetone. Table 2N-2 on page 139 lists properties of NMR solvents.

How to Prepare a Sample for ^{13}C NMR Spectroscopy

Weigh out 20–40 mg of sample into a clean, tared vial. Add about 0.75 mL of a deuterated solvent, usually $CDCl_3$, and mix to dissolve. Using a Pasteur pipet, transfer the dissolved sample to a Pasteur filter pipet that is placed in a clean, dry 5-mm NMR tube. Using a ruler, adjust the liquid column length to the height specified by the instructor by adding pure solvent. Carefully cap the NMR tube and shake to thoroughly mix.

General Approach to Determining an Unknown Structure

^{13}C NMR spectra are normally used in conjunction with IR and ^{1}H NMR spectra. Taken together, the structures of most simple organic compounds can be deduced. For some simple organic compounds, proton-decoupled or off-resonance decoupled ^{13}C NMR spectra may be sufficient to determine the structure of the compound. A general procedure of seven steps for structure determination follows.

1. **Determine the elements of unsaturation (unsaturation number or degrees of unsaturation) from the formula, if known.** This number will indicate the number of double bonds, triple bonds, and/or rings that are in the molecule.

$$U = \frac{(2C + 2) - H}{2}$$

2. **Examine the IR spectrum (if available).** The IR spectrum is helpful in determining which functional groups are present in the molecule.
3. **Examine the ^{1}H NMR spectrum (if available).** The integration, splitting patterns, and chemical shifts of the protons can provide important information about the general structure of the compound.
4. **Examine the proton-decoupled ^{13}C NMR spectrum.** Determine the number of magnetically unique carbon atoms from the number of signals in the spectrum. Remember that the height of the peak in the proton-decoupled ^{13}C NMR spectrum does not usually correlate with the number of carbons in the compound; however, a very small peak is frequently an indication of a quaternary carbon.
5. **Examine the off-resonance decoupled ^{13}C NMR spectrum (if available).** Determine the number hydrogens attached to each carbon atom. (Alternately, this information can be obtained from the ^{1}H NMR spectrum.)
6. **Use the ^{13}C Chemical Shift Table (Table 2N-3).** Correlate the chemical shift with the hybridization of the carbon and proximity to electronegative atoms.
7. **Assemble the fragments so that each atom has the correct number of bonds.**

An example will illustrate this approach.

Example: Figure 2N-18 shows the proton noise-decoupled spectrum of an unknown organic molecule with the molecular formula $C_{11}H_{15}NO$. The three signals at $\delta 77.5$ are due to the deuterochloroform. The multiplicities were obtained from the off-resonance

**Chemical Shifts
and Multiplicities:**
a. 13.2q
b. 47.3t
c. 112.1d
d. 128.1s
e. 133.4d
f. 152.6s
g. 191.2d

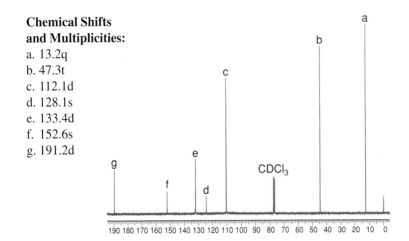

Figure 2N-18

^{13}C NMR spectrum of an unknown compound

SOURCE: Reprinted with permission of Aldrich Chemical.

spectrum (s = singlet, d = doublet, t= triplet, q = quartet). The IR and ^{1}H NMR spectra of this compound are not available. From the noise-decoupled and off-resonance decoupled ^{13}C NMR alone, determine the structure of this compound.

Solution

First, calculate the elements of unsaturation. The unsaturation number is 5. There is a good possibility that the compound is aromatic.

Next, count the number of signals in the ^{13}C NMR spectrum. Frequently there may be signals in the spectrum due to the solvents, such as a singlet for TMS at $\delta 0$ and three peaks of equal height at $\delta 77.5$ for CDCl$_3$. This spectrum shows seven singlets, representing seven different types of carbon in the molecule. These peaks are labeled a–g. Notice that there are eleven carbons in the formula but only seven unique types of carbon.

Much information may be ascertained from the chemical shifts of the signals. Two of the peaks are upfield. Peaks "a" and "b" must be sp^3 hybridized. Peak "b" is further downfield, so the carbon must be bonded to an electronegative atom like nitrogen, oxygen, or a halogen. Since no halogen atom is present, this carbon must be directly bonded to the nitrogen or oxygen atom. There are four signals between $\delta 110$ and $\delta 150$, which is the aromatic region. Having four magnetically different carbon atoms would seem to indicate a *para-* disubstituted aromatic ring. Peak "g" is far downfield, indicative of a carbonyl (C=O) group. Therefore the carbon represented by peak "b" must be attached to a nitrogen atom. It might be possible to infer the structure of the molecule from this spectrum alone.

Now consider the data from the off-resonance decoupled spectrum. Since peak "a" is split into a quartet, there must be three hydrogens bonded to the carbon. This is a methyl group (CH$_3$–). Peak "b" is split into a triplet; hence it is a CH$_2$– group. Peaks "c" and "e" are aromatic and both have one hydrogen attached; peaks "d" and "f" are singlets with no hydrogens attached. This must be the point of attachment of the substituents. Peak "g" is a doublet, indicating that one hydrogen is bonded to the carbonyl. This must then be an aldehyde functionality. The fragments are shown here.

Fragments "a" and "b" obviously combine to form an ethyl group. The formula shows two more carbon atoms. It is apparent that there are actually two ethyl groups in the compound (in order to satisfy the valency of the nitrogen atom). The identity of the organic compound is 4-(N,N-diethylamino)benzaldehyde. (For further information on assigning carbons labeled "c," "d," "e," and "f," consult one of the references at the end of Technique N.)

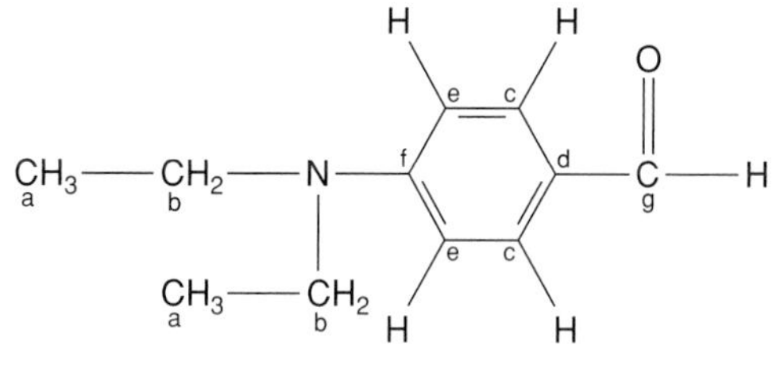

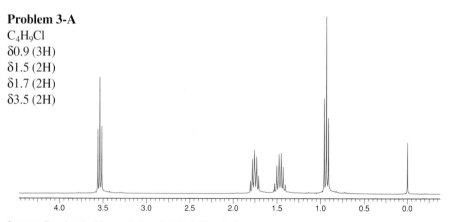

Exercise N.1: Recording a ^{1}H NMR Spectrum

Prepare a sample for ^{1}H NMR, as directed by the instructor. Record and analyze the spectrum. Prepare a detailed report, interpreting the spectrum.

Exercise N.2: Recording a ^{13}C NMR Spectrum

Prepare a sample for ^{13}C NMR, as directed by the instructor. Record and analyze the spectrum. Prepare a detailed report, interpreting the spectrum.

Exercise N.3: ^{1}H NMR Spectral Problems

From the spectral information and molecular formula, propose structures that are consistent with the data for 3-A through 3-G. The spectra are 300 MHz FT-NMR ^{1}H NMR spectra. The NMR solvent is CDCl$_3$ containing TMS.

Problem 3-A
C$_4$H$_9$Cl
δ0.9 (3H)
δ1.5 (2H)
δ1.7 (2H)
δ3.5 (2H)

SOURCE: Reprinted with permission of Aldrich Chemical.

Problem 3-B
C$_4$H$_9$Cl
δ1.0 (6H)
δ1.9 (1H)
δ3.4 (2H)

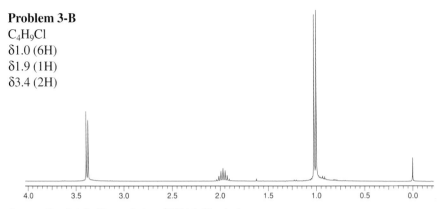

SOURCE: Reprinted with permission of Aldrich Chemical.

Problem 3-C
C$_7$H$_{16}$
δ0.9 (12H)
δ1.0 (2H)
δ1.6 (2H)

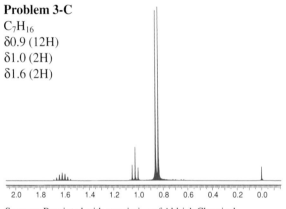

SOURCE: Reprinted with permission of Aldrich Chemical.

Problem 3-D
C$_5$H$_{10}$O
δ1.1 (6H)
δ2.5 (4H)

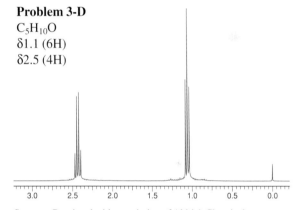

SOURCE: Reprinted with permission of Aldrich Chemical.

Problem 3-E
C$_8$H$_8$O
δ2.5 (3H)
δ7.2−7.7 (4H)
δ10.0 (1H)

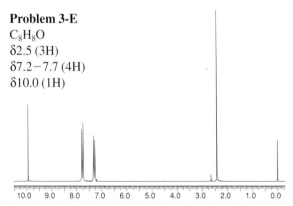

SOURCE: Reprinted with permission of Aldrich Chemical.

Problem 3-F
C$_6$H$_{14}$O
δ0.9 (6H)
δ1.6 (4H)
δ3.4 (4H)

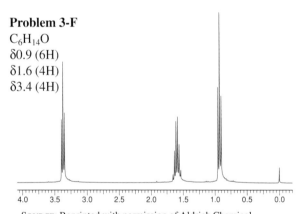

SOURCE: Reprinted with permission of Aldrich Chemical.

Problem 3-G.   From the data given below, determine the structure of each of the following unknown compounds:

1. C$_3$H$_7$Br: δ1.71 doublet (6H); δ4.29 heptet (1H)
2. C$_4$H$_9$Cl: δ1.62 singlet
3. C$_4$H$_8$O: δ1.1 triplet (3H); δ2.1 singlet (3H); δ2.5 quartet (2H)
4. C$_4$H$_{10}$O$_2$: δ3.4 singlet (6H); δ3.6 singlet (4H)
5. C$_8$H$_{10}$O: δ1.5 triplet (3H); δ4.1 quartet (2H); δ7.0–7.8 multiplet (5H)

Exercise N.4: ^{13}C NMR Spectral Problems

Predict the structure of the organic compounds that give each of the spectra labeled 4-A through 4-C. Note that the three signals at δ77.5 are due to solvent, deuterochloroform.

Problem 4-A

$C_6H_{14}O$

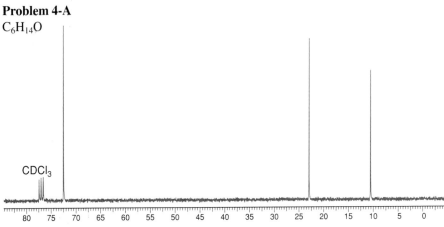

SOURCE: Reprinted with permission of Aldrich Chemical.

Problem 4-B

$C_7H_{12}O$

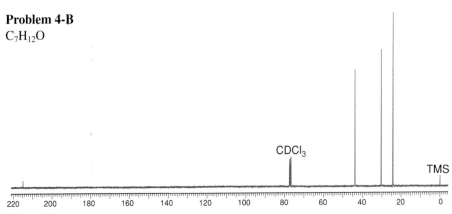

SOURCE: Reprinted with permission of Aldrich Chemical.

Problem 4-C

C_8H_{10}

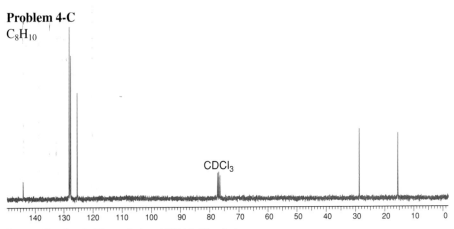

SOURCE: Reprinted with permission of Aldrich Chemical.

Exercise N.5: Spectral Identification Using ^{1}H and ^{13}C NMR

Predict the structure of the organic compounds that give the ^{1}H and ^{13}C NMR spectra labeled 5-A through 5-C.

Problem 5-A
$C_{11}H_{14}O_2$
$\delta 1.0$ (3H)
$\delta 1.5$ (2H)
$\delta 1.7$ (2H)
$\delta 4.4$ (2H)
$\delta 7.4-8.1$ (5H)

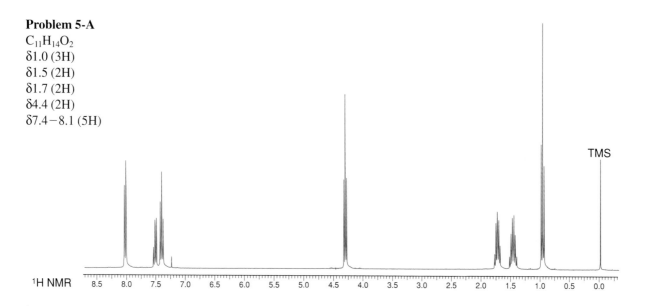

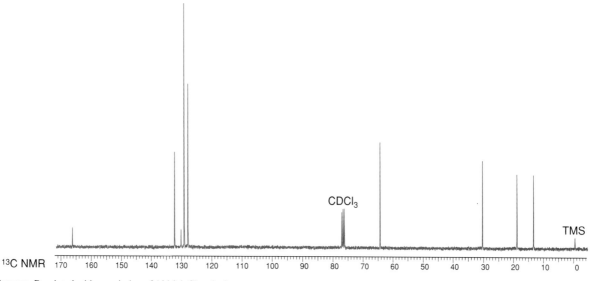

Problem 5-B
C_7H_8O
$\delta 3.7$ (3H)
$\delta 7-7.3$ (5H)

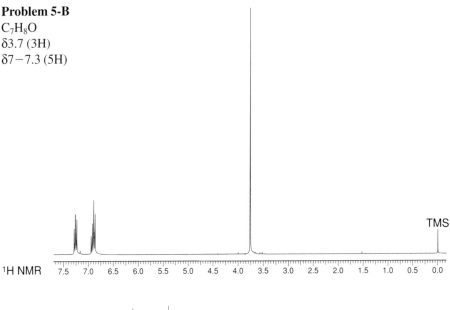

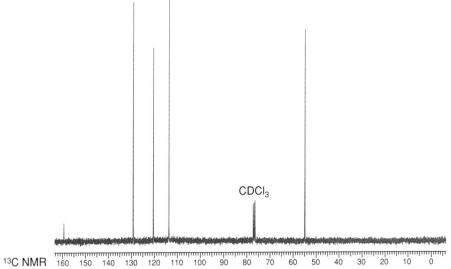

SOURCE: Reprinted with permission of Aldrich Chemical.

Questions

1. Are the effects on chemical shifts of attached electronegative atoms additive? Explain.

2. Explain how to differentiate between acetic acid and methanol using ^{1}H NMR spectroscopy.

3. Explain why protons on alcohols show singlets in the CW ^{1}H NMR spectra.

4. How could ^{1}H NMR spectroscopy be used to differentiate between 2-butanone and methyl propanoate?

Problem 5-C

C_4H_6O

$\delta 2.2$ (3H)

$\delta 5.9-6.3$ (3H)

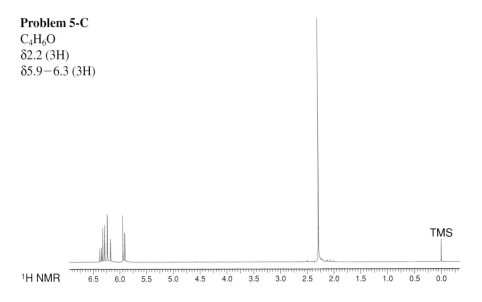

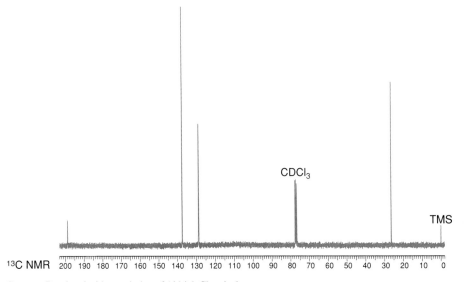

SOURCE: Reprinted with permission of Aldrich Chemical.

5. The ^{1}H NMR spectrum of cyclohexane shows one singlet. If the sample is cooled to –80°C, the spectrum changes. Predict the spectrum of the cooled cyclohexane and explain why the spectrum changes as the solution is cooled.

6. What is the purpose of TMS?

7. Predict the chemical shift of the protons in methyllithium (CH_3Li).

8. A specific proton in an organic compound has a chemical shift of $\delta 3.4$ in a 60 MHz NMR spectrum. What will be the chemical shift if the spectrum is recorded using a 90 MHz NMR instrument?

9. The coupling between two protons in an alkene is 10.8 Hz at 60 MHz. What will be the value of the coupling constant between these same two protons at 200 MHz?

References

Abraham, R. J., Fisher, J., and Loftus, P. *Introduction to NMR Spectroscopy.* New York: Wiley, 1992.

Breitmaier, E. *Structure Elucidation by NMR in Organic Chemistry.* Chichester, England: Wiley, 2002.

Kemp, W. *Organic Spectroscopy.* 3d ed. New York: W. H. Freeman and Company, 1991.

Lambert, J. B., Shurvell, H. F., Lightner, D., and Cooks, R. G. *Organic Structural Spectroscopy.* Upper Saddle River, NJ: Prentice-Hall, Inc., 1998.

Pavia, D., Lampman, G., and Kriz, G. *Introduction to Spectroscopy: A Guide for Students of Organic Chemistry.* 3rd ed. Pacific Grove: Brooks/Cole, 2001.

Silverstein, R. M., and Webster, F. X. *Spectrometric Identification of Organic Compounds.* 6th ed. New York: Wiley, 1998.

Technique O: Ultraviolet and Visible Spectroscopy

Introduction

IR and NMR spectroscopy both play an extremely important role in structure identification in the organic laboratory. Ultraviolet and visible (UV-vis) spectroscopy plays a supporting role. The wavelength ranges for UV and visible spectroscopy are shown in Figure 2O-1.

Conjugated unsaturated compounds, such as dienes, trienes, and aromatic compounds, absorb UV light. Extensively conjugated molecules, such as methyl orange, β-carotene, and β-chlorophyll, are colored and absorb visible light. Most saturated organic molecules, such as hexane and ethanol, do not absorb UV (>200 nm) or visible light. Alkenes absorb UV light, but the wavelength of maximum absorption (λ_{max}) falls below the normal UV region (< 200 nm). Carbonyl compounds, such as aldehydes and ketones, also absorb weakly in the UV region.

UV spectra can be useful in the analysis of conjugated molecules and visible spectra can be useful in the analysis of colored compounds. Spectra may be used qualitatively to determine absorption maxima and minima or quantitatively to measure concentrations of substances. UV spectroscopy is generally of more use in organic chemistry than visible spectroscopy because many more organic compounds absorb UV light than absorb visible light. An advantage of UV-vis spectroscopy is that very small amounts of sample are required—much smaller than the amounts needed for IR and NMR spectroscopy.

X ray	Vacuum UV	UV	Visible	Near IR	IR	Far IR and Microwave
Short λ	< 200 nm	200–400 nm	400–800 nm	0.8–2.5 μ	2.5–25 μ	Long λ
		UV	Visible		(1μ = 1000 nm)	

Figure 2O-1   Relationship of UV and visible light to other forms of electromagnetic radiation

Theory

Certain compounds absorb UV and visible radiation because an outer π or nonbonding electron is excited from its ground state to an excited state. For conjugated dienes, trienes, and aromatics, an electron from the highest lying π molecular orbital is excited to a π^* molecular orbital. This effect is illustrated in Figure 2O-2 using molecular orbital theory for the π molecular orbitals of 1,3-butadiene. The theory is also used for comparison with the π molecular orbitals of a simple alkene, ethene.

The highest occupied and lowest unoccupied orbitals are labeled HOMO and LUMO, respectively. The energy difference between the HOMO and LUMO is greater for ethene than for 1,3-butadiene. Therefore the energy required for excitation is greater for ethene (164 kcal/mole at 171 nm) than for 1,3-butadiene (129 kcal/mole at 217 nm). The greater the number of conjugated double bonds, the closer the energies of the LUMO and the HOMO. More extensively conjugated substances are colored and have HOMO and LUMO energies that are very close. This requires less energetic, longer wavelength light for excitation. The wavelengths required for excitation fall in the visible region for colored organic compounds such as Methyl orange dyes, β-carotene, and β-chlorophyll.

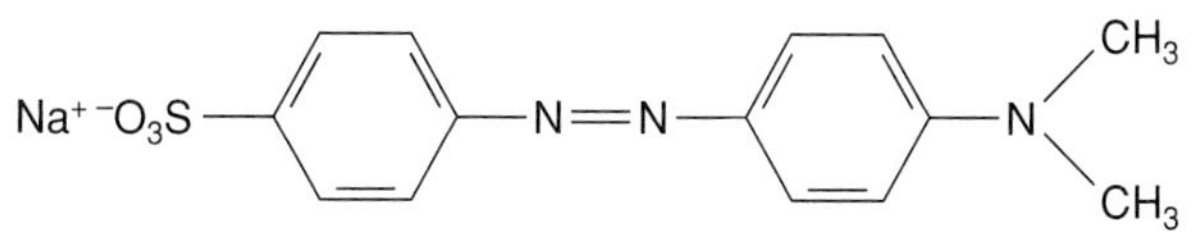

Methyl orange

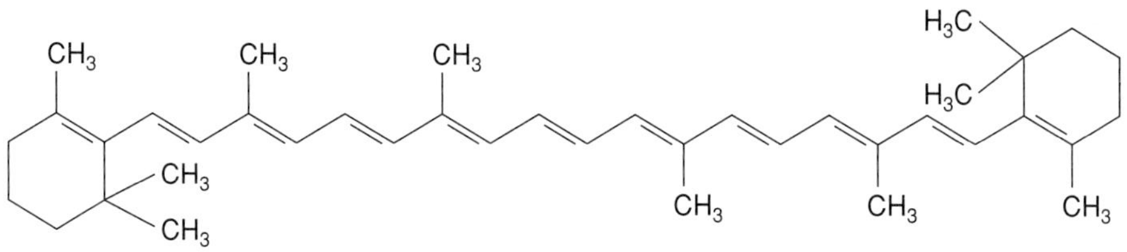

β-Carotene

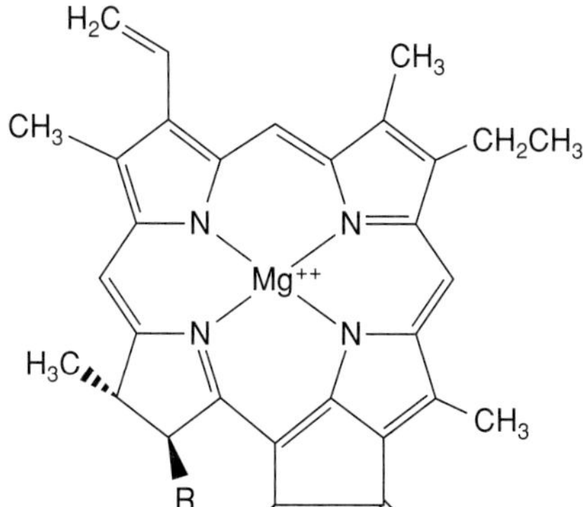

R = phytylpropanoyl = $C_{24}H_{43}O_2$

β-Chlorophyll

UV spectra are plots of absorbance or molar absorptivity versus wavelength (200–360 nm). Compounds having extended conjugation have λ_{max} values at longer

Figure 20-2

Molecular orbital diagrams for ethene and 1,3-butadiene

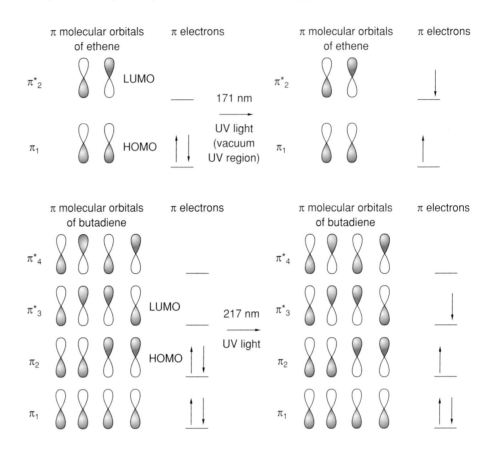

wavelengths than less conjugated compounds. Aromatic compounds, such as benzonitrile and toluene, absorb in the UV region due to their conjugated structures. UV spectra consist of very broad peaks and valleys because the electronic states responsible for UV absorption have many vibrational and rotational states. Electronic excitation may start and end at any of several different energy levels, resulting in a wavelength range of excitation. Examples of typical UV spectra are shown in Figure 20-3.

Uses of UV-Visible Spectroscopy

Qualitative Analysis

UV spectroscopy can be used to distinguish between conjugated and nonconjugated unsaturated compounds. Some examples of λ_{max} values for unsaturated hydrocarbons are given below.

Compound	λ_{max} (nm)	Number of conjugated double bonds
1-pentene	177	0
1,4-pentadiene	178	0
1,3-butadiene	217	2
1,3-pentadiene	224	2
trans-1,3,5-hexatriene	253, 263, 274	3
benzene	256	3
toluene	261	3

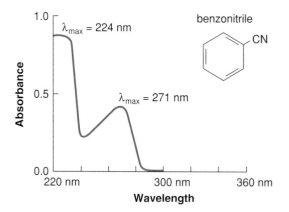

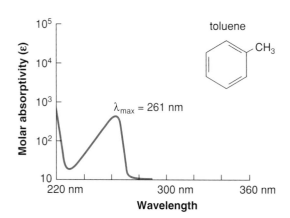

Figure 20-3

Examples of UV spectra

   Substitution patterns of conjugated dienes and trienes can be predicted using the Woodward-Fieser rules (Wade 2003, pp. 1211–1215). Alkylated, conjugated dienes and aromatic compounds have λ_{max} values at longer wavelengths than 1,3-butadiene.

Quantitative Analysis

Beer's Law states that the intensity of absorption is proportional to the concentration of a dilute solution of an absorbing compound, where A = absorbance, c = concentration (M), d = width of the sample cell (usually 1 cm), and ε= molar absorptivity (L/mole-cm):

$$A = \varepsilon cd$$

Beer's Law

Molar absorptivity (ε) is a measure of how effectively a compound absorbs radiation at a particular wavelength. Molar absorptivity values at specific wavelengths are physical constants that are available in spectral catalogs (Weast 1973). The magnitude of ε values depends upon the type of excitation. Transitions that are inefficient, such as the n → π^* transition of carbonyl groups, have low ε values. (An n → π^* transition is excitation of a nonbonding electron to an antibonding π^* orbital.) The $\pi \to \pi^*$ transition of the π bond of alkenes is very efficient. These transitions have large ε values.

Applications of Beer's Law

Example 1: A sample of 1,3-cyclohexadiene has a molar absorptivity of 10,000 at the λ_{max} of 259 nm. The measured absorbance reading is 0.35 and the width of the sample tube is 1.0 cm. The concentration of the sample can be determined from the equation $A = \varepsilon cd$. For this sample, c is calculated to be 3.5×10^{-5} M.

Example 2: UV spectroscopy can be used to monitor reaction rates. Because absorbance is proportional to concentration, UV and visible spectroscopy can be used to monitor changes of absorbing species during a reaction. A plot of 1/absorbance versus time for the second-order reduction of an azo dye at 520 nm is given below. The plot of 1/A versus time should give a straight line for a second-order reaction as shown in Figure 2O-4.

In Example 2, the concentration of azo dye decreases with time as the dye is reduced. The intensity of absorption decreases at the λ_{max} of the azo dye linearly as a function of 1/A. In this way UV and visible spectroscopy can be used as a convenient tool for kinetics studies.

How to Operate a Spectronic 20

1. Turn the instrument power on and allow the instrument to warm up for several minutes.
2. Set the dial to the desired wavelength for analysis. Use the left knob to set % transmittance to zero.
3. Prepare a cuvette or test tube containing a blank. The blank is usually a solution containing everything but the absorbing species. Place the container in the sample compartment and adjust the % transmittance to 100% using the right knob.
4. Remove the tube containing the blank and replace it with a test tube containing the solution to be analyzed. If the same container is used for the blank and the sample, rinse out the container several times with the sample solution.
5. Record the % transmittance and calculate absorbance.

$$A = 2 - \log\%T$$

A	1/A	Time (minutes)
.83	1.20	0
.81	1.23	1
.80	1.25	2
.76	1.31	4
.73	1.37	6
.70	1.42	8
.65	1.53	12
.58	1.71	16
.54	1.82	20

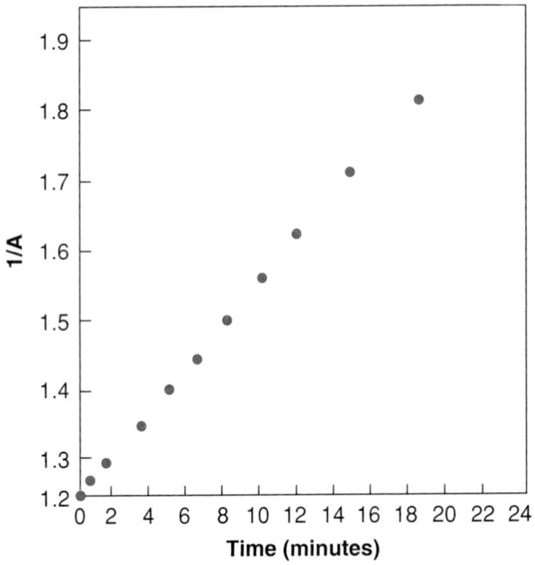

Figure 2O-4   Second-order rate plot for data from kinetics experiment monitored using a Spectronic 20

6. Repeat steps 3–5 as necessary for the same sample and for diluted samples.
7. Turn off the instrument power unless the instrument is to be used again soon.

How to Operate a UV-Visible Spectrometer

1. Turn the instrument power on. If necessary, set the instrument for either UV or visible use. In the UV, a deuterium lamp is used and should be turned on. A tungsten lamp is used for analyses in the visible region.
2. Using a matched pair of fused silica cuvettes, fill each cuvette with solvent being used for the experiment.
3. Zero the baseline and scan the wavelength region of interest. In the UV, the region to be scanned is generally from 200 to 400 nm. Use 400–800 nm for a scan in the visible region. There should be little or no absorption observed throughout the scan. A trace of the baseline should be flat.
4. Remove the cuvette from the sample compartment, but leave the cuvette containing solvent in the reference compartment. Empty the sample cuvette and add some of the solution to be analyzed and rinse. Repeat this twice and then fill the cuvette with the solution to be analyzed. Return the cuvette to the sample compartment.
5. Scan the region of interest and be sure that the desired absorbance readings are achieved. If the absorbance reading is lower than 0.30, try increasing the concentration of the solution being analyzed and repeat the scan. If the solution is too concentrated, the absorbance reading will be too high and the sample solution should be diluted. For quantitative work, the absorbance reading at the λ_{max} should be between 0.3 and 0.8 absorbance units.
6. Obtain a recording of the spectrum. Be sure to label the spectrum with name, date, sample identification, solvent, and concentration of the sample, if known.
7. Remove all solutions from the cuvettes. Be careful, as the matched cells are expensive. Return the cuvettes to their storage container and turn off the instrument power.

Questions

1. The energy of UV light is 70 kcal/mole at 400 nm and 140 kcal/mole at 200 nm. Why is UV radiation damaging to the eyes and skin? (Hint: Think about the energy required to break a C−C bond.)
2. Which is more energetic: 235 nm or 325 nm light? Explain.
3. Why is UV spectroscopy used more frequently in organic chemistry than visible spectroscopy?
4. Using the structure of methyl orange as an example, draw resonance forms to show how conjugation extends throughout the molecule.
5. Depict the π molecular orbitals for 1,3,5-hexatriene. Show the electron occupancy of each orbital. Label the LUMO and HOMO. Is the expected wavelength for UV excitation at a longer or shorter wavelength than that for 1,3-butadiene?
6. Calculate the molar absorptivity (ε) of benzene at 260 nm for a solution of benzene in ethanol where the concentration of benzene is 3.3×10^{-3} M and the absorbance reading is 0.70 using a 1-cm cell.
7. Is toluene a good solvent for UV spectroscopy? Explain.

References

Pavia, D., Lampman, G., and Kriz, G. S. *Introduction to Spectroscopy.* 3rd ed. Pacific Grove: Brooks/Cole, 2001.

Wade, L. G. Jr. *Organic Chemistry.* 5th ed. Upper Saddle River, NJ: Prentice-Hall, 2003.

Weast, R. C., ed. *Atlas of Spectral Data and Physical Constants for Organic Compounds.* Boca Raton, FL: CRC Press, 1973.

Technique P: Mass Spectrometry

Principles of Mass Spectrometry

In mass spectrometry (MS), an organic molecule is bombarded with an electron beam in an evacuated chamber. This causes the molecule to lose an electron, forming a molecular ion $M^{+\bullet}$ (a radical cation). The energy required is usually sufficient to cause the molecular ion to fragment into carbocations and radicals. When carbocations are formed, an equal number of free radicals are also formed, but these are not normally detected. Only the positively charged species are usually detected in MS. The ionization and fragmentation processes take place in the gas phase in a very highly evacuated environment (10^{-5} torr). This process, called electron impact (EI), is shown here for methane.

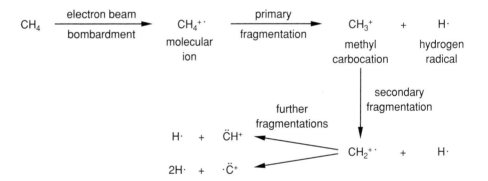

The $CH_4^{+\bullet}$ radical cation is called the molecular ion. Generally, only the molecular ions and cations from primary fragmentation are important in MS. (There are some exceptions.) In MS, the methyl carbocation is formed by primary fragmentation of $CH_4^{+\bullet}$. For methane, a plot of the mass to charge ratio (m/z) versus relative abundance of fragments shows that the molecular ion at m/z 16 and the primary fragment at m/z 15 are the most abundant ions. Because the molecular masses of CH_4 and $CH_4^{+\bullet}$ differ only by the mass of one electron, the position on the m/z scale for CH_4^+ gives the molecular mass of the compound. Since methane (CH_4) is neutral, it is not detected in mass spectrometry. Fragments resulting from secondary and higher-order fragmentations are not very abundant and are generally not observed as major fragments in MS. However, the mass spectrum of a very simple molecule, such as CH_4, demonstrates the principle of successive fragmentations. The mass spectrum of methane is shown below as a bar graph. The largest peak is called the base peak. In this case, the base peak and the molecular ion peak are the same.

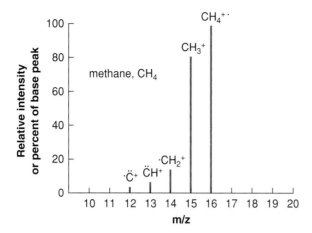

Simple alkanes show a peak corresponding to the molecular ion, $M^{+\bullet}$. For ethane, this peak occurs at m/z 30. For propane, the molecular ion peak is m/z 44. The molecular ion of propane can lose a hydrogen atom to form an isopropyl carbocation of m/z 43. The primary propyl carbocation is also possible, but less likely, because it is less stable than the secondary carbocation. The predominant fragmentation affords an ethyl carbocation of m/z 29 (base peak) and a methyl radical. Formation of the alkyl radical via carbon-carbon cleavage is preferable to formation of a hydrogen atom (radical).

$$CH_3CH_2CH_3 \xrightarrow{70\ eV} \left[CH_3CH_2CH_3 \right]^{+\bullet}$$

m/z 44

$$\xrightarrow{-H\bullet} CH_3\overset{+}{C}HCH_3 \quad \text{m/z 43}$$

$$\xrightarrow{-CH_3\bullet} CH_3CH_2^+ \quad \text{m/z 29}$$

The mass spectrum of butane shows the $M^{+\bullet}$ at m/z 58 and fragment peaks at m/z 43 and 29, corresponding to the isopropyl carbocation (after rearrangement from the initially formed propyl carbocation) and ethyl carbocation. For isobutane, loss of methyl radical furnishes the isopropyl carbocation directly at m/z 43 and this is the expected fragmentation. Loss of hydrogen radical to give a *tert*-butyl carbocation is unlikely due to the easier bond breaking of a $C-C$ (85 k cal/mol) versus a $C-H$ (95 k cal/mol) bond.

$$\left[(CH_3)_2CHCH_3\right]^{+\bullet} \longrightarrow CH_3\overset{+}{C}HCH_3 + CH_3\bullet$$

m/z 43

Mass spectra of higher-molecular-weight alkanes also show abundant C_3 and C_4 fragments. Spectra of more complex molecules show clusters of bars, indicating masses of the more stable fragments. For example, an isopropyl cation at m/z 43 can lose a hydrogen atom to form a propenyl cation at m/z 42 and the propenyl cation can lose a hydrogen radical to form an allyl cation at m/z 41. In MS, just as in solution, the more stable fragments are those that are the more stable carbocations. However, because of the high energies involved, numerous carbocations can be formed in MS that are not typically formed in solution. Also, losses of hydrogen atoms and other neutral species can occur for carbocation fragments in MS while this is not observed for carbocations in solution.

In addition to breaking apart to give a fragment and a free radical, molecular ions may also lose a stable neutral molecule, such as ethylene. Loss of a neutral molecule from the molecular ion results in a new radical ion, which for compounds containing carbon and hydrogen or carbon, hydrogen, and oxygen, has an even-numbered mass. Loss of ethylene is observed for cyclic alkanes such as cyclohexane. The mass spectrum of cyclohexane is shown below. Peaks are observed for positive fragments, but not for radicals or neutral molecules.

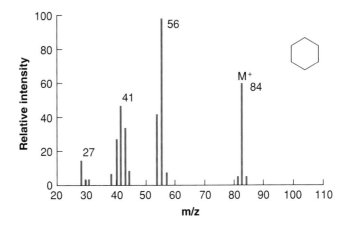

Note that the molecular weight of cyclohexane is an even number (84). Molecular masses of molecules without nitrogen or that have an even number of nitrogens are even numbered. In the example above, cyclohexane has a mass of 84. The molecular ion peak, $M^{+\bullet}$, has the identical mass of 84. Odd-numbered fragments (41 and 27) are carbocations. However, the even-numbered peak at m/z 56 is not a carbocation. It is a radical cation that has a mass that is 28 mass units less than the mass of cyclohexane due to loss of ethylene. The largest peak (base peak) is observed at m/z 56. This peak cannot be a carbocation because the mass of the fragment is even and carbocations such as CH_3^+, $C_2H_5^+$, etc., have odd-numbered masses. The fragment at m/z 56 must result by loss of a neutral molecule (ethylene) from the molecular ion.

$$\left[\bigcirc \right]^{+\bullet} \longrightarrow \left[\square \right]^{+\bullet} \quad + \quad CH_2{=}CH_2$$

m/z 84   m/z 56 ·
                              $C_4H_8^+$

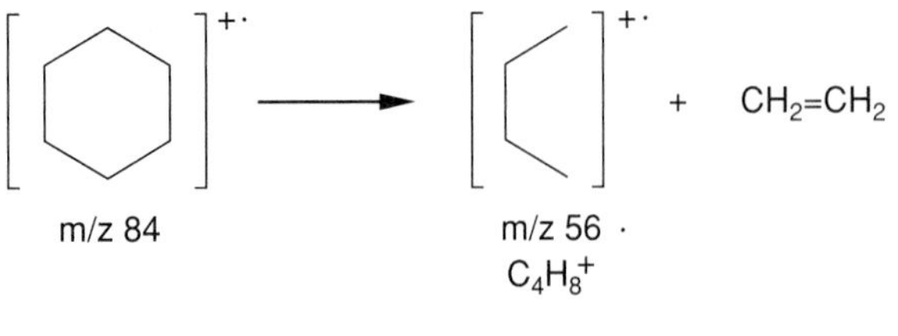

The fragment at m/z 41 is due to an alkenyl carbocation. There is no simple cleavage pattern to get from the radical cation m/z 56 ion to the m/z 27 fragment. The latter fragment is a result of secondary fragmentation. These fragmentations are depicted here.

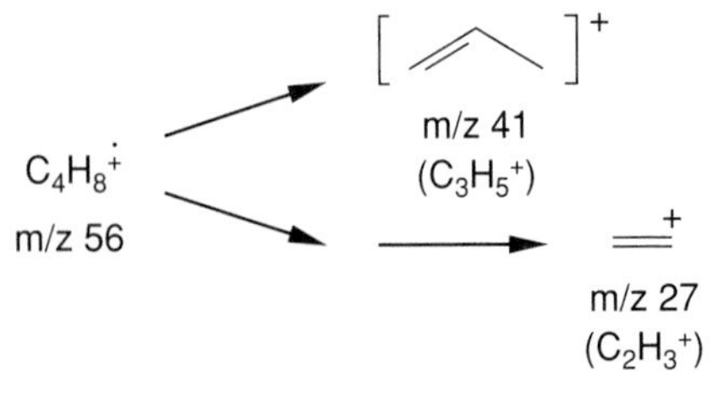

$$C_4H_8^{\dot{+}} \qquad \left[\diagup\!\!\!\diagup \right]^{+}$$

               m/z 56   m/z 41
                                $(C_3H_5^+)$

                                       m/z 27
                                       $(C_2H_3^+)$

Fragmentation Patterns

It is important to understand where peaks occur (at which m/z values) and why some peaks are large and others are small. The base peak is the largest peak and corresponds to a stable, positively charged fragment or molecule. Odd-numbered fragments of compounds that contain no nitrogen or an even number of nitrogens will show large peaks if they correspond to stable carbocations. The abundance varies for each compound and depends on the mass of the compound, the skeletal structure (straight chain or branched), and the types of functional groups that are present. Some of the more important fragmentation pathways for representative functional classes of compounds are presented in this section.

The principal objective is to become familiar with basic patterns of fragmentation so that the main fragments in a mass spectrum will be identifiable. Cleavage generally favors formation of the most stable cations and radicals, although the radicals are not seen in MS. Primary carbocations have very little stability, and tertiary carbocations are usually the most stable fragments. Stabilization of the radical (formed in producing the carbocation) is also important. Fragmentation also depends upon the relative bond strengths of bonds being broken. Breaking a C−H bond requires more energy than breaking a C−C bond. Allylic and benzylic C−H bonds are broken more easily than other C−H bonds. (Refer to the table of bond dissociation energies in Table 2P-1.)

Alkanes

Of the common cleavage patterns, simple cleavage is observed for alkanes. In simple cleavage, a carbon-carbon bond is broken from the molecular ion. Some carbon-carbon bonds are more likely to break than others. For 3-methyloctane, cleavage is more likely to occur near the branching carbon because the resulting carbocations are secondary rather than primary. In this case there are two possibilities, because cleavage may occur on either side of the branching carbon (C−3). (A third possibility for breaking a tertiary C−H bond is less likely.) The base peak in the MS of 3-methyloctane is at m/z 99. Other peaks occur at m/z 57 and m/z 113.

Table 2P-1 Bond Dissociation Energies[*]

Bond	kJ/mol	kcal/mol
CH_3-H	435	104
CH_3CH_2-H	410	98
$(CH_3)_2CH-H$	397	95
$(CH_3)_3C-H$	380	91
C_6H_5-H	460	110
$CH_2=CH-H$	452	108
$CH_2=CH-CH_2-H$	372	89
$C_6H_5CH_2-H$	356	85

[*]Bond dissociation energies refer to the single bond indicated in structural formula for each compound.

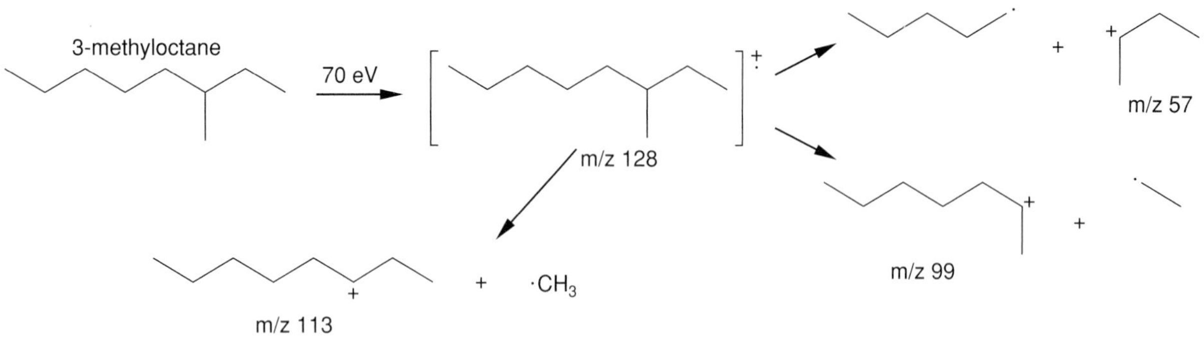

3-methyloctane

70 eV

m/z 128

m/z 57

m/z 113   +   ·CH₃

m/z 99

The odd electron can also remain with the secondary carbon, giving m/z 71 and 29 peaks. Although observable in the MS of 3-methyloctane, these peaks are much smaller.

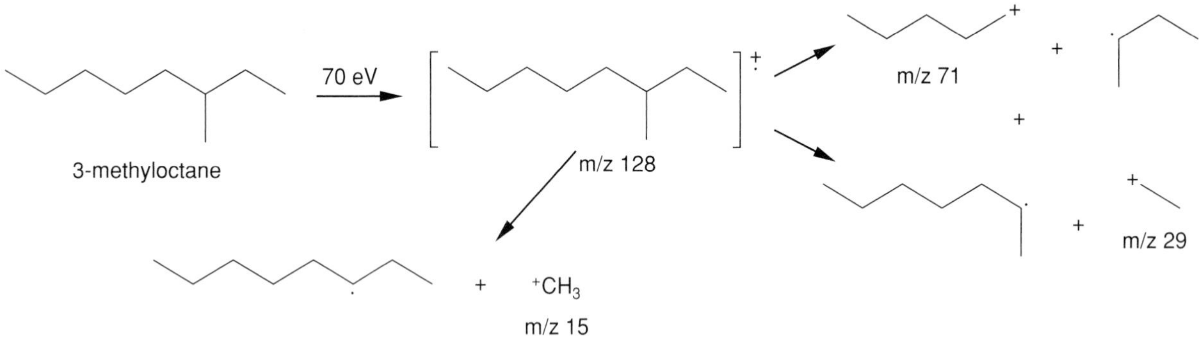

3-methyloctane

70 eV

m/z 128

m/z 71

m/z 29

+   ⁺CH₃

m/z 15

Alkenes

Alkenes often fragment to yield allylic ions. For example, the base peak in the mass spectrum of 1-heptene is m/z 41 due to formation of the allyl cation.

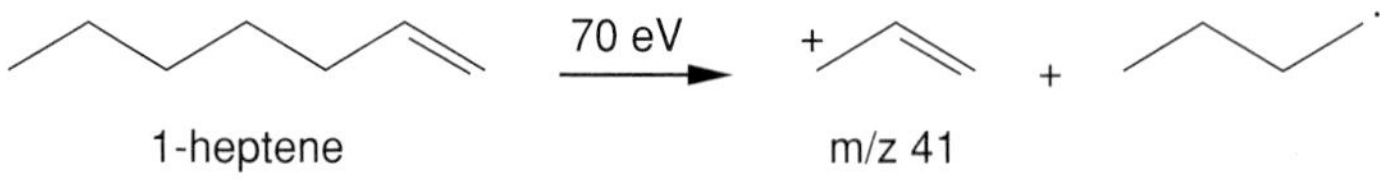

1-heptene   70 eV   m/z 41

Cyclic alkanes and alkenes often split out stable neutral molecules such as ethylene or substituted ethylenes. The resulting cationic fragment is the radical cation of an alkene or a diene. For example, cyclohexene is known to give ethylene as shown below. This is similar to a retro Diels-Alder reaction.

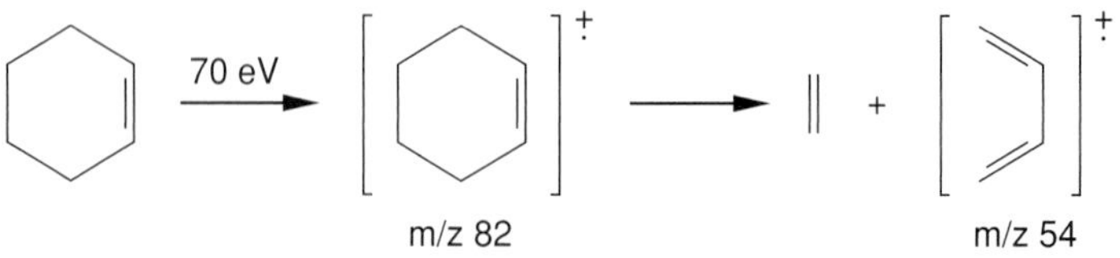

70 eV

m/z 82   m/z 54

Aromatic Compounds

Aromatic compounds that contain alkyl side chains often cleave to yield benzylic carbocations as the main fragmentation path. The benzylic ions initially formed rearrange to the more stable tropylium ions.

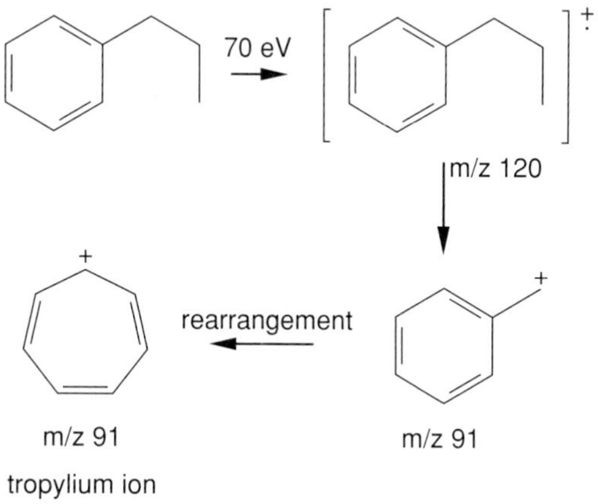

m/z 91
tropylium ion

m/z 91

Compounds with Heteroatoms

Molecules containing oxygen, nitrogen, halogens, or other heteroatoms may undergo a similar type of cleavage, called α-cleavage, where the driving force is formation of resonance-stabilized cations. An example is 4-heptanone, where α-cleavage yields a four-carbon acylium ion and a propyl radical.

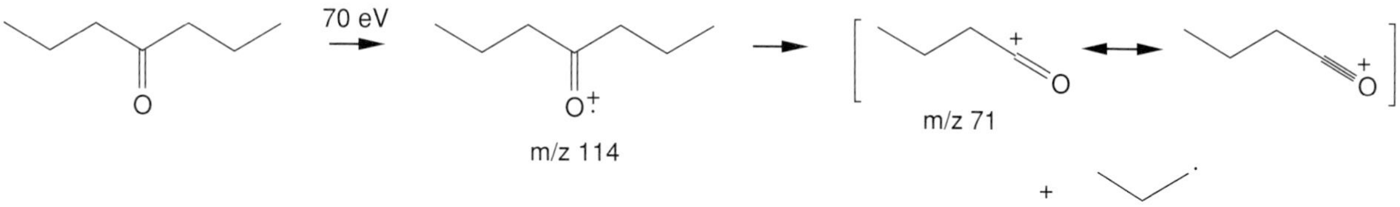

Another example of a simple α-cleavage involving an ether is shown below.

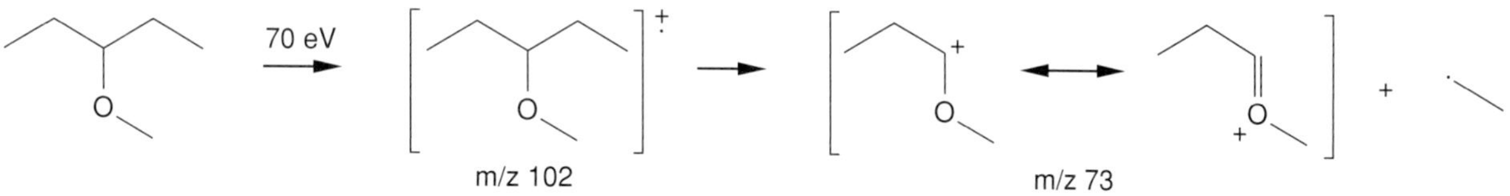

Compounds with More Than One Heteroatom

Cleavage will generally occur at a more substituted carbon bearing the heteroatom as shown. This is what would be predicted based on the relative stabilities of the carbocations formed.

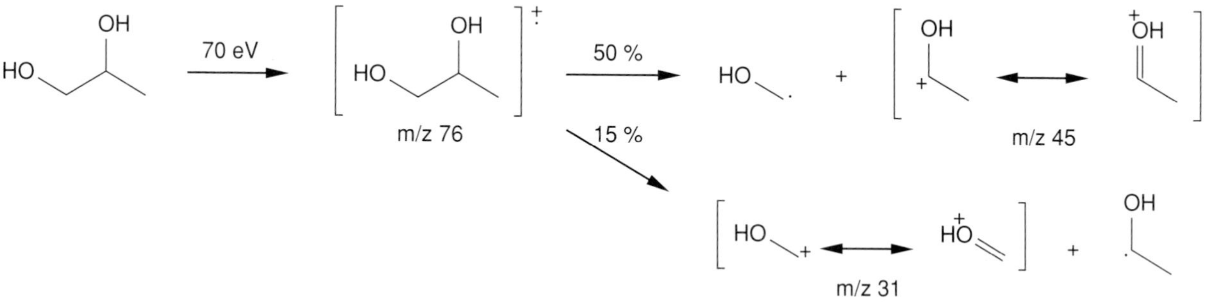

Among different heteroatoms, less electronegative atoms exert a greater stabilizing effect than more electronegative atoms. This effect may be seen from the example below. Note that cleavage is favored to give the ion that contains the nitrogen atom. This is what would be predicted based on the electronegativities of the atoms.

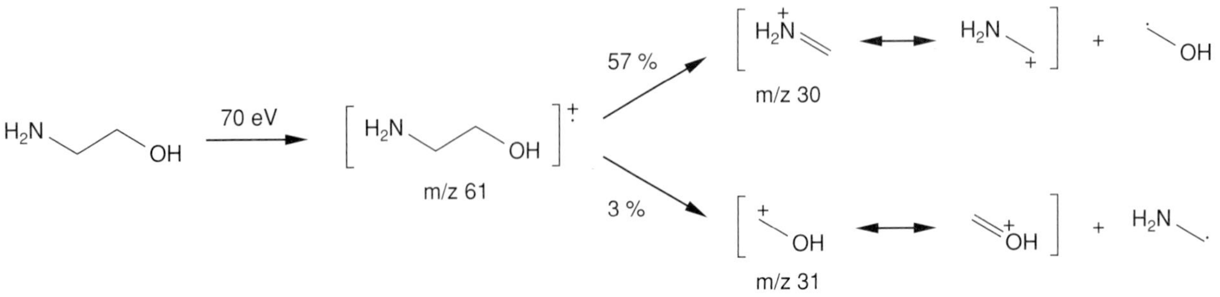

Carbonyl Compounds

Carbonyl compounds, such as aldehydes, ketones, and esters, may also split off neutral molecules from the molecular ion. In this case, not only is a neutral molecule expelled, but there is also a transfer of an atom via rearrangement. The rearrangement is called a McLafferty rearrangement. In order to give the rearrangement there must be at least a three-carbon chain attached to the carbonyl and the γ-carbon must have at least one hydrogen. It is the γ-hydrogen that is transferred during rearrangement. The resulting radical cation, which has an even mass number, can also give further fragmentation.

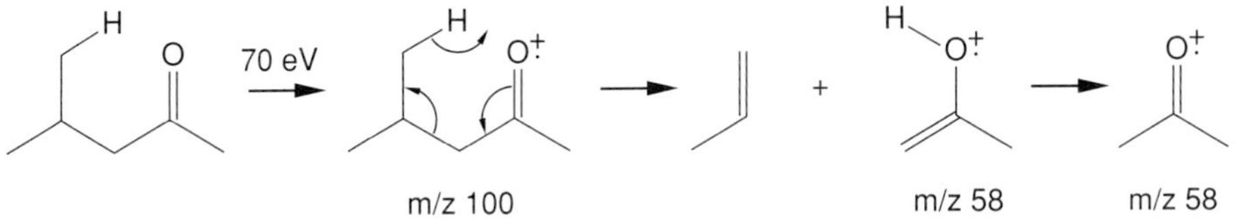

The ester below lacks enough carbons to give a McLafferty rearrangement, but it has a large side chain that gives rise to a related rearrangement.

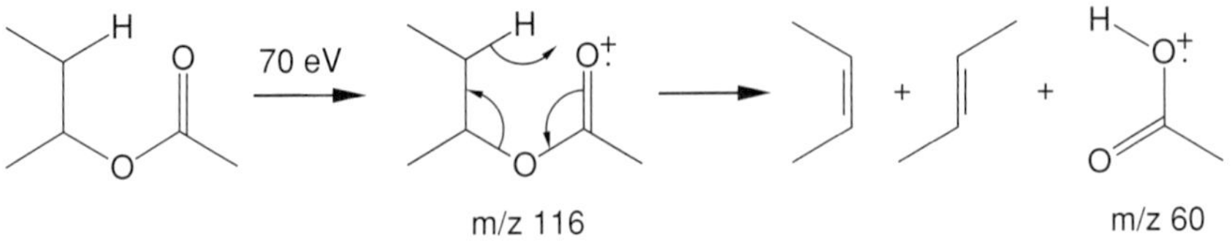

Molecular Ion Peak

A principal use of MS is to determine molecular mass. This is effective for most compounds provided that a particular compound gives a molecular ion peak. In high resolution MS (HRMS), the mass of the molecular ion peak is measured precisely to four decimal places. This value may be compared with tables of masses (Silverstein and Webster 1998). For example, both $C_2H_3N_2$ and C_3H_5N have masses of 55, but the exact mass of $C_2H_3N_2$ is 55.0297 and that of C_3H_5N is 55.0422. This difference is easily determined in HRMS. Not all compounds give large molecular ion peaks; some branched alkanes and alcohols do not show a molecular ion peak at all. Generally, more highly branched compounds are likely to fragment more easily and show smaller molecular ion peaks. On the other hand, cyclic compounds, aromatics, normal chain compounds, and alkenes generally show large molecular ion peaks. Occasionally, the molecular ion will also be the base peak (largest peak in the spectrum), but more often a stable fragment will be the base peak.

Isotopes

MS is particularly valuable for compounds that contain chlorine and bromine atoms. Substances that contain chlorine or bromine atoms give large molecular ion peaks and

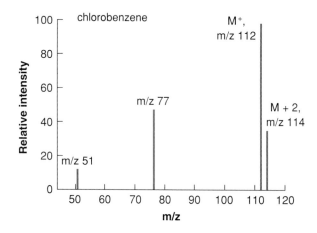

very sizable peaks (called M + 2 peaks) two mass units higher than the molecular ion peaks. For example, chlorobenzene shows a molecular ion peak at m/z 112 and an M + 2 peak at m/z 114. This latter peak is given by molecules (25% of the total) that contain ^{37}Cl rather than ^{35}Cl. The relative abundance of this peak is about one-third of the molecular ion abundance, since the relative abundances of ^{35}Cl and ^{37}Cl are roughly 75:25.

The MS of a monobrominated compound is similarly distinctive. The bromine isotopes of 79 and 81 are of equal abundance. Therefore, a compound containing one bromine atom will show M and M + 2 peaks that are of equal size. For example, bromobenzene shows two peaks of equal size (equal abundance) at m/z 156 and 158.

Compounds with more than one bromine or chlorine give additional isotope peaks (M + 4, M + 6, etc.). The m/z values and the abundance of these isotope peaks indicate the type and number of chlorine and bromine atoms present. Fluorine and iodine atoms have only single isotopes in natural abundance.

Gas Chromatography/Mass Spectrometry (GC/MS)

Gas chromatography/mass spectrometry (GC/MS) is a very important and useful instrumental method of analysis for organic substances. Applying this technique, a multicomponent mixture of volatile organic compounds is injected into a GC column through the injection port. This procedure is the same as for ordinary GC analysis (see Technique J). As the individual compounds are eluted, they are analyzed by a mass selective detector. Mass spectra are computed in less than a second and these may be analyzed during or after the GC run is completed. There are many applications. For example, the organic extract of a sample of water from a stream or a sample of drinking water may be analyzed for the presence of pesticides and herbicides.

Strategy for Solving Structural Problems Using MS, IR, and NMR

In practice, structural identification is usually done using NMR and IR. The IR spectrum is used to determine the types of functional groups that are present. Interpretation of the NMR spectrum gives specific structural information about the nature of the carbon skeleton and the types of attached protons. Sometimes it is necessary to get additional information and this is where MS may be useful. It is also conceivable that in certain situations, IR and NMR data may not be available, either because the compound in question is a constituent of a complex mixture or because there was not enough sample to obtain a good IR or NMR spectrum. One very favorable feature of MS is that only very small samples are needed for analysis. MS is also very useful for determining the molecular mass of the compound (although some compounds, such as alcohols and highly branched compounds, may not show a molecular ion peak) and for determining whether the compound contains bromine or chlorine. MS can also be helpful for identifying whether the compound contains an odd number of nitrogen atoms. To use MS for structural determination, follow the guidelines below.

Mass Spectrometric Analysis Guidelines

1. Consult your instructor for directions on how to prepare solutions for MS and GC/MS analyses.
2. Determine the molecular mass if possible. Remember that some alcohols and compounds that are highly branched may show very small or nonexistent molecular ion peaks.
3. Determine whether the compound contains bromine or chlorine or contains an odd number of nitrogens.
4. Map the major fragmentations observed by subtracting the masses of each fragment from the molecular mass. Refer to Tables 2P-2 for masses of fragment ions, 2P-3 for masses of fragment radicals, and 2P-4 for masses of neutral molecules expelled from molecular ions. Stabilized fragment ions, such as tertiary carbocations, allylic ions, acylium ions, and tropylium ions are seen most frequently as major fragment ions.
5. Propose fragmentation patterns to explain the presence of each fragment and the relative abundance of each fragment. It is usually possible to propose a likely structure or structures for an unknown.

Worked Examples of Spectral Analysis

MS of Unknown A (Refer to spectrum on p. 169.)

1. *Molecular weight:* The molecular ion is at m/z 182. An almost equally abundant peak is at m/z 184 (M + 2 peak).
2. *Analysis of nitrogen or halogens:* Presence of one bromine atom in the molecule is inferred because there are two peaks of approximately equal intensity in the area of the molecular ion. One peak is at m/z 184 and the other peak is at m/z 182. Two peaks are observed because bromine has two almost equally abundant isomers, ^{79}Br and ^{81}Br. The molecular ion is taken to be the peak with the lower mass, which contains ^{79}Br. The molecular mass is even, an indication that no nitrogen atoms or an even number of nitrogen atoms are present.

Table 2P-2 Typical Mass Spectrometric Fragment Ions

m/z	Cation
27	$C_2H_3^+$
29	$C_2H_5^+$
30	$CH_2NH_2^+$
31	CH_2OH^+, CH_3O^+
35	Cl^+
39	$C_3H_3^+$
41	$C_3H_5^+$
43	$C_3H_7^+$, $CH_3C{=}O^+$
53	$C_4H_5^+$
55	$C_4H_7^+$
57	$C_4H_9^+$
59	$CH_3OC{=}O^+$
65	$C_5H_5^+$
67	$C_5H_7^+$
69	$C_5H_9^+$
71	$C_5H_{11}^+$, $C_3H_7C{=}O^+$
75	$CH_3CH_2OC{=}O^+$
77	$C_6H_5^+$
79	Br^+
85	$C_6H_{13}^+$, $C_4H_9C{=}O^+$
91	$C_6H_5CH_2^+$, $C_7H_7^+$
99	$C_7H_{15}^+$
105	$C_6H_5C{=}O^+$
107	$C_6H_5CH_2O^+$, $HOC_6H_4CH_2^+$
127	I^+

Table 2P-3 Typical Mass Spectrometric Radicals

Mass	Radical
15	CH_3
17	OH
27	$CH_2{=}CH$
29	CH_3CH_2
30	$CH_2{=}NH_2$
31	CH_3O, CH_2OH
35	Cl
41	$CH_2{=}CHCH_2$
43	C_3H_7, $CH_3C{=}O$
45	CH_3CH_2O
49	CH_2Cl
57	C_4H_9
59	$CH_3OC{=}O$
69	C_5H_9

Table 2P-4 Typical Mass Spectrometric Neutral Molecules Formed

Mass	Molecule
18	H_2O
26	$H{-}C{\equiv}C{-}H$
27	HCN
28	$CH_2{=}CH_2$
32	CH_3OH
36	HCl
42	$CH_2{=}C{=}O$, $CH_3CH{=}CH_2$
54	$CH_2{=}CHCH{=}CH_2$
78	C_6H_6
80	HBr
128	HI

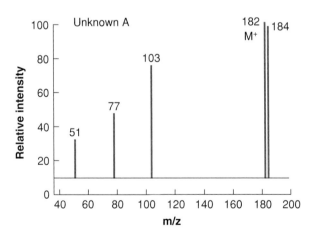

3. *Determination of fragment ions:* A large fragment ion at m/z 103 is indicative of loss of a bromine radical (having mass of either 79 or 81) giving a phenylethenyl cation. Also prominent are peaks at m/z 77, corresponding to the phenyl cation, and m/z 51, corresponding to secondary fragmentation of the phenyl cation. The phenyl cation loses the neutral molecule, acetylene (MW = 26), to form a cation containing four carbons and having m/z 51. These fragmentations are illustrated here.

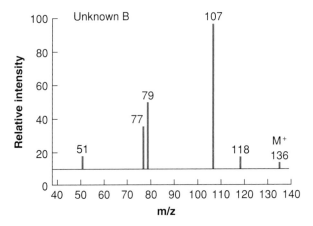

4. From the data available, unknown A appears to be either (Z) or (E)-1-bromo-2-phenylethene (β-bromostyrene). The distinction between Z and E isomers can be made by measuring coupling constants for the alkene protons in the ^{1}H NMR spectra.

(E)-1-bromo-2-phenylethene   (Z)-1-bromo-2-phenylethene

MS of Unknown B

1. *Molecular weight:* The molecular ion peak indicates a molecular weight of 136.
2. *Analysis of nitrogen or halogens:* The molecular mass is even, an indication that no nitrogen atoms or an even number of nitrogens are present.
3. *Determination of fragment ions:* Alcohols frequently show very small or nonexistent molecular ion peaks. This is because the molecular ion easily loses water to give an M-18 fragment. In this case that fragment appears at m/z 118. The ion at m/z 107 is due to loss of an ethyl radical (mass = 29) from the molecular ion. The 107 peak is the base peak in this spectrum and it is due to a benzylic ion that is stabilized by an adjacent −OH group. The fragment at m/z 77 is due to a phenyl cation, formed from the molecular ion. The peak at m/z 51 is a result of secondary fragmentation of the phenyl cation, resulting in the loss of acetylene. The peak at m/z 79 is also a

result of secondary fragmentation and it may result from loss of CO from the 107 peak. It is usually not possible to assign identities to all fragments or to ascribe all fragmentation pathways with certainty. These fragmentations are summarized here.

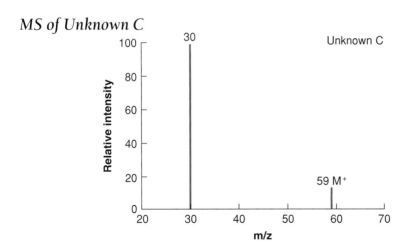

4. The likely structure of Unknown B is 1-phenyl-1-propanol.

1-phenyl-1-propanol

MS of Unknown C

1. *Molecular weight:* The molecular weight is 59.
2. *Analysis for nitrogen or halogens:* The odd-numbered molecular ion peak indicates the presence of a nitrogen atom.
3. *Determination of fragment ions:* Loss of a 29 radical indicates cleavage to form an ethyl radical and a resonance stabilized iminium ion fragment ion ($CH_2=NH_2^+$) of m/z 30.
4. Unknown C is 1-aminopropane ($CH_3CH_2CH_2NH_2$).

Exercise P.1: Solving Problems in Mass Spectrometry

In this exercise, you will learn to apply the basic principles of mass spectrometry, the important modes of fragmentation of organic compounds, and the methods of interpretation of mass spectra of simple organic compounds.

1. For each of the spectra labeled a–f below, determine the molecular mass. Explain how it is possible to tell if nitrogen, chlorine, or bromine is present.

2. For each of the spectra labeled a–f, identify the major features and suggest identities of major fragments. It will not generally be possible to do this for all fragment

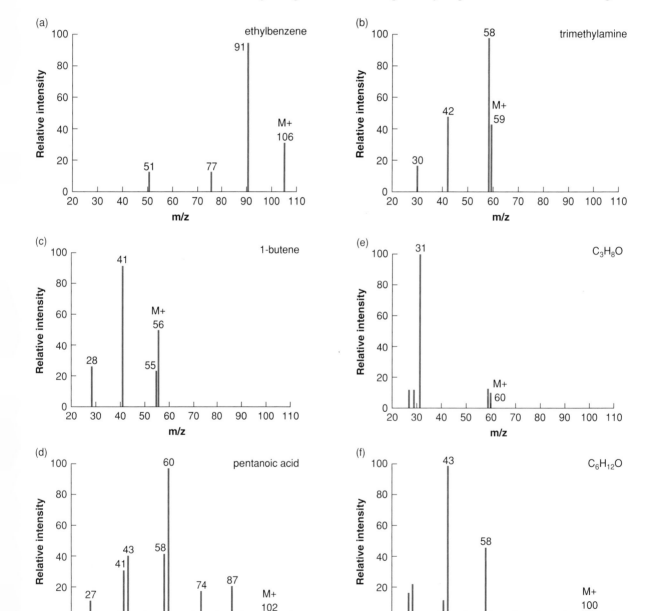

peaks. For spectra where only molecular formulas are furnished, calculate the elements of unsaturation (see Technique N) before proceeding with the MS analysis.

Questions

1. What ion fragments are expected from the mass spectrum of ethane?

2. Is 2-methylpropane expected to show a large ion fragment at m/z 29? Explain.

3. What is special about the stability of an ion fragment of m/z 41?

4. For 3-methyloctane, which end of the molecule is more likely to undergo fragmentation? Explain.

5. What major ion fragment is expected from the mass spectrum of acetone? Explain using structural drawings.

6. Illustrate the cleavage pattern for:

    a. ethylene glycol.

    b. ethylenediamine.

7. What major ion fragment is expected from the mass spectrum of ethylbenzene?

8. Predict the mass spectrum of 2-chloropropane. Explain.

9. Illustrate the McLafferty rearrangement for:

    a. 2-hexanone.

    b. propyl acetate.

10. *Library project:* Locate a journal article by either J. B. Fenn or K. Tanaka. Write a brief summary highlighting the significance of the work. Fenn and Tanaka received the 2002 Nobel Prize in chemistry along with K. Wuthrich.

References

Lambert, J. B., Shurvell, H. F., Lightner, D., and Cooks, R. G. *Organic Structural Spectroscopy.* Upper Saddle River, NJ: Prentice-Hall, Inc., 1998.

Pavia, D., Lampman, G., and Kriz, G. *Introduction to Spectroscopy: A Guide for Students of Organic Chemistry.* 3rd ed. Pacific Grove: Brooks/Cole, 2001.

Silverstein, R. M., and Webster, F. X. *Spectrometric Identification of Organic Compounds.* 6th ed. New York: Wiley, 1998.

Technique Q: Molecular Modeling

Introduction

Chemists have sought ways to make perception of organic molecules more visual and ways to make the presentation of chemistry more exact and scientific. Theoretical chemists have successfully made each of these objectives possible by developing molecular modeling software. Most desktop computers are now fast enough to be able to handle the mathematical calculations required for molecular modeling.

Molecular modeling encompasses both the two-dimensional projection of molecules and the three-dimensional assembly of molecules and their manipulation. For computer modeling, projections can be presented in various ways including ball-and-stick, tube, bond-line, space-filling, and other projections. Molecular modeling of projected molecules enables the interpretation and use of estimates of physical data, such as energies, dipole moments, and electron densities based on force-field or quantum mechanical calculations. In this technique, we will discuss aspects of each of these features of molecular modeling and present exercises in calculating minimized energies of alternative conformations of simple organic molecules, predicting the relative stability of organic isomers, and calculating a number of physical properties.

Most students of organic chemistry have ready access to a molecular model kit. The next section on molecular modeling is a brief discussion of molecular model kits and their importance in learning organic chemistry. Use of these kits is encouraged in lecture and laboratory work.

Organic Molecular Models

Use of molecular model kits provides an excellent vehicle to help students visualize molecules and the spatial relationships between bonded and nonbonded atoms within molecules. Particularly important is the ability to see complex molecules in three dimensions, to physically rotate the models around single bonds, and to appreciate the very limited rotation of bonds within rings or multiple bonds. A common application is to visualize different conformations of rings, such as substituted cyclohexanes.

An important use of molecular models is assisting in the understanding of stereochemical relationships. Elements of symmetry can often be seen more easily than from two-dimensional drawings. Building models can also help the student understand relationships between structures. Model kits can be used to determine if structures are either superimposable or whether they are enantiomers or diastereomers.

Drawing Organic Molecules Using Computer Software

It is often necessary to be able to draw an organic structure prior to doing modeling studies. The structure may be represented as a bond-line drawing, a framework model showing bonds but not atoms, or a ball-and-stick model. Organic molecules can be drawn using commercial software. Several drawing programs are available for use on desktop computers. Among those that are primarily designed for drawing organic molecules are ACD/ChemSketch (Advanced Chemistry Development), ChemDraw (CambridgeSoft Corp.), ISIS/Draw (MDL Information Systems Inc.), and ChemWindow (Bio-Rad Laboratories). Structural drawings can be readily pasted or inserted into word processing and presentation documents. These documents are useful in preparing laboratory reports.

Molecules may be also represented as space-filling drawings or as electrostatic potential map drawings. By representing molecules in these ways, a clearer picture emerges of the electron clouds within molecules. Calculations based upon models such as these fall within the realm of molecular modeling.

Molecular Modeling Using Computer Software

Until the recent era of fast personal computers, molecular modeling on the computer was confined to research laboratories equipped with expensive computers, with results reported in research articles in scientific journals. Calculations are now possible by anyone who owns a desktop computer equipped with molecular modeling software. Some molecular modeling programs employed in modeling studies, calculating energies, and estimating various physical properties are CAChe (CAChe Scientific, Inc.), CHARMM

(M. Karplus, Harvard), Chem3D (CambridgeSoft), Discover (Molecular Simulations, Inc.), MacroModel (Schrodinger), HyperChem (Hypercube, Inc.), Spartan (Wavefunction, Inc.), and SYBYL (Tripos, Inc.).

These programs and others like them facilitate learning in organic chemistry and learning in the organic laboratory. When we can model on a molecular basis what is happening in a reaction flask, it helps us to conceptualize and visualize physical processes such as distillation and chemical processes such as organic reactions and the mechanisms of these processes.

Two separate approaches are used in molecular modeling. One uses a subatomic approach that is based upon quantum mechanics and quantum mechanical approximations. The other treats atoms in a molecule as if they were connected by springs to certain other atoms in the molecule. This approach is faster and is known as molecular mechanics. In some cases, both approaches are used for different aspects of the calculations.

Sample calculations and exercises are included for the two types of modeling that follow. Students are encouraged to work through the sample calculations using computer software of their own and then to try the exercises at the end of each section. Additional exercises are included as critical thinking questions at the end of a number of the experiments. Molecular mechanics is discussed in the next section.

Molecular Mechanics

The potential energy of a molecule can be represented by the total contributions of energies of all molecular components. These contributions include bonding and nonbonding interactions.

$$PE = E = \Sigma E \text{ bonding interactions} + \Sigma E \text{ nonbonding interactions}$$

The bonding contributions are stretching energies of bonds summed over all bonds, bending and scissoring energies summed over all bonds, and torsional energies summed over all bonds. The nonbonding interactions summed over all bonds are electrostatic field effects through space and van der Waals forces between atoms that are nonbonded. The nonbonding forces may be attractive or repulsive.

$$E = \Sigma E \text{ stretching} + \Sigma E \text{ bending} + \Sigma E \text{ torsional} + \Sigma E \text{ field} + \Sigma E \text{ van der Waals}$$

Calculations of the force field begins with an estimate of the "look" of the molecule. This is usually a drawing of a molecule that you would make using one of the structural drawing programs. Each of the terms in the equation is then calculated using one of the modeling programs. The programs may include different levels of sophistication. For example, simple force-field calculations are done by considering the bonds as simple harmonic oscillators. More advanced force fields may take into consideration potentials beyond simple harmonic and may consider effects due to electronegativity differences of atoms and other effects. The more advanced force fields may require more calculation time, but they can give better agreement between calculated and experimental results. Among many applicable force fields are those known as MM2, MM3, and MM4; SYBYL and MMFF.

There are a few issues to keep in mind concerning calculated energies. The calculated energies can only be used as comparisons with one another. They are not absolute energies. The energy comparisons are roughly equivalent to enthalpy comparisons of molecules in the gas phase. Molecular mechanics can be used as an aid in evaluating relative energies of different conformations in conformational analysis. Entropy contributions are ignored in these calculations.

Sample Calculation 1: Bond Angles

Bond angles can be estimated using force-field calculations. To use 3-bromopropene as an example, if the initial drawing of the structure is drawn in one of the calculation software programs, the bond angles appear as 109° around the sp^3 carbon (C3) and as 102° around the sp^2 carbons (C1 and C2). Calculate the bond angles in the lowest energy conformation.

Solution: Use molecular mechanics to measure the minimum energy geometry. The H−C1−C2 bond angle is calculated to be 123.2° and the C2−C3−Br bond angle is calculated to be 114.8° for 3-bromopropene.

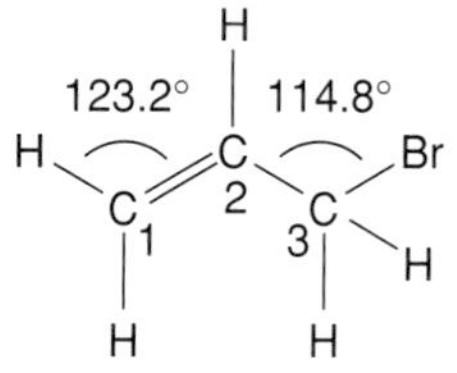

Sample Calculation 2: Dihedral Angles

Molecular mechanics calculations can furnish data on preferred dihedral angles and relative energies of conformations of molecules. Calculate the dihedral angle of the preferred conformation of 3-bromopropene. Calculate the relative energies of the preferred conformation and the conformation having a dihedral angle of 0°.

Solution: The dihedral angle for C1−C2−C3−Br of the energy-minimized (preferred conformation) of 3-bromopropene using a molecular mechanics MMFF calculation is 118.34°. The energies of different conformations of the molecule such as eclipsed or other conformations may also be calculated by constraining the C1−C2−C3−Br dihedral angle. The relative energy of the conformation having a dihedral angle of 0° is 3.56 kcal/mol higher than that of the preferred conformation. Conformations of 3-bromopropene and the designated dihedral angles are shown below.

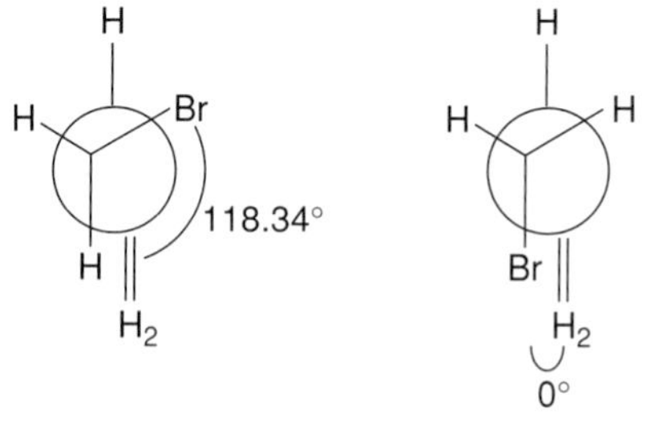

Sample Calculation 3: Conformational Energies

Relative energies of different conformations of molecules can be calculated, tabulated, and plotted. Determine the relative energies for ten different conformations of 1,2-difluoroethane? (The number of conformations selected can be set in the software program.

Solution: A tabulation of calculated conformational energies of 1,2-difluoroethane from 180° (anti conformation) to 0° (fluoro-fluoro eclipsed conformation) is shown on the next page.

1,2-Difluoroethane conformation; Dihedral angle	Relative energy (kcal/mol)
180° (anti)	0.00
160°	0.70
140°	2.16
120° (fluoro-hydrogen eclipsing)	2.86
100°	2.00
80°	0.89
60°	1.48
40°	4.23
20°	7.57
0° (fluoro-fluoro eclipsing)	9.06

Sample Calculation 4: Gauche, Anti, and Eclipsed Forms

Calculate the relative energies of the gauche and anti conformations of butane. How much more stable is the anti conformation than the eclipsed conformation having two eclipsed methyl groups? Compare your calculations with the experimental values given in your lecture text.

Solution: After drawing in the structure of butane, the program allows you to constrain the C2−C3 bond angle to 180° for the anti conformer. The calculated energy using SYBYL is 0.989 kcal/mol. Similarly, the energies for the 60° (gauche) and 0° (methyl-methyl eclipsed) conformers are 1.606 kcal/mol and 6.884 kcal/mol respectively. Now setting the anti conformation to 0 kcal/mol and recalculating gives a gauche energy of 0.617 kcal/mol and methyl-methyl eclipsed energy of 5.895 kcal/mol. These values compare favorably with the experimental values.

Relative Energies of Butane Conformers

Conformation	Calculated energy (kcal/mol)	Experimental energy (kcal/mol)
Eclipsed	5.895	Varies from 4 to 6 kcal/mol
Gauche	0.617	About 0.8
Anti	0	0

A different application of molecular mechanics concerns the interactions of molecules in a mixture. If organic molecules are mixed with water molecules, molecular mechanics predicts that the water molecules will attract one another and be at close proximity. This can be illustrated by drawing in two water molecules and three ethane molecules on the same screen (using one of the software programs). Energy optimization brings the water molecules together.

It is interesting to see how molecular mechanics calculations compare with experimental data. Being able to estimate the properties of molecules through calculation helps us to validate our theories of how atoms are connected in molecules and how molecules look and behave on a microscopic level.

Exercises Q1–3 are introductory examples that a student may encounter during the first part of an organic chemistry course. Additional assignments are included in several experiments in this text, including Experiment 5.1, 6.1, 8.2, 9.1 and subsequent experiments.

Exercises involving molecular mechanics calculations for the determination of molecular properties such as equilibrium geometries, strain energies, vibrational frequencies, and those involving calculations of energies of different conformers, their bond distances, and angles, may be found as questions in the Critical Thinking sections of several experiments in this text, including Experiments 3.7, 7.2 and later experiments.

Quantum Mechanics

The main advantage of molecular mechanics calculations is their relative speed and the ability to perform calculations on molecules having a thousand or more atoms. In the second approach for calculating molecular properties, quantum mechanical modeling is considered. This approach requires more calculation time and is limited to smaller molecules, usually those having one thousand or fewer atoms. However, this limitation is not severe for many practicing experimentalists.

Modeling using quantum mechanical approximations depicts a molecule as consisting of electron clouds containing embedded pointlike positively charged nuclei. Although they may require more time than molecular mechanics calculations, quantum mechanics calculations are generally used to model various possible transition states and to predict their relative stabilities. Quantum mechanics calculations can be used to predict product distributions for reactions that may give multiple products.

Hartree-Fock calculations are based upon the Schrodinger wave equation, but assume motionless nuclei (Born-Oppenheimer approximation) and use an approximation known as the linear combination of atomic orbitals (LCAO). Although the assumptions and resulting approximations are considerable, it is still possible to derive useful qualitative and quantitative information from the calculations. Hartree-Fock models are used in calculations of transition-state energies for molecules having fewer than one hundred atoms. Calculations of equilibrium energies and physical properties, such as geometry and dipole moments, may be done using Hartree-Fock calculations.

Semiempirical calculations make assumptions beyond those used in the Hartree-Fock modeling. Possible effects due to electrons in filled quantum shells are ignored and further approximations are made. Nonetheless, semiempirical calculations are useful for determining heats of formation, equilibrium geometries of molecules, and geometries of transition states during reactions of molecules.

Other calculation methods include Møller-Plesset (MP2, MP3, etc.) and Density Functional (DFT) methods. DFT can be used much like Hartree-Fock, but the MP2 and MP3 methods require more calculation time.

Sample Calculation 5

Calculate the relative energies of the gauche and anti conformations of butane. How much more stable is the anti conformation than the eclipsed conformation having two eclipsed methyl groups? Compare calculations with the experimental values given in the lecture text and with the molecular mechanics values in Calculation 4.

Solution: After drawing the structure of butane, the program allows you to constrain the C2−C3 bond angle to 180° for the anti conformer. The calculated energy using SYBYL is 0.975 kcal/mol. Similarly, the energies for the 60° (gauche) and 0° (methyl-methyl eclipsed) conformers are 1.587 kcal/mol and 6.682 kcal/mol, respectively. Now setting the anti conformation to 0 kcal/mol and recalculating gives a gauche energy of 0.612 kcal/mol and methyl-methyl eclipsed energy of 5.707 kcal/mol. Compare these values with the values in Calculation 4.

Relative Energies of Butane Conformers

Conformation	Molecular mechanics (kcal/mol)	Hartree-Fock (kcal/mol)
Eclipsed	5.895	5.707
Gauche	0.617	0.612
Anti	0	0

For further practice repeat the calculations for propane and 2-methylbutane. Compare these results to those obtained in molecular mechanics Exercises Q.2 and Q.3.

Sample Calculation 6

Calculate the equilibrium geometry for formaldehyde, acetone, and benzaldehyde. Determine the dipole moment and net atomic charges on various atoms. Compare the relative order of the calculated dipole moments and compare with values reported in the CRC.

Solution: After drawing in the structures of each aldehyde or ketone, minimize the energy of each and calculate the charge on each atom and calculate the dipole moment of each compound using 3-21G* Hartree-Fock calculation. Calculated dipole moments compare favorably on a relative basis with experimental values.

Compound	Calculated dipole moment (D)	Experimental dipole moment (D)
Formaldehyde	2.64	2.3
Acetone	3.12	2.9
Benzaldehyde	3.38	3.0

The net charges on the carbonyl carbons are calculated to be 0.471 for formaldehyde, 0.830 for acetone, and 0.525 for benzaldehyde, whereas the carbonyl oxygens each have partial negative charges according to the calculations.

Using Molecular Mechanics with Quantum Mechanics

Calculations using quantum mechanics often start with a molecular geometry based upon a relatively fast molecular mechanics calculation. As a result, both approaches may be used for different parts of a calculation. For example, establishing the lowest-energy conformation required for a thermal rearrangement reaction, such as the Claisen rearrangement of allyl phenyl ether, can be done by first applying molecular mechanics to determine possible conformations of the molecule. The rearrangement calls for a chairlike transition state. Energies of the different possible conformations using AM1 geometries can be calculated. The energy of the most favorable chairlike conformation for reaction may be compared with the energy of the lowest energy conformation possible to calculate the energy required to get from the lowest energy conformation to the most favorable conformation for reaction. Applying molecular mechanics before AM1 facilitates the quantum mechanics calculation and shortens the overall calculation time.

allyl phenyl ether   transition state   2-allylphenol

Claisen rearrangement of allyl phenyl ether

Generally, when calculating relative energies and preferred conformations of a molecule such as 3-bromopropene in Sample Calculation 2, molecular mechanics MMFF and quantum mechanical Hartree-Fock and semiempirical AM1 and PM3 calculations are done at the same time (back-to-back). MMFF calculations are usually fast and Hartree-Fock calculations take longer, but otherwise modern software programs allow any and all of these calculations. To further pursue the relative energies of conformations of 3-bromopropene, let us calculate the preferred conformations of 3-bromopropene using both molecular mechanics and quantum mechanics calculations.

Sample Calculation 7

Calculate the dihedral angle of the preferred conformation of 3-bromopropene and the relative energy of the preferred conformation compared to the conformation having a dihedral angle of $0°$.

Solution: From Calculation 2, the dihedral angle for $C1-C2-C3-Br$ of the energy-minimized (preferred conformation) of 3-bromopropene using a molecular mechanics MMFF calculation is $118.34°$. A summary of results from molecular mechanics and quantum mechanics calculations is shown here. (Note that 1 Hartree = 627.51 kcal/mol.)

Calculation method	Dihedral angle of preferred conformation	Energy of preferred conformation relative to the $0°$ conformation
MMFF	$118.34°$	-3.561 kcal/mol
AM1	$111.18°$	-1.414 kcal/mol
PM3	$120.31°$	-1.736 kcal/mol
Hartree-Fock 3.21G*	$117.70°$	-1.004 kcal/mol

Exercises involving quantum mechanics calculations for the determination of molecular properties such as equilibrium geometries, heats of formation and physical properties, charge densities in molecules, and other calculations may be found in the Critical Thinking sections of several experiments in this text, including Experiments 5.1, 6.1, 8.2, 12.2, 12.4 and other experiments.

Exercises

Part I Molecular Mechanics Exercises

Exercise Q.1: Conformational Analysis of Butane and Other Molecules

Calculate the relative energies for ten different conformations of butane (about the $C2-C3$ axis). Use the calculation software to plot energy (kcal/mol) vs dihedral angle. Alternatively, do calculations on a different molecule such as 1,2-dichloroethane, 1,2-dibromoethane, or another molecule assigned by the instructor.

Exercise Q.2: Conformational Analysis of Propane

Set the number of propane conformers desired using the calculation software and calculate the relative energies of the different conformers. Tabulate and compare with experimental data. Calculate the relative energies of propane conformers. Compare the calculated energies with the experimental values: eclipsed conformation, 2.90 kcal/mol; staggered conformation, 0 kcal/mol.

Exercise Q.3: Conformational Analysis of 2-Methylbutane

Calculate the relative energies of 2-methylbutane conformers. Compare the calculated energies with the predicted values based upon experiments: methyl-methyl eclipsed conformation, ~6.4 kcal/mol; hydrogen-methyl eclipsed conformations, 3.8 kcal/mol; dimethyl-methyl gauche conformation, 1.6 kcal/mol; methyl-methyl gauche conformation, 0.8 kcal/mol.

Part II Quantum Mechanics Exercises

Exercise Q.4: Identifying Reactive Sites

Determine the most reactive site for propanal and propanone with nucleophiles. This calculation can be done by determining the net charges. Reactive sites may also be visualized using orbital representations of electrostatic potential maps where relatively electron deficient regions are shown in blue and electron-rich regions are shown in red. For the lowest unoccupied molecular orbital (LUMO) of aldehydes and ketones, the electron density will show as a blue color over the carbonyl carbon and as a red color over the oxygen.

Exercise Q.5: Heats of Formation and Dipole Moments

Calculate the heats of formation and dipole moments for 1-butene, *cis*-2-butene, and *trans*-2-butene. Compare with the reported gas phase heats of formation (CRC, 2001, 5-25 to 5-60); 1-butene: 0.1 kcal/mol; *cis*-2-butene: –7.1 kcal/mol; *trans*-2-butene: –11.4 kcal/mol. Dipole moments are calculated for specific conformations. Compare the calculated values with experimental dipole moments (CRC, 2001, 9-44 to 9-50); 1-butene (cis): 0.438 Debyes (D); 1-butene (skew): 0.359 D; *cis*-2-butene: 0.253 D; *trans*-2-butene: 0 D.

Exercise Q.6: LUMO Energies of Alkenes

Compare the LUMO energies of the alkene π bonds of ethyl acrylate ($CH_2=CHCO_2CH_2CH_3$) and diethyl fumarate ($CH_3CH_2O_2CCH=CHCO_2CH_2CH_3$). The LUMO of lower energy (more negative value) generally reacts faster with electrons in the highest occupied molecular orbital of an electron pair donor molecule, such as in a Diels-Alder reaction (see Experiment 9.1). Use an AM1 calculation of initial geometry after selecting a Hartree-Fock single-point energy calculation. Which would be predicted to be more reactive, ethyl acrylate or diethyl fumarate?

Exercise Q.7: Conformational Analysis of 3-Fluoropropene

Use molecular mechanics MMFF and quantum mechanical Hartree-Fock and semi-empirical AM1 and PM3 calculations to calculate the preferred conformation of 3-fluoropropene. Compare the results with those obtained for 3-bromopropene. Do the calculations enable insight into molecular modeling by calculation vs using molecular models?

Reference

Lide, D. R., *CRC Handbook of Chemistry and Physics,* 82nd edition. CRC Press, Boca Raton, FL 2001.

Chapter 3

Applications Using Laboratory Resources and Techniques

One of the tasks required for all experiments is gathering of physical data for each of the chemicals used in the experiments. Physical data are available for many organic compounds from common reference works, such as the *Handbook of Chemistry and Physics* (CRC Press, Inc.). The scavenger hunt (Experiment 3.1) provides some practice in locating physical data for specific organic compounds and a review of stoichiometric calculations. Other experiments in Chapter 3 serve as a framework for the elaboration of techniques detailed in Chapter 1.

In Experiment 3.2, measurements of boiling point and refractive index are used to identify liquid unknowns. Solvents are the focus of other experiments in Chapter 3. Solvents occupy a central role in organic chemistry because most reactions are best carried out in solution and liquid solvents provide the most convenient means of mixing reactants. Solvents play important roles in recrystallization (Experiment 3.4), extraction (Experiments 3.5 and 3.6), and chromatography (Experiment 3.7).

Solvent properties that are important for recrystallization are explored in Experiment 3.4. Distributing different solutes between two solvents that do not mix well can serve as a separation technique, such as the separation of caffeine from coffee or NoDoz in Experiment 3.6. For solvents that form separate layers in intimate contact, a desired compound can often be drawn by extraction into one of the solvents, leaving other substances (impurities) behind in the other solvent.

Chromatography is another technique where solvents play a major role. Chromatography, such as thin-layer chromatography in Experiment 3.7, is used to separate compounds from a mixture (see also Technique K). In Experiment 3.7, thin-layer chromatography is used to separate three compounds of differing polarity.

Experiment 3.1: Scavenger Hunt: *Introduction to Chemical Data Reference Books and Calculations*

There are over ten million known organic compounds. Properties of several thousand of these can be found in chemical data reference books. In this experiment, you will learn:

◆ which reference books to use to find data for specific compounds.

◆ where to locate the appropriate data in the various reference books.

◆ what the symbols and abbreviations found in reference books mean.

◆ how to locate information on material safety data sheets (MSDS).

◆ how to use the information to perform simple calculations of molar quantities, limiting reactants, and yields.

Background

Understanding how to use chemical handbooks and reference books is pivotal to a student's success in organic chemistry. Whether finding the melting point of a product or obtaining safety information about a reagent used in the laboratory, it will be necessary to use these reference books throughout the course. Learning to use these books efficiently will decrease the amount of time spent preparing for laboratory work.

The most common chemistry reference books are the *Chemical Rubber Company's Handbook of Chemistry and Physics* (CRC Press, Inc.) and *The Merck Index* (Merck & Co., Inc.), but other books, such as *Lange's Handbook of Chemistry* (McGraw-Hill), *Handbook of Tables for Organic Compound Identification* (CRC Press, Inc.), and chemical supply catalogs, such as those of Aldrich Chemical Company, Lancaster Synthesis, Inc., and Fluka Chemical Company, can be used depending upon the information desired.

Handbook of Chemistry and Physics (CRC)

The CRC has much information about all fields of chemistry and physics, such as thermodynamic properties of elements, dissociation constants of acids and bases, and ionization potentials. Topics are listed alphabetically in the index in the back of the book. Of most interest to organic students is the section "Physical Constants of Organic Compounds." Physical properties such as molecular weight, boiling point, melting point, density, refractive index, and solubilities are tabulated for selected organic compounds. Other physical data such as crystalline form, color, and specific rotation may also be included.

The data can best be understood by reading "Explanation of Table of Physical Constants of Organic Compounds" and "Symbols and Abbreviations," which can be found in the pages immediately preceding the tables. For example, the entry for acetaldehyde gives the refractive index (n_D) as 1.4409_D^{20} and the solubility as w, al. From "Explanation of Table of Physical Constants of Organic Compounds," it can be seen that the refractive index is reported for the D line of sodium and that the "20" refers to the temperature at which the refractive index was measured. From the list of abbreviations and symbols, it can be seen that the compound is soluble in water and ethyl alcohol. Before starting the assignment that follows, refer to the section "physical constants of organic compounds," find these pages, and see what other information is available on organic compounds.

To find the physical properties of a given organic compound in the CRC, it is important to have a correct name. Organic compounds are listed alphabetically by their International Union of Pure and Applied Chemistry (IUPAC) names under the name of the parent compound. For example, the IUPAC name of a compound with the structural

PHYSICAL CONSTANTS OF ORGANIC COMPOUNDS (Continued)

No.	Name, Synonyms, and Formula	Mol. wt.	Color, crystalline form, specific rotation and λ_{max} (log ε)	b.p. °C	m.p. °C	Density	n_D	Solubility	Ref.
11963	1-Propanol, 2-methyl or Isobutyl alcohol.............. $(CH_3)_2CHCH_2OH$	74.12	....	108	−108	$0.8018^{20/4}$	1.3955^{20}	w, al, eth, ace	B1[4], 158

SOURCE: Reprinted with permission from Chemical Rubber Company's *Handbook of Chemistry and Physics,* 74th ed., 3–416, 1993. Copyright CRC Press, Boca Raton, Florida.

formula $(CH_3)_2CHCH_2OH$ is 2-methyl-1-propanol. This compound is indexed in the CRC as **1-propanol, 2-methyl.** The CRC entry for this compound is shown above.

The ease of use of the CRC is governed by being able to locate the parent name correctly and then to locate the substituent(s) under the parent. A comprehensive guide to naming compounds by the IUPAC system is given before the section on physical constants of organic compounds. Alternately, the compound can be located from its molecular formula in the Formula Index for Organic Compounds. However, since there are generally many organic compounds with the same molecular formula, this tends to be a tedious way to locate a compound. For example, the molecular formula of 2-methyl-1-propanol is $C_4H_{10}O$. The 63rd edition of the CRC lists reference numbers to twelve different compounds with this same molecular formula.

The Merck Index

Another reference book frequently used in the organic lab is *The Merck Index,* which is an alphabetical compilation of common names of compounds, much like a dictionary. The major focus of *The Merck Index* is on pharmaceutical, medicinal, and natural products, although some common organic compounds are also included. In addition to physical properties such as molecular weight, boiling point, melting point, refractive index, density, color, crystalline structure, solubility data, and recrystallization solvents, *The Merck Index* gives the IUPAC name, Chemical Abstracts name and alternative names, Chemical Abstracts reference numbers, toxicity and biological hazard information, biological and pharmacological information, and one or more references to the synthesis of the compound. A section at the beginning of *The Merck Index* shows a sample abstract and gives abbreviations and selected definitions of the terms used in the abstract. Refer to these pages to interpret the physical data given in the abstract.

To look up a compound in *The Merck Index,* go to the index at the back of the book and find the name of the compound alphabetically. The index cross references the various names of a compound, so this is the quickest way to find the given compound. For example, the compound with the structural formula $(CH_3)_2CHCH_2OH$ is cross referenced under **2-methyl-1-propanol** (the IUPAC name) and under **isobutyl alcohol** (a common name). Beside each entry is an abstract number. The abstracts are numbered and arranged sequentially. Refer to the specific abstract to find information on the given compound. *The Merck Index* entry for this compound is shown here.

5146. Isobutyl Alcohol. *2-Methyl-1-propanol;* isopropylcarbinol; 1-hydroxymethylpropane; fermentation butyl alcohol. $C_4H_{10}O$; mol wt 74.12. C 64.82%, H 13.60%, O 21.58%. $(CH_3)_2CHCH_2OH$. Present in fusel oil; also produced by fermentation of carbohydrates: Baraud, Genevois, *Compt. Rend.* **247,** 2479 (1958); Sukhodol, Chatskii, *Spirt. Prom.* **28,** 35 (1962), *C.A.* **57,** 5124e (1962). Prepn: Wender *et al., J. Am. Chem Soc.* **71,** 4160 (1949); Schreyer, **U.S.** pat. **2,564,130** (1951 to du Pont); Pistor, **U.S.** pat. **2,753,366;** Harrer, Ruhl, **Ger.** pat. **1,011,865;** Himmler, Schiller, **U.S.** pat. **2,787,628** (1956, 1957, 1957 all to BASF). Toxicity study: Smyth *et al., Arch. Ind. Hyg. Occup. Med.* **10,** 61 (1954).

Colorless, refractive liq; flammable; odor like that of amyl alcohol, but weaker. d^{15} 0.806. bp 108°. mp −108°. Flash pt, closed cup: 82°F (28° C). n_D^{15} 1.3976. Sol in about 20 parts water; misc with alcohol, ether. LD_{50} orally in rats: 2.46 g/kg (Smyth).

Caution: Potential symptoms of overexposure are irritation of eyes and throat; headache, drowsiness; skin irritation, skin cracking. *See NIOSH Pocket Guide to Chemical Hazards* (DHHS/NIOSH 90-117, 1990) p 132.

Use: Manuf esters for fruit flavoring essences; solvent in paint, varnish removers

SOURCE: *The Merck Index,* 12th ed., 1996.

Chemical Catalogs

A third source of chemical information comes from the catalogs of chemical suppliers. Although the information in these catalogs is limited, physical properties for a number of common organic, inorganic, and biochemical compounds sold by the companies are given, and the catalogs are usually easy to use. In addition to molecular weight, melting point, boiling point, refractive index, density, and specific rotation, other information such as safety hazards, disposal information, purity, and cost are included. The physical properties are for the chemicals as supplied by the company and may be slightly different from those for the pure compounds.

Compounds are listed alphabetically under common or IUPAC names. Thus $(CH_3)_2CHCH_2OH$ is found under **2-methyl-1-propanol** in the Aldrich catalog. Two Aldrich catalog entries for this compound are shown here.

27,046-6   **2-Methyl-1-propanol,** 99.5%, HPLC grade [*78-83-1*] (isobutyl alcohol)..............................**100mL**
★ $(CH_3)_2CHCH_2OH$ FW 74.12 mp -108° n_D^{20} 1.3960 d 0.803 Fp 82°F(27°C)   **1L**
     Beil. **1,**373 *Merck Index* **13,**5146 *FT-NMR* **1**(1),170A *FT-IR* **1**(1),114C *SI* 14,B,7   **6x1L**
     Safety **2,**2422A *R&S* **1**(1),123N *RTECS#* NP9625000 *FLAMMABLE LIQUID IRRITANT*   **2L**
     Glass distilled   **4x2L**
     (*Packaged under nitrogen, 100mL, 1L, and 2L bottles are Sealed for Quality*^{TM})   **4x4L**

     Max. U.V. Abs. (1 cm. cell - vs. H_2O)

λ(nm)	400-330	260	250	245	230	219
A		0.01	0.03	0.05	0.10	0.50	1.0

     Water <0.05%   Evapn. residue <0.0005%

29,482-9   **2-Methyl-1-propanol,** anhydrous, 99.5%, [*78-83-1*] (isobutyl alcohol)............................ **100mL**
★ $(CH_3)_2CHCH_2OH$ **1L**
     Water <0.005% (100 ml unit)   **6x1L**
     (*Packaged under nitrogen in Sure/Seal*^{TM} *bottles*)   **2L**
     Water <0.003%   Evapn. residue <0.0005%   **4x2L**

SOURCE: Aldrich Catalog, 2003–2004.

The material safety data sheet (MSDS) is a good source of information about safety and other hazardous materials information (Appendix D). Required by law for every chemical, the MSDS provides important information about toxicity, permissible levels of exposure, health consequences of exposure, protocols for safe use, and disposal. Each supplier of chemicals provides MSDS sheets for each chemical distributed. One particularly important section of an MSDS is the Health Hazard Data section. This delivers critical information about how the chemical enters the body, the routes of entry (such as inhalation, ingestion, and skin contact), and the appropriate personal protective equipment (PPE) required to prevent exposure (such as gloves, respirator, etc.). A MSDS also lists the permissible exposure limit (PEL), which refers to the amount of exposure to a chemical that is considered safe. PELs are given as a time-weighted average exposure over eight hours. The lower the PEL, the more hazardous the chemical. For example, hydrochloric acid has a PEL of 5, while ethyl acetate has a PEL of 400. Further information about an MSDS and a sample MSDS can be found in Appendix D.

Online Sources

In addition to reference texts, there are some web-based search engines, such as ChemFinder (http://www.chemfinder.com). Here, physical properties are found by typing the name of the compound and searching for the information. For example, typing either **2-methyl-1-propanol** or **isobutyl alcohol** in the search box brings up the structure, formula, other names, and physical properties of the compound. Information about health, safety, and MSDS is available by clicking on the relevant links. Other sources of online chemical information can be found using the Combined Chemical Dictionary (CCD) Online or Merck Index Online. While ChemFinder is free, these latter programs must be purchased.

Other Sources

Other reference books such as *Lange's Handbook of Chemistry* and *The Chemist's Companion* can also be used to find specific information. In addition, there are also reference books in which spectral information about organic compounds is tabulated and others in which melting points of derivatives can be found. Your lab instructor may indicate the use of other reference books, depending upon the information desired.

Part A: Scavenger Hunt in the Chemical Reference Books

Use the reference sources described above to provide answers for each of the following questions. Cite the source.

1. The entry in the CRC for "acetaldehyde, bromo, diethylacetal" gives the physical properties listed below. Refer to the Explanation of Table of Physical Constants of Organic Compounds and the Abbreviation and Symbols sections to answer the questions:
    a. bp = 180.6^{18}: What does the superscript 18 signify?
    b. d = $1.280^{20/4}$: What does the superscript 20/4 signify?
    c. $n_D = 1.4376^{20}$: What does the subscript D signify?
    d. sol. = **al, eth**: List two solvents in which this compound is soluble.
2. Cinnamaldehyde is a constituent of Ceylon and Chinese cinnamon oils.
    a. What is the molecular weight of cinnamaldehyde?
    b. What is the structure and formula of cinnamaldehyde?
    c. Cite a reference for the preparation of cinnamaldehyde.
    d. What is the density of cinnamaldehyde?

    e.   What is the boiling point of cinnamaldehyde at 760 torr?
    f.   What is the refractive index of cinnamaldehyde?
3. What is a common name of 2-methyl-2-propanol? Is this compound a liquid, a solid, or a gas at 25°C?
4. Give a name (common or IUPAC) for any compound having the molecular formula $C_6H_4N_2O_6$.
5. List three solvents in which 2,6-dinitrobenzaldehyde is soluble.
6. What is the cost of 500 mg of glycylglycylglycylglycylglycine?
7. Diethyl ether is frequently used in extractions.
    a.   What is the flash point for diethyl ether?
    b.   What fire extinguishing media should be used to extinguish a diethyl ether fire?
    c.   What is the LD_{50} for diethyl ether? Explain what this means.
8. Cite the PEL values for acetone and bromine. Which compound is more hazardous?

Part B: Scavenger Hunt and Stoichiometric Calculations

The next part of the exercise includes examples of simple calculations required in the organic lab. Show all work in your lab notebook.

1. Consider the reaction of 5.0 g of propane with 10.0 mL of molecular bromine to form 2,2-dibromopropane and hydrogen bromide. The **balanced** reaction is shown below. Look up necessary information in a reference book and use the information to answer the following questions:

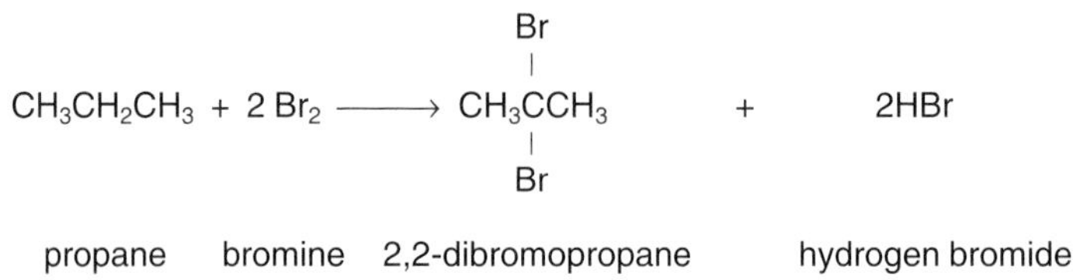

$$CH_3CH_2CH_3 + 2\,Br_2 \longrightarrow CH_3\underset{\underset{Br}{|}}{\overset{\overset{Br}{|}}{C}}CH_3 \quad + \quad 2HBr$$

    propane   bromine   2,2-dibromopropane   hydrogen bromide

    a.   Look up the molecular weight of propane and calculate the moles of propane present before reaction.
    b.   Look up the density and molecular weight of bromine and calculate the moles of molecular bromine present before reaction.
    c.   Determine whether propane or bromine is the limiting reactant.
    d.   Calculate the mass (g) remaining of the excess reactant after the reaction is complete.
    e.   Calculate the theoretical yield (g) of 2,2-dibromopropane.
    f.   Suppose that 15.58 g of 2,2-dibromopropane were actually formed in the reaction. Calculate the percent yield of 2,2-dibromopropane.
    g.   Look up the density of 2,2-dibromopropane and calculate the volume of 15.58 g of 2,2-dibromopropane.
2. Calculate the number of mmol of HCl in 3.0 μL of 12 M HCl.

References

Aldrich Handbook of Fine Chemicals and Laboratory Equipment, Aldrich Chemical Co., Inc., Milwaukee, WI, 2003–2004.

Budavari, S., ed. *The Merck Index: An Encyclopedia of Drugs, Chemicals and Biologicals.* 12th ed. Whitehouse Station, NJ: Merck and Co., Inc.,1996; O'Neil, M. J. et al. ed., 13th ed., 2001.

Dean, J. A., ed. *Lange's Handbook of Chemistry*, 14th ed., New York: McGraw-Hill, 1992.

Gordon, A. J., and Ford, R. A. *The Chemist's Companion: A Handbook of Practical Data, Techniques and References.* New York: Wiley, 1973.

Lide, D. R., Jr. *Handbook of Chemistry and Physics.* 82nd ed. Boca Raton, FL: CRC Press, 2001.

Rappoport, Z. *Handbook of Tables for Organic Compound Identification.* 3rd ed. Cleveland, OH: Chemical Rubber Co., 1967.

Experiment 3.2: Identification of Organic Liquids by Physical Properties

The purpose of this experiment is to determine the identity of an unknown organic liquid using a combination of measurements of physical properties. These include refractive index and boiling point. In this experiment, you will:

♦   measure the refractive index of a pure liquid.

♦   determine the boiling point of a pure liquid.

♦   apply correction factors for temperature and for barometric pressure.

Techniques

Technique B	Measurement of liquids
Technique D	Boiling point
Technique E	Refractive index

Background

By measuring the boiling point and refractive index of a simple organic liquid and comparing the experimental data with the physical properties of known liquids, it is often possible to determine the identity of the liquid. Careful and accurate measurements of the boiling point and refractive index should enable you to successfully identify the unknown liquid or at least to narrow down the list of possibilities.

The boiling point of a pure liquid can be measured by placing the liquid in a small-diameter test tube or Craig tube and heating slowly until the liquid boils. Alternatively, a microscale procedure can be used with a glass bell and capillary tube. The observed boiling point must be corrected if the ambient pressure differs from 1 atmosphere. Overheating or too rapid heating will superheat the liquid, causing the measured temperature to be too high.

The refractive index of a liquid can be measured very accurately. The observed refractive index is both temperature and wavelength dependent. To standardize the measurements, the values of the refractive index are measured with respect to the yellow D line in the sodium emission spectrum (589.3 nm). A mathematical correction must be applied if the observed temperature is not 20°C. This equation is shown below. Note that t is the ambient temperature in °C:

$$n_D^{20} = n_{obs} + 0.00045 \, (t - 20.0)$$

By comparing the values for the corrected physical properties with tables of physical properties of known liquids, the identity of the unknown can often be determined.

Prelab Assignment

1. Read the procedure for using the Abbe refractometer (Technique D).
2. The boiling point and refractive index of an unknown liquid were measured in a laboratory where the ambient pressure and temperature were 700 torr and 22°C, respectively. The observed boiling point of the unknown liquid was 105°C; the observed refractive index was 1.4959. Calculate the corrected boiling point and the corrected refractive index for this unknown liquid.
3. How does altitude affect the observed boiling point of a liquid?

Experimental Procedure

S a f e t y   F i r s t !

Always wear eye protection in the laboratory.

1. Wear eye protection at all times in the laboratory.
2. Assume that all organic liquids are harmful. Wear gloves and don't breathe the vapors.
3. Never heat a closed system.

General Directions

Obtain a 1-mL sample of an unknown liquid from the instructor. Record the identification number of the unknown. Measure and record the ambient pressure and temperature.

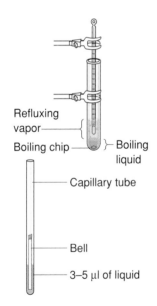

Refluxing vapor

Boiling chip —)— Boiling liquid

Capillary tube

Bell

3–5 µl of liquid

Part A: Boiling Point Determination

Pipet 0.3–0.5 mL of the liquid into a small diameter test tube or Craig tube that contains a boiling chip. Clamp the tube in a heat block or sand bath. Suspend the thermometer so that the bottom of the thermometer bulb is 0.5 cm above the level of the liquid and clamp in place. Position the thermometer so that it does not contact the sides of the tube. Gently heat the liquid to boiling. Continue to heat slowly until the refluxing vapor forms a ring of condensate about 1–2 cm above the top of the bulb. When the temperature stabilizes for at least one minute, record the temperature. Correct the measured boiling point for atmospheric pressure using a nomograph. (See also Figure 1D-1.)

Alternate procedure: The boiling point of a liquid may also be measured using a capillary tube and bell. The instructor will demonstrate how to make and load the glass bell. Heat the capillary tube fairly rapidly until a steady stream of bubbles emerges from the glass bell. Turn down the heat. Watch carefully until the last bubble collapses and liquid enters the glass bell. Note the temperature. The temperature at which the liquid reenters the bell is the observed boiling point. Correct the measured boiling point for atmospheric pressure using a nomograph.

Part B: Refractive Index Determination

Place two drops of the purified liquid between the plates of an Abbe refractometer. Measure and record the refractive index of the liquid. Also record the ambient temperature. If this differs from 20°C, apply a correction factor.

Results and Conclusions for Parts A and B

1. Report the ambient temperature, the barometric pressure, and the identification number of the unknown liquid.
2. Calculate the corrected boiling point of the liquid.
3. Calculate the corrected refractive index.
4. Compare the corrected boiling point and corrected refractive index of the unknown liquid with the physical properties of known liquids from Table 3.2-1. Determine the identity of the unknown or list more than one compound if a clear distinction cannot be made between two or more compounds. Justify your conclusion.
5. Calculate the percent error for each of the measured physical properties (experimental values) from the reported physical properties of the identified liquid (literature values). Percent error is given as:

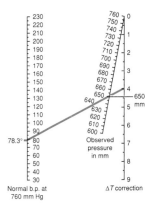

$$\% \text{ error} = \left| \frac{\text{literature value} - \text{experimental value}}{\text{literature value}} \right| \times 100$$

Cleanup & Disposal

Dispose of the glass capillary tubes and melting point capillary tubes in a glass collection container. Return all remaining liquid unknown in the original labeled vial.

Table 3.2-1   Physical Properties of Selected Organic Liquids

Organic compound	Boiling point (°C)	Refractive index (n_D^{20})
2,3-Dimethylbutane	58.0	1.3750
3-Methylpentane	63.3	1.3765
Tetrahydrofuran	67.0	1.4050
Hexane	69.0	1.3751
Ethyl acetate	77.1	1.3723
2,2-Dimethylpentane	79.2	1.3822
2-Butanone	79.6	1.3788
2,4-Dimethylpentane	80.5	1.3815
Cyclohexane	80.7	1.4266
Acetonitrile	81.6	1.3442
2-Methyl-2-propanol	82.3	1.3878
2-Propanol	82.4	1.3776
2,3-Dimethylpentane	89.8	1.3919
2-Methylhexane	90.0	1.3848
3-Methylhexane	92.0	1.3885
1-Propanol	97.4	1.3850
Heptane	98.4	1.3878
2-Butanol	99.5	1.3978
3-Pentanone	101.7	1.3924
2-Pentanone	102.0	1.3895
Toluene	110.6	1.4961
1-Butanol	117.2	1.3993
Ethylbenzene	136.2	1.4959
p-Xylene	138.3	1.4958

Critical Thinking Questions

1. Which physical property is more useful for determining the identity of the liquid? Explain.

2. Explain how nonvolatile impurities affect the vapor pressure and the boiling point of a liquid.

3. Will an impurity affect the refractive index of a liquid? Explain.

4. Why does temperature affect the refractive index?

5. After correcting for ambient pressure and temperature, a student determines that the corrected boiling point for an unknown liquid is 80°C and the corrected refractive index is 1.4260. Use Table 3.2-1 to determine the identity of the unknown.

6. A student measured the refractive index of an unknown liquid as 1.3815. The ambient temperature in the lab was 24°C. The student erroneously reported the corrected refractive index as 1.3995 (when the corrected value should have been 1.3833). What mistake in the calculation led to this error?

Experiment 3.3: Relationships Between Structure and Physical Properties

The structure of an organic compound greatly influences physical properties such as boiling point, melting point, and refractive index. In this experiment, you will make measurements of the physical properties of alkanes and other functionalized organic compounds. From the data, you will determine:

♦ the boiling point of a liquid.

♦ the refractive index of a liquid.

♦ the melting point of a solid.

♦ the relationship between structure and boiling point.

♦ the relationship between functionality and boiling point.

♦ the relationship between structure and refractive index.

Techniques

Technique C	Melting point
Technique D	Boiling point
Technique E	Index of refraction

Background

The boiling point of a liquid depends upon the strength of interactions between the molecules. The stronger the intermolecular forces the higher the boiling point of the liquid. The weakest intermolecular attractions are London forces. These are transient interactions. Dipole-dipole attractions are stronger intermolecular attractions. These occur between molecules that have a permanent dipole. The strongest intermolecular attractions are hydrogen bonds. These occur between molecules that contain hydrogen bonded to a strongly electronegative atom such as oxygen, nitrogen, or fluorine.

In the first part of this experiment, the boiling points of unbranched alkanes and branched alkanes will be compared to determine the effect that chain length and branching have on boiling point. The unbranched alkanes used in this experiment have the formula $CH_3(CH_2)_nCH_3$ where n = 3 for pentane, n = 4 for hexane, n = 5 for heptane, n = 6 for octane, n = 7 for nonane, and n = 8 for decane. The branched alkanes used in this experiment are 2-methylpentane and 2,2-dimethylbutane, which have the structures shown here:

$$CH_3$$
$$|$$
$$CH_3CCH_2CH_2CH_3 \qquad\qquad CH_3CCH_2CH_3$$
$$| \qquad\qquad\qquad\qquad\qquad |$$
$$H \qquad\qquad\qquad\qquad CH_3$$

$$\text{2-methylpentane} \qquad \text{2,2,-dimethylbutane}$$

In the second part of this experiment, the influence of functional groups on boiling point will be determined using 2-pentanone, pentanal, 1-pentanol, and butanoic acid. Boiling points of the compounds will be compared with the boiling point of hexane. All of these compounds have similar molecular weights. The structures of the functionalized compounds are shown below:

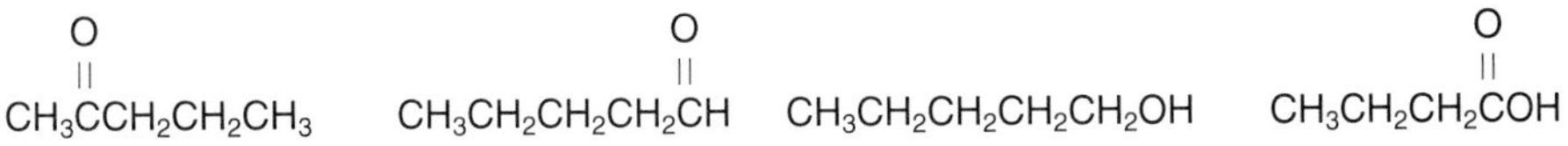

$$O \qquad\qquad\qquad O \qquad\qquad\qquad\qquad\qquad\qquad\qquad\qquad O$$
$$|| \qquad\qquad\qquad || \qquad\qquad\qquad\qquad\qquad\qquad\qquad\qquad ||$$
$$CH_3CCH_2CH_2CH_3 \quad CH_3CH_2CH_2CH_2CH \quad CH_3CH_2CH_2CH_2CH_2OH \quad CH_3CH_2CH_2COH$$

$$\text{2-pentanone} \qquad\qquad \text{pentanal} \qquad\qquad \text{1-pentanol} \qquad\qquad \text{butanoic acid}$$

The melting point of a molecular compound is determined by the ability of the molecules to pack together and the strength of attractions between molecules. Ionic compounds, with electrostatic attractions, have extremely high melting points. Covalent compounds have significantly lower melting points. The melting point is strongly influenced by symmetry. By comparing the melting points of pairs of compounds, the student can determine how symmetry affects the melting point. The structures of these compounds are shown below:

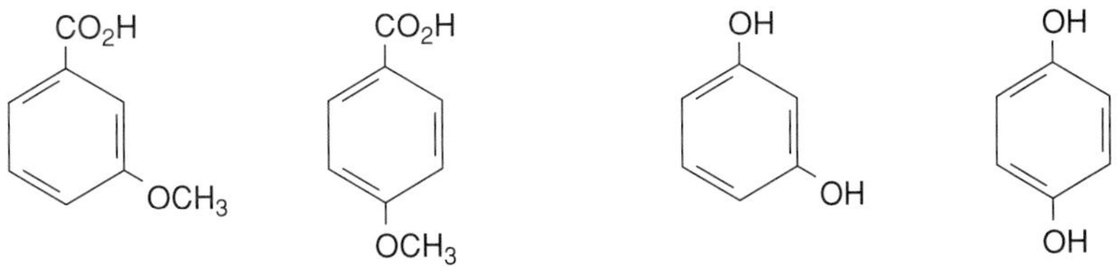

m-anisic acid   *p*-anisic acid   resorcinol   hydroquinone

The refractive index of liquids can be measured very accurately. When a refractive index is known, refractometry is an easy method for determining the purity of liquids. The refractive index is both temperature and wavelength dependent. To standardize the measurements, the values of the refractive index are measured with respect to the yellow D line in the sodium emission spectrum (589.3 nm) and a mathematical correction for the actual temperature is performed. In this experiment, the effect of chain length and branching on refractive index will be determined.

Prelab Assignment

1. Straight-chain alkanes have greater surface area than branched chain alkanes of the same molecular weight. Rank 2-methylpentane, hexane, and 2,2-dimethylbutane in terms of increasing surface area.

2. Classify each of the compounds listed in Table 3.3-3 as polar, moderately polar, or nonpolar and indicate what type of intermolecular attractions (London dispersion, dipole-dipole, or hydrogen-bonding) would be expected to predominate.

3. Examine the structures for planes of symmetry and determine which compound in each pair of compounds has a greater degree of symmetry: *m*-anisic acid or *p*-anisic acid; resorcinol or hydroquinone.

Experimental Procedure

1. Wear goggles at all times in the laboratory.
2. Avoid skin contact and inhalation of all compounds. Wear gloves.

In this experiment, each student will work on individual samples and then pool the results with the rest of the class. Each student should determine the boiling point and refractive index of a liquid and the melting point of a solid. Obtain a liquid and a solid sample from the laboratory instructor. The liquid samples may be branched alkanes, unbranched alkanes, or other types of organic compounds. Record the name of the samples and draw their structures. Prepare tables in the lab notebook similar to Tables 3.3-1 to 3.3-4.

Part A: Determination of the Effect of Structure on Boiling Point

This procedure works well for liquids that boil higher than 50°C. Add 0.3–0.5 mL of the liquid to a small diameter test tube or Craig tube that contains a boiling chip. Clamp the tube in a heat block or sand bath. Suspend the thermometer so that the bottom of the thermometer bulb is 0.5 cm above the level of the liquid and clamp in place. Position the thermometer so that it does not contact the sides of the tube (see Figure 1D-3). Gently heat the liquid to boiling. Continue to heat slowly until the refluxing vapor forms a ring of condensate about 1–2 cm above the top of the bulb. When the temperature stabilizes for at least one minute, record the temperature. Record the boiling point in the appropriate table and correct for atmospheric pressure, if necessary.

Part B: Determination of the Effect of Structure on Refractive Index

Measure the refractive index of the liquid. Apply a temperature correction if the ambient temperature differs from 20°C. Record the corrected refractive index in the appropriate table.

Part C: Determination of the Effect of Structure on Melting Point

Prepare a capillary tube with a few crystals of the solid. Record the temperature at which the crystals start to melt and the temperature at which the last crystal melts. Record the melting point range in the appropriate table.

Cleanup & Disposal

Dispose of the glass capillary tubes in a glass collection container. Pour any remaining liquid unknown into a labeled container for recycling. Return any remaining solid unknown to the instructor.

Data Tables

Table 3.3-1 Physical Properties of Straight-chain Alkanes

Compound	Number of carbons	Boiling point (observed)	Boiling point (corrected)	Refractive index (observed)	Refractive index (corrected)
Pentane	5				
Hexane	6				
Heptane	7				
Octane	8				
Nonane	9				
Decane	10				

Table 3.3-2 Physical Properties of Branched Alkanes

Compound	Number of carbons	Boiling point (observed)	Boiling point (corrected)	Refractive index (observed)	Refractive index (corrected)
Hexane	6				
2-Methylpentane	6				
2,2-Dimethylbutane	6				

Table 3.3-3 Boiling Point of Functionalized Compounds

Compound	Molecular mass (g/mol)	Boiling point (observed)	Boiling point (corrected)
Hexane	86.17		
2-Pentanone	86.13		
Pentanal	86.13		
1-Pentanol	88.15		
Butanoic acid	88.10		

Table 3.3-4   Melting Point and Symmetry of Isomeric Compounds

Compound	Melting point	Compound	Melting point
m-Anisic acid		Resorcinol	
p-Anisic acid		Hydroquinone	

Results and Conclusions

1. Look up the physical properties of the assigned compounds in the CRC or *The Merck Index*. Calculate the percent error in the boiling point, refractive index, and melting point.
2. Make a graph of boiling point (corrected) versus chain length (number of carbons in chain) for the unbranched alkanes (Table 3.3-1). Summarize the effect that chain length has on boiling point.
3. Compare the boiling points of the branched alkanes (Table 3.3-2) to the boiling point of hexane. Summarize the effect that branching has on boiling point. Explain the observed differences in boiling points for the unbranched and branched alkanes in terms of intermolecular attractions.
4. Arrange the compounds in Table 3.3-3 in order of increasing boiling point. Explain how functional groups affect intermolecular attractions.
5. Explain the relationship between symmetry of a compound and its melting point. Rationalize this relationship.
6. Compare the refractive indices of hexane, 2-methylpentane, and 2,2-dimethylbutane. Do isomers have identical or different refractive indices?
7. Compare the refractive indices of the straight-chain alkanes. Is there a linear relationship between chain length and refractive index? Make a graph of refractive index (corrected) versus chain length (number of carbons in chain) for the unbranched alkanes (Table 3.3-1). Summarize the effect that chain length has on refractive index.

Critical Thinking Questions

1. How would you expect the boiling point of an amine (RNH_2) to differ from that of a similar molecular weight alcohol (ROH)? Explain.
2. Ethanol (CH_3CH_2OH) has a boiling point of 78.3°C. Dimethyl ether (CH_3OCH_3), an isomer of ethanol, boils at −24.8°C. Explain this difference.
3. Draw all isomers having the molecular formula $C_4H_{10}O$. Match each isomer to the correct boiling point: 31, 34, 39, 83, 100, 108, and 118°C. Justify.
4. Trimethylamine and propylamine are isomers. They have the same molecular weight but boil at different temperatures. One boils at 3°C; the other boils at 49°C. Assign a structure to each boiling point. Explain.
5. 2-Hydroxybenzoic acid (also called salicylic acid) melts at 158–160°C, while 3-hydroxybenzoic acid has a much higher melting point, 201–203°C. The melting point of 4-hydroxybenzoic acid is similar to that of the 3-hydroxybenzoic acid (215°C). Look carefully at the structures of these compounds and propose a reason why the melting point of 2-hydroxybenzoic acid is so much lower than the other two isomers.

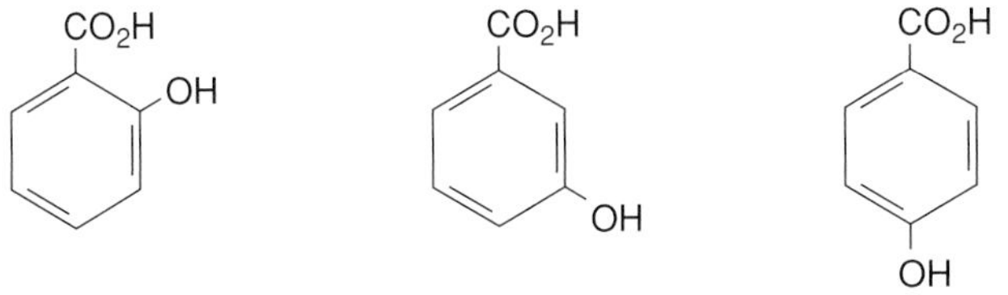

2-hydroxybenzoic acid   3-hydroxybenzoic acid   4-hydroxybenzoic acid
(salicylic acid)

Experiment 3.4: Properties of Solvents and Recrystallization of Organic Solids

Recrystallization is one of the most common techniques to purify solids. In this experiment, you will:

◆ determine the solubility of an organic solid in various solvents.

◆ select a suitable solvent or solvent pair to recrystallize the unknown.

Techniques

Technique C	Melting point
Technique F	Recrystallization and vacuum filtration

Background

In the first part of this exercise, the solubilities of three organic solutes of varying polarities will be tested in five different solvents. The solutes are *trans*-stilbene, 9-fluorenone, and benzoic acid. The structures are shown here.

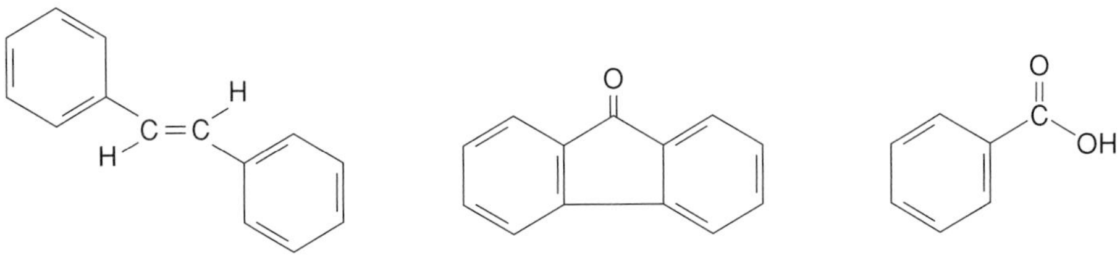

trans-stilbene   9-fluorenone   benzoic acid

     The solubilities of these solutes will be tested in hexane, toluene, acetone, ethanol, and water.

Prelab Assignment

1. Classify *trans*-stilbene, 9-fluorenone, and benzoic acid as polar, nonpolar, or moderately polar.
2. Draw the structures of ethanol, acetone, toluene, hexane, and water. Classify each solvent as polar, nonpolar, or moderately polar.
3. Prepare a solubility table similar to Table 3.4-1.

Table 3.4-1   Solubilities of Selected Organic Solutes in Five Different Solvents

	Hexane		Toluene		Acetone		Ethanol		Water	
	cold	hot	cold	hot	cold	hot	cold	hot	cold	hot
trans-Stilbene										
9-Fluorenone										
Benzoic acid										

Experimental Procedure

1. Wear eye protection at all times in the laboratory.
2. All organic compounds and solvents should be considered hazardous; wear gloves and avoid breathing the vapors. Work under the hood or in a well-ventilated area.

Part A: Selecting an Appropriate Solvent

Obtain 15 clean, dry 12 × 75-mm test tubes. Five of the test tubes will be used for the *trans*-stilbene, five will be used for the 9-fluorenone, and five will be used for the benzoic acid. Label each of the test tubes with the name of a compound and the solvent in which it is to be tested. To the test tubes, add 10–20 mg of the appropriate compounds. It is not necessary to weigh out each sample if a 10–20 mg sample in a test tube is available for comparison. Be careful not to exceed 20 mg or it may be difficult to evaluate solubility. To each test tube, add 10 drops (or 0.5 mL) of the appropriate solvent. Swirl to mix. Note whether the solid dissolves in the given solvent. If the compound dissolves, it is soluble. Mark an "s" in the appropriate column in the table. If the compound does not dissolve, it is insoluble in the given solvent. Mark an "i" in the appropriate column in the table. If it appears that some of the solid has dissolved in the solvent, add another 10 drops of the solvent and notice whether more of the solid dissolves in the additional solvent. This would be recorded as "p" (for partially soluble) in the table. Remove the test tubes in which all of the solid dissolved.

To test for solubility in hot solvent, heat the remaining test tubes in a warm sand bath under the hood for several minutes. Watch the test tubes carefully to make certain that the solvents do not boil away. If the compound dissolves in the hot solvent, mark an "s" under the appropriate column. If the compound does not dissolve in the hot solvent, add another 10 drops of the solvent and reheat. If the compound dissolves in the additional solvent, mark a "p" in the appropriate column. If the compound still does not dissolve, mark an "i" in the appropriate column.

Results and Conclusions for Part A

1. Rank *trans*-stilbene, 9-fluorenone, and benzoic acid from least polar to most polar.
2. Rank the solvents tested from least polar to most polar.

3. What conclusions can be drawn from this experiment about solubilities of polar, moderately polar, and nonpolar compounds in various solvents?
4. List the best recrystallization solvent (or solvents) for each of the three solids: *trans*-stilbene, 9-fluorenone, and benzoic acid.

Part B: Microscale Recrystallization of an Organic Solid

Show the results from Part A to the laboratory instructor. The instructor will assign one of the compounds to be recrystallized. Use the results in Table 3.4-1 to select an appropriate solvent for recrystallization. (Remember that the solvent chosen should have a boiling point below the melting point of the compound. The solvent should dissolve the compound when hot but not when cold.) Using the results of Part A, recrystallize the solid using a single solvent or a solvent pair.

Single-Solvent Recrystallization

Weigh out 200 mg of the assigned compound and put this in a 10-mL Erlenmeyer flask. In another small Erlenmeyer flask containing a boiling chip, heat the selected solvent. With a Pasteur pipet, add a small portion of the hot solvent to the solid. Heat the flask on a sand bath, swirling to dissolve. Add hot solvent in small portions, swirling after each addition, until the crystals have dissolved. If necessary, use a filter pipet to separate the solution from solid impurities. Remove the flask from the heat, and allow it to cool to room temperature. Crystallization should occur as the solution cools to room temperature. Place the solution in an ice bath for 5 to 10 minutes to complete the crystallization process.

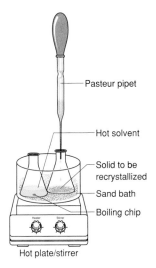

Pasteur pipet

Hot solvent

Solid to be recrystallized

Sand bath

Boiling chip

Hot plate/stirrer

If crystals form too quickly, reheat the solution, adding slightly more hot solvent, and let the solution cool slowly. If no crystals appear after keeping the solution in an ice bath, scratch the side of the Erlenmeyer flask with a glass rod, add a seed crystal, or put the solution back on the hot plate to boil away some of the excess solvent. When crystallization is complete and the mixture is cold, filter the crystals through a Hirsch funnel and wash with a small portion of ice-cold solvent. Collect and air dry the crystals. Weigh the dry crystals and calculate the percent recovery.

Solvent-Pair Recrystallization

If a satisfactory solvent was not found, try a solvent-pair recrystallization. The boiling point of each of the solvents should be lower than the melting point of the crystals. To serve as a solvent pair, the two solvents must be miscible (soluble in each other). To determine miscibility, add 10 drops of each solvent to a test tube; shake well. If two layers form, the solvents are immiscible and may not be used for a solvent pair recrystallization. If only one layer forms, the solvents are miscible and may be used. Examples of commonly used solvent pairs are ethanol-water and acetone-water.

Weigh out 200 mg of the assigned compound and add it, along with a boiling chip, to a 10-mL Erlenmeyer flask. Add a small amount of the hot solvent in which the compound is soluble. Swirl the flask on a hot plate, adding just enough hot solvent to dissolve the crystals. Add the second solvent dropwise and swirl after each addition. Continue to add this hot solvent until the solution just starts to turn cloudy. Then add a drop or two of the first solvent to clarify the solution. Cool to room temperature. Complete the crystallization process by placing the flask in an ice bath. Filter using a Hirsch funnel. Wash with a small portion of the ice-cold solvent in which the crystals were insoluble. Weigh the dry crystals and calculate the percent recovery.

Characterization for Part B

Melting Point: Determine the melting point of the recrystallized solid. The reported melting points are *trans*-stilbene (124°C), 9-fluorenone (84°C), and benzoic acid (122°C).

Results and Conclusions for Part B

1. Calculate the percent recovery for the recrystallization process. Explain why it is not 100%.
2. Explain and evaluate the effectiveness of the recrystallization solvent in terms of percent recovery and purity of the recrystallized solid.
3. Suggest other solvents or solvent pairs that might have been used for this recrystallization.

Cleanup & Disposal

Place the solvents used for recrystallization in a container labeled "nonhalogenated organic solvent waste." Aqueous solutions can be washed down the drain with water.

Critical Thinking Questions *(The harder one is marked with a ❖.)*

1. List the main criteria for selecting a recrystallization solvent.
2. When is it necessary to use a solvent-pair recrystallization?
3. Why should the recrystallization solvent have a fairly low boiling point?
❖ 4. Will the following pairs of solvents be suitable for doing a solvent-pair recrystallization? Explain.
   a. ethanol (bp 78.5°C) and water
   b. methylene chloride (bp 40°C) and water
   c. dimethylformamide (bp 153°C) and diethyl ether (bp 37°C)
5. If a solute is soluble in cold solvent, is it necessary to test the solubility of the solute in the same solvent when hot? Explain.
6. Arrange the following solvents in order of increasing polarity: ethanol, ethyl acetate, petroleum ether, toluene, and acetone.
7. Methylene chloride (CH_2Cl_2) is polar, whereas carbon tetrachloride (CCl_4) is nonpolar. Explain.
8. Carbon disulfide (CS_2) is sometimes used as a recrystallization solvent. Will this solvent dissolve polar or nonpolar compounds? Explain.

Experiment 3.5: Separations Based upon Acidity and Basicity

Extraction is a technique in which a solute is transferred from one solvent to another. In this experiment, you will investigate acid-base extraction. You will:

♦ determine the solubilities of an organic acid, an organic base, and a neutral organic compound.

♦ design a flow scheme to separate an organic acid, an organic base, and a neutral compound.

♦ use microscale extraction techniques to separate and isolate each component of a mixture of naphthalene, benzoic acid, and ethyl 4-aminobenzoate.

♦ use miniscale extraction techniques to separate and isolate a mixture of benzoic acid and ethyl 4-aminobenzoate.

Techniques

Technique C   Melting point

Technique F   Vacuum filtration

Technique I   Drying and extraction

Background

A water-insoluble, acidic organic compound such as a carboxylic acid or phenol can be easily separated from neutral and basic organic compounds by conversion to a water-soluble salt.

$$\underset{\substack{\text{a water-insoluble}\\\text{carboxylic acid}}}{\text{RCO-H}} + \text{NaOH} \longrightarrow \underset{\substack{\text{a water-soluble}\\\text{carboxylate salt}}}{\text{RCO}^-\text{Na}^+} + H_2O$$

Neutral and basic organic compounds remain in the organic layer. The two layers can then be separated. Addition of HCl to the aqueous layer regenerates the water-insoluble carboxylic acid, which can then be filtered or extracted into an organic solvent:

$$\underset{\substack{\text{a water-soluble}\\\text{carboxylate salt}}}{\text{RCO}^-\text{Na}^+} + \text{HCl} \longrightarrow \underset{\substack{\text{a water-insoluble}\\\text{carboxylic acid}}}{\text{RCO-H}} + \text{NaCl}$$

A similar scheme can be used to separate a basic compound, such as a water-insoluble amine, from neutral or acidic organic compounds by conversion of the amine to a water-soluble salt:

$$\underset{\substack{\text{a water-insoluble}\\\text{amine}}}{\text{RNH}_2} + \text{HCl} \longrightarrow \underset{\substack{\text{a water-soluble}\\\text{ammonium salt}}}{\text{RNH}_3^+ \text{Cl}^-}$$

Neutral compounds and acidic organic compounds remain in the organic solvent, where they can be removed. Addition of sodium hydroxide to the aqueous layer regenerates the amine, which is now insoluble in the aqueous solution. The amine can be filtered or extracted into an organic solvent.

$$\underset{\substack{\text{a water-soluble}\\\text{ammonium salt}}}{\text{RNH}_3^+ \text{Cl}^-} + \text{NaOH} \longrightarrow \underset{\substack{\text{a water-insoluble}\\\text{amine}}}{\text{RNH}_2} + \text{NaCl}$$

The neutral compound remains in the organic solvent, where it can be recovered by drying the solution to remove traces of water, filtering off the drying agent, and evaporating the solvent.

In this exercise, the solubilities of an organic acid (benzoic acid), an organic base (ethyl 4-aminobenzoate), a neutral compound (naphthalene), and the organic salts (ethyl 4-aminobenzoate hydrochloride and sodium benzoate) will be tested in methylene chloride and water.

From the solubilities, you will construct a flow scheme outlining the separation of naphthalene, benzoic acid, and ethyl 4-aminobenzoate. In Part B, you will use the flow scheme to separate a mixture of naphthalene, benzoic acid, and ethyl 4-aminobenzoate in microscale. In Part C, you will use the flow scheme to separate a mixture of benzoic acid and ethyl 4-aminobenzoate in miniscale.

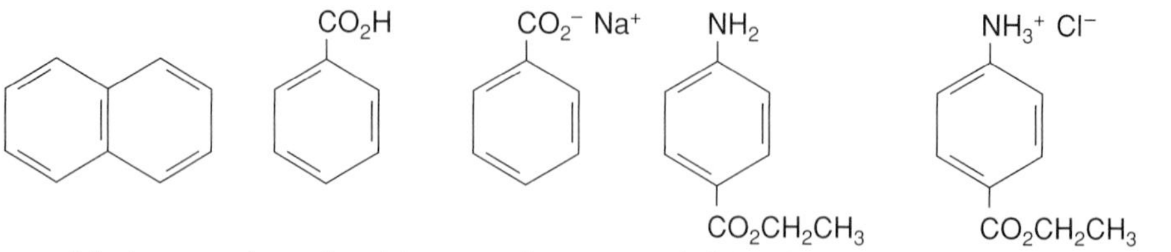

naphthalene | benzoic acid | sodium benzoate | ethyl 4-aminobenzoate | ethyl 4-aminobenzoate hydrochloride

The instructor may substitute other compounds for those shown here.

Prelab Assignment

1. Read Technique I on the theory and technique of extraction and do all assigned problems.
2. Construct a solubility table similar to Table 3.5-1 in the experimental section.
3. Identify the conjugate acid/conjugate base pairs for the structures above.
4. Write the reaction (if any) and give the products for the reaction of each pair of reagents below. If no reaction occurs, write NR. Indicate whether the product will be water-soluble or water-insoluble.
    a. benzoic acid with NaOH.
    b. sodium benzoate with HCl.
    c. ethyl 4-aminobenzoate with HCl.
    d. ethyl 4-aminobenzoate hydrochloride with NaOH.
    e. naphthalene and NaOH.
    f. ethyl 4-aminobenzoate with NaOH.
5. Determine whether each of the five compounds is predominantly ionically or covalently bonded. Based upon this answer, indicate whether the compound would be expected to be more soluble in water or more soluble in methylene chloride.

Experimental Procedure

Safety First!

Always wear eye protection in the laboratory.

1. Wear eye protection at all times in the laboratory.
2. Wear gloves when handling reagents in this experiment.
3. Methylene chloride is a toxic irritant and a suspected carcinogen. Do not breathe the vapors. Work under the hood or in a well-ventilated area.
4. NaOH and HCl are corrosive and toxic and can cause burns.

Part A: Determination of Solubilities

Obtain 20 small, dry test tubes or a spot plate. Place approximately 10–20 mg of benzoic acid into four of the test tubes or wells; place 10–20-mg of sodium benzoate into four other test tubes or wells. Repeat, using 10–20-mg samples of the other solutes. It is

Table 3.5-1   Solubility Table

	HCl	**NaOH**	**CH$_2$Cl$_2$**	**Water**
Naphthalene				
Benzoic acid				
Sodium benzoate				
Ethyl 4-aminobenzoate				
Ethyl 4-aminobenzoate hydrochloride				

not necessary to weigh out these quantities; use the amount that fits on the tip of a small spatula. Test the solubility of each of the compounds in 3M HCl, 3M NaOH, methylene chloride, and water by adding about 10 drops (or 0.5 mL) of each solvent to the appropriate test tube or well. Note which compounds dissolve and/or react. If a portion, but not all, of the compound appears to dissolve in the solvent, add another 10 drops and see if more of the compound dissolves. Record the results in your table. Mark an "s" if the compound is soluble, an "i" if the compound is insoluble, and a "p" if the compound is partially soluble.

Results and Conclusions for Part A

Based upon the results of the solubility tests, propose a flow scheme for separating naphthalene, benzoic acid, and ethyl 4-aminobenzoate based upon the technique of extraction. Prepare an outline for this procedure, indicating clearly how the compounds will be separated and how each will be recovered.

Part B: Microscale Separation of Naphthalene, Benzoic Acid, and Ethyl 4-Aminobenzoate

Develop a flow scheme that shows clearly how naphthalene, benzoic acid, and ethyl 4-aminobenzoate could be separated from each other and recovered. The flow scheme may be developed using the results from Part A, or it may be developed using answers from Prelab Assignment question 5.

Obtain a 200-mg sample of a mixture of naphthalene, benzoic acid, and ethyl 4-aminobenzoate. Dissolve the mixture in 2.5 mL of methylene chloride in a 5-mL conical vial. Follow the flow scheme to separate and recover the three solid components. Weigh each of the dried solids.

Characterization for Part B

Melting Point: Determine the melting point of each of the isolated solids. The reported melting points are naphthalene (80°C), benzoic acid (122°C), and ethyl 4-aminobenzoate (91–92°C).

Results and Conclusions for Part B

1. Calculate the composition of the unknown mixture (in weight percent).
2. Evaluate the purity of each isolated product by comparing the observed melting point with the literature value.

Part C: Miniscale Separation of Benzoic Acid and Ethyl 4-Aminobenzoate

Develop a flow scheme that shows clearly how benzoic acid and ethyl 4-aminobenzoate could be separated from each other and recovered. The flow scheme may be developed using the results from Part A, or it may be developed using answers from Prelab Assignment question 5.

Give the flow scheme to the laboratory instructor for approval. If directed to do so by the instructor, include naphthalene in the flow scheme and devise a plan for removing the methylene chloride by evaporation or distillation.

Obtain a 0.5-g sample of a mixture of benzoic acid and ethyl 4-aminobenzoate. Dissolve the mixture in 10 mL of methylene chloride in a 50-mL Erlenmeyer flask. Follow the flow scheme to separate and recover the two components. Weigh each of the dried solids.

Characterization for Part C

Melting Point: Determine the melting point of each of the isolated solids. The reported melting points are benzoic acid (122°C) and ethyl 4-aminobenzoate (91–92°C).

Results and Conclusions for Part C

1. Calculate the composition of the unknown mixture (in weight percent).
2. Evaluate the purity of each isolated product by comparing the observed melting point with the literature value.
3. Explain how naphthalene could be separated from a mixture of naphthalene, benzoic acid, and ethyl 4-aminobenzoate and explain how the methylene chloride should be removed from naphthalene.

Cleanup & Disposal

Place methylene chloride solutions in a container labeled "halogenated organic solvent waste." Aqueous solutions should be neutralized and washed down the drain with water.

Critical Thinking Questions

1. Describe three properties of good extraction solvents.
2. Diethyl ether and methylene chloride are generally not suitable for use in recrystallization, but are frequently used as extraction solvents. Explain.
3. Acetone and isopropyl alcohol are generally not suitable for use in extraction, but are frequently used as recrystallization solvents. Explain.
4. After separating layers in an extraction, it is necessary to "dry" the organic layer with a substance that absorbs water. Explain why this is necessary.
5. Suggest two methods for determining which layer in an extraction is the aqueous layer and which layer is the organic layer.
6. An organic compound has a distribution coefficient of 1.5 in methylene chloride and water. If 50 mg of this compound is dissolved in 15 mL of water, in which case will more of the compound be extracted into the methylene chloride: one extraction with 15 mL of methylene chloride or three extractions using 5 mL of methylene chloride each time? Show all work. Refer to Appendix B if necessary.

7. Phenol and benzoic acid are both weak acids that are soluble in diethyl ether and insoluble in water. Benzoic acid is soluble in both 10% NaOH and 10% NaHCO₃. Phenol is soluble only in the stronger base, 10% NaOH. From this information, write a flow scheme for separating and recovering phenol and benzoic acid.

8. When oxidizing benzyl alcohol to benzaldehyde, a common side product is benzoic acid, caused by overoxidation of benzaldehyde. Explain how benzaldehyde might be separated from the benzoic acid impurity. Benzyl alcohol, benzaldehyde, and benzoic acid are all soluble in ether and only slightly soluble in water.

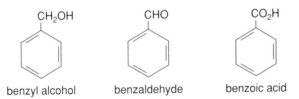

| benzyl alcohol | benzaldehyde | benzoic acid |

9. Sometimes when extracting a polar organic solute into an organic solvent, brine (a saturated solution of sodium chloride) is used in place of water. What is the advantage of using brine rather than water?

10. Which of the following pairs of solvents would be suitable for extraction? Explain.

    a. ethanol and water

    b. methylene chloride and diethyl ether

    c. acetone and water

    d. methylene chloride and acetone

    e. diethyl ether and water

Experiment 3.6: Isolation of a Natural Product

Natural products are naturally occurring organic compounds that are found in (or produced by) living organisms. Many useful and important organic compounds may be extracted from plants, such as caffeine from tea. In this experiment, you will:

◆   perform a miniscale solid-liquid extraction.

◆   perform a miniscale liquid-liquid extraction.

◆   perform a hot gravity filtration.

Techniques

Technique C	Melting point
Technique F	Recrystallization, hot gravity filtration, and vacuum filtration
Technique I	Extraction

Background

The technique of extraction is one of the oldest and most useful procedures in organic chemistry. Brewing tea with a tea bag, making a cup of drip coffee, adding a bay leaf to spaghetti sauce—all of these are examples of solid-liquid extraction. In this technique, the solid is washed with a solvent in which a particular component is soluble. The compound dissolves in the solvent, leaving an insoluble residue behind. In Part A, Part B, and Part C of this experiment, caffeine is isolated from tea leaves, instant coffee, and NoDoz tablets respectively. In Part D of this experiment, cholesterol is isolated from simulated gallstones. The structures of caffeine and cholesterol are shown below:

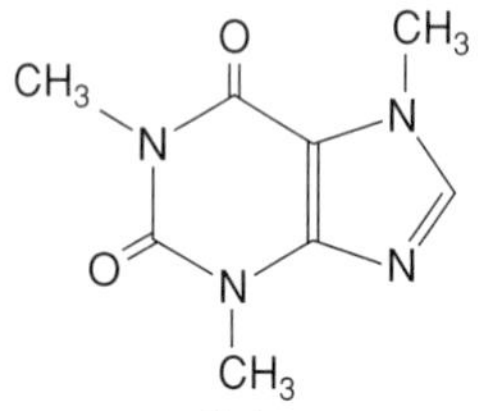

caffeine

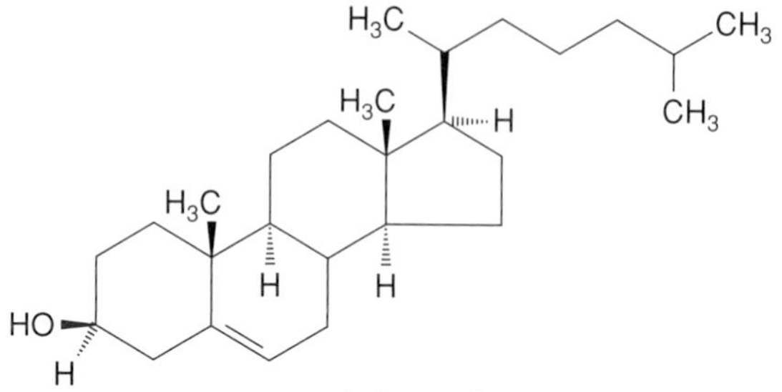

cholesterol

Isolating a natural component from a complex mixture can be difficult, depending upon the composition of the mixture and the physical and chemical properties of the components. Tea leaves are composed of a variety of components, the major component of which is cellulose. Caffeine accounts for only about 5% of the mass of the tea leaves. However, the extreme water-solubility of caffeine makes it relatively easy to separate the caffeine from the water-insoluble cellulose by washing the tea leaves with hot water and filtering off the cellulose. Other water-soluble components, such as tannins, will dissolve in hot water with the caffeine. Tannins are acidic and can be separated from the caffeine by the addition of a base such as sodium carbonate. The water-soluble sodium carboxylate salts of the tannins are insoluble in methylene chloride. This solvent can be used to extract the caffeine (which is more soluble in methylene chloride than in water) from the salts.

Removing caffeine from coffee works the same way: the coffee beans are rinsed with a solvent that dissolves and removes the caffeine from the flavorants. The first process for decaffeinated coffee, invented in 1905, used benzene, which is now considered a carcinogen. More recent decaffeination techniques are extraction with methylene chloride, water processing, and supercritical carbon dioxide decaffeination. The latter two methods are preferred, as they leave no chemical residue on the coffee beans. In order to be called decaffeinated, a coffee must have at least 97% of its caffeine removed. A regular 5-oz cup of coffee contains from 70 to 155 mg of caffeine; the same amount of decaffeinated coffee contains less than 5 mg of caffeine.

Caffeine is a central nervous system stimulant that causes alertness and sleeplessness in mild quantities. Excessive intake of caffeine (more than 600 mg a day) causes a condition called caffeinism, which is characteristized by chronic insomnia, ringing in the ears, persistent anxiety, stomach upset, and mild delirium. NoDoz tablets contain a large amount of caffeine (about 100 mg of caffeine per tablet), with the remainder being inorganic binder. Because the caffeine and the binder have different physical properties, caffeine is easily isolated by dissolution in methylene chloride and filtration of the insoluble binder.

Cholesterol, another natural product, is a steroid produced by the body to regulate passage across cell membranes. Cholesterol is a precursor of steroid hormones. Excessively high levels of cholesterol can result in atherosclerosis or hardening of the arterial walls. Cholesterol can also crystallize in the gallbladder as a large solid mass called a gallstone. If they become large enough, gallstones can block flow through the bile duct. Gallstones are about 75% cholesterol. In Part D, cholesterol will be isolated from simulated gallstones. Simulated gallstones are used because use of actual gallstones may pose a health hazard.

Cholesterol is insoluble in water, but soluble in organic solvents such as 2-propanol, acetone, 2-butanone, and diethyl ether. Cholesterol can be isolated by dissolving the simulated gallstones in a solvent such as 2-propanol, filtering off insoluble bilirubin pigments (which give bile its characteristic yellow-brown color), and removing the solvent. The crude cholesterol can be purified by recrystallization or sublimation.

Prelab Assignment

1. Design a flow scheme for the isolation of caffeine from tea leaves (Part A), the isolation of caffeine from instant coffee (Part B), the isolation of caffeine from NoDoz (Part C), or the isolation of cholesterol from gallstones (Part D).
2. In the extraction of tea with methylene chloride and water, will methylene chloride be the upper or lower layer? How could this be verified?
3. What is an emulsion? Give an example.

Experimental Procedure

1. Always wear eye protection in the laboratory.
2. Methylene chloride is a toxic irritant and a suspected carcinogen. Perform all operations using this solvent under the hood or in a well-ventilated area.

Part A: Miniscale Extraction of Caffeine from Tea Leaves

Obtain two tea bags from the instructor. Record the brand of the tea. Carefully open each tea bag and weigh the tea leaves. Record the mass of the tea leaves. Replace the tea leaves in the tea bags and staple the tea bags tightly shut. Place the tea bags in a 125-mL Erlenmeyer flask. To the flask add 20 mL of distilled water and 4 g sodium carbonate and stir. Place a watch glass on top of the flask; heat the flask using a sand bath. Boil gently for 20 minutes. Since the tea bag has a tendency to float, it may be necessary to press it back down into the water. Do this with a glass stirring rod—**gently.** Remove the flask from the heat and cool to room temperature. With a Pasteur pipet, transfer the aqueous solution to a centrifuge tube or small flask. With the glass rod, gently roll the tea bag against the side of the flask to squeeze out the water. Get as much water out of the tea bag as possible without breaking it. Add this water to the centrifuge tube.

Add 4 mL of methylene chloride to the aqueous solution in the centrifuge tube. Cap the tube and shake and invert to mix the layers. To prevent pressure buildup, vent by frequently releasing the cap. With a pipet, remove the lower organic layer with a clean, dry pipet and place in an Erlenmeyer flask. Add another 4-mL aliquot of methylene chloride to the aqueous layer in the centrifuge tube. Shake the tube, venting frequently, and allow the layers to separate. Transfer the organic layer (bottom layer) to the Erlenmeyer flask containing the first extract using a clean, dry Pasteur pipet.

To remove traces of water from the organic layer, add small amounts of anhydrous sodium sulfate (about the tip of a small spatula) until no more clumping is observed. Let stand for about 10 minutes. Use a filter pipet to transfer the dried solution to a clean, dry Erlenmeyer flask and add a boiling chip. In the hood, evaporate the methylene chloride

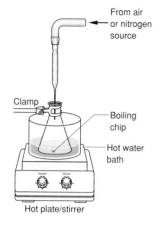

From air
or nitrogen
source

Clamp

Boiling
chip

Hot water
bath

Hot plate/stirrer

by warming the flask on a sand bath, steam bath, or water bath. When the methylene chloride has evaporated, a layer of off-white crystals will remain.

Caffeine can be recrystallized in an Erlenmeyer flask using 2-propanol. Place the crude crystals in a small Erlenmeyer flask. Add hot 2-propanol until the crystals dissolve. Remove the solution from the sand bath. Cool to room temperature, then place in an ice bath to insure complete crystallization. Collect the crystals by vacuum filtration. Wash with ice-cold petroleum ether. Air dry. Weigh the crystals.

Characterization for Parts A, B, and C

Melting Point: Because caffeine sublimes below its melting point, the melting point should be measured using a sealed capillary tube. The melting point of pure caffeine is 238°C.

Results and Conclusions for Part A

1. Calculate the percentage of caffeine in tea leaves.
2. Explain why sodium carbonate was added to the mixture of tea and water.
3. Assume that one tea bag makes one 8-oz cup of tea. How many cups of the assigned tea can you drink and stay under the recommended limit of 500 mg of caffeine a day?

Part B: Miniscale Extraction of Caffeine from Instant Coffee

Weigh approximately 4 g of instant coffee. Record the brand of coffee used and the exact weight in your lab notebook. Transfer the instant coffee to a 125-mL Erlenmeyer flask fitted with a stir bar. Add 4 g of sodium carbonate and 50 mL of distilled water to the flask. Heat, with stirring, for 5–10 minutes. Let cool to room temperature. Pour into a 125-mL separatory funnel. Pour 10 mL of methylene chloride into the separatory funnel. Do not shake vigorously, as an emulsion is likely to form. Instead, gently swirl the two layers for a few minutes to allow transfer of the caffeine from the water into the methylene chloride. Allow the layers to settle, then drain off the lower organic layer into a clean Erlenmeyer flask. Add a fresh 10-mL portion of methylene chloride to the separatory funnel, and repeat the extraction. Combine the methylene chloride layers. Add anhydrous sodium sulfate to the methylene chloride solution until no more clumping is observed. Let stand for 5–10 minutes. Filter the dried solution into a tared suction flask containing a spin bar. Wash the drying agent with a fresh 0.5-mL portion of methylene chloride and add this to the flask. Evaporate the solvent under vacuum on a water or sand bath under the hood to leave a yellowish solid. Weigh the crude product. Recrystallize from 95% ethanol, using about 5 mL/g. Collect the product by suction filtration, using a Hirsch funnel. Rinse the crystals with a minimum amount of ice-cold 95% ethanol. After the crystals are dry, weigh the purified caffeine. Characterize the product as in Part A.

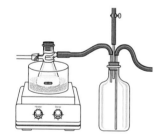

Results and Conclusions for Part B

1. Calculate the mass percent of caffeine in the assigned brand of instant coffee.
2. Compare the amount of caffeine in the different brands of instant coffee.
3. Ingesting a moderate amount of caffeine (under 500 mg per day) is generally considered harmless. Assuming that coffee is your only source of caffeine, calculate the maximum number of 5-oz cups of the assigned brand of coffee you could drink to stay under 500 mg.

Part C: Miniscale Isolation of Caffeine from NoDoz

Weigh one NoDoz tablet. Grind the tablet with a mortar and pestle, and transfer to a 100-mL beaker containing a boiling stone. Add 40 mL of water. Boil the solution on a hot plate for 5–10 minutes, then let the solution cool to room temperature. Gravity filter the cooled suspension into a clean Erlenmeyer flask to remove the insoluble binder. Pour the aqueous solution into a 125-mL separatory funnel and extract with 5–10 mL of methylene chloride. Do not shake the separatory funnel vigorously, as an emulsion is likely to form. Instead, gently swirl the layers to transfer the caffeine from the water into the methylene chloride. Let the layers separate, then drain the lower methylene chloride layer into a clean Erlenmeyer flask. Add a second 7–8 mL portion of methylene chloride to the separatory funnel and repeat the extraction. Combine the methylene chloride layers. Add anhydrous sodium sulfate until no further clumping occurs. Gravity filter the dried solution into a clean, tared Erlenmeyer flask that contains a boiling stone. Evaporate the solvent under the hood using a steam bath. Weigh the crude caffeine. Recrystallize from 95% ethanol (5 mL/g). Suction filter the crystals using a Hirsch funnel. Rinse with a minimum amount of ice-cold 95% ethanol. Weigh the crystals when they are dry. Characterize the product as in Part A.

Results and Conclusions for Part C

1. Calculate the mass percent of caffeine in the NoDoz tablet.
2. Calculate the percent recovery of caffeine, assuming that the tablet contained 100 mg of caffeine.
3. Explain where in the workup procedure product might have been lost.

Part D: Miniscale Isolation of Cholesterol from Simulated Gallstones

Place 2 g of simulated gallstones into a 50-mL Erlenmeyer flask. Add 15 mL of 2-propanol. Heat the mixture on a sand bath for 5–10 minutes to dissolve the gallstones. Filter off any undissolved solids using a hot gravity filtration. Add 10 mL of methanol to the filtrate. Add a spatula-tip full of decolorizing carbon. Heat and swirl the mixture, then remove the solids by gravity filtration with a prewarmed funnel and filter paper into a 50-mL Erlenmeyer flask. Heat the solution to boiling and add 1 mL of hot water. Swirl to dissolve any solid that has formed. Remove the flask from the heat source and let cool to room temperature. Cholesterol will crystallize out as colorless plates. Cool the solution in an ice bath until crystallization is complete. Isolate the crystals by vacuum filtration and wash with a small amount of ice-cold methanol. Recrystallize from methanol. Weigh the dry crystals.

Characterization for Part D

Melting Point: Determine the melting point of the solid. Pure cholesterol melts at 148–150°C.

Results and Conclusions for Part D

Calculate the weight percent of pure cholesterol obtained from the simulated gallstones.

Cleanup & Disposal

Place methylene chloride in a container labeled "halogenated organic solvent waste." Aqueous solutions can be washed down the drain with water. Organic solvents used for recrystallization should be placed in a container labeled "nonhalogenated organic solvent waste."

Critical Thinking Questions *(The harder one is marked with a ❖.)*

1. A student neglected to add sodium carbonate when extracting the tea leaves with hot water. Yet a very high yield of caffeine was obtained. The crystals melted at 202–214 °C. Explain.

2. The weight of the purified caffeine is often much lower than the weight of the crude caffeine. Explain why the percent recovery is frequently low.

3. Cholesterol is purified using a hot gravity filtration. Why must the beaker and filter paper be kept hot?

4. Is it better to do two extractions with 10 mL of methylene chloride or one extraction with 20 mL of methylene chloride? Explain.

5. Why is decolorizing carbon (charcoal) used in the purification of cholesterol?

6. Identify all of the functional groups present in cholesterol.

❖ 7. Look up the structure of sugar (sucrose). What problems might be encountered in trying to extract caffeine from a soft drink that contains sugar?

8. *Library Project:* Explain the relationship between excess cholesterol and heart disease.

Experiment 3.7: Solvent and Polarity Effects in Thin-layer Chromatography (TLC)

Thin-layer chromatography (TLC) is a useful technique for analyzing purity and identifying components of a mixture. In this experiment, you will:

- determine the factors that affect the rate of elution of organic compounds.

- determine the relationship between polarities of compounds and their rates of elution (R_f).

- determine the relationship between solvent polarity and rate of elution (R_f).

- select a solvent or pair of solvents to separate a mixture of organic compounds.

Technique

Technique K   Thin-layer chromatography

Background

Thin-layer chromatography (TLC) is the separation of nonvolatile components of a mixture by differential elution on an absorbent. It is one of the most useful techniques for estimating the purity of a compound or for analyzing mixtures of nonvolatile compounds. TLC is rapid, and requires only a very small amount of sample. This makes it particularly valuable for microscale experiments.

There are several different types of TLC. The one used most frequently in the organic chemistry lab is adsorption chromatography. In adsorption TLC, the nature of the adsorbent and solvent (the eluent) are important in separating components of a mixture. The adsorbent (called the stationary phase) is usually silica gel or alumina. Typically it is coated onto a plastic or glass plate. Silica gel and alumina are both polar and bind polar compounds more strongly than they bind nonpolar compounds. The solvent (or solution, if a mixture of solvents) is called the mobile or liquid phase.

TLC solvents that are volatile and have low viscosities are advantageous because they elute rapidly up the TLC plate. Volatile solvents also evaporate rapidly, so that a developed TLC plate can be analyzed soon after the chromatography is complete. Solvents most commonly used for TLC are those that are inexpensive and nontoxic. Pentane, hexane, petroleum ether (hydrocarbon mixture), diethyl ether, acetone, ethyl acetate, and ethanol are common TLC solvents.

In TLC, a solution of the compound is spotted near the bottom of the TLC plate and the plate is placed in a chamber containing the solvent (called the "eluent"). The distance traveled by the compound will depend upon the relative affinity of the compound for the adsorbent versus the eluent. In general, compounds of differing polarities have differing relative affinities for the adsorbent and eluent. The rate of elution (R_f) is defined as the solute distance divided by the solvent distance.

The purpose of this experiment is to analyze the effect of polarity of solvents on the rate of elution of *trans*-stilbene, 9-fluorenone, and benzoic acid. The structures are shown here:

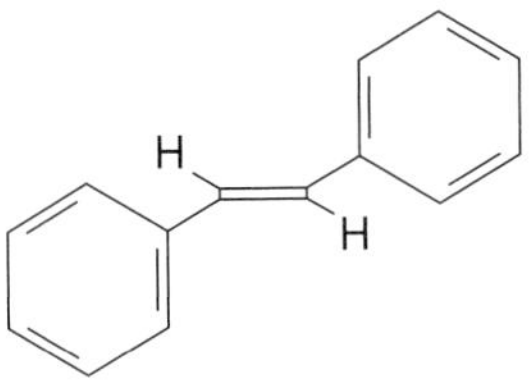

| *trans*-stilbene | 9-fluorenone | benzoic acid |

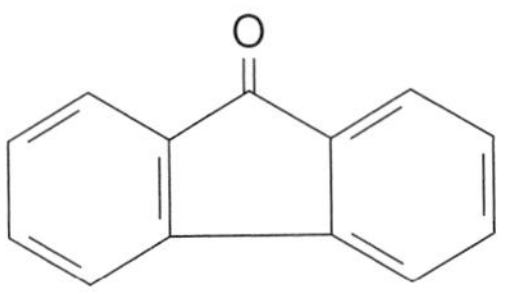

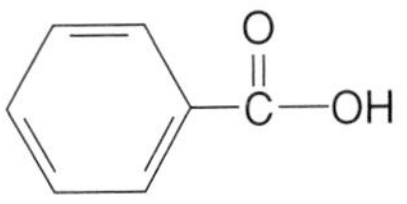

One of these compounds is quite polar, one is nonpolar, and one is moderately polar. These compounds will be spotted on silica gel TLC plates, which will then be developed in a series of solvents. These solvents are hexane, acetone, ethyl acetate, and ethanol. You will measure the R_f values of each of the three compounds in all of the solvent systems. You also will determine how the polarity of the organic compound and the polarity of the solvent system affects the rate of elution. Based on the results, a solvent or mixed solvent system will be chosen to separate a mixture of the three compounds. Mixed solvent systems can be prepared that have polarities falling in between those of the pure solvents.

Prelab Assignment

1. Rank *trans*-stilbene, 9-fluorenone, and benzoic acid in terms of increasing polarity.
2. Look up the dielectric constants for the solvents used in this experiment. Rank the solvents in terms of increasing dielectric constant.

3. Draw the structures of each of the solvents used in this experiment. Classify each solvent as polar, nonpolar, or moderately polar. Rank the solvents in order of increasing polarity.

4. Prepare a data table (like Table 3.7-1) to record the R_f values for each of the compounds in each of the solvents.

Experimental Procedure

1. All organic compounds and solvents should be considered hazardous; wear gloves and avoid breathing the vapors.

2. Wear goggles at all times.

3. When using the ultraviolet lamp, use proper eye protection to avoid exposing eyes to ultraviolet light. Do not look into the light.

Part A: Determining the Effect of Polarity on Elution

Work individually, in pairs, or in larger groups as directed by the instructor. Obtain four 400-mL clean, dry beakers or jars. Line each container with a piece of filter paper. (This helps keep the beaker saturated with vapors and speeds development time.) Pour 10 mL of hexane into an appropriately labeled beaker. Place a piece of aluminum foil tightly over the top of the beaker. Prepare three additional TLC setups, using acetone, ethyl acetate, and ethanol or other solvents assigned by the instructor.

Obtain four silica gel TLC plates. Silica Gel F plates work well because they are coated with a fluorescent indicator, which facilitates visualization of spots on the developed chromatogram. With a soft lead pencil (not a mechanical pencil) draw a very light line approximately 1 cm from the bottom of the plate. Be careful not to chip the silica gel or the plate will not develop evenly. Dissolve 1–2 mg of *trans*-stilbene, 9-fluorenone, and benzoic acid in 10 drops of methylene chloride in separate labeled vials. With a drawn capillary tube or spotter, apply a small amount of the solution of *trans*-stilbene to the plate. Touch the spotter to the plate very briefly. The solution will flow by capillary action. In the same manner, spot the solutions of 9-fluorenone and benzoic acid. View the spots under ultraviolet light. (**Caution: Use proper eye protec-**

Table 3.7-1 R_f Values for *trans*-Stilbene, 9-Fluorenone, and Benzoic Acid in Various Solvents

Solvents	*trans*-**Stilbene**	**9-Fluorenone**	**Benzoic acid**
Hexane			
Ethyl acetate			
Acetone			
Ethanol			

tion to avoid exposing eyes to ultraviolet light.) If the spots are not visible under the UV light, the concentration on the plate is too low. Apply a second spot directly over the first and recheck. Repeat this procedure for each of the TLC plates.

Place a TLC plate in each of the containers (see Figure 3.7-1a). Make certain that the level of the solvent is below the level of the spots on the plate. Prop the plate against the inside of the container. Be careful not to slosh the solvent around in order to avoid uneven development. Cover the container. The solvent will rise by capillary action. Watch carefully. When the solvent has traveled approximately 0.5 mm from the top of the plate, remove the plate from the container and immediately mark the position of the solvent front with a pencil. (The solvent evaporates, so the solvent front must be marked while it is still visible.) Let the plates air dry. View the plate under ultraviolet light. (Caution!) Silica Gel F plates are coated with a fluorescent indicator. Compounds will appear as dark spots under UV light. Circle the position of each of the compounds on the plate. These compounds may also be observed by placing the plate in an iodine chamber. Most organic compounds give orange-brown spots with iodine as reversible, colored complexes are formed. Measure the distance the compound traveled from the origin and the distance to the solvent front (see Figure 3.7-1b). Calculate R_f values for each of the compounds in the given solvents, where the R_f value is the distance the compound traveled divided by the distance the solvent traveled. Record the R_f values in Table 3.7-1.

Results and Conclusions for Part A

1. A solvent that gives a high R_f value is said to have high eluting ability. Rank the solvents in terms of eluting ability of *trans*-stilbene. (The solvent with the highest eluting power gives the highest R_f value and is listed as 1.) Repeat with 9-fluorenone and benzoic acid. How does the polarity of the solvent affect its eluting ability? Explain.

Rank	*trans*-Stilbene	9-Fluorenone	Benzoic acid
1			
2			
3			
4			

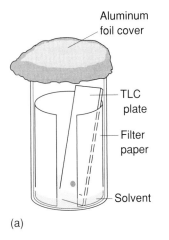

(a)

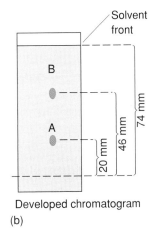

Developed chromatogram

(b)

Figure 3.7-1 Thin-layer chromatography chamber and chromatogram

2. For each of the solvents used, rank the compounds (*trans*-stilbene, 9-fluorenone, and benzoic acid) in terms of increasing elution (increasing R_f value). Does the order of elution of the three compounds vary according to the solvent polarity? Explain.

Solvent	Compound with lowest R_f value	Compound with intermediate R_f value	Compound with highest R_f value
Hexane			
Acetone			
Ethyl acetate			
Ethanol			

3. Do polar or nonpolar compounds travel faster on silica gel? Explain.
4. Do polar or nonpolar solvents have greater eluting ability? Explain.

Part B: Separation and Identification of Components of a Mixture of trans-Stilbene, 9-Fluorenone, and Benzoic Acid

Based on the results from Part A, choose a solvent or mixed solvent system that will allow separation of a mixture of *trans*-stilbene, 9-fluorenone, and benzoic acid. Obtain a vial from the instructor and record the number of the unknown. The vial will contain a mixture of one, two, or all three of the organic compounds. Dissolve 1–2 mg of the solid in 10 drops of methylene chloride. Spot a TLC plate with this solution. Also spot solutions of *trans*-stilbene, 9-fluorenone, and benzoic acid to use as standards. Develop the TLC in the appropriate solvent system. Visualize under UV light (Caution!) and circle the position of each component. Calculate an R_f value for each spot.

Results and Conclusions for Part B

1. Identify the components of the unknown mixture. Justify the choice of solvent for separating a mixture of *trans*-stilbene, 9-fluorenone, and benzoic acid. Were there other solvents or mixed solvents that would have separated these components? Explain.
2. Suppose that the solvents used in the TLC chamber were not dry, but contained a small amount of water. How would the resulting chromatogram be expected to differ from the one obtained in this experiment? Would it still be possible to identify all of the components of the mixture? Explain based upon your results.

Cleanup & Disposal

Pour the TLC solvents into the appropriately labeled bottles for recovery.

Critical Thinking Questions

1. Methyl alcohol, cyclohexane, isopropyl alcohol, and acetic acid are other solvents that are routinely used for TLC. Look up the dielectric constants and predict where these solvents would fit into the order of eluting ability.

CH_3OH CH_3CHCH_3 O

                                          OH $CH_3C - OH$

methanol   cyclohexane   2-propanol   acetic acid

2. Examine the structures of the following functionalized compounds and then rank in order of increasing elution sequence: benzoic acid, acetophenone, benzyl alcohol, benzylamine, benzylamine hydrochloride, and toluene.

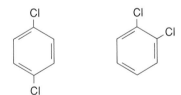

benzoic acid   acetophenone   benzyl alcohol   benzylamine   benzylamine hydrochloride   toluene

3. A compound has an R_f value of 0.98 in ethyl acetate but an R_f value of 0.05 in hexane. Suggest a solvent or solvent system that might be better.

4. In order to identify a compound using TLC, a sample of the pure compound must be spotted on the same plate as the unknown. Why is this necessary?

5. Suppose a TLC plate shows only one spot at an R_f value of 0.99. Does this mean that the sample is pure? Explain.

6. Suppose a TLC plate shows only one spot at an R_f value of 0.01. Does this mean that the sample is pure? Explain.

7. What is the purpose of placing a piece of filter paper in the developing chamber?

8. Why must the TLC chamber be covered with a lid or aluminum foil?

9. *trans*-1,2-Dibenzoylethene and the *cis*-isomer can be separated by TLC. Which compound is more polar? Which compound should have the higher R_f value?

10. *p*-Dichlorobenzene and *o*-dichlorobenzene can be separated by TLC. Which compound is more polar? Which compound travels faster in hexane? Explain.

p-dichlorobenzene   *o*-dichlorobenzene

11. *p*-Dihydroxybenzene and *o*-dihydroxybenzene can be separated by TLC using silica gel plates and hexane/ethyl acetate as the developing solvent. The *ortho* isomer has a significantly higher R_f value than the *para* isomer. Suggest a reason for the difference in R_f values.

p-dihydroxybenzene   *o*-dihydroxybenzene

12. What would be the appearance of the developed TLC plate if too much compound were spotted? What would be the appearance of the developed TLC plate if too little compound were spotted?

13. Molecular modeling assignment: Calculate the dipole moment of each of the compounds and solvents used in this experiment. Draw the structure of each compound, minimize the energy, and then perform a single point energy calculation using molecular mechanics (MMFF), semiempirical (AM1), or Hartree-Fock calculations (3-21G*).

Experiment 3.8: Purification and Analysis of a Liquid Mixture: *Simple and Fractional Distillation*

Distillation is a method of purification of volatile organic liquids. Simple and fractional distillation are both used as methods of separating components of a mixture. In this experiment, you will:

◆ perform a microscale or miniscale simple distillation to separate a mixture of hexane and a higher-boiling organic liquid.

◆ perform a microscale fractional distillation to separate a mixture of hexane and a higher-boiling organic liquid.

◆ compare the efficiencies of simple and fractional miniscale distillation.

Techniques

Technique E	Refractive index
Technique G	Simple distillation and fractional distillation

Background

Distillation is the main technique used for the purification of organic liquids. Distillation or fractional distillation is used throughout the text as a method of purification of liquids prepared in miniscale experiments. The goal of this experiment is to compare results from simple distillation and fractional distillation for a series of binary liquid mixtures that differ in boiling points from 13 to 105°C. In Part A of this experiment, each student or pair of students will distill a different mixture of hexane and another higher-boiling liquid. The student will measure the volume of distillate collected and the boiling point at which the distillate is collected throughout the distillation and will prepare a graph of temperature versus volume of distillate. The student will determine the boiling temperature and the purity of the initial distillate using a measurement of refractive index. This will be repeated for the final few drops of distillate. From pooled class data, the student will evaluate the effectiveness of simple distillation for the pairs of liquids distilled.

In Part B of this experiment, the same mixture will be distilled using a fractionating column. The composition of the initial and final distillate will be analyzed by measurement of the refractive index. A graph of temperature versus mL distilled will be constructed and compared with the diagram constructed in Part A. The purity of initial and final distillates obtained by each technique will also be compared.

Prelab Assignment

1. Look up and record the boiling point and refractive index of the assigned pair of liquids. Each pair will include hexane and one other liquid, which will be one of the following: cyclohexane, heptane, octane, nonane, decane, toluene, *p*-xylene, or other hydrocarbon assigned by the instructor. Record the difference in boiling points for the assigned pair of liquids.
2. At the boiling point of hexane, which of the components of the mixture has a higher vapor pressure? Explain.
3. Suppose that the initial distillate is observed to have a refractive index that is significantly different from that of hexane. What would this indicate? Would the same behavior be expected for the fractional distillation of the same mixture?
4. What main advantage and disadvantage can you predict about fractional distillation?
5. Why is it necessary to stir or add a boiling chip to the mixture being distilled?

Experimental Procedure

1. Wear eye protection at all times in the lab.
2. Hexane and the other organic liquids are flammable and irritants and may be toxic. Wear gloves when handling these reagents and work in a well-ventilated area.
3. Never heat a closed system. Always make sure that the system is open to the atmosphere.
4. Never distill a flask to dryness. Always stop the distillation and remove the source of heat while liquid remains in the distilling flask.

Part A: Microscale Distillation

Pipet 3.0 mL of hexane and 3.0 mL of an organic liquid assigned by the instructor into a 10-mL round-bottom flask that contains a spin bar or boiling chip. Fit the flask with a still head, water condenser, and bent adapter as shown. Use a 5-mL vial as a receiver. Place a thermometer so that the bulb of the thermometer is centered just below the take-off region of the still head. Gently turn on the water to the condenser. Turn on the heat source and slowly raise the temperature until vapors can be seen rising in the still head. Maintain the temperature of the heat source as drops of distillate begin to collect in the vial. Allow the first 1–2 drops to pass into the vial. Collect the next 4–5 drops in a separate vial labeled "Fraction 1" and cap the vial. Measure the refractive index of the liquid as time allows. Record the temperature and the volume of liquid collected throughout the distillation. Take readings every 0.2 mL or so. Continue the distillation until 5 mL of distillate have been collected. Disconnect the collection vial. In a second vial, labeled as "Fraction 2," collect 4–5 drops of distillate and cap the vial. Turn off the heat source and allow the apparatus to cool. Measure the refractive index of the liquids in the vials. Disassemble the apparatus. Transfer the distillate to a container in the hood marked "organic distillate."

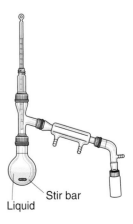

Stir bar

Liquid

Figure 3.8-1

Microscale simple distillation.

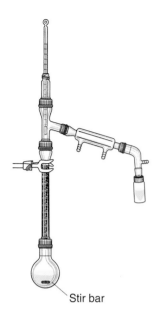

Stir bar

Part B: Microscale Fractional Distillation

Pipet 3.0 mL of hexane and 3.0 mL of an organic liquid assigned by the instructor into a 10-mL round-bottom flask that contains a spin bar or boiling chip. Fit the flask with air condenser packed with copper wool and a still head and an adapter as shown. Wrap the air condenser and bottom of the still head with aluminum foil or other insulation. Use a 5-mL vial as the collection vial. Turn on the heat source and raise the temperature until vapors can be seen rising in the still head. A higher setting is necessary for the heat source than for the simple distillation. Increase the temperature of the heat source slightly as drops of distillate begin to collect in the vial. Allow the first 1–2 drops to pass into the vial. Collect the next 4–5 drops in a separate vial labeled "Fraction 1" and cap the vial. Measure the refractive index of the liquid as time allows. Record the temperature and the volume of liquid collected throughout the distillation. Take readings every 0.2 mL or so. About halfway through the distillation, it may be necessary to turn up the heat source substantially to enable distillation of the second liquid. Continue the distillation until 4.5–5 mL of distillate have been collected. Disconnect the collection vial and save for disposal. Collect 4–5 drops of distillate in a vial labeled "Fraction 2" and cap the vial. Turn off the heat source and allow the apparatus to cool. Measure the refractive index of the liquids in the vials. Disassemble the apparatus. Transfer the distillate to a container in the hood marked "organic distillate."

Part C: Miniscale Distillation

Work individually or in pairs, as directed by the laboratory instructor. Lay out all of the glassware needed to perform a simple distillation. The glassware should be clean, dry, and free of star cracks. Calibrate the thermometer if necessary.

Measure out 10 mL of hexane and 10 mL of a second assigned liquid and add each to a 50-mL round-bottom flask containing a boiling chip or a magnetic stir bar if using a magnetic stirrer. Assemble the apparatus for simple distillation as shown in Figure 1G-8. The bulb of the thermometer must extend below the neck of the still head. Any other placement of the thermometer may result in erroneous temperature readings. Be certain that the tip of the bent adapter fits well into the neck of the collection flask (a 25-mL graduated cylinder) to minimize evaporation of the distillate. Turn on the water (gently) to the condenser. Turn on the heat source and heat the liquids until the vapors reach the takeoff region of the still head. Maintain or slightly increase the rate of heating to achieve and maintain distillation. Discard the first 5 drops of distillate, which may contain water or volatile impurities. Collect the next 4–5 drops of distillate in a vial labeled "Fraction 1" and immediately cap and save for measurement of the refractive index. Maintain the distillation rate at about 1–2 drops per second. Record the temperature versus volume of distillate (in mL) during the entire distillation. Take readings every 1 mL or more frequently if the temperature is changing rapidly. As the distillation proceeds, the more volatile component will be depleted and the pot will become enriched in the higher-boiling liquid. It will be necessary to turn up the heat to continue the distillation at a suitable rate. Distill until only 1–2 mL of liquid remains in the pot. Collect the last 4–5 drops of distillate in a vial labeled "Fraction 2" for measurement of the refractive index. Remove the heat source and let the apparatus cool to room temperature. Disassemble the distillation apparatus and clean the glassware. Transfer the distillate to a container in the hood marked for organic distillate.

Part D: Miniscale Fractional Distillation

Work individually or in pairs, as directed by the laboratory instructor. Lay out all of the glassware needed to perform a fractional distillation. The glassware should be clean, dry, and free of star cracks. Calibrate the thermometer if necessary.

Measure out 10 mL of hexane and 10 mL of a second assigned liquid and add each to a 50-mL round-bottom flask containing a boiling chip or a magnetic stir bar if using a magnetic stirrer. Assemble the apparatus for fractional distillation as shown in Figure 1H-2. For the fractionating column, use either a Vigreux column or a packed column as shown. Wrap the column with aluminum foil or other insulation. The bulb of the thermometer must extend below the neck of the still head. Any other placement of the thermometer may result in erroneous temperature readings. Be certain that the tip of the bent adapter fits well into the neck of the collection flask (a 25-mL graduated cylinder) to minimize evaporation of the distillate. Turn on the water (gently) to the condenser. Turn on the heat source and heat the liquids until the vapors reach the takeoff region of the still head. Maintain or slightly increase the rate of heating to achieve and maintain distillation. Discard the first 5 drops of distillate, which may contain water or volatile impurities. Collect the next 4–5 drops of distillate in a vial labeled "Fraction 1" and immediately cap and save for measurement of the refractive index. Maintain the distillation rate at about 1–2 drops per second. Record the temperature versus volume of distillate (in mL) during the entire distillation. Take readings after every 1 mL of distillate or more frequently if the temperature is changing rapidly. As the distillation proceeds, the more volatile component will be depleted and the pot will become enriched in the higher-boiling liquid. It will be necessary to turn up the heat to continue the distillation at a suitable rate. Distill until only 1–2 mL of liquid remains in the pot. Collect the last 4–5 drops of distillate in a vial labeled "Fraction 2" for measurement of the refractive index. Remove the heat source and let the apparatus cool to room temperature. Disassemble the distillation apparatus and clean the glassware. Transfer the distillate to a container in the hood marked for organic distillate.

Vigreux column

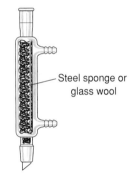

Steel sponge or glass wool

Packed column

Results and Conclusions

1. Correct the boiling points to standard pressure.
2. Measure the refractive index of each standard. Determine the refractive index of the distillate collected in each of the vials. Correct the readings for temperature.
3. From the refractive index measurements, estimate the relative purity of the first and last portions of distillate.
4. Plot corrected boiling point temperature versus volume of distillate (in mL). On this graph mark the temperatures at which pure hexane and the other pure liquid would be predicted to distill if pure. On the basis of this graph and the values measured for the refractive index of the two samples of distillate, estimate the volumes (if any) of pure hexane and pure assigned liquid.
5. Examine the boiling point versus distillate graphs of the simple distillation of the other binary mixtures. Note which pairs of liquids were efficiently separated by simple distillation. For these liquids, note the differences in boiling points between the liquids.
6. Compare the simple distillation and the fractional distillation of the same binary mixture. Which distillation technique was more effective in separating the two liquids? Explain.

7. Examine the boiling point versus distillate graphs of the fractional distillation of the other binary mixtures. Note which pairs of liquids were efficiently separated by simple distillation. For these liquids, note the differences in boiling points between the liquids.

8. Use Raoult's Law (Technique G) to explain how and why the temperature changes as the distillation proceeds.

Cleanup & Disposal

Place the organic distillate in an appropriately labeled container in the hood.

Critical Thinking Questions

1. Compare and contrast the techniques of microscale and miniscale distillation in terms of efficiency of distillation. What are the advantages of each? What are the disadvantages of each?

2. What is the purpose of adding a spin vane, boiling chip, or magnetic stir bar to the distilling flask?

3. Explain why a liquid should not be distilled to dryness.

4. How will distillation be affected by:
   a. distilling too rapidly?
   b. distilling too slowly?
   c. adding too much liquid to the flask or vial?
   d. forgetting to add a boiling chip?

5. Distillation assemblies in this experiment are left open to the atmosphere. Why is this important?

6. Why must boiling points be corrected for barometric pressure? Explain.

Chapter 4

Alcohols and Alkyl Halides

Alcohols and alkyl halides comprise two of the most common functional classes in organic chemistry. Experiment 4.1 details a clear and efficient conversion of an alcohol to an alkyl chloride using HCl and purification by simple distillation. Free radical chlorination using chlorine gas is not a reaction that is feasible for most undergraduate laboratories. Liquid sulfuryl chloride is used in place of chlorine. The free radical nature of the mechanism is emphasized in Experiment 4.2. The alkane to be halogenated has two different types of hydrogen (primary and tertiary) and a comparison of the relative amounts of chlorinated products obtained gives a measure of relative reactivities of the hydrogens. This is best determined by using gas chromatography to separate and quantify the products.

Experiment 4.1: Synthesis of an Alkyl Halide from an Alcohol

In this experiment, you will investigate the reactions and conditions used for the conversion of alcohols to alkyl halides as well as the mechanism involved. You will:

◆ synthesize an alkyl halide from an alcohol.

◆ observe a reaction occurring via an S_N1 mechanism.

Techniques

Technique D	Micro boiling point
Technique G	Microscale distillation
Technique I	Extraction and drying

Background and Mechanism

Tertiary alcohols can be easily converted to alkyl chlorides by the addition of concentrated hydrochloric acid to the alcohol. In this experiment, concentrated hydrochloric acid is used to prepare 2-chloro-2-methylbutane from 2-methyl-2-butanol.

2-methyl-2-butanol   2-chloro-2-methylbutane

The mechanism of this S_N1 reaction has three steps. The first is a rapid (and reversible) protonation of the alcohol, followed by a much slower and rate-determining loss of water to give a tertiary carbocation. In the last step the carbocation rapidly combines with a halide ion to form the alkyl halide, which separates from the aqueous layer due to the fact that the alkyl halide is insoluble in water.

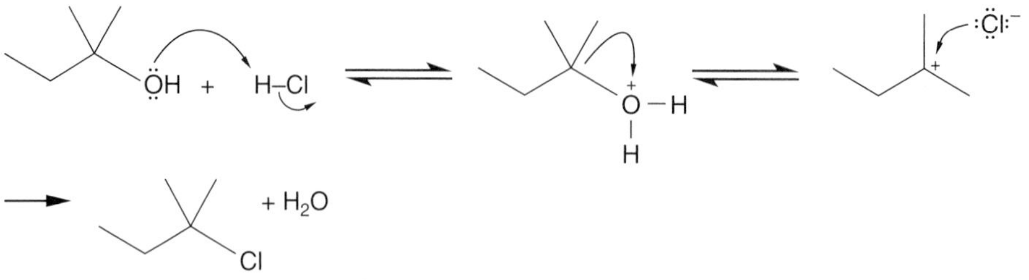

$$\text{(reaction mechanism shown)}$$

There are several advantages to this experiment. It can be accomplished quickly without heating the reaction mixture. Since no molecular rearrangement such as a hydride or methyl shift occurs, a single substitution product is expected. Elimination affords minor alkenyl side products, but these are easily removed during purification of the product.

Prelab Assignment

1. Draw the structures of all alkenyl side products that could be formed by dehydration of 2-methyl-2-butanol by an E1 mechanism.
2. Propose a mechanism for the formation of the alkene side products.
3. Record the physical properties of each alkene side product that could form in this reaction.
4. Write a flow scheme for the workup procedure of the reaction mixture for this experiment.
5. Explain where the alkenyl side products are removed in the experimental procedure.
6. What is the purpose of using sodium bicarbonate in the workup procedure of this reaction? Write an equation for the reaction.

Experimental Procedure

Safety First!

Always wear eye protection in the laboratory.

1. Wear eye protection at all times in the laboratory.
2. HCl is corrosive and toxic and causes burns. Wear gloves when handling this reagent.
3. Use caution when neutralizing acids with sodium bicarbonate. Vent frequently to prevent buildup of gas pressure.

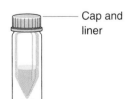
Cap and liner

Microscale Synthesis of 2-Chloro-2-methylbutane from 2-Methyl-2-butanol

Place 1.0 mL of 2-methyl-2-butanol into a 5-mL conical vial or centrifuge tube. Add 2.5 mL of concentrated hydrochloric acid to the vial and shake vigorously (capped tightly with liner inserted in the cap) for about one minute. Carefully loosen the cap to vent.

Repeat the shaking and venting process three times. The product will separate as an upper layer when the mixture is allowed to stand.

Wear gloves

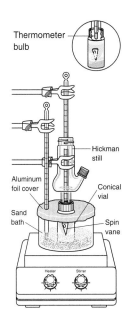

The alkyl halide product is readily hydrolyzed back to the alcohol so it is important to proceed directly with the next steps. Remove the lower (aqueous) layer and transfer it to another small vial or Erlenmeyer flask. Wash the organic layer with 1 mL of water, shaking vigorously several times. Upon standing, the layers will separate. Remove the lower (aqueous) layer. Add 1 mL of saturated sodium bicarbonate solution to the organic layer. Mix gently with a spatula or stirring rod and note the evolution of gas. Cap the vial and shake it gently for one minute. Vent frequently. Allow the layers to separate. Remove the lower (aqueous) layer. Wash the organic layer with 1 mL of water.

After separation of the layers and removal of the lower (aqueous) layer, dry the organic layer over anhydrous sodium sulfate. Add the minimum amount of drying agent, but add enough so that some is free flowing (not clumped together). The dried liquid product should be clear. Using a Pasteur filter pipet, transfer the alkyl halide to a 3-mL conical vial for distillation, being careful not to draw up any drying agent. Add a spin vane and distill using a Hickman still and water-jacketed condenser. Position the thermometer so that the bottom of the bulb is level with the bottom of the cap. Record the boiling point at which the product is collected during the distillation. Transfer the product to a clean, tared vial. Weigh the product.

Characterization

Boiling Point: Determine the micro boiling point of 2-chloro-2-methylbutane. The literature value is 86°C.

Refractive Index: Determine the refractive index of the product. The literature value (n_D) is 1.4055^{20}.

Gas Chromatography: Determine the purity of the product via GC analysis. Inject 1–5 μL of the product via syringe onto a column. Record the gas chromatogram. Inject standards of 2-chloro-2-methylbutane, 2-methyl-2-butanol, and alkenes, if available. Determine the relative amount of product obtained in the reaction.

Infrared Spectroscopy (optional): Record the IR spectrum of the product. (The product is volatile.) Interpret the spectrum noting the location of the bands that confirm the presence of the C−Cl stretching (750–650 cm^{-1}) and the sp^3 C−H stretching (3000–2800 cm^{-1}).

NMR Spectroscopy (optional): Record the NMR spectrum of the product in CDCl$_3$ and interpret it noting the chemical shift positions of the absorptions and the areas using integration values. Only three signals are expected for the product, but they may be close together. The product can be distinguished from the starting material by the absence of a singlet for the hydroxyl proton of the alcohol.

Results and Conclusions

1. Determine the percent yield of the product.
2. Evaluate the purity of the product based on the boiling point, refractive index, or other characterization method.
3. Measuring the boiling point during a microscale distillation with a Hickman still is generally not as accurate as doing a micro boiling point determination. Explain.

Cleanup & Disposal

The neutralized aqueous solutions may be washed down the drain with water.

Critical Thinking Questions

1. Sodium hydroxide can often be used in place of sodium bicarbonate but it cannot be used in this procedure because of a complicating side reaction. Write a reaction that might occur if NaOH were present.

2. If the reaction is carried out in 80% aqueous ethanol, a new product is formed. It is ethyl *tert*-pentyl ether. Write a mechanism for its formation.

3. Careful examination of the reaction mixture from question 2 reveals a small amount of another ether, with the formula $C_{10}H_{22}O$. Propose a structure and suggest a mechanism for its formation.

4. Rearrangement of the intermediate carbocation in this experiment is not observed. Explain why there is no rearrangement.

5. The reaction of *tert*-pentyl alcohol with HNO_3 does not produce *tert*-pentyl nitrate. Explain why this reaction fails and predict the structure of the product formed in this reaction.

6. *Molecular modeling assignment:* Construct a butyl carbocation, a 2-butyl carbocation, and a 2-methyl-2-propyl carbocation, making certain to include a +1 charge on each carbocation. Show electron density with charge density mapped on top. Determine whether the charge is localized on one carbon or delocalized. Analyze the electron density maps to justify the order of carbocation stability. Calculate the heats of formation of the carbocations, both with and without solvation effects.

Experiment 4.2: Selectivity of Free Radical Chlorination of 2,3-Dimethylbutane

In this experiment, you will investigate the relative reactivity of different types of carbon-hydrogen bonds through free radical halogenation of an alkane. You will:

◆   synthesize a mixture of chloroalkanes from an alkane via a free radical reaction.

◆   analyze the ratios of chloroalkanes via gas chromatography.

◆   determine the relative reactivity of primary versus tertiary carbon-hydrogen bonds toward free radical chlorination.

Techniques

Technique G	Reflux
Technique I	Extraction and drying
Technique J	Gas-liquid chromatography

Background and Mechanism

Alkanes are not reactive toward most reagents. A notable exception is the free radical halogenation of alkanes. Halogenation may be initiated by either heat or light. The overall reaction for monohalogenation is shown below, where X is Br, Cl, or F. An excess of the alkane must be used to avoid polyhalogenation.

$$R{-}H + X_2 \xrightarrow[\text{or } \Delta]{hv} R{-}X + HX$$
(excess)

The monohalogenation reaction works well only for chlorine and bromine. The reaction with fluorine is too exothermic and gives polyfluorination. The reaction with iodine, on the other hand, is too endothermic and leads to no significant product formation.

The first step in the mechanism of free radical halogenation is thermal cleavage of the free radical initiator, 2,2′-azobisisobutyronitrile (AIBN), to form two stabilized radicals:

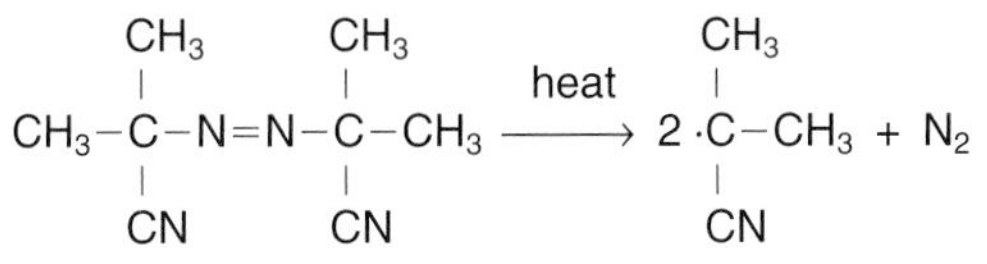

```
        CH3      CH3                    CH3
         |        |          heat        |
  CH3 - C - N = N - C - CH3  ------->  2 ·C - CH3  +  N2
         |        |                     |
         CN       CN                    CN

        AIBN                       isobutyronitrile radical
```

In the next step, isobutyronitrile radical abstracts a chlorine atom from sulfuryl chloride to yield a chlorosulfonyl radical:

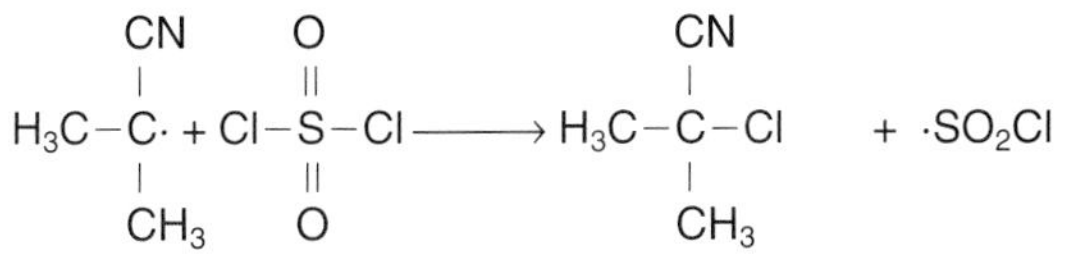

```
       CN      O                      CN
        |      ||                      |
  H3C - C·  + Cl - S - Cl  -------> H3C - C - Cl   +  ·SO2Cl
        |      ||                      |
        CH3    O                       CH3

      sulfuryl chloride            chlorosulfonyl radical
```

The chlorosulfonyl radical then decomposes to sulfur dioxide and chlorine radical:

$$\cdot SO_2Cl \longrightarrow SO_2 \;+\; Cl\cdot$$

chlorine radical

The next phase of the mechanism involves propagation steps in which the chlorine radical abstracts a hydrogen atom from the alkane, forming an alkyl radical. The alkyl radical abstracts a chlorine atom from sulfuryl chloride to form a chloroalkane, and the chlorosulfonyl radical then decomposes to form another chlorine radical. This is illustrated here for methane.

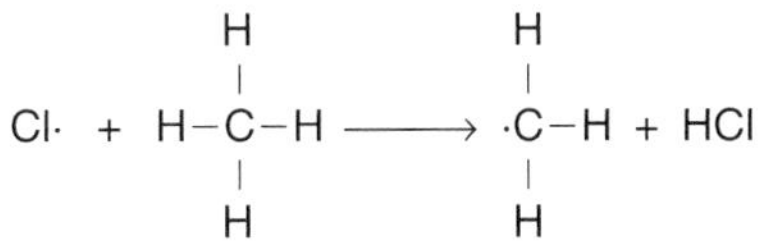

```
              H                    H
              |                    |
  Cl·  +  H - C - H  ------->  ·C - H  + HCl
              |                    |
              H                    H

           methane            methyl radical
```

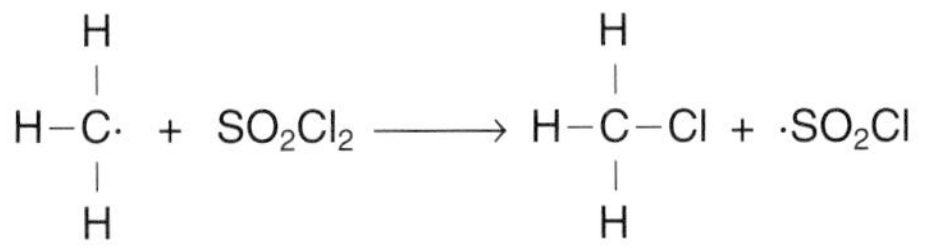

```
      H                             H
      |                             |
  H - C·  +  SO2Cl2  ------->   H - C - Cl  +  ·SO2Cl
      |                             |
      H                             H

  methyl radical               chloromethane
```

$$\cdot SO_2Cl \longrightarrow SO_2 + Cl\cdot$$

The chain reaction will continue as long as there are adequate concentrations of the alkane and sulfuryl chloride. When these reagents are depleted, the chance of two radicals recombining increases. The termination step in the mechanism occurs when any two radicals recombine.

In this experiment, the free radical chlorination of 2,3-dimethylbutane will be investigated and the relative amounts of monochlorinated products will be determined by gas chromatography.

$$\text{2,3-dimethylbutane} \xrightarrow[\text{SO}_2\text{Cl}_2]{\text{AIBN}} \text{monochlorinated products}$$

2,3-dimethylbutane

Two different monochlorinated products of 2,3-dimethylbutane can be formed. The product distribution is governed by two factors: the differing reactivities of hydrogens attached to the primary and tertiary carbons and the number of each type of hydrogen. In general, reactivity factors can be found by dividing the mass percent of each monochlorinated product by the number of the primary, secondary, or tertiary hydrogens that would yield that product.

As an illustration, the free radical chlorination of propane at 25°C produces two monochlorinated products: 1-chloropropane, which can be made by substitution of any of six primary hydrogens, and 2-chloropropane, which can be made by substitution of either of two secondary hydrogens.

$$CH_3CH_2CH_3 \ + \ Cl_2 \ \xrightarrow[25°C]{h\nu} \ CH_3CH_2CH_2Cl \ + \ \underset{\underset{Cl}{|}}{CH_3CHCH_3}$$

1-chloropropane   2-chloropropane

43%   57%

The reactivity factors at 25°C for the primary and secondary hydrogens can then be calculated as follows (reactivity = relative percent/number of hydrogens):

Reactivity of primary hydrogen at 25°C = 43% / 6 = 7.2,

Reactivity of secondary hydrogen at 25°C = 57% / 2 = 28.5

Relative reactivity of secondary hydrogen to primary hydrogen is 28.5/7.2 or 3.9 to 1.

It is also possible to get some dichlorinated and trichlorinated products in addition to the monochlorinated products. During analysis, the polychlorinated products are eluted from the gas chromatography column much later than the monochlorinated products.

Prelab Assignment

1. Draw the structures of all monochlorinated products that could be formed in this reaction. Assign the IUPAC name for each compound. Look up and record the boiling point of each compound.
2. For the chlorination of 2,3-dimethylbutane, predict the percentages of monochlorinated products that would be expected on a purely statistical basis.
3. Write the mechanism for the thermal decomposition of AIBN to produce isobutyronitrile radicals and nitrogen, showing the direction of electron flow with arrows.
4. Calculate the quantity (mmol) of each reagent used in this experiment. Which reagent is limiting?

Experimental Procedure

S a f e t y   F i r s t !

Always wear eye protection in the laboratory.

1. Wear eye protection at all times in the laboratory.
2. Sulfuryl chloride is corrosive and toxic. It is also a lachrymator (a substance that causes tearing of the eyes). Use gloves and handle it only in the hood. Avoid its

vapors. Any contact with the skin should be washed immediately with copious amounts of water.

3.   Small amounts of gaseous SO_2 and gaseous HCl are formed in this experiment. This experiment should be conducted in a hood or at the bench with a gas trap.

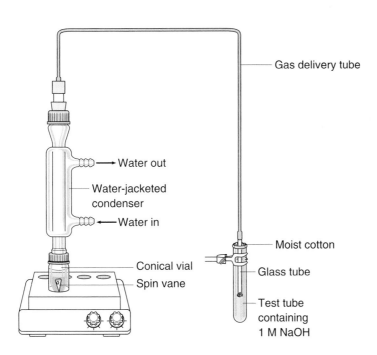

Figure 4.1-1   Operational Setup for Experiment 4.2

Microscale Chlorination of 2,3-Dimethylbutane with Sulfuryl Chloride

Place 1.5 mL of 2,3-dimethylbutane and 30 mg of 2,2′-azobisisobutyronitrile (AIBN) in a dry 5-mL conical vial equipped with a spin vane. Add 150 μL of sulfuryl chloride to the vial. Attach a water-jacketed condenser to the reaction vial. To the top of the condenser attach gas outlet tubing. The tip of the tubing should extend into a large test tube that has a piece of damp cotton placed into its mouth or a small beaker. The tubing should be slightly below the surface of 5–10 mL of a 1 M NaOH solution. This will neutralize the HCl and SO_2 gases evolved. Turn on the cold water to the water-jacketed condenser and reflux the reaction mixture for 30 minutes in a hot sand bath or heat block.

Hood

Remove the gas outlet tubing from the water-jacketed condenser to prevent water from being drawn back in from the trap and cool to room temperature. Remove the spin vane. Add 1.0 mL of water to the vial. Mix the two layers well. Draw off the aqueous (bottom) layer. Wash the organic layer with 1.0 mL of 5% $NaHCO_3$ solution. Cap the vial; shake cautiously and vent frequently. Draw off the aqueous layer. Wash the organic layer with 1.0 mL of water. Cap, shake, and vent. Draw off the aqueous layer. Dry the organic layer over sodium sulfate. Cap the vial and store until ready for gas-liquid chromatographic analysis.

Inject 0.1–1.0 μL of the sample into the gas chromatograph. Adjust the attenuation for analysis of the two chloroalkane products. Inject samples of authentic monochlorinated products to determine the identity of the chlorinated products. If standards of the pure monochlorinated products are not available, determine the identity of each peak based on relative boiling points of the products.

Since there was no purification by distillation, there may be a considerable amount of 2,3-dimethylbutane remaining, which should elute first from the column. There may also be some dichlorinated products formed in this reaction, which will have longer retention times.

Results and Conclusions

1. Identify each monochlorinated product in the chromatogram, based on the retention times of authentic samples or the boiling points of the alkane and chlorinated products.
2. Measure the area of the peak of each monochlorinated product. Determine the relative percentages of the monochlorinated products formed in this reaction.
3. Determine the relative reactivities of primary hydrogens and tertiary hydrogens toward free radical chlorination. Compare the values with the values given in your lecture textbook.
4. Write a detailed mechanism of the formation of the major product of this reaction.
5. Write the structures of all dichlorinated products that could possibly be formed in this reaction and name them.
6. Draw a potential energy diagram for the propagation steps in the formation of the major monochlorinated product.

Cleanup & Disposal

The neutralized aqueous extracts can be washed down the drain with water. The material collected in the gas trap should be neutralized and washed down the drain.

Critical Thinking Questions

1. Write the structures of the expected monochlorinated products from free radical chlorination of 2,4-dimethylpentane.

2. Would there be a difference in product ratios if a brominating reagent were used rather than a chlorinating agent? Explain. What product ratio might be expected if fluorine were used?

3. Other chlorinating agents such as molecular chlorine could be used in place of SO_2Cl_2. What major experimental and safety problems are associated with using chlorine?

4. Explain the chemistry occurring in the gas trap. Remember that both gases evolved in this process require the presence of the trap.

5. The relative reactivity factors of free radical bromination of primary, secondary, and tertiary hydrogens are 1, 82, and 1640, respectively.

   a. Explain why free radical bromination is more selective than free radical chlorination, using energy diagrams where appropriate.
   b. Calculate the expected product distribution of the free radical bromination of 2-methylbutane.

Chapter 5

Synthesis of Alkenes

The two simplest methods for the preparation of alkenes are introduced in Chapter 5. These methods are dehydration and dehydrohalogenation. Alkenes can be made by either acid-catalyzed dehydration of alcohols in Experiment 5.1 or base-induced dehydrohalogenation of alkyl halides in Experiment 5.2. A third commonly used method is the Wittig synthesis found in Chapter 17.

Dehydration follows an E1 mechanism that proceeds via a carbocation intermediate. Dehydrohalogenation with strong base occurs by a concerted E2 mechanism. Since both experiments can produce the same volatile alkenes, the two methods of preparation may be directly compared when both experiments are chosen.

Experiment 5.1: Alkenes Via Acid-catalyzed Dehydration of Alcohols

In this experiment, you will investigate alcohol dehydration to form alkenes. You will

- convert an alcohol to a mixture of alkenes.
- study a reaction occurring via an E1-like mechanism.
- evaluate the relative stabilities of alkenes produced by the reaction.
- determine the product distribution.
- determine the identity of an unknown alcohol based upon the product distribution.

Techniques

Technique G	Reflux and distillation
Technique H	Steam distillation
Technique I	Extraction and drying
Technique J	Gas-liquid chromatography
Technique M	Infrared spectroscopy

Background and Mechanism

Elimination by alcohol dehydration is readily accomplished by heating in the presence of an acid catalyst such as sulfuric acid or phosphoric acid. Sulfuric acid can be used for dehydration reactions, but phosphoric acid is a milder acid catalyst that results in higher yields of alkenes and fewer side products.

The first step in the mechanism of this reaction is a rapid and reversible protonation of the alcohol by the acid, followed in the second step by a rate-determining loss of water to form a tertiary carbocation. This process is illustrated here for 3-methyl-3-pentanol.

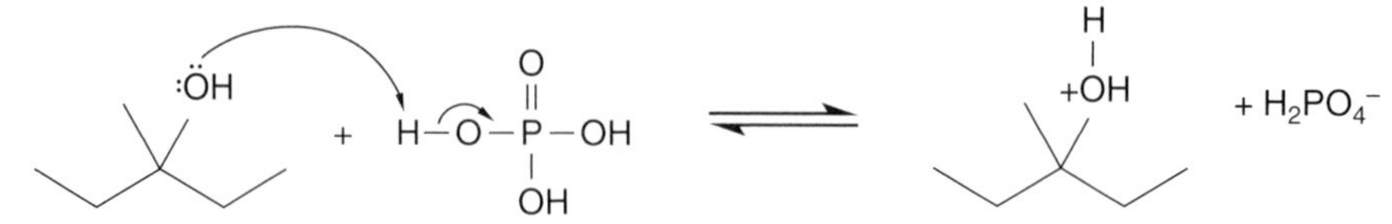

3-methyl-3-pentanol

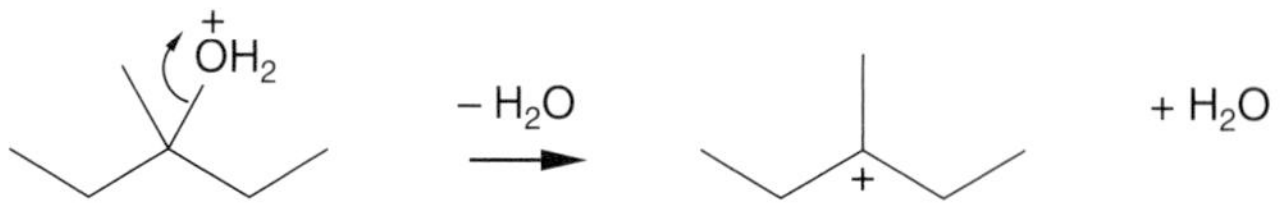

In the third step, the carbocation loses a proton to give a mixture of alkenes.

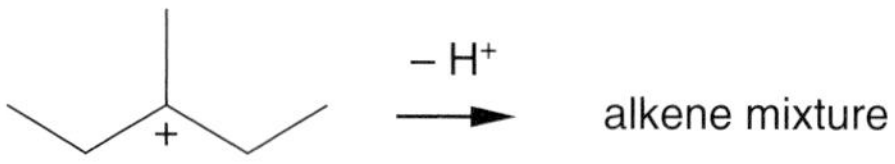

In this experiment, there are several different alkenes that can be formed. However, one of the alkenes often predominates.

In Part A of this experiment, the alcohol to be dehydrated is 3-methyl-3-pentanol, while in Part B, the alcohol will be 3,3-dimethyl-2-butanol. The product compositions will be evaluated by gas chromatography.

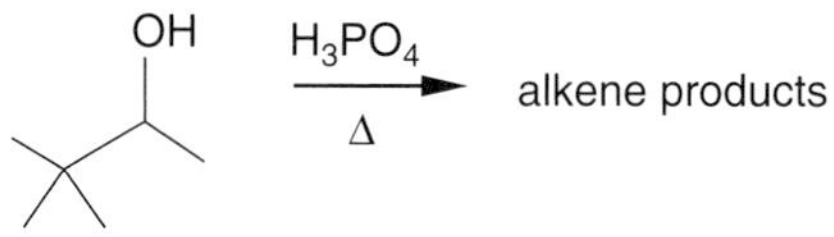

In Part C, you will dehydrate an unknown alcohol. The two possible unknown alcohols are 2-methyl-1-phenyl-2-propanol or 2-phenyl-2-butanol. Based upon the number of alkenes formed in the reaction, you will identify the structure of the unknown alcohol. You will then identify the predominant alkene through analysis of the product mixture by gas chromatography (GC) or gas chromatography-mass spectrometry (GC-MS).

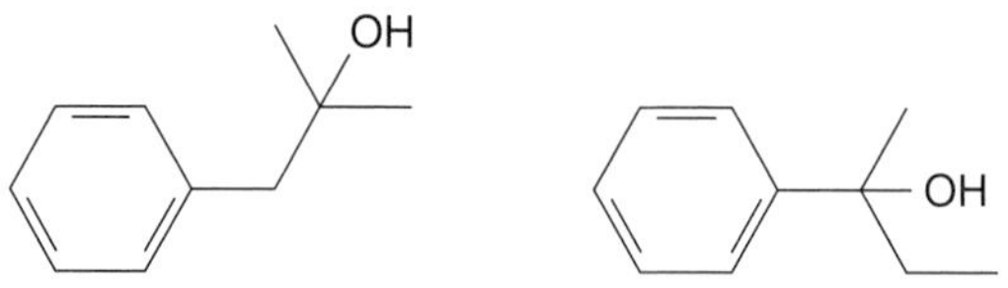

2-methyl-1-phenyl-2-propanol   2-phenyl-2-butanol

Prelab Assignment

1. Parts A and B: Draw the structures of all of the alkenes that could be formed during the assigned experiment. Provide an IUPAC name for each possible product.
2. Predict which alkene should be the major product.
3. Complete the mechanism, showing the loss of a proton that generates each alkene that could be formed from the assigned reaction.
4. Part C: Draw the structures of the alkenes that could be formed from each of the two possible alcohols. Provide an IUPAC name for each possible product.

5. Look up the boiling points of the possible products for the assigned reaction and predict the order of elution from the GC column.
6. Write a flow scheme for the workup procedure for this experiment.

Experimental Procedure

1. Wear eye protection at all times in the laboratory.
2. Phosphoric acid is corrosive. Wear gloves when handling this reagent.

Part A: Microscale Synthesis of Alkenes Via the Acid-catalyzed Dehydration of 3-Methyl-3-pentanol

Place 1.25 mL of 3-methyl-3-pentanol in a 5-mL conical vial containing a spin vane or boiling chip. Add 0.25 mL of 85% phosphoric acid to the alcohol. Fit the reaction vial with a Hickman still and place a water-jacketed condenser on top for cooling. Place the vial in a sand bath or heat block and turn on the water to the condenser. Adjust the temperature to between 100 and 105°C. Heat the reaction mixture for 30 minutes or until the collar of the Hickman still is filled with the liquid alkene mixture. The alkene products are very volatile (bp = 65–68°C). Transfer the liquid product from the collar of the still to a vial and dry over $CaCl_2$. Cap the vial during the drying process. Use a filter pipet to transfer the dried liquid into a clean, tared vial and weigh. Determine the product distribution by gas chromatography.

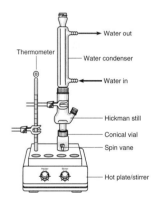

Thermometer — Water out — Water condenser — Water in — Hickman still — Conical vial — Spin vane — Hot plate/stirrer

Characterization for Part A

Because of the extreme volatility of the alkene products, many of the common characterization techniques such as micro boiling point, refractive index, and infrared spectroscopy are not suitable for characterization.

Gas-liquid Chromatography: Determine the product distribution via GC analysis. Inject 1–5 µL of the product onto a nonpolar column. Record the chromatogram. Inject standards of 3-methyl-3-pentanol and each of the alkenes, if available.

Results and Conclusions for Part A

1. Determine the percent yield of alkenes. For this calculation, consider the combined yield of alkenes since they are isomers and have the same molecular weight.
2. Determine the product distribution of the alkenes from analysis of the gas chromatogram.
3. Is the predominant alkene obtained in this experiment the one that was expected? Explain.

Part B: Miniscale Synthesis of Alkenes Via the Acid-catalyzed Dehydration of 3,3-Dimethyl-2-butanol

Construct a simple distillation setup using a 25-mL round-bottom flask. Place 6.0 mL of 3,3-dimethyl-2-butanol in the 25-mL round-bottom flask fitted with a spin bar. Add 6.0 mL of 85% phosphoric acid to the flask and begin stirring. Turn the water on to

Refer to Figue 1G-8.

the condenser and begin heating. Due to the extreme volatility of the products, collect the distillate in a graduated cylinder set inside a beaker filled with an ice-water mixture. The distillate will be cloudy and should separate into two layers. Do not let the temperature rise above 75°C. After the product mixture has been distilled, turn off the heat, and let the distillation apparatus cool down. Clean the distillation glassware and rinse with acetone. Set aside the glassware to dry. Transfer the distillate to a small separatory funnel. Drain the lower aqueous layer into a beaker. Wash the organic layer with about 20 mL of saturated sodium chloride solution (brine). Drain the lower aqueous layer into the beaker and decant the organic layer into a clean 10-mL Erlenmeyer flask. Dry with anhydrous magnesium sulfate. Gravity filter the clear solution into a dry 10-mL round-bottom flask. Reassemble the apparatus for distillation. Cool the receiving flask in an ice-water bath, as alkenes are very volatile. Weigh a 10-mL sample vial with cap. As product distills, transfer it to the tared vial. Reweigh the vial and distillate. Keep the vial tightly capped while waiting to do the analysis.

Characterization for Part B

Gas Chromatography: Determine the product distribution via GC analysis. Inject 1–5 μL of the product onto a nonpolar column. Record the chromatogram. Inject standards of the starting material (3,3-dimethyl-2-butanol) and 3,3-dimethyl-1-butene.

Infrared Spectroscopy: Place a few drops of the product on a salt plate and quickly record the IR spectrum.

Results and Conclusions for Part B

1. From the gas chromatogram, indicate the extent of conversion from alcohol to product.
2. Examine the IR spectrum for evidence of sp^2 C$-$H stretching (around 3100 cm^{-1}) and C=C stretching (around 1600 cm^{-1}). Are these absorption bands prominent in the spectrum? Knowing that the major product of this reaction is an alkene, what conclusions can you draw about the possible structure of that alkene?
3. Write the structure of the major product formed in this reaction. Was the major product 3,3-dimethyl-1-butene? Explain.
4. Propose a mechanism to account for the formation of the observed major product.
5. Was the major product obtained the one that is the most thermodynamically stable? Explain.

Part C: Microscale Synthesis of Alkenes Via the Acid-catalyzed Dehydration of an Unknown Alcohol

Each student will receive a vial containing 1.5 mL of an unknown alcohol, which will be either 2-methyl-1-phenyl-2-propanol or 2-phenyl-2-butanol. Transfer the alcohol to a 5-mL conical vial containing a spin vane. Add 0.25 mL of 85% phosphoric acid to the alcohol. Fit the reaction vial with a Hickman still and place a thermometer in the heat block. Place the vial in a heat block and start stirring. Adjust the temperature to approximately 100–105°C and heat for 30 minutes. As the solution is heating, water vapor condensation may appear in the Hickman still. After 30 minutes, remove the thermometer from the heat block and place in the Hickman still. Gradually increase the temperature of the heat block until distillate begins to appear in the Hickman still. The alkenes will distill around 170–180°C. Measure the temperature at which distillation occurs. Transfer the liquid products from the collar of the still to a clean vial.

Dry the distillate in the vial over CaCl$_2$. While the product is drying, tare a clean vial. Transfer the dried liquid into the clean, tared vial, using a filter pipet. Reweigh the

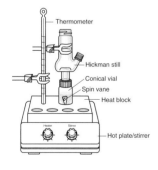

Thermometer

Hickman still

Conical vial

Spin vane

Heat block

Hot plate/stirrer

vial and determine the mass of the alkenes. Inject the mixture onto a GC or a GC-MS and analyze the results.

Characterization for Part C

Gas Chromatography: Inject 1–5 μL of the product onto a nonpolar column. Record the chromatogram. Inject standards of the possible alkene products, if available.

Optional Analysis with Gas Chromatography-Mass Spectrometry (GC-MS): If available, analyze a small portion of the reaction mixture. Analyze the spectrum and identify the products formed during the reaction.

Results and Conclusions for Part C

1. Based upon the number of peaks evident in the chromatogram, determine the identity of the unknown alcohol.
2. Based on boiling points and/or retention times, identify the alkene products formed in this reaction.
3. Determine the product composition of the alkenes.
4. Identify the major alkene formed in the reaction. Is the predominant alkene obtained in this experiment the one that is the most thermodynamically stable? Explain.
5. Determine the percent yield of the alkenes. For this calculation, consider the combined yield of alkenes since they are isomers and have the same molecular weight.
6. Write a detailed mechanism of the dehydration of the assigned alcohol to form the major alkene.

Cleanup & Disposal

Neutralize the acidic pot residue with sodium bicarbonate solution and wash down the drain with water.

Critical Thinking Questions

1. Why is elimination favored at high temperatures?
2. Rearrangement of the carbon skeleton is not important in this reaction. Explain.
3. Why is elimination favored over substitution with phosphoric acid?
4. If alkenes are not distilled away from the acid, polymerization may occur. Write a mechanism to show this process. Follow it through the incorporation of three monomeric units.
5. What would be the expected results if HBr were substituted for H_3PO_4 in this reaction? Write the structure(s) of the expected product(s).
6. Provide a mechanism to account for the fact that acid-catalyzed dehydration of 2,2-dimethylcyclohexanol and of 2,3-dimethylcyclohexanol (mixed isomers) produces the same major product.
7. Explain why the distillate in the dehydration of 3,3-dimethyl-2-butanol was cloudy.
8. *Molecular modeling assignment:* Construct models of the alkene products in the assigned reaction. Optimize the geometry using the semiempirical AM1 molecular orbital method. Record the heat of formation of each optimized structure. Which alkene isomer is lowest in energy? Do the molecular modeling calculations confirm or refute the experimental findings? Explain.

9. *Molecular modeling assignment:* Construct models of the carbocations involved in the assigned reaction. Optimize the geometry using the semiempirical AM1 molecular orbital method. Record the heat of formation of each optimized structure. Which carbocation is the lowest in energy? Is this finding supported by your experimental results? Explain.

Experiment 5.2: Alkenes Via Base-induced Elimination of Alkyl Halides

In this experiment, you will investigate the structural, mechanistic, and stereochemical aspects of dehydrohalogenation reactions. You will:

♦   synthesize alkenes from an alkyl halide.

♦   study a reaction occurring by an E2 mechanism.

♦   analyze the product distribution of alkenes.

Techniques

Technique G	Reflux and distillation
Technique I	Drying
Technique J	Gas-liquid chromatography

Background and Mechanism

Dehydrohalogenation of alkyl halides occurs upon heating in the presence of a strong base. The reaction rate is dependent on the concentrations of both base and alkyl halide. In this one-step process, called β-elimination, bond-making and bond-breaking occur simultaneously. The mechanism is illustrated below for 2-chlorobutane.

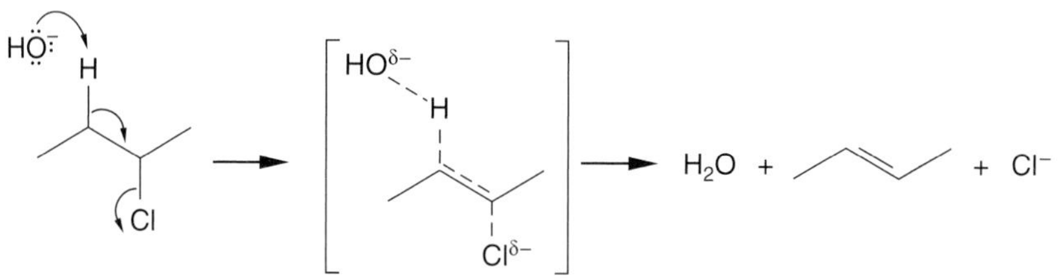

If all the β-hydrogens are not equivalent, it is possible to form more than one alkene. In this experiment, a strong base will be used to cause dehydrohalogenation of 3-chloro-3-methylpentane to produce a mixture of alkenes. The product distribution will be determined and compared with the mixture resulting from the E1 reaction of an alcohol (see Experiment 5.1).

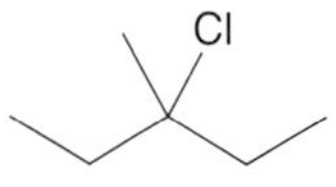

3-chloro-3-methylpentane + KOH $\xrightarrow{\Delta}$ mixture of alkenes + KCl + H_2O

Prelab Assignment

1. Write the structures of each alkene that could be formed from E2 elimination of 3-chloro-3-methylpentane and provide the IUPAC name.
2. Predict which alkene should predominate. Explain.
3. Write a mechanism for the dehydrohalogenation reaction of 3-chloro-3-methylpentane with KOH.

Experimental Procedure

1. Wear eye protection at all times in the laboratory.
2. Potassium hydroxide is very corrosive and toxic. Always wear gloves when handling this reagent. In the event of skin contact, immediately wash the affected area with cold water and notify the instructor immediately.

Microscale Synthesis of Alkenes via Dehydrohalogenation of 3-Chloro-3-methylpentane

Place 0.42 g of potassium hydroxide and 3 mL of absolute ethanol in a dry, 5-mL conical vial equipped with a spin vane. Fit the vial with a Hickman still and place a reflux condenser on top of the still. Turn on the cold water to the reflux condenser. Heat the mixture, with stirring, until the potassium hydroxide dissolves. Cool the solution in a water bath. Add 680 µL of 3-chloro-3-methylpentane. Using a sand bath or heat block, gently reflux this mixture, with stirring, for 45 minutes. The temperature of the reaction mixture should be maintained between 90 and 95°C. Any solid present may cause bumping. The alkene product mixture will condense in the collar of the Hickman still. The products are very volatile (bp 65–68°C). Remove the liquid product mixture from the collar of the Hickman still with a pipet. Transfer the alkenes into a capped vial and dry over $MgSO_4$. Transfer the dried alkenes with a filter pipet to a tightly capped, tared vial. Weigh the product. Analyze by gas chromatography and determine the product distribution.

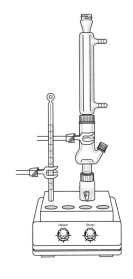

Characterization

Because of the extreme volatility of the products, many of the common characterization techniques such as micro boiling point, refractive index, and infrared spectroscopy are not suitable in this experiment.

Gas-liquid Chromatography: Determine the product distribution via GC analysis. Inject 1–5 µL of the product onto a nonpolar GC column. Record the chromatogram. Inject standards of 3-chloro-3-methylpentane and the alkenes, if available.

Results and Conclusions

1. Determine the percent yield of the mixture of alkenes. For this calculation, consider the combined yield of alkenes since they are isomers and have the same molecular weight.
2. Determine the product distribution of the alkenes from analysis of the gas chromatogram.
3. Using a Newman projection, explain the formation of the major product.

4. Was there any evidence of unreacted alkyl halide? Explain.
5. Compare the results of this experiment with the results of the acid-catalyzed dehydration of 3-methyl-3-pentanol (see Experiment 5.1). Explain any differences.

Cleanup & Disposal

Carefully neutralize the ethanolic KOH solution with aqueous HCl and wash down the drain with water.

Critical Thinking Questions

1. How would the product distribution change if potassium *tert*-butoxide (a bulky base) were substituted for potassium hydroxide? Explain.

2. A possible side product of the reaction is an ether. Predict its structure and the mechanism by which it is formed.

3. Give two simple qualitative tests to confirm the presence of an alkene.

4. Is rearrangement of the carbon skeleton in this experiment a possibility? Is it a probability? Explain the answers in mechanistic terms.

5. The order of reactivity of alkyl halides in E2 elimination reactions is 3°>2°>1°. Explain this order in mechanistic terms.

6. In general, what experimental factors favor an E2 reaction over an S_N2 reaction?

Chapter 6

Alkene Addition Reactions

A ddition reactions are the most common reactions of alkenes. In Experiment 6.1, the hydrogenation of alkenes is explored along with partial hydrogenation of a natural oil. Hydration of alkenes using aqueous sulfuric acid via electrophilic addition yields alcohols, as illustrated in Experiment 6.2. An alternative procedure for preparing alcohols is the hydroboration-oxidation method of Experiment 6.3, which proceeds by anti-Markovnikov addition of water to the carbon-carbon double bond. In Experiment 6.4, alkenes are transformed into polymers using a free radical process.

In each of the reactions in Chapter 6, the π bond of the carbon-carbon double bond is broken and new bonds are formed at each of these carbons. The double bond serves as a convenient vehicle for the production of new functional groups and as a method for joining simple molecules to form long, continuous chains known as polymers.

Experiment 6.1: Catalytic Hydrogenation of Alkenes

The conversion of alkenes to alkanes is investigated using an organic hydrogen transfer agent in the presence of a catalyst. The reaction can be selective, allowing reduction of alkenyl and alkynyl groups in the presence of other functional groups such as esters. In this experiment, you will:

♦ perform a catalytic hydrogenation on an alkene.

♦ convert an unsaturated triglyceride to a partially saturated fatty acid triester.

♦ use IR spectroscopy and refractive index to determine the identity of the product.

Techniques

Technique C	Index of refraction
Technique F	Filtration and recrystallization
Technique G	Reflux
Technique M	Infrared spectroscopy
Technique N	Nuclear magnetic resonance spectroscopy

Background

It is possible to do catalytic hydrogenation safely in a student laboratory environment using an enclosed hydrogen atmosphere (Landgrebe, 1995). However, a more convenient technique is to avoid the use of hydrogen gas by using a method of hydrogenation known as catalytic transfer hydrogenation (CTH) (Brieger and Nestrick, 1974; Johnstone et al., 1985; De et al., 1984; Hansen, 1997; Bunjes et al., 1997). The technique works best by employing a molecule that can readily donate the equivalent of a hydrogen molecule. Hydrogen is donated by cyclic alkenes that upon donation achieve aromatization. Compounds such as cyclohexene, 1-methylcyclohexene, and 2,3-dihydrofuran work well as "donors." These compounds form the aromatic compounds benzene, toluene, and furan respectively after donating hydrogen. "Acceptor" molecules are other alkenes that receive hydrogen from the donor molecules. The acceptor molecules are hydrogenated via addition of hydrogen to a double bond. When using cyclohexene as donor, the best acceptor molecules are monosubstituted alkenes such as 1-hexene or 1-decene. Disubstituted alkenes are hydrogenated more slowly. CTH reactions are selective: Alkenes can be reduced in the presence of a number of other functional groups such as ketones, carboxylic acids, and esters. In Part A, you will react cyclohexene and 1-decene. In this reaction, illustrated below, cyclohexene serves as the donor and 1-decene serves as the acceptor.

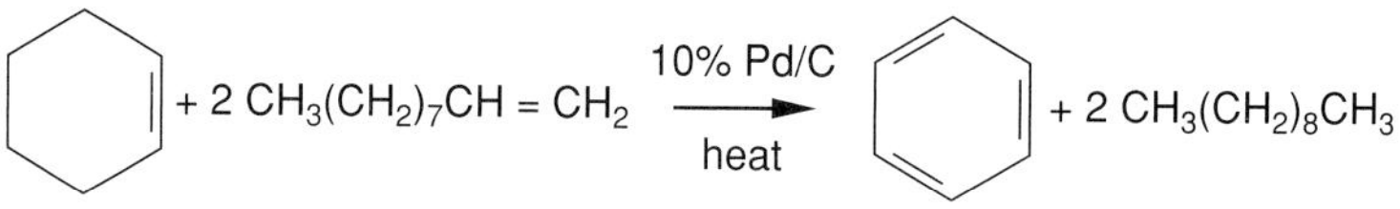

In Part B of this experiment, allylbenzene will be used as the acceptor. You will determine whether hydrogenation of the isolated double bond is possible without affecting the aromatic ring.

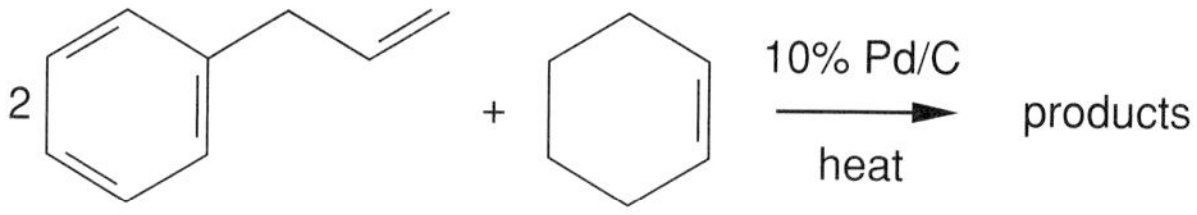

In Part C, hydrogenation of an unsaturated fat (oil) to a saturated fat is investigated. Unsaturated triglycerides, such as olive oil, contain large amounts of oleic acid esters. These fats are partially hydrogenated commercially to produce fats that contain fewer double bonds. These "partially hydrogenated vegetable oils" melt over a wide temperature range because they are mixtures. These substances are higher melting than unsaturated oils and are therefore more suitable for making candies than are unsaturated oils. Olive oil contains (in esterified form) 7% palmitic acid (the straight-chain C_{16} acid), 2% stearic acid (the straight-chain C_{18} acid), 84% oleic acid (an enoic C_{18} acid), and 5% linoleic acid (a dienoic C_{18} acid). Olive oil has a tendency to solidify upon hydrogenation or partial hydrogenation to form a buttery mass. Testing with bromine in methylene chloride serves to distinguish the product from olive oil. The physical characteristics of the product are also distinguishable from olive oil.

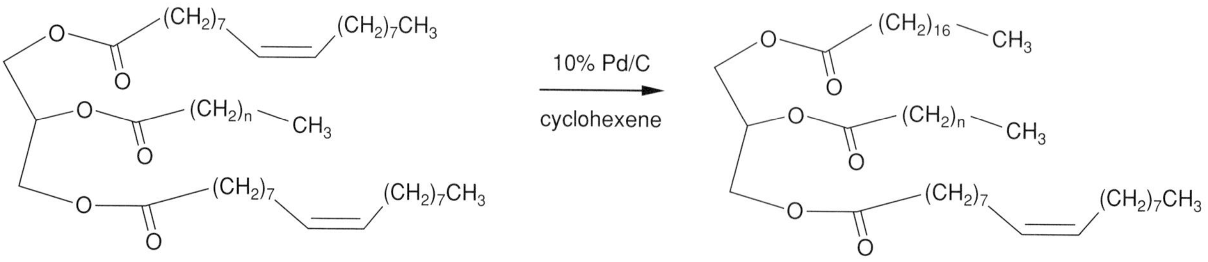

olive oil, an unsaturated triglyceride
(a representative structural formula)

partially hydrogenated olive oil

Prelab Assignment

1. Calculate the quantities of reagents required for this experiment.
2. Draw the products of the reaction of cyclohexene with allyl benzene if (a) the isolated alkenyl group is reduced only; (b) the aromatic ring is fully reduced but not the alkenyl group; and (c) both alkenyl group and aromatic ring are reduced.
3. Predict the IR spectrum for each of the possible products in Part B. What bands would be significant in determining whether the isolated alkene is reduced or the aromatic ring?
4. Predict the NMR spectrum of each of the possible products in Part B. What signals would be significant in determining the structure of the product?

Experimental Procedure

1. Wear safety goggles at all times in the laboratory.
2. Wear gloves when handling reagents and solvents in this experiment. Notify the instructor if any chemicals get on your skin.
3. Avoid breathing solvent vapors in this experiment. Evaporate solvents using a fume hood. Cyclohexene is an irritant, and benzene, formed in small quantities in this experiment, is a cancer suspect agent.

Part A: Microscale Hydrogenation of 1-Decene

To a 5-mL conical vial containing a spin vane, add 2.0 mmol of 1-decene, 24 mmol of cyclohexene, and 40–50 mg of 10% Pd on charcoal. Fit the vial with a water condenser. Reflux the mixture on a sand bath or heat block for 20 minutes. Cool the mixture to room temperature. Prepare a Pasteur filter pipet, and add a small portion of Celite. Transfer a portion of the mixture to the filter pipet. Place a plastic bulb on the filter pipet and force the filtered solution out of the pipet and into a clean, tared suction flask containing a stir bar or a boiling chip. The black catalyst should remain in the pipet. Repeat until all of the mixture has been filtered. Add 2.0 mL of petroleum ether to the pipet in portions, again using the pipet bulb to push the solvent through the Celite and out into the tared suction flask. Stopper and connect the suction flask to a vacuum aspirator and swirl to allow any remaining solvent to evaporate. The flask will become cool and a small amount of liquid should remain at the bottom of the flask. Allow the flask to warm to room temperature and weigh the flask.

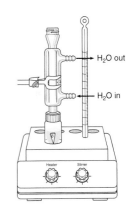

Characterization for Part A

Bromine in Methylene Chloride Test: Use this test to determine how much unsaturation is present in the unknown sample. Add 10 drops of a 1.0 M bromine in methylene chloride solution to a 10-cm test tube. Using a pipet, add dropwise a sample of the starting alkene until the red color of the bromine disappears. Swirl the tube between added drops. Note the amount of alkene required. Repeat using another 10 drops of 1.0 M bromine solution for a sample of the product. Record the results.

Refractive Index: Determine the refractive index (n_D^{20}) of the product and compare to the reported values for decane, 1.4110.

Infrared Spectroscopy (optional): Transfer a drop of the liquid product to a salt plate. Obtain the IR spectrum and compare it with the IR spectrum of the starting

alkene. The absorption present in the alkene spectrum at 3100 cm^{-1} (vinyl C—H stretching) and 1640 cm^{-1} (C=C stretching) should be absent in the IR spectrum of the product from Part A.

Results and Conclusions for Part A

1. Calculate the percent yield of product.
2. From the IR, determine whether the reaction went to completion.
3. Determine the purity of the product using the refractive index.

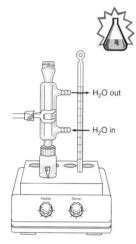

Part B: Microscale Hydrogenation of Allylbenzene

To a 5-mL conical vial containing a spin vane, add 2.0 mmol of allylbenzene, 24 mmol of cyclohexene, and 40–50 mg of 10% Pd on charcoal. Fit the vial with a water condenser. Reflux the mixture on a sand bath or heat block for 20 minutes. Cool the mixture to room temperature. Prepare a Pasteur filter pipet, and add a small portion of Celite. Transfer a portion of the mixture to the filter pipet. Place a plastic bulb on the filter pipet and force the filtered solution out of the pipet and into a clean, tared suction flask containing a stir bar or a boiling chip. The black catalyst should remain in the pipet. Repeat until all of the mixture has been filtered. Add 2.0 mL of petroleum ether to the pipet in portions, again using the pipet bulb to push the solvent through the Celite and out into the tared suction flask. Stopper and connect the suction flask to a vacuum aspirator and stir to allow any remaining solvent to evaporate. The flask will become cool and a small amount of liquid should remain at the bottom of the flask. Allow the flask to warm to room temperature and weigh the flask.

Characterization for Part B

Potassium Permanganate Test: To one test tube, add 3 drops of alkene. To a second test tube, add 3 drops of product from the hydrogenation. To each tube add 2–3 drops of 1% KMnO$_4$ solution. Swirl the tubes and record the results.

Infrared Spectroscopy: Transfer a drop of the liquid product to a salt plate. Obtain the IR spectrum and compare it with the IR spectrum of the starting alkene. Absorption present in the alkene spectrum at 3100 cm^{-1} (C—H stretching) and 1640 cm^{-1} (C=C stretching) should be used to compare absorptions in the same regions for the product.

Refractive Index: Measure the refractive index of the product. The refractive index of propylbenzene is 1.4910. The refractive index of allylcyclohexane is 1.4500.

NMR Spectroscopy: Dissolve a sample of the product in CDCl$_3$. Filter into a clean, dry NMR tube. Obtain the NMR spectrum and interpret it, noting the chemical shift positions and the integration values.

Results and Conclusions for Part B

1. From the IR, determine the identity of the hydrogenation product of allylbenzene.
2. Determine the percent yield of product.
3. After determining the product based upon IR evidence, verify your prediction by measuring the refractive index (n^{20}) of the product and compare it to the reported value.
4. Was the product obtained the one expected, based upon thermodynamic considerations? Explain.
5. *Molecular modeling assignment:* Construct a model of allylbenzene. Minimize the geometry and calculate the heat of formation of this compound. Then construct models of propyl benzene and allylcyclohexane and repeat the calculations. Calculate the difference in energies for the formation of the two products. Use the calculations to support or refute the formation of the observed product. Explain.

Part C: Microscale Partial Hydrogenation of Olive Oil

To a 5-mL conical vial containing a spin vane, add 600 mg of olive oil, 12 mmol of cyclohexene, and 40–50 mg of 10% Pd on charcoal. Fit the vial with a water condenser. Reflux the mixture on a sand bath or heat block for 50 minutes. Cool the mixture to room temperature. Prepare a Pasteur filter pipet, and add a small portion of Celite. Transfer a portion of the mixture to the filter pipet. Place a plastic bulb on the filter pipet and force the filtered solution out of the pipet and into a clean, tared suction flask containing a boiling chip. The black catalyst should remain in the pipet. Repeat until all of the mixture has been filtered. Add 1.0 mL of petroleum ether to the pipet in portions, again using the pipet bulb to push the solvent through the Celite and out into a tared suction flask containing a stir bar. Reduce the volume of the solvent by warming briefly on a heat block or sand bath. Connect the suction flask to a vacuum aspirator and allow any remaining solvent to evaporate. The flask will become cool and a milky solid should appear at the bottom of the flask. Allow the flask to warm to room temperature and weigh the flask. The white solid should remain in the form of a buttery mass.

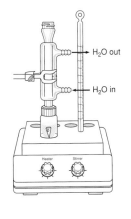

Characterization for Part C

Bromine in Methylene Chloride Test: Use this test to determine how much unsaturation is present in the unknown sample. Add 10 drops of a 1.0 M bromine in methylene chloride solution to a 10-cm test tube. Using a pipet, add dropwise a sample of warmed liquid olive oil until the red color of the bromine disappears. Swirl the tube after each addition. Note the amount of olive oil required. Repeat using another 10 drops of 1.0 M bromine solution for a sample of the warmed liquid product. Record the results.

Hood!

    Infrared Spectroscopy: Transfer a drop of the liquid olive oil or product to a salt plate. Obtain the IR spectra of both olive oil and product and compare. Absorption in the spectrum of the product at 3020 cm^{-1} (sp^2 C−H stretching) and 1620 cm^{-1} (C=C stretching) should be absent if the olive oil has been completely hydrogenated. Also, because *cis*-alkenes absorb strongly in the 650–720 cm^{-1} region, this absorption is observed for olive oil, but should be absent in the product.

Results and Conclusions

1. Determine the percent yield of product.
2. Is there any evidence of any starting material present in the product? Explain.
3. Could any product be lost during the workup procedure? Explain.
4. Was the oil partially or fully hydrogenated? Explain.

Cleanup & Disposal

Place the filter pipet containing residual catalyst in a container in the hood labeled "Filter pipets containing Pd catalyst."

Critical Thinking Questions

1. Suggest a reason that oils are intentionally partially hydrogenated commercially rather than being totally hydrogenated.
2. Predict the product of catalytic hydrogenation of 1,2-dimethylcyclohexene using Pd/C in a hydrogen atmosphere.
3. Predict the product of catalytic transfer hydrogenation of methyl oleate with deuterated cyclohexene (C_6D_{10}).

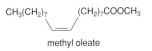

$CH_3(CH_2)_7$ $(CH_2)_7COOCH_3$

methyl oleate

4. What are the similarities between heterogeneous catalytic hydrogenation and cat-alytic transfer hydrogenation?

5. Suggest a reason why cyclohexene rather than cyclohexadiene is used as the hydrogen donor.

References

Brieger, G., and Nestrick, T. J. "Catalytic Transfer Hydrogenation." *Chem. Rev.* 74 (1974): 567.

Bunjes, A., Eilks, I., Pahlke, M., and Ralle, B. J. "On the Disproportionation of Cyclohexene and Related Compounds." *Chem. Educ.* 74 (1997): 1323.

De, S., Gambhir, G., and Krishnamurthy, H. G. "A Simple and Safe Catalytic Hydrogenation of 4-Vinylbenzoic Acid." *J. Chem. Educ.* 71 (1994): 992.

Hansen, R. W. "Catalytic Transfer Hydrogenation Reactions for Undergraduate Practical Programs." *J. Chem. Educ.* 74 (1997): 430.

Johnstone, R. A. W., Wilby, A. H., and Entwistle, I. D. "Heterogeneous Catalytic Transfer Hydrogenation and Its Relation to Other Methods for Reduction of Organic Compounds" *Chem. Rev.* 85 (1985): 129.

Landgrebe, J. A. "Selective and Quantitative Catalytic Hydrogenation: Safe, Inexpensive Experiment for Large Classes." *J. Chem. Educ.* 72 (1995): A220.

Experiment 6.2: Hydration of Alkenes

The purpose of this experiment is to investigate the mercuric ion or acid-catalyzed hydration of alkenes and the regioselectivity and stereoselectivity of addition. You will:

♦   analyze gas chromatographic results to determine the nature of the product(s) and the regiochemistry of the mercuric ion-catalyzed reaction.

♦   evaluate stereochemical results based upon melting points and determine the stereoselectivity of an acid-catalyzed hydration reaction.

Techniques

Technique E	Refractive index
Technique F	Recrystallization and sublimation
Technique I	Extraction
Technique J	Gas-liquid chromatography
Technique M	Infrared spectroscopy (optional)

Background and Mechanism

Alkenes react with aqueous sulfuric acid to give alcohols by electrophilic addition. If at least one of the carbons of the carbon-carbon double bond is disubstituted, addition occurs quickly and rearrangements are not observed.

In Part A of this experiment, the regiochemistry of the mercuric ion-catalyzed addition of water to a double bond will be investigated by reacting 2-ethyl-1-butene with mercuric acetate in aqueous solution (oxymercuration) and reduction of the organomercury intermediate with sodium borohydride (demercuration). The boiling points of the two possible alcohols are sufficiently distinct to permit a definitive analysis by gas chromatography.

Therefore, the nature of the product(s) and orientation of the addition will be determined using gas chromatography. An alternate method of analysis is to measure the refractive index of the product.

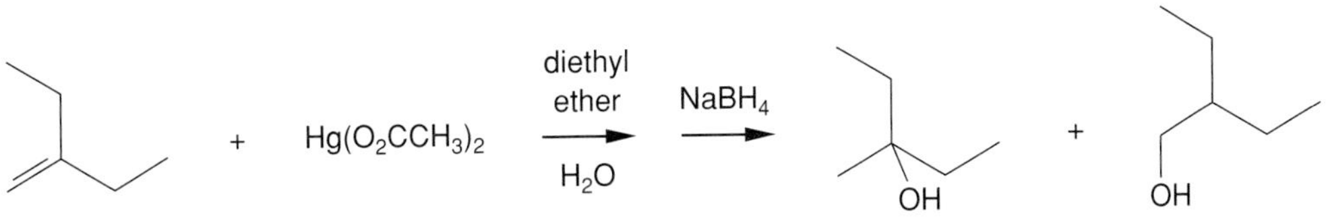

2-ethyl-1-butene

diethyl ether   NaBH$_4$

Hg(O$_2$CCH$_3$)$_2$ H$_2$O

3-methyl-3-pentanol
bp 123°C

2-ethyl-1-butanol
bp 146°C

In Parts B and C, dilute acid treatment of the bicyclic alkene, norbornene, will be investigated to determine which of two stereoisomeric products is formed. In this case, the regiochemistry of addition is the same for both products, but the stereoselectivity of addition could result in either product depending upon the mechanism of addition. Hydration could produce either *exo*-norborneol or *endo*-norborneol or both products.

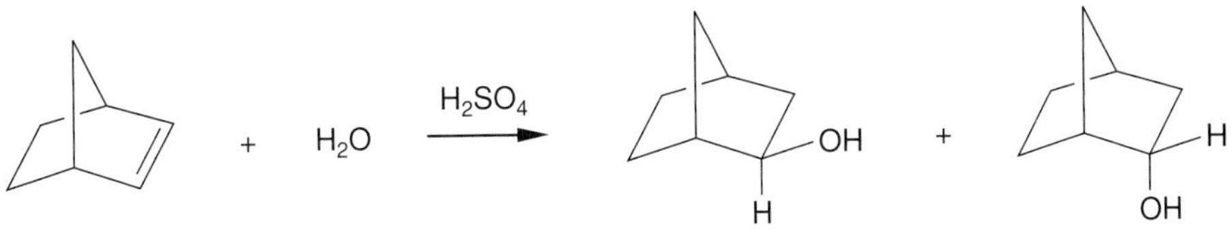

+   H$_2$O   H$_2$SO$_4$

exo-norborneol
mp 124–126°C

endo-norborneol
mp 149–151°C

In this instance, the melting behavior of the two alcohols is distinctive. Isolation of the product and determination of the melting point permit a definitive analysis of the product to determine if the hydroxyl group is on the "top side" of the molecule (*exo*-norborneol) or the "bottom side" (*endo*-norborneol).

Prelab Assignment

1. Determine the limiting reagent in your experiment.
2. Write a flow scheme for the experiment and workup procedure.

Experimental Procedure

Safety First!

Always wear eye protection in the laboratory.

1. Wear eye protection at all times in the laboratory.
2. Sulfuric acid is a corrosive oxidizer. Mercuric acetate is a highly toxic and corrosive solid. Sodium borohydride is a corrosive solid. Wear gloves to avoid skin contact with these compounds and their solutions.
3. Work in a hood or in a well-ventilated area. Do not breathe alkene vapors. Low molecular weight alkenes have disagreeable odors. Measure out the alkenes in the hood and cap the vials in order to prevent alkene vapors from getting into the air.
4. Methylene chloride is a toxic irritant and a suspected carcinogen. Diethyl ether is a toxic liquid. Avoid breathing vapors of these solvents.

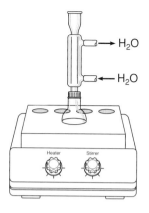

See Cleanup and
Disposal for handling
waste.

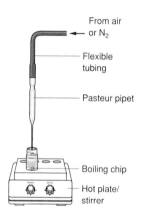

Part A: Microscale Hydration of 2-Ethyl-1-butene

Add 2.5 mL of water to a 10-mL round-bottom flask containing a stir bar. Place the flask in a heat block or sand bath at room temperature and carefully add 0.64 g of mercuric acetate with stirring. Stir the mixture until the solid has dissolved. Add 1.0 mL of diethyl ether. Attach a reflux condenser to the flask and adjust the flow of cold water through the condenser. Using a calibrated pipet, measure out 290 μL of 2-ethyl-1-butene in a conical vial and add 1.0 mL of diethyl ether. Use a Pasteur pipet to transfer this solution in small portions to the stirred aqueous solution down through the condenser. Stir vigorously for 45 minutes. Remove the condenser. Add 1.0 mL of 6 M NaOH to the mixture using a Pasteur pipet, followed by a solution of 0.50 g of sodium borohydride in 2.0 mL of 3 M NaOH. Stir the mixture vigorously for 15 minutes. A mercury precipitate should form at the bottom of the flask. Remove the apparatus from the heat block or sand bath and remove the condenser. Allow the precipitate to settle.

Use a Pasteur filter pipet to transfer the supernatant liquid (both layers) to a centrifuge tube. Set aside the mercury and mercuric salts for proper disposal. Transfer the lower aqueous layer to a second centrifuge tube. Place the diethyl ether layer in a 5-mL conical vial. Add 1 mL of diethyl ether to the aqueous solution in the centrifuge tube. Cap and swirl gently. Carefully remove most of the ether layer and add it to the ether solution in the conical vial. Repeat the extraction with a second portion of 1 mL of diethyl ether. Add the ether layer to the ether solution in the conical vial and add a small portion of anhydrous sodium sulfate. After a few minutes use a Pasteur filter pipet to transfer the solution to a tared 5-mL conical vial containing a boiling chip. If doing IR characterization (below), put a few drops on a salt plate and run the IR spectrum now. Evaporate the diethyl ether using a heat block, a sand bath, or water bath (40–45°C). Weigh the vial to determine the weight of the product. Keep the vial capped during analysis because the product is volatile.

Characterization for Part A

Gas-liquid Chromatography: Using a 10-μL syringe, remove a 1-μL aliquot and inject it onto a moderately polar or polar GC column. One or more peaks may be recorded on the chromatogram including a peak for any remaining solvent and peaks for any remaining starting material and product(s). Authentic samples of potential products (if available) and starting material should be used as standards.

Refractive Index: Use a few drops of product to determine the refractive index and apply the appropriate temperature correction. 3-Methyl-3-pentanol has an RI (n_D) of 1.4190^{20}, and 2-ethyl-1-butanol has an RI (n_D) of 1.4220^{20}.

Infrared Spectroscopy (optional): Transfer 2–3 drops of the diethyl ether solution containing the product(s) to a salt plate, and press a second plate against the first to spread out the liquid. Remove the plates from one another briefly and press together a second time. The liquid remaining should now be free or nearly free of diethyl ether. Obtain the IR spectrum. Compare the spectrum to authentic 3-methyl-3-propanol (a tertiary alcohol) and 2-ethyl-1-butanol (a primary alcohol). 3-Methyl-3-pentanol shows strong $C-O$ absorption at 1380, 970, and 900 cm^{-1}; these bands are not present in 2-ethyl-1-butanol.

Results and Conclusions for Part A

1. Identify each of the peaks by comparison of GC retention times of observed peaks with standards.
2. Measure the areas of the peaks in the chromatogram to calculate the mass percent. The correction factors for the alcohol and the starting alkene are approximately equal.
3. Determine the extent of conversion of 2-ethyl-1-butene to alcohol.
4. Write a detailed mechanism for the formation of the observed product(s).
5. Using your mechanism from above, explain the predominance of the observed product.

Part B: Microscale Hydration of Norbornene

Pipet 0.25 mL of water into a 5-mL vial equipped with a spin vane. Cool the vial in an ice bath. Carefully add 0.5 mL of concentrated H_2SO_4 dropwise using a Pasteur pipet. Remove the ice bath. Transfer 150 mg of norbornene to the vial and stir for 10 minutes or until the solid has dissolved.

Transfer the solution to a centrifuge tube. Rinse the vial with 1 mL of water and swirl briefly. Transfer the liquid in the vial to the centrifuge tube. Cool the tube in an ice bath and carefully add 3.5 mL of 6 M NaOH dropwise to the mixture using a Pasteur pipet. Swirl the mixture and check the pH using moist blue litmus paper. The paper should not turn red. If it turns red, add additional base in 0.1-mL increments and check the pH using moist blue litmus paper after each increment has been added. Remove the tube from the ice bath. Make certain that the solution is at room temperature before proceeding.

Add 1.5 mL of methylene chloride and cap the vial. Swirl gently and vent. Repeat. Shake the vial and vent. Remove the cap and transfer the organic layer to a 5-mL vial. Add another 1.5 mL of methylene chloride to the aqueous solution and repeat the extraction. Separate the layers and combine the second methylene chloride extract with the first. Wash the organic layer with 1 mL of water by capping the vial and shaking gently. Transfer the organic layer to an Erlenmeyer flask. Add a small amount of anhydrous sodium sulfate to the combined methylene chloride extracts and swirl. Let stand for 5–10 minutes. Using a Pasteur filter pipet, transfer the solution to a clean, tared 5-mL conical vial containing a boiling chip. Evaporate the solvent using a warm sand bath or heat block. Allow the flask to stand at room temperature for at least 10 minutes and then weigh the flask to determine the yield of product. Purify the product by sublimation or recrystallize from an ethanol/water solvent pair.

Hood!

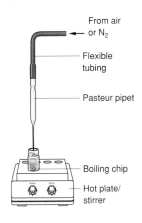

Part C: Miniscale Hydration of Norbornene

Pipet 0.75 mL of water into a 25-mL Erlenmeyer flask equipped with a spin bar. Cool the flask in an ice bath. Carefully add 1.5 mL of concentrated H_2SO_4 dropwise using a calibrated pipet. Allow to stir briefly and then remove the ice bath. Transfer 450 mg of norbornene to the flask and stir for 15 minutes or until the solid has dissolved.

Cool the flask in an ice bath and carefully add 11 mL of 6 M NaOH dropwise with stirring to the mixture using a calibrated pipet. Check the pH using moist blue litmus paper. The paper should not turn red. If it turns red, add additional base in 0.2-mL increments and check the pH using moist blue litmus paper after each increment has been added. Make certain that the solution is at room temperature before proceeding. Remove the stir bar from the flask.

Transfer the solution to a separatory funnel. Rinse the flask with 5 mL of water and swirl. Transfer to the separatory funnel. Add 5 mL of methylene chloride to the flask and swirl. Transfer to the separatory funnel. Swirl and then stopper the funnel. Shake carefully with frequent venting. Remove the stopper and draw off the methylene chloride layer into a Erlenmeyer flask. Add another 2 mL of methylene chloride to the aqueous solution remaining in the separatory funnel and repeat the extraction. Separate the layers and combine the second methylene chloride extract with the first. Transfer the aqueous solution in the separatory funnel to a flask for disposal. Add anhydrous sodium sulfate to the combined methylene chloride extracts and swirl. Let stand for 5–10 minutes. Transfer the methylene chloride solution to a clean, tared suction flask containing a boiling chip. Evaporate the solvent using a warm sand bath or water bath. Allow the flask to stand at room temperature for at least 10 minutes and then weigh the flask to determine the yield of product. Purify the product by sublimation or recrystallization from an ethanol/water solvent pair.

Hood!

Characterization for Parts B and C

Melting Point: Determine the melting point of the product. The reported melting point of *exo*-norborneol is 124–126°C and the reported melting point of *endo*-norborneol is 149–151°C.

Infrared Spectroscopy (optional): Obtain the IR spectrum of the product as a Nujol mull or KBr pellet. Correlate absorptions with those expected for *exo*-norborneol and *endo*-norborneol. The carbon-oxygen stretching vibration for the exo isomer is a strong absorption at 1000 cm^{-1}, whereas the same absorption for the endo isomer occurs at 1030 cm^{-1}.

Results and Conclusions for Parts B and C

1. Determine the percent yield of product.
2. What was the stereochemical outcome of the reaction? What evidence permitted a definitive assignment of stereochemistry?
3. Assemble a molecular model of norbornene and use it to explain why the observed product was obtained rather than the isomer.

Cleanup & Disposal

Place solutions and solid residue that may contain mercury or mercuric ion in a special container in the hood. Neutralized aqueous solutions can be washed down the drain with water. Sodium sulfate may be dissolved in water and washed down the drain.

Critical Thinking Questions *(The harder ones are marked with a ❖.)*

1. Predict the major product of hydration when 3-methyl-1-hexene is treated with
   a. aqueous mercuric acetate, followed by sodium borohydride in aqueous base.
   b. dilute sulfuric acid.
2. Predict the major product of hydration when 2-methyl-1-hexene is treated with aqueous mercuric acetate, followed by sodium borohydride in aqueous base.
3. The norbornyl carbocation does not rearrange to a tertiary carbocation even though there is a tertiary carbon adjacent to the secondary carbon bearing the positive charge. Suggest a possible explanation.
4. Oxymercuration-demercuration is often used as a method to give Markovnikov addition of water and alcohols to alkenes. What is a major advantage of the oxymercuration-demercuration method compared to aqueous sulfuric acid?
❖ 5. Norbornene is a strained alkene. It is very reactive toward hydration in aqueous acid. Explain how strain contributes to the high reactivity.

Experiment 6.3: Preparation of Alcohols from Alkenes by Hydroboration-oxidation

In this experiment, you will investigate conversion of alkenes to alcohols via an anti-Markovnikov addition mechanism. You will:

- perform a regioselective synthesis of an alcohol from an alkene.
- work with air-sensitive solutions.

Techniques

Technique G	Distillation and reflux
Technique I	Extraction and drying
Technique J	Gas-liquid chromatography
Technique M	Infrared spectroscopy (optional)
Technique N	Nuclear magnetic resonance spectroscopy (optional)

Background and Mechanism

There are three important methods for converting alkenes to alcohols. The first of these, and the oldest, is the acid-catalyzed addition of water (see Experiment 6.2). The mechanism involves converting alkenes to alcohols via carbocations, which are subject to rearrangements. The second method is oxymercuration-demercuration (see Experiment 6.2). The major product expected is the Markovnikov addition adduct; rearrangements are not observed.

acid-catalyzed addition

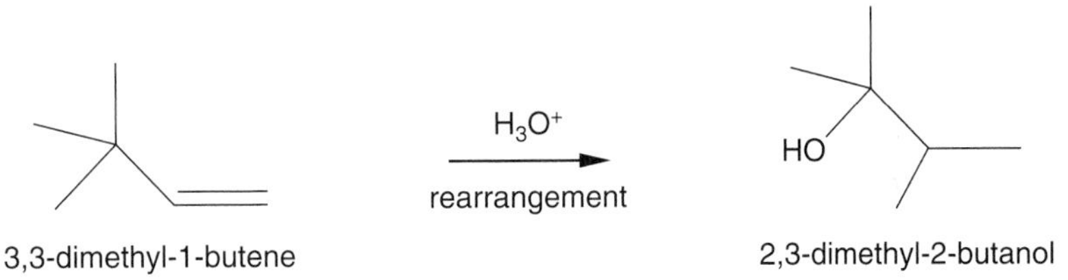

3,3-dimethyl-1-butene H_3O^+ rearrangement   HO   2,3-dimethyl-2-butanol

oxymercuration-demercuration

3,3-dimethyl-1-butene $Hg(OAc)_2$ $NaBH_4$ H_2O/THF   HO   3,3-dimethyl-2-butanol

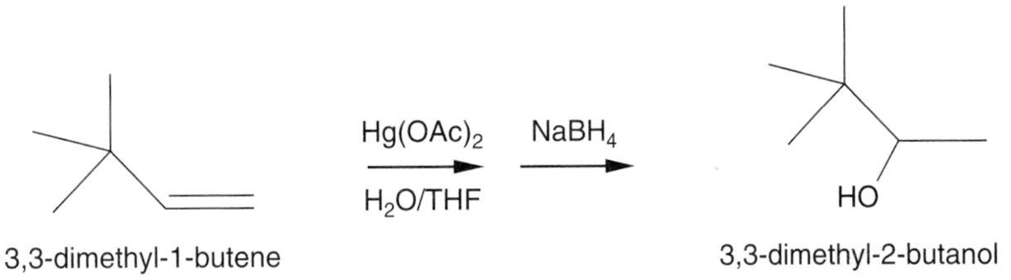

A third method is hydroboration-oxidation. The mechanism involves anti-Markovnikov addition and the stereochemistry of addition is "syn." (Syn means that both the hydrogen and hydroxyl are added to the same side of the double bond.) In this experiment, 3,3-dimethyl-1-butene will undergo hydroboration-oxidation to yield 3,3-dimethyl-1-butanol.

hydroboration-oxidation

3,3-dimethyl-1-butene $(BH_3)_2$ THF H_2O_2 OH^- OH   3,3-dimethyl-1-butanol

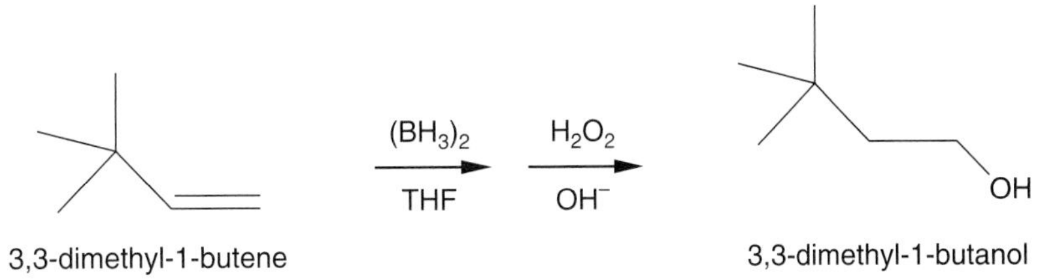

The starting material is a very volatile liquid that can be easily removed along with the solvent, leaving the product for analysis.

The mechanism of hydroboration is shown here.

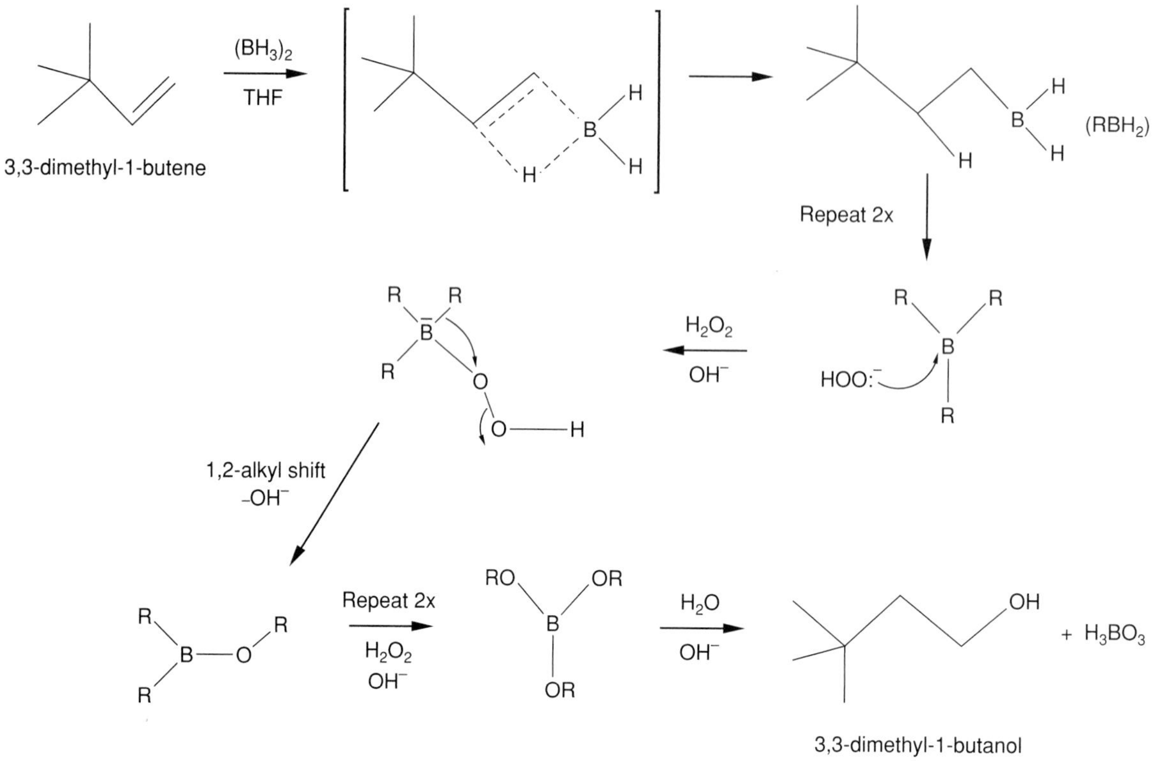

3,3-dimethyl-1-butene

(RBH$_2$)

Repeat 2x

H$_2$O$_2$ / OH$^-$

HOO:$^-$

1,2-alkyl shift
–OH$^-$

Repeat 2x
H$_2$O$_2$
OH$^-$

H$_2$O
OH$^-$

3,3-dimethyl-1-butanol

+ H$_3$BO$_3$

R = (CH$_3$)$_3$CCH$_2$CH$_2$–

Prelab Assignment

1. Explain the meaning of each dotted line in the transition state for the mechanism of hydroboration-oxidation of 3,3-dimethyl-1-butene.
2. Which part of the reaction mechanism covers hydroboration and which covers oxidation? Explain.
3. What is the limiting reagent for the reaction?
4. Prepare a flow scheme for the workup procedure of the reaction mixture.

Experimental Procedure

S a f e t y   F i r s t !

Always wear eye protection in the laboratory.

1. Wear eye protection at all times in the laboratory.
2. Wear gloves when handling strong bases. Avoid skin contact with basic solutions. If base contacts the skin, immediately wash the affected area with copious amounts of cool water.
3. Work in the hood or in a well-ventilated area during this experiment.
4. Diborane is highly toxic. Dispense it in a hood. Avoid skin contact and inhalation of vapors.
5. Hydrogen peroxide is an oxidant. Wear gloves when handling this reagent.

Part A: Microscale Hydroboration-oxidation of 3,3-Dimethyl-1-butene

Dry all glassware in an oven prior to the experiment. To a 5-mL conical vial equipped with spin vane add 150 µL of 3,3-dimethyl-1-butene and 1 mL of dry THF. Swirl to mix. Attach a Claisen adapter fitted with a calcium chloride drying tube on the side neck and a plastic cap or septum (or a disposable rubber septum) on the center neck. Add dropwise 300 µL of 1 M diborane/THF solution through the septum in the Claisen adapter using a syringe. The addition should take about 1 minute. Stir the solution for 45 minutes at room temperature.

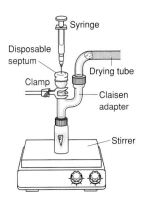

Remove the Claisen adapter from the vial and add 15 drops of 2 M NaOH. Using a syringe, add dropwise 0.75 mL of 30% hydrogen peroxide with stirring. Place a water-jacketed condenser on top of the vial. Heat the mixture for 30 minutes at reflux using a sand bath or heat block. Remove the apparatus from the heat source and remove the condenser. Cool the reaction mixture to room temperature. Reduce the volume of the solution to about 2.0 mL using a warm sand bath. Cool to room temperature. Cool the vial using ice or cold water and remove the spin vane.

Carefully add 500 µL of 3 M HCl. Add 1.5 mL of methylene chloride. Cap the vial and shake. Vent the vial by carefully loosening the cap. Draw off the lower (organic) layer and save this solution in a separate vial. Add 1.5 mL of methylene chloride to the reaction vial and repeat the extraction. Remove the methylene chloride extract and combine it with the first extract. Dry the combined methylene chloride solution over sodium sulfate and transfer the solution using a flilter pipet into a tared, conical vial containing a boiling chip. Evaporate the solution to a small volume in a hood using a warm (~40°C) sand bath, leaving the product as an oily residue. (Be careful not to also boil away the product!) Reweigh the vial containing the product.

Hood!

Part B: Miniscale Hydroboration-oxidation of 3,3-Dimethyl-1-butene

To a 50-mL round-bottom flask equipped with stir bar add 2.00 mL of 3,3-dimethyl-1-butene and 7 mL of dry THF. Swirl to mix. Attach a Claisen adapter fitted with a calcium chloride drying tube on the side neck and a septum on the center neck. Add dropwise 4.0 mL of 1 M diborane/THF solution through the septum in the Claisen adapter using a syringe. The addition should take about 5 minutes. Stir the solution for 45 minutes at room temperature.

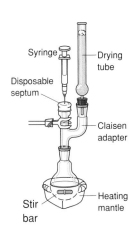

Remove the Claisen adapter from the flask and add 10.0 mL of 2 M NaOH. Using a syringe, add dropwise 10.0 mL of 30% hydrogen peroxide with stirring. Attach a water-jacketed condenser to the flask. Heat the mixture with stirring for 30 minutes at reflux using a sand bath or heating mantle. Remove the apparatus from the heat source and remove the condenser. Cool the reaction mixture to room temperature and remove the stir bar. Cool the flask using ice or cold water.

Carefully add 7.0 mL of 3 M HCl. Transfer the solution to a 125-mL separatory funnel. Add 5 mL of diethyl ether. Stopper and shake. Vent frequently to relieve pressure. Draw off the lower (aqueous) layer and save. Pour the ether solution into a 25-mL Erlenmeyer flask. Put the aqueous layer back into the separatory funnel and repeat the extraction using a second 5-mL portion of diethyl ether. Drain off the aqueous extract and save for disposal. Combine the second diethyl ether extract with the first extract and dry using anhydrous sodium sulfate. Transfer the solution to a small suction flask and add a boiling chip. Connect the suction flask to an aspirator with vacuum tubing and draw a vacuum on the flask for a few minutes or until the diethyl ether has evaporated. Swirl the flask occasionally as necessary to avoid frothing. Transfer the remaining liquid to a conical vial containing a boiling chip and set up for distillation using a Hickman still. Collect the distilled product (bp 143°C) from the lip of the Hickman still at intervals and transfer to a tared, capped vial. Reweigh the vial containing the product.

Characterization for Parts A and B

Boiling Point: Determine the micro boiling point and correct for atmospheric pressure. The reported boiling point of 3,3-dimethyl-1-butanol is 143°C.

Gas-liquid Chromatography: Dissolve a drop of product in 5 drops of reagent-grade ether and draw up 1–5 µL of solution in a syringe. Inject the sample through the injection port and record the chromatogram. If the product consists of more than one component, identify each using standard alcohol samples. Determine the relative amounts of each component.

Infrared Spectroscopy (optional): Run an IR spectrum of the product. Look for the absence of C=C stretching absorption near 1650 cm^{-1} and the presence of a broad OH absorption centered at 3400–3300 cm^{-1} and C$-$O absorption at 1100–1050 cm^{-1}.

NMR Spectroscopy (optional): If there is product left after determining the boiling point and obtaining an IR spectrum, add CDCl$_3$ and transfer the solution into a clean, dry NMR tube using a filter pipet. Record and interpret the integrated NMR spectrum.

Results and Conclusions (for Parts A and B)

1. Calculate the percent yield of 3,3-dimethyl-1-butanol.
2. Did the hydroboration reaction go to completion? How might the yield be improved?
3. How can you be sure that the main product is not the Markovnikov addition product?

Cleanup & Disposal

Place the aqueous washings in the sink and wash them down the drain. Be sure that there are no residues of diborane solution or peroxide solution left on the lab bench. Place any recovered methylene chloride solutions in the container labeled "halogenated solvent waste" in the hood.

Critical Thinking Questions *(The harder ones are marked with a ❖.)*

1. Predict the major product for the reaction of 3-methyl-1-butene with
   a. H$^+$, H$_2$O.
   b. Hg(OAc)$_2$, H$_2$O; NaBH$_4$.
2. If the GC analysis gives overlapping peaks, what can be done to improve resolution?
3. Write the structure of the product expected from the hydroboration-oxidation of 2-deuteropropene.
❖ 4. What type of chromatography would be useful for qualitatively analyzing the product if 2-phenyl-1-pentene were the starting material in this experiment? Explain.
5. Why must water be excluded from the hydroboration reaction?
❖ 6. When borane is allowed to react with two moles of an optically active alkene, the resulting dialkylborane can react with another alkene, such as *cis*-2-butene, followed by oxidation to give optically active 2-butanol. This is called enantioselective oxidation. Report on this reaction by consulting the journal article by H. C. Brown and coworkers, "Hydroboration. XVIII . . . A Convenient Synthesis of Optically Active Alcohols and Olefins of High Optical Purity and Established Configuration" [*J. Amer. Chem. Soc.* 86 (1964): 397].

Experiment 6.4: Addition Polymers: *Preparation of Polystyrene and Polymethyl Methacrylate*

In this experiment, you will study the reactions and conditions used for the conversion of alkenyl monomers to polymers. You will:

♦   synthesize polystyrene and polymethyl methacrylate.

♦   study a reaction occurring by a free radical chain mechanism.

Techniques

Technique F	Vacuum filtration
Technique G	Reflux
Technique M	Infrared spectroscopy

Background

Polymers are very large natural or synthetic substances, which have linked sequences of repeating units. Some common natural polymers are hair, horns, and cellulose. Most natural polymers are condensation polymers, formed by condensing monomeric units by splitting out water or other small molecules. Nylon, rayon, latex rubber, and many plastics are synthetic polymers. Preparation of synthetic Nylon 66 is an example of condensation polymerization.

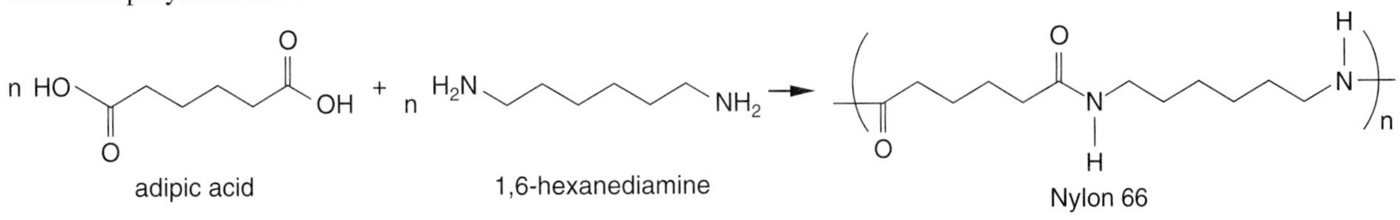

adipic acid   1,6-hexanediamine   Nylon 66

Another important type of polymerization is addition polymerization in which units are added together to form long chains. An example is preparation of polypropylene, which is used in many common consumer products, such as luggage and carpeting.

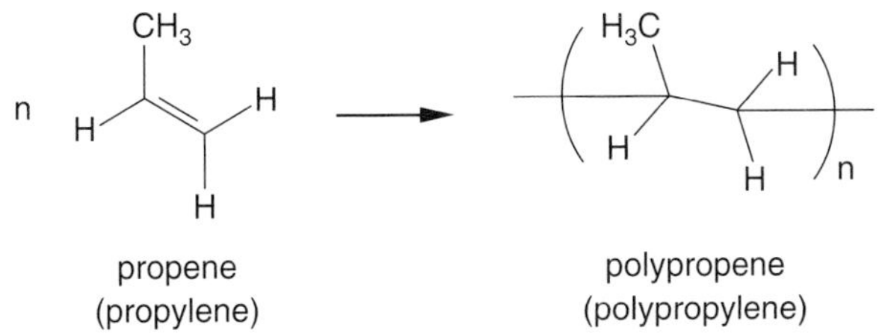

propene   polypropene
(propylene)   (polypropylene)

This experiment will focus on addition polymerization using the alkenyl monomers styrene and methyl methacrylate. Polystyrene is used in packaging and for many household items.

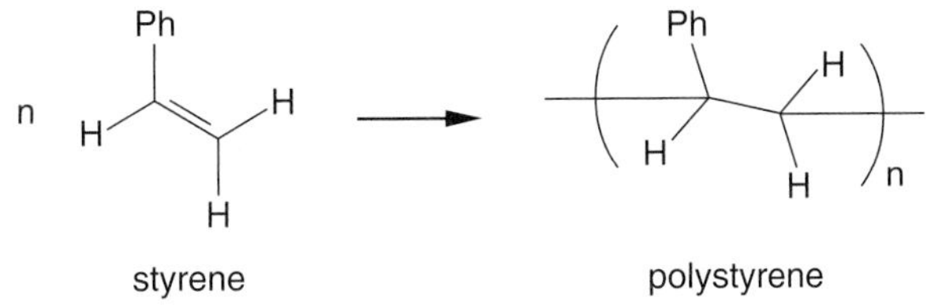

styrene   polystyrene

Polymethyl methacrylate is more commonly known as Lucite or Plexiglas.

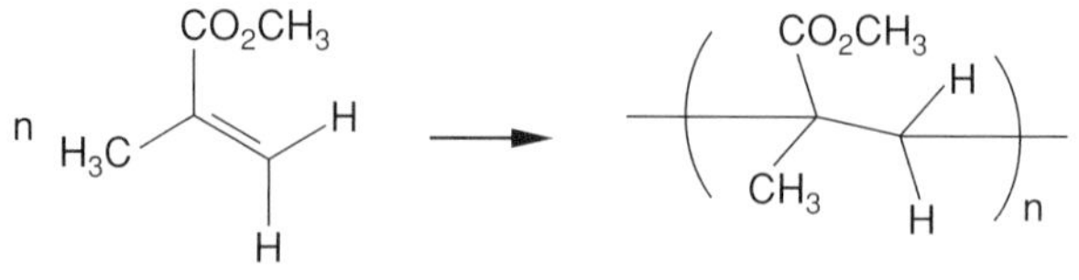

methyl methacrylate   polymethyl methacrylate

In this experiment, two different experimental techniques will be used: bulk polymerization and solution polymerization. Bulk polymerization involves mixing the monomer or monomers with an initiator. For example, a free radical initiator is added to a liquid monomer and light or heat is used to activate the initiator and start the polymerization reaction. You may suspend an object such as a coin in the reaction medium and obtain a product with the suspended object embedded in the transparent polymeric product. In the solution polymerization technique, the monomer and initiator are dissolved in a solvent. In this experiment, styrene and *tert*-butylperoxybenzoate are dissolved in xylene.

In order to prevent premature polymerization, commercial monomers normally contain free radical inhibitors, which must be removed before starting the reaction. Inhibitors are often phenolic substances. The inhibitor can often be removed by washing the monomer with base or by freshly distilling the monomer.

Mechanism of Polymerization

Free radical initiation is the first step in the polymerization reaction. The free radical initiators typically used are organic peroxides such as *tert*-butylperoxybenzoate and azo compounds such as 2,2′-azobisisobutyronitrile (AIBN). Peroxides and azo compounds readily cleave under mild heating (50–100°C) to generate radicals. These radicals, designated In·, initiate a free radical chain process. After initiation, the second step is chain propagation in which a radical adds to a monomer to give a new radical. This process continues and the polymer grows in size until a chain termination process (combination of two radicals) occurs or until the monomer is consumed. A mechanistic scheme for the free radical chain growth reaction is shown below:

Initiation

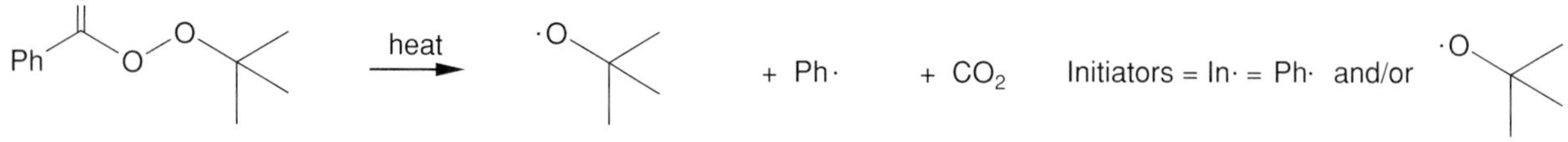

Chain propagation

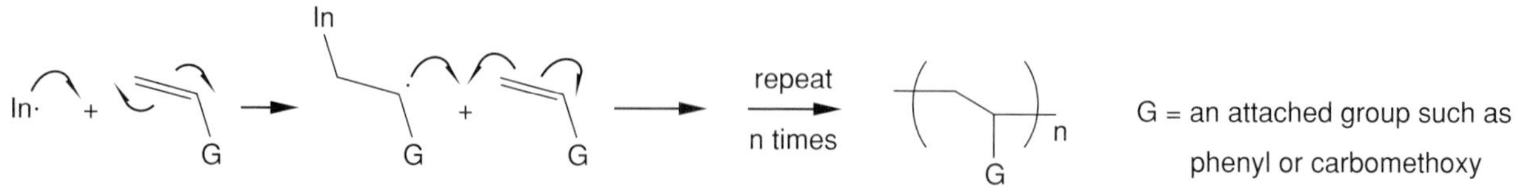

The polymer produced has a head-to-tail orientation where the head is the end of the vinyl group that has the group (G) attached (e.g., phenyl for styrene and ester for the methacrylate). This orientation gives the most stable radical (e.g., a benzyl radical for styrene). Head-to-head (or tail-to-tail) polymers are possible but much less common.

Prelab Assignment

1. Write the general formula for polystyrene. Draw a section of this polymer containing three monomeric units.

2. Write the general formula for polymethyl methacrylate. Draw a section of this polymer containing three monomeric units.

Experimental Procedure

1. Wear eye protection at all times.

2. Wear gloves when handling all of the reagents in this experiment.

3. Work under the hood or in a well-ventilated area.

4. The reagents used in this experiment are irritants. In addition, *tert*-butylperoxybenzoate is an oxidizer; styrene is flammable and toxic; and methyl methacrylate is a lachrymator. Avoid contact with these substances and avoid breathing the vapors.

Part A: Miniscale Polymerization of Styrene (Bulk Method)

Place 2.0 mL of styrene (with the commercial inhibitor removed) into a small soft-glass test tube. Add 7 drops of *tert*-butylperoxybenzoate. Swirl gently. Use a cotton plug or a loose-fitting cork to stopper the tube. Place the tube in a heat block or sand bath at 80–100°C. Heat the tube for 75 minutes. Remove it carefully from the heat and cool the tube in an ice bath. The solid mass of polystyrene should separate from the glass and be removed for examination. If it does not separate on cooling, it may be necessary to carefully break the tube to remove the product. Weigh the product and calculate the percent yield. Test the solubility of the polystyrene in acetone by placing a piece of the polystyrene in a small test tube and adding 0.50 mL of acetone. Swirl and note whether the polymer dissolves. Repeat the solubility test using methanol and then toluene. Record observations.

Part B: Microscale Polymerization of Styrene (Solution Method)

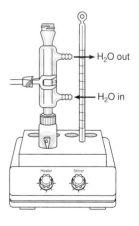

Place 1.0 mL of styrene (with the commercial inhibitor removed) into a 5-mL conical vial equipped with a spin vane and a water-jacketed condenser. Add 2 mL of xylene and 3 drops of *tert*-butylperoxybenzoate. Reflux with stirring for 20 minutes, using a warm sand bath or heat block. Cool to room temperature. Slowly add the solution to a 50-mL beaker containing 25 mL of methanol to precipitate the polystyrene. Collect the precipitate by decantation or vacuum filtration. Transfer the polymer to a clean beaker and add 15 mL of fresh methanol. Stir vigorously with a glass rod until the polymer is no longer sticky. Use additional methanol, if necessary. Vacuum filter and wash with cold methanol. Weigh the product after it is dry.

Part C: Miniscale Polymerization of Methyl Methacrylate (Bulk Method)

This reaction requires more than one lab period to polymerize. In a soft-glass test tube, place 2.0 mL of freshly distilled monomeric methyl methacrylate. Add 7 drops of *tert*-butylperoxybenzoate. Swirl gently. Stopper the tube and set it in direct sunlight until the next lab period. A sunlamp may be used if there is not sufficient sunlight. Follow the instructor's instructions regarding recovery of the solid polymer. This may require breaking the test tube under carefully controlled conditions. Weigh the product and calculate the percent yield. Test the solubility of the polymethyl methacrylate in acetone, cyclohexane, and methanol.

Characterization for Parts A, B, and C

Infrared Spectroscopy: Obtain an IR spectrum of each product isolated. Dissolve a piece of the polymer in one of the solvents in which it is soluble. Put 1–2 drops of the solution on a polyethylene film card and record the spectrum. Interpret the spectra noting the key bands characteristic of the functional groups present.

Results and Conclusions for Parts A, B, and C

1. Determine the percent yield of each polymer obtained.
2. Compare the IR spectrum obtained from the polystyrene in this experiment with that of a spectrum of polystyrene reference sample. This is usually available next to the IR instrument as a thin film for calibration of the band positions. Interpret the results.
3. Draw the structure of the intermediate radical for each polymer prepared in this experiment. Explain why head-to-tail orientation is observed in the polymerization of styrene and methyl methacrylate.
4. Write a mechanism for the formation of each polymer prepared in this experiment. Carry through the incorporation of two monomer units.

Cleanup & Disposal

Place solutions of methanol, cyclohexane, toluene, and acetone in a container labeled "nonhalogenated organic solvent waste."

Critical Thinking Questions *(The harder ones are marked with a ❖.)*

1. Draw the structure for the polymer derived from each of the following monomers: (a) vinyl chloride, (b) propene, and (c) acrylonitrile.
2. Draw the structure of the polymer derived from 1,3-butadiene as a monomer. Is there more than one possibility? Which one would be expected to predominate and why?
❖ 3. Suggest a method for the preparation of polystyrene foam.
4. Styrene can be polymerized via acid catalysis as well as by free radical initiation. Show the mechanism for polymerization under acid conditions.
5. Teflon is a polymer made from tetrafluoroethylene. Draw the structure of Teflon.
6. Copolymerization occurs when two or more different monomers are polymerized together. Propose a structure for the copolymer formed from styrene and acrylonitrile.
7. What would be the effect on polymer chain length if the amount of peroxide initiator were doubled?
❖ 8. Explain why the term "molecular weight" has a different meaning for polymers than for simple nonpolymeric molecules.

Chapter 7

Stereochemistry

S tereochemistry is encountered throughout organic chemistry and throughout this text. As an introduction to the importance of this topic, the preparation of fumaric acid by isomerization of maleic acid and addition of molecular bromine to fumaric acid are illustrated in Experiment 7.1. These reactions could potentially give rise to different stereoisomers. Being able to control the stereochemical outcome of reactions is of fundamental importance in organic chemistry.

Molecular models are very useful for visualizing organic chemicals. Models of molecules can be assembled and examined in particular conformations that can be compared with other more stable or less stable conformations. An example could be comparison of the chair, half-chair, skew-boat, and boat forms of cyclohexane. Reasons for differing stabilities can be appreciated by observing proximities of neighboring bonds, atoms, and groups of atoms. Rotational properties of single bonds in acyclic organic compounds may be examined. Properties of mirror-image forms of chiral molecules can be directly compared. Molecular models are generally helpful in perceiving stereochemical relationships within a molecule and between isomers, particularly stereoisomers. Being able to see these relationships and to represent three-dimensional compounds as two-dimensional projections is the focus of Experiment 29.5. Molecular modeling on the computer is also very useful. An exercise on computer modeling is included in Experiment 29.5.

Experiment 7.1: Stereochemistry of Alkenes and Derivatives

In this experiment, you will study isomerization about the π bond in an alkene and stereospecific addition of bromine to the alkene. You will:

- study the interconversion of geometric isomers.

- study the mechanism of cleavage of π bonds by electrophiles and regeneration of π bonds to give the more stable product.

- study the differences in physical properties of a pair of geometric isomers.

- determine the stereochemistry of addition of bromine to an alkene.

- investigate the mechanism of isomerization of dimethyl maleate to dimethyl fumarate.

Techniques

Technique F	Vacuum filtration and recrystallization
Technique G	Reflux
Technique K	Thin-layer chromatography
Technique M	Infrared spectroscopy (optional)

Background and Mechanism

Rotation about π bonds is restricted due to the requirement of parallel p-orbital overlap. In the presence of an acid, the π bond is broken. Rotation about the sigma (σ) bond occurs readily. Loss of a proton can lead to the formation of either the cis or the trans isomer. Under equilibrium conditions, the thermodynamically favored product predominates. In Part A of this experiment, maleic acid will be isomerized to fumaric acid with concentrated hydrochloric acid. The mechanism for this reaction is shown here.

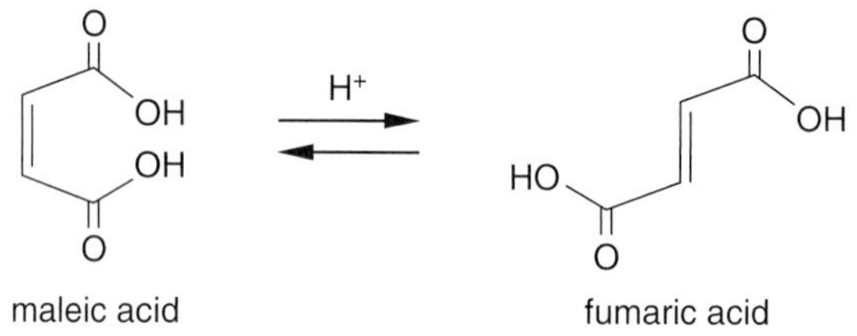

maleic acid   fumaric acid

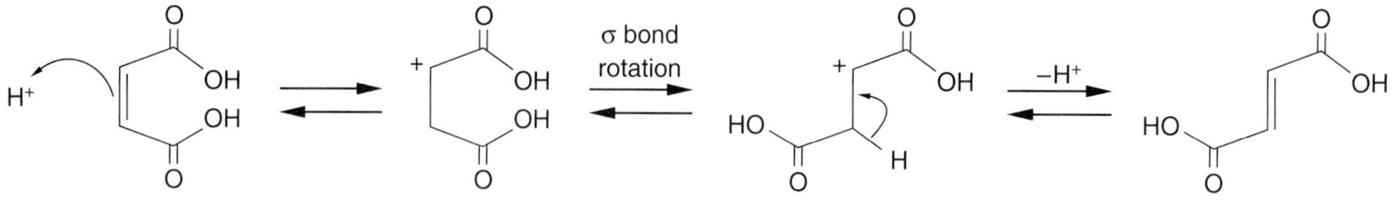

In Part B of this experiment, you will react fumaric acid with molecular bromine to form a 2,3-dibromosuccinic acid:

fumaric acid   2,3-dibromosuccinic acid

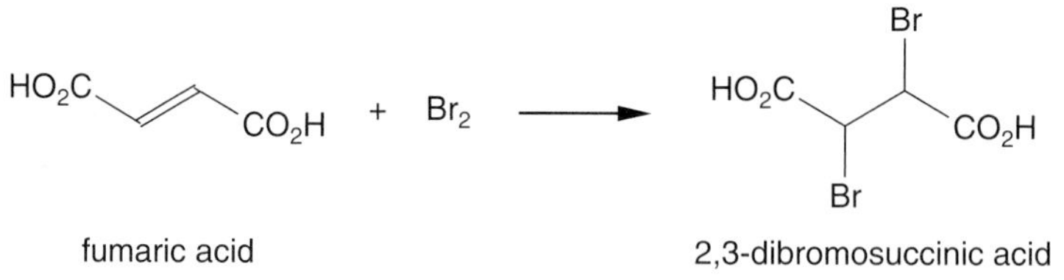

The first step in the mechanism of the electrophilic addition of bromine to an alkene is the formation of a cyclic bromonium ion:

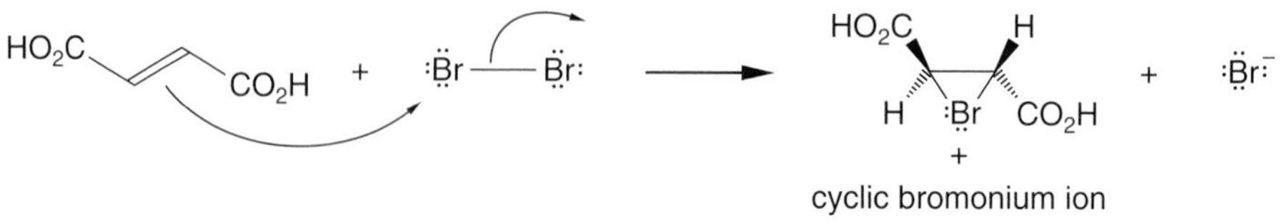

cyclic bromonium ion

Nucleophilic attack by bromide anion causes the cyclic bromonium ion ring to open. The net result is the addition of two bromine atoms to the double bond. The stereochemistry of this reaction will be examined in Part B of this experiment. You will characterize and identify the product formed in the reaction and can therefore deduce the stereochemistry of bromine addition to alkenes.

In Part C of this experiment, dimethyl maleate will be isomerized in the presence of bromine to dimethyl fumarate. You will investigate the mechanism of this isomerization by varying the reaction conditions and by monitoring the extent and outcome of the reaction.

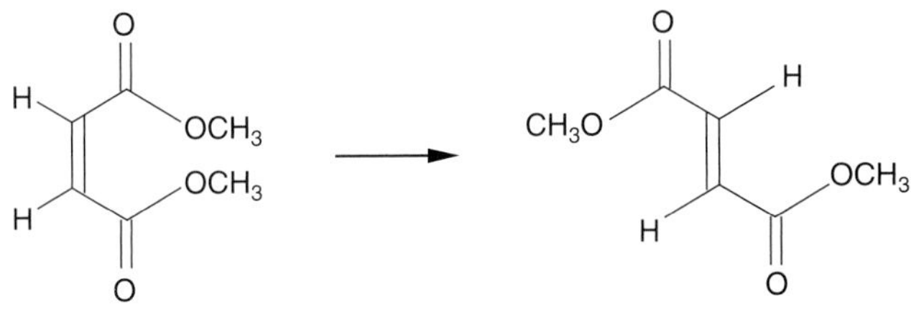

dimethyl maleate   dimethyl fumarate

The isomerization reaction will be conducted under two different sets of conditions: the dimethyl maleate will be mixed with bromine and kept either in the light or in the dark. By following the reaction with TLC, you will propose a mechanism to account for the observed product(s) and any intermediates.

Prelab Assignment

1. Predict which alkene will predominate at equilibrium. Explain.
2. Predict the stereochemistry of the products formed if the addition of bromine to fumaric acid occurs via:
   a. syn addition.
   b. anti addition.
   c. stereorandom addition.
3. The stock solution of bromine for Part B is prepared by dissolving 7.75 g of Br_2 and 6.25 g of KBr in water and diluting to 25.00 mL in a volumetric flask. Calculate the molarity of Br_2 in the stock solution. (KBr is present to facilitate the reaction.)
4. The isomerization of dimethyl maleate to dimethyl fumarate can potentially occur by a free radical mechanism, formation of a bromonium cation, or addition of bromine to the double bond via a carbocation to form a dibromide, followed by debromination. Write detailed mechanistic steps for each of these possibilities.
5. Which compound (dimethyl maleate or dimethyl fumarate) would be expected to have a higher R_f value on TLC? Explain.

Experimental Procedure

1. Wear eye protection at all times in the laboratory.
2. HCl is corrosive and toxic and can cause severe burns when in contact with skin. If this occurs, immediately rinse the affected area with cold water. Cautiously neutralize any acid spills on the lab bench with sodium bicarbonate. Notify the lab instructor immediately.
3. Bromine is a highly toxic oxidizer. Keep bromine solutions under the hood and avoid breathing the vapors. Always wear gloves when handling this reagent

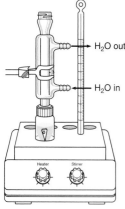

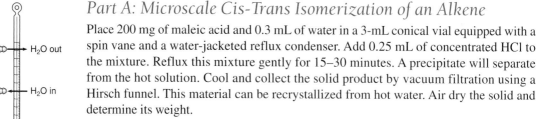

Part A: Microscale Cis-Trans Isomerization of an Alkene

Place 200 mg of maleic acid and 0.3 mL of water in a 3-mL conical vial equipped with a spin vane and a water-jacketed reflux condenser. Add 0.25 mL of concentrated HCl to the mixture. Reflux this mixture gently for 15–30 minutes. A precipitate will separate from the hot solution. Cool and collect the solid product by vacuum filtration using a Hirsch funnel. This material can be recrystallized from hot water. Air dry the solid and determine its weight.

Characterization for Part A

Melting Point: Determine the melting point of the product. Pure maleic acid melts at 130.5°C; fumaric acid has a melting point of 287°C, but sublimes at 200°C. If the solid hasn't melted by 150°C, assume it is fumaric acid. Do a mixed melting point of the solid with maleic acid.

   Thin-layer Chromatography: Dissolve 1–2 mg of the product in methylene chloride and spot on a TLC plate. Spot standards of fumaric acid and maleic acid. Develop the chromatogram in ethyl acetate/absolute ethanol (1:1).

   Infrared Spectroscopy (optional): Obtain an IR spectrum of the product as a Nujol mull or KBr pellet. Also obtain an IR spectrum of the starting material. Note differences between the spectra, particularly around 870 cm^{-1} and 1600 cm^{-1}.

Results and Conclusions for Part A

1. Was the major product of this reaction the cis or the trans isomer? Explain.
2. Determine the percent yield of each isomer. (The weight of the other isomer is the difference between the mass of the product and the mass of the starting material.)
3. Explain the significance of the mixed melting point determination.
4. Explain why formation of one geometric isomer is favored over another.

Part B: Microscale Addition of Bromine to Fumaric Acid

To a 5-mL conical vial containing a spin vane, add 100 mg of fumaric acid and 1.25 mL of a stock solution of bromine. (The stock solution was prepared by dissolving 7.75 g of Br$_2$ and 6.25 g of KBr in water and diluting to 25.00 mL in a volumetric flask.) Swirl to mix. Attach an air condenser. Heat the solution on a warm sand bath or heat block. After a few minutes, the color will turn orange. After 10 minutes a white precipitate will form and the color of the mixture will turn light yellow. If the solution becomes colorless as it is being heated, add a few drops more of the bromine solution. Continue heating for 15 minutes. Remove the solution from the heat and cool in an ice bath. Collect the product by vacuum filtration on a Hirsch funnel, washing the crystals with several small portions of ice-cold water. See Cleanup and Disposal for handling of filtrate. The product may be recrystallized from water. Air dry the crystals or dry in an oven set at 100–110°C. Weigh the solid.

Characterization for Part B

Melting Point: Determine the melting point of the product. Pure *racemic*-2,3-dibromo-succinic acid melts at 166–167°C; *meso*-2,3-dibromosuccinic acid melts at 255–256°C. If the product hasn't melted by 175°C, assume that it is the meso product.

Results and Conclusions for Part B

1. Calculate the percent yield of the product.
2. Determine the stereochemistry of the 2,3-dibromosuccinic acid.
3. Explain the stereochemistry of bromine addition.

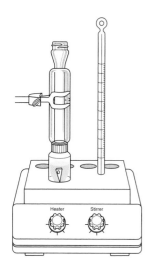

Hood!

Part C: Determining the Mechanism of the Isomerization of Dimethyl Maleate to Dimethyl Fumarate (Miniscale)

Each student will perform two reactions. (Alternately, the instructor may assign students to work in pairs and share data.) Obtain a 20 cm × 20 cm TLC plate and carefully cut it in half. Draw a line in soft pencil 1 cm from the bottom and mark places for eight spots. Spot standard solutions of dimethyl maleate, dimethyl fumarate, and 2,3-dibromosuccinate (if available). The other solutions to be spotted are the reaction mixture when initially mixed, then the reaction mixture after 2 minutes, 5 minutes, and 20 minutes. The final spot will be for the isolated product.

Obtain two 10-mL Erlenmeyer flasks. One will be used for the reaction mixture in the light and the other will be used for the dark reaction. Label the Erlenmeyer flasks accordingly. Into each of the Erlenmeyer flasks, add 4 mL of methylene chloride and 0.75 mL of dimethyl maleate. Swirl to mix. Into one of the Erlenmeyer flasks, add 10 drops (0.5 mL) of 0.5 M Br_2 in methylene chloride and swirl to mix. Immediately spot an aliquot of the reaction mixture on the TLC plate. Repeat with the other Erlenmeyer flask, spotting an aliquot of this reaction mixture on the second TLC plate.

Stopper the flasks. Place one Erlenmeyer flask in the dark and the other one in front of an incandescent light. After 2 minutes, spot each of the reaction mixtures on the appropriate TLC plate. Repeat after 5 minutes and after 20 minutes.

Add a boiling chip to any flask that indicates dimethyl fumarate has been produced and reduce the volume by half on an aluminum heat block or steam bath. Chill in an ice bath and add 6 mL of ice-cold hexanes. Cool until crystallization is complete. Suction filter the crystals, washing with small amounts of cold hexanes, and air dry. If no product crystallizes after adding the cold hexanes, set the reaction flask aside. Spot solutions of the isolated crystals (dissolved in methylene chloride) or the uncrystallized reaction solution.

Develop the plates in a 1:3 ethyl acetate: hexanes mixture in TLC chambers. Develop the plates until the solvent front is 1–2 cm from the top of the plates. Remove the plates, air dry, and view the spots under UV light. Weigh the crystals and obtain a melting point.

Characterization for Part C

Melting Point: Determine the melting point of the isolated product. Dimethyl fumarate melts at 103–104°C.

Results and Conclusions for Part C

1. Calculate the percent yield of product.
2. Evaluate each of the reaction conditions: Did the isomerization occur under both light and dark conditions? Explain the implications for the mechanism. Does this finding rule out any of the proposed mechanisms? Explain.
3. From the TLC plate, determine whether intermediates or other products were formed in this reaction. What implications does this finding hold for the proposed mechanism?
4. Propose a detailed, step-by-step mechanism for the isomerization of dimethyl maleate in the presence of bromine. Explain how each of the experimental findings supports this mechanism.
5. At the end of the 20-minute time period, was the reaction complete or did some unreacted dimethyl maleate remain? Were the crystals obtained pure or were they a mixture of dimethyl maleate and dimethyl fumarate? Explain why the addition of cold hexanes facilitated the isolation of pure dimethyl fumarate.
6. Explain why dimethyl fumarate is a solid while its geometric isomer (dimethyl maleate) is a liquid.

7. Write a balanced reaction to show how aqueous sodium bisulfite (NaHSO$_3$) reduces bromine to bromide (and itself forms sodium sulfate).
8. Explain the role of bromine in this experiment.

Cleanup & Disposal

The aqueous residue from Part A should be carefully neutralized with 10% sodium bicarbonate and poured down the drain with running water. The filtrates from Part B and Part C should be neutralized with aqueous sodium bisulfite to reduce any remaining bromine. Wash aqueous solutions down the drain with water. Place recovered chlorinated solvents in a container labeled "halogenated waste."

Critical Thinking Questions *(The harder one is marked with a ❖.)*

1. Maleic acid and fumaric acid have very different melting points. One of the geometric isomers shows intramolecular hydrogen bonding while the other shows intermolecular hydrogen bonding. Draw structures to explain this observation and explain how this affects the melting point.
2. HCl is known to give addition reactions to carbon-carbon double bonds. Why is this behavior not observed in this reaction?
❖ 3. Fumaric acid is much less soluble in water than maleic acid. Could this be a driving force for the isomerization? Explain.
4. Are the observed TLC results consistent with the predicted results based on polarities of the components? Explain.
5. Predict the structure of the product expected from addition of molecular bromine to maleic acid.
6. *Molecular modeling assignment:* Construct models of maleic and fumaric acids. Determine the net dipole for each of the compounds. How does this affect the properties of the compound? Explain.
7. *Molecular modeling assignment:* Construct models of maleic and fumaric acids and calculate the energy of the most stable conformation. Which compound is more thermodynamically stable? Explain.

Chapter 8

Introduction to Nucleophilic Substitution Reactions

One important question about nucleophilic substitution is "What are the differences between unimolecular and bimolecular substitution?" This question is addressed in Experiment 8.1, which is concerned with relative rates of unimolecular and bimolecular substitution for a variety of alkyl halides. Previously, an example of a unimolecular substitution reaction (S_N1) was observed in Experiment 4.1. Further examples of nucleophilic aliphatic substitution are given in Experiments 16.1 and 24.1.

Aliphatic nucleophilic substitution via the S_N2 pathway provides an excellent synthetic procedure for preparing many primary and some secondary alkyl derivatives. An example is the synthesis of 1-bromobutane from 1-butanol in Experiment 8.2.

Experiment 8.1: Relative Rates of Nucleophilic Substitution Reactions

In this experiment, the relative rates of substitution reactions for different alkyl halides will be examined. From the results, you will determine:

- the preferred structure of alkyl halides undergoing S_N2 reactions.

- the preferred structure of alkyl halides undergoing S_N1 reactions.

- the effect of the nature of the leaving group on the rates of S_N1 and S_N2 reactions.

- the effect of the size of the nucleophile in S_N2 reactions.

- the effect of solvent polarity on the rate of S_N1 reactions.

- the rate laws for S_N2 and S_N1 reactions.

Background

S_N2 Reactions of Alkyl Halides

The mechanism of an S_N2 reaction involves a one-step bimolecular displacement of the leaving group by a nucleophile with inversion of configuration:

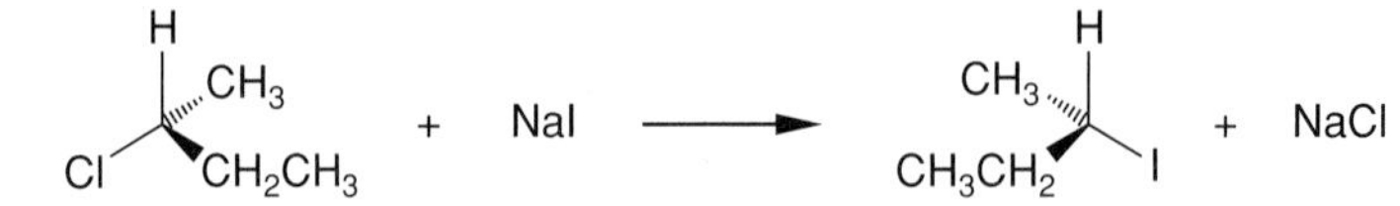

(R)-2-chlorobutane   (S)-2-iodobutane

In Part A of the experiment, you will examine factors that affect the relative rates of the S_N2 reaction of an alkyl bromide or chloride with a solution of sodium iodide in acetone. You will study how alkyl halide structure and the nature of the leaving group affect the rate of an S_N2 reaction. The effect of β-branching on the rate of the reaction will also be examined. Finally, the rate law for an S_N2 reaction will be deduced by varying the concentration of alkyl halide and nucleophile and noting the change in the rate of the reaction.

The nucleophile in each reaction will be iodide, in the form of sodium iodide, which is soluble in acetone. Sodium chloride and sodium bromide are insoluble in acetone. If a reaction occurs, the iodide ion will displace the leaving group and NaBr or NaCl will precipitate from solution. The formation of a precipitate provides visual verification that the reaction has occurred. The rate at which the solution turns cloudy will indicate the rate of reaction.

$$R{-}X + NaI \longrightarrow RI + NaX\downarrow$$

S_N1 Reactions of Alkyl Halides

In Part B of the experiment, you will examine factors that affect the relative rates of the S_N1 reaction of an alkyl halide with a solution of silver nitrate in ethanol. The mechanism of an S_N1 reaction involves initial formation of a carbocation, followed by rapid reaction with the nucleophile, ethanol:

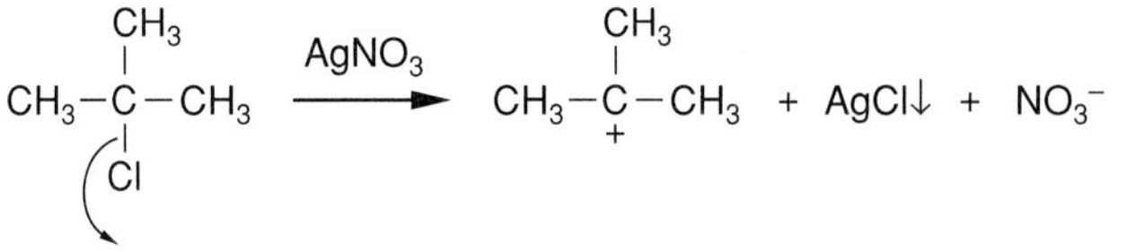

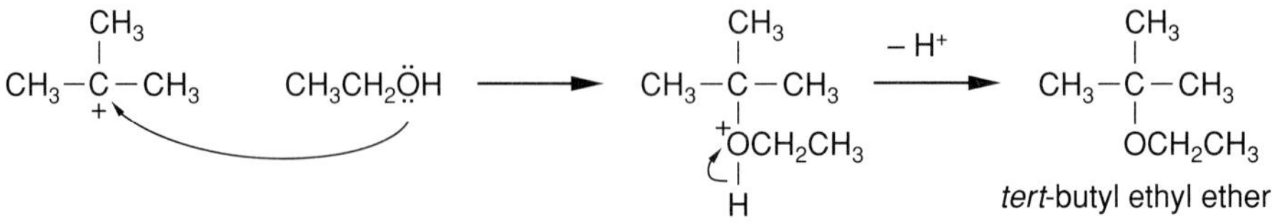

tert-butyl ethyl ether

Because nitrate is a much weaker nucleophile than ethanol, alkyl nitrates are not formed in the reaction. Insoluble silver halide is also formed in this reaction. The rate at which the solution turns cloudy will indicate the rate of reaction. The overall reaction is shown below:

$$R{-}X + CH_3CH_2OH \xrightarrow{AgNO_3} R{-}OCH_2CH_3 + HNO_3 + AgX\downarrow$$

You will study how alkyl halide structure, nature of the leaving group, and solvent polarity affect the rate of an S_N1 reaction.

Prelab Assignment

1. Draw structural formulas for 1-bromobutane, 2-bromobutane, 2-bromo-2-methylpropane, 1-bromo-2-methylpropane, and 1-bromo-2,2-dimethylpropane. Classify each compound as a primary halide, secondary halide, or tertiary halide.
2. Draw structural formulas for acetone and ethanol. Which solvent is more polar? Explain.
3. A student wishes to determine the rate law for the reaction A + B → C. The student doubles the concentration of A while holding the concentration of B constant and finds that the rate of the reaction doubles. The student then doubles the concentration of B while holding the concentration of A constant and finds that the rate is unchanged. Write the rate law for this reaction. (Refer to a general chemistry text, if necessary.)

Experimental Procedure

Safety First!

Always wear eye protection in the laboratory.

1. Wear eye protection at all times in the laboratory.
2. Alkyl halides are toxic. Some alkyl bromides are lachrymators and suspected carcinogens. Work under the hood! Wear gloves and avoid skin contact. Do not breathe alkyl halide vapors.
3. Silver nitrate will stain hands and clothing. Avoid contact with this reagent.

Part A: Determination of Factors Affecting the Relative Rates of S_N2 Reactions

All test tubes must be dry.

Effect of Structure of the Alkyl Halide

Measure 2 mL of 15% sodium iodide in acetone into each of three clean, dry 10-cm test tubes. Add 2 drops of 1-bromobutane (butyl bromide) to the first test tube; add 2 drops of 2-bromobutane (*sec*-butyl bromide) to the second test tube; and add 2 drops of 2-methyl-2-bromopropane (*tert*-butyl bromide) to the third test tube. Stopper the tubes with corks. (Do not use rubber stoppers.) Shake the tubes to mix. Observe closely during the first 15–20 minutes, then at intervals throughout the lab period. Observe the test tubes for a sign of cloudiness or precipitation. Record your observations.

Steric Effects

Measure 1 mL of 15% sodium iodide in acetone into each of three clean, dry 10-cm test tubes. Into one tube, add 2 drops of 1-bromobutane; into the second test tube, add 2 drops of 1-bromo-2-methylpropane (isobutyl bromide); into the third test tube, add 2 drops of 1-bromo-2,2-dimethylpropane (neopentyl bromide). Stopper the tubes and shake. Observe closely. Record your observations.

Effect of the Leaving Group

Measure 1 mL of 15% sodium iodide in acetone into each of two clean, dry 10-cm test tubes. Add 2 drops of 1-bromobutane into one test tube and add 2 drops of 1-chlorobutane into the other test tube. Stopper and shake the tubes. Observe closely. Record your observations.

Determination of the Rate Law

Measure 1.0 mL of a 15% solution of sodium iodide in acetone into two clean, dry test tubes. At the same time (or do one at a time, marking the time carefully!), add 0.1 mL (or 2 drops) of 1.0 M 1-bromobutane to one test tube and add 0.1 mL (or 2 drops) of 2.0 M 1-bromobutane to the other. Record your observations and compare the rates of the two reactions.

Measure 1.0 mL of a 1.0 M solution of 1-bromobutane into two clean, dry test tubes. Into one test tube, add 0.1 mL (or 2 drops) of a 7.5% solution of sodium iodide in acetone. At the same time, add 0.1 mL (or 2 drops) of a 15% solution of sodium iodide in acetone into the other test tube. (Alternately, use one test tube at a time, noting carefully the amount of time for precipitate formation.) Record your observations and compare the rates of the two reactions.

Results and Conclusion for Part A

1. Which alkyl bromide reacted fastest with sodium iodide in acetone: 1-bromobutane, 2-bromobutane, or 2-bromo-2-methylpropane? Which alkyl bromide reacted slowest? Explain how the structure of the alkyl halide affects the rate of an S_N2 reaction.
2. Which alkyl bromide reacted fastest with sodium iodide in acetone: 1-bromobutane, 1-bromo-2-methylpropane, or 1-bromo-2,2-dimethylpropane (neopentyl bromide)? All of these are primary halides. Why was there a difference in reactivity?
3. Which halide reacted faster with sodium iodide in acetone: 1-bromobutane or 1-chlorobutane? Explain how the nature of the leaving group affects the rate of an S_N2 reaction.
4. Write balanced equations for all substitution reactions that took place between the alkyl halides and NaI.
5. How did the following changes affect the rate of the reaction with 1-bromobutane with NaI: doubling the concentration of 1-bromobutane? Doubling the concentration of NaI?
6. Write the generalized rate expression for an S_N2 reaction.
7. List the factors studied in this lab that affect the rate of an S_N2 reaction.

All test tubes must be dry.

Part B: Determination of Factors Affecting the Relative Rates of S_N1 Reactions

Effect of Structure of the Alkyl Halide

Measure 2 mL of a 0.1 M solution of silver nitrate in absolute ethanol into each of three clean, **dry** 10-cm test tubes. Add 1 drop of 1-bromobutane into the first test tube, 1 drop of 2-bromobutane into the second, and 1 drop of 2-bromo-2-methylpropane to the third. Stopper and shake the tubes continually. Watch closely for evidence of cloudiness or the formation of a white precipitate. Record all observations.

Effect of the Leaving Group

Measure 2 mL of a 0.1 M solution of silver nitrate in absolute ethanol into two clean, dry test tubes. Add 1 drop of 2-bromo-2-methylpropane into one test tube. At the same time, add 1 drop of 2-chloro-2-methylpropane into the other test tube. Stopper and shake. Record your observations.

Effect of Solvent Polarity

Measure 2 mL of a 0.1 M solution of silver nitrate in absolute ethanol into a dry test tube. Into a second dry test tube, measure 2 mL of a 0.1 M silver nitrate solution in 5% ethanol/95% acetone. At the very same time (get a partner to help!) add 1 drop of 2-chloro-2-methylpropane to each of the test tubes. Stopper and shake. Observe. Record your observations.

Determination of the Rate Law

To a clean, dry test tube, add 0.5 mL of 0.1 M 2-chloro-2-methylpropane in ethanol. To another clean, dry test tube, add 0.5 mL of 0.2 M 2-chloro-2-methylpropane in ethanol. At the same time, add 1.0 mL of 0.1 M silver nitrate solution in ethanol to each of the two test tubes. (Alternately, do one at a time and carefully note the time it takes for precipitation.) Record your observations and compare the rates for the two reactions.

   To a clean, dry test tube, add 1.0 mL of 0.1 M silver nitrate in ethanol. To another clean, dry test tube, add 0.5 mL of 0.1 M silver nitrate solution in absolute ethanol and 0.5 mL absolute ethanol (to ensure that the volumes are the same in the two test tubes). At the same time, add 1.0 mL of 0.1 M 2-chloro-2-methylpropane in ethanol to each of the two test tubes. (Alternately, work with one test tube at a time and carefully note the time it takes for precipitation.) Record your observations and compare the rates for the two reactions.

Results and Conclusions for Part B

1. Which reacted fastest with silver nitrate in absolute ethanol: 1-bromobutane, 2-bromobutane, or 2-bromo-2-methylpropane? Which reacted slowest? Explain how the structure of the alkyl halide affects the rate of an S_N1 reaction.

2. Which alkyl halide reacted faster: 2-chloro-2-methylpropane or 2-bromo-2-methylpropane? Explain how and why the nature of the leaving group affects the rate of an S_N1 reaction. Would 2-iodo-2-methylpropane react faster or slower than the other alkyl halides? Explain.

3. In which of the two solvents did 2-chloro-2-methylpropane react faster: ethanol or the ethanol/acetone mixture? Explain how the polarity of the solvent affects the rate of an S_N1 reaction.

4. How did changing the concentration of 2-chloro-2-methylpropane affect the rate of the reaction of 2-chloro-2-methylpropane with ethanol and silver nitrate? Determine the rate law for an S_N1 reaction.

5. List the factors that affect the rates of S_N1 reactions.

C l e a n u p   &   D i s p o s a l

Place the contents of the test tubes from Part A (S_N2 reactions) into a container labeled "halogenated organic solvent waste." Place all solutions from Part B (S_N1 reactions) into a separate container labeled "silver waste," so that the silver can be recovered.

1-Bromoadamantane

Critical Thinking Questions *(The harder ones are marked with a ❖.)*

1. What determines whether 2-bromobutane undergoes S_N1 and/or S_N2 reactions?

❖ 2. 1-Bromoadamantane is a tertiary halide, yet it is 10,000 times slower than *tert*-butyl bromide when reacting with silver nitrate in ethanol. Consider the structure of 1-bromoadamantane and explain this observation. (Hint: Consider the geometry of a carbocation. It might be helpful to build a model of 1-bromoadamantane.)

3. On the same graph, draw reaction coordinate diagrams for the nucleophilic substitution reactions of sodium iodide with:

    a.   1-bromobutane.

    b.   2-bromobutane.

4. On the same graph, draw reaction coordinate diagrams for the nucleophilic substitution reactions of ethanol with:

    a.   2-bromobutane.

    b.   2-bromo-2-methylpropane.

❖ 5. Benzyl bromide reacts rapidly with sodium iodide in acetone to yield a white precipitate. Benzyl bromide also reacts rapidly with ethanol and silver nitrate. Bromobenzene does not react with sodium iodide or with ethanol. Explain these observations.

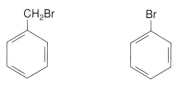

benzyl bromide   bromobenzene

❖ 6. Allyl bromide is a primary alkyl halide, yet it undergoes rapid reaction with silver nitrate in ethanol. Explain. The structure of allyl bromide is shown here.

$$H_2C=CHCH_2Br$$

7. Are alkyl fluorides good substrates in S_N1 or S_N2 reactions? Explain.

8. What are the characteristics of a good leaving group?

9. Why doesn't the nitrate ion react to give S_N1 substitution products?

❖ 10. A student measures the dry weight of AgBr produced in the S_N1 reaction of 2-bromo-2-methylpropane and calculates the moles of AgBr formed. The student proposes that the amount (mol) of *tert*-butyl ethyl ether formed via S_N1 reaction will be the same as the amount (mol) of AgBr formed. Critique this conclusion.

11. *Molecular modeling assignment:* SCN^- is an ambient nucleophile, capable of bonding at either the sulfur or the nitrogen end of the molecule. Construct a model of this molecule and examine the electrostatic potential map of this anion. Which atom is the most electron-rich? What product would be obtained from the reaction of ethyl iodide with this nucleophile?

12. *Molecular modeling assignment:* 1-bromo-3-chloropropane can potentially react with a nucleophile at two different sites. Construct a model of this molecule and examine the electrostatic potential map of this electrophile. Which site has the greater positive charge? Now examine the lowest occupied molecular orbital (LUMO) of the electrophile. Which site is most likely for nucleophilic attack? What product would most likely be obtained from the reaction of 1-bromo-3-chloropropane with sodium azide (NaN_3)?

Experiment 8.2: Nucleophilic Aliphatic Substitution: *Synthesis of 1-Bromobutane*

In this experiment, you will study nucleophilic substitution of an aliphatic alcohol. You will:

◆   carry out a bimolecular nucleophilic aliphatic substitution reaction (S_N2).

◆   purify the product by simple distillation.

◆   analyze the purity of the product by gas-liquid chromatography (GC).

Techniques

Technique E	Refractive index
Technique G	Distillation and reflux
Technique H	Steam distillation
Technique I	Extraction and drying
Technique J	Gas-liquid chromatography
Technique M	Infrared spectroscopy (optional)
Technique N	Nuclear magnetic resonance spectroscopy (optional)

Background and Mechanism

Primary alcohols react with hydrogen bromide to form alkyl bromides. Nucleophilic aliphatic substitution reactions of unbranched, primary alcohols occur by attack of a nucleophile on the protonated alcohol. The direct displacement process is an example of an S_N2 reaction. Vigorous heating (reflux) must be used to cause the reaction to reach completion. In this experiment, 1-butanol will be converted to 1-bromobutane by reaction with sulfuric acid and sodium bromide.

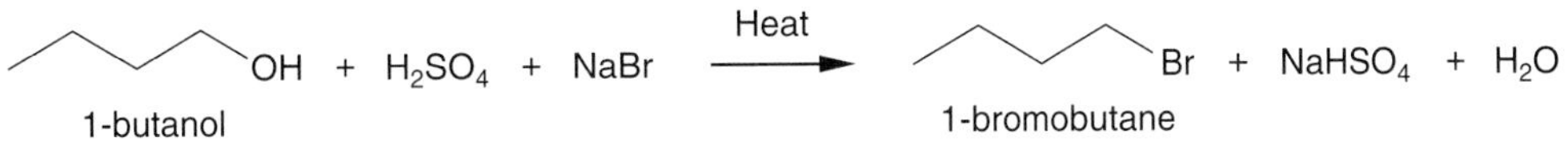

The effective reagent for the reaction is gaseous hydrogen bromide, generated *in situ* by the reaction of sulfuric acid with sodium bromide:

$$H_2SO_4 + NaBr \longrightarrow HBr + NaHSO_4$$

Protonation of 1-butanol by hydrogen bromide, followed by nucleophilic attack by bromide ion on the protonated alcohol, gives 1-bromobutane. (Protonation of 1-butanol converts the OH group to a satisfactory leaving group.)

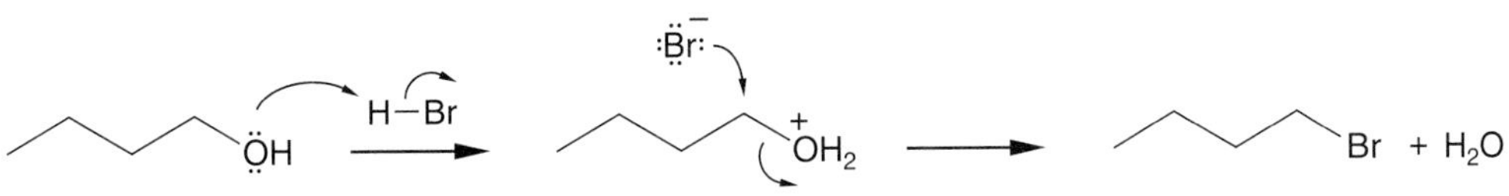

Purification of 1-bromobutane by distillation effectively removes it from most of the unreacted 1-butanol and impurities that are formed as side products, chiefly dibutyl ether. GC analysis is an excellent method to evaluate purity of the product. Testing for traces of 1-butanol in the product may be done by evaluating the infrared spectrum of the distillate.

Prelab Assignment

1. Calculate the number of mmoles of 1-butanol, sulfuric acid, and sodium bromide used in the procedure.
2. Write a mechanism to explain the formation of dibutyl ether as a by-product.
3. What is the role of water in the reaction mixture?
4. In this experiment, it is important to determine which layer is aqueous following extraction. Describe how to do this.
5. Write a flow scheme for the workup procedure in this experiment.

Experimental Procedure

1. Wear eye protection at all times in the laboratory.
2. Wear gloves while handling reagents for this experiment.
3. Work in a hood or a well-ventilated area for this experiment.
4. Rigorously avoid contact with sulfuric acid. It is corrosive and can cause severe burns. Keep reagents under the hood.
5. Any spills should be cleaned up immediately. Notify the instructor immediately about any contact with reagents or any spills.

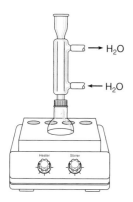

Part A: Microscale Synthesis of 1-Bromobutane

Add 1.75 g of sodium bromide and 1.25 mL of water to a 10-mL round-bottom flask fitted with a spin bar. Stir for a few minutes until most of the sodium bromide has dissolved. Add 1.25 mL of 1-butanol to the flask. Place the flask in an ice bath and continue stirring. Carefully measure out 1.3 mL of concentrated sulfuric acid into a clean, dry 3-mL conical vial. Clamp the vial to a ring stand in the hood. Using a Pasteur pipet, cautiously add sulfuric acid dropwise to the flask. After the addition is complete, remove the flask from the ice bath and place it in a heat block or sand bath for reflux. Attach a water-cooled condenser to the flask and heat the resulting solution at reflux with stirring for 1 hour. Allow the mixture to cool until the layers separate. Transfer the mixture to a 5-mL conical vial.

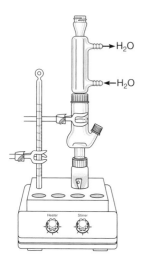

Use a Pasteur pipet to transfer the top layer to a conical vial. Add 1.5 mL of 80% sulfuric acid to the vial. Cap and shake to mix, being careful to vent and not spill the contents. After allowing the layers to separate, transfer the bottom aqueous layer to a test tube for proper disposal. Add 2.0 mL of water to the vial. Cap and shake and then allow the layers to separate. Transfer the bottom, organic layer to another conical vial. (Check to be sure that the bottom layer is indeed the organic layer.) Set aside the aqueous layer for proper disposal. Add 2.0 mL of a saturated sodium bicarbonate solution to the conical vial. Swirl carefully to allow bubbles of carbon dioxide to escape. Cap the tube and shake carefully with frequent venting. Transfer the bottom, organic layer to a dry conical vial and set aside the aqueous layer for proper disposal. Add a small amount of anhydrous sodium sulfate to the organic layer in the vial and allow the vial to stand with occasional swirling for five minutes. The liquid should be clear and some of the drying agent should not be clumped. Add a little more drying agent if necessary.

Use a filter pipet to transfer the solution to a clean, dry 3-mL conical vial and add a spin vane. Fit the vial with a Hickman still and a water condenser. Distill the liquid and collect distillate in the lip of the Hickman still. Transfer aliquots from the lip of the Hickman still to a tared sample vial as necessary. Do not collect any distillate that boils above 105°C. Keep the product vial tightly capped. Determine the weight of the product.

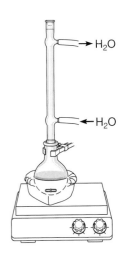

Part B: Miniscale Synthesis of 1-Bromobutane

Add 12.5 g of sodium bromide and 12 mL of water to a 100-mL round-bottom flask fitted with a stir bar. Clamp the flask to a ring stand. Stir for a few minutes until most of the sodium bromide has dissolved. Add 10 mL of 1-butanol to the flask. Place in an ice bath and continue stirring. Carefully measure out 11.0 mL of concentrated sulfuric acid into a clean, dry 50-mL Erlenmeyer flask. Clamp the flask to a ring stand in the hood and place in a separate ice bath. Cautiously transfer the sulfuric acid in small portions to the round-bottom flask over a period of several minutes. After the addition is complete, remove the round-bottom flask from the ice bath and fit with a heating mantle and a water-cooled condenser. (If instructed, fit the top of the condenser with a drying tube containing solid sodium bicarbonate.) Reflux the resulting solution with stirring for 45 minutes. Allow the solution to cool for a few minutes. Remove the condenser from the flask and insert a still head and thermometer. Attach the condenser to the still head for distillation. Use a 50-mL round-bottom flask for collection. See Figure 1G-1 for setup. Distill until 20–25 mL of distillate have been collected or until no more water-insoluble liquid appears in the distillate. The distillate will consist of two layers.

Pour the distillate into a 125-mL separatory funnel. Add approximately 50 mL of water to the graduated cylinder, swirl, and pour into the separatory funnel. Stopper and shake to thoroughly mix the layers. After allowing the layers to separate, draw off the bottom, organic layer into a 125-mL Erlenmeyer flask. Pour the upper, aqueous layer into a beaker and set aside for proper disposal. Add 5 mL of water to the beaker and check to see if mixing occurs. If so, the layer in the beaker is indeed the aqueous layer.

Transfer the organic layer from the Erlenmeyer flask to the separatory funnel and add 25 mL of water. Stopper and shake and allow the layers to separate. After allowing the layers to separate, draw off the bottom, organic layer into a 125-mL Erlenmeyer flask. Pour the upper, aqueous layer into a beaker and set aside for proper disposal. Check to be sure that the solution in the beaker is indeed aqueous. Transfer the organic layer from the Erlenmeyer flask to the separatory funnel and add 25 mL of a saturated sodium bicarbonate solution to the vial. Swirl carefully to allow bubbles of carbon dioxide to escape. Stopper and carefully shake with frequent venting and then allow the layers to separate. Draw off the bottom, organic layer into a 125-mL Erlenmeyer flask. Pour the upper, aqueous layer into a beaker and set aside for proper disposal. Add anhydrous sodium sulfate to the organic layer in the Erlenmeyer flask and allow to stand with occasional swirling. The liquid should be clear and some of the drying agent should not be clumped. Add more drying agent if necessary.

Filter the dried liquid into a clean, dry 50-mL round-bottom flask and add a spin bar. Place the flask in a heating mantle and fit with a still head, thermometer, water condenser, and tared receiving vessel. See Figure 1G-1 for setup. Distill the liquid until most of the liquid in the distilling flask has been distilled (about 105°C). Weigh the receiving vessel containing the product and cap the vessel. Determine the weight of the product.

Characterization for Parts A and B

Boiling Point: Measure the boiling point and correct for atmospheric pressure. The reported boiling point of 1-bromobutane is 102°C.

Gas Chromatography: Dissolve a drop of product in 5 drops of reagent-grade diethyl ether and draw up 1–5 μL of solution into a syringe. Inject the sample through the injection port and record the chromatogram. Determine the relative amounts of each component and identify each using standard samples of 1-butanol, 1-bromobutane, and dibutyl ether.

Refractive Index: Record the refractive index of the product and correct for temperature. The reported refractive index of 1-bromobutane is 1.4390^{20}.

Infrared Spectroscopy (optional): Run an IR spectrum of the product. Interpret the spectrum. Look for the absence of OH stretching at $3400–3300 \text{ cm}^{-1}$.

NMR Spectroscopy (optional): Dissolve a small portion of the sample in $CDCl_3$ and transfer the solution into a clean, dry NMR tube using a filter pipet. Record and interpret the integrated ^1H NMR spectrum. The singlet for an alcoholic OH should not be present, and there should be one triplet for a methyl group, indicating a reasonably pure product. Interpret the spectrum.

Results and Conclusions for Parts A and B

1. Determine the percent yield of the product.
2. Calculate the percent purity of the product as determined by GC analysis.
3. Can the product be clearly differentiated from the starting material by IR and/or NMR spectroscopy? Explain.

Cleanup & Disposal

Neutralize acidic extracts with dilute base. Wash the neutralized, aqueous extracts down the drain with running water.

Critical Thinking Questions

1. Draw the structures for products from reaction of 1-butanol with each of the following:
   a. HCl
   b. HI
   c. H_2SO_4

2. A student neglected to add sulfuric acid to the reaction mixture. What is the expected outcome? Explain.

3. An alternate reagent for this experiment is phosphorus tribromide. List health and safety considerations for phosphorus tribromide.

4. What happens to the $NaHSO_4$ formed as a by-product in this experiment?

5. A possible side product from the experiment is dibutyl ether. Where in the procedure is this compound removed?

6. What products are expected when 2-butanol is substituted for 1-butanol in this experiment?

7. Why is it necessary to do two distillations in this experiment?

8. Why is HBr generated *in situ* from NaBr and H_2SO_4 rather than just using concentrated HBr?

9. What problems might occur when using the reaction conditions in this experiment to convert allyl alcohol to allyl bromide? Explain.

10. A student is asked to react 3-methyl-2-butanol with HBr. Predict the products and explain why the experiment will not be as successful for production of an alkyl bromide as the reaction of 1-butanol with HBr.

Chapter 9

Dienes and Conjugation

Conjugated dienes are often thought of as containing a single functional group, consisting of four carbons connected by both a σ-bonding network and a π-bonding network. The π-bonding network reacts as a unit, rather than as two separate π bonds. For example, 1,3-butadiene can react with HBr by conjugate addition (1,4-addition) to yield a mixture of *cis-* and *trans-*1-bromo-2-butenes.

The Diels-Alder reaction is an important ring-forming reaction of conjugated dienes. There are only a few important synthetic methods for forming new rings. The Diels-Alder reaction is one of the best methods for making six-membered rings. A conjugated diene is required as a reactant for this reaction. Reaction with an activated alkene or alkyne by a process known as cycloaddition yields six-membered ring compounds as in Experiment 9.1A. If the diene is cyclic, then the Diels-Alder product is bicyclic, as illustrated in Experiment 9.1B.

Experiment 9.1 Dienes and the Diels-Alder Reaction

In this experiment, you will investigate the Diels-Alder reaction and the synthesis of six-membered rings. You will:

- perform a Diels-Alder cycloaddition reaction.
- study the stereochemistry of the Diels-Alder reaction.
- use molecular modeling to predict the structure of a Diels-Alder adduct.

Techniques

Technique F	Vacuum filtration and recrystallization
Technique G	Reflux
Technique K	Thin-layer chromatography
Technique M	Infrared spectroscopy
Technique N	Nuclear magnetic resonance spectroscopy
Technique Q	Molecular modeling

Background and Mechanism

The Diels-Alder reaction provides a synthesis of six-membered ring compounds (Diels and Alder, 1928). The discovery and subsequent development of this reaction won a

Nobel Prize for Otto Diels and Kurt Alder in 1950. Today this reaction is used as a tool to prepare a wide variety of compounds containing six-membered rings, including natural products and materials of medicinal value.

An example of the use of the Diels-Alder reaction is the reaction of 2-methoxy-5-methyl-2,5-cyclohexadiene-1,4-dione with 1,3-butadiene to form a Diels-Alder adduct, developed as a key initial step for the synthesis of cholesterol reported by R. B. Woodward and coworkers in 1952.

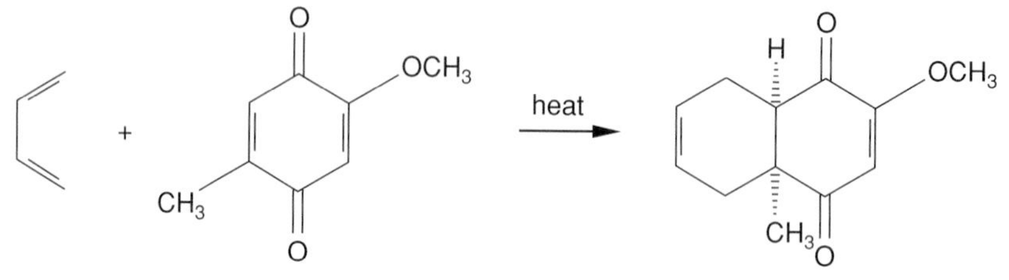

Diels-Alder adduct

Diels-Alder reactions have been used in numerous other syntheses of natural product such as vitamin B_{12} and longifolene. A family of chlorinated insecticides is based upon Diels-Alder reactions of hexachlorocyclopentadiene with various dienophiles such as 2,5-bicycloheptadiene for the preparation of Aldrin.

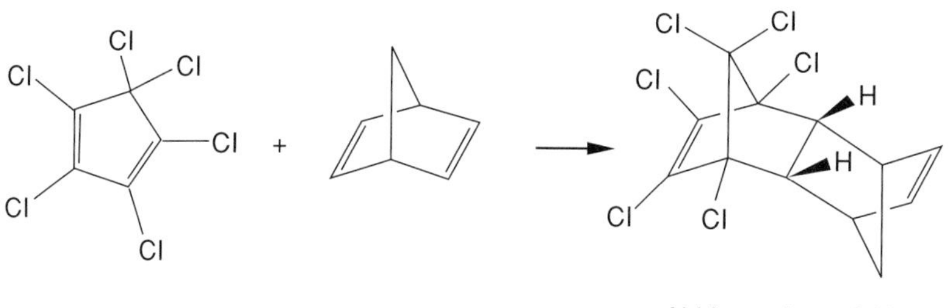

Aldrin, an insecticide

Unfortunately, Aldrin has been implicated as being a carcinogen in male rats. Aldrin is a neurotoxin and accumulates in mammalian tissue. Other chlorinated insecticides are suspected carcinogens or are known carcinogens and all have been removed from the market.

The two components required for the Diels-Alder reaction are a conjugated diene and a dienophile. The dienophile ("diene lover") must contain at least one π bond. The diene must be in the s-cis conformation for the proper overlap of the terminal p-orbitals of the diene with the p-orbitals of the dienophile; the molecule cannot react via a Diels-Alder reaction from the s-trans conformation. The Diels-Alder reaction is a [4 + 2] π cycloaddition reaction, reflecting the total number of π electrons (6) in the two reacting components that are directly involved in the reaction. The Diels-Alder reaction proceeds by way of a concerted, single-step mechanism:

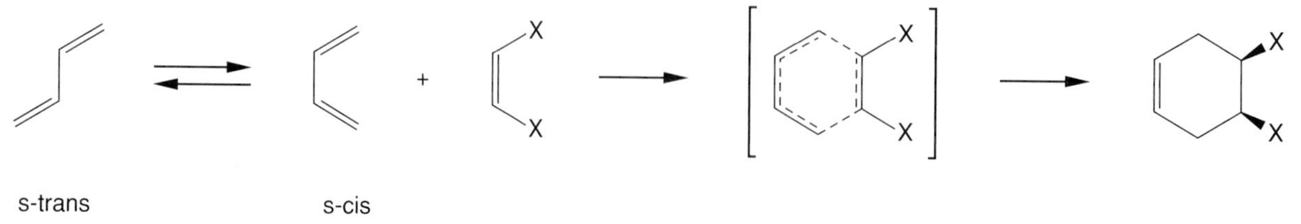

s-trans   s-cis

X = electron-withdrawing group

The reaction is stereospecific because the stereochemistry of the reacting components is maintained; that is, trans substituents on the dienophile are trans in the Diels-
Alder adduct and cis substituents are cis in the Diels-Alder adduct. The Diels-Alder
reaction proceeds best if the dienophile contains electron-withdrawing groups, such as
C=O, or CN, attached to the π bond and if the diene contains electron-donating groups,
such as alkyl groups.

A feature of Diels-Alder reactions of cyclic dienes with electron deficient alkenes
is the possibility of forming either exo or endo product. For example, in the reaction
below, endo and exo bicyclic products are possible. The endo product has favorable overlap in the transition state of the reaction and is often the preferred product (Jarret et al.,
2001; Lee, 1992). However, the exo product is more stable thermodynamically and may
be formed under equilibrium conditions. The kinetically controlled endo product undergoes reverse reaction faster than the thermodynamically controlled exo product, allowing the gradual buildup of the exo product.

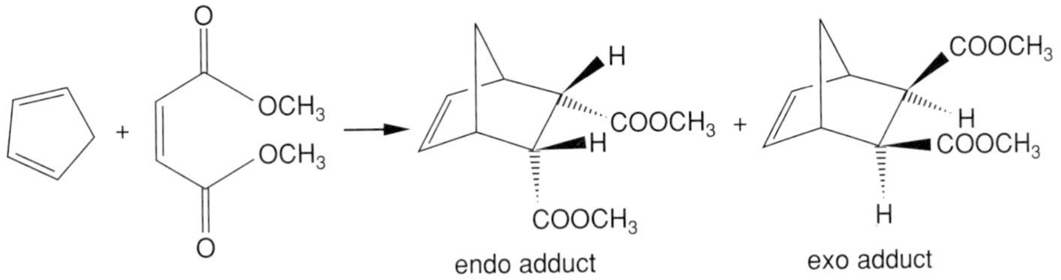

endo adduct   exo adduct

For discovery experiments in Parts C or D, you will prepare the Diels-Alder adduct
and use the melting point to determine the stereochemistry of the product to see which
of the isomers is formed under the reaction conditions used.

In Part A of this experiment, the diene is 1,3-butadiene, which is generated by heating 3-sulfolene. Sulfur dioxide is a by-product in this reaction.

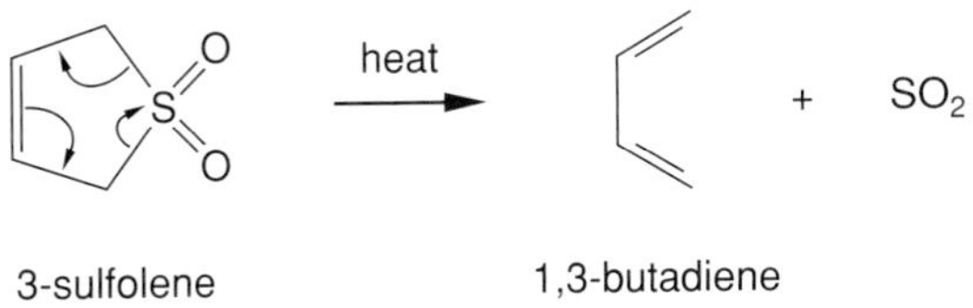

3-sulfolene   1,3-butadiene

The 1,3-butadiene reacts rapidly with maleic anhydride, which is quite reactive due to
the presence of electron-withdrawing carbonyl groups.

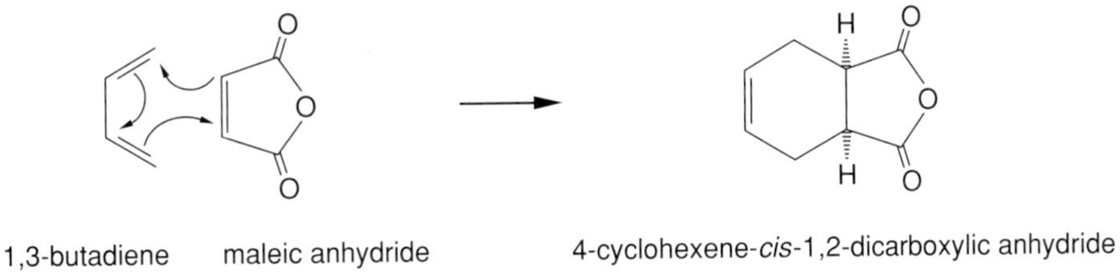

1,3-butadiene   maleic anhydride   4-cyclohexene-*cis*-1,2-dicarboxylic anhydride

For the similar miniscale version of this experiment in Part B, the diene is 2,3-dimethyl-1,3-butadiene and the dienophile is maleic anhydride. Once again, the cis product is expected because of the stereospecificity of the Diels-Alder reaction.

cis adduct

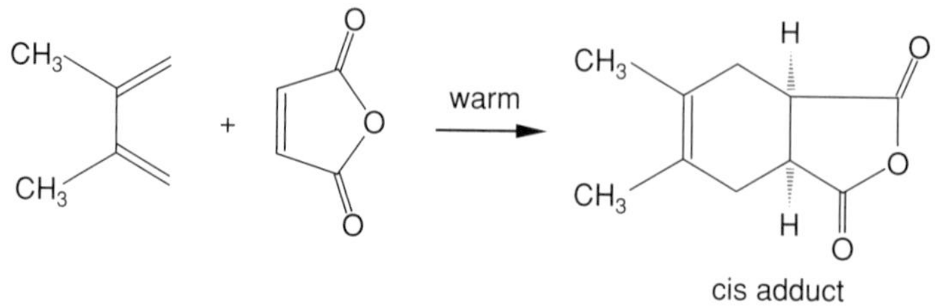

In Part C of this experiment, the diene is 1,3-cyclopentadiene. This diene is unstable and must be prepared soon before performing the Diels-Alder reaction by "cracking" the cyclopentadiene dimer (a reverse Diels-Alder reaction). Cyclopentadiene should be used soon after it is collected, but it can be stored in the refrigerator overnight (or longer in the freezer) before being used for a Diels-Alder experiment.

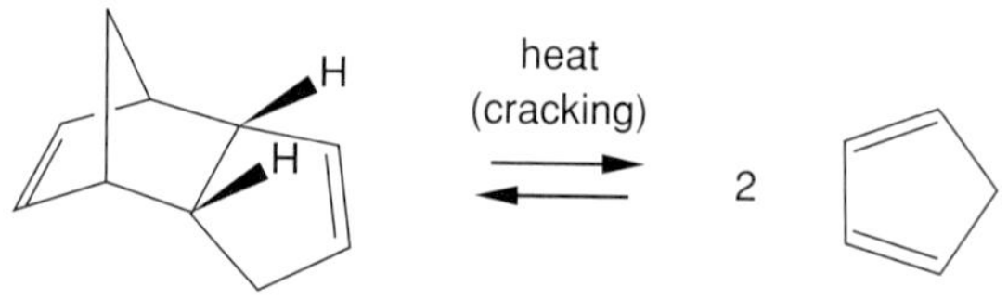

Cyclopentadiene reacts readily with dienophiles, such as maleic anhydride, at room temperature. It is locked into the s-*cis* conformation by virtue of the small ring size. The methylene group acts as a rate-enhancing alkyl substituent on the diene system. Cyclopentadiene must be kept cold after it is collected and it must be used soon after its generation or it will dimerize and return to the dicyclopentadiene starting material. The instructor will prepare cyclopentadiene for the class just prior to the laboratory period.

In Parts C and D, the dienophile is maleic anhydride. You are asked to prepare the Diels-Alder adduct and to determine whether the adduct is an endo adduct or an exo adduct based upon the melting point. IR and ^{1}H NMR spectroscopy can be used to show that a C=C and an anhydride group are present in the product.

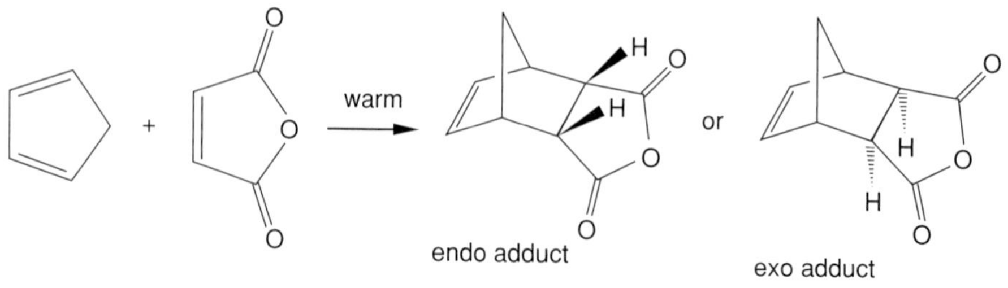

endo adduct   exo adduct

In Part E, two possible Diels-Alder adducts can be formed. Maleic anhydride is again used as a dienophile in a Diels-Alder reaction with anthracene. You will determine the structure of the adduct and provide a rationale for its formation.

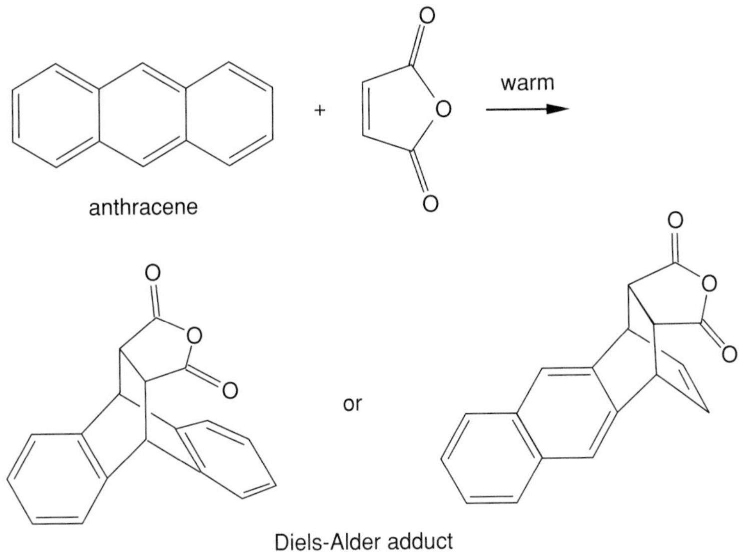

anthracene

warm

Diels-Alder adduct

Prelab Assignment

1. Write a mechanism for the formation of 1,3-butadiene from 3-sulfolene or cyclopentadiene from dicyclopentadiene.
2. Why is it important to avoid water in these reactions? Write a reaction showing the results of the presence of water.
3. Write a reaction to show why freshly distilled cyclopentadiene must be reacted right away with maleic anhydride, rather than allowing it to stand until the next lab period.
4. (Part E) Using a molecular modeling program on the computer, construct both possible Diels-Alder adducts and use molecular mechanics to calculate energies of each. Use the results to predict the expected thermodynamic product.
5. (Part E) Examine the structures of the two possible Diels-Alder adducts and indicate how the compounds could be differentiated using ^{1}H NMR spectroscopy and/or IR spectroscopy.

Experimental Procedure

1. Wear eye protection at all times in the laboratory.
2. Cyclopentadiene and dicyclopentadiene have unpleasant odors, are flammable, and have very low PELs. They should be dispensed and used in a hood. Do not use a flame in or near this experiment. Avoid breathing these chemicals.
3. Sulfur dioxide gas is generated in Part A. Carry out this reaction under the hood.

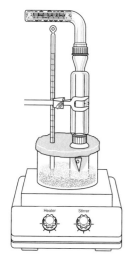

Part A: Microscale Reaction of 1,3-Butadiene with Maleic Anhydride

Place 200 mg of 3-sulfolene and 90 mg of maleic anhydride in a 5-mL conical vial containing 250 μL of dry xylene. The vial should be equipped with an air condenser and spin vane. Place a drying tube filled with $CaCl_2$ on the condenser. This must not be a closed system since a gas is generated! Heat the reaction mixture to reflux, with stirring, in a sand bath or heat block at 190°C for 25 minutes. Cool the mixture to room temperature. Add 0.5 mL of toluene and then add petroleum ether (0.5–0.75 mL) dropwise until slight cloudiness remains. Reheat the mixture until it is essentially clear and then cool in an ice bath. Collect the white product by vacuum filtration using a Hirsch funnel and wash it on the filter with 1 mL of cold petroleum ether. Determine the percent yield of 4-cyclohexene-*cis*-1,2-dicarboxylic anhydride.

Characterization for Part A

Melting Point: Determine the melting point of the product. The reported melting point of 4-cyclohexene-*cis*-1,2-dicarboxylic anhydride is 103–104°C. The reported melting point of 4-cyclohexene-*trans*-1,2-dicarboxylic anhydride is 130°C.

Infrared Spectroscopy (optional): Obtain an IR spectrum of the product using either KBr or the Nujol mull technique and interpret the spectrum. The product should show two carbonyl peaks at 1880–1780 cm^{-1}.

NMR Spectroscopy (optional): Obtain an NMR spectrum of the product and interpret the spectrum. The product should show absorption due to alkenyl protons around δ6.

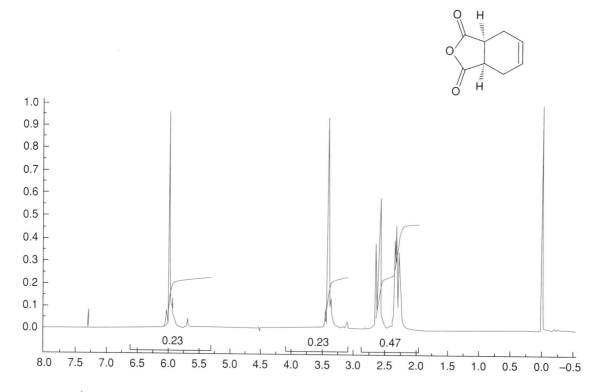

Figure 9.1-1 ^{1}H NMR spectrum of 4-cyclohexene-*cis*-1,2-dicarboxylic anhydride

Part B: Miniscale Reaction of 2,3-Dimethyl-1,3-butadiene with Maleic Anhydride

Under the hood, pipet 1.0 mL of 2,3-dimethyl-1,3-butadiene into a 10-mL or 25-mL round-bottom flask containing 0.88 g of powdered maleic anhydride. Fit the flask with an air condenser and drying tube containing $CaCl_2$ or other drying agent. The flask will become warm soon after mixing indicating that a reaction is taking place. After about 20 minutes, the flask should have cooled to room temperature. Remove the air condenser and drying tube. Break up the solid mass in the flask with a spatula. Add 5 mL of ice-cold water and use the spatula to mix the solids with the water. Perform a vacuum filtration to collect the solid on a Hirsch funnel. Wash the solid with two additional 5-mL portions of ice-cold water. Then wash with two 2-mL portions of ice-cold tetrahydrofuran. Air dry the solid and weigh the product. Recrystallize from hexanes.

Characterization for Part B

Melting Point: Determine the melting point of the product. The reported melting point of 4,5-dimethyl-4-cyclohexene-*cis*-1,2-dicarboxylic acid anhydride (*cis*-1,2,3,6-tetrahydro-4,5-dimethylphthalic anhydride) is 78–79°C.

Infrared Spectroscopy: Obtain the IR spectrum of the recrystallized product using either the KBr pellet or Nujol mull technique and interpret the spectrum. The product should show two carbonyl peaks at 1880–1780 cm^{-1}.

NMR Spectroscopy: Record the ^1H NMR spectrum of the product and interpret the spectrum. The product should not show absorption due to alkenyl protons.

Results and Conclusions for Parts A and B

1. Determine the percent yield of the product.
2. Was the product obtained the cis isomer? Justify your answer.
3. How were IR and/or NMR spectroscopy useful in helping to identify structural features of the adduct?

Part C: Microscale Reaction of Cyclopentadiene with Maleic Anhydride

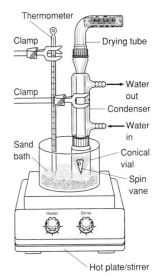

Cyclopentadiene will be prepared by the instructor just prior to the laboratory period. Place 200 mg of maleic anhydride in a dry 5-mL conical vial. Equip the vial with a water-jacketed condenser or air condenser and a drying tube. Dissolve the anhydride in 0.75 mL of dry ethyl acetate by warming in a sand bath or heat block. To this solution add 0.75 mL of dry hexanes. Cool the resulting mixture in an ice bath. Add 0.20 mL of freshly distilled cyclopentadiene and swirl the solution. After the initial exothermic reaction is complete, a white solid will separate. Warm the mixture until the product is dissolved. Cool the mixture and allow it to stand undisturbed. Collect the white crystals of norbornene-*cis*-5,6-dicarboxylic anhydride by suction filtration using a Hirsch funnel and determine the yield.

Part D: Miniscale Reaction of Cyclopentadiene with Maleic Anhydride

Cyclopentadiene will be prepared by the instructor just prior to the laboratory period. Place 2.0 g of powdered maleic anhydride in a 50-mL Erlenmeyer flask. Add 5 mL of dry ethyl acetate to partially dissolve the anhydride. Warm the flask using a sand bath until all of the solids have dissolved. Add 5 mL of dry hexanes to the solution and cool the solution in an ice bath. Slowly add 2.0 mL of freshly distilled cyclopentadiene and swirl to mix. The solution will become warm. Soon a white solid will begin to precipitate. Cool the mixture in an ice bath and allow to stand undisturbed. Collect the solid product of nor-bornene-*cis*-5,6-dicarboxylic anhydride by suction filtration using a Büchner funnel and allow to air dry. Weigh the dry product.

Characterization for Parts C and D

Melting Point: Determine the melting point of the product. The reported melting point of *endo*-norbornene-*cis*-5,6-dicarboxylic anhydride is 164–165°C. The reported melting point of *exo*-norbornene-*cis*-5,6-dicarboxylic anhydride is 142–143°C.

Infrared Spectroscopy: Obtain the IR spectrum of the product using either the KBr pellet or Nujol mull technique and interpret the spectrum. The product should show two carbonyl peaks at 1850–1750 cm^{-1}.

NMR Spectroscopy: Record the ^{1}H NMR spectrum of the product and interpret the spectrum. The product should show absorption due to alkenyl protons around δ6.3.

Results and Conclusions for Parts C and D

1. Determine the percent yield of the product.
2. Write the structure of the product obtained, showing clearly the stereochemistry.
3. Explain, using molecular orbital theory, why the observed product was obtained.
4. Has a reaction clearly taken place and is the product clearly different from the starting materials? Explain based upon evidence obtained in the characterization of the product.
5. How were IR/NMR spectroscopy useful in identifying structural features of the adduct?
6. Could TLC be used to follow either of these reactions (Parts A and B)? If so, explain what would be observed.

Cleanup & Disposal

Place all liquid organic solutions in a container labeled "nonhalogenated organic solvent waste."

Part E: Miniscale Reaction of Anthracene with Maleic Anhydride

Add 0.89 g of anthracene and 0.49 g of powdered maleic anhydride to a 25-mL round-bottom flask containing a stir bar (or a boiling stone). Add 12 ml of dry xylenes and fit the flask with a water-jacketed or an air condenser and a drying tube. Position the flask on a heat block or place in a sand bath or heating mantle and reflux with stirring for about 30 min. Cool to room temperature. Some solids may appear upon cooling. Disassemble the apparatus and place the flask in an ice bath in order to allow more solid to accumulate. Suction filter the product using a Büchner funnel. Wash the solid on the filter with 1–2 mL of cold xylenes and allow to air dry on the filter. Scrape the product onto a new piece of filter paper and allow to air dry. When thoroughly dry, determine the weight of the product.

Cotton
Drying agent
Drying tube
Cotton
Rubber or cork stopper
Water out
Condenser
Water in
Clamp
Round-bottom flask
Heating mantle
Stir bar
Magnetic stirrer

Miniscale

Characterization for Part E

Melting Point (optional): Determine the melting point of the product. The reported melting point of the Diels-Alder adduct is 261–262°C. (The melting point is quite high. The instructor will have special instructions for this melting point determination.)

Thin-layer Chromatography: Spot aliquots of solutions of the Diels-Alder product and anthracene on a silica gel TLC plate and develop in hexanes. View spots of the developed chromatogram under a UV lamp.

Infrared Spectroscopy: Obtain the IR spectrum of the product using either the KBr pellet or Nujol mull technique and interpret the spectrum. The product should show two carbonyl peaks at 1880–1780 cm^{-1}.

NMR Spectroscopy: Record and interpret the ^{1}H NMR spectrum of the product.

Results and Conclusions for Part E

1. Determine the percent yield of the product assuming a 1:1 molar reaction between reactants.
2. Give the structure of the product obtained in this reaction.
3. Explain how the presence or absence of alkenyl absorption in the ^{1}H NMR spectrum is useful in determining the structure of the product.
4. Does the structure of the product make sense in terms of the aromaticity of anthracene and the product? Explain.
5. Compare your experimental results with the predictions from molecular modeling calculations.

Critical Thinking Questions

1. Draw the structure for the product from the reaction of 1,3-cyclohexadiene and maleic acid (*cis*-2-butenedioic acid).
2. What is the product expected from the reaction of 1,3-butadiene and 3-sulfolene? Why is this not a major concern as a by-product?
3. In Part B, the dimer dicyclopentadiene is a possible contaminant of cyclopentadiene. It is also a diene. Can dicyclopentadiene undergo reaction with cyclopentadiene? Can dicyclopentadiene act as a diene and undergo reaction with the anhydride? Show structures for any possible products from these reactions.
4. Explain why reverse Diels-Alder reactions are possible at high temperatures.
5. Predict the Diels-Alder product of each of the following reactions:

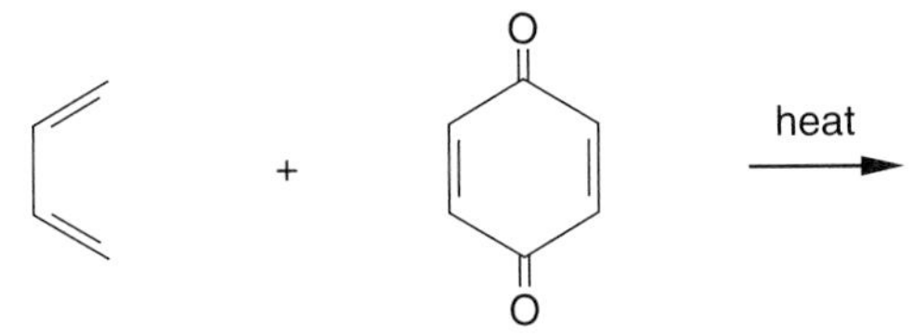

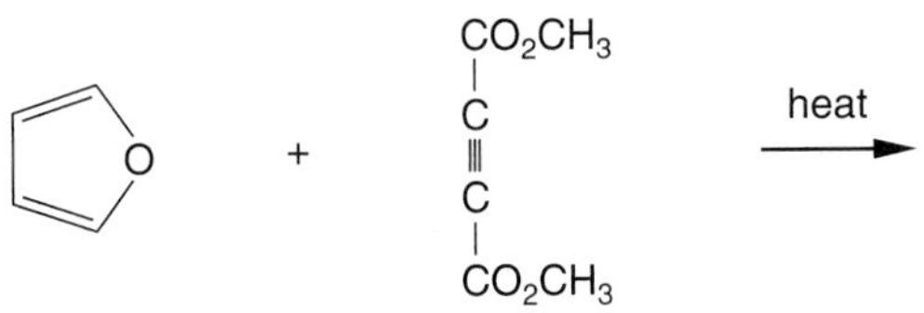

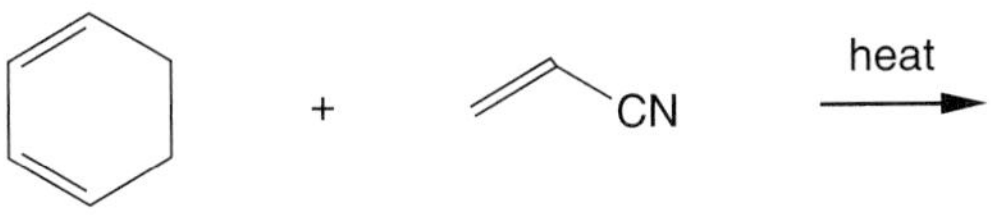

6. Explain the principal synthetic value of the Diels-Alder reaction.
7. Explain why the compound shown here fails to react as a Diels-Alder diene.

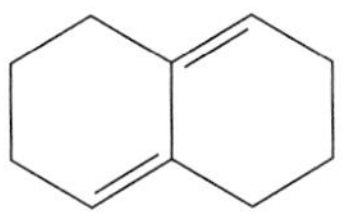

8. The Diels-Alder reaction proceeds well in nonpolar solvents. Why are polar solvents not required?

9. The dienophile used by Woodward and coworkers as part of the total synthesis of cholesterol might have reacted at either of the alkene double bonds. Justify formation of the observed Diels-Alder adduct.

10. *Molecular modeling assignment:* Calculate the heats of formation of the endo and exo products of reaction of furan and maleimide using AM1 and geometry optimization. Which product is more stable? Examine each molecule using the different drawing options. Look for steric effects using the space-filling option. Explain what is meant by the terms kinetic product and thermodynamic product.

11. *Molecular modeling assignment:* Construct the LUMO of maleimide and the HOMO of furan using AM1. Call up both MOs on the same screen and determine if better overlap occurs for exo or endo product formation. If available, use software to calculate the transaction state energies leading to the two products. According to the calculation, should the exo product or endo product be formed more easily?

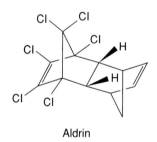

Aldrin

12. *Library project:* Do a literature search to find out more about the toxicity of insecticides such as Aldrin. Results of one study are available on the Internet (http://ntp-server.niehs.nih.gov/htdocs/LT-studies/TR021.html). Suggest a reason that the insecticide Aldrin accumulates in animal tissue.

References

Diels, O., and Alder, K. "Syntheses in the Hydroaromatic Series, I. Addition of "Diene" Hydrocarbons." *Ann. Chem.* 460, (1928): 98.

Jarret, R. M., New, J., Hurley, R., and Gillooly, L. "Looking Beyond the Endo Rule in a Diels-Alder Discovery Lab." *J. Chem. Educ.* 78 (2001): 1262.

Lee, M. "The Microscale Synthesis and the Structure Determination of Endo-9-methoxycarbonyl-3-oxatricyclo[4.2.1.0^{4,5}]-2-nonanone." *J. Chem Educ.* 69 (1992): A172–A174.

Woodward, R. B., Sondheimer, F., Taub, D., Heusler, K., and McLamore, W. M. "The Total Synthesis of Steroids." *J. Am. Chem. Soc.* 74 (1952): 4223.

Chapter 10

Qualitative Organic Analysis I

In qualitative organic analysis, physical and chemical properties of an unknown sample are obtained and evaluated in order to identify the structure. Experiment 10.1 illustrates simple chemical test reactions and measurements of physical properties that can be used to elucidate structures of alkenes, dienes, alkynes, and alkyl halides. These classes of compounds were encountered during the earlier part of the course. The purpose of Experiment 10.1 is to identify the structure of an unknown organic liquid by determining physical, chemical, and spectroscopic properties of the substance.

In Qualitative Organic Analysis II (Chapter 28), similar testing is used to elucidate structures of alcohols, ketones, and other functional molecules. In Experiment 25.1, qualitative organic analysis will be used to test for sugars and sugar derivatives.

Experiment 10.1: Qualitative Analysis of Alkyl Halides, Alkenes, Dienes, and Alkynes

The purpose of this experiment is to identify the structure of an unknown compound by determining several physical, chemical, and spectroscopic properties of the substance. You will:

♦ purify an unknown liquid.

♦ measure the physical properties of an unknown compound.

♦ use infrared (IR) spectroscopy to characterize an unknown compound.

♦ use ultraviolet (UV) spectroscopy to analyze and characterize an organic compound.

Techniques

Technique E	Refractive index
Technique G	Distillation and reflux
Technique M	Infrared spectroscopy
Technique N	Nuclear magnetic resonance spectroscopy (optional)
Technique O	Ultraviolet and visible spectroscopy

Background

There are several properties of compounds that can be used to identify unknown organic compounds. In this experiment, all unknown samples are liquids, so this discussion will apply to analysis of liquids. Chief among important physical properties of liquids are density, refractive index, boiling point, and solubility. As a general class of compounds, alkyl halides may be distinguished from unsaturated compounds on the basis of insolubility of alkyl halides in concentrated sulfuric acid. Whereas alkyl chlorides, bromides, and iodides are insoluble in concentrated sulfuric acid, alkenes, dienes, and alkynes are soluble.

It is not usually possible, however, to conclusively determine the structure of a compound from the knowledge of these properties alone. It is sometimes difficult to distinguish between compounds having similar physical properties, such as the compounds shown below. Such compounds can be differentiated by their chemical properties. A sodium fusion reaction, followed by analysis of the halide ion produced by the reaction, allows classification of an unknown as a bromide, chloride, or iodide.

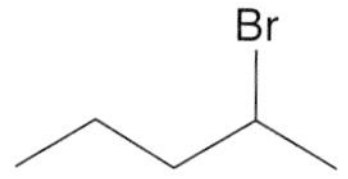

2-bromopentane
bp 116–117°C
n_D 1.4413[20]

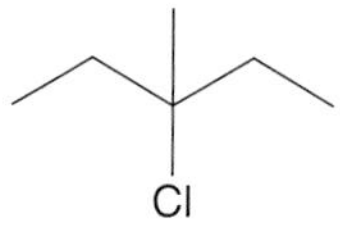

3-chloro-3-methylpentane
bp 115–116°C
n_D 1.4211[20]

Differences in chemical reactivity of compounds having the same functional group can be used to gain structural information. Once it has been established that the compound is an alkyl bromide, for example, chemical test reactions can be used to determine if the compound is a primary, secondary, or tertiary alkyl bromide. The rate of an S_N2 reaction of a secondary alkyl bromide is slower than that of a primary alkyl bromide. This kind of reaction is easily carried out in the laboratory. A second test is the reaction of alkyl halides with silver nitrate in ethanol. This S_N1 reaction proceeds faster for tertiary halides and slower for secondary halides. In the presence of this reagent, a terminal alkyne forms a precipitate via the formation of a silver acetylide, whereas an internal alkyne does not form a precipitate.

Spectroscopic analysis is an excellent tool for organic structure identification. Infrared spectroscopy is very useful in determining the presence or absence of functional groups. IR can generally be used to determine if a carbon-carbon double bond is present. It can also be used to gain further structural information about functional groups in a compound, such as whether an alkene is at the end of a carbon chain or is internal, whether a disubstituted carbon-carbon double bond has cis or trans geometry, whether or not a substance contains methyl groups, or whether an alkyne is internal or terminal. Although there are differences between the spectra of various alkyl halides, there are no definitive characteristic differences between C—Cl and C—Br absorptions in the IR. IR cannot be used to distinguish between these functional classes, nor can it distinguish between primary, secondary, or tertiary alkyl halides. Chemical tests are useful for these determinations, however.

In this experiment, all three approaches to identification will be used to identify an unknown organic liquid. An additional spectroscopic method, ultraviolet (UV) spectroscopy, can be used in certain instances to test for conjugation in dienes. The wavelength of maximum UV absorption is a function of the extent of conjugation in a molecule, and this can be used as an aid in structure identification, as in the case of 1,3-cyclohexadiene versus 1,4-cyclohexadiene. The 1,3-isomer is a conjugated diene that absorbs in the UV region at 256 nm, whereas the 1,4-isomer absorbs only in the vacuum UV region below 200 nm.

1,3-cyclohexadiene
λ_{max} = 256 nm
bp 80°C
n_D 1.4740
d 0.841 g/mL

1,4-cyclohexadiene
λ_{max} = 184 nm
bp 88°C
n_D 1.4720
d 0.847 g/mL

As an optional technique, the instructor may include ^{1}H NMR spectroscopy as a method of analysis of the unknown liquid.

Types of Possible Unknown Compounds

The unknown liquid sample may be an alkyl bromide, chloride, or iodide, an alkene, a diene, or an alkyne. Within a particular functional class, there may be different types of compounds. For example, alkyl halides may be primary, secondary, or tertiary. Alkynes may be terminal or internal. Dienes may be isolated (nonconjugated) or conjugated. Alkenes may be mono-, di-, tri-, or tetrasubstituted. Within each class, substances may be cyclic or acyclic. Lists of all possible unknowns are given in Tables 10.1-1, 10.1-2, 10.1-3, and 10.1-4 at the end of this experiment.

Tests

Boiling Point Determination: The boiling point should be determined using either the microscale or miniscale procedure.

Density: The density of a liquid is easily measured by transferring a specific volume of liquid unknown to a tared vial using a graduated pipet.

Refractive Index: The refractive index should be measured and the reading corrected to 20°C.

Solubility: Solubility in concentrated sulfuric acid is used to distinguish alkyl halides from compounds containing double and triple bonds.

Infrared Spectroscopy: The IR spectrum of a pure liquid unknown should be useful for all unknowns. A chart of useful IR absorption ranges for compounds used in this experiment can be found in the back inside cover of this text.

Ultraviolet Spectroscopy: This spectroscopic method should be applied only when attempting to determine if a diene is conjugated or not conjugated.

Sodium Iodide in Acetone Test (S_N2): This chemical test will differentiate between primary, secondary, and tertiary alkyl chlorides and bromides.

Silver Nitrate in Ethanol Test (S_N1): This chemical test will differentiate between primary, secondary, and tertiary alkyl halides. It can also be used to distinguish between terminal and internal alkynes.

Bromine in Methylene Chloride Test: This test can be used to determine the extent of unsaturation present in the unknown sample. It can be used to distinguish alkenes from dienes, for example. A specified amount of alkene is used to titrate a fixed amount of 1.0 M bromine solution, assuming an approximate molecular weight of the unknown of 100 g/mol. This test can also be used to identify internal alkynes, which can sometimes be difficult to identify from IR spectra.

Potassium Permanganate Test: This test may be used in place of or in addition to the bromine in methylene chloride test. Aqueous potassium permanganate (1%) is used for this test.

Sodium Fusion Test: This test is used on alkyl halides to determine the type of halogen present. Reaction of an alkyl halide with molten sodium converts halogenated compounds to sodium halides that can be tested for the nature of the halogen in the alkyl halide.

Prelab Assignment

1. How do the alkyl halides differ in physical properties from the other functional classes of compounds in this experiment?
2. How can a *trans*-alkene be differentiated from a *cis*-alkene using IR spectroscopy?
3. How can an internal alkyne be differentiated from a terminal alkyne using IR spectroscopy?
4. How can a diene be differentiated from an alkene in this experiment?
5. Summarize the IR data that can be used to differentiate between alkenes, such as terminal alkenes, *cis*- and *trans*-disubstituted alkenes, and more highly substituted alkenes.
6. Develop a flow scheme that can be followed to identify any of the possible unknowns. The compound may be an alkyl chloride, bromide, or iodide, an alkene, a diene, or an alkyne. Have the plan approved by the laboratory instructor before proceeding to do laboratory work on this experiment.
7. If a liquid unknown boiled near 140°C, had an observed density of 1.50 g/mL, was insoluble in concentrated sulfuric acid and gave a positive silver nitrate test, what is a likely structure of the unknown? Use tables found at the end of this experiment. How might IR spectroscopy be of use in solving this unknown?
8. Suppose that an unknown liquid gave a measured boiling range of 116–120°C and a measured density of 0.90 g/mL. The unknown dissolved in concentrated sulfuric acid. A silver nitrate test was negative. The bromine in methylene chloride test was positive, but the unknown required about half the number of drops of reagent required for a known alkene. Suggest a likely structure for the unknown using the tables found at the end of this experiment.

Experimental Procedure

S a f e t y   F i r s t !

Always wear eye protection in the laboratory.

1. Wear eye protection at all times in the laboratory.
2. Wear gloves to avoid skin contact with organic liquids. Do not breathe liquid vapors. Low molecular weight alkenes and alkynes have disagreeable odors. Many alkyl halides are suspected carcinogens and all should be handled with care. Avoid skin contact and inhalation of vapors of all compounds used in this experiment.

Hood!

3. Work in the hood or in a well-ventilated area for this experiment. Keep all liquid unknowns under the hood at all times. Sulfuric acid and nitric acid are strong acids. Avoid skin contact. Wear gloves when handling these reagents. Avoid skin contact with solutions containing bromine, potassium permanganate, and silver nitrate.
4. Sodium is a very reactive metal. Be very careful when handling sodium. Wear gloves. Do not allow sodium to come in contact with water.
5. Many of the compounds used in this experiment are flammable. Make certain that there are no solvents near the Bunsen burner if doing sodium fusion.

Procedure for Qualitative Organic Analysis

Obtain 3 mL of an unknown liquid sample from the instructor. Distill the liquid (to give about 2 mL of distillate) using a Hickman still, if instructed to do so by the instructor. Carry out a micro or mini boiling point determination and correct the observed boiling point to the normal boiling point. Also, measure the density and refractive index of the liquid. Correct the refractive index reading to 20°C.

Following your flow scheme, perform additional chemical and spectroscopic tests as necessary to be able to correctly identify the unknown. It may not be necessary to carry out all tests, but remember that it may be necessary to be able to answer questions about each test.

Solubility Test: Use this test to determine if the unknown is an alkyl halide. Place 2 drops of the unknown liquid in a small, dry test tube or vial. Add 5 drops of concentrated sulfuric acid. Solubility is indicated by a homogeneous solution, color change, or evolution of heat or gas. Alkenes, dienes, and alkynes are soluble, but alkyl halides are not.

Caution: Strong acid!

Sodium Fusion Test: Use this test for alkyl halides to determine the type of halogen present. Run this test in an area of the lab where there are no organic vapors present. Obtain a small, disposable glass test tube that is made of soft glass (non-Pyrex tube). Grip the clean, dry tube with test tube holder and add a small, pea-size piece of sodium metal to the tube. Heat the tube using a Bunsen burner so that the sodium melts. Carefully add 2 drops of unknown liquid to the tube. Try to add the liquid so that it does not contact the side of the tube. Heat the tube gently until there is no more reaction and then heat strongly for about 2 minutes while the tube is red hot. Cool the tube all the way to room temperature and then carefully add methanol dropwise to react with remaining sodium metal. Heat the tube gently until the methanol has evaporated and then heat more strongly until the tube is red hot. Place the red hot tube into a small beaker containing 20 mL of deionized water. The tube will break upon contact with the water. If the tube does not break, break it using a heavy glass rod. Heat to boiling and gravity filter into a small Erlenmeyer flask giving a clear solution.

Caution: Heat gently at first. Point opening of tube away from people.

Halide test A: Transfer 0.5 mL of the aqueous solution from sodium fusion to a clean test tube and acidify the solution with 3M nitric acid. Add 2 drops of 0.1 M silver nitrate solution to precipitate white silver chloride, light-yellow silver bromide, or yellow silver iodide.

Halide test B: Transfer 0.5 mL of the aqueous solution from sodium fusion to a clean test tube and add 5 drops of 1% $KMnO_4$ solution. Stopper and shake the tube. Add about 15 mg of oxalic acid (or enough to discharge the purple color due to excess permanganate). Add 0.5 mL of methylene chloride to the tube. Stopper and shake the tube. Allow the layers to separate and observe the color of the bottom layer. A clear layer indicates chloride ion, a brown layer indicates bromide, and a purple layer indicates iodide.

Sodium Iodide in Acetone Test: Use this chemical test if the unknown appears to be an alkyl bromide or chloride. Measure 2 mL of 15% sodium iodide in acetone into a clean, dry 10-cm test tube. For purposes of comparison, similarly prepare three additional tubes. Add 2 drops of liquid unknown to the first tube. If your unknown is an alkyl chloride, add 2 drops of each of the following alkyl chlorides respectively to the remaining tubes: 1-chloropropane, 2-chloropropane, and 2-methyl-2- chloropropane. If your unknown is an alkyl bromide, add the corresponding alkyl bromides. Stopper and shake the tubes and observe them over a period of 15 to 45 minutes. Record any observed cloudiness or precipitation in any of the tubes, as well as the time of formation in each case. Compare the behavior of the unknown liquid with that of the known samples.

Bromine in Methylene Chloride Test: Use this test to determine how much unsaturation is present in the unknown sample. Add 10 drops of a 1.0 M bromine in methylene chloride solution to a 10-cm test tube. Using a pipet, add a sample of a known alkene, such as 1-heptene, dropwise until the red color of the bromine disappears. Swirl the tube between added drops. Note the amount of alkene required. Repeat using another 10 drops of 1.0 M bromine solution for a sample of the liquid unknown. Record the results. If the unknown liquid requires the same amount to discharge the bromine, the unknown is an alkene. If only half the amount of unknown liquid is required, the unknown sample is either a diene or an alkyne.

Mix vigorously after each addition.

Potassium Permanganate Test: Use this test instead of or in addition to the Bromine in Methylene Chloride Test. Add 10 drops of a freshly prepared 1% $KMnO_4$ solution to a 10-cm test tube. Using a pipet, add a sample of a known alkene, such as 1-heptene, dropwise until the purple color disappears. Swirl the tube between added drops. Note the number of drops of alkene required. Repeat using another 10 drops of 1% $KMnO_4$ solution for a sample of the liquid unknown. Record the results. If the unknown liquid requires the same amount to discharge the purple color, the unknown is an alkene. If only half the amount of unknown liquid is required, the unknown sample is either a diene or an alkyne.

Silver Nitrate in Ethanol Test: Use this chemical test if the unknown appears to be an alkyl halide or an alkyne. Measure 2 mL of a 0.1 M solution of silver nitrate in ethanol into each of six clean, dry 10-cm test tubes. Add a drop of liquid unknown to the first tube and a drop of each of the following alkyl chlorides, bromides, or iodides, respectively, to the remaining tubes: 1-halopropane, 2- halopropane, 2-methyl-2-halopropane. If the unknown appears to be an alkyne, then test 1-hexyne and 2-hexyne. Stopper and shake the tubes and watch closely for evidence of cloudiness or precipitate formation. Record observations and the corresponding times. Compare the behavior of the unknown liquid with that of the known samples.

Infrared Spectroscopy: Obtain the IR spectrum of a drop or two of the neat liquid. Assign as many of the peaks in the spectrum as possible.

Ultraviolet Spectroscopy: Use this technique to determine whether the unknown is a conjugated diene. Dissolve a drop of unknown liquid in 5 mL of 95% ethanol. Dissolve a drop of this solution in 5 mL of 95% ethanol. Place 1 mL of the second solution in one of a matched pair of cuvettes and fill with 95% ethanol. Place pure 95% ethanol in the other cuvette. Place the cuvette containing only solvent in the reference beam of the sample compartment of the UV spectrophotometer. Place the solution containing the unknown liquid in the sample beam of the sample compartment. Obtain a UV spectrum by scanning the wavelength region between 200 and 300 nm. If the spectrum is too intense, try diluting the sample until the recorder pen remains on scale. Record the wavelength of maximum absorption (λ_{max}).

Nuclear Magnetic Resonance Spectroscopy (optional):   Record the ^{1}H NMR spectrum of the liquid unknown. Interpret the spectrum in as much detail as possible. Refer to Technique N for directions on how to prepare a sample for ^{1}H NMR spectroscopy.

Results and Conclusions

1. Summarize the results obtained for each test and the conclusions reached for each test.
2. Give the structure of the compound that best fits the data for the unknown liquid.

3. Report the corrected boiling point, the corrected refractive index, and the density of the unknown liquid.

4. Calculate the percent error between the measured boiling point and the reported boiling point of the compound. Similarly, calculate the percent error in refractive index and density. Which property has the lowest percent error?

5. Summarize the IR data that support the structure assignment.

6. If NMR was used, explain how the recorded NMR spectrum was used to solve the structure of the unknown.

7. Write equations for all positive chemical tests.

Cleanup & Disposal

Place all silver salt solutions from the silver nitrate tests into a container labeled "recovered silver salts." Place all mixtures from the sodium iodide test in a container labeled "recovered sodium iodide solutions." Place all other solutions that contain halogens or that contain halogenated hydrocarbons in a container labeled "halogenated hydrocarbons." Place nonhalogenated hydrocarbons in a container labeled "nonhalogenated hydrocarbons."

Critical Thinking Questions

1. Suppose an instructor issued an alkane as an unknown. Suggest a scheme to distinguish an alkane from the other classes of compounds.

2. How could allyl bromide ($CH_2{=}CHCH_2Br$) be identified if it were an unknown in this experiment?

3. Explain how 1-butene can be distinguished from the isomers of 2-butene in this experiment.

4. An unknown liquid (A) showed no IR absorption between 1500 and 2500 cm^{-1}. Similarly, there was no absorption above 3000 cm^{-1}. Compound A reacted with alcoholic silver nitrate to give cloudiness after several minutes and it also reacted with sodium iodide in acetone to give a precipitate after several minutes. What conclusion can be made about the identity of compound A?

5. Suppose that unknown solids were issued as unknowns in this experiment. What changes in the experiment can be anticipated?

6. A liquid unknown showed IR absorption at 2125 cm^{-1}. What conclusions can be drawn about the structure?

7. Suppose that an unknown liquid is a primary alkyl iodide. Explain how this compound can be identified.

8. A student has concluded that an unknown sample is a conjugated diene that contains a terminal double bond. Cite the relevant tests that should have been done and the expected results that would have led to this conclusion.

9. Explain how the structure of an internal alkyne can be determined using tests given in this experiment. How might NMR be useful?

Table 10.1-1   Alkyl Halides

Halide	Boiling point (°C)	Density g/mL	Refractive index (n_D^{20})
A. Chloride			
2-Chloro-2-methylpropane	51	0.851	1.3840
2-Chlorobutane	68	0.873	1.3971
1-Chloro-2-methylpropane	69	0.881	1.3970
1-Chlorobutane	77	0.886	1.4021
1-Chloro-2,2-dimethylpropane	85	0.866	1.4030
2-Chloro-2-methylbutane	85	0.865	1.4040
2-Chloropentane	97	0.873	1.4069
1-Chloro-3-methylbutane	99	0.870	1.4084
1-Chloropentane	106	0.882	1.4127
Chlorocyclopentane	114	1.005	1.4510
1-Chlorohexane	134	0.879	1.4199
Chlorocyclohexane	142	1.000	1.4626
B. Bromide			
2-Bromopropane	59	1.314	1.4251
1-Bromopropane	71	1.354	1.4343
2-Bromobutane	91	1.259	1.4366
1-Bromo-2-methylpropane	91	1.260	1.4350
1-Bromobutane	101	1.276	1.4401
1-Bromo-2,2-dimethylpropane	109	1.199	1.4370
2-Bromopentane	117	1.212	1.4413
3-Bromopentane	118	1.211	1.4441
1-Bromo-3-methylbutane	119	1.261	1.4400
1-Bromopentane	129	1.219	1.4447
Bromocyclopentane	137	1.387	1.4886
C. Iodide			
Iodoethane	73	1.950	1.5130
2-Iodopropane	89	1.703	1.5028
1-Iodopropane	102	1.749	1.5058
2-Iodobutane	118	1.592	1.4991
1-Iodo-2-methylpropane	119	1.600	1.4940
2-Iodo-2-methylbutane	128	1.494	1.4981
1-Iodobutane	129	1.616	1.5001
3-Iodopentane	142	1.518	1.4974
2-Iodopentane	142	1.510	1.4961
1-Iodo-3-methylbutane	147	1.512	1.4939
1-Iodopentane	155	1.512	1.4959

Table 10.1-2 Alkenes

Alkene	Boiling point (°C)	Density g/mL	Refractive index (n_D^{20})
2,3-Dimethyl-1-butene	56	.680	1.3890
1-Hexene	66	.673	1.3858
trans-3-Hexene	67	.677	1.3940
2-Hexene (mixture of cis and trans)	68	.681	1.3928
2,3-Dimethyl-2-butene	73	.708	1.4120
1-Methylcyclopentene	73	.780	1.4310
Cyclohexene	84	.809	1.4492
1-Heptene	94	.697	1.3998
2,3,3-Trimethyl-1-butene	101	.715	1.4082
2,4,4-Trimethyl-1-pentene	101–102	.708	1.4080
4-Methylcyclohexene	101–102	.799	1.4410
2,4,4-Trimethyl-2-pentene	104	.720	1.4160
1-Methylcyclohexene	110–111	.813	1.4500
Cycloheptene	115	.823	1.4580
2,3,4-Trimethyl-2-pentene	116	.743	1.4280
1-Octene	121	.716	1.4088
1,4-Dimethyl-1-cyclohexene	128	.801	1.4470

Table 10.1-3 Dienes

Diene	Boiling point (°C)	Density g/mL	Refractive index
2-Methyl-1,4-pentadiene	56	.692	1.4060
1,5-Hexadiene	59	.690	1.4010
trans-2-Methyl-1,3-pentadiene	75–76	.718	1.4460
1,3-Cyclohexadiene	80	.841	1.4740
2,4-Hexadiene	81	.715	1.4493
1,4-Cyclohexadiene	88–89	.847	1.4720
2,4-Dimethyl-1,3-pentadiene	94	.744	1.4410
2,5-Dimethyl-1,5-hexadiene	114	.742	1.4290
1-Methyl-1,4-cyclohexadiene	114–115	.838	1.4710
1,3-Cycloheptadiene	120–121	.868	1.4980
2,5-Dimethyl-2,4-hexadiene	132–134	.773	1.4760

Table 10.1-4 Alkynes

Alkyne	Boiling point (°C)	Density g/mL	Refractive index
1-Hexyne	70	.712	1.3990
3-Hexyne	81	.724	1.4115
2-Hexyne	84–85	.731	1.4140
1-Heptyne	100	.750	1.4180
1-Octyne	132	.747	1.4172

Chapter 11

Reactions of Aromatic Side Chains

T he inherent stability of aromatic rings is evidenced by their resistance to oxidation by powerful oxidants, such as $KMnO_4$. This is not the case with alkyl side chains, however. Reactions of alkyl substituents take place at the benzylic carbon, where intermediates may be especially stabilized as benzylic radicals or benzylic carbocations. In Experiment 11.1, methyl substituents are oxidized by $KMnO_4$ to give carboxyl substituents. The aromatic ring is unaffected by the oxidant. Another oxidation of an aromatic side chain that leaves the aromatic ring intact is the hypochlorite oxidation of *trans*-4-phenyl-3-buten-2-one in Experiment 19.2. The mild oxidizing agent causes oxidative cleavage only next to the carbonyl group.

Experiment 11.1: Benzylic Oxidation: *Benzoic Acid from Toluene; A Phthalic Acid from an Unknown Xylene*

In this experiment, you will oxidize all aromatic alkyl side chains present on an aromatic ring to carboxyl groups. You will:

♦ oxidize an alkylbenzene or dialkylbenzene using alkaline permanganate.

♦ use spectroscopic analysis to determine the structure of an unknown xylene derivative.

Techniques

Technique F	Filtration and recrystallization
Technique G	Reflux
Technique M	Infrared spectroscopy
Technique N	Nuclear magnetic resonance spectroscopy

Background

A convenient method of preparing aromatic acids is oxidation of a side chain attached to an aromatic ring using hot, alkaline potassium permanganate. If the side chain is an alkyl group, the benzylic carbon must have at least one attached hydrogen. Tertiary benzylic groups will not react. In Part A of this experiment, toluene is oxidized to benzoic acid.

The purple oxidant is reduced to water-insoluble, brown manganese dioxide. The organic product is water-soluble sodium benzoate. After filtration of the MnO_2, solid benzoic acid may be recovered by acidification of the filtrate.

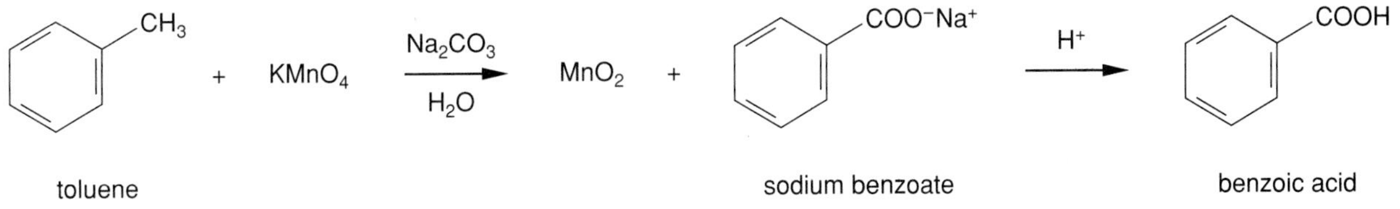

If more than one alkyl group is present, all alkyl groups with at least one benzylic hydrogen are subject to oxidation. For example, *p*-xylene gives terephthalic acid, as shown below.

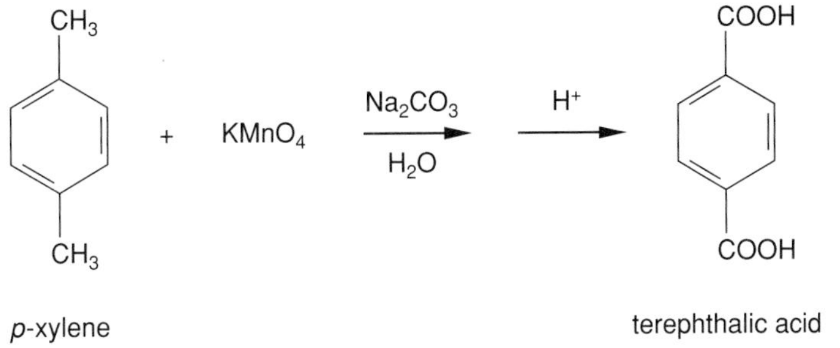

Benzylic oxidation has been widely used in structure determination of aromatic natural products. The aromatic carboxylic acids are easily characterized as solids and it is often possible to determine the number and placement of alkyl side chains on aromatic rings of unknown compounds. For example, basic permanganate was used to elucidate the structure of the A ring of griseofulvin, an antibiotic from *Penicillium griseofulvum,* in 1952. In this example, it should be noted that cleavage also occurs next to carbonyl groups, but that halogen and methoxy ring substituents are unaffected by treatment with $KMnO_4$.

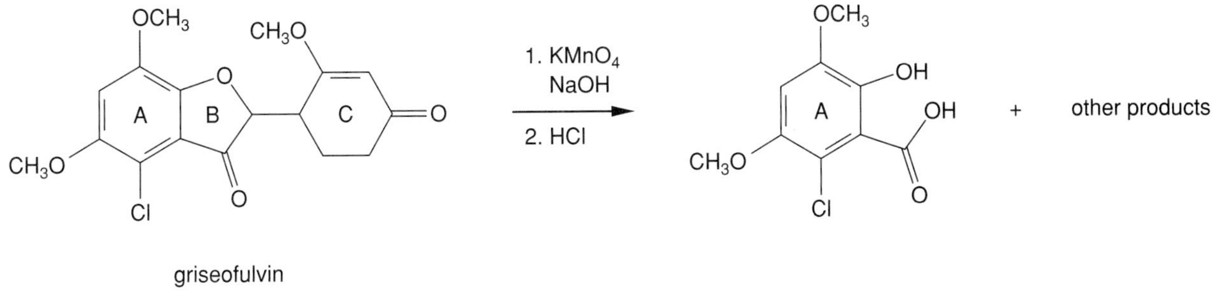

In Part B of this experiment, the unknown compound is either *o-, m-,* or *p*-xylene, a far cry from the complexity of griseofulvin, but nevertheless still useful in showing the oxidative method for determining the pattern of substituents on an aromatic ring.

The mechanism of oxidation is not well understood, but benzylic free radical or radical ion intermediates are thought to be involved. Other strong oxidants, such as nitric acid and chromic acid, may also be used. Ring nitration may occur as a side reaction if nitric acid is used as the oxidant.

A detergent, such as Tide, may be added to the reaction mixture to help mix the liquid phases. This is beneficial because the aromatic liquids are not soluble in water, which contains the oxidant.

Spectroscopic analysis is particularly useful in Part B because differentiation between the possible products requires a method other than determining the melting point. Infrared spectroscopy can be used and comparison can be made with Figures 11.1-1, 11.1-2, and 11.1-3. Use of nuclear magnetic resonance (NMR) spectroscopy is even more effective in distinguishing between the possible isomeric products. Refer to ^{1}H NMR spectra in Figures 11.1-4, 11.1-5, and 11.1-6, and ^{13}C NMR spectra in Figures 11.1-7, 11.1-8, and 11.1-9.

Prelab Assignment

1. Write a balanced equation for the reaction in this experiment. What ratio of reactants is required according to the balanced equation?
2. Write a flow scheme for the workup procedure.
3. (Part A) Write the formula of any detergent. Explain how detergents function as phase-transfer agents.
4. (Part B) Draw the structures of the three possible xylene unknowns and their oxidation products. Look up the properties of the three possible phthalic acid products. Note differences in the IR and NMR spectra between the isomers.

Experimental Procedure

S a f e t y   F i r s t !

Always wear eye protection in the laboratory.

1. Wear eye protection at all times in the laboratory.
2. Wear gloves when handling reagents in this experiment.
3. Work in the hood or in a well-ventilated area for this experiment.
4. Potassium permanganate is a strong oxidant. Avoid contact as it will stain skin and clothing. Wash the affected area thoroughly with soap and water. Notify the instructor if any $KMnO_4$ comes in contact with your skin.
5. Toluene and the xylenes are toxic liquids and the phthalic acids are mild irritants. Avoid skin contact and inhalation of the vapors.
6. Concentrated hydrochloric acid is corrosive and toxic. Be careful when handling this reagent.

Part A: Microscale Oxidation of Toluene to Benzoic Acid

Caution: The mixture has a tendency to bump, due to precipitation of MnO_2 as the oxidation proceeds.

Place 825 mg of potassium permanganate, 100 mg of sodium carbonate, and 3 mL of water in a 10-mL round-bottom flask along with a spin bar. Attach a reflux condenser to the flask. Dissolve the reagents by gentle warming and stirring. Cool slightly and add 250 µL of toluene and 10 mg of Tide or a similar detergent. Reflux the mixture with stirring in a sand bath or heat block for 45 minutes. Bumping can be minimized by stirring and not overheating. Most of the purple color should be gone by the end of the reflux period. Turn off the flow of water and remove the reflux condenser. While the mixture is still warm, add sodium bisulfite until the purple color is gone. Filter the MnO_2 by vacuum filtration using a Hirsch funnel containing Celite (filter aid). Rinse the flask with

0.5 mL of water and pour through the filter. Add the rinse to the filtrate. Dispose of the Celite/MnO_2 mixture on the filter according to directions of the instructor. If necessary, destroy any excess permanganate in the filtrate by slowly adding sodium bisulfite until the solution is colorless (pink to brown to colorless). Filter again if a brown precipitate forms. Do not add too much sodium bisulfite. Cool the filtrate in ice and acidify with 1 mL of concentrated HCl. (The pH should be about 2.) Collect the crystals by vacuum filtration using a Hirsch funnel and wash the crystals with 1 mL of cold water. Recrystallize from water. Dry the crystals. Weigh the product and determine the melting point.

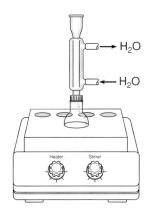

Characterization for Part A

Melting Point: Compare the melting point to the literature value of 122°C.

  Infrared Spectroscopy: Obtain an IR spectrum using either KBr or Nujol and interpret the spectrum noting the key bands characteristic of the structure of the product. Carboxylic acids show broad OH absorption from 3500 to 2500 cm^{-1} and carbonyl absorption near 1700 cm^{-1}.

  NMR Spectroscopy (optional): Obtain ^1H and ^{13}C NMR spectra of the product in $CDCl_3$. Interpret the spectra. The acidic proton of carboxylic acids is expected to show a signal at δ10–14 in the ^1H NMR spectrum. The ^{13}C NMR spectrum should give five signals.

Results and Conclusions for Part A

1. Determine the percent yield of the product.
2. Write a balanced equation for the reaction involving sodium bisulfite. Bisulfite is oxidized to sulfate and $KMnO_4$ is reduced to MnO_2.
3. Why is addition of hydrochloric acid necessary prior to isolation of the product?

Part B: Microscale Oxidation of a Xylene to a Phthalic Acid

Obtain a vial containing an unknown xylene. Record the number of your unknown. Place 825 mg of potassium permanganate, 100 mg of sodium carbonate, and 3 mL of water in a 10-mL round-bottom flask, along with a spin bar. Attach a reflux condenser to the flask. Dissolve the reagents by gentle warming and stirring. Cool slightly and add 125 μL of an unknown xylene and 10 mg of Tide or similar detergent. Reflux the mixture with stirring in a sand bath or heat block for 45 minutes. Bumping can be minimized by stirring and not overheating. Most of the purple color should be gone by the end of the reflux period leaving a dark, gunky residue. Turn off the flow of water and remove the reflux condenser. While the mixture is still warm, add sodium bisulfite until the purple color is gone. Filter the MnO_2 by vacuum filtration using a Hirsch funnel containing Celite (filter aid). Rinse the flask with 0.5 mL of water and pour through the filter. Dispose of the Celite/MnO_2 mixture on the filter according to directions of the instructor. If necessary, destroy any excess permanganate in the filtrate by slowly adding sodium bisulfite until the solution is colorless (pink to brown to colorless). Filter again if a brown precipitate forms. Do not add too much sodium bisulfite. Cool the filtrate in ice and acidify with concentrated HCl until the pH is about 2 (approximately 1 mL). Collect the white crystals by vacuum filtration using a Hirsch funnel. Wash the crystals with 1 mL of cold water. Recrystallize from ethanol or ethanol-water. Dry thoroughly. Weigh the product and determine the melting point if possible.

Caution: The mixture has a tendency to bump, due to precipitation of MnO_2 as the oxidation proceeds.

Characterization for Part B

Melting Point: The recorded melting point of phthalic acid is 210°C. Isophthalic acid and terephthalic acid each melt above 300°C.

Infrared Spectroscopy: Obtain an IR spectrum using either KBr or Nujol and interpret the spectrum noting the key bands characteristic of the structure of the product. Carboxylic acids show broad –OH absorption from 3500–2500 cm^{-1} and carbonyl absorption near 1700 cm^{-1}. The substitution pattern in the 900–600 cm^{-1} region can help to clarify the structure of the product. The out-of-plane bending region can be used to differentiate between the ortho-, meta-, and para-isomers. Ortho-disubstituted rings have one strong absorption band near 750 cm^{-1}. Meta-disubstituted rings show two strong bands around 690 cm^{-1} and 800 cm^{-1}, with a third less-intense band occurring near 890 cm^{-1}. One strong band is observed near 830 cm^{-1} for para-disubstituted rings.

NMR Spectroscopy: Obtain ^{1}H and ^{13}C NMR spectra of the product in CDCl$_3$. Interpret the spectra and compare with spectra in the accompanying figures. The acidic protons of phthalic acids are expected to show a signal at δ10–14 in the ^{1}H NMR spectrum. The pattern of the aromatic signals can help distinguish between the isomeric phthalic acids.

Figure 11.1-1

IR spectrum of phthalic acid (Nujol)

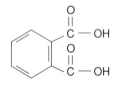

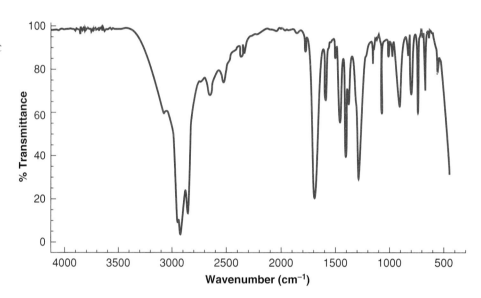

Figure 11.1-2

IR spectrum of isophthalic acid (Nujol)

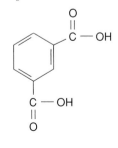

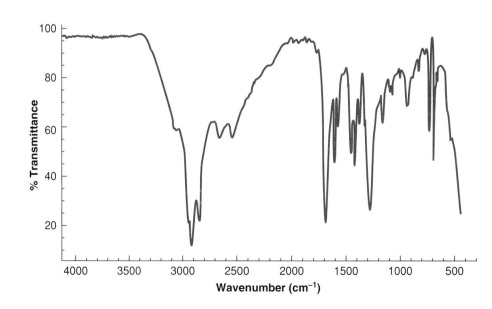

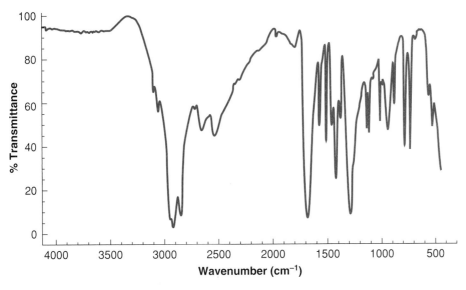

Figure 11.1-3

IR spectrum of tere-phthalic acid (Nujol)

Results and Conclusions for Part B

1. Determine the percent yield of the product.
2. From the melting point and spectroscopic data, identify the structure of the phthalic acid isomer and deduce the structure of the unknown xylene.
3. Explain the role of sodium bisulfite in the procedure.
4. How does acidification with hydrochloric acid reduce the solubility of the product?

Cleanup & Disposal

Recovered solid MnO_2 should be placed in a labeled container. The aqueous acidic filtrate can be washed down the drain with running water. Residues of MnO_2 on glassware may be difficult to remove. If this is a problem, a 10% sodium bisulfite solution can be used to rinse glassware to remove the residue.

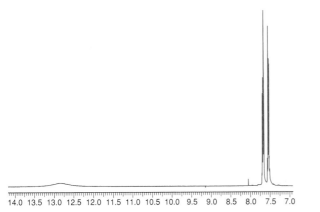

Figure 11.1-4 ^{1}H NMR spectrum of phthalic acid (CDCl$_3$)

SOURCE: Reprinted with permission of Aldrich Chemical.

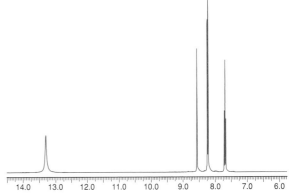

Figure 11.1-5 ^{1}H NMR spectrum of isophthalic acid (CDCl$_3$)

SOURCE: Reprinted with permission of Aldrich Chemical.

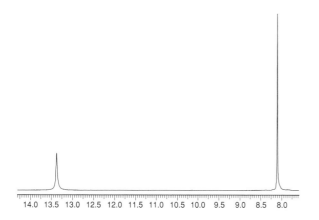

Figure 11.1-6 ^{1}H NMR spectrum of terephthalic acid (CDCl$_3$)

SOURCE: Reprinted with permission of Aldrich Chemical.

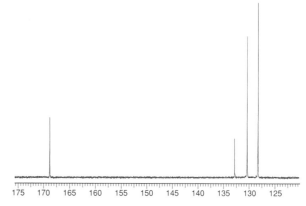

Figure 11.1-7 ^{13}C NMR spectrum of phthalic acid (CDCl$_3$)

SOURCE: Reprinted with permission of Aldrich Chemical.

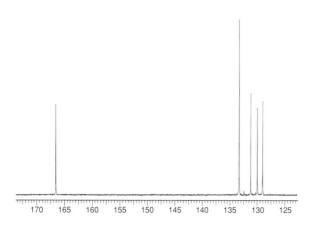

Figure 11.1-8 ^{13}C NMR spectrum of isophthalic acid (CDCl$_3$)

SOURCE: Reprinted with permission of Aldrich Chemical.

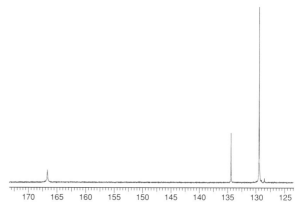

Figure 11.1-9 ^{13}C NMR spectrum of terephthalic acid (CDCl$_3$)

SOURCE: Reprinted with permission of Aldrich Chemical.

Critical Thinking Questions

1. Write a balanced equation for the oxidation of *p*-xylene using chromic acid (K$_2$Cr$_2$O$_7$ and H$_2$SO$_4$). Chromic acid is reduced to Cr^{+3}.

2. Write a balanced equation for the oxidation of ethylbenzene in alkaline permanganate solution. Note that K$_2$CO$_3$ is a by-product.

3. Given the options of alkaline permanganate or chromic acid for the oxidation of arenes (alkylaromatics), why are alkaline conditions preferred?

4. Devise a synthesis of 4-nitrobenzoic acid from toluene. Give all important reagents and reaction conditions. Two separate synthetic steps are required.

5. A student attempted to prepare 4-chlorobenzoic acid by oxidizing toluene and then chlorinating benzoic acid. The compound she isolated was not 4-chlorobenzoic

acid. What was the product and why was it obtained? Suggest a suitable route to the desired product.

6. Phthalic acid (benzene-1,2-dicarboxylic acid) is a commercially important material that is used in the preparation of plastics. Give the structures of two different aromatic compounds that could be used to synthesize phthalic acid.

7. 1,2,3,4-Tetrahydronaphthalene has the structure shown here. Write the structure of the $KMnO_4$ oxidation product.

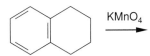

8. Alkenyl, alkynyl, and acyl groups attached directly to aromatic rings may also be oxidized to carboxyl groups. Predict the expected aromatic products of alkaline permanganate oxidation of each of the following compounds: styrene, phenylacetylene, acetophenone, and isopropylbenzene.

Chapter 12

Electrophilic Aromatic Substitution

A romatic compounds are much more stable than alkenes and alkynes. For example, many of the constituents of coal are aromatic compounds that have been formed over thousands of years. Electrophilic reagents will not generally give addition reactions to aromatic rings as they do to alkenes and alkynes. However, aromatic compounds will undergo electrophilic substitution reactions during which charged complexes are formed and products ultimately result from the complexes. The aromatic ring is regenerated in going from the complex to the product.

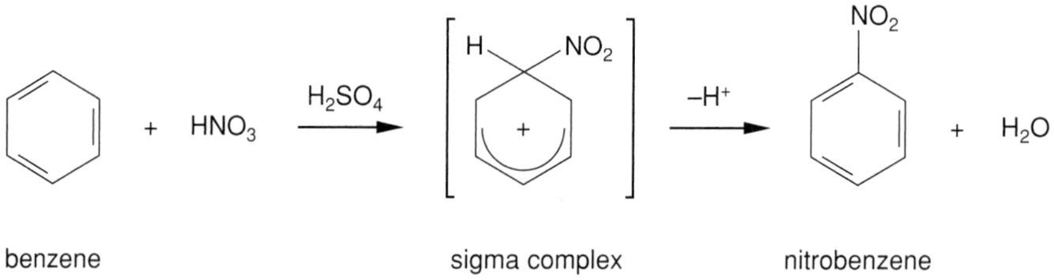

| benzene | sigma complex | nitrobenzene |

Nitration of benzene was commonly performed in organic chemistry laboratories worldwide for many years until it was discovered that nitrobenzene is really quite toxic and that it can be readily absorbed through the skin. Benzene is a cancer suspect agent. Other aromatic compounds can be safely nitrated, such as methyl benzoate (Experiment 12.2).

One of the more fascinating features of electrophilic aromatic substitution is the influence of substituents on the aromatic ring. Substituents that are already present give directions (like traffic police) to incoming groups telling them where to link up. These directive effects of substituents are important in the aromatic substitution experiments in this chapter. Ester groups exert a particular directive effect in the nitration of methyl benzoate (Experiment 12.2) and phenyl and acetamido substituents are observed to have specific directive effects toward acylation (Experiments 12.3 and 12.4). Substituents also influence the rate of aromatic substitution, such as aromatic bromination (Experiment 12.1).

Experiment 12.1: Activating and Deactivating Effects of Aromatic Substituents: *Relative Rates of Bromination*

The purpose of this experiment is to determine the effect of substituents on the rate of electrophilic aromatic bromination. In this experiment, you will:

♦ classify the relative bromination rates of substituted benzenes compared to toluene.

♦ correlate the rate of bromination with the electronic properties of the substituent.

♦ interpret the mechanism of electrophilic aromatic bromination.

♦ classify aromatic substituents as electron-donating or electron-withdrawing.

Background and Mechanism

The electrophilic bromination of aromatic compounds is an example of electrophilic aromatic substitution. The electrophile in this case is molecular bromine:

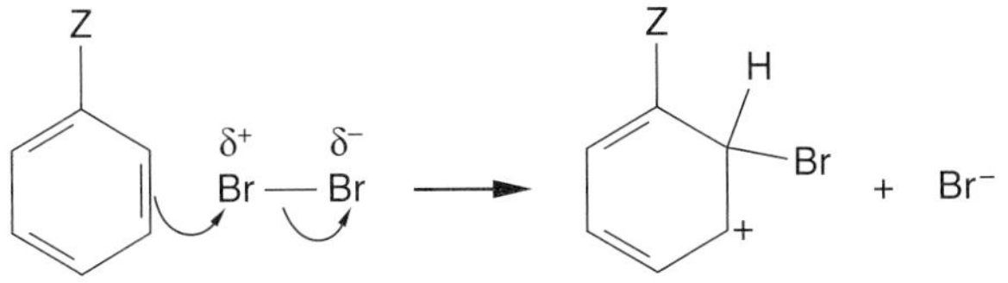

The rate-determining step of the reaction is the addition of π electrons from the aromatic ring to the electrophile (bromine) to form a resonance-stabilized intermediate. This is illustrated for ortho- and meta-addition of the electrophile.

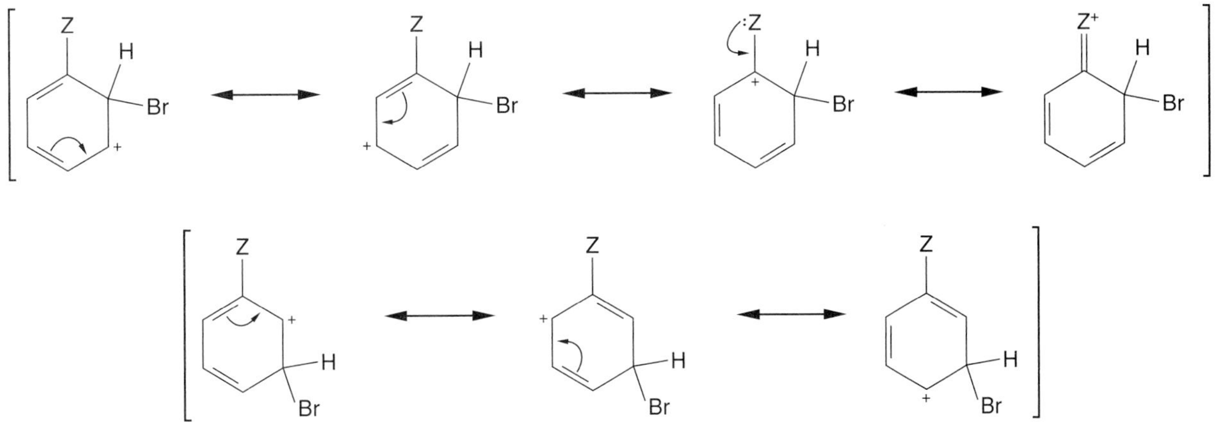

The final step is the regeneration of aromaticity by the rapid abstraction of the proton by a base such as water.

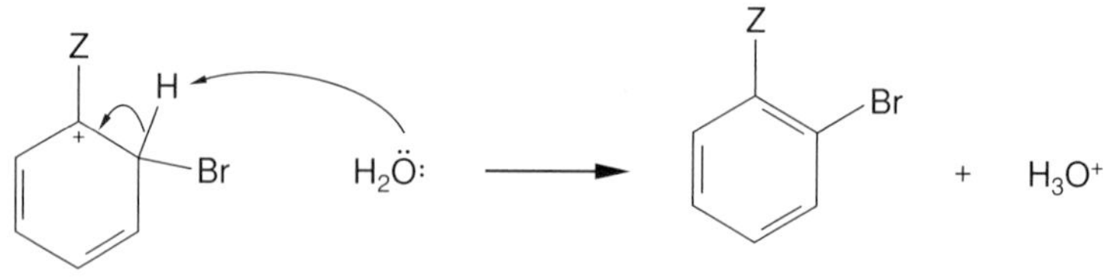

ortho product

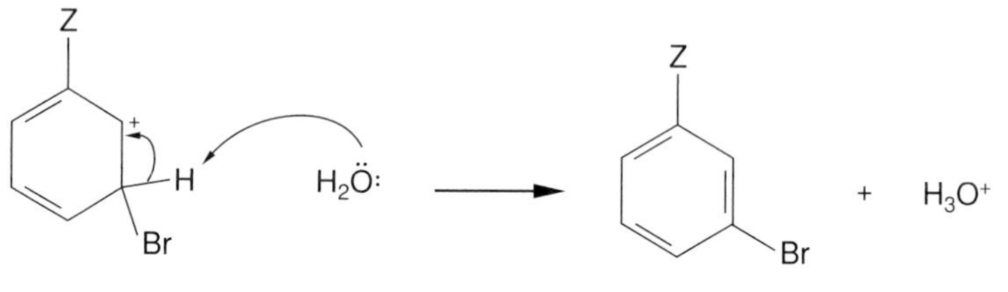

meta product

The rate of this reaction is dependent upon the nature of the substituent (Z) on the aromatic ring. Most substituents that are ortho, para-directors increase the rate of the reaction as compared to benzene, because they lower the energy of activation leading to the resonance-stabilized intermediate. Substituents that are meta-directors decrease the rate of the reaction because they raise the energy of activation leading to the resonance-stabilized intermediate.

The purpose of this experiment is to determine how each substituent affects the rate of electrophilic aromatic bromination and to analyze the similarities between substituents that increase the rate (and are ortho, para-directors) and similarities between substituents that decrease the rate (and are meta-directors).

The rate at which this reaction proceeds is easily determined by measuring the time it takes for the reddish color of bromine to disappear. The brominated aromatic compounds are colorless. The influence of the substituent on the rate of electrophilic aromatic bromination will be examined by comparing the rates of bromination of a series of aromatic compounds. Ideally, the rates should be compared with benzene, since it is unsubstituted. Unfortunately, benzene is hazardous, so it will not be used in this experiment. Toluene will be used as the reference. You will observe the rate of decolorization of toluene and compare the rates of various substituted aromatic compounds as occurring faster, slower, or about the same as that of toluene. It is possible that toluene might also undergo free radical reaction with bromine to generate benzyl bromide. Therefore, the test tube containing the toluene solution should be kept away from light during the reaction.

In addition to the nature of the substituent, other factors influence the rate at which the reaction occurs. These include temperature, concentration of reactants, solvent polarity, nature of catalyst, and the presence of water. To ensure that the only factor affecting the rate in this experiment is the nature of the substituent on the aromatic ring, the reaction will be run in a water bath so that the temperature remains constant during the reaction. The solvent, 15 M acetic acid, is 90% acetic acid and 10% water. This is a polar solvent with a high (and constant) percentage of water. Using the same concentration of reactants each time will ensure that the only variable in the experiment is the nature of the substituent.

The compounds to be tested in this experiment are shown here. The instructor may assign other aromatic compounds.

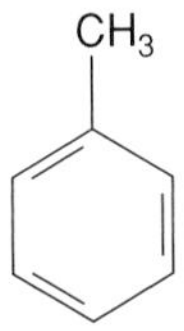
CH3

toluene

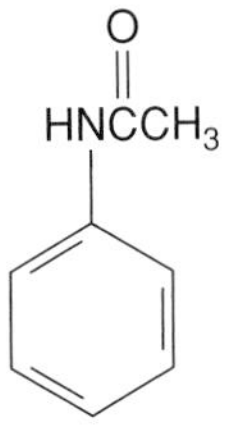
O
HNCCH3

acetanilide

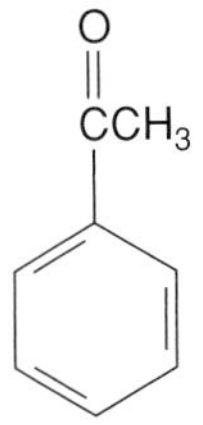
O
CCH3

acetophenone

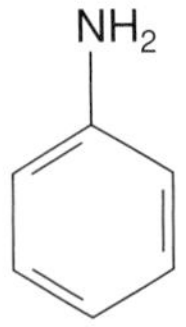
NH2

aniline

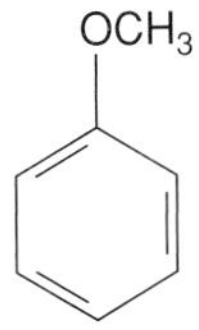
OCH3

anisole

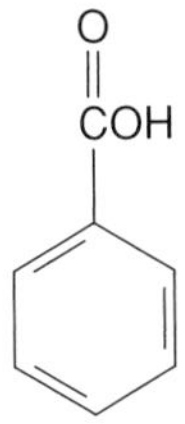
O
COH

benzoic acid

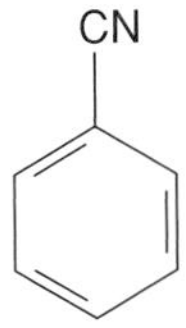
CN

benzonitrile

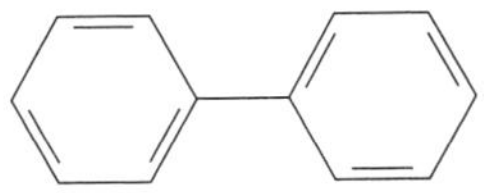

biphenyl

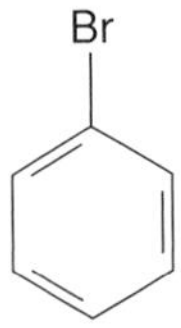
Br

bromobenzene

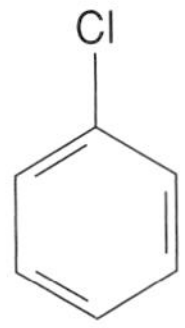
Cl

chlorobenzene

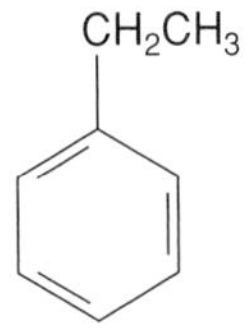
CH2CH3

ethylbenzene

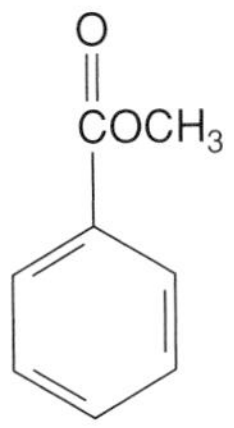
O
COCH3

methyl benzoate

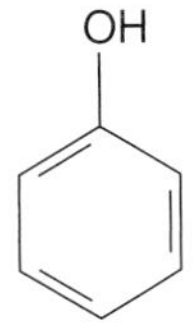
OH

phenol

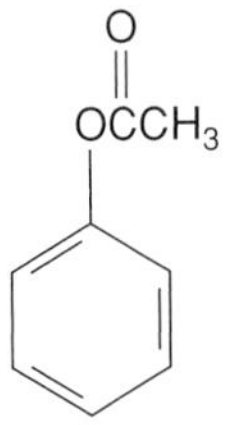
O
OCCH3

phenyl acetate

Prelab Assignment

1. The methyl group is an ortho, para-director. Write the mechanism for the para-bromination of toluene.
2. Construct a table to record the reaction rates for toluene and each of the substituted aromatic compounds tested in the lab. The table should have two columns, one for the compound tested and one to record the rate of the reaction as compared to toluene. In the table, list all of the compounds to be tested.

Experimental Procedure

Safety First!

Always wear eye protection in the laboratory.

1. Wear eye protection at all times in the laboratory.
2. Wear gloves when handling all reagents in this experiment.
3. Bromine is a highly toxic oxidizer. Keep bromine solutions under the hood and avoid breathing the vapors. The solutions used in this experiment are dilute; nevertheless, extreme care must be taken when working with bromine solutions. Inform the instructor of any spills.

4. Some students are sensitive to halogenated aromatic compounds. Avoid skin contact with all aromatic compounds. Wash carefully if your skin itches or tingles.

5. Phenol is highly toxic and corrosive. Wear gloves when handling solutions of phenol.

Microscale Bromination of Substituted Benzene Derivatives

Do all work under the hood!

Prepare a table in your lab notebook. Record all results in the table. Prepare an ice-water bath and a water bath set to approximately 70°C. The water level of the water bath must be higher than the volume of liquid in the test tubes. (Alternately, an 800-mL beaker and a 250-mL beaker can be used. The 250-mL beaker should be inverted inside the 800-mL beaker and the beaker filled with water. Let stand for 15–20 minutes to come to thermal equilibrium.) Do all work under the hood and keep solutions away from direct sunlight.

Obtain a test tube rack, fourteen 15×125 mm test tubes and 12 cork stoppers. The individual test tubes will fit around the 250-mL beaker without falling. Label the test tubes with the names of the aromatic compounds to be tested.

In the first test tube, put 1.0 mL of 0.1 M toluene. In the other test tubes, add 1.0 mL of 0.1 M solutions of the assigned substituted aromatic compounds to the appropriately labeled test tube. Chill the solutions in an ice bath for 5–10 minutes. Working under the hood, add 1.5 mL of bromine in acetic acid all at once to each test tube and cork the tube. Rapidly swirl to mix the contents thoroughly. Return the test tubes to the ice bath for 5 minutes. Record any tubes that react (the red color of the bromine disappears) and remove these tubes from the ice bath. The solution may be colorless or it may have a slight yellow tinge.

Remove the ice bath and let the solutions stand at room temperature for 5 minutes. Record any reactions that occur and their relative order and remove these test tubes. Place the remaining test tubes in the 70°C water bath, noting color changes as they occur. Continue to monitor the reactions for one hour or until all but one test tube has reacted.

In the table, estimate the rate of the reaction with respect to toluene. If a color change occurs while in an ice bath, record the results as "very fast." If a color change occurs at room temperature, but much faster than toluene, record the results as "fast." If the color change occurs about the same time as toluene, record the results as "same time as toluene." If a color change occurs slower than toluene, record the results as "slow." And if the solution does not change color within one hour, record the results as "very slow."

Results and Conclusions

1. From the data, separate the compounds into the following groups depending upon their relative reactivity as compared to toluene: compounds that react very rapidly, compounds that react fast, compounds that react about the same, compounds that react slowly, and compounds that react very slowly. Substituents that react very rapidly are said to be "strongly activating." Substituents that react the same or faster than toluene are said to be "activating." Substituents that react slower are said to be "deactivating," and those that react very slowly (if at all) are said to be "strongly deactivating." Classify each compound as strongly activating, activating, deactivating, or strongly deactivating.

2. For each compound within each group, draw complete Lewis structures of the compounds, showing lone pairs of electrons. Indicate the partial charge on the atom directly attached to the aromatic ring. Within each group, examine the structures of the compounds for similarities. What general trait describes compounds

that have activating substituents? What general trait describes compounds that have deactivating substituents? What types of compounds exhibit comparable reactivity to toluene?

3. How do the structures of the aromatic substituents differ between those that are activating and those that are strongly activating? Why does the change in structure change the reactivity?

4. How do the structures of the aromatic substituents differ between those that are deactivating and those that are strongly deactivating? Why does the change in structure change the reactivity?

5. Do compounds that exhibit decreased reactivity toward electrophilic aromatic bromination contain electron-withdrawing or electron-donating groups? Explain.

6. Do compounds that exhibit decreased reactivity toward electrophilic aromatic bromination contain electron-withdrawing or electron-donating groups? Explain.

7. Draw a potential energy diagram to show the effect that aromatic compounds containing electron-donating groups have on the rate of electrophilic aromatic substitution. On the same diagram, show the effect that electron-withdrawing substituents have on the rate.

8. Knowing that activating substituents direct ortho, para and that deactivating substituents direct meta, draw structures on the monobrominated products formed by reaction of each compound with bromine.

Cleanup & Disposal

Neutralize any excess bromine by cautious addition of 10% aqueous sodium bisulfite. When neutralized, the solution should be colorless. Wash the neutralized aqueous solution down the drain with water. Place all organic solutions in a container labeled "halogenated organic solvent waste."

Critical Thinking Questions *(The harder ones are marked with a ❖.)*

1. Explain the observed order of reactivity of compounds tested in terms of resonance and inductive effects.

2. Explain how the electronic properties of an attached group can affect the rate of electrophilic aromatic substitution and the orientation.

3. Explain why aromatic compounds undergo electrophilic aromatic substitution rather than addition.

4. The reactivities of phenol and aniline toward bromination change when the reaction is performed in sulfuric acid rather than acetic acid. Explain this observation.

5. Most substituents that are activating toward electrophilic aromatic bromination are also ortho, para-directors, while substituents that are deactivating are meta-directors. Iodobenzene is less reactive than benzene, yet is still an ortho, para-director. Explain.

6. Predict the major monobrominated product for the reaction of each compound with bromine:

(a) $H_2N-\!\!\!\bigcirc\!\!\!-NHCCH_3$ (b)

❖ 7. β-Naphthol has the structure shown below. Use resonance forms of brominated intermediates to predict the major product obtained from electrophilic aromatic bromination.

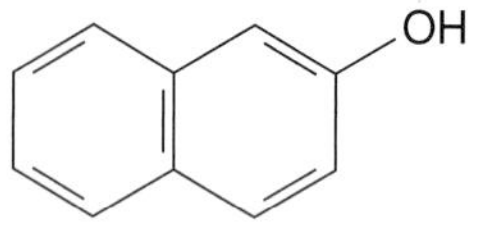

β-naphthol

❖ 8. Some aromatic compounds also undergo nucleophilic aromatic substitution. Propose a mechanism for this reaction and indicate what types of substituents (electron-donating or electron-withdrawing) would be expected to be the most reactive.

9. In the cleanup after this experiment, molecular bromine is treated with sodium bisulfite to discharge the red color. What type of a reaction does this illustrate? The products of this reaction are sodium bromide and sodium sulfate. Write a balanced chemical equation for this reaction.

Experiment 12.2: Nitration of Methyl Benzoate or an Unknown Aromatic Compound

The nitration of methyl benzoate is an example of an electrophilic aromatic substitution reaction. In this experiment, you will:

♦ perform an electrophilic aromatic substitution reaction.

♦ predict the effect that an electron-withdrawing group will have on the orientation of substitution.

♦ determine the identity of the product and write a mechanism to account for the formation of the observed product.

Techniques

Technique F	Vacuum filtration and recrystallization
Technique M	Infrared spectroscopy
Technique N	Nuclear magnetic resonance spectroscopy

Background and Mechanism

Aromatic compounds undergo electrophilic aromatic substitution reactions. In Parts A and B of this experiment, the electrophile is the nitronium ion (NO_2^+), the aromatic compound is methyl benzoate, and the substitution is the replacement of a hydrogen with a nitro group.

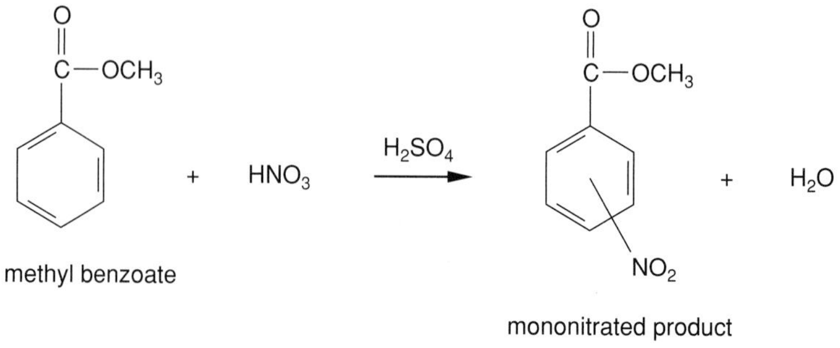

methyl benzoate

mononitrated product

The mechanism is illustrated below for benzene. The first step is formation of the nitronium ion by an acid-base reaction between H_2SO_4 and HNO_3. HNO_3 accepts a proton from H_2SO_4, then loses water to generate the nitronium ion:

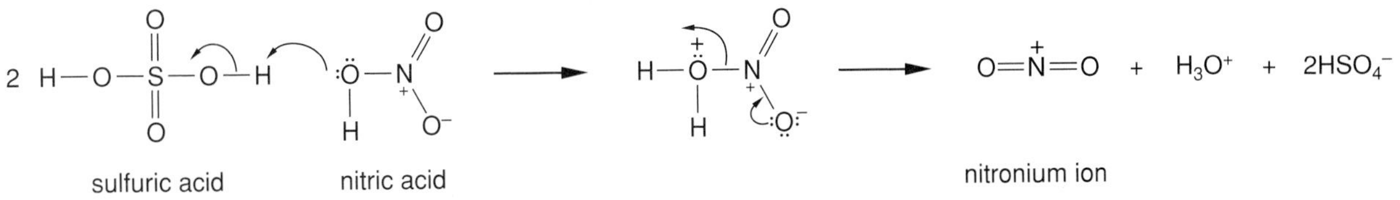

sulfuric acid   nitric acid   nitronium ion

The next step in the reaction (the rate-limiting step) involves reaction of the nitronium ion with the aromatic ring to form a sigma complex:

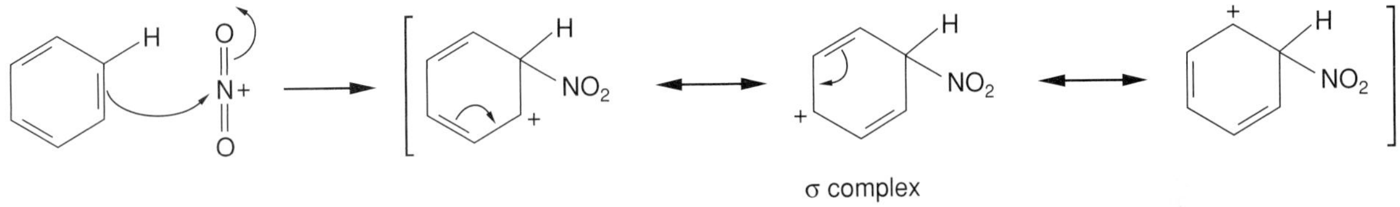

σ complex

Even though the intermediate sigma complex is resonance-stabilized, it has lost aromaticity due to the formation of the sp^3-hybridized carbon. Loss of a proton regenerates the aromatic ring.

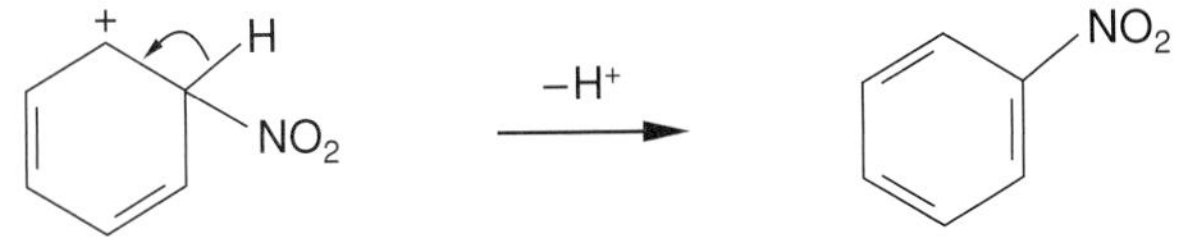

The product is less likely to undergo nitration than the starting material. (Why?) Polynitrated products are generally not formed unless the temperature is increased.

In Parts A and B, the reaction of methyl benzoate with HNO_3 and H_2SO_4 can form three possible methyl nitrobenzoates. You will determine the identity of the mononitrated product formed in this reaction.

In Part C of this experiment, you will nitrate an unknown aromatic compound. Based upon the melting point of the recrystallized product, you will identify the compound and propose a mechanism to account for its formation. For a related experiment, see McElveen et al., 1999.

Prelab Assignment

1. Determine the number of mmoles of HNO_3, H_2SO_4, and methyl benzoate that will be used in this experiment. Concentrated HNO_3 is 16 M; concentrated H_2SO_4 is 18 M.
2. Determine the limiting reagent in this experiment.
3. Write a flow scheme for the workup procedure for this reaction. Be sure to indicate where in the reaction workup excess nitric and sulfuric acid are removed and where any excess methyl benzoate is removed.

Experimental Procedure

1.   Wear eye protection at all times in the laboratory.

2.   Wear gloves when handling all of the reagents in this experiment.

3.   Concentrated sulfuric acid and nitric acid are extremely caustic and can cause severe burns. Avoid skin contact. In case of skin contact, immediately rinse the affected area with cold water. If acid spills on the lab bench, cautiously neutralize the spill with sodium bicarbonate. Be sure to notify the lab instructor.

4.   Some individuals are sensitive toward nitrated aromatic compounds. Avoid contact with the product of this reaction. If tingling or swelling is observed, rinse off the affected area with cool water and notify the instructor.

Part A: Microscale Nitration of Methyl Benzoate

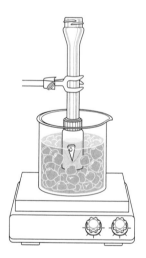

Prepare an ice bath in a 100-mL beaker on a magnetic stir plate. To a clean, dry 5-mL conical vial, pipet 0.20 mL of methyl benzoate. To the methyl benzoate, add 0.45 mL of concentrated (18 M) sulfuric acid and a spin vane. Attach an air condenser to the conical vial. Carefully clamp the apparatus so that the vial is immersed in the ice bath. Begin stirring. Into a clean, dry 3-mL conical vial, add 0.15 mL of concentrated (18 M) sulfuric acid and 0.15 mL of concentrated (16 M) nitric acid. Cool the mixture in a second ice bath. With a Pasteur pipet, slowly add the cooled sulfuric acid–nitric acid solution to the stirred solution of methyl benzoate through the air condenser. The addition should take approximately 15 minutes. Adding the nitrating agent more rapidly increases the amount of by-products and decreases the yield.

When all the nitrating agent has been added, carefully remove the ice bath. Continue stirring until the reaction has come to room temperature. Then let the reaction flask stand undisturbed for 15 minutes. Place 2 g of ice in a 30-mL beaker. With a Pasteur pipet, transfer the reaction mixture to the beaker. Rinse the reaction vial with 1 mL of cold water. When the ice has melted, vacuum filter the crude crystals using a Hirsch funnel. Wash the crystals with two 1.0-mL portions of cold water and finally with two 0.3-mL portions of cold methanol. The crude product can be recrystallized from methanol. Record the mass of the dry product and calculate the yield.

Part B: Miniscale Nitration of Methyl Benzoate

Caution!

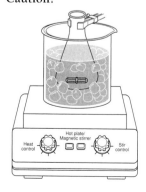

Prepare an ice bath in a 100-mL beaker on a magnetic stir plate. To a clean, dry 25-mL Erlenmeyer flask, add 1.0 mL of methyl benzoate via pipet. Calculate the mass of methyl benzoate. To the methyl benzoate, add 2.25 mL of concentrated (18 M) sulfuric acid and a spin bar. Carefully clamp the flask so that it is immersed in the ice bath and begin stirring. Prepare the nitrating solution by adding 0.75 mL of concentrated (18 M) sulfuric acid and 0.75 mL of concentrated (16 M) nitric acid to a clean, dry vial. Cool the mixture in a second ice bath. With a Pasteur pipet, slowly add the cooled sulfuric acid–nitric acid solution to the stirred solution of methyl benzoate. The addition should take approximately 15 minutes. Adding the nitrating agent more rapidly will increase the amount of by-products and decrease the yield.

After all of the nitrating agent has been added, carefully remove the ice bath. Continue stirring until the reaction has come to room temperature. Then let the reaction flask stand undisturbed for 15 minutes. Place 10 g of ice in a 100-mL beaker. With a Pasteur pipet, carefully transfer the reaction mixture to the beaker. Rinse the flask with 5 mL of cold water. When the ice has melted, vacuum filter the crude crystals using a Büchner funnel. Wash the crystals with two 5-mL portions of cold water and finally with two 1.5-mL portions of cold methanol. The crude product can be recrystallized from methanol. Record the mass of the dry product and calculate the yield.

Characterization for Parts A and B

Melting Point: Determine the melting point of the recrystallized product. The melting points of the three isomeric methyl nitrobenzoates are methyl 2-nitrobenzoate (–13°C), methyl 3-nitrobenzoate (78°C), and methyl 4-nitrobenzoate (95°C).

Infrared Spectroscopy: Obtain an IR spectrum using either the KBr or Nujol method. Use the overtone bands (2000–1667 cm^{-1}) and out-of-plane bending vibrations (900–690 cm^{-1}) to determine the substitution pattern of the product.

NMR Spectroscopy: Obtain a ^{1}H NMR spectrum of the product in CDCl$_3$ containing TMS. Interpret the spectrum, examining closely the aromatic absorption around δ7. The ^{1}H NMR spectra of the three isomeric methyl nitrobenzoates are shown in Figure 12.2-1.

Results and Conclusions for Parts A and B

1. What is the major product formed in this reaction? Justify your reasoning.
2. Write the mechanism for the formation of the observed product.
3. Calculate the percent yield of the product.
4. Explain, using appropriate resonance structures, why the other mononitrated isomers were not formed to any significant extent in this reaction.
5. Explain why polynitration is not favored in this experimental procedure.

Part C: Miniscale Nitration of an Unknown Aromatic Compound

Prepare an ice-salt bath in a 100- or 150-mL beaker on a magnetic stir plate. The temperature of the ice-salt bath should be –8°C. Place 1 g of the aromatic compound in a 25-mL Erlenmeyer flask and add 2.25 mL of concentrated (18 M) sulfuric acid and a stir bar. Carefully place the flask in the icebath so it stays immersed and begin stirring. Prepare the nitrating solution by adding 0.75 mL of concentrated (18 M) sulfuric acid and 0.75 mL of concentrated (16 M) nitric acid to a clean, dry vial. Cool this mixture in the ice bath. With a Pasteur pipet, slowly add the cooled nitrating solution to the stirring solution containing the aromatic compound. The addition should take approximately 15 minutes. Adding the nitrating agent more rapidly will increase the amount of by-products and decrease the yield.

After all the nitrating agent has been added, carefully remove the icebath. Continue stirring until the reaction has come to room temperature. Then let the reaction flask stand undisturbed for 15 minutes. Place 10 g of ice in a 50-mL beaker. With a Pasteur pipet, carefully transfer the reaction mixture to the beaker containing the ice. Rinse the flask with 5 mL of cold water. When the ice has melted, vacuum filter the crude crystals using a Büchner funnel. Wash the crystals with two 5-mL portions of cold water. Recrystallize the product. To select the recrystallization solvent, test small amounts of the crude product in methanol, 2-propanol, or other solvents. Alternatively, the instructor may suggest an appropriate recrystallization solvent for the specific unknown. Let the crystals dry thoroughly before measuring the melting point. Record the mass of the dry product and calculate the percent yield. Determine the identity of the product based on the melting point. Melting points of the possible products are detailed in Table 12.2-1.

Figure 12.2-1

^{1}H NMR spectra of methyl 2-nitrobenzoate, methyl 3-nitrobenzoate, and methyl 4-nitrobenzoate

SOURCE: Reprinted with permission of Aldrich Chemical.

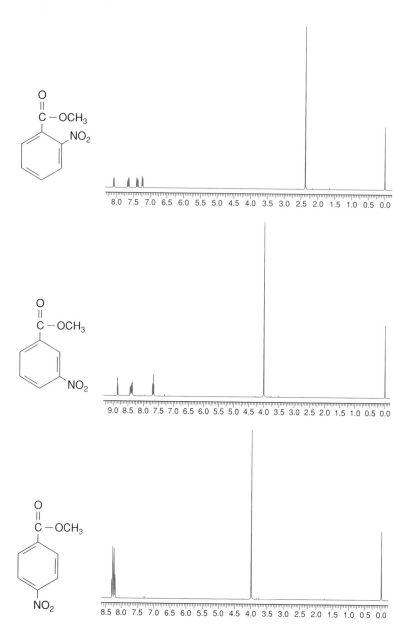

Results and Conclusions for Part C

1. Based upon the observed melting point, identify the major product formed in this reaction.
2. Write the mechanism to account for the formation of the observed product.
3. Explain, using resonance structures of the starting compound, why the observed product was obtained rather than either of the other isomers.
4. Calculate the percent yield of the product.
5. Explain why polynitration is not favored in this experimental procedure.

Table 12.2-1: Melting Points of Products

Starting compound	Melting points of possible nitration products		
	ortho-Product (°C)	*meta*-Product (°C)	*para*-Product (°C)
Methyl benzoate	−13	78	95
Acetanilide	94	156	217
Benzaldehyde	44	59	104
Benzamide	178	143	201
Benzonitrile	109	118	149
Benzoic acid	148	142	239

Cleanup & Disposal

Carefully neutralize acidic filtrates with 10% sodium bicarbonate solution. Wash the neutralized aqueous solutions down the drain with water. Place the filtrate from recrystallization in a container labeled "nonhalogenated organic solvent waste."

Critical Thinking Questions *(The harder one is marked with a ❖.)*

1. Why must the nitric acid/sulfuric acid solution be added slowly to the methyl benzoate?

2. If the temperature of the reaction is too high, dinitration can occur. Explain why higher temperatures are necessary for dinitration. Predict the structure of your dinitrated product.

3. What is the purpose of washing the crystals with water and with ice-cold methanol?

4. How could the procedure be modified to produce dinitrated product?

5. What is the purpose of pouring the reaction mixture over ice?

❖ 6. A student neglected to cool the methyl benzoate and the nitrating agent. In addition, the student added the nitrating agent all at once. The major product, which contained two nitro groups, dissolved when the crystals were washed with methanol. In order to characterize this by-product, an alert student suggested trying to dissolve the by-product in aqueous $NaHCO_3$. The by-product readily dissolved. Suggest a possible structure for this compound. Explain how it might have been formed and suggest a method of removing this by-product.

7. Predict the major mononitrated product for each of the following compounds:
   a. phenyl benzoate.
   b. biphenyl.
   c. phenyl acetate.

8. A possible contaminant in the reaction mixture is a nitrobenzoic acid. How is this by-product formed? Suggest a method for removing the by-product.

9. Explain why toluene and trifluoromethylbenzene give different mononitrated products.

10. A student elected to synthesize methyl *meta*-nitrobenzoate by the following reaction scheme. Predict the outcome of this sequence and explain.

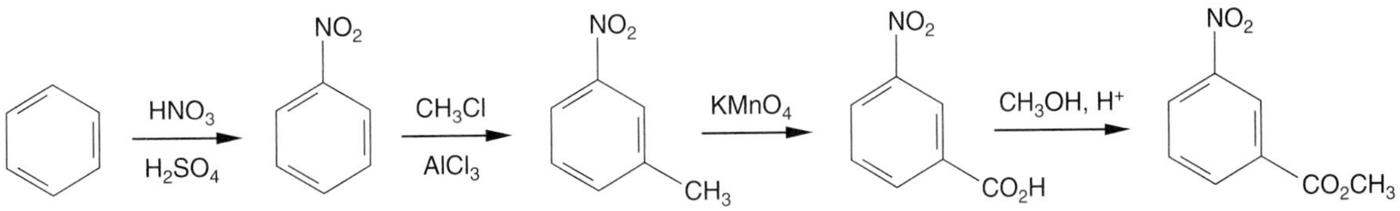

11. Devise a reasonable synthesis of methyl *p*-nitrobenzoate from benzene.

12. *Library project:* Polynitrated compounds are often explosive. TNT (trinitrotoluene) is probably the most well-known polynitrated explosive, but there are others. Use the chemical literature to find the structure of another explosive polynitrated compound. Give the complete reference citation.

13. *Library project:* In this experiment the nitrating agent used was HNO_3/H_2SO_4. Use the chemical literature to find other possible nitrating agents. Cite the complete references.

14. *Molecular modeling assignment:* Construct a model of the starting reagent and examine the highest occupied molecular orbital (HOMO). Where are the possible sites of electrophilic attack? Which site is most probable? Draw resonance forms for the intermediates formed in the assigned reaction and calculate the energy leading to its formation. Draw resonance forms for the other possible products and calculate those energies. Compare the energies.

Reference

McElveen, S. R., Gavaardinas, K., Stamberger, J. A., and Mohan, R. S. The Discovery-Oriented Approach to Organic Chemistry 1. Nitration of Unknown Organic Compounds. *J. Chem. Educ.* 75 (1999): 535.

Experiment 12.3: Friedel-Crafts Acylation Reactions

The purpose of Part A of this experiment is to illustrate Friedel-Crafts acylation where a single isolable product may be obtained in good overall yield. In Part B, several acylation products may be visualized using TLC analysis. Part C of the experiment illustrates the use of column chromatography to separate the colored reactant and colored products. You will:

*   carry out a Friedel-Crafts acylation reaction.

*   study the directive and steric effects of phenyl substituents in electrophilic aromatic substitution.

*   apply TLC techniques to assess reaction progress and outcome.

*   use a Craig tube for recrystallization.

*   purify the product using column chromatography.

Techniques

Technique F	Filtration and recrystallization
Technique G	Reflux
Technique I	Extraction and drying
Technique K	Thin-layer and column chromatography
Technique M	Infrared spectroscopy
Technique N	Nuclear magnetic resonance spectroscopy

Background

Acylation of biphenyl using acetyl chloride and aluminum chloride is a convenient, practical illustration of Friedel-Crafts acylation. A phenyl substituent is an ortho, para-directing group. Because of the large size of the phenyl group, the principal acylation product is the para isomer.

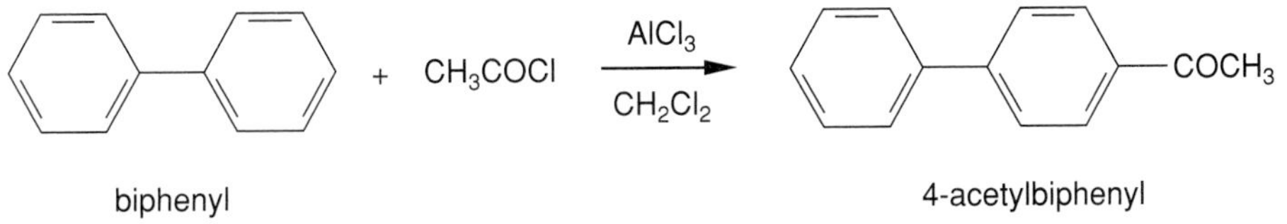

biphenyl $+$ CH_3COCl $\xrightarrow[CH_2Cl_2]{AlCl_3}$ 4-acetylbiphenyl

Phenanthrene has multiple sites for acylation and you will determine which sites are acylated using thin-layer chromatography (TLC).

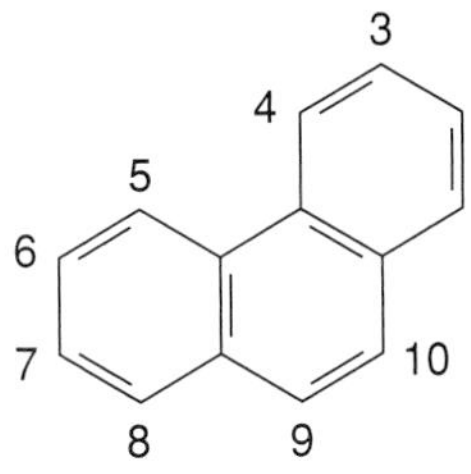

phenanthrene

Ferrocene is an aromatic organometallic compound. Ferrocene consists of activated dicyclopentadienyl rings, which readily undergo acylation reactions. Benzoylation yields highly colored products that can be visualized directly on TLC and in column chromatography. In this part of the experiment, you have the option of making mostly monobenzoylated or mostly dibenzoylated product, depending upon the amount of benzoyl chloride that is used in the synthesis.

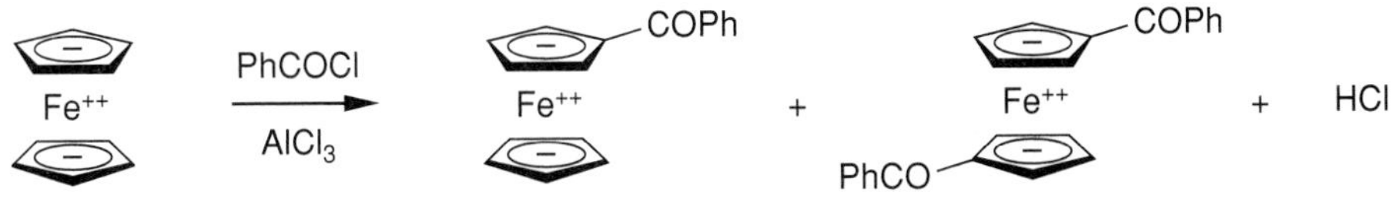

Prelab Assignment

1. Calculate the masses of each reagent used in the reaction and determine the limiting reagent.
2. Design a flow scheme for the workup procedure used in this experiment.
3. (Part A) Write a reaction mechanism for the formation of 4-acetylbiphenyl.
4. (Part A) Explain why 4-acetylbiphenyl is the main product expected.
5. (Part B) Draw structures of all possible monoacylated phenanthrenes.
6. (Part B) Write resonance forms of the intermediate formed in the acetylation of 3-acetylphenanthrene. There are seven resonance forms.
7. (Part C) Explain the high reactivity of ferrocene toward electrophilic substitution.
8. (Part C) Write the mechanism of benzoylation of ferrocene by benzoyl chloride and $AlCl_3$.

Experimental Procedure

1. Wear eye protection at all times in the laboratory.

2. Wear gloves when transfering reagents in this experiment. Use care when adding glass wool to a chromatography column because small glass fibers can penetrate the skin.

3. Acetyl chloride and benzoyl chloride are lachrymators that must be measured out carefully in a fume hood. Methylene chloride is a toxic irritant and a suspected carcinogen. Avoid skin contact and breathing the vapors.

4. Aluminum chloride reacts vigorously with water and with water vapor in the air to form noxious HCl. All glassware must be completely dry before starting the experiment. Aluminum chloride should be measured out quickly. Avoid skin contact.

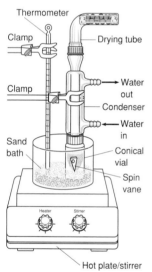

Part A: Microscale Procedure for Acetylation of Biphenyl

All glassware to be used for the reaction must be dried in an oven or with a flame before starting the experiment. Cool all glassware to room temperature after drying. Fit a 5-mL vial containing a spin vane with a water condenser and a drying tube that contains fresh granular drying agent (CaCl$_2$). Add 1.25 mmol of anhydrous aluminum chloride and 2 mL of methylene chloride to the vial. Next, add 0.5 mmol of powdered biphenyl. Add 50 µL of acetyl chloride by automatic pipet. Stir the mixture for a few minutes at room temperature. Then stir and heat the mixture at reflux for 10 minutes using a sand bath or heat block. Cool to room temperature. Place the reaction vial in an ice bath. Add 1 mL of 3 M HCl dropwise to the vial. Remove the spin vane. Cap the vial and shake the contents with venting. Any remaining aluminum chloride should dissolve in the aqueous layer. Draw off the bottom layer with a Pasteur pipet and save. Remove and discard the aqueous layer remaining in the vial. Place the organic layer back in the vial. Add 1 mL of 1 M NaOH solution to the vial. Cap the vial and shake, with venting. Remove the organic layer and transfer to another vial. Draw off and discard the aqueous layer. Add 1 mL of water to the vial containing the organic layer. Repeat the washing procedure. Draw off the organic layer and place in an Erlenmeyer flask. Dry the organic layer over anhydrous sodium sulfate. Transfer the dry organic solution to a clean, dry 10-mL Erlenmeyer flask using a filter pipet. Evaporate the methylene chloride from the filtrate to give the crude solid product. Recrystallize from ethanol using either a Craig tube or an Erlenmeyer flask followed by filtration using a Hirsch funnel. Weigh the recrystallized product.

Characterization for Part A

Melting Point: The reported melting point of 4-acetylbiphenyl is 120–121°C.

Infrared Spectroscopy: Obtain an IR spectrum of the product using a KBr pellet or Nujol mull. The spectrum should show strong carbonyl absorption at 1700 cm^{-1} and both mono- and disubstituted aromatic absorption bands.

NMR Spectroscopy: Dissolve the product in CDCl$_3$ and obtain the proton NMR spectrum. The product should show a methyl singlet at 2.6. Interpret the signals observed in the aromatic region. Also run the ^{13}C NMR spectrum, if possible.

Thin-layer Chromatography: Spot solutions of samples to be analyzed, including a standard solution of 4-acetylbiphenyl (if available) and biphenyl. Develop the TLC in 19:1 methylene chloride-ethyl acetate.

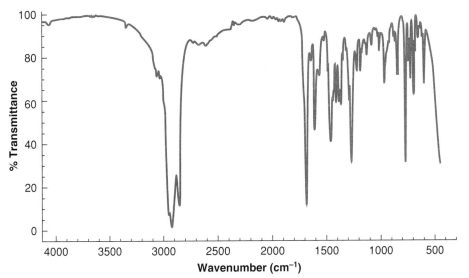

Figure 12.3-1

IR spectrum of 4-acetylbiphenyl (Nujol)

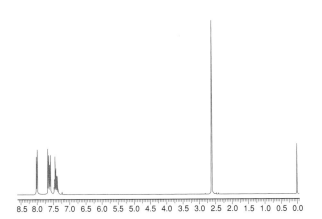

Figure 12.3-2 ^{1}H NMR spectrum of
4-acetylbiphenyl (CDCl$_3$)

SOURCE: Reprinted with permission of Aldrich Chemical.

Figure 12.3-3 ^{13}C NMR spectrum of
4-acetylbiphenyl (CDCl$_3$)

SOURCE: Reprinted with permission of Aldrich Chemical.

Results and Conclusions for Part A

1. Calculate the percent yield of product.
2. Was 4-acetylbiphenyl the major product obtained? Justify.
3. Based upon the yield obtained of acetylated product, how successful was this preparation? At what points in the procedure might product have been lost?

Part B: Microscale Procedure for Acetylation of Phenanthrene with TLC Analysis

All glassware to be used for the reaction must be dried in an oven or flame before starting the experiment. Cool all glassware to room temperature after drying. Fit a 5-mL vial containing a spin vane with a water-jacketed condenser and a drying tube that contains fresh granular drying agent (CaCl$_2$). Add 1.25 mmol of anhydrous aluminum chloride

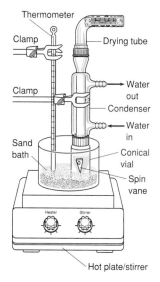

Thermometer

Clamp — Drying tube

Clamp

Water out
Condenser
Water in

Sand bath

Conical vial
Spin vane

Heater Stirrer

Hot plate/stirrer

and 2 mL of methylene chloride to the vial. Next, add 0.5 mmol of powdered phenanthrene. Add 50 μL of acetyl chloride by automatic pipet. Stir the mixture for a few minutes at room temperature. Stir and heat the mixture at reflux for 10 minutes using a sand bath or heat block. Allow to cool to room temperature. Place the reaction vial in an ice bath. Add 1 mL of 3 M HCl dropwise to the vial. Cap the vial and shake the contents with venting. Any remaining aluminum chloride should dissolve in the aqueous layer. Draw off the bottom layer with a Pasteur pipet and save. Remove and discard the aqueous layer remaining in the vial. Place the organic layer back in the vial. Add 1 mL of 1 M NaOH solution to the vial. Cap the vial and shake, with venting. Remove the organic layer and transfer to another vial. Draw off and discard the aqueous layer. Add 1 mL of water to the vial containing the organic layer. Repeat the washing procedure. Draw off the organic layer and place in an Erlenmeyer flask. Dry the organic layer over anhydrous sodium sulfate.

While the solution is drying, spot a TLC plate with standard solutions of phenanthrene, 2-acetylphenanthrene, 3-acetylphenanthrene, and 9-acetylphenanthrene. Tare a conical vial containing a boiling chip and transfer the dry organic solution using a filter pipet. Spot an aliquot of the solution on the TLC plate next to the standard compounds. Develop the TLC plate in 19:1 methylene chloride/ ethyl acetate. Record the R_f values for each component on the TLC plate. Evaporate the methylene chloride under the hood to give an oily mixture of products. Record the mass of the product mixture.

Characterization for Part B

Infrared Spectroscopy: Obtain an IR spectrum of the product mixture. The spectrum should show strong carbonyl absorption at 1700 cm^{-1} and aromatic absorption bands. The spectrum may be compared with recorded spectra in the literature to help determine which isomer is the major product.

Results and Conclusions for Part B

1. Calculate the percent yield of product mixture based upon the weight of oil obtained.
2. Based on the TLC data, determine the structures of the products formed in this reaction. Justify.
3. Which isomer is the major product? Explain why this isomer predominates.
4. At what points in the procedure might product have been lost?

Part C: Microscale Benzoylation of Ferrocene with Column Chromatography

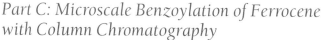

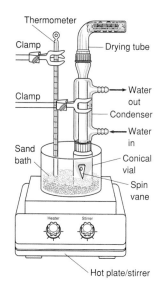

All glassware to be used for the reaction must be dried in an oven or a flame before starting the experiment. Cool all glassware to room temperature after drying. Fit a 5-mL vial containing a spin vane with a water condenser and a drying tube that contains fresh granular drying agent (CaCl$_2$). Add 0.6 mmol of anhydrous aluminum chloride and 2 mL of methylene chloride to the vial. Next, add 0.5 mmol of ferrocene. According to your instructor's directions add 40–80 μL of benzoyl chloride by automatic pipet. (Addition of 80 μL favors dibenzoylation.) Stir the mixture for a few minutes at room temperature. Stir and heat the mixture at reflux for 5 minutes using a sand bath or heat block and allow to cool to room temperature. Place the reaction vial in an ice bath. Add 1 mL of 3 M HCl dropwise to the vial. Cap the vial and shake the contents with venting. Any remaining aluminum chloride should dissolve in the aqueous layer. Draw off the bottom layer with a Pasteur pipet and save. Remove and discard the aqueous layer remaining in the vial. Place the organic layer back in the vial. Add 1 mL of 1 M NaOH solution to the vial. Cap the vial and shake, with venting. Remove the organic layer and transfer to another vial. Draw off and discard the aqueous layer. Add 1 mL of water to the vial containing the organic layer.

Repeat the washing procedure. Draw off the organic layer and place in an Erlenmeyer flask. Dry the organic layer over anhydrous sodium sulfate. Transfer the dry organic solution to a clean, dry conical vial using a filter pipet. Evaporate all but about 0.5 mL of the methylene chloride and cork the flask. Run a silica gel TLC of a sample of the solution.

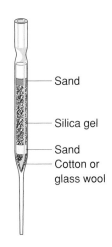

Prepare a pipet for chromatography by first adding cotton or glass wool, then sand, silica gel, and finally a top layer of sand. The pipet should be about two-thirds full. Pour petroleum ether through the column until solvent begins to elute from the bottom of the column. Drain solvent from the column until the level in the reservoir is even with the top of the sand. Using a pipet, apply the methylene chloride solution of benzoylated ferrocene evenly to the top of the column. Drain solvent from the column until the level is even with the top of the sand. Carefully add a few mL of petroleum ether while minimally affecting the layer of sand. Allow a few mL of solvent to elute from the column. Do not allow any portion of the column to go dry. Now add more petroleum ether to the column. Elute solvent from the column. Collect fractions using tared Erlenmeyer flasks. A yellow band of any remaining ferrocene should be collected in the first flask. Next, add 1:1 petroleum ether/methylene chloride to elute the red band consisting of benzoylferrocene into a second flask. The last purplish band of dibenzoylferrocene may be eluted into a third flask using methylene chloride. Evaporate the solvents on a warm sand bath or water bath in a hood. Use a stream of dry air or nitrogen to hasten the evaporation. Weigh the flasks with the solid residues and determine the mass of each compound.

Characterization for Part C

Melting Point: The reported melting points are: benzoylferrocene, 111–112°C; 1,1'-dibenzoylferrocene, 105–106°C; ferrocene, 172–173°C.

Infrared Spectroscopy: Obtain an IR spectrum of the product using a KBr pellet or Nujol mull. The spectrum should show strong carbonyl absorption at 1700 cm^{-1} and aromatic absorption bands.

NMR Spectroscopy: Dissolve each of the products in $CDCl_3$ and obtain the ^1H NMR spectra. Compare the spectra of benzoyl- and 1,1'-dibenzoylferrocene with the NMR of ferrocene.

Results and Conclusions for Part C

1. Calculate the percent yield of benzoylferrocene and 1,1'-dibenzoylferrocene.
2. If three equivalents of benzoyl chloride were used, what product would be obtained?

C l e a n u p & D i s p o s a l

Discard aqueous solutions down the drain. Neutralize and clean up any spills of aqueous acid or base before leaving the work area. Transfer solutions of petroleum ether to a bottle in the hood for nonhalogenated hydrocarbons. Place any remaining methylene chloride solutions in a container labeled "halogenated waste."

Critical Thinking Questions *(The harder ones are marked with a ❖.)*

1. One difference between Friedel-Crafts alkylation and acylation is that a stoichiometric amount of $AlCl_3$ is required for the acylation reaction. Explain.

2. Suggest a reason why 3 M HCl was used instead of aqueous base in the workup of the crude reaction mixtures from acetylation.

❖ 3. 2-Acetylbiphenyl is a possible acylation product in Part A of this experiment, starting with biphenyl. Where would this product be removed in the workup procedure?

4. 3-Acetylbiphenyl is a low-melting solid (mp 35°C). Is it necessary to be concerned about removal of this substance during purification of the product from Part A? Explain.

5. Predict the reaction product for the Clemmensen reduction of 4-acetylbiphenyl. This product can also be made through direct Friedel-Crafts alkylation of biphenyl with ethyl bromide and $AlCl_3$. Discuss disadvantages of the alkylation method.

❖ 6. Predict the expected product of acetylation of 4-bromobiphenyl. Explain.

7. Predict the structure of the major diacetylated product expected from biphenyl.

8. Why does benzoylation not occur twice on the same ring of ferrocene?

9. Account for the fact that ferrocene is eluted by the nonpolar solvent, petroleum ether, during column chromatography.

10. What feature of the benzoylation of ferrocene makes it particularly attractive for a column chromatography experiment?

Experiment 12.4: Aromatic Bromination

This experiment illustrates aromatic substitution and directive effects of substituents. In this experiment, you will:

♦ brominate an activated aromatic compound and isolate the solid product.

♦ evaluate the products expected from bromination of activated aromatic rings.

♦ evaluate the directive effects for rings that have more than one substituent.

♦ design an experiment to brominate an acetanilide derivative.

Techniques

Technique F	Recrystallization
Technique I	Extraction
Technique K	Thin-layer, column and high-performance liquid chromatography
Technique M	Infrared spectroscopy
Technique N	Nuclear magnetic resonance spectroscopy

Background and Mechanism

Aromatic bromination is a common example of electrophilic aromatic substitution. Molecular bromine is used as the reagent in bromination reactions. A Lewis acid catalyst, such as $FeBr_3$, is required for those rings that lack activating groups. In this experiment, an activating acetamido group is present as a substituent in the starting material. Therefore, catalysis by a Lewis acid is unnecessary. Bromination of acetanilide is accomplished using molecular bromine in acetic acid.

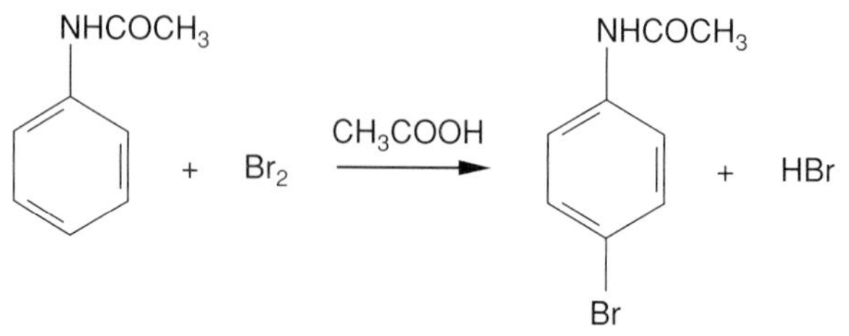

Bromination of *p*-methylacetanilide in Part B illustrates the relative activating strengths of two different activating groups. Both substituents (acetamido and methyl) are ortho/para-directing, activating groups. There are two possible monobrominated products. The acetamido substituent is more powerfully activating than an alkyl group, so the major product is predicted to be 2-bromo-4-methylacetanilide.

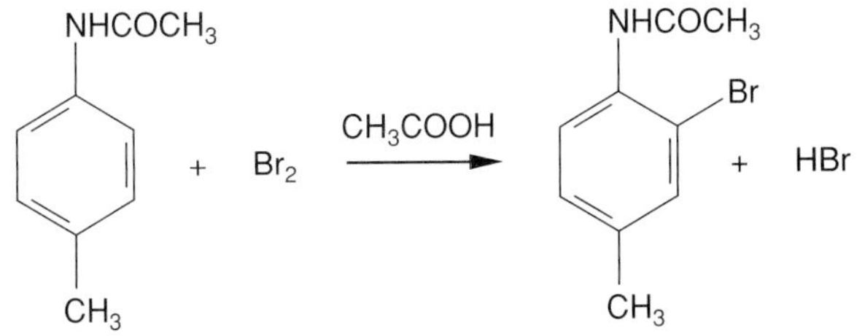

In Part C of this experiment, you will design a procedure to brominate an assigned aromatic compound, which will be one of the following compounds: *o*-chloroacetanilide, *m*-chloroacetanilide, *p*-chloroacetanilide, or another substituted acetanilide as designated by the instructor. A number of factors must be considered when planning the procedure, such as reactivity of the compound toward bromination, possible side products, physical properties of the expected products, and numerous other factors. Based upon these factors, you will synthesize, purify, and characterize a mono-brominated product from the assigned compound. In order to modify the procedure appropriately, you must analyze directive effects to predict the possible product(s) and consider reactivity and physical properties of reagents and product(s).

Prelab Assignment

1. (For Parts A and B) Calculate the mass of acetanilide or another aromatic compound used in the experiment.
2. (For Parts A and B) Provide a detailed mechanism for the reaction used in the designated procedure.
3. (For Part C) Develop a procedure to brominate the assigned acetanilide.

Experimental Procedure

1. Wear eye protection at all times in the laboratory.
2. **Bromine is an extremely hazardous oxidant and irritant.** Wear gloves when handling solutions containing molecular bromine. Bromine solutions should be dispensed in a hood. If any bromine solution comes in contact with the skin, if vapors are inhaled, or if any bromine solution is spilled in the hood, inform the instructor immediately.
3. Acetic acid is corrosive. Use gloves when handling acetic acid and the solution of bromine in acetic acid.

Part A: Microscale Bromination of Acetanilide

Weigh out acetanilide (1.60 mmol) and place it in a 5.0-mL reaction vial (or 10-mL Erlenmeyer) containing a spin vane or a stir bar. Add 2.0 mL of acetic acid to the vessel and swirl the mixture until all of the solid has dissolved. Add a few more drops of acetic acid if the acetanilide does not dissolve completely. Turn on the stirrer. Add 0.5 mL of freshly prepared 4.1 M bromine in acetic acid solution (1:4 v/v of liquid bromine/glacial acetic acid). Cap the vial or cork the Erlenmeyer as soon as the bromine solution has been added. Soon after the addition, a solid product should begin to precipitate. Stir vigorously at room temperature for 15 minutes.

Work in the hood when working with bromine.

Pour the mixture into a 25-mL Erlenmeyer containing 15 mL of water. Add a solution of 30% sodium thiosulfate dropwise to discharge the reddish-orange color of bromine. Cool the mixture in an ice bath for a few minutes and then collect the solid product on a Hirsch funnel by vacuum filtration. Continue to apply vacuum. Rinse the container with 2 mL of cold water and pour the washings into the Hirsch funnel. If the crystals are yellow, wash with several drops of 30% thiosulfate solution. Wash the crystals with 2 mL of ice-cold water. Continue to apply the vacuum to the system for 10–15 minutes to partially dry the crystals. Recrystallize the crude product using an ethanol/water solvent pair. Dissolve the solid in 2–3 mL of hot 95% ethanol and add hot water until the cloud point is reached. Add a few drops of ethanol until the solution is clear. Set the solution aside undisturbed until crystals form. Cool in an ice bath and collect the product using a suction flask and a Hirsch funnel. Allow the product to air dry to constant weight. (This might require keeping the crystals in your locker until the next period.) Record the yield of product and its melting point.

Characterization for Part A

Melting Point: The reported melting point of *p*-bromoacetanilide is 168°C.

Infrared Spectroscopy: Record the IR spectrum using either a Nujol mull or a KBr pellet. The spectrum of *p*-bromoacetanilide should show N−H absorption at 3400–3200 cm^{-1} and C=O absorption at 1680 cm^{-1}. Examine the spectrum for the characteristic absorption bands common to *p*-disubstituted benzene derivatives. The IR spectrum of *p*-bromoacetanilide is shown in Figure 12.4-1.

NMR Spectroscopy: The ^{1}H NMR spectrum of the product should be run using CDCl$_3$. The spectrum should show an N-H signal near δ8.0, aromatic protons signal near δ7.2 showing the pattern typical of *p*-disubstituted benzenes, and a signal for methyl protons at δ2.2.

Figure 12.4-1

IR spectrum of
p-bromoacetanilide
(Nujol)

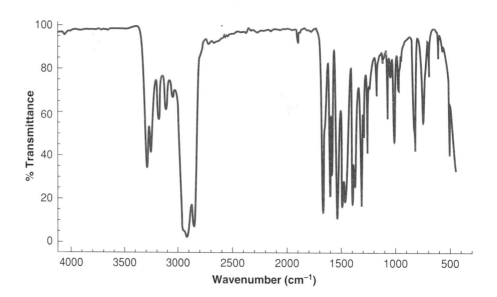

Thin-layer Chromatography: Spot ethanolic solutions of acetanilide, *p*-bromo-acetanilide, and the isolated product on three separate plates. Elute with ethyl acetate, methylene chloride, and diethyl ether respectively. (Use alternate solvents if directed by your instructor.) Measure R_f values for each component on each plate.

Results and Conclusions for Part A

1. Calculate the percent yield of the product.
2. Which TLC solvent was the most effective in distinguishing between product and reactant?
3. Did the quality and quantity of the product isolated match the R_f data about the reaction from TLC?
4. Sodium thiosulfate reduces excess molecular bromine to form sodium bromide and sodium sulfate. Write a balanced equation for the reaction.

Part B: Microscale Bromination of p-Methylacetanilide with TLC and Column Chromatography or HPLC Analysis

Weigh out *p*-methylacetanilide (0.40 mmol) and place it in a 5.0-mL reaction vial containing a spin vane. Add 2.0 mL of acetic acid to the vial and swirl the mixture until all of the solid has dissolved. Add a few more drops of acetic acid if the *p*-methylacetanilide does not dissolve completely. Start the stirrer. Use a Pasteur pipet to add 10 drops of 4.1 M bromine in acetic acid solution. Cap the vial as soon as the bromine solution has been added. Soon after the addition, an oily semisolid product should form. Stir the mixture vigorously at room temperature for 15 minutes.

Work in the hood when working with bromine.

Add 1.0 mL of water and several drops of a solution of 30% sodium thiosulfate dropwise to discharge the reddish-orange color of bromine. The color should disappear, leaving a precipitate at the bottom of the vial. Add 1 mL of methylene chloride, remove the spin vane, and cap the vial. Shake with venting. Draw off the organic layer and place in a separate 5-mL vial. Extract the aqueous layer with another 1 mL of methylene chloride. Draw off the methylene chloride and combine with the first extract. Set aside the aqueous layer to be discarded. Wash the organic layer with two 2-mL portions of 1 M NaOH, followed by one 1-mL portion of water. Dry the organic layer over sodium sulfate. Draw off the dried solution using a filter pipet and transfer the solution to a 3-mL vial. The reaction products can be analyzed qualitatively using TLC. Alternatively, the product mixture can be analyzed by high-pressure liquid chromatography (HPLC).

Purification by column chromatography. Apply the remaining methylene chloride solution dropwise to the top of a small silica gel column. Elute the column using methylene chloride. Collect several 2-mL fractions. Spot aliquots of each fraction on TLC plates and develop using ethyl acetate-hexane (4.1). Pool fractions that have identical R_f values into small, tared Erlenmeyer flasks. Evaporate the solvent from each pooled fraction and weigh each flask to determine the yield of each fraction. Determine the melting point for each fraction. Alternate purification by recrystallization: Evaporate the solution and recrystallize the residue from ethanol-water, but overnight refrigeration may be required.

HPLC analysis. Withdraw a 0.1 mL aliquot of the methylene chloride solution and place it in an Erlenmeyer flask. Evaporate the solvent. Dissolve the residue in HPLC-grade acetonitrile (5 ml) and gravity filter the solution. Inject a 20-μL sample onto a 250 mm (length) × 4.6 mm (inside diameter) or similar column packed with ODS (octadecylsilane, C-18). Elute with 70% acetonitrile/water. Detection is by UV analysis at 254 nm. Similarly prepare and inject standard solutions of *p*-methylacetanilide and 2-bromo-4-methylacetanilide (if available).

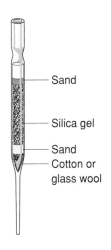

Sand

Silica gel

Sand
Cotton or
glass wool

Characterization for Part B

Melting Point: The melting point of 2-bromo-4-methylacetanilide is 117°C.

Infrared Spectroscopy: Obtain the IR spectrum using a Nujol mull or KBr pellet. The product is expected to show N-H absorption at 3400–3200 cm^{-1} and C=O absorption near 1680 cm^{-1}.

NMR Spectroscopy: Obtain the ^{1}H NMR spectrum of the product mixture using CDCl$_3$ solvent. The spectrum is expected to show an N-H signal near δ8.0, a signal for aromatic protons near δ7.2, signals for acetamido methyl protons near δ2.2, and signals for aromatic methyl protons near δ2.3. Overlapping peaks can be expected for the isomeric products.

Thin-layer Chromatography (optional): Spot solutions of *p*-methylacetanilide, 2-bromo-4-methylacetanilide (if available), and the product on three separate plates. The chromatogram should be developed using a solution of 4:1 ethyl acetate-hexane, methylene chloride, and diethyl ether, respectively. Measure R$_f$ values for each component on each plate.

Results and Conclusions for Part B

1. Determine the number of products formed in this reaction. What is the major product?
2. Is TLC an effective method for identifying the products formed in this reaction?
3. Which compound is eluted first by column chromatography? What is its structure?
4. Based upon the HPLC chromatogram of the mixture, identify peaks based upon standards and calculate relative quantities of UV-absorbing components identified as reactant and products.
5. What conclusions can be reached based upon the HPLC analysis? Are there any extra peaks in the HPLC chromatogram that might indicate a side product? Explain.

Part C: Experimental Design of a Procedure for Bromination of an Acetanilide Derivative

Each student will be assigned one of the following aromatic compounds: *o*-chloroacetanilide, *m*-chloroacetanilide, *p*-chloroacetanilide, *m*-methylacetanilide, or another assigned acetanilide derivative. Based upon considerations of the factors enumerated below, you will design a procedure to brominate 300 mg of the assigned starting material. Procedures may be adapted from the procedure described in Part A (Microscale Bromination of Acetanilide) or from the chemical literature. Write the procedure and present it to the instructor for approval before starting the experiment. Carry out the procedure and analyze the product. Calculate the percent yield. If directed to do so by the laboratory instructor, modify the revised procedure to improve the yield and purity of the product. Write a detailed report indicating the results that were obtained and providing a rationale for the experimental design, including reaction, purification, and characterization.

When designing the procedure, you should consider many factors, including those presented below:

1. Is the assigned compound expected to be more or less reactive toward bromination than acetanilide? Would this reaction likely require more or less time than the reaction of acetanilide? Explain.
2. Based upon directive and steric effects, draw the structures of the expected major product. Why would this product be expected to predominate?
3. Draw the structures of expected side products in this reaction. Which of these side products will be significant?

4. Look up the physical properties of all of the expected products and side products. Will these be solid or liquid at room temperature?
5. What safety precautions are important in the procedure?
6. Calculate the number of mmoles of the assigned compound assuming that you start with 300 mg of the compound.
7. Calculate the volumes of acetic acid and 4.1M bromine solution that should be used in the reaction.
8. Calculate the volume of water that should be used in the workup procedure.
9. Determine the appropriate size of glassware that should be used.
10. Decide how to analyze the purity of the product.
11. Decide how to characterize your product.
12. Would IR or NMR spectroscopy be helpful in characterizing your product? If so, indicate what you would look for in the spectra. If these spectroscopic techniques would not be helpful, explain why.
13. Is hazardous waste a problem with any of the reaction products or by-products? How should used solvents and any remaining solutions be handled for disposal?

Cleanup & Disposal

Pour aqueous solutions that contain no free bromine (no reddish-orange color) down the drain. Collect acetonitrile solutions in a bottle labeled "recovered acetonitrile." Pour methylene chloride solutions into a bottle labeled "halogenated organic solvent waste."

Critical Thinking Questions *(The harder ones are marked with a ❖.)*

1. Can aqueous sodium bicarbonate be substituted for sodium thiosulfate in this experiment? Why or why not?
2. Can IR be used to definitively identify the bromination reaction product(s) in this experiment? Explain.
3. Why is it necessary to wash the crude *p*-bromoacetanilide with water?
4. Why is it not necessary to include Fe or $FeBr_3$ as a catalyst for the bromination reactions in this experiment?
5. ❖ Would *o*-bromoacetanilide be expected to have a larger or a smaller R_f than *p*-bromoacetanilide? Explain.
6. The HPLC analysis uses a reversed-phase stationary phase. Is it possible to predict which of the brominated isomers will elute first in the HPLC experiment? Explain.
7. What role does the NaOH wash play in the procedure leading to the HPLC analysis?
8. ❖ What method or methods can be used to help identify peaks on the HPLC chromatogram?
9. What changes in procedure (if any) are required if phenyl acetate is substituted for acetanilide in Part A of this experiment? Explain.
10. Aniline is more highly activated toward electrophilic aromatic substitution than acetanilide. Bromination of aniline affords 2,4,6-tribromoaniline. To prepare

p-bromoaniline from aniline, aniline is first converted to acetanilide. Show a synthesis of *p*-bromoaniline starting with aniline.

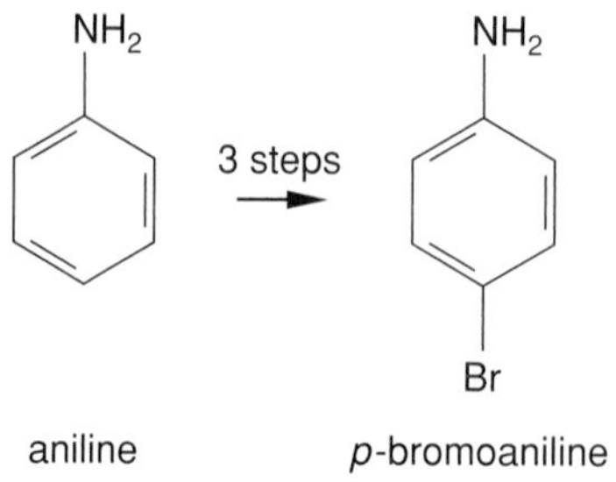

aniline   *p*-bromoaniline

11. *Molecular modeling assignment:* Construct a model of the intermediate formed from *para*-bromination of acetanilide in which the positive charge is localized on the carbon bearing the acetamido group. Calculate the energy of this intermediate. Now construct comparable models of the intermediates formed from *para*-bromination of *o*-nitroacetanilide and of *m*-nitroacetanilide. Calculate energies. What conclusions can be drawn by comparing the energies of the three intermediates?

Chapter 13

Combined Spectroscopy and Advanced Spectroscopy

Spectroscopic methods are often used in combination because each method furnishes specific information that may be necessary to solve a structural problem. Data from infrared spectroscopy and nuclear magnetic resonance spectroscopy are frequently sufficient to enable determination of the structures of relatively simple organic compounds. Use of combined IR and ^{1}H NMR is illustrated in Experiment 13.1.

Solving structures of more complex molecules may require additional spectroscopic techniques, such as mass spectrometry (MS) and ultraviolet/visible spectroscopy (UV-vis). MS is important in organic chemistry and also in biochemistry and polymer chemistry. MS furnishes structural information that is often very helpful and significantly different from that obtainable from other spectroscopic methods. MS furnishes the molecular mass of a compound directly or indirectly. The fragmentation of molecules in the mass spectrometer follows certain characteristic patterns, which can be cataloged and often analyzed. Also, most spectrometers are equipped with spectral libraries that can be used to match the spectrum of an unknown with those in the library so that direct comparisons may be made. It is not uncommon for computers associated with mass (and IR) spectrometers to have data cataloged from thousands of compounds. UV spectroscopy is similarly useful for conjugated compounds that absorb UV radiation. The wavelength of maximum absorbance indicates the degree of conjugation present. Experiment 13.2 illustrates the combined use of NMR, IR, MS, and UV for identifying the structures of unknown compounds.

Experiment 13.1: Infrared and Nuclear Magnetic Resonance Spectroscopy of Alcohols, Ethers, and Phenols

The purpose of this experiment is to identify alcohols, ethers, and phenols using infrared (IR) or proton nuclear magnetic resonance (^{1}H NMR) spectroscopy or both. You will:

- identify features of IR spectra of alcohols and phenols due to $O-H$ and $C-O$ groups.

- identify features of IR spectra of ethers due to the $C-O$ group.

- locate differences in IR spectra of primary, secondary, and tertiary alcohols and phenols.

- interpret ^{1}H NMR spectra of alcohols, ethers, and phenols.

- use IR and ^{1}H NMR together to solve structural problems.

- obtain IR spectra of solids as Nujol mulls and KBr pellets.

- run ^{1}H NMR spectra of solids in solution.

- identify features of the IR spectra of *o*-, *m*-, and *p*-substituted phenols.

Techniques

Technique M	Infrared spectroscopy
Technique N	Nuclear magnetic resonance spectroscopy

Background

IR and NMR spectroscopy are extremely useful for the analysis of pure organic compounds. IR spectroscopy furnishes information about functional groups that may be present or absent in the compound being analyzed and the substitution patterns in alkenes and aromatic rings. NMR spectroscopy provides information about the nature of the carbon skeleton. Refer to a spectroscopy book for a detailed treatment of IR and NMR spectroscopy (Lambert et al., 1998; Pavia et al., 1996; Silverstein and Webster, 1998; Kemp, 1991).

Infrared Spectroscopy of Alcohols, Ethers, and Phenols

The IR spectrum of an alcohol is characterized by an intense, broad absorption in the region of 3600–3200 cm^{-1}. This applies to all common alcohols of low to moderate molecular weight. The broad peak results from intermolecular hydrogen bonding. (IR spectra of gas phase or very dilute solutions of alcohols show sharp absorption at 3650–3580 cm^{-1} for a non-hydrogen-bonded O$-$H.) The C$-$O bond in alcohols shows a strong absorption band at 1250–1000 cm^{-1}. The wide variation in the position of this absorption band may be used to determine the nature of the alcohol. Among the simple, acyclic alcohols, primary alcohols show C$-$O absorption at 1085–1025 cm^{-1}, secondary alcohols at 1205–1085 cm^{-1}, and tertiary alcohols at 1205–1125 cm^{-1}. There is some overlap in absorption ranges for cyclic and unsaturated alcohols.

1-butanol	2-butanol	2-methyl-2-propanol
OH 3300 cm^{-1} (broad)	OH 3300 cm^{-1} (broad)	OH 3300 cm^{-1} (broad)
C–O 1080 cm^{-1}	C–O 1105 cm^{-1}	C–O 1190 cm^{-1}

IR spectra of phenols show an O$-$H absorption centered at 3300 cm^{-1} and a C$-$O absorption at 1260–1180 cm^{-1}. A carbon-oxygen absorption due to C$-$O$-$H bending may also be observed at 1390–1330 cm^{-1}. The spectra of phenols also show absorptions due to aromatic C=C bonds in the region of 1625–1425 cm^{-1}. Substituted aromatic rings show absorptions in the region of 900–700 cm^{-1}, which are not present in aliphatic alcohols. The IR spectra of isopropyl alcohol and *p*-cresol are shown in Figures 13.1-1 and 13.1-2.

Both spectra show broad O$-$H stretching absorption at 3600–3200 cm^{-1}. Other than the C$-$O absorption at 1200–1150 cm^{-1} and the C$-$H absorptions at 3000–2890 cm^{-1} and 1500–1350 cm^{-1}, the spectrum of the aliphatic alcohol is rather featureless compared to the "busy" spectrum of the phenol derivative. The aromatic compound shows

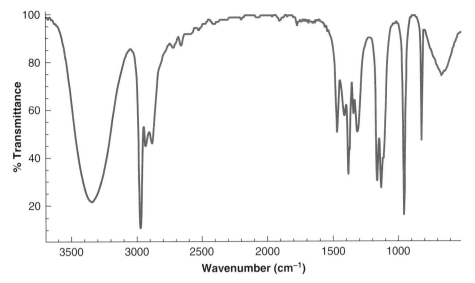

Figure 13.1-1

IR spectrum of
isopropyl alcohol

Source: ©BIO-RAD
Laboratories, Sadtler Division.

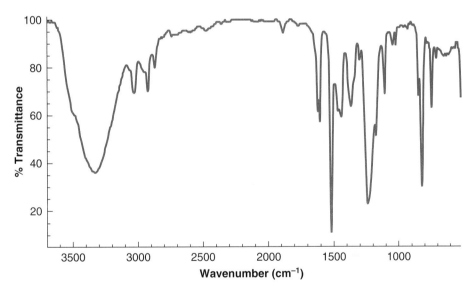

Figure 13.1-2

IR spectrum of *p*-cresol

Source: ©BIO-RAD
Laboratories, Sadtler Division.

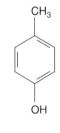

both a greater number of absorptions and absorptions of greater intensity (lower transmittance) from 1600–700 cm^{-1} than the aliphatic compound.

IR spectra of aliphatic ethers show C$-$O$-$C asymmetric stretching at 1085–1150 cm^{-1}. The usual position is about 1125 cm^{-1}. A strong absorption band around 1125 cm^{-1} in the absence of absorption due to alcohols or carbonyl compounds is indicative of an ether. Branching at the α-carbon of an ether causes the absorption at 1125 cm^{-1} to be split into two peaks. Alkyl aryl ethers show two C$-$O$-$C absorptions: an asymmetric stretch at 1275–1200 cm^{-1} and a symmetric stretch at 1075–1020 cm^{-1}. Vinyl ethers, like alkyl aryl ethers, show two absorptions in the same regions. Further examples of IR spectra are available in spectral collections (Pouchert, 1981).

NMR Spectroscopy of Alcohols, Ethers, and Phenols

The most significant aspect of the ^{1}H NMR spectra of oxygen-containing compounds such as alcohols, ethers, and phenols is that the chemical shift values of protons located on carbons attached directly to oxygen are in the $\delta 3.5$–4.0 range. In addition, the signals for the O$-$H protons of alcohols and phenols give variable chemical shift values. For 60 or 90 MHz spectra, the chemical shift value for hydroxyl protons is $\delta 1$–5 and is concentration dependent. Similarly, the O$-$H protons of phenols have chemical shifts of $\delta 4.5$–7.5. At lower concentrations, the signal appears upfield. At higher concentrations the signal is observed downfield, closer to $\delta 5$, due to intermolecular hydrogen bonding.

High-field FT-NMR spectrometers require use of dilute solutions of samples. Therefore, for high-field instruments, it is less common to observe the concentration effects seen in more concentrated solutions. The chemical shift of hydroxyl protons is usually $\delta 1$–2 for the dilute solutions used to obtain high-field spectra.

An important feature of 60 or 90 MHz NMR spectra of alcohols is that splitting of the O$-$H signal is generally not observed. This is due to rapid exchange of O$-$H protons caused by trace impurities of acids or bases. Highly purified alcohols show split signals and the n + 1 rule is followed. Thus, the hydroxyl proton of pure ethanol shows a triplet. Most ethanol samples of ordinary purity show O$-$H peaks as singlets. If there is doubt as to whether a peak is due to an O$-$H or not, the concentration of the solution can be changed to see if the suspected peak changes position. After a spectrum has been measured in CDCl$_3$ solvent, a drop or two of D$_2$O can be added and, after shaking the capped tube, the spectrum can be run again to see if the suspected O$-$H peak has diminished in size due to hydrogen-deuterium exchange. (The OD signal is not seen because ^{2}D NMR spectra are observed at different frequencies than ^{1}H NMR spectra.)

Splitting of O$-$H signals in high-field FT NMR spectra is often observed because the concentration of the alcohol sample in CDCl$_3$ is very small. For example, refer to Compound 9 in question 9 in the set of spectra in this experiment. The O$-$H signal appears as a triplet at $\delta 3.20$ because of an attached CH$_2$ group. Further examples of ^{1}H NMR spectra are available in spectral collections (Pouchert, 1983; Ault, 1980).

Part A: Experimental Procedure for IR Spectroscopy

Obtain from the instructor a sample of a solid unknown. The sample will be an alcohol, an ether, or a phenol (or other functional classes as directed by the instructor). Prepare a Nujol mull or a KBr pellet of the unknown and run the spectrum. Consider the peaks observed in the spectrum and choose the best match from the following compounds: 2-naphthol, 4-nitrobenzyl alcohol, isoborneol, hexadecanol, *o*-nitrophenol, *m*-nitrophenol, *p*-nitrophenol, and 4-bromophenyl ether. Justify your choice.

Part B: Experimental Procedure for IR and NMR Spectroscopy

Obtain a sample of an unknown from the instructor. The sample may be a solid or a liquid alcohol, ether, or a phenol. Obtain the IR and ^{1}H NMR spectra for the unknown sample. If the sample is a solid, the CDCl$_3$ solution must be filtered prior to obtaining the NMR spectrum. Carefully analyze the spectra and propose a likely structure or structures that fit the data.

Part C: Exercises for IR and NMR Spectroscopy

Propose a structure or structures consistent with each of the spectra or sets of spectra below. Include a rationale to justify a choice or choices for each compound.

1. Compound 1 gave the IR spectrum shown in Figure 13.1-3. The molecular formula of compound 1 is C_6H_6O.

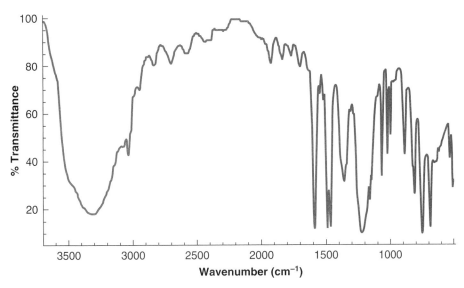

Figure 13.1-3

IR spectrum of compound 1

SOURCE: ©BIO-RAD Laboratories, Sadtler Division.

2. Compound 2 gave the IR spectrum shown in Figure 13.1-4. The molecular formula of compound 2 is $C_{12}H_{10}O$.

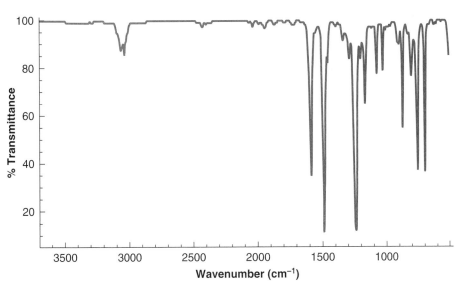

Figure 13.1-4

IR spectrum of compound 2

SOURCE: ©BIO-RAD Laboratories, Sadtler Division.

3. Compound 3 gave the ^{1}H NMR spectrum shown in Figure 13.1-5. The molecular formula of compound 3 is $C_5H_{12}O_2$.

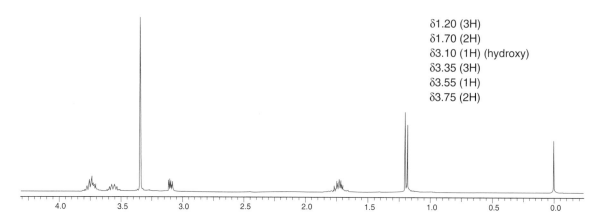

δ1.20 (3H)
δ1.70 (2H)
δ3.10 (1H) (hydroxy)
δ3.35 (3H)
δ3.55 (1H)
δ3.75 (2H)

Figure 13.1-5 ^{1}H NMR spectrum of compound 3

SOURCE: Reprinted with permission of Aldrich Chemical.

4. Compound 4 gave the ^{1}H NMR spectrum shown in Figure 13.1-6. The molecular formula of compound 4 is C_4H_8O.

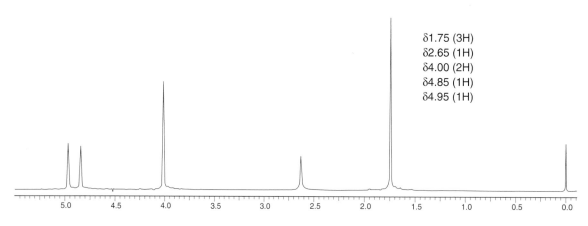

δ1.75 (3H)
δ2.65 (1H)
δ4.00 (2H)
δ4.85 (1H)
δ4.95 (1H)

Figure 13.1-6 ^{1}H NMR spectrum of compound 4

SOURCE: Reprinted with permission of Aldrich Chemical.

5. Compound 5 gave the ¹H NMR spectra shown in Figures 13.1-7a and 13.1-7b. The molecular formula of compound 5 is $C_8H_{10}O$.

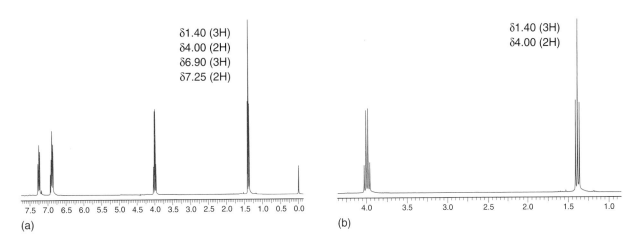

δ1.40 (3H)
δ4.00 (2H)
δ6.90 (3H)
δ7.25 (2H)

(a)

δ1.40 (3H)
δ4.00 (2H)

(b)

Figure 13.1-7a ¹H NMR spectrum of compound 5

SOURCE: Reprinted with permission of Aldrich Chemical.

Figure 13.1-7b ¹H NMR spectrum of compound 5 (expanded)

SOURCE: Reprinted with permission of Aldrich Chemical.

6. Compound 6 gave the ¹H NMR spectrum shown in Figure 13.1-8. The molecular formula of compound 6 is C_3H_4O.

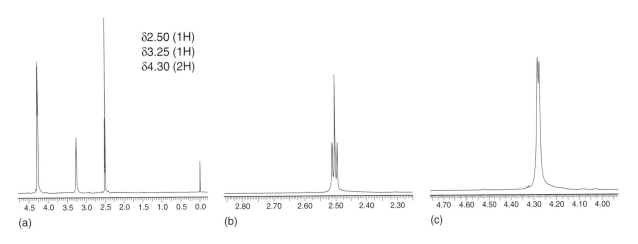

δ2.50 (1H)
δ3.25 (1H)
δ4.30 (2H)

(a)

(b)

(c)

Figure 13.1-8 (a) ¹H NMR spectra of compound 6; (b) and (c) expanded regions

SOURCE: Reprinted with permission of Aldrich Chemical.

7. Compound 7 gave the IR spectrum and the ^{1}H NMR spectrum shown in Figures 13.1-9 and 13.1-10. The molecular formula of compound 7 is C_3H_6O.

Figure 13.1-9

IR spectrum of compound 7

SOURCE: ©BIO-RAD Laboratories, Sadtler Division.

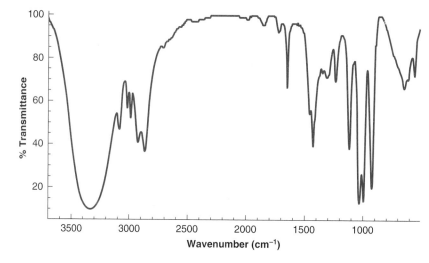

Figure 13.1-10

^{1}H NMR spectrum of compound 7

SOURCE: Reprinted with permission of Aldrich Chemical.

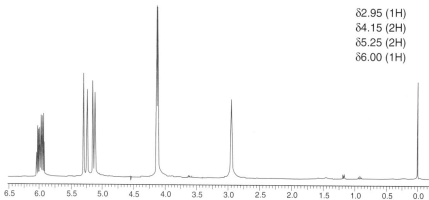

δ2.95 (1H)
δ4.15 (2H)
δ5.25 (2H)
δ6.00 (1H)

8. Compound 8 gave the IR spectrum and the ^{1}H NMR spectrum shown in Figures 13.1-11 and 13.1-12.

Figure 13.1-11

IR spectrum of compound 8

SOURCE: ©BIO-RAD Laboratories, Sadtler Division.

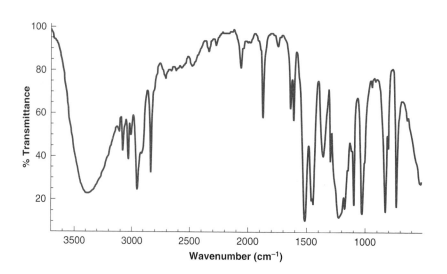

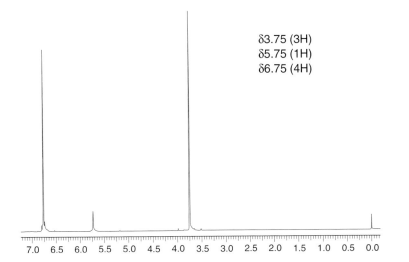

δ3.75 (3H)
δ5.75 (1H)
δ6.75 (4H)

Figure 13.1-12

^{1}H NMR spectrum
of compound 8

SOURCE: Reprinted with
permission of Aldrich Chemical.

9. Compound 9 gave the IR spectrum and the ^{1}H NMR spectrum shown in Figures
13.1-13 and 13.1-14.

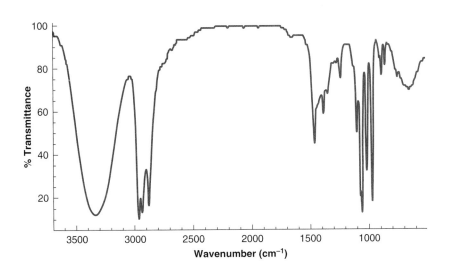

Figure 13.1-13

IR spectrum of
compound 9

SOURCE: ©BIO-RAD
Laboratories, Sadtler Division.

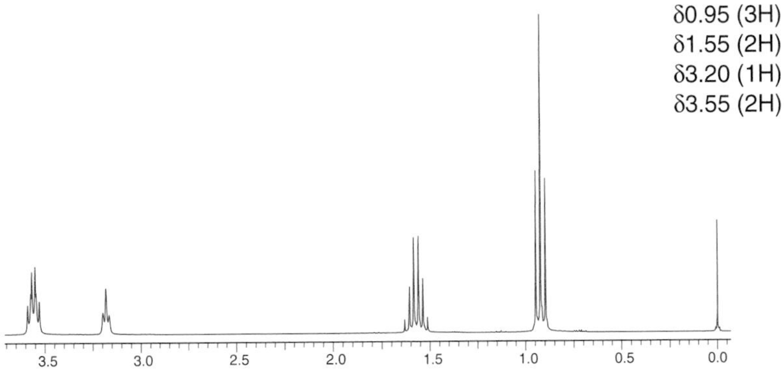

δ0.95 (3H)
δ1.55 (2H)
δ3.20 (1H)
δ3.55 (2H)

Figure 13.1-14

^{1}H NMR spectrum
of compound 9

SOURCE: Reprinted with
permission of Aldrich Chemical.

Critical Thinking Questions *(The harder ones are marked with a ❖.)*

1. The ^{1}H NMR spectrum of a concentrated solution of an organic compound having the molecular formula $C_3H_6O_2$ showed three singlets: $\delta 2.1$ (3H), $\delta 4.5$ (2H), and $\delta 4.9$ (1H). Suggest a structure for the compound.
2. How can the ^{1}H NMR signal for the alkyne C$-$H be distinguished from the O$-$H signal for propargyl alcohol (HC$\equiv$CCH$_2$OH)?
3. How does the ^{1}H NMR spectrum of highly pure 1-propanol differ from the NMR spectrum of 1-propanol that contains acidic or basic impurities?
4. How much of a sample is "enough" to obtain a good ^{1}H NMR spectrum? Does it matter which type of spectrometer is being used?
❖ 5. It is not always possible to deduce the structure of a compound given only the IR and ^{1}H NMR spectra. The molecular formula is also a very useful item. How can information be obtained about molecular formulas?
❖ 6. A student has prepared a Nujol mull of a solid to run an IR spectrum. The spectrum is obtained, but the baseline absorbance is very high throughout the spectrum and particularly in the 4000–3000 cm^{-1} region. What should be done to obtain a better spectrum?
7. Suppose a student has obtained an IR spectrum of a sample of a solid alcohol. The only peaks visible in the spectrum are due to Nujol. Even the expected O$-$H peak is not observed. What action should the student take to get a good spectrum?
8. An IR spectrum is obtained using "cloudy" salt plates. The spectrum obtained has a few unexpected peaks. Unfortunately, there are no other salt plates available. What action should be taken in order to get meaningful results?
9. Explain the proper techniques for cleaning salt plates and NMR tubes.
❖ 10. *Library project:* Explain why it is necessary to measure NMR spectra of solutions in spinning sample tubes. Explain the difference in spectra expected for ethanol in a spinning versus a nonspinning tube.

References

Ault, A., and Ault, M. R. *A Handy and Systematic Catalog of NMR Spectra.* Mill Valley, CA: University Science Books, 1980.

Kemp, W. *Organic Spectroscopy.* 3rd ed. New York: W. H. Freeman Co., 1991.

Lambert, J. B., Shurvell, H. F., Lightner, D., and Cooks, R. G. *Organic Structural Spectroscopy.* Upper Saddle River, NJ: Prentice-Hall, Inc., 1998.

Pavia, D. L., Lampman. G. M., and Kriz, G. S. *Introduction to Spectroscopy.* 2nd ed. Ft. Worth, TX: Saunders, 1996.

Pouchert, C. J. *Aldrich Library of NMR Spectra, Volumes 1 and 2.* Milwaukee, WI: Aldrich Chemical Co., 1983.

Pouchert, C. J. *Aldrich Library of Infrared Spectra.* 3rd ed. Milwaukee, WI: Aldrich Chemical Co., 1981.

Silverstein, R. M., and Webster, F. X. *Spectrometric Identification of Organic Compounds.* 6th ed. New York: Wiley, 1998.

Experiment 13.2: Combined Spectral Analysis:
Infrared, Ultraviolet, and Nuclear Magnetic Resonance Spectroscopy and Mass Spectrometry

The objectives of this experiment are to interpret infrared (IR), ultraviolet (UV), and nuclear magnetic resonance (NMR) spectra and mass spectrometry (MS) data and to use these interpretations to determine the structures of unknown compounds. You will:

- interpret IR, ^{1}H NMR, and ^{13}C NMR spectra.

- use IR, NMR, and MS together to solve structural problems.

- use UV spectroscopy as an aid to solve unknown structures.

- obtain spectra of solid and liquid samples.

Techniques

Technique M	Infrared spectroscopy
Technique N	Nuclear magnetic resonance spectroscopy
Technique O	Ultraviolet spectroscopy
Technique P	Mass spectrometry

Background

Previous exercises and experiments on IR and NMR spectroscopy have emphasized basic techniques and spectral interpretation. Substances with complex structures may require contributions of several sources of spectroscopic information.

Solved Examples of the Use of Combined Spectroscopy

Structures of some organic unknowns can be solved from their NMR spectra alone or from their IR and NMR spectra. In certain others, additional information from ^{13}C NMR, UV spectroscopy, or MS is necessary. Sometimes ^{13}C NMR and MS data are used simply to verify that the correct structural assignment has been made. In other cases, it is critical to use these data to establish a definitive identification.

Spectra for Unknown A

The following are UV data: λ_{max} (274 nm); log ε (1.2), solvent (methanol).

Solved Unknown A

Since many structures can be identified using only IR and ^{1}H NMR spectra, such as those shown in Figures 13.2-1 and 13.2-2, examine these spectra first.

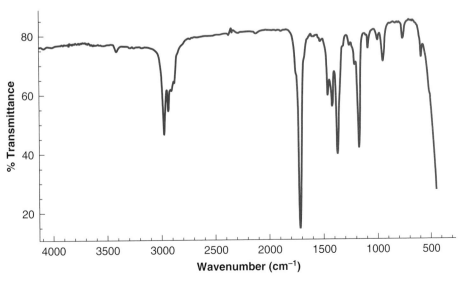

Figure 13.2-1

IR spectrum for unknown A (neat)

Figure 13.2-2

^{1}H NMR spectrum for unknown A (CDCl$_3$)

SOURCE: Reprinted with permission of Aldrich Chemical.

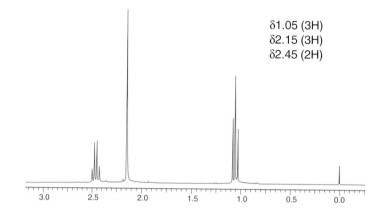

δ1.05 (3H)
δ2.15 (3H)
δ2.45 (2H)

Step 1. Determine which functional groups are indicated by the IR and NMR spectra. There is strong absorption at 1720 cm^{-1}, indicative of a carbonyl group of an aldehyde or ketone. Because there are no IR bands at 2800–2700 cm^{-1} where aldehyde C—H stretching would be expected and because there are no NMR signals at δ9–10, the compound appears to be a ketone.

Step 2. Determine whether the compound is aliphatic or aromatic. If aliphatic, determine whether the compound is saturated or unsaturated. The IR spectrum has relatively few absorptions, a pattern that is characteristic of an aliphatic compound. There is no absorption characteristic of alkenyl C=C or aromatic C=C groups. There are no NMR signals in the δ5–6 or δ7–8 regions, indicating no alkenyl or aromatic protons. The compound is aliphatic.

Step 3. Assign the structure unambiguously by further examination of the NMR spectrum and if necessary, the IR data. There are three signals. The sharp singlet at δ2.15 is likely a methyl group because the signal integrates to three hydrogens. The value of the chemical shift suggests that the methyl group is adjacent to a carbonyl group. The other signals indicate an ethyl group. The methylene carbon is attached to the carbonyl, causing a downfield shift to about δ2.45. The integrations for the triplet and quartet are consistent with the methyl and methylene groups, respectively, of an ethyl group. Based upon the evidence, the compound is predicted to be 2-butanone.

Step 4. Evaluate the ^{13}C NMR spectrum shown in Figure 13.2-3. Mass spectral data in Figure 13.2-4 and UV data below add support to the structural assignment. The m/z of the parent ion of the MS is 72. The base peak is at m/z 43, consistent with a (CH$_3$C=O)$^+$ fragment. The ^{13}C NMR shows four signals, indicating four different carbons. One is far downfield, consistent with a carbonyl carbon. This information supports the structural assignment of 2-butanone. The UV spectrum has a λ$_{max}$ at 274 nm (ε = 16), consistent with the presence of a carbonyl group. The evidence supports assignment of the structure as 2-butanone.

2-butanone

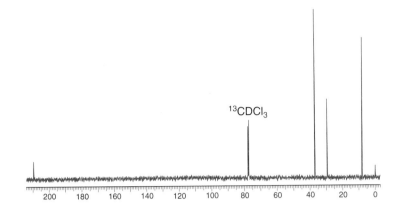

m/z	% abundance
29	35
43	100
57	20
72 (M)	38

Figure 13.2-3 ¹³C NMR spectrum for unknown A (CDCl₃)

SOURCE: Reprinted with permission of Aldrich Chemical.

Figure 13.2-4 MS data for unknown A

Spectra for Unknown B

The following are UV data: no absorption >200 nm.

Solved Unknown B

IR and NMR spectra shown in Figures 13.2-5 and 13.2-6 are considered first to see whether the structure can be solved using only this spectral information. The same general approach used to solve unknown A will be used for unknown B. Data are tabulated below.

Step 1. Determine which functional groups are present. The IR and NMR data are of primary importance. Note that only certain IR bands are assigned.

IR Data:	3030 cm⁻¹	alkene C−H stretching
	2950–2850 cm⁻¹	alkane C−H stretching
	1660 cm⁻¹	C=C stretching
	715 cm⁻¹	alkene C−H bending for *cis*-disubstituted alkene
NMR Data:	δ5.6	alkene protons

Step 2. Determine whether the compound is aliphatic or aromatic. If aliphatic, determine if there is evidence of unsaturation. The compound is not aromatic according to the NMR data (no signal in the δ7–8 region, indicating no aromatic protons). The presence of a C=C is strongly suggested (1660 cm⁻¹, 715 cm⁻¹).

Step 3. Determine if the structure can be unambiguously assigned by further examination of the NMR spectrum and, if necessary, the IR data. Additional NMR data are summarized below from Figure 13.2-6.

Additional NMR Data:	δ1.6	relative area = 2; methylene protons
	δ2.0	relative area = 2; methylene protons

As noted above, the signal in the NMR spectrum at δ5.6 is characteristic of vinyl protons. There appear to be three distinct signals. Each signal has very small splitting. A cyclic structure seems likely because there are no methyl groups present. From the IR data, there

Figure 13.2-5

IR spectrum for unknown B (neat)

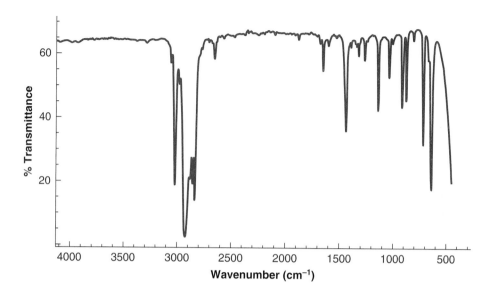

Figure 13.2-6

^{1}H NMR spectrum for unknown B ($CDCl_3$)

SOURCE: Reprinted with permission of Aldrich Chemical.

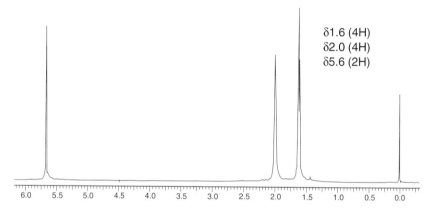

δ1.6 (4H)
δ2.0 (4H)
δ5.6 (2H)

is a characteristic absorption of *cis*-disubstituted alkenes at 715 cm^{-1}. This conclusion is consistent with the structure of cyclohexene.

Step 4. Evaluate ^{13}C NMR, UV, and mass spectra to lend support to the structural assignment. The mass spectrum shows a molecular ion peak at m/z 82. The ^{13}C NMR spectrum shown in Figure 13.2-7 has three signals; two are assignable to aliphatic carbons and one to alkenyl carbons. This supports assignment of the structure as cyclohexene. If this information had been taken into consideration along with the IR and ^{1}H NMR data initially, it might have taken less time to reach a definitive conclusion. The ^{1}H NMR appears not to be as useful as it was for unknown A because the coupling constants for unknown B are small and the connectivities of nonequivalent, adjacent protons are not as obvious. In such a case, evidence from MS and ^{13}C NMR can be more useful. MS gives the molecular mass (82), but interpretation of the fragment ions in Figure 13.2-8 is complex for this compound. The UV spectrum of cyclohexene shows no wavelength maximum >200 nm (e.g., no UV spectrum), consistent with the properties expected of an alkene.

cyclohexene

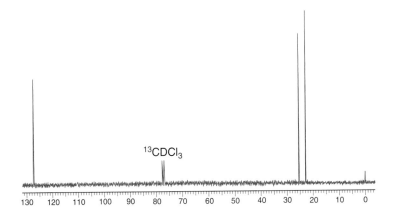

m/z	% abundance
28	30
39	42
54	60
67	100
82 (M)	70

Figure 13.2-7 ^{13}C NMR spectrum for unknown B (CDCl$_3$)

SOURCE: Reprinted with permission of Aldrich Chemical.

Figure 13.2-8 MS data for unknown B

Spectra for Unknown C

The following are UV data: no absorption >200 nm.

Solved Unknown C

IR and ^{1}H NMR spectra shown in Figures 13.2-9 and 13.2-10 are considered first to see whether the structure can be solved using only this spectral information.

Step 1. Determine which functional groups are present. IR absorptions are few, with the band at 1070 cm^{-1} being assignable to C–O stretching. The IR evidence is suggestive of an ether. The NMR spectrum is also simple, showing just two sets of triplets of equal area at $\delta 1.85$ and $\delta 3.75$.

Step 2. Determine whether the compound is aliphatic or aromatic. If aliphatic, determine if there is evidence of unsaturation. The IR spectrum shows no absorption for

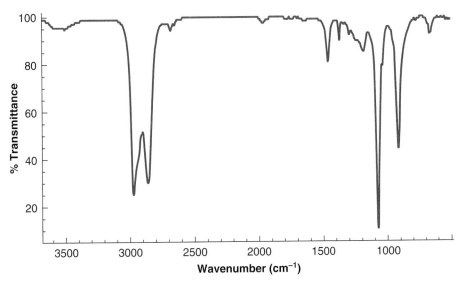

Figure 13.2-9

IR spectrum for unknown C

SOURCE: ©BIO-RAD Laboratories, Sadtler Division.

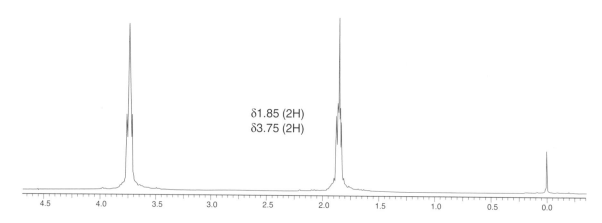

δ1.85 (2H)
δ3.75 (2H)

Figure 13.2-10   ¹H NMR spectrum for unknown C (CDCl₃)
SOURCE: Reprinted with permission of Aldrich Chemical.

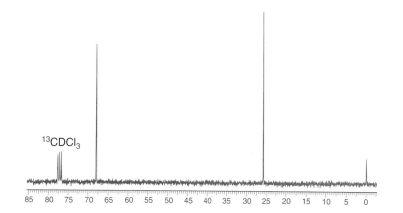

¹³CDCl₃

Figure 13.2-11   ¹³C NMR spectrum for unknown C (CDCl₃)
SOURCE: Reprinted with permission of Aldrich Chemical.

m/z	% abundance
27	32
28	40
39	22
41	55
42	100
43	25
71	40
72 (M)	45

Figure 13.2-12   MS data for
unknown C

alkenyl C=C stretching or aromatic C=C stretching. The NMR spectrum shows only two signals at δ1.85 and δ3.75. There are no signals in the δ5–8 region, ruling out aromatic protons and alkenyl protons.

Step 3.   Determine if the structure can be unambiguously assigned by further examination of the NMR spectrum and, if necessary, the IR data. The NMR spectrum shows no signal near δ1–2, ruling out the presence of skeletal methyl groups. The NMR spectrum shows two signals of equal area, one for protons attached to a carbon bonded to oxygen. The signals are two triplets, suggesting $-O-CH_2-CH_2-$. It is likely that the unknown is a cyclic ether, possibly tetrahydrofuran.

Step 4.   Evaluate ¹³C NMR, UV, and mass spectra to support the structural assignment. Only two signals appear in the ¹³C NMR spectrum shown in Figure 13.2-11. One of these must be next to an oxygen, because of the chemical shift of δ68.0. The molecular ion in the mass spectrum appears at m/z 72. From the data you can propose structures having molecular mass of 72. The likely structure suggested in step 3 is tetrahydrofuran and this is in agreement with the ^{13}C NMR and mass spectral data given in Figure 13.2-12.

tetrahydrofuran (THF)

Spectra for Unknown D

The following are UV data: λ_{max} (266 nm); log ε (3.6), solvent (methanol).

Solved Unknown D

The approach used here is similar to that used in the examples above, starting with IR and ^{1}H NMR spectra shown in Figures 13.2-13 and 13.2-14.

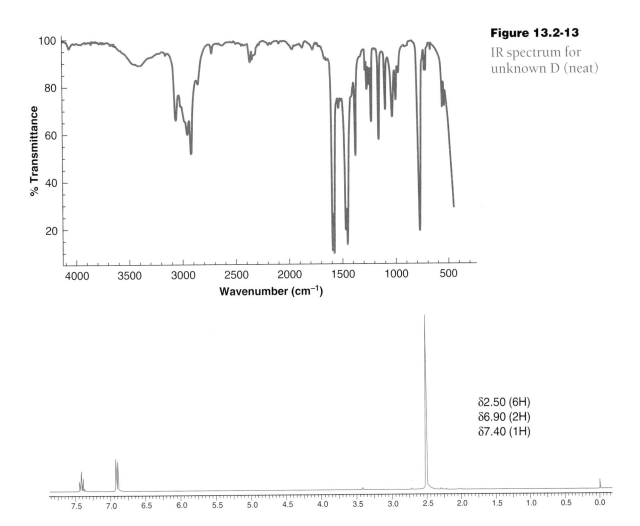

Figure 13.2-13

IR spectrum for unknown D (neat)

δ2.50 (6H)
δ6.90 (2H)
δ7.40 (1H)

Figure 13.2-14 ^{1}H NMR spectrum for unknown D (CDCl$_3$)

SOURCE: Reprinted with permission of Aldrich Chemical.

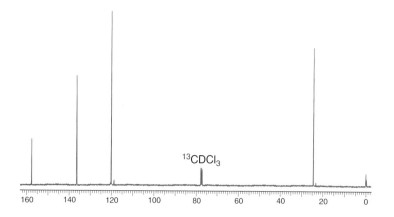

m/z	% abundance
39	38
65	24
66	25
92	22
107 (M)	100

Figure 13.2-15 ^{13}C NMR spectrum for unknown D (CDCl$_3$)

SOURCE: Reprinted with permission of Aldrich Chemical.

Figure 13.2-16 MS data for unknown D

Step 1.   Determine which functional groups are present. There is a broad, weak absorption at ~3500 cm^{-1}. However, since there is no C—O absorption apparent at 1200–1000 cm^{-1}, this absorption could be due to a trace of water. There is no absorption in the carbonyl region.

Step 2.   Determine if the compound is aliphatic or aromatic. If aliphatic, determine if there is evidence of unsaturation. The ^{1}H NMR shows signals in the δ7–8 region, indicating aromatic protons. This is supported by IR absorption at 1600 cm^{-1}.

Step 3.   Determine if the structure can be unambiguously assigned by further examination of the ^{1}H NMR spectrum and if necessary, the IR data. The relative ratio of peak areas in the ^{1}H NMR is important. The relative areas of aliphatic to aromatic protons is 2:1, suggesting either two methylene groups and two aromatic protons or two methyl groups and three aromatic protons. Two methyl groups are more likely because it is difficult to envision a structure having an aromatic ring and just two methylene groups. The chemical shift of a methyl group attached directly to an aromatic ring generally appears at δ2.3–2.5. The signal observed at δ2.50 in Figure 13.2-14 is consistent with the presence of two methyl groups.

Step 4.   Evaluate the ^{13}C NMR spectrum shown in Figure 13.2-15, UV and mass spectra to make the structural assignment. The particularly revealing feature about the spectral data is the odd-numbered molecular ion peak in the mass spectrum at m/z 107 as shown in Figure 13.2-16. This corresponds to a structure having an odd number of nitrogens. A possible structure can now be constructed. If there are two methyl groups, then four aromatic protons will be present if the ring is a benzene ring. If a nitrogen is substituted for one of the ring carbons, then the structure is a pyridine containing two methyl substituents. This is in agreement with the ^{13}C NMR spectrum, which has only three signals in the aromatic region. One of these signals is shifted downfield to δ157.5 by an adjacent nitrogen. The molecular mass matches the value of the molecular ion at m/z 107 given by the MS data. The UV absorption data are indicative of conjugation such as in aromatic compounds.

Step 5.   Distinguish between possible isomeric structures. It remains to decide the substitution pattern of the methyl groups on the ring. The NMR spectrum shows a distorted triplet and a distorted doublet. This is consistent with two equivalent protons coupled to a third, as in 2,6-dimethylpyridine. The symmetry of this compound is consistent with having only three signals in the aromatic region of the ^{13}C NMR spectrum.

2,6-dimethylpyridine

Experimental Section: Acquisition of Spectra of Unknown Substances

An unknown compound will be assigned for analysis. The unknown may be a solid or liquid. There are two steps to follow in obtaining spectra of the unknown: purification of the compound and preparation of solutions for spectral analysis.

Part A: Purification of the Unknown

Unless directed otherwise, the unknown should be purified prior to obtaining spectra. A solid unknown should have a sharp melting point. Solid unknowns should be recrystallized at least once before proceeding. Impurities in the sample could cause errors in spectral interpretation. Liquids should be distilled prior to analysis.

Part B: Preparation of the Sample and Spectroscopic Analysis

Infrared Spectroscopy: IR spectra of solid unknowns should be run as Nujol mulls or as KBr pellets. IR spectra of liquid samples should be run for the neat liquid on salt plates.

NMR Spectroscopy (High field): ^{1}H NMR spectra of solid samples should be run in CDCl$_3$ containing TMS unless directed otherwise. Dissolve about 5 mg of solid unknown in about 0.7 mL of solvent. All of the solid should dissolve. Transfer the solution by filter pipet to a clean, dry NMR tube. Add more solvent as necessary to adjust the level of liquid correctly. Cap the tube. For liquid unknowns, add 1 drop of unknown to an NMR tube using a Pasteur pipet. About 0.6 mL of solvent should be added. Adjust the level of liquid to the correct level. Cap the tube.

^{13}C NMR Spectroscopy: Similar preparations are required for ^{13}C NMR spectroscopy using CDCl$_3$ solvent containing TMS. It is best to use more concentrated solutions (20–50 mg per 0.6 mL) because multiple scans are required for ^{13}C NMR spectroscopy. For some instruments, both ^{1}H and ^{13}C NMR spectra can be run using the same sample if 20–30 mg of sample are used for both ^{1}H and ^{13}C NMR spectra.

NMR Spectroscopy (60–90 MHz): ^{1}H NMR spectra of solid samples should be run in CDCl$_3$ containing TMS unless directed otherwise. Dissolve at least 30 mg of solid unknown in about 0.5–0.75 mL of solvent. All of the solid should dissolve. Transfer the solution by filter pipet to a clean, dry NMR tube. Add more solvent as necessary to adjust the level of liquid correctly. Cap the tube. For liquid unknowns, add 3–5 drops of unknown to an NMR tube using a Pasteur pipet. About 0.5–0.75 mL of solvent should be added. Adjust the level of liquid to the correct level. Cap the tube.

Mass Spectrometry: For mass spectrometry using a GC/MS instrument, dissolve about 1–2 mg of sample in about 5 mL of methylene chloride. Add 1 drop of this solution to 5 mL of the solvent in a separate container. Add 1 drop of the solution to another 5 mL of the solvent. Use this solution for injection onto the GC column.

UV Spectroscopy: To obtain a UV spectrum, a very dilute solution of unknown in ethanol must be used. Dissolve 1–2 mg or 1–2 μL of unknown in 2–3 mL of ethanol. Pipet 2 drops of the solution into 5 mL of ethanol and transfer a portion of the solution to a cuvette to take the UV spectrum. Dilute further as necessary.

Cleanup & Disposal

The NMR tubes containing the $CDCl_3$ solutions should be emptied into a container labeled "used $CDCl_3$" or "halogenated waste." Remove the plastic cap carefully from the tube because the glass is thin. NMR tubes should be rinsed with acetone and allowed to dry. The tubes should be returned to the instructor in a clean, dry condition. Clean all glassware used for the mass spectral sample. Extra solution should be returned to a bottle labeled "halogenated solvent waste." Ethanol solutions should be poured down the drain.

Results and Conclusions for Parts A and B

1. Prepare a table listing the main features of each spectrum for the unknown.
2. Make an entry next to each listing in the table indicating the significance of the entry.
3. Analyze the data from the table in a stepwise fashion as illustrated in each of the solved examples. Propose a structure or structures for the unknown.

Part C: Spectroscopic Exercises

Propose a structure or structures consistent with each of the spectra or sets of spectra shown on the following pages. In addition to proposing structures, include a rationale to justify a choice or choices for each compound.

Problem 1:

UV: wavelength (ε_{max}): 268 (101), 264 (158), 262 (147), 257 (194), 252 (153)

^{13}C NMR(in ppm): 20.7, 66.1, 128.1, 128.2, 128.4, 136.2, 170

MS:

m/z	% abundance
43	71
91	68
108	100
150 (M)	28

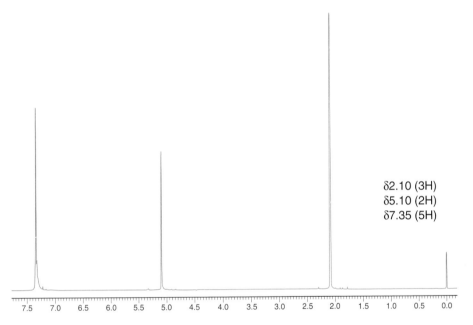

^{1}H NMR spectrum
(CDCl$_3$)

SOURCE: Reprinted with permission of Aldrich Chemical.

δ2.10 (3H)
δ5.10 (2H)
δ7.35 (5H)

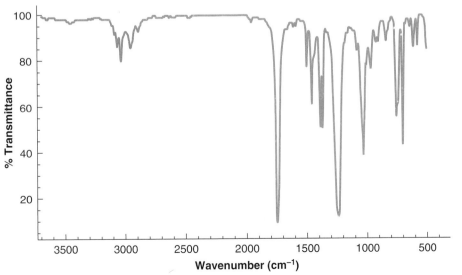

IR spectrum

SOURCE: ©BIO-RAD Laboratories, Sadtler Division.

Problem 2:

UV: wavelength (ε_{max}): 292 (20)

^{13}C NMR(in ppm): 25.1, 27.1, 41.9, 211.3

MS:

m/z	% abundance
42	81
55	100
70	20
83	7
98 (M)	30

^{1}H NMR spectrum (CDCl$_3$)

SOURCE: Reprinted with permission of Aldrich Chemical.

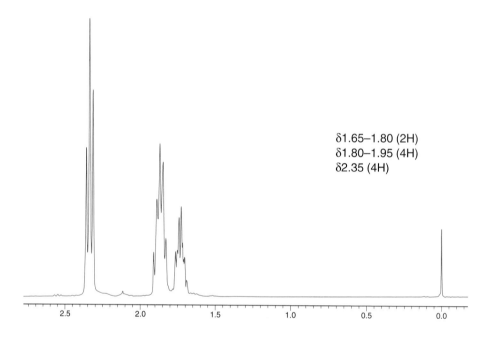

δ1.65–1.80 (2H)
δ1.80–1.95 (4H)
δ2.35 (4H)

IR spectrum

SOURCE: ©BIO-RAD Laboratories, Sadtler Division.

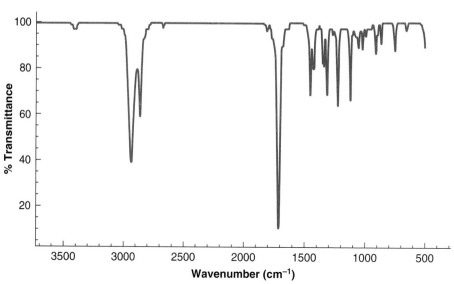

Problem 3:

UV: wavelength (ε_{max}): no absorption peaks >205

^{13}C NMR(in ppm): 24.2, 25.5, 35.6, 70.5

MS:

m/z	% abundance
57	100
67	68
82	53
100 (M)	2.4

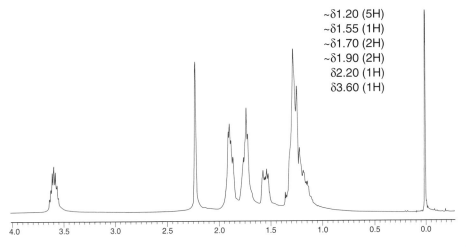

~δ1.20 (5H)
~δ1.55 (1H)
~δ1.70 (2H)
~δ1.90 (2H)
δ2.20 (1H)
δ3.60 (1H)

^{1}H NMR spectrum (CDCl$_3$)

SOURCE: Reprinted with permission of Aldrich Chemical.

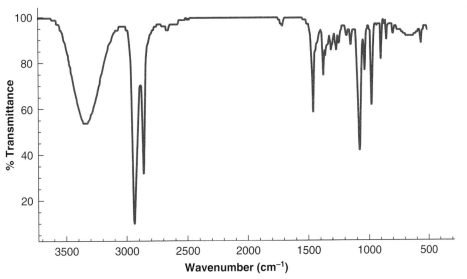

IR spectrum

SOURCE: ©BIO-RAD Laboratories, Sadtler Division.

Problem 4:

UV: no absorption peaks > 205

^{13}C NMR(in ppm): 11.3, 26.8, 44.1

MS:

m/z	% abundance
30	100
59 (M)	7.8

^{1}H NMR spectrum (CDCl$_3$)

SOURCE: Reprinted with permission of Aldrich Chemical.

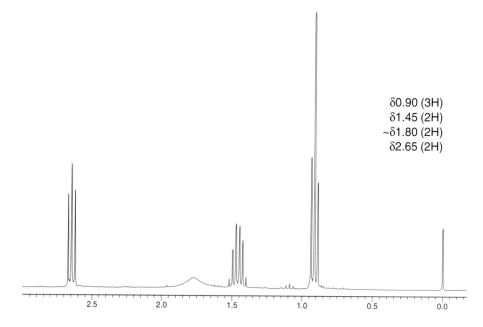

δ0.90 (3H)
δ1.45 (2H)
~δ1.80 (2H)
δ2.65 (2H)

IR spectrum

SOURCE: ©BIO-RAD Laboratories, Sadtler Division.

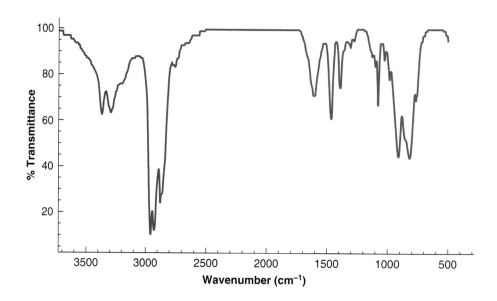

Problem 5:

UV (data unavailable)

^{13}C NMR(in ppm): 30.7, 113.4, 121.6, 128.9, 147.8

MS:

m/z	% abundance
28	100
106	15
140	25
141 (M)	42
143 (M+2)	13

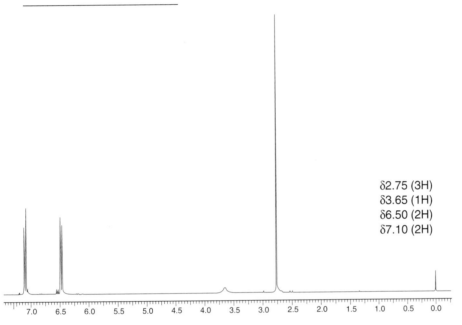

^{1}H NMR spectrum
(CDCl$_3$)

SOURCE: Reprinted with
permission of Aldrich Chemical.

δ2.75 (3H)
δ3.65 (1H)
δ6.50 (2H)
δ7.10 (2H)

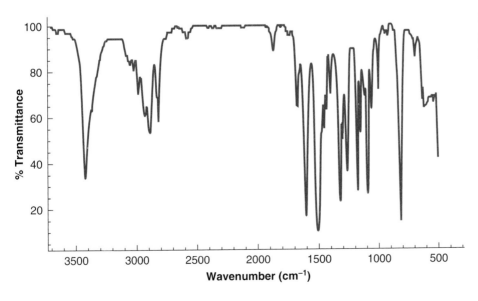

IR spectrum

SOURCE: ©BIO-RAD
Laboratories, Sadtler Division.

Problem 6:

UV: wavelength (ε_{max}): 210 (11,500), 315 (14)

^{13}C NMR(in ppm): 138.1, 139.5, 195.3

MS:

m/z	% abundance
27	100
56 (M)	65

^{1}H NMR spectrum
(CDCl$_3$)

SOURCE: Reprinted with
permission of Aldrich Chemical.

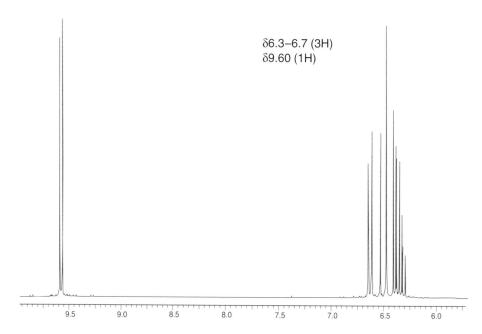

δ6.3–6.7 (3H)
δ9.60 (1H)

IR spectrum

SOURCE: ©BIO-RAD
Laboratories, Sadtler Division.

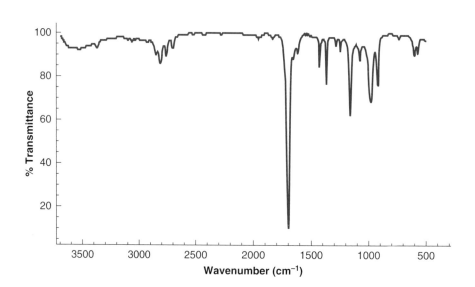

Problem 7:

UV: no absorption peaks > 220 nm

^{13}C NMR(in ppm): 22.9, 28.9, 29.3, 34.6, 69.3, 176.2

MS:

m/z	% abundance
42	76
55	100
114 (M)	13

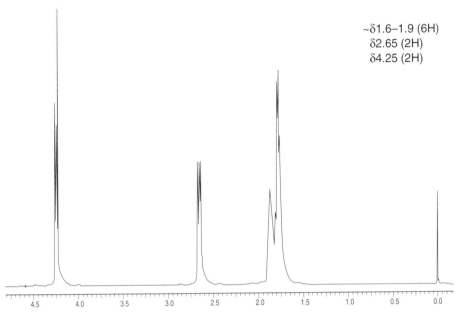

~δ1.6–1.9 (6H)
δ2.65 (2H)
δ4.25 (2H)

^{1}H NMR spectrum
(CDCl$_3$)

SOURCE: Reprinted with
permission of Aldrich Chemical.

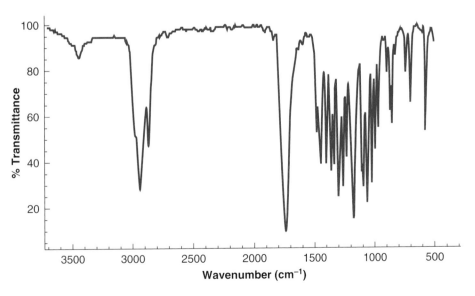

IR spectrum

SOURCE: ©BIO-RAD
Laboratories, Sadtler Division.

Critical Thinking Questions

1. An unknown is believed to be an alcohol because the IR spectrum showed broad absorption at 3200–3500 cm^{-1} and no absorption in the 1700–1750 cm^{-1} region. The ^{1}H NMR spectrum showed a number of signals in the δ1–5 region. What can be done experimentally to identify the OH proton in the NMR spectrum?

2. The compounds below cannot be identified from IR and ^{1}H NMR spectra alone. Explain. Suggest how other types of spectroscopy might be useful in solving the structure.
    a. 4-undecanone
    b. 1-bromo-1-chlorocyclohexane
    c. (E,E)-4-ethyl-2,4-heptadiene

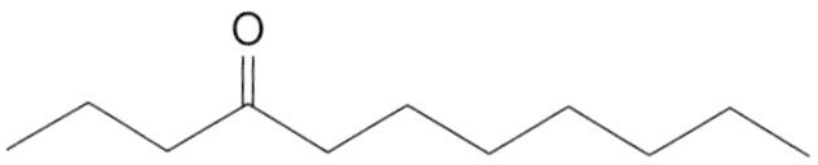

4-undecanone

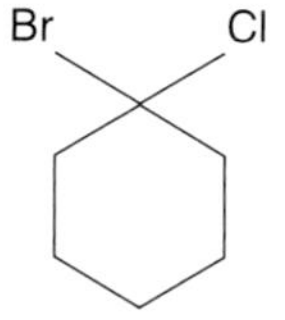

1-bromo-1-chlorocyclohexane

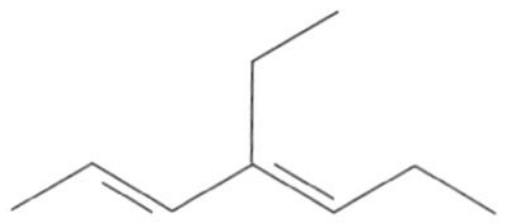

(E,E)-4-ethyl-2,4-heptadiene

Chapter 14

Organometallics

ompounds containing metal-carbon bonds such as Grignard reagents are known as organometallics. Organometallic compounds have been known since before 1900, but new methodologies are being developed that use a variety of metals for organic synthesis. A classical method for making new carbon-carbon bonds uses reactants containing magnesium-carbon bonds, known as Grignard reagents. Grignard reagents react with a variety of classes of organic compounds such as aldehydes and ketones, epoxides, esters, acid halides, and nitriles. Reactions of Grignard reagents with ketones and esters are explored in Experiment 14.1. A more novel methodology using indium intermediates is illustrated in Experiment 14.2.

Experiment 14.1: Grignard Synthesis: *Preparation of Triphenylmethanol and 3-Methyl-2-phenyl-2-butanol*

In this experiment, you will study reactions of organometallic compounds. You will:

♦ synthesize a Grignard reagent and use it to prepare a tertiary alcohol.

♦ use anhydrous conditions for a moisture-sensitive reaction.

♦ purify a product using column chromatography.

Techniques

Technique F	Recrystallization and filtration
Technique G	Reflux
Technique I	Extraction
Technique K	Column chromatography and thin-layer chromatography
Technique M	Infrared spectroscopy
Technique N	Nuclear magnetic resonance spectroscopy
Technique O	Ultraviolet spectroscopy

Background

Victor Grignard, a French chemist at the turn of the twentieth century, discovered that organic halides and magnesium react to form organomagnesium halides. In 1912 he was awarded the Nobel Prize for this discovery. Grignard reagents, which are examples of an important class of compounds called organometallics, may be formed by reacting elemental magnesium with primary, secondary, and tertiary alkyl halides or aryl and vinyl halides.

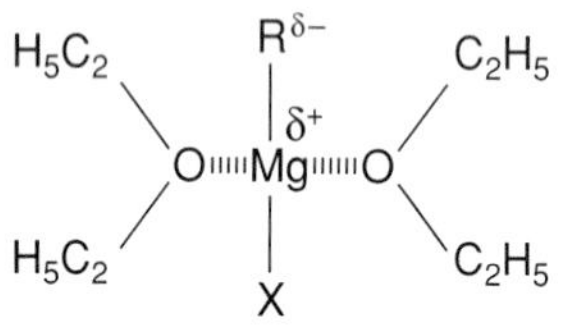

The solvent plays a key role in the Grignard reaction. Anhydrous ether solvents such as diethyl ether or tetrahydrofuran (THF) are most commonly used. These solvents form soluble complexes with the Grignard reagent. Diethyl ether also protects the Grignard reagent from reaction with oxygen. It is important to use anhydrous conditions when preparing Grignard reagents. Since Grignard reagents are strong bases, they react readily with any compound having an acidic proton. Water, alcohols, and carboxylic acids all react with a Grignard reagent and destroy the reagent. Sometimes, reaction of magnesium and the organic halide is difficult to initiate. Iodine crystals can be added if necessary to initiate reaction of the magnesium metal.

The carbon-magnesium covalent bonds in Grignard reagents are polar, with partial negative charge on the carbon atoms of the carbon-magnesium bonds. These carbons are nucleophilic. Grignard reagents are widely used in the formation of new carbon-carbon bonds, created by nucleophilic attack of carbon-magnesium bonds of Grignard reagents on the electrophilic carbon of polarized π-bonded functional groups such as carbonyl groups found in aldehydes, ketones, and esters. Reaction of Grignard reagents with carbonyl groups initially gives alkoxides and subsequently, after hydrolysis, alcohols.

In Part A of this experiment, the Grignard reagent (phenylmagnesium bromide) is reacted with benzophenone to afford a tertiary alcohol, triphenylmethanol.

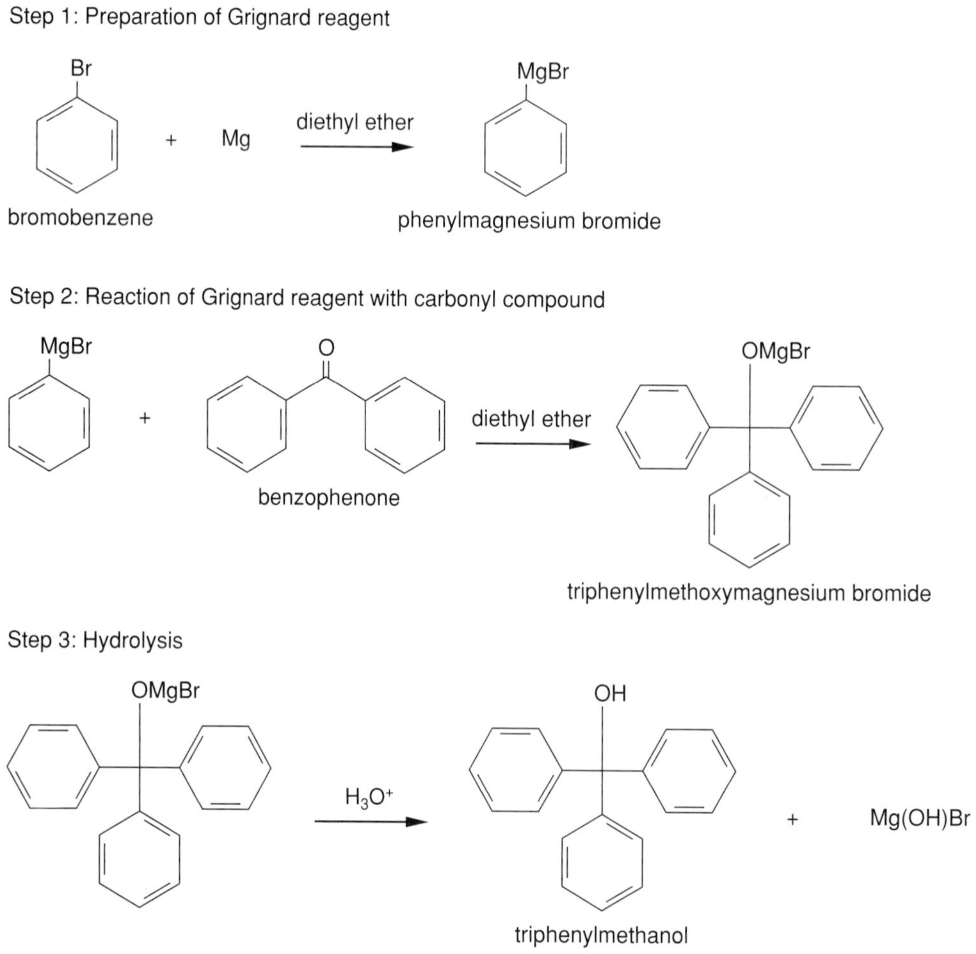

Complex of diethyl ether with a Grignard reagent.

Step 1: Preparation of Grignard reagent

bromobenzene   phenylmagnesium bromide

Step 2: Reaction of Grignard reagent with carbonyl compound

benzophenone   triphenylmethoxymagnesium bromide

Step 3: Hydrolysis

triphenylmethanol

In Part B, the Grignard reagent is reacted with ethyl benzoate to also afford triphenyl-methanol. In Part B, the ratio of Grignard to ethyl benzoate is 2:1. Benzophenone is the initial product after one Grignard addition. Benzophenone immediately undergoes a second Grignard reaction to afford triphenylmethanol.

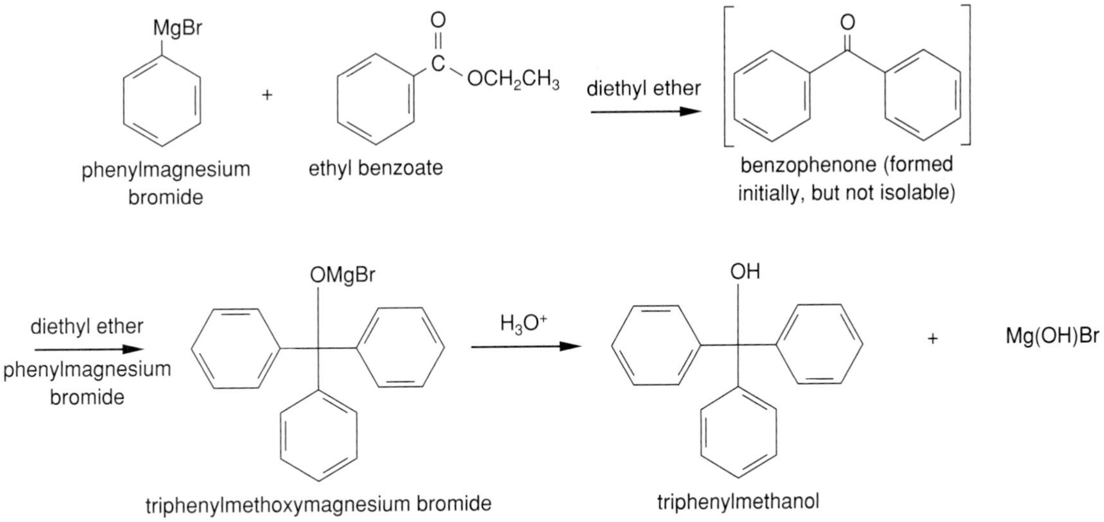

Coupling between the Grignard reagent and unreacted aryl halide is a significant side reaction. The coupling product in this reaction is biphenyl, which must be removed in a purification step.

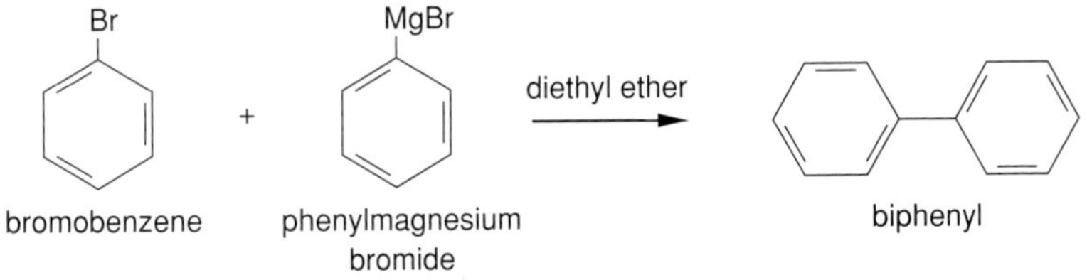

In Part C of this experiment, the Grignard reagent, prepared from magnesium and 2-bromopropane, is reacted with acetophenone, to form 3-methyl-2-phenyl-2-butanol, a liquid that can be purified by column chromatography.

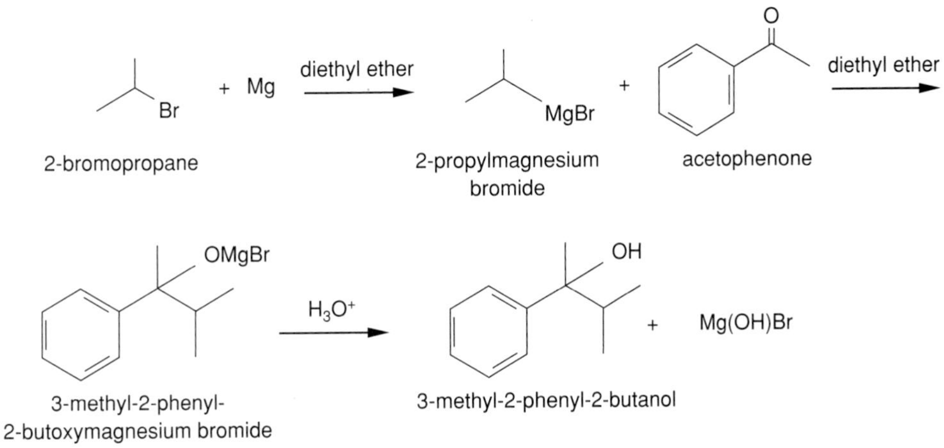

Prelab Assignment

1. Write a flow scheme for the synthesis of the assigned product.
2. Write a detailed mechanism for reaction of the Grignard reagent with the ketone or ester to be used in your experiment.
3. List the items of glassware that must be dried before starting this experiment.

Experimental Procedure

1. Always wear eye protection in the laboratory.
2. Diethyl ether and petroleum ether are extremely volatile and flammable. Make certain that there are no flames anywhere in the lab when these solvents are being used. Use in a well-ventilated area and avoid breathing the vapors. Petroleum ether may be harmful if inhaled or absorbed through the skin. Avoid prolonged contact or inhalation.
3. Bromobenzene, 2-bromopropane, and ethyl benzoate are irritants. Avoid contact with these liquids and do not breathe their vapors.

Part A: Microscale Synthesis of Triphenylmethanol from Benzophenone

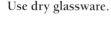

Use dry glassware.

All glassware must be dried prior to starting this experiment. Glassware can be dried in an oven at 110°C for at least 1 hour prior to use. Do not put plastic parts in the oven. Dry a 10-mL round bottom flask, a 5-mL conical vial, a water-jacketed condenser, a drying tube, and a Claisen adapter. A heat gun may be used instead to heat glassware. Allow the glassware to cool to room temperature. Assemble glassware as pictured.

Weigh 96 mg of magnesium turnings and place in a dry 10-mL round bottom flask containing a stir bar. Add a tiny crystal of iodine to the flask.

In a dry 5-mL conical vial, dissolve 0.42 mL of dry bromobenzene in 2.0 mL of anhydrous diethyl ether from a freshly opened can. Cap the vial. Draw this solution into a dry 3-cc syringe and add it in small portions through the septum into the 10-mL round-bottom flask containing the magnesium and iodine. Cloudiness and bubbles at the surface of the metal are indicators that the reaction has begun. This may take 5 minutes. If no reaction is observed at the end of 5 minutes, warm the flask in the palm of your hand or try adding another crystal of iodine. If the reaction still hasn't started, consult the instructor.

Once the reaction has started, turn on the stirrer and add the remainder of the bromobenzene/diethyl ether solution dropwise through the septum using a syringe over 5 to 10 minutes. This reaction is exothermic. Rinse the 5-mL conical vial with 1.0 mL of anhydrous diethyl ether and add the rinse through the septum via syringe to the reaction vial. Reflux the reaction mixture gently, with stirring, in a sand bath or heat block for 15 minutes after the addition of bromobenzene is complete. The final Grignard mixture should be cloudy and most of the magnesium should be gone. Cool to room temperature and add more anhydrous diethyl ether as necessary through the septum using a dry syringe to maintain the original volume. The Grignard reagent must be used immediately for the next step.

Dissolve 730 mg of benzophenone in 2.0 mL of anhydrous diethyl ether in the same 5-mL conical vial used above. Add the benzophenone solution dropwise via syringe through the septum to the stirred Grignard solution. Adjust the rate of addition so that the mixture refluxes gently. Rinse the 5-mL vial with 1.0 mL anhydrous diethyl ether and add the rinse to the 10-mL round-bottom flask. After all of the benzophenone has been added, reflux the mixture, with stirring, for 15 minutes. A red-pink precipitate should form. Cool the solution to room temperature. Add additional diethyl ether as necessary to maintain the overall volume. If time permits, go on to the next step. Otherwise, if there is not enough time to complete the procedure, disassemble the apparatus, cap the reaction vial, and seal it with parafilm until the next period.

Use dry glassware.

If necessary, add additional diethyl ether to replace any solvent lost by evaporation. The diethyl ether need not be anhydrous. (Why?) Remove the stir bar. Carefully add 3 M hydrochloric acid dropwise to the solution. Have an ice bath available to cool the flask if the reaction becomes too exothermic. Use a stirring rod or spatula to break up the precipitate. Continue to add 3 M HCl until the pH of the aqueous layer (the bottom layer) is 2–3. Add more diethyl ether as necessary to dissolve all of the product. Transfer to a centrifuge tube .

Draw up a drop of the bottom layer to test the pH.

The reaction mixture should separate into two layers as the alkoxide salt dissolves. The product is in the diethyl ether (upper) layer. Remove the lower aqueous layer to a clean vial and transfer the organic layer to a clean vial. Wash the aqueous layer with two 1.0-mL portions of diethyl ether. Combine the diethyl ether layers and wash with 1.0 mL of water. Remove the lower aqueous layer. Transfer the diethyl ether solution to a clean, dry Erlenmeyer flask or conical vial and dry over anhydrous sodium sulfate for 10 minutes. Using a filter pipet, transfer the dried solution to a tared 25-mL Erlenmeyer flask that contains a boiling chip. Rinse the drying agent with an additional 2 mL of diethyl ether. If doing TLC, spot a sample of this solution on the TLC plate. Carefully evaporate the diethyl ether under the hood to leave an off-white solid, which is a mixture of triphenylmethanol and biphenyl.

Add 2.0 mL of petroleum ether (bp 30–60°C). Stir the mixture thoroughly and vacuum filter using a Hirsch funnel. Rinse the crystals of triphenylmethanol with small amounts of petroleum ether.

Recrystallize the crude product from isopropyl alcohol. This will require approximately 3.0 mL of isopropyl alcohol. Weigh the purified triphenylmethanol. If instructed to do so, isolate the biphenyl side product from the petroleum ether filtrate by evaporating the solvent under the hood using a warm sand bath.

Part B: Miniscale Synthesis of Triphenylmethanol from Ethyl Benzoate

All glassware must be dried prior to starting this experiment. Glassware can be dried in an oven at 110°C for at least 1 hour prior to use. Do not put plastic parts in the oven. Dry a 50-mL round-bottom flask, a stir bar, a condenser, a drying tube, a 125-mL addition funnel, a Claisen adapter, a graduated cylinder, and several small Erlenmeyer flasks. A heat gun may be used instead to heat glassware. Allow the glassware to cool to room temperature before beginning the experiment.

Weigh 960 mg of magnesium turnings and place in a dry, round-bottom flask that contains a stir bar. Assemble glassware as pictured here. (It may be necessary to attach a second clamp to secure the apparatus.) In an addition (separatory) funnel, place 4.18 mL (6.28 g) of dry bromobenzene and 20 mL of anhydrous diethyl ether from a freshly opened can. Swirl to mix. Place a small paper strip between the stopper and funnel to avoid building up a vacuum as liquid is drained from the funnel. Prepare an ice bath to cool the solution if the reaction becomes too vigorous. Add approximately 1 mL of the bromobenzene/diethyl ether solution to the round-bottom flask and turn on the cooling water to the condenser.

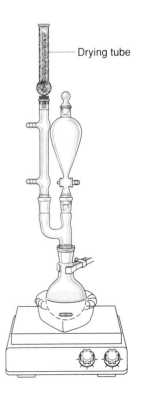

Drying tube

Cloudiness and bubbles at the surface of the metal are indicators that the reaction has begun. This may take 5 minutes. If no reaction is observed at the end of 5 minutes, add a small crystal of iodine. If the reaction still doesn't start, consult the instructor.

Once the reaction has started, turn on the stirrer. Add the remainder of the bromo-benzene solution dropwise to maintain a steady reflux. The addition should take approximately 45 minutes. Rinse the addition funnel with 3 mL of anhydrous diethyl ether and add the rinse to the round-bottom flask. Fit the flask with a heating mantle and reflux the reaction mixture gently for 15 minutes. The final Grignard reaction mixture should be cloudy and most of the magnesium metal should be gone. Cool to room temperature and add more anhydrous diethyl ether as necessary to maintain the original volume. The Grignard reagent must be used immediately in the next step.

Dissolve 1.41 mL (1.48 g, 10.8 mmol) of ethyl benzoate in 10 mL of anhydrous diethyl ether in the addition funnel. Add the ethyl benzoate solution dropwise to the stirred Grignard solution at a rate to keep the reaction solution at gentle reflux. Have an ice bath available to cool the solution if the reaction becomes too vigorous. The addition will take approximately 20 minutes. After all of the ethyl benzoate has been added, reflux the mixture, with stirring, for 15–30 minutes, or let stand until the next lab period if there is not enough time to complete the procedure. This is a good stopping point. Turn off the stirrer, disassemble the apparatus, stopper the flask, and seal it with parafilm until the next lab period. The flask should be nested inside a beaker in your drawer so that it doesn't tip over.

Replace any diethyl ether lost by evaporation. The diethyl ether need not be anhydrous. Remove the stir bar. Pour the reaction mixture into a 50-mL Erlenmeyer flask that contains 8 g of ice. Rinse the round-bottom flask with additional diethyl ether and add the rinse to the Erlenmeyer flask. While stirring, carefully add 3 M hydrochloric acid dropwise to the solution. Have an ice bath available if the reaction becomes too exothermic. Use a stirring rod or spatula to break up the solid. Continue to add 3 M HCl until the pH of the aqueous layer (the bottom layer) is just acidic to litmus. Don't overacidify. Add more diethyl ether as necessary to dissolve the product.

Carefully pour the reaction mixture into a separatory funnel, leaving any unreacted magnesium in the flask. Rinse the flask with 5 mL of diethyl ether and add the rinse to the separatory funnel. The reaction mixture should separate into two layers as the alkoxide salt dissolves. The product is in the diethyl ether (upper) layer. Separate the layers. Wash the lower aqueous layer with 5 mL of diethyl ether. Combine the diethyl ether layers and wash with 5 mL of water. Remove the lower aqueous layer. Transfer the diethyl ether solution to a clean, dry Erlenmeyer flask and dry over anhydrous sodium sulfate for 10 minutes. Decant or gravity filter the dried solution into a tared 50-mL Erlenmeyer flask that contains a boiling chip. Rinse the drying agent with an additional 1 mL of diethyl ether. If doing TLC, spot a sample of this solution on the TLC plate. Carefully evaporate the diethyl ether on a warm sand bath under the hood to leave an off-white solid, which is a mixture of triphenylmethanol and biphenyl.

Add 10 mL of petroleum ether (30–60°C). Stir the mixture thoroughly and vacuum filter using a Hirsch funnel or a Büchner funnel. Rinse the crystals of triphenylmethanol with small amounts of petroleum ether.

Recrystallize the crude product from isopropyl alcohol. This will require approximately 10–15 mL of isopropyl alcohol. Weigh the purified triphenylmethanol. If instructed to do so, isolate the biphenyl side product from the petroleum ether filtrate by evaporating the solvent under the hood using a warm sand bath.

Characterization for Parts A and B

Melting Point: Determine the melting points of triphenylmethanol and biphenyl. The reported melting points are 164°C for triphenylmethanol and 72°C for biphenyl.

Thin-layer Chromatography: Dissolve a few crystals of the product in a small amount of methylene chloride and apply a small spot to a silica gel plate. Also spot samples of authentic triphenylmethanol and biphenyl as standards. Develop the chromatogram with methylene chloride.

Infrared Spectroscopy: Run an IR spectrum of the product in either KBr or Nujol. Interpret the spectrum. The spectrum should show a broad OH stretch around 3450 cm^{-1}. The IR spectrum of triphenylmethanol is shown in Figure 14.1-1.

NMR Spectroscopy: Run the ^{1}H NMR spectrum of the product and interpret the spectrum. The spectrum should show strong absorption around $\delta 7$. The NMR spectrum of triphenylmethanol is shown in Figure 14.1-2.

UV Spectroscopy: Obtain the UV spectrum of the product in methanol. The maximum wavelength should be around 258 nm.

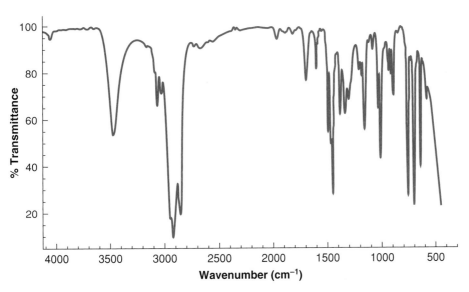

Figure 14.1-1

IR Spectrum of triphenylmethanol (Nujol)

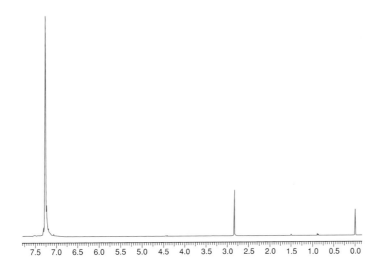

Figure 14.1-2

^{1}H NMR spectrum of triphenylmethanol (CDCl$_3$)

SOURCE: Reprinted with permission of Adrich Chemical.

Results and Conclusions for Parts A and B

1. Determine the percent yield of triphenylmethanol.
2. Explain the role of iodine in the preparation of phenylmagnesium bromide.
3. Why does some magnesium remain unreacted at the end of the preparation of the Grignard reagent?
4. Write balanced equations to show what happens to excess magnesium metal after preparation of phenylmagnesium bromide.

Part C: Microscale Synthesis of 3-Methyl-2-phenyl-2-butanol

Use dry glassware.

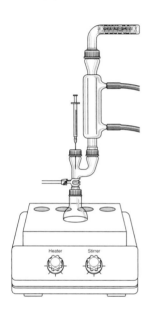

All glassware must be dried prior to starting this experiment. Glassware can be dried in an oven at 110°C for at least 1 hour prior to use. Do not put plastic parts in the oven. Dry a 10-mL round-bottom flask, two 3-mL conical vials, a jacketed condenser, a drying tube, and a Claisen adapter. A heat gun may be used instead to heat glassware. Allow the glassware to cool to room temperature.

Weigh 192 mg of magnesium turnings and place the turnings and a crystal of iodine in a dry 10-mL round-bottom flask containing a spin bar. Immediately cap the flask until ready to use. Assemble glassware as pictured. To a dry 3-mL conical vial, add 0.76 mL of 2-bromopropane and 1.2 mL of anhydrous diethyl ether from a freshly opened can. Cap the vial and swirl to mix the contents. Draw this solution into a dry syringe and add a portion of it dropwise through the septum into the 10-mL round-bottom flask containing the magnesium and iodine. Cloudiness and bubbles at the surface of the metal are indicators that the reaction has begun. Turn on the stirrer and add the rest of the solution. The addition should take approximately 5 minutes. During the addition, the mixture may bubble or change color. Rinse the 3-mL conical vial with 0.3 mL of anhydrous diethyl ether and add the rinse through the septum via syringe to the reaction flask. As soon as all of the 2-bromopropane solution has been added, reflux under low heat for 40 minutes. At the end of the reflux period, remove the reaction flask from the heat and cool in an ice bath. If necessary, add additional anhydrous diethyl ether to replace solvent lost by evaporation. Some unreacted magnesium may still be present.

The Grignard reagent must be used immediately in the next step. While it is cooling, prepare a solution of 0.66 mL of acetophenone and 1.5 mL of anhydrous diethyl ether in a dry 3-mL conical vial. Cap the vial and swirl to mix. Draw up the acetophenone solution in a dry syringe and slowly add it to the cooled, stirred mixture of the Grignard reagent. The addition should take approximately 5 minutes. Rinse the 3-mL vial with 0.3 mL of anhydrous diethyl ether and add the rinse to the 10-mL round-bottom flask. After the addition is complete, reflux the solution under low heat using a sand bath or heat block for 20 minutes. Cool to room temperature. If time permits, go on to the next step. If there is not enough time to complete the procedure, disassemble the apparatus, cap the reaction flask and seal it with parafilm until the next period.

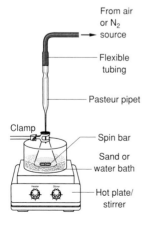

From air or N$_2$ source

Flexible tubing

Pasteur pipet

Clamp

Spin bar

Sand or water bath

Hot plate/ stirrer

If necessary, replace any diethyl ether lost by evaporation. The diethyl ether need not be anhydrous. Remove the spin bar. Transfer to a centrifuge tube. Add 2.5 mL of 2 M NH$_4$Cl. Cap the tube and shake, venting frequently. The solution should form two layers. After the layers have separated, remove the lower aqueous layer to a clean vial and transfer the organic layer to a clean vial. Wash the aqueous layer with two 0.5-mL portions of diethyl ether. Combine the diethyl ether layers and dry over anhydrous sodium sulfate for 10 minutes. Using a filter pipet, transfer the dried solution to a tared 25-mL Erlenmeyer flask that contains a boiling chip or stir bar. Rinse the drying agent with an additional 1 mL of diethyl ether. Carefully evaporate the diethyl ether under the hood, using a warm sand bath.

Purify the product by column chromatography using a silica gel column. Prepare a column by putting a glass wool plug in the bottom of the column, applying a thin layer of sand, and filling the column with 15 g of silica gel. The eluent is 93% hexanes: 7% ethyl acetate. Pour the solvent system through the column, being careful not to disturb the silica gel. Drain the solvent until the level is at the top of the silica gel. As the solvent drains, prepare a solution of the crude product in 0.5 mL of methylene chloride. With a clean pipet, apply the solution to the top of the silica gel. Rinse the flask with a few drops of methylene chloride and add to the column. Pour a thin layer of sand on top of the sample.

Carefully add the eluent to the column and collect fractions in test tubes. Change test tubes after every 4–5 mL. Spot each fraction on a TLC plate, along with standards of acetophenone and 3-methyl-2-phenyl-2-butanol, if available. Visualize TLC spots with a UV lamp. Collect 10–12 fractions or more if not all of the product has eluted by the 12th fraction. Combine fractions containing only 3-methyl-2-phenyl-2-butanol in a tared Erlenmeyer flask containing a boiling stone. Evaporate the solvent under the hood on a warm sand bath or an aluminum heat block. Weigh the purified product.

Caution: Use proper eye protection to avoid exposing eyes to UV light.

Characterization for Part C

Infrared Spectroscopy: Run an IR spectrum of the product. Interpret the spectrum. The spectrum should show a broad OH stretch around 3300 cm^{-1}.

NMR Spectroscopy: Run an NMR spectrum of the product in CDCl$_3$ and interpret the spectrum. The spectrum should show a pair of doublets for the two diastereotopic methyl groups, a singlet for the C-1 methyl group, a heptet centered around $\delta 1.7$, a singlet OH peak, and aromatic absorption around $\delta 7$.

Results and Conclusions for Part C

1. Determine the percent yield of 3-methyl-2-phenyl-2-butanol.
2. Explain the role of iodine in the preparation of the Grignard reagent.
3. Why is there some unreacted magnesium at the end of the preparation of the Grignard reagent?
4. Write balanced equations to show what happens to any excess magnesium metal after preparation of the Grignard reagent.
5. A possible side product in this reaction is 2,3-dimethylbutane. Propose a mechanism for its formation and explain where in the procedure it would be eliminated.
6. Explain why 2 M NH$_4$Cl was used in this procedure as the proton source rather than HCl or H$_2$SO$_4$.

Cleanup & Disposal

Neutralize the combined acidic aqueous layers with sodium bicarbonate and wash down the drain with water. Place methylene chloride solutions in a container labeled "halogenated organic solvent waste." Place petroleum ether solutions and the chromatography solvent into a container labeled "nonhalogenated organic waste solvent."

Critical Thinking Questions *(The harder ones are marked with a ❖.)*

1. In addition to the synthetic pathway used in this experiment, devise two other synthetic pathways using Grignard reagents for the preparation of 3-methyl-2-phenyl-2-butanol.
2. Give the product, after acid hydrolysis, of the reaction of methyl benzoate with two moles of methylmagnesium iodide. Suggest an alternate Grignard synthesis of this compound.

3. Grignard reagents react with carbon dioxide. Give the product, after acid hydrolysis, of the reaction of phenyl magnesium bromide with carbon dioxide.

4. What would be the product of a Grignard reagent with an acid chloride such as acetyl chloride? How many moles of the Grignard would react per mol of acid halide?

5. What would be the product if D_2O were added to a Grignard reagent such as phenylmagnesium bromide? Is this type of product in Grignard reactions of any value? If so, why?

6. What would be the product if D_2O/D_3O^+ were added to $R_3COMgBr$?

7. The preparation of phenylmagnesium bromide from bromobenzene and magnesium metal is an example of an oxidation-reduction reaction. What is oxidized and what is reduced?

❖ 8. The structural nature of a Grignard reagent has been studied extensively, but is still the subject of some controversy. Evidence suggests the presence of a dialkylmagnesium species, such as diphenylmagnesium. Write a reaction showing formation of diphenyl-magnesium and the other product that must also be present as a result of this reaction.

9. Give the product of the reaction of a Grignard with a nitrile such as acetonitrile, followed by acid hydrolysis.

10. Is a Grignard reagent covalent or ionic? Explain.

❖ 11. Explain why Grignard reagents react readily with C=O groups but not with simple alkenes (C=C).

12. *Molecular modeling assignment:* Construct a model of methylmagnesium bromide. Minimize energy. Examine the electrostatic potential map for the molecule, noting carefully the charged regions. Now construct a model of the solvated complex (methylmagnesium bromide-diethyl ether complex) and compare the charges and reactivity. Summarize how solvation affects the reactivity of the Grignard reagent.

Experiment 14.2: Using Indium Intermediates: *Reaction of Allyl Bromide with an Aldehyde*

In this experiment, you will study the reaction of an alkyl halide with an aldehyde in the presence of a transition metal in aqueous solution. You will:

- carry out a reaction of an alkyl halide with a transition metal and an aldehyde.
- analyze the purity of the product by infrared (IR) spectroscopy and gas chromatography (GC).
- compare the results expected for a Grignard synthesis with the results obtained in this experiment.

Techniques

Technique I	Extraction
Technique M	Infrared spectroscopy
Technique N	Nuclear magnetic resonance spectroscopy

Background

The U.S. Congress passed the Pollution Prevention Act of 1990 establishing a national policy to prevent or reduce pollution wherever possible. The Environmental Protection Agency embarked upon a program of Green Chemistry aimed at making chemical processes used throughout the nation more environmentally friendly and less polluting. The stated goal of the EPA program is: "To promote innovative chemical technologies that reduce or eliminate the use or generation of hazardous substances in the design, manufacture, and use of chemical products." (http://www.epa.gov/greenchemistry)

This initiative is being done in cooperation with numerous agencies and organizations including the American Chemical Society, National Science Foundation, and National Research Council. Methods for reducing pollution are several. Recycling of chemicals used in industrial processes is one technique designed to reduce pollution. Another technique is to replace noxious or carcinogenic chemical reagents with reagents that are less hazardous. For example, if an alcohol can be oxidized on an industrial scale using molecular oxygen and a catalyst instead of aqueous chromic acid, this would eliminate the need for using and having to dispose of a hazardous, carcinogenic chemical. If the quantities of organic solvents used in industrial organic processes can be reduced, then lesser quantities of such solvents need to be manufactured and shipped to users and lesser quantities of waste solvents need to be processed. The challenge is for researchers to develop new methods and techniques that address these pollution-reducing goals.

In this experiment, the use of a transition metal and an aqueous solvent will be examined as an alternative to using the traditional Grignard conditions, which must be maintained in a scrupulously anhydrous manner. The objective of this experiment is to carry out the reaction of 3-bromopropene with *o*-chlorobenzaldehyde to determine if the reaction occurs in a manner analogous to a Grignard synthesis or whether a different result is obtained (Li and Chan, 1991; Li, 1996).

3-bromopropene   *o*-chlorobenzaldehyde

The reaction uses powdered indium metal in an aqueous environment. Indium and nearby metals in the Periodic Table, such as lead and bismuth, are "soft" metals (relatively large and polarizable) that are not particularly reactive with water, especially when compared to Group I and Group II metals. Indium is not a hazardous chemical. It has a moderately high first ionization potential and can be considered to be relatively reactive, although not necessarily with "hard" oxygen-containing compounds such as water (Li and Chan, 1997).

Grignard reagents and their reactions with a variety of carbonyl compounds and epoxides provide extremely valuable avenues for chemists to build larger structures containing functional groups at predictable and key positions in molecules. Can reactions done in aqueous solution using indium metal also be effective? Some additional examples of reactions of indium-promoted couplings using allylic bromides have been reported (Paquette et al., 1997; Paquette et al., 1998).

Prelab Assignment

1. Calculate the number of mmoles of 3-bromopropene, indium, and *o*-chlorobenzaldehyde used in the procedure.
2. If indium were to behave like magnesium in this experiment, predict the structure of the expected organic product and propose a mechanism to explain its formation.

3. Explain how the procedure for this experiment is environmentally friendly.
4. Write a flow scheme for the workup procedure in this experiment.

Experimental Procedure

1. Wear eye protection at all times in the laboratory.
2. Wear gloves while handling reagents for this experiment.
3. 3-Bromopropene is a volatile chemical and a lachrymator. Do not breathe the vapors.
4. Work in a hood when working on this experiment.
5. Any spills should be cleaned up immediately. Notify the instructor immediately about any contact with reagents or any spills.

Miniscale Reaction of 3-Bromopropene with o-Chlorobenzaldehyde and Indium

Add 0.43 g of indium metal powder to 0.35 mL of 3-bromopropene in a 10-mL Erlenmeyer flask fitted with a spin bar. Place a cork loosely on top of the flask and stir the mixture for 20 minutes at room temperature. Add 0.30 mL of *o*-chlorobenzaldehyde to the mixture and continue stirring for an additional 30 minutes. Add 2.0 mL of concentrated HCl to the mixture and stir for 5–10 minutes. Transfer the mixture to a centrifuge tube and add 2 mL of methylene chloride. Cap the tube and shake to mix the layers with venting. Allow the layers to separate. Transfer the bottom layer to a dry Erlenmeyer flask. Repeat using a second 2-mL portion of methylene chloride and combine the organic layers. Dry over anhydrous sodium sulfate with occasional swirling for about 10 minutes. Transfer the dried organic solution to a tared Erlenmeyer flask containing a boiling chip.

Hood!

Heat the vial in a heat block or sand bath at 40°C to remove the solvent, and then at about 75° for a few minutes to remove any remaining 3-bromopropene. Cool the vial and weigh. Cap the vial for use in subsequent analyses.

Characterization

Thin-layer Chromatography: Spot the product and *o*-chlorobenzaldehyde on a TLC plate. Develop the plate using methylene chloride or diethyl ether as eluent. Visualize spots under a UV lamp.

Infrared Spectroscopy: Run an IR spectrum of the product and interpret the results.

Nuclear Magnetic Resonance Spectroscopy: Dissolve a small portion of the sample in $CDCl_3$ and transfer the solution into a clean, dry NMR tube using a filter pipet. Record and interpret the integrated ^1H NMR spectrum.

Results and Conclusions

1. Determine the percent yield of the product.
2. Based upon the evidence from IR and/or NMR spectroscopic analysis, propose a structure for the product.
3. Can the product be clearly differentiated from the starting materials by IR spectroscopy and NMR spectroscopy? Explain.
4. Estimate the percent purity of the product as determined by IR or NMR analysis.
5. Were the results of this experiment the same or different from results expected from a Grignard experiment using the same organic halide and aldehyde?

C l e a n u p & D i s p o s a l

Wash the aqueous extracts down the drain with running water. Clean the vials and flasks that contained indium carefully. Indium mixtures tend to bind up and get sticky if not washed immediately.

Critical Thinking Questions *(The harder ones are marked with a ❖.)*

1. Cite as many differences as you can between the indium reaction in this experiment and a similar Grignard synthesis carried out under anhydrous conditions.

2. Suggest other metals that might be substituted for indium in this experiment.

3. Suggest the structure of a possible reaction intermediate in this experiment.

❖ 4. Differentiate between "hard" and "soft" reactants. Describe the nature of the reactants in this experiment in terms of being "hard" or "soft." (You may wish to consult a text on inorganic chemistry when answering this question.)

5. What products are expected when *o*-chlorobenzaldehyde is replaced by benzaldehyde in this experiment?

6. It has been observed that 1-bromopropane fails to react with indium powder and an aldehyde in aqueous solution. Explain the difference in reactivity governing the failure of 1-bromopropane to react.

❖ 7. Might 3-bromo-1-butene be expected to react with indium and an aldehyde as in this experiment? Why or why not? What about using 4-bromo-1-butene instead?

8. Can the rate of the reaction in this experiment be followed by using TLC? Explain in detail how this might be accomplished.

9. Explain what differences in procedure if any might be recommended if 3-chloropropene were to be substituted for 3-bromopropene in this experiment.

10. *Library project.* Search for the various types of Green Chemistry experiments or processes that have been proposed. Give the advantages of each over traditional methods.

11. Explore the nature and properties of indium. One leading website for indium is http://www.resource-world.net/In.htm.

12. Suggest some changes for your laboratory that might reduce pollution and foster the concept of Green Chemistry.

References

Li, C-J. Aqueous Barbier-Grignard Type Reactions: Scope, Mechanism and Synthetic Applications. *Tetrahedron* 52 (1996): 5643.

Li, C-J., and Chan, T-H. *Organic Reactions in Aqueous Media.* New York: Wiley-Interscience, 1997.

Li, C-J., and Chan, T-H. Organometallic Reactions in Aqueous Media with Indium. *Tetrahedron Letters* 32 (1991): 7017.

Paquette, L. A., Bennett, G. D., Chhatriwalla, A., and Isaac, M. B. Factors Influencing 1,4-Asymmetric Induction during Indium-Promoted Coupling of Oxygen-Substituted Allylic Bromides to Aldehydes in Aqueous Solution. *J. Org. Chem.* 62 (1997): 3370.

Paquette, L. A., Bennett, G. D., Isaac, M., and Chhatriwalla, A. B. Effective 1,4-Asymmetric C—C/C—O Stereoinduction in Indium-Promoted Coupling Reactions of Aldehydes to Protected and Unprotected [1-(Bromomethyl)vinyl] Alkanols. The Status of Intramolecular Chelation within Functionalized Allylindium Reagents. *J. Org. Chem.* 63 (1998): 1836.

Chapter 15

Alcohols and Diols

Alcohols are among the most readily available and useful classes of organic compounds. Simple alcohols are often used as reactants to synthesize other organic compounds because many alcohols are relatively cheap and they are relatively nonhazardous, allowing easy shipment.

Alcohols undergo numerous reactions and in the process are converted to other functional compounds, such as alkenes. Acid-catalyzed dehydration of 3-methyl-3-pentanol affords alkenes, as illustrated in Experiment 5.1. An alcohol is used as a reactant in Experiment 4.1 in the preparation of 2-chloro-2-methylbutane.

Preparation of alcohols in the laboratory is often accomplished by hydration of alkenes or by reduction of aldehydes or ketones. In Experiment 6.2, hydrolysis and mercuration-demercuration reactions of alkenes are explored. In Experiment 15.1, the chemical reducing agent sodium borohydride is used to reduce benzil, a diketone. Alcohols can also be prepared as the products of carbon-carbon condensation reactions as in Experiment 14.1.

Experiment 15.1: Stereoselective Reduction of Ketones with Sodium Borohydride

There are a number of reagents that reduce carbonyl groups. In this experiment, one of these reagents will be examined. You will:

- reduce a ketone to an alcohol.

- determine the stereochemistry of the product.

Techniques

Technique F	Vacuum filtration
Technique K	Thin-layer chromatography
Technique M	Infrared spectroscopy
Technique N	Nuclear magnetic resonance spectroscopy

Background

Sodium borohydride ($NaBH_4$) is a mild reducing agent that will selectively reduce aldehydes and ketones, but not esters, alkenes, carboxylic acids, or amides. The mechanism of this reaction involves transfer of hydride from the borohydride ion to the carbonyl carbon and protonation of the resulting alkoxide by the solvent, usually ethanol or methanol. In Part A of this experiment, benzil will be reduced with sodium borohydride to hydrobenzoin. The addition of the first equivalent of hydride forms an α-hydroxyketone (benzoin); the second equivalent of hydride forms the diol (hydrobenzoin). The addition of hydride to the top or to the bottom face of benzoin produces different stereoisomers. The product can be characterized from the melting point or through thin-layer chromatography.

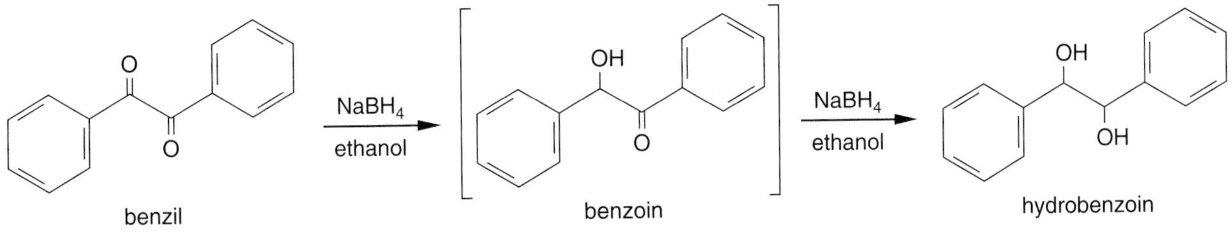

In Part B of the experiment, camphor will be reduced to borneol or isoborneol. The predominant mode of addition (exo or endo) by the borohydride can be determined by examining the ratio of these two alcohols.

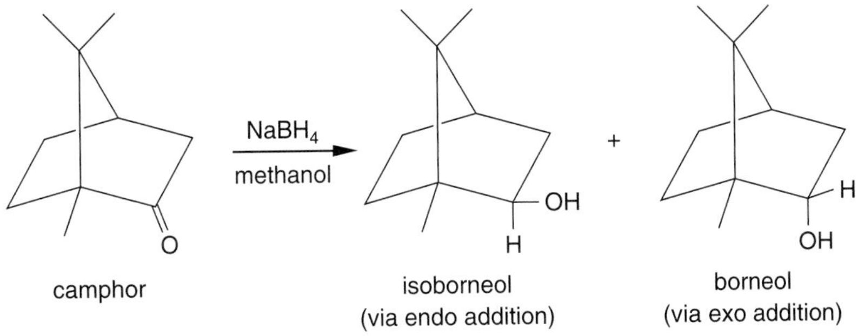

If the borohydride adds to the carbonyl from the bottom face of the camphor (endo), the product is isoborneol; if addition occurs from the top face (exo), the product is borneol. The product mixture can be characterized by thin-layer chromatography or nuclear magnetic resonance spectroscopy.

Prelab Assignment

1. Write balanced equations for the reduction of benzil or camphor, assuming that each mol of $NaBH_4$ provides four moles of hydride.
2. (Part A) Give the structures of all of the possible stereoisomers produced by the reduction of benzil to hydrobenzoin. Draw Fischer projections of the molecules. Designate each stereogenic center as R or S. Indicate whether the molecule is chiral or achiral. Designate the stereochemical relationship between the isomers.
3. (Part B) Assign R or S configurations to each stereogenic center of isoborneol and borneol. Give the relationship between these two stereoisomers.

Experimental Procedure

1. Wear eye protection at all times in the laboratory.

2. Sodium borohydride should be used under the hood. Wear gloves.

3. When using the ultraviolet lamp, use proper eye protection to avoid exposing eyes to ultraviolet light. Do not look into the lamp.

Part A: Microscale Reduction of Benzil

Weigh out 200 mg of benzil and add it to a 10-mL Erlenmeyer flask. Add 2 mL of 95% ethanol. Swirl to mix. Add 75 mg of sodium borohydride to the solution of benzil. Let stand 10 minutes with occasional swirling. The yellow color will disappear as benzil is reduced. Add 2 mL of water slowly as foaming may occur. Heat the solution to boiling. The solution should be clear and colorless. If necessary, filter off any insoluble impurities, using a Pasteur filter pipet. Transfer the hot clear solution to a clean Erlenmeyer flask using a filter pipet. Add an additional 2 mL of hot water. Allow the solution to crystallize undisturbed to form shiny thin plates. Cool in an ice bath and then vacuum filter the crystals, using a Hirsch funnel. Wash the crystals with a minimum amount of cold water. Weigh the product.

Characterization for Part A

Melting Point: Determine the melting point of the product. *meso*-Hydrobenzoin has a reported melting point of 137°C; (±)-hydrobenzoin has a reported melting point of 120°C.

Thin-layer Chromatography: Dissolve 1–2 mg of the product in about 0.5 mL of ethyl acetate and apply a small spot to a silica gel plate. Also spot samples of authentic *meso*- and (±)-hydrobenzoin dissolved in ethyl acetate. The plate may be developed using methylene chloride/ethyl acetate (9:1). Visualize the developed chromatogram in an iodine chamber or under UV light.

Use eye protection when using UV light.

Infrared Spectroscopy: Record the IR spectrum of the product as a Nujol mull or KBr pellet and interpret the spectrum, noting key bands characteristic of functional groups present. Compare the spectrum with the IR spectra of benzil and hydrobenzoin, if available. The IR spectra of *meso*-hydrobenzoin and (±)-hydrobenzoin are shown in Figures 15.1-1 and 15.1-2, respectively.

NMR Spectroscopy: Record the ^{1}H NMR spectrum of the product in CDCl$_3$ solution. The alkyl C–H signals are different for *meso*-hydrobenzoin ($\delta 4.72$) and (±)-hydrobenzoin ($\delta 4.57$).

Results and Conclusions for Part A

1. Determine the structure of the major stereoisomer(s) formed in this reaction.
2. Calculate the percent yield of the reaction.
3. Is the product mixture expected to be optically active? Explain.

Part B: Microscale Reduction of (1R)-(+)-Camphor

Add 100 mg of (1R)-(+)-camphor to a 5-mL conical vial containing a spin vane. Add 0.5 mL of methanol. Stir to dissolve. Cool the solution in an ice bath. Weigh out 60 mg of

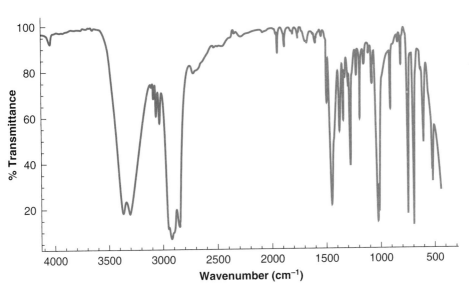

Figure 15.1-1

IR spectrum of *meso*-hydrobenzoin (Nujol)

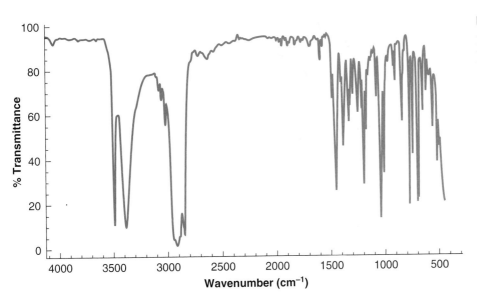

Figure 15.1-2

IR spectrum of (±)-hydrobenzoin (Nujol)

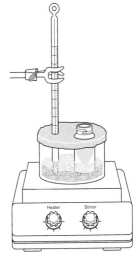

sodium borohydride. Cautiously add the sodium borohydride in three batches to the stirred cooled solution of camphor. Let the reaction subside between additions. After all of the sodium borohydride has been added, heat the solution to boiling using a sand bath. Cool to room temperature. Cautiously add 4 mL of ice-cold water with stirring to the grayish suspension. Remove the spin vane. Allow the mixture to stand for several minutes to solidify. Suction filter the white precipitate with a Hirsch funnel. Transfer the air-dried crystals into a small Erlenmeyer flask. Dissolve the crystals in 5 mL of diethyl ether. Dry the solution with anhydrous sodium sulfate. Let stand 5 minutes. Use a filter pipet to transfer the liquid to a clean, dry, tared 25-mL Erlenmeyer flask. Rinse the sodium sulfate with 1 mL of diethyl ether and add the rinse to the tared flask. Evaporate the solvent using a sand bath under the hood to leave an off-white solid. Air dry the crystals. (If desired, the product may be further purified by vacuum sublimation.) Weigh the product. Characterize the product using one or more of the suggested methods.

Characterization for Part B

Melting Point: Determine the melting point of the product. The reported melting point of borneol is 207–208°C; the reported melting point of isoborneol is 212–214°C. Also do a mixed melting point determination with your product mixed with a sample of either pure borneol or isoborneol.

Use eye protection when using UV light.

Thin-layer Chromatography: Dissolve a few crystals of the product in a small amount of methylene chloride and apply a small spot to a silica gel plate. Also spot samples of authentic borneol, isoborneol, and (1R)-(+)-camphor. Develop the plate using ethyl acetate. Visualize the developed chromatogram in an iodine chamber or under UV light.

NMR Spectroscopy: Prepare a sample in CDCl$_3$ and record the ^{1}H NMR spectrum. The spectra of borneol and isoborneol are distinctive. The peak at δ3.6 corresponds to the proton adjacent to the OH group in isoborneol; the comparable proton in borneol has a chemical shift of δ4.0. The integrated ^{1}H NMR spectra of isoborneol and borneol are shown in Figure 15.1-3 and Figure 15.1-4, respectively.

Results and Conclusions for Part B

1. Determine the structure of the major product formed in the reaction. Is exo or endo the preferred mode of addition? Build a model of this compound and use it to explain the observed stereochemistry.
2. Calculate the percent yield of the product.
3. Is the product mixture expected to be optically active? Explain.
4. What results would be expected if racemic camphor were used in place of optically active camphor? Explain.

Figure 15.1-3

^{1}H NMR spectrum of isoborneol (CDCl$_3$)

SOURCE: Reprinted with permission of Aldrich Chemical.

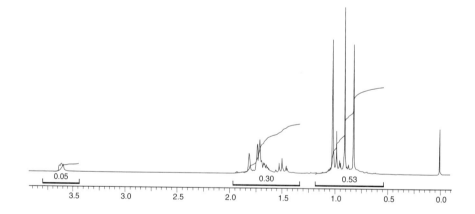

Figure 15.1-4

^{1}H NMR spectrum of borneol (CDCl$_3$)

SOURCE: Reprinted with permission of Aldrich Chemical.

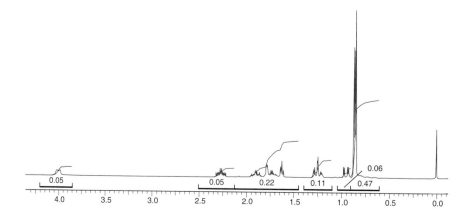

Cleanup & Disposal

Solutions of sodium borohydride should be neutralized with dilute HCl. Use caution since hydrogen gas may be evolved. Once neutralized, the aqueous solutions may be washed down the drain with water. Place TLC solvents into a container labeled "nonhalogenated organic solvent waste." Place methylene chloride from the TLC characterization into a container labeled "halogenated organic solvent waste."

Critical Thinking Questions

1. Why is it important to use a large excess of sodium borohydride when doing a reduction in aqueous ethanol? (Hint: Consider what reaction might occur between water and sodium borohydride.)

2. The reduction of benzil to hydrobenzoin can be followed by TLC or IR or by watching the disappearance of the yellow color. Explain.

3. Draw the structure of the reduction products of the reaction of the assigned ketone with:
    a. $NaBD_4$ in CH_3CH_2OH and H_2O.
    b. $NaBH_4$ in CH_3CH_2OD and D_2O.
    c. $NaBH_4$ in CD_3CD_2OH and H_2O.

4. Predict the products of the sodium borohydride reduction of each of the molecules below. Show all stereoisomers expected. Label each stereogenic center as R or S.

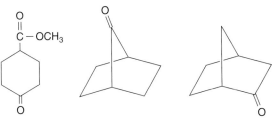

5. Draw the structure of the intermediate hydroxyketone formed by the transfer of one hydride to benzil. Is this intermediate optically active? Explain.

6. What is the purpose of adding water to the reaction mixture after the reaction between sodium borohydride and the ketone is complete?

Experiment 15.2: Experimental Design for an Alcohol Oxidation

Oxidation is one of the most versatile methods of functional group transformation. Many different methods of oxidation have been developed to convert primary alcohols to aldehydes or carboxylic acids and secondary alcohols to ketones. In this experiment, the student will design and carry out a procedure to do one or more of the following:

♦   oxidize a secondary alcohol to a ketone.

♦   oxidize a primary alcohol to an aldehyde.

♦   oxidize a primary allylic alcohol to an α, β-unsubstituted aldehyde.

♦   oxidize a primary alcohol to a carboxylic acid.

Background

Many different methods of oxidation have been developed to transform primary alcohols into aldehydes or carboxylic acids, secondary alcohols into ketones, and alkyl benzenes into benzoic acid derivatives. Oxidation of alcohols is important because of the commercial availability of many alcohols as starting materials. Many of the common methods use oxidizing agents of chromium or manganese that are inexpensive and readily available.

A primary alcohol can be converted either to an aldehyde or to a carboxylic acid, depending upon the oxidizing agent:

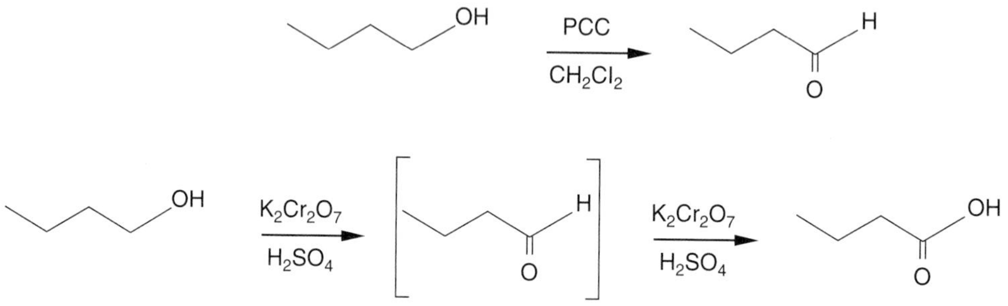

The oxidation first produces the aldehyde, which can easily be oxidized further to a carboxylic acid. Most oxidizing agents react so rapidly with the aldehyde that isolation of the aldehyde product is not possible. PCC (Pyridinium chlorochromate) is said to be a "selective" oxidizing agent, because it will not oxidize the aldehyde further to the carboxylic acid in methylene chloride solutions. PCC is a complex formed from pyridine, chromium trioxide, and HCl. Other oxidizing agents that permit the isolation of the aldehyde are PDC (pyridinium dichromate) and Collins reagent. PDC is a complex formed from two pyridine molecules and chromic acid. The Collins reagent is a complex formed from two pyridine molecules and chromium trioxide. In addition to these reagents, primary alcohols that are allylic or benzylic can also be selectively oxidized to α, β-unsaturated aldehydes by MnO_2 without affecting the carbon-carbon double bond.

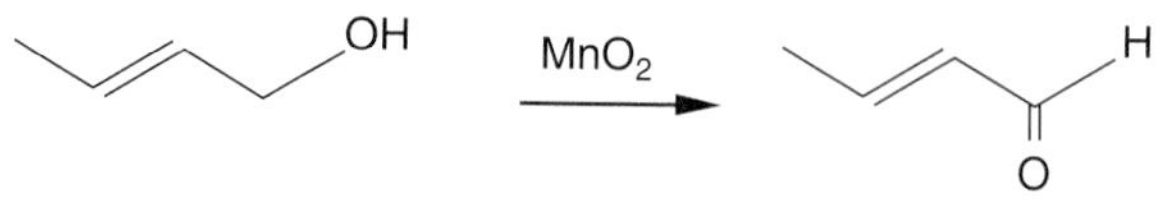

Secondary alcohols are cleanly converted to ketones with Jones reagent (an acetone solution of chromic acid and sulfuric acid) or with one of a large number of other oxidizing agents, since there is little danger of overoxidation. PCC, Collins reagent, and the Jones reagent are only a few of the many different oxidizing agents that can be used. Tertiary alcohols are not oxidized except under very harsh conditions required for carbon-carbon bond cleavage.

You will be assigned one or more of the oxidative transformations listed here, although others may be assigned by the instructor.

4-hydroxy-3-methoxybenzyl alcohol → vanillin

4-hydroxy-3-methoxybenzyl alcohol → vanillic acid

2-nitrobenzyl alcohol → 2-nitrobenzaldehyde

3-nitrobenzyl alcohol → 3-nitrobenzaldehyde

4-nitrobenzyl alcohol → 4-nitrobenzaldehyde

2-naphthalenemethanol → 2-naphthaldehyde

benzoin → benzil

9-hydroxyfluorene → 9-fluorenone

benzhydrol → benzophenone

cholesterol → cholest-5-en-3-one

4-phenyl-3-buten-2-ol → 4-phenyl-3-buten-2-one

coniferyl alcohol (4-hydroxy-3-methoxycinnamyl alcohol) → coniferyl aldehyde
   (4-hydroxy-3-methoxycinnamaldehyde)

p-chlorobenzyl alcohol → *p*-chlorobenzoic acid

The selection of an appropriate oxidizing agent will depend upon selectivity of the reagent, ease of use, toxicity, and cost considerations. It is important to consider the structure of the entire compound to be oxidized when choosing an oxidizing agent. Are there acid-labile functional groups present? If so, the strongly acidic conditions of the Jones reagent may not be appropriate and the oxidation should be carried out under neutral or basic conditions. Are other functional groups present (such as double bonds or aldehydes) that may react with the oxidizing agent? If so, an oxidizing agent must be chosen that will not react with the other functional groups. The oxidation procedure chosen should be relatively easy to perform so that the reaction can be completed in one lab period.

The procedure should be adapted for the physical properties of the alcohol and product. Scale the amounts of reagents to use approximately 100 mg of the alcohol. Write a detailed procedure for the oxidation of the assigned alcohol, including the masses and/or volumes of all reagents to be used in the experiment and a detailed workup procedure for isolating and characterizing the product. Address safety concerns and how the reagents will be disposed of after the experiment. Note the reference upon which the procedure is based. To make certain that specific reagents are available, the procedure must be approved by the instructor before beginning to work in the lab.

Prelab Assignment

1. Find one or more procedures in the library to accomplish the assigned oxidation. It is unlikely that a procedure will be found for oxidizing the exact assigned alcohol, but a procedure should be found for oxidizing a similar type of alcohol. *Organic Reactions, Organic Synthesis, Advanced Organic Chemistry (March), Vogel's Practical Organic Chemistry (Vogel), and Comprehensive Organic Chemistry (Barton)* are only a few of the secondary sources that might be helpful in locating primary sources. Record the specific reference(s).

2. Look up the physical properties of the assigned reactant and product. Are the compounds liquids or solids? Based upon the properties, determine how the product can best be isolated from any unreacted starting material.

3. What possible by-products could be obtained in this reaction? Look up the properties of the by-products and determine how the by-products can best be removed from the desired product.

4. Consider the hazards of all of the reagents and solvents involved and decide what safety precautions must be considered when carrying out the assigned transformations.

5. Adapt the literature procedure to the assigned reaction and scale the procedure to start with 100 mg of the assigned alcohol. Write detailed directions for the reaction, workup, and characterization of the product.

6. Look up the costs of all of the reagents and solvents in a chemical supply company. Estimate the cost of performing the experiment in the laboratory.

Experimental Procedure

Present the procedure to the instructor at least one week prior to performing the experiment in order to gather the necessary supplies. After the procedure has been approved, gather the reagents, and conduct the experiment to isolate and characterize the product.

Results and Conclusions

1. Rationalize the selection of the oxidizing agent based on selectivity, ease of use, cost, and any other considerations.
2. Determine the percent yield obtained in the reaction.
3. Evaluate the purity of the product obtained in the reaction.
4. Write a flow scheme for the oxidation workup procedure used in this experiment.
5. Give the structures of any possible impurities and explain where in the procedure these impurities are separated from the product.

Critical Thinking Questions

1. Indicate whether or not the following proposed oxidations will yield the desired product. If the oxidation is not feasible, explain why. If possible, suggest a suitable oxidant for the transformation.

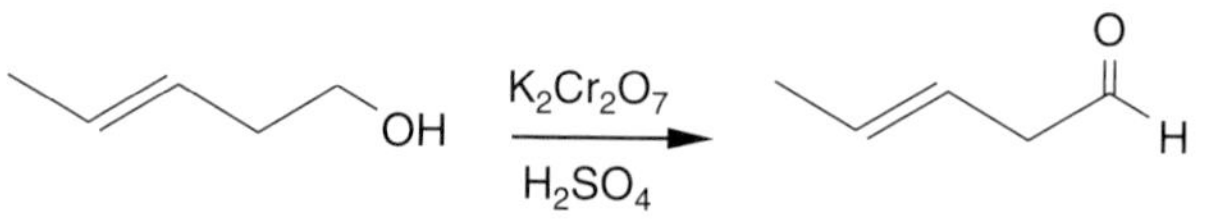

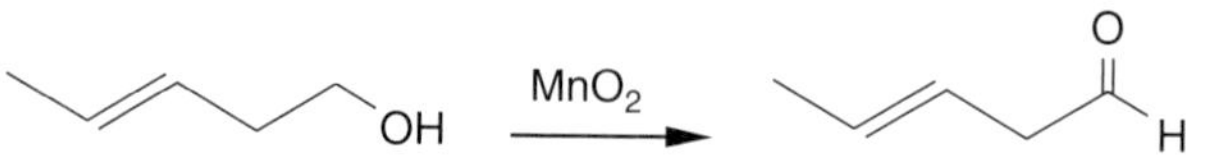

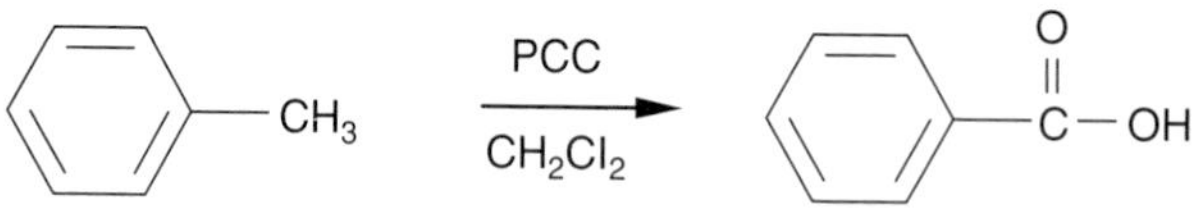

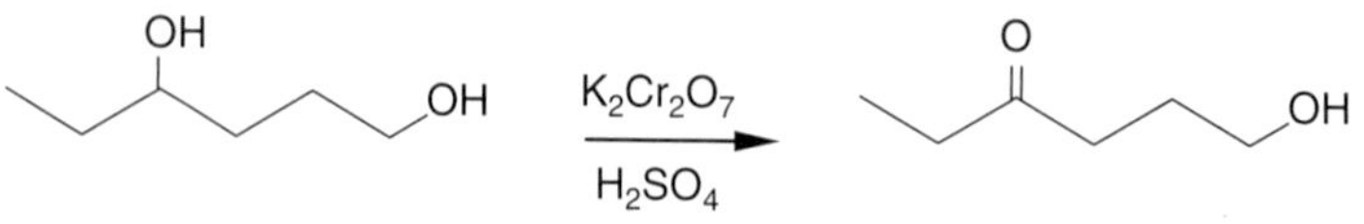

2. Write balanced equations for the reaction shown below:
$$CH_3CH_2CH_2CH_2OH + K_2Cr_2O_7 + H_2SO_4 \rightarrow CH_3CH_2CH_2CO_2H + Cr_2(SO_4)_3$$
3. Explain how the following techniques could differentiate between the compounds:
   a. ^{13}C NMR: 1-propanol and propanal.
   b. IR: 1-butanol, butanal, and butanoic acid.
   c. ^1H NMR: 2-propanol and 2-propanone (acetone).
   d. IR: toluene and benzoic acid.
   e. ^{13}C NMR: toluene and benzoic acid.

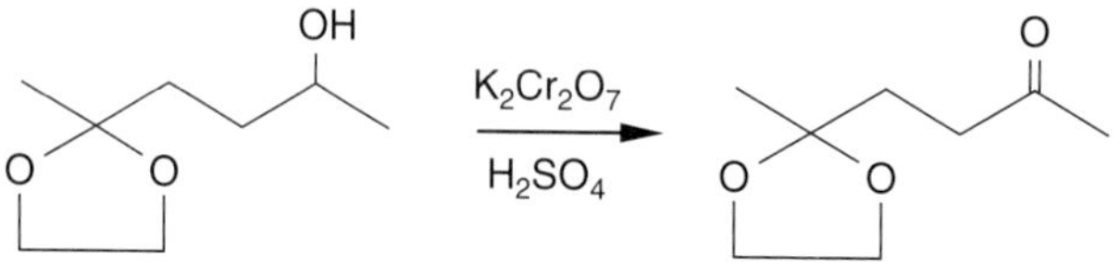

4. *Library project:* Clay has been used as a solid support in polymer-bound oxidation reactions. Find a reference in the literature. Give the complete citation to the reference.
5. *Library project:* Microbes can be very selective oxidizing agents. *Rhizopus nigricans* can transform progesterone into 11-α-hydroxyprogesterone. Find a procedure in the literature for accomplishing this transformation.

Experiment 15.3: Photochemical Oxidation of Benzyl Alcohol

In this experiment, you will investigate an oxidation reaction occurring by a free radical mechanism. You will:

◆   carry out a reaction that uses light as an initiator.

◆   perform GC separation of the starting material and product.

◆   use thin-layer chromatography to identify an oxidation product.

Techniques

Technique J	Gas chromatography
Technique K	Thin-layer chromatography
Technique M	IR spectroscopy
Technique P	Mass spectrometry

Background

Most organic students are probably familiar with free radical halogenation of alkanes using chlorine or bromine and light. This experiment is one that also goes by a free radical chain mechanism, initiated by light. Short-chain alcohols such as benzyl alcohol can be oxidized by N-iodosuccinimide (NIS) in chlorobenzene solvent in the presence of light to give a single product (Beebe et al., 1983). You will synthesize the product and use thin-layer chromatography or gas chromatography-mass spectrometry to identify the oxidation product. Based upon the product obtained, you will propose a mechanism for the reaction.

The reaction, which requires two moles of the NIS reagent for each mole of benzyl alcohol, is shown below.

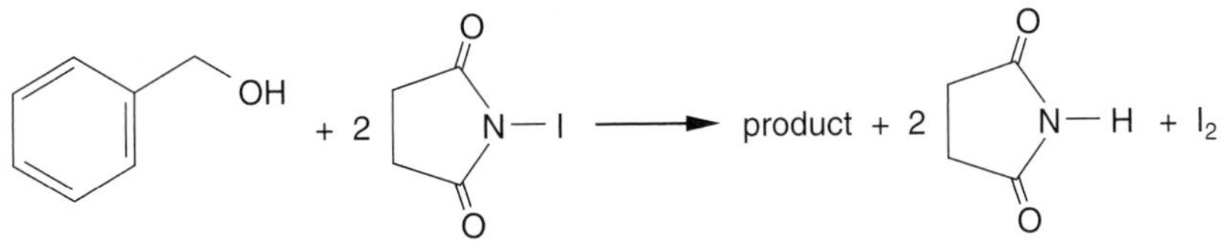

$$\text{(benzyl alcohol, } C_6H_5CH_2OH) + 2 \text{ (N-iodosuccinimide)} \longrightarrow \text{product} + 2 \text{ (succinimide, N—H)} + I_2$$

Prelab Assignment

1. Draw the structures of the two possible oxidation products of benzyl alcohol.
2. Look up the physical characteristics of the two possible products, benzyl alcohol and chlorobenzene. Recognizing that unreacted benzyl alcohol and chlorobenzene will still be present in the reaction mixture, estimate the order of elution on a nonpolar GC column.

3. Prepare a flow scheme for the assigned experiment.

4. Molecular iodine is produced during the course of the experiments. What should happen to the color of the reaction mixture as the reactions proceed?

Experimental Procedure

1. Chlorobenzene may cause respiratory irritation. Use gloves when measuring out this liquid and avoid breathing the vapors.

2. A bright light is used to catalyze the reaction. The lamp will be hot and can burn the skin. Be careful to avoid looking directly at the lamp.

3. N-Iodosuccinimide is a skin irritant. Wear gloves when handling this substance.

4. Wear safety goggles at all times in the laboratory.

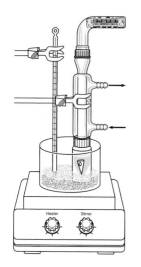

Microscale Experimental Procedure for the Oxidation of Benzyl Alcohol with N-Iodosuccinimide

Into a 5-mL conical vial, add 150 μL of benzyl alcohol via pipet and 3 mL of chlorobenzene. Swirl to mix the liquids. Add 700 mg of solid N-iodosuccinimide and a spin vane to the vial. Attach a water-jacketed condenser with drying tube to the vial and stir the solution for 60 minutes about 6–8 inches away from a sunlamp. Two or three setups may be grouped to allow irradiation from a single sunlamp. After the 60-minute period, turn off the stirrer and remove the condenser. Allow any precipitate to settle and transfer the solution to a conical vial using a Pasteur pipet. Wash the solution with 1 mL of 5% aqueous sodium thiosulfate solution. Remove the organic layer to a clean flask and dry over anhydrous sodium sulfate. Use a filter pipet to transfer the dried solution to a clean vial. Use a syringe to draw up the dry organic solution and inject about 2–3 μL of the sample onto a GC column according to directions given by the instructor. The sample will contain unreacted benzyl alcohol (bp 205°C) and chlorobenzene (bp 135°C) in addition to the oxidation product.

Characterization

Gas Chromatography (or GC-Mass Spectrometry): Analyze an aliquot of the reaction mixture using GC on a semi-polar column.

Thin-layer Chromatography: Spot standard solutions of benzyl alcohol, chlorobenzene, and the two possible oxidation products on a silica gel plate containing a fluorescent indicator. Spot the product mixture. Develop the chromatogram in 1:1 hexanes: ethyl acetate. Visualize under a UV lamp.

Infrared Spectroscopy: IR analysis can be done by capturing an aliquot at the exit port of the GC (with thermal conductivity detector). Interpret the spectrum in terms of the peaks expected for the product and those bands of the starting material that are expected to be absent in the product. Absorption in the 3200–3600 cm^{-1} region indicates the presence of OH absorption, characteristic of the starting alcohol.

Results and Conclusions

1. Based upon the GC data or the IR spectrum and/or the results of thin-layer chromatography, determine the identity of the oxidation product.
2. Did the reaction go to completion? Could additional irradiation time have given a higher yield of product? Explain.
3. Propose a mechanism for the formation of the observed product.
4. Identify the element being oxidized and the element being reduced for the reaction in this experiment.

Cleanup & Disposal

Place the aqueous washings in the sink and wash them down the drain. Be sure that the area around the balances is clean and that no residue of N-iodosuccinimide remains on or near the balances. Place used halogenated solvent in a container labeled "halogenated organic solvent waste."

Critical Thinking Questions

1. Is there any evidence for the formation of molecular iodine during the reaction?
2. This reaction does not proceed without light initiation. Look up the N-I bond strength and propose a mechanism by which the reaction is initiated.
3. Write an equation for the reaction that takes place when iodine reacts with sodium thiosulfite.
4. Predict the product of reaction when *o*-methylbenzyl alcohol is irradiated in the presence of N-iodosuccinimide.
5. Explain what is meant by the term "homolytic process."
6. *Library project:* Investigate other oxidative reactions induced by NIS. For example, see Beebe and Howard, 1969, Beebe et al., 1972, and Beebe et al., 1974.

References

Beebe, T. R., Adkins, M., Kwok, P., Roehm, R. Reaction of N-Iodosuccinimide with Tertiary Alcohols. *J. Org. Chem.* 37(1972): 4220.

Beebe, T. R, Adkins, R. L., Bogardus, C. C., Champney, B., Hii, P. S., Reinking, P., Shadday, J., Weatherford III, W. D., Webb, M. W., and Yates, S. W. Primary Alcohol Oxidation with N-Iodosuccinimide. *J. Org. Chem.* 48(1983): 3126.

Beebe, T. R., Howard, F. M. Oxidation of 1-Phenylethanol by the Succinimidyl Radical. *J. Amer. Chem. Soc.* 91(1969): 3379.

Beebe, T. R., Lin, A. L., Miller, R. D. Reaction of N-Iodosuccinimide with Secondary Alcohols. *J. Org. Chem.* 39(1974): 722.

Chapter 16

Ethers

E thers are often used as solvents in organic chemistry. They may be readily synthesized via S_N2 reactions, as illustrated for the synthesis of aryl ethers in Experiment 16.1. Synthesis of ethers from alkoxides and alkyl halides is called the Williamson ether synthesis. Aryl ethers can also be prepared by nucleophilic aromatic substitution on aromatic rings containing strongly electron-withdrawing groups, such as 2,4-dinitrobromobenzene and p-nitrofluorobenzene. Nucleophilic aromatic substitution reactions are explored in Experiment 23.1. A nucleophilic aliphatic substitution and elimination puzzle involving ethers is investigated in Experiment 16.2.

Experiment 16.1: Ether Synthesis by S_N2 Displacement

In this experiment, you will investigate an S_N2 reaction where the nucleophile is an alkoxide or a phenoxide and the product is an ether. You will:

♦   follow the progress of a reaction using thin-layer chromatography (TLC).

♦   perform a useful synthetic application of an S_N2 reaction.

♦   investigate the acidic properties of phenols.

♦   learn how to plan an experiment as a prelab assignment.

Techniques

Technique G	Reflux
Technique I	Extraction
Technique K	Thin-layer and column chromatography
Technique M	Infrared spectroscopy
Technique N	Nuclear magnetic resonance spectroscopy

Introduction

In the S_N2 synthesis of ethers, an alkoxide or phenoxide reacts with a methyl halide or primary alkyl halide by nucleophilic displacement. Such nucleophilic substitution reactions are very common in the laboratory because the starting materials and reagents are relatively cheap and available, the reaction conditions are mild and the reactions can be completed quickly and very often in good yield.

Elimination side reactions can be avoided or minimized by using methyl and primary halides, particularly ones that lack β-hydrogens. In Part A of this experiment, the starting materials are halogenated phenoxides and methyl or primary alkyl tosylates. Tosylates are used because of their greater reactivity.

X = F, Br, Cl, I; R = 1° alkyl or methyl; OTs = *p*-toluenesulfonyl = tosyl

The reaction can be monitored by TLC since the aromatic phenoxide can be detected under a UV lamp. The product, an alkyl halophenyl ether, is a high-boiling liquid (an oil), which can be isolated and purified using column chromatography.

For TLC analysis of the starting material and product of this reaction, small aliquots of the reaction mixture are removed at convenient intervals and spotted on a TLC plate. Development of the chromatogram in a moderately polar solvent will show a fast-moving spot corresponding to the product. The ionic phenoxide starting material travels at a much slower rate than the product so that reaction progress can be monitored using TLC.

In Part B of the experiment, the reaction of potassium *tert*-butoxide with benzyl bromide affords benzyl *tert*-butyl ether. The mechanism is S_N2 and there is no interference from an E2 reaction.

benzyl *tert*-butyl ether

Prelab Assignment

1. Will E2 reactions compete with S_N2 for the reaction of methyl or ethyl tosylate with potassium phenoxide? Explain.
2. Prepare a flow scheme for the experiment.

Prelab Preparation for Part A

Draw structural formulas of three or four sodium or potassium alkoxides. Draw structural formulas of three different halophenols. Name each structure. Consult a chemical catalog (Aldrich, Acros, Fluka, Lancaster, etc.) to determine which of the compounds are commercially available. From these compounds and from any others that become obvious during the search of the catalog, choose a halophenol and an alkyl tosylate to use for the experiment. Factors to consider are commercial availability of the compounds, costs, and relative reactivities. (Be aware that achieving an optimum reaction rate is important, while keeping costs down.) It will also be necessary to choose a solvent. Prepare a plan and a backup plan and get approval from the instructor before the laboratory period. Be prepared to support both plans.

1. Wear eye protection at all times in the laboratory.

2. Wear gloves when handling phenols and strong bases. Avoid skin contact with basic solutions. If base contacts the skin, wash the affected area with copious amounts of cool water. Many phenols are skin irritants.

3. Work in the hood or in a well-ventilated area. Alkyl tosylates should be handled with care as they are cancer suspect agents and toxic irritants. Halogenated phenols are toxic and corrosive. Benzyl bromide is a lachrymator. Do not breathe the vapors. Use gloves when weighing out these chemicals.

4. Be careful to wear proper eye protection when using an ultraviolet lamp. Do not look directly into the light.

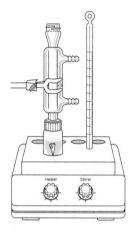

Use proper eye
protection to prevent
passage of UV light into
the eyes.

Part A: Microscale Preparation of an Alkyl Halophenyl Ether

Into a dry, 5-mL conical vial containing a spin vane, add 1 mmol of a halophenol and 2 mL of absolute ethanol. Stir to dissolve the solid. Add 1 mmol of KOH from a stock solution of standardized 1.0 M KOH in ethanol. Add 1 mmol of alkyl tosylate to the vial. Use a spatula to free up the solid if necessary. Attach a water-jacketed condenser (lightly greased) to the vial and reflux with stirring for 10 minutes using a heat block or a hot sand bath. Remove the apparatus from the heat source and allow the reaction mixture to cool for a minute. Remove the condenser and remove an aliquot of liquid using a micropipet. Spot the sample on a TLC plate. Also, spot samples of standards of the starting materials. Repeat the refluxing/spotting process three more times over a period of one hour. Develop the TLC plate using methylene chloride as the eluent. After 1 hour, cool the reaction mixture to room temperature. Add 1 mL of 1 M NaOH (aqueous) and reflux for an additional 10 minutes. Remove the apparatus from the heat source and cool to room temperature. Remove the condenser. Observe the developed TLC plate under the ultraviolet lamp.

Carefully reduce the volume of the mixture in a hood to 0.5–1.0 mL using a sand bath or heat block and a stream of nitrogen or air. Cool to room temperature. Add 1.5 mL of methylene chloride and 1.5 mL of water. Cap the vial and shake. Vent the vial by carefully loosening the cap. Draw off the bottom (organic) layer and place in a separate vial. Add 1.5 mL of fresh methylene chloride to the aqueous layer and perform a second extraction. Draw off the lower organic layer and combine the two organic extracts in a 5-mL vial and add 1.5 mL of water. Cap the vial and shake. Vent the vial by slowly loosening the cap. Draw off the organic layer into a clean conical vial. Dry the methylene chloride solution over sodium sulfate and transfer the solution into a tared vial containing a boiling chip. Evaporate the methylene chloride using a sand bath or heat block in a hood to give an oily residue. Reweigh the vial containing the crude product. The product may be purified by column chromatography.

Column Chromatography

The oil from the procedure above contains both product and unreacted starting materials. The contaminants may be removed by column chromatography using petroleum ether as the eluting solvent. Add several drops of 50% methylene chloride/50% petroleum ether

to dissolve the oil and apply the solution to the top of a silica gel column. Add 20% methylene chloride/80% petroleum ether to the column continuously and collect 10–15 fractions of about 2–3 mL each. Use TLC analysis of the fractions to determine which fractions contain product. Pool the fractions that contain the product as the only UV-absorbing compound in a tared Erlenmeyer flask containing a boiling chip. Concentrate the solution to give an oil. Determine the weight of the purified product and run an IR spectrum of the purified product.

Characterization for Part A

Infrared Spectroscopy: Run an IR spectrum of the product (neat). Interpret the spectrum in terms of the peaks expected for the product. Ethers show strong C−O absorption in the 1200–1000 cm^{-1} region.

NMR Spectroscopy: Dissolve a portion of the product in CDCl$_3$. Transfer the solution into a clean, dry NMR tube using a filter pipet. Record the ^{1}H NMR spectrum and analyze the spectrum. Note the alkyl protons and the ratio of aromatic/aliphatic protons.

Results and Conclusions for Part A

1. Calculate the percent yield of product.
2. Is TLC a good technique to monitor the progress of the reaction? Explain.
3. Judging from the TLC results, would additional heating (beyond 1 hour) give a higher yield of product? Explain.

Cleanup & Disposal

Neutralize all aqueous washings with dilute HCl and wash them down the drain. Be sure that there are no residues of alkoxide or phenoxide left on the lab bench. Be sure that the area around the balances is clean and that no residue of phenols or alkyl tosylate remains on or near the balances. Place any residual halogenated solvent into a container labeled "halogenated organic solvent waste."

Prelab Preparation for Part B

Devise a procedure for preparation of 3 mmoles of benzyl *tert*-butyl ether. Starting materials are benzyl bromide and potassium *tert*-butoxide. Use acetonitrile as the solvent. Assume a yield of 50% of the product. Use a threefold excess of *tert*-butoxide and a 30-minute reflux. Prepare a plan and get approval from the instructor prior to the lab period. Include the following items in the plan: apparatus, amounts of reagents and limiting reagent, amount of solvent, safety precautions, reaction time, workup procedure, purification, and characterization of the product. Assume that TLC will be useful, but that column chromatography is not necessary. The product is a high-boiling oil that is not to be distilled. Use the oil for characterization.

Part B: Miniscale Preparation of Benzyl tert-Butyl Ether

Each student will develop his or her own procedure for this experiment as directed in the section on prelab preparation.

Characterization for Part B

Each student will develop his or her own characterization plan for this experiment as directed in the prelab assignment.

Results and Conclusions for Part B

1. Calculate the percent yield of product.
2. How useful is TLC as a technique to monitor the progress of this reaction? Explain.
3. Did the reaction go to completion? Explain.
4. What changes in the experimental procedure you used for preparation of benzyl *tert*-butyl ether, if any, would you recommend?
5. Explain the NMR pattern for the aromatic protons of the product.

Cleanup & Disposal

Devise a cleanup and disposal plan for Part B of the experiment. Have it checked by your instructor before starting the experiment.

Critical Thinking Questions *(The harder ones are marked with a ❖.)*

1. A student proposes to prepare benzyl *tert*-butyl ether from *tert*-butyl bromide and potassium benzyloxide. Predict the outcome of the reaction.
2. Check the physical constants and hazards for methyl bromide. Would methyl bromide be a suitable choice as a starting material for the synthesis in Part A?
3. Predict the main product of reaction from Part B if wet potassium *tert*-butoxide is used as a reagent.
4. Why is it important that equimolar amounts of base and phenol be used in Part A?
❖ 5. What difference in results is predicted if isopropyl iodide were used in place of an alkyl tosylate in Part A?
❖ 6. Must the alkyl group of the alcohol be the same as the alkyl group of the tosyl ester in Part A? Explain.
7. Is there any way to show that inversion of configuration occurred during the reaction in Part A or Part B?
8. Comment on the prelab preparation required for Part B compared to having the experimental procedure available as in most of the previous experiments.

Experiment 16.2: Nucleophilic Aliphatic Substitution
Puzzle: *Substitution Versus Elimination*

In this experiment, you will study a nucleophilic substitution reaction designed to form a cyclic product. You will:

- carry out a bimolecular nucleophilic aliphatic substitution reaction (S_N2).

- purify the product by column chromatography.

- propose an explanation for the observed results.

Techniques

Technique G	Distillation and reflux
Technique I	Extraction
Technique K	Column chromatography
Technique N	Nuclear magnetic resonance spectroscopy
Technique P	Mass spectrometry

Background Information

Alcohols react with sodium metal or with sodium hydride to form alkoxide bases with evolution of hydrogen gas. The alcohols may be of any type. In this experiment, hydrobenzoin will be reacted with sodium hydride to form the alkoxide with evolution of molecular hydrogen.

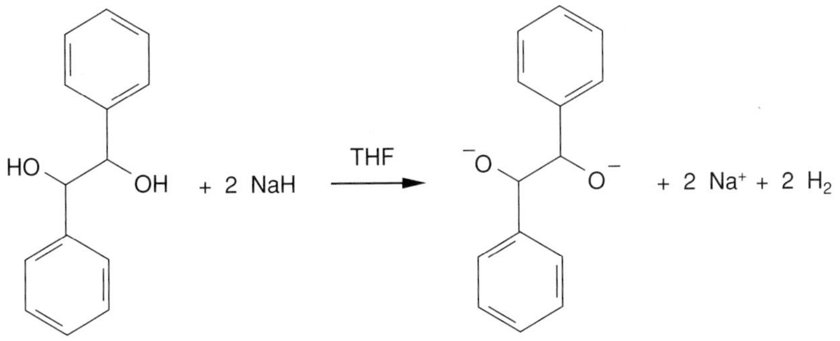

Nucleophilic aliphatic substitution reactions of unbranched, primary alkyl halides occur by attack of a nucleophilic alkoxide base on the carbon bearing the halogen. The direct displacement process is an example of an S_N2 reaction. Mild heating is used to cause the reaction to reach completion. In theory, reacting the dianion of *meso*-hydrobenzoin with 1,3-dibromopropane should result in the synthesis of the cyclic ether shown here via two S_N2 reactions. In this experiment, you will react the dianion of *meso*-hydrobenzoin with 1,3-dibromopropane by an S_N2 reaction in an attempt to synthesize the cyclic ether below.

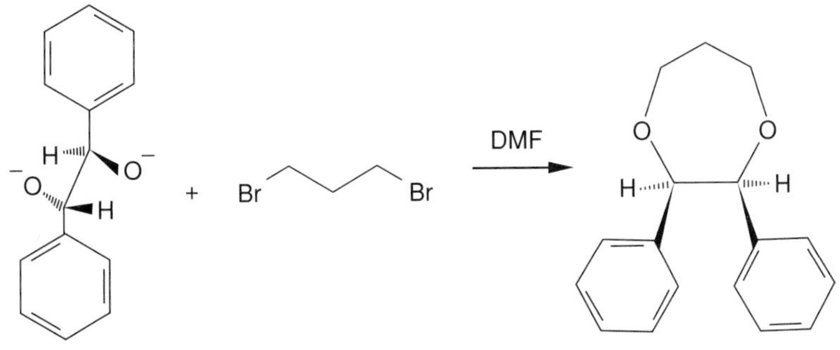

meso-hydrobenzoin dianion   target molecule

Unfortunately, this reaction does not produce the target molecule shown. Instead, another product is formed (Jha and Joshi, 2002). Your task in this experiment is to run the reaction, isolate the major product, and identify the product.

The experimental conditions require the use of dry tetrahydrofuran (THF), dry N,N-dimethylformamide (DMF), and dry sodium hydride. The sodium hydride is shipped as a dispersion in mineral oil. Column chromatography must be used to separate the major product from other products and by-products. Success of the experiment depends on care in ensuring that all reagents and reaction conditions meet these requirements.

Prelab Preparation

1. Write a mechanism for the formation of the target molecule.
2. What by-product is produced?
3. Write a flow scheme for the workup procedure in this experiment.
4. The sodium hydride is packaged as a 60% dispersion in mineral oil. Where in the procedure is the mineral oil to be removed?
5. Calculate the mmol of NaH if 80 mg of a 60% dispersion of NaH in mineral oil is used.

Experimental Procedure

Safety First!

Always wear eye protection in the laboratory.

1. Wear eye protection at all times in the laboratory.
2. Wear gloves while handling reagents for this experiment.
3. Work in a hood or a well-ventilated area for this experiment. Keep reagents under the hood. Diethyl ether is very volatile and flammable. Be careful to use this solvent in the hood and away from flames.
4. Any spills should be cleaned up immediately. Notify the instructor immediately about any contact with reagents or any spills.

Miniscale Reaction of Hydrobenzoin with Sodium Hydride and 1,3-Dibromopropane

To a stirred suspension of 80 mg of NaH (60% dispersion in mineral oil) in 1 mL of dry THF in a 10-mL round-bottom flask equipped with a magnetic stir bar, add a solution of 210 mg of *meso*-hydrobenzoin in 1.5 mL of dry THF. Stir the reaction mixture until no more hydrogen is evolved (about 20 min). To the suspension add 220 mg of 1,3-dibromopropane dissolved in 1 mL of dry DMF. Stir the reaction mixture for about an hour. Quench the reaction by adding 1 mL of 1M HCl. Dilute the reaction mixture by adding 5 mL of water. Transfer the mixture to a centrifuge tube and add 5 mL of diethyl ether. Mix the layers thoroughly, being careful to vent the tube.

Remove the lower layer using a Pasteur pipet. Add 5 mL of brine to the ether layer in the tube. Mix the layers thoroughly and remove the brine layer with a Pasteur pipet. Repeat the brine wash using a second 5 mL of brine. Transfer the ether solution to an Erlenmeyer flask and add a small amount of anhydrous sodium sulfate until no clumping is observed. Reduce the volume of the solution to about 2 mL and apply it to the top of a silica gel column. Elute with 20% ethyl acetate/hexanes and collect 2-mL fractions. The second component should be the major one. Follow the progress of the chromatography by spotting aliquots and analyzing by TLC. After an initial side product, the main product should elute quickly. Combine the appropriate fractions and evaporate the solvent. It is not necessary to continue the chromatography after the main fraction has been eluted.

Characterization

Infrared Spectroscopy: Run an IR spectrum of the oily product. Look for features that are inconsistent with the expected IR spectrum of the target molecule. List the major absorptions present in the product. Predict what functional groups are present in the product.

Nuclear Magnetic Resonance Spectroscopy: Dissolve 5–10 mg of the product in CDCl$_3$ and transfer the solution into a clean, dry NMR tube using a filter pipet. Record the integrated ^{1}H NMR spectrum.

Gas Chromatography/Mass Spectrometry: Dissolve a drop of the oily product in 1 mL of reagent-grade ethyl acetate. Add a drop of that solution to about 1 mL of ethyl acetate and draw up 1–2 μL of solution into a syringe. Inject the sample through the injection port of the GC instrument onto a semi-polar or nonpolar column and record the chromatogram. Determine the relative amounts of each component. Use MW data and fragmentation patterns in the mass spectra to verify the structure of the compound.

Results and Conclusions

1. IR spectroscopy and NMR spectroscopy are used to identify the structure of the product. Predict the major absorptions and peaks expected in the IR and NMR spectra of the target molecule. Summarize the differences between the product and those anticipated for the target.
2. Given the IR and NMR evidence, propose a structure for the product.
3. Determine the percent yield of the product.
4. Calculate the percent purity of the product as determined by GC analysis.
5. Propose a mechanism to account for the formation of the product.

Cleanup & Disposal

Neutralize acidic extracts with dilute base. Wash the neutralized, aqueous extracts down the drain with running water.

Critical Thinking Questions

1. Suppose that the starting material was the methyl ether of *meso*-hydrobenzoin shown below. The observed product from this reaction is shown. Is the formation of this product consistent with the mechanism of the reaction?

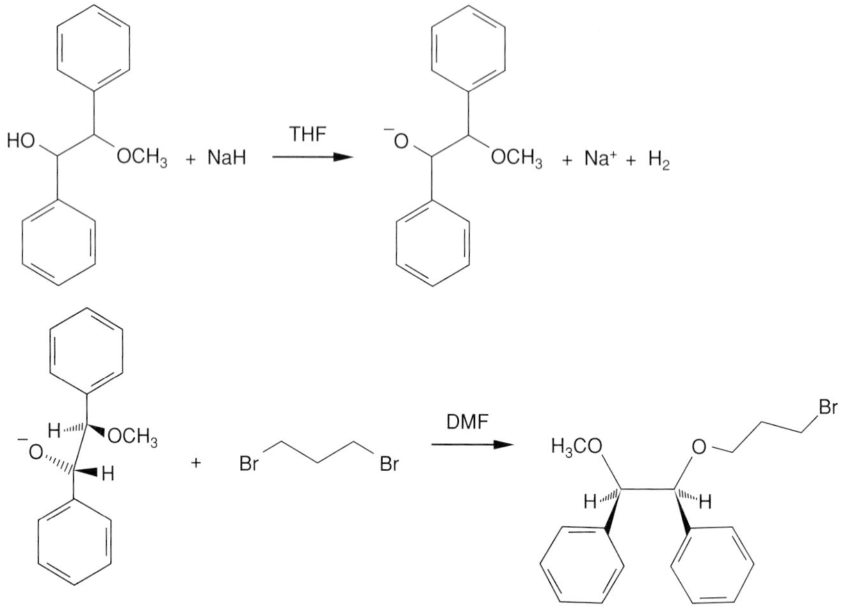

2. Brine is used in the workup of the reaction mixture. What is brine and why is it used here?

3. An alternate reagent for this experiment is sodium metal in place of NaH. Can Na metal be used in the place of NaH? Think about the physical nature of Na and how the reaction with *meso*-hydrobenzoin would take place.

4. Where in the procedure is the THF removed? Where is the DMF removed? Explain.

5. Give a reason or reasons for the failure of the reaction below.

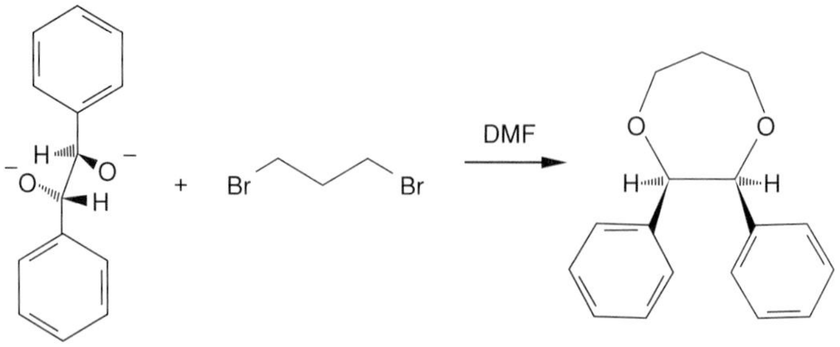

meso-hydrobenzoin dianion   target molecule

Reference

Jha, S. C., and Joshi, N. N, Intramolecular Dehydrohalogenation during Base-Mediated Reaction of Diols with Dihaloalkanes. *J. Org. Chem.* 67(2002): 3897.

Chapter 17

Aldehydes and Ketones

Several aldehydes and ketones are important natural products. Steroidal ketones, such as progesterone and testosterone, are crucial in human physiology as hormones.

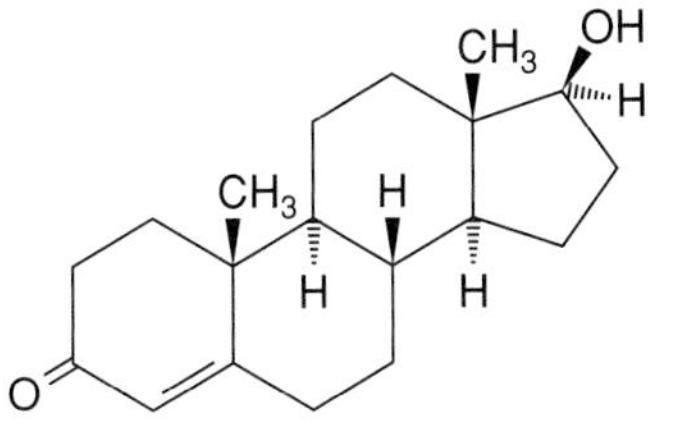

testosterone, a male hormone 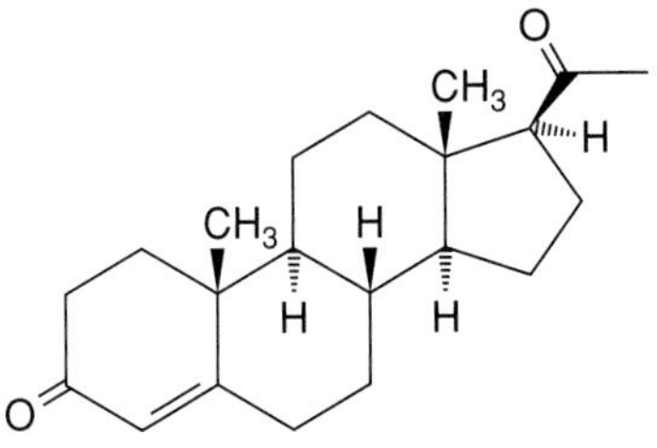   progesterone, a female hormone

Many ketones have great industrial importance. For example, acetone is one of the most common organic solvents. Aldehydes, such as *trans*-cinnamaldehyde, and ketones, such as muscone, are used as flavorants and perfumes.

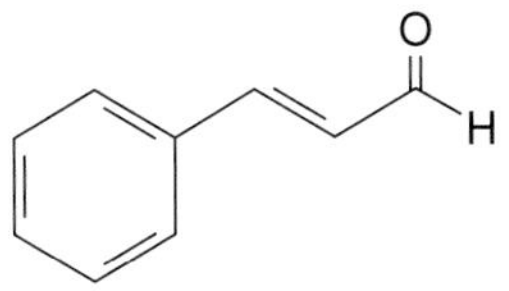

trans-cinnamaldehyde 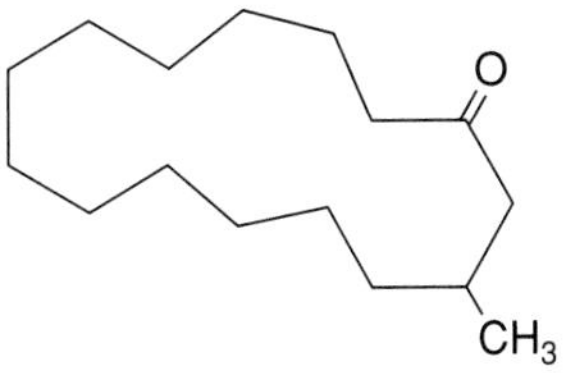   muscone

Aldehydes and ketones are often prepared in the laboratory by direct oxidation of alcohols. Aromatic ketones can be prepared by Friedel-Crafts acylation as illustrated in Experiment 12.3. An interesting rearrangement reaction, known as the pinacol rearrangement, can be used to prepare certain ketones (Experiment 17.3).

Aldehydes and ketones are convenient starting materials for preparing other compounds. One example of the use of ketones in synthesis is the preparation of a heterocycle, illustrated by the preparation of 2,5-dimethyl-1-phenylpyrrole or of an unknown pyrrole in Experiment 21.2. The syntheses of pyrazoles and pyrimidines are illustrated in Chapter 22.

Perhaps the most well-known organic reaction of aldehydes and ketones is the Grignard synthesis of alcohols from aldehydes or ketones and a Grignard reagent

(Experiment 14.1). The reaction is a principal method for making new carbon-carbon bonds. A second, more recent method is the Wittig synthesis of alkenes, starting with an aldehyde or ketone and a Wittig reagent (see Experiment 17.1). The Wittig synthesis and the aldol condensation (Experiment 18.1) both utilize resonance-stabilized carbanions as key intermediates. Nucleophilic additions of cyanide and other carbanions to carbonyl groups are well known, as are addition-elimination reactions, such as those with hydroxyl-amine, phenylhydrazine, ammonia and amines, and semicarbazide. Examples of these reactions will be encountered in organic qualitative analysis (see Experiments 28.2).

The carbonyl group in aldehydes and ketones occupies a key niche among functional groups. Carbonyl groups may be efficiently reduced to alcohols using sodium borohydride as in Experiment 15.1. Aldehydes can be selectively oxidized by a number of oxidants, including moist silver oxide in Experiment 24.1 and in qualitative organic analysis using the Tollens test in Chapter 28. Although more difficult to oxidize than aldehydes, ketones react with peroxides to give esters by a rearrangement reaction known as the Baeyer-Villiger oxidation in Experiment 20.2.

Many of the starting materials and products in the latter half of this text involve a ketone or a ketone derivative. These compounds are extremely important in synthesis, both as starting materials and intermediates. The experiments in this chapter represent three different types of reactions of ketones.

Experiment 17.1: Stereoselective Synthesis of Alkenes

The Wittig and Horner-Emmons reactions convert aldehydes and ketones to alkenes. The geometry of the alkene formed may vary depending upon the synthetic method used. In this experiment, you will:

♦   condense an aldehyde and a Wittig reagent to synthesize an alkene.

♦   condense an aldehyde with diethylbenzyl phosphonate to synthesize an alkene.

♦   determine the stereochemistry of the alkene using melting points, infrared spectroscopy, and/or ^{1}H NMR spectroscopy.

Techniques

Technique F	Recrystallization and filtration
Technique I	Extraction
Technique K	Thin-layer chromatography
Technique M	Infrared spectroscopy
Technique N	Nuclear magnetic resonance spectroscopy

Background

The Wittig reaction is one of the most useful reactions in synthetic organic chemistry because it transforms carbon-oxygen double bonds to carbon-carbon double bonds. Both aldehydes and ketones can be converted to alkenes. In Part A of this experiment, you will convert 9-anthraldehyde to *trans*-9-(2-phenylethenyl)anthracene using 50% NaOH and benzyltriphenylphosphonium chloride. The overall reaction is shown here.

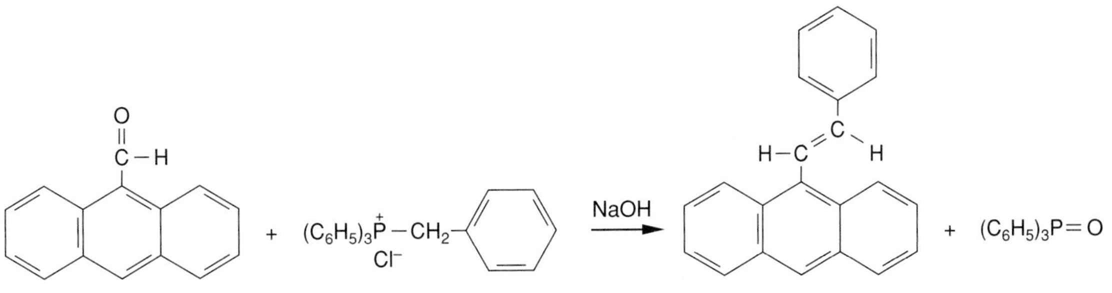

9-anthraldehyde   benzyltriphenylphosphonium chloride   *trans*-9-(2-phenylethenyl)-anthracene   triphenylphosphine oxide

The α-protons of the benzyl group of benzyltriphenylphosphonium chloride are relatively acidic (pK_a = 20). Sodium hydroxide removes an α-hydrogen from benzyltriphenylphosphonium chloride and forms a resonance-stabilized intermediate known as an ylide.

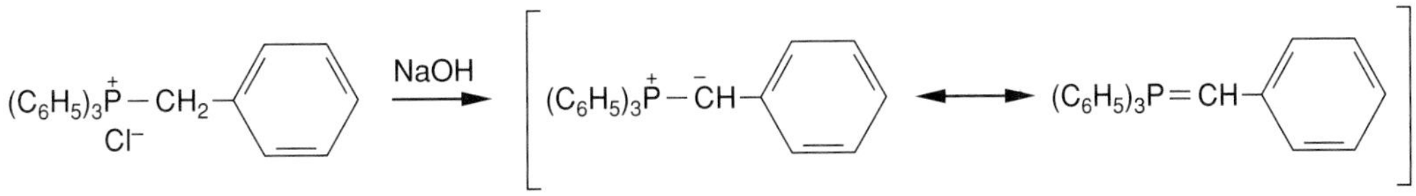

benzyltriphenylphosphonium chloride

benzylidenetriphenylphosphorane (a phosphorus ylide or Wittig reagent)

The ylide then reacts with 9-anthraldehyde to give an alkene:

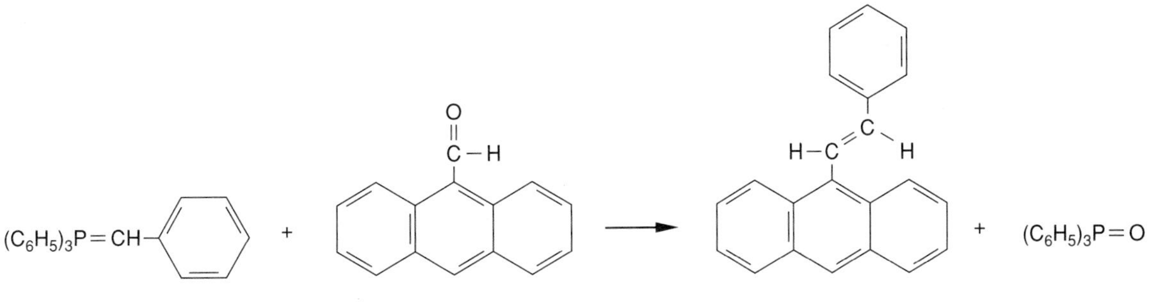

trans-9-(2-phenylethenyl)anthracene

Isolation of the product, followed by recrystallization, gives a moderate to high yield of the trans isomer exclusively. The geometry about the double bond can be substantiated by NMR.

The Wittig reaction produces the trans-isomer in this instance because this isomer has less steric hindrance (Silversmith, 1986). In the Horner-Emmons reaction, in this experiment diethyl benzylphosphonate and benzaldehyde can theoretically produce the cis isomer, the trans isomer, or a mixture of the geometric isomers of stilbene.

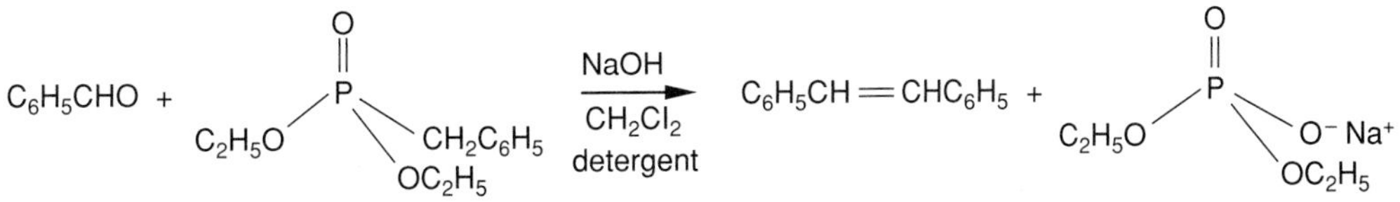

benzaldehyde diethyl benzylphosphonate   *cis*-or *trans*-stilbene   sodium diethyl phosphate

You will analyze the resulting product(s) via thin-layer chromatography, melting point, infrared spectroscopy, and/or ^{1}H NMR spectroscopy and determine the stereochemistry of the product produced in the Horner-Emmons reaction.

Prelab Assignment

1. Write a detailed mechanism for (a) the formation of the ylide derived from benzyltriphenylphosphonium chloride and sodium hydroxide; and (b) the reaction of the ylide with 9-anthraldehyde to produce the product. Use your lecture text, if necessary.
2. Explain how ^{1}H NMR spectroscopy can be used to characterize the *trans* geometry of the double bond of the alkene formed in the reaction.
3. Explain how infrared spectroscopy could be used to differentiate between *cis*- and *trans*-alkenes. Tabulate specific absorbance frequencies.
4. Write a balanced reaction for the acid-base reaction of diethylbenzyl phosphonate with sodium hydroxide. Which hydrogens are the most acidic? Explain by writing resonance structures.
5. Propose a detailed mechanism for the reaction of the anion of diethyl benzylphosphonate with benzaldehyde.
6. Write a flow scheme for the workup procedure for this experiment, showing where by-products would be separated from the product.
7. Which isomer—*cis*-stilbene or *trans*-stilbene—would be expected to have a higher R_f value in thin-layer chromatography? Explain.

Experimental Procedure

1. Wear eye protection at all times in the laboratory.
2. 50% NaOH is very caustic! Wear gloves and goggles at all times. If NaOH comes in contact with your skin, wash the affected area with copious amounts of cool water. If burning or tingling persists, seek medical attention.
3. Methylene chloride is a toxic irritant and a suspected carcinogen. Use under the hood or in a well-ventilated area. Do not breathe its vapors.
4. Never look directly at the ultraviolet light.

Part A: Microscale Wittig Synthesis of trans-9-(2-Phenylethenyl)anthracene

Obtain a clean, dry 5-mL conical vial and a spin vane or a 10-mL Erlenmeyer flask and a stir bar. To the vial or flask add 100 mg of 9-anthraldehyde, 190 mg of benzyltriphenylphosphonium chloride and 1.5 mL of methylene chloride. Stir vigorously. Cautiously add 5 drops of 50% NaOH dropwise, being careful to avoid splattering. Stir for 30 minutes. Add 1.0 mL each of methylene chloride and water and remove the spin vane; cap and shake vigorously. Remove the lower organic layer to a clean Erlenmeyer flask. Extract the aqueous layer remaining in the conical vial with another 1-mL portion of methylene chloride. Combine the organic layers and dry with anhydrous sodium sulfate. With a filter pipet, transfer the dry organic solution to a 10-mL Erlenmeyer flask that contains a boiling chip or stir bar. Rinse the drying agent with an additional 1 mL of methylene chloride. Spot the crude product on a silica gel TLC plate. Evaporate the solvent under the hood using a warm sand bath. Recrystallize the crude yellow crystals from 2-propanol. Filter the crystals using a Hirsch funnel. Air dry the yellow crystals and weigh.

Part B: Miniscale Wittig Synthesis of trans-9-(2-Phenylethenyl)anthracene

To a clean, dry 25-mL Erlenmeyer flask containing a spin bar, add 0.50 g of 9-anthraldehyde, 0.95 g of benzyltriphenylphosphonium chloride, and 7.5 mL of methylene chloride. Stir vigorously. Cautiously add 1.25 mL of 50% NaOH dropwise, being careful to avoid splattering. Stir for 30 minutes. Add 5 mL each of methylene chloride and water and pour the solution into separatory funnel. Shake, venting often. Allow the layers to separate, then remove the lower organic layer to a clean Erlenmeyer flask. Extract the aqueous layer with another 5-mL portion of methylene chloride. Combine the organic layers and dry with anhydrous sodium sulfate. Decant or gravity filter the dry organic solution into a 25-mL suction flask that contains a boiling chip or stir bar. Save a small aliquot of the crude product for TLC. Evaporate the solvent under vacuum in the hood using a warm sand bath. Recrystallize the crude yellow crystals from 2-propanol. Filter the crystals using a Büchner funnel. Air dry the yellow crystals and weigh.

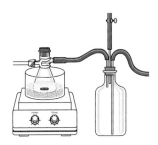

Characterization for Parts A and B

Melting Point: Determine the melting point of the product and compare it to the literature value. The reported melting point of *trans*-9-(2-phenylethenyl)anthracene is 131–132°C.

Thin-layer Chromatography: Dissolve a few crystals of the crude product (before recrystallization) in a small amount of methylene chloride and spot on a silica gel plate. Dissolve a few crystals of the recrystallized product in methylene chloride and spot it beside the crude solution. Also spot samples of the starting materials, authentic 9-(*trans*-2-phenylethenyl)anthracene (if available), and triphenylphosphine oxide. The chromatogram may be developed using methylene chloride. Visualize the developed chromatogram under UV light.

Use eye protection when using UV light.

Infrared Spectroscopy: Obtain an IR spectrum using the Nujol mull or KBr pellet technique. The product should show absorptions typical of alkenes and contain no carbonyl absorption. Compare the IR spectrum with that of the starting material.

NMR Spectroscopy: Dissolve a sample of the product in $CDCl_3$. Record the ^{1}H NMR spectrum. The vinylic protons should appear as a pair of doublets. One doublet appears at $\delta 6.82$; the other doublet appears at $\delta 7.82$ and is embedded in the signal of the aromatic protons ($\delta 7.3–8.5$).

Results and Conclusions for Parts A and B

1. Determine the percent yield of *trans*-9-(2-phenylethenyl)anthracene.
2. Identify each proton in the ^{1}H NMR spectrum and measure the coupling constant between the vinylic protons. Verify that only the trans isomer is produced in the reaction.
3. Suggest a reason why only the trans alkene is formed in this reaction even though the Wittig reaction usually gives mixtures of cis and trans isomers.
4. Summarize your TLC results.

Part C: Microscale Horner-Emmons Reaction of Diethyl Benzylphosphonate and Benzaldehyde

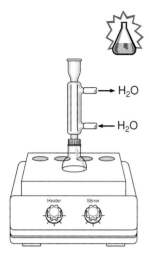

In a 10-mL round-bottom flask equipped with a stir bar, add 3 drops of Aliquot 336, 100 µL of benzaldehyde, 200 µL of diethyl benzylphosphonate, and 2.0 mL of hexanes. Stirring vigorously, cautiously add 2.0 mL of 40% NaOH. Attach a water-jacketed reflux condenser that has been lightly greased. Stirring vigorously, heat to reflux for 1 hour, then let cool to room temperature.

Add approximately 1 mL of methylene chloride to dissolve any crystals that formed during the reaction. Transfer the solution to a centrifuge tube, rinsing the flask with an additional 0.5 mL of methylene chloride. As you add the additional methylene chloride, note which layer increases in volume. Depending upon the density of the aqueous layer, the methylene chloride may be the top or the bottom layer. If the methylene chloride layer is not the bottom layer, add an additional 1 mL of methylene chloride and an additional 1 ml of water to the centrifuge to ensure that the organic layer is on the bottom.

Cap and shake the centrifuge tube, venting often. Let the layers separate. With a filter pipet, remove the lower organic layer to a clean beaker. Transfer the aqueous layer to another beaker and set aside. Return the organic layer to the centrifuge tube and add 1 mL of water. Shake. Allow the layers to fully separate. Remove the bottom organic layer to the flask, then remove the aqueous layer to the beaker. Return the methylene chloride solution to the centrifuge tube and repeat the extraction with a second 1 mL of water. Dry the combined methylene chloride solution with anhydrous sodium sulfate. Use a filter pipet to transfer the dried solution to a 10-mL tared Erlenmeyer flask containing a boiling chip. Spot an aliquot of the solution on a TLC plate. Place the flask on an aluminum heat block and evaporate the solvent. After all of the solvent has evaporated, remove the flask from the hot plate and let cool to room temperature. There should be no odor of methylene chloride present. The properties of the product will differ depending upon whether the *cis-* or the *trans-*alkene is the major product. Examine the flask carefully to see whether it contains an off-white solid (indicating the *trans-*product) or an oil (indicating the *cis-*product). Follow the appropriate procedure to isolate the product:

If the product is a solid: Recrystallize the crude sample from hot 95% ethanol. Suction filter the crystals, rinsing with a small amount of ice-cold 95% ethanol. Weigh the dried crystals and record the melting point.

If the product is an oil: Weigh the flask, liquid, and boiling chip. No further purification is required.

Characterization for Part C

Physical properties: Determine the physical properties of the product and compare them to the literature values for the *cis-* and the *trans-*stilbene. The reported melting point of *trans-*stilbene is 124°C and the reported boiling point at 760 torr is 306–307°C. The reported melting point of *cis-*stilbene is –5°C and the reported boiling point is 135°C at 10 torr. The reported refractive index of *cis-*stilbene is 1.6220.

Infrared Spectroscopy: Obtain an IR spectrum. The product should show absorptions typical of alkenes and contain no carbonyl absorption. Compare the IR spectrum with that of the starting material.

NMR Spectroscopy: Dissolve about 8 mg of the sample in $CDCl_3$. Record the ^1H NMR spectrum. Spectra of authentic samples as shown in Figures 17.1-1 and 17.1-2.

Thin-layer Chromatography: Dissolve a small portion of the product in methylene chloride and spot it beside the crude solution. Also spot samples of the starting materials

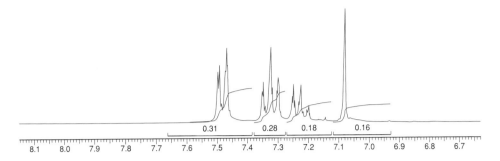

Figure 17.1-1 ^{1}H NMR spectrum of *trans*-stilbene (CDCl$_3$)

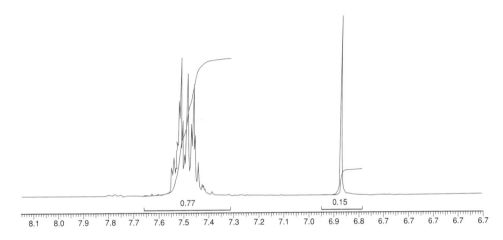

Figure 17.1-2 ^{1}H NMR spectrum of *cis*-stilbene (CDCl$_3$)

and authentic *cis*- and *trans*-stilbenes on silica gel plates with fluorescent indicator. The chromatography plates may be developed using hexanes. Visualize the developed chromatogram in an iodine chamber or under UV light.

Use eye protection when using UV light.

Results and Conclusions for Part C

1. Determine the percent yield of the product.
2. Based upon the physical properties, TLC data, or IR spectroscopic data, determine the identity of the major product. Justify. Based upon these results, evaluate the stereoselectivity of the Horner-Emmons reaction.
3. Calculate R$_f$ values for the product(s) and standards. Justify the order of elution. Based upon the TLC results, how effective was the recrystallization procedure?
4. Identify regions in the IR that indicate the observed stereochemistry.
5. Suggest a reason why a specific stereoisomer was formed in this reaction rather than the other alternatives.
6. Propose a mechanism for this reaction that explains the formation of the observed product.
7. Construct a detailed flow scheme for the workup procedure in this experiment.

Cleanup & Disposal

Place all remaining methylene chloride solutions in a container labeled "halogenated organic solvent waste." Place 2-propanol solutions in a container labeled "nonhalogenated organic solvent waste." Aqueous solutions may be washed down the drain with water after being neutralized with 3 M HCl.

Critical Thinking Questions *(The harder one is marked with a ❖.)*

1. Benzyltriphenylphosphonium chloride is prepared via an S_N2 reaction of triphenylphosphine (Ph_3P) and benzyl chloride ($PhCH_2Cl$). Write an equation for this reaction. Which species acts as the nucleophile? Which species acts as the electrophile?

2. A by-product of the Wittig reaction is triphenylphosphine oxide. At what step in the experimental procedure is this compound removed from the product?

3. Which of the two resonance structures of the ylide is expected to be the major contributor? Explain.

4. Give the expected products of each of the following Wittig reactions:

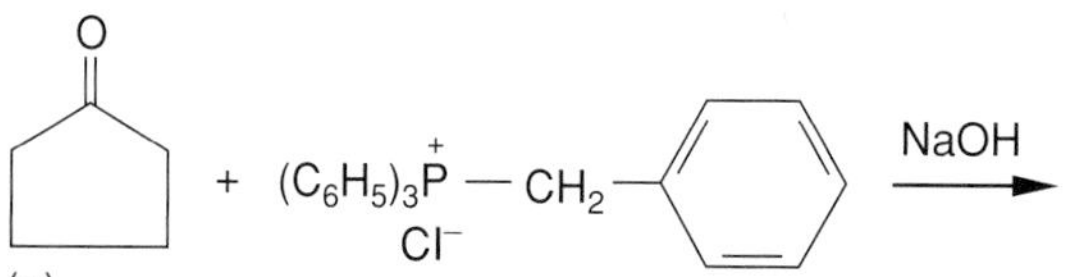

(a)

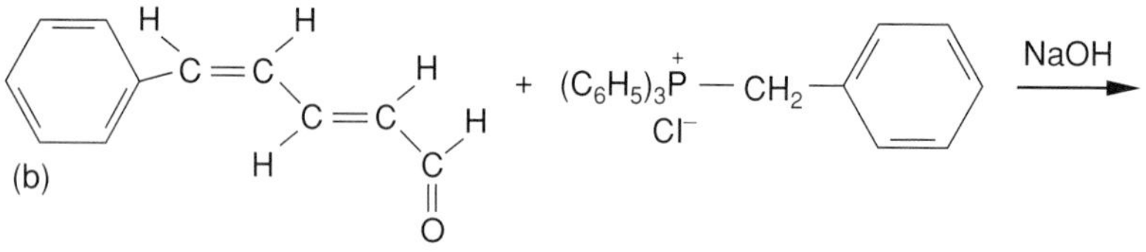

(b)

5. Devise an efficient synthesis of each of the following alkenes using a carbonyl compound and an alkylphosphonium chloride:

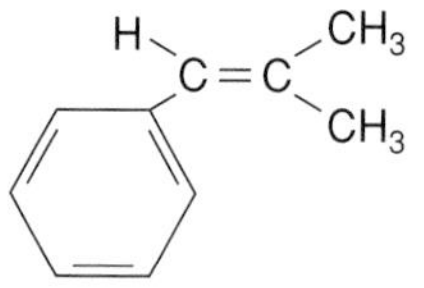

(a) 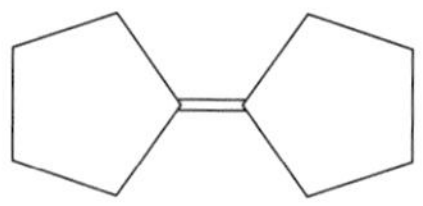   (b)

6. Which of the following alkyl halides could not be used to prepare an alkyltriphenylphosphonium bromide (Wittig reagent)? Explain.

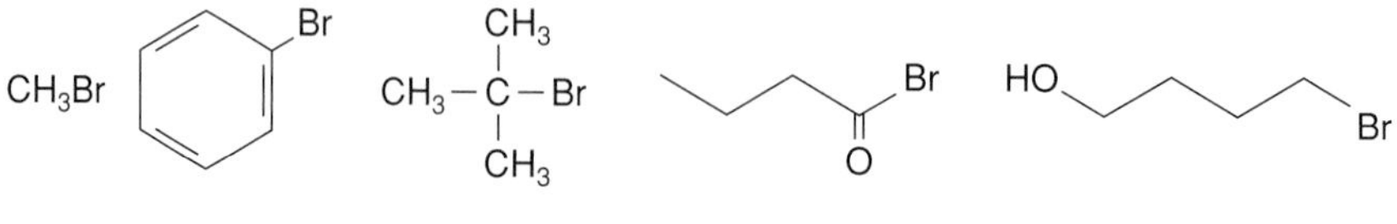

7. Could each of the following compounds be synthesized using a Wittig reaction? If the answer is "yes," give the structures of the carbonyl compound and the

alkyltriphenylphosphonium chloride that would give the product. If the answer is "no," explain why.

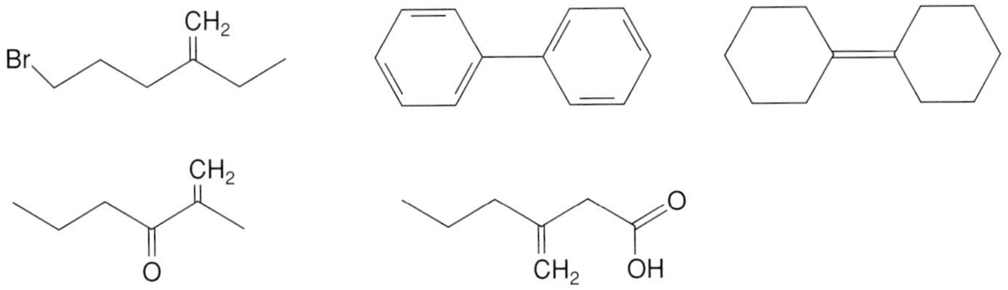

❖ 8. Benzyltriphenylphosphonium chloride can be converted to an ylide with sodium hydroxide, whereas methylcyclohexyltriphenylphosphonium chloride requires *n*-butyllithium. Explain why the methylene protons of the benzyl group are more acidic than the methyl protons of the methylcyclohexyl group.

9. *Library project:* Georg Wittig was awarded a Nobel Prize for developing the reaction that bears his name. Locate the original article in which this work was published. Give the complete citation to the reference. Identify the well-known organic chemist who shared the Nobel Prize with Georg Wittig.

10. Diethyl benzylphosphonate can be synthesized from triethyl phosphite and benzyl chloride. Write a balanced equation and explain what is happening in the reaction.

11. Explain why the Horner-Emmons reaction requires 40% NaOH rather than the stronger bases required in many Wittig-type reactions.

12. *Library project:* Aliquot 336 is a detergent that can act as a phase-transfer agent. Look up the properties of detergents and propose a reason for its inclusion in the Horner-Emmons reaction.

13. *Molecular modeling assignment*: Construct models of *cis*-stilbene and *trans*-stilbene. Minimize energies to get the conformations with the lowest potential energies. Examine the torsional angles for each of the models. Do the calculations support the experimental findings? Explain.

Reference

Silversmith, E. A Wittig Reaction That Gives Only One Stereoisomer. *J. Chem. Educ.* 63 (1986): 645–647.

Experiment 17.2: Conversion of Cyclohexanone to Caprolactam

In this experiment, you will learn about the synthesis of oximes (ketone derivatives) and the Beckmann rearrangement of oximes to amides. You will:

◆ carry out a two-step synthetic sequence.

◆ prepare an oxime as a ketone derivative.

◆ perform a Beckmann rearrangement of an oxime to a cyclic amide, caprolactam.

Techniques

Technique F	Vacuum filtration and recrystallization
Technique I	Extraction and drying
Technique M	Infrared spectroscopy
Technique N	Nuclear magnetic resonance spectroscopy

Background

In this reaction sequence, cyclohexanone is converted in two steps to ε-caprolactam (caprolactam).

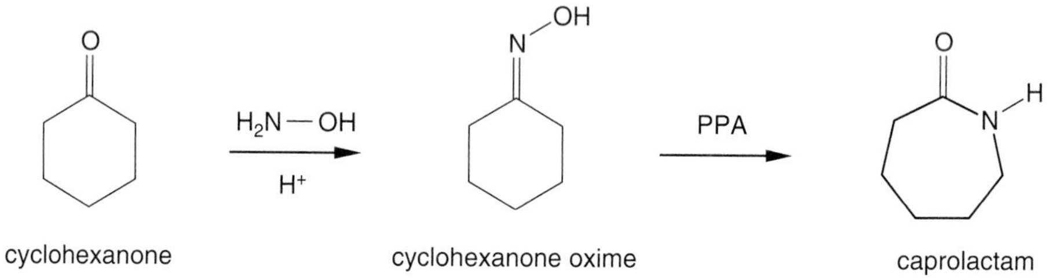

cyclohexanone   cyclohexanone oxime   caprolactam

In the first step of this experiment, cyclohexanone is converted to cyclohexanone oxime. In the second step, the oxime is transformed into caprolactam as the result of a molecular rearrangement using polyphosphoric acid (PPA). This type of rearrangement is known as the Beckmann rearrangement.

The mechanism of the Beckmann rearrangement is shown here. The first step is protonation of the hydroxyl group, followed by migration of the carbon-carbon bond anti to the hydroxyl. The resulting ion reacts with water and then tautomerizes to an amide. Cyclic ketones such as cyclohexanone form cyclic amides, known as lactams.

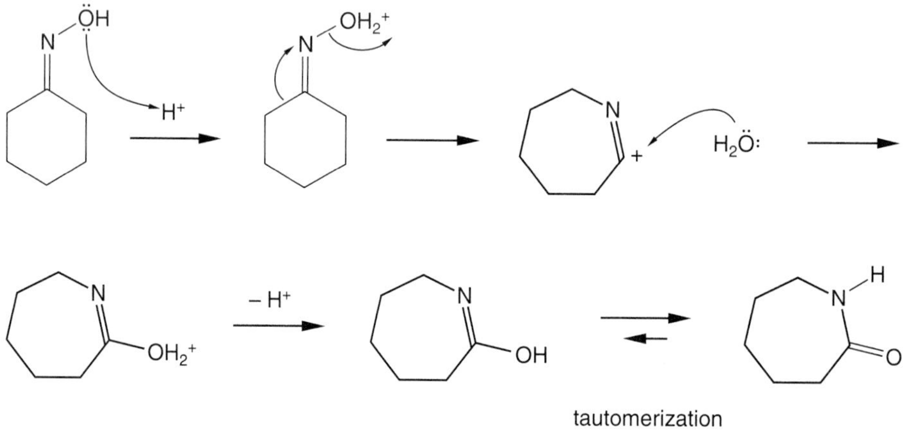

tautomerization

Prelab Assignment

1. Calculate the amount (mmol) of cyclohexanone and hydroxylamine to be used in this experiment. Which is the limiting reagent?
2. Write a flow scheme for the preparation and workup of cyclohexanone oxime.
3. Write a mechanism for the formation of cyclohexanone oxime.
4. The caprolactam produced in Part B of this experiment is called ε-caprolactam. Explain the meaning of the ε (epsilon).

Experimental Procedure

1. Wear eye protection at all times in the laboratory.

2. Polyphosphoric acid is corrosive and caprolactam is a potential skin irritant. Avoid contact with these reagents and wear gloves.

3. Methylene chloride is a toxic irritant and a suspected carcinogen. Use under a hood or in a well-ventilated area.

Part A: Microscale Conversion of Cyclohexanone to Caprolactam

A.1: Preparation of Cyclohexanone Oxime

Dissolve 110 mg of hydroxylamine hydrochloride in 1.5 mL of water in a 5-mL conical vial. With stirring, add 0.16 mL (0.15 g) of cyclohexanone followed by dropwise addition of a solution of 85 mg of sodium carbonate in 1.3 mL of water. Continue stirring for 10 minutes after the addition. Cool the reaction mixture in an ice bath and collect the oxime product by vacuum filtration on a Hirsch funnel. Wash the product with 0.1–0.2 mL of cold water. Dry the product for a few minutes on the funnel, pressing out as much water as possible. The crude product should be sufficiently pure for use in the next step. Save a few mg for characterization. If instructed to do so, recrystallize the crude oxime from water or ethanol/water solvent pair.

A.2: Beckmann Rearrangement of Cyclohexanone Oxime to Caprolactam

Use a hood for this reaction. Place 0.3 mL (0.6 g) of polyphosphoric acid in a 5-mL conical vial. Add the oxime (about 40 mg) prepared in Part A.1. Use gloves in this next operation. With great caution, use a thermometer to mix the viscous mixture while it is being heated on a sand bath or heat block. The temperature is important! The reaction temperature should rise to 130°C over a 15-minute period (approximately 7°/minute). When the desired temperature is reached (130°C), stop heating and carefully remove the vial from the hot plate, using tongs or thermal gloves. Cool the mixture to about 100°C. Carefully add 2 mL of ice water.

Adjust the pH of the reaction mixture by adding sodium carbonate until the pH reaches 6 (slightly acidic). Extract the aqueous reaction mixture with two 0.5-mL portions of methylene chloride. Combine the organic extracts and wash with 0.3 mL of water. Transfer the organic layer to a dry vial and dry over anhydrous sodium sulfate. Use a filter pipet to transfer the dried solution to a clean Erlenmeyer flask. Add a boiling chip and heat the solution on a sand bath in the hood to evaporate the methylene chloride. Recrystallize the crude residue from hexanes or petroleum ether. Weigh the recrystallized product. Characterize the product as shown for Part B.

Part B: Miniscale Conversion of Cyclohexanone to Caprolactam

B.1: Preparation of Cyclohexanone Oxime

Dissolve 1.1 g of hydroxylamine hydrochloride in 15 mL of water in a 50-mL Erlenmeyer flask or beaker. With stirring, add 1.6 mL (1.5 g) of cyclohexanone followed by dropwise addition of a solution of 0.85 g of sodium carbonate in 13 mL of water. Continue stirring for 10 minutes after the addition. Cool the reaction mixture in an ice bath and collect the oxime product by vacuum filtration on a Büchner funnel. Wash the product with 1–2 mL of cold water. Dry the product a few minutes on the funnel, pressing out as much water as possible. The crude product should be sufficiently pure for use in the next step. Save a few mg for characterization. If instructed to do so, recrystallize the crude oxime from water or ethanol/water solvent pair.

B.2: Beckmann Rearrangement of Cyclohexanone Oxime to Caprolactam

Use a hood for this reaction. Place 3 mL (6 g) of polyphosphoric acid in a 25-mL beaker or Erlenmeyer flask. Add the oxime (about 400 mg) prepared in Part B.1. Use gloves in this next operation. With great caution, use a thermometer to mix the viscous mixture while it is being heated on a hot plate. The temperature is important! The reaction temperature should rise to 130°C over a 15-minute period (~ 7°/minute). When the desired temperature is reached (130°C), stop heating and carefully remove the beaker or flask from the hot plate, using tongs or thermal gloves. Cool the mixture to about 100°C. Carefully pour the reaction mixture into a 50-mL beaker containing 18 mL of ice water. Rinse the original reaction flask with 2 mL of water and add this rinse to the product mixture.

Adjust the pH of the reaction mixture by adding sodium carbonate until the pH reaches 6 (slightly acidic). Pour the mixture into a separatory funnel. Extract the aqueous reaction mixture with two 6-mL portions of methylene chloride. Combine the organic extracts and wash with 3 mL of water. Dry the organic extract over anhydrous sodium sulfate. Gravity filter the solution into a dry Erlenmeyer flask containing a boiling stone. Heat the solution on a sand bath in the hood and evaporate the methylene chloride. Recrystallize the crude residue from hexanes or petroleum ether. Weigh the recrystallized product.

Characterization for Parts A and B

Melting Point: Determine the melting point of the product of each reaction. The reported melting points are 91°C for the oxime and 72°C for caprolactam.

Infrared Spectroscopy: Obtain an IR spectrum of each product isolated using either the KBr or the Nujol mull technique. Interpret spectra noting key bands characteristic of the functional groups present. Caprolactam should show a strong N–H absorption around 3200 cm^{-1} and a carbonyl stretch for an amide around 1690 cm^{-1}. The IR spectrum of caprolactam is shown in Figure 17.2-1.

Figure 17.2-1

IR spectrum of caprolactam (Nujol)

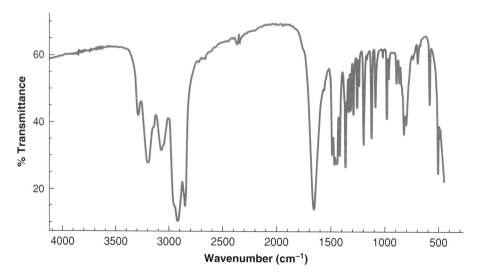

NMR Spectroscopy: Obtain an NMR spectrum of each of the products obtained. Interpret the spectra noting chemical shift positions and the integration as they support the anticipated structures. The ^{1}H NMR and the ^{13}C NMR spectra of caprolactam are shown in Figure 17.2-2 through Figure 17.2-4, respectively. The ^{13}C NMR spectrum of cyclohexanone oxime is shown in Figure 17.2-5 for comparison.

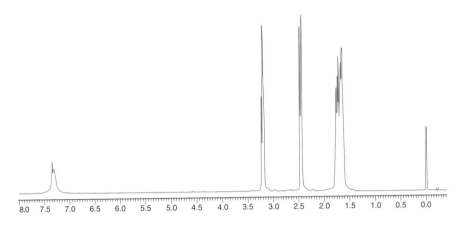

Figure 17.2-2
^{1}H NMR spectrum of caprolactam (CDCl$_3$)

SOURCE: Reprinted with permission of Aldrich Chemical.

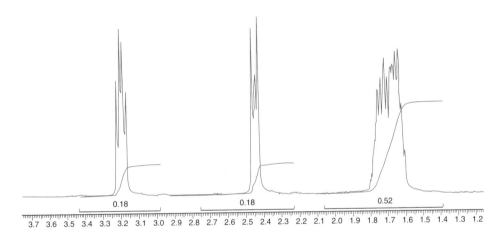

Figure 17.2-3 Expanded ^{1}H NMR spectrum of caprolactam (CDCl$_3$)

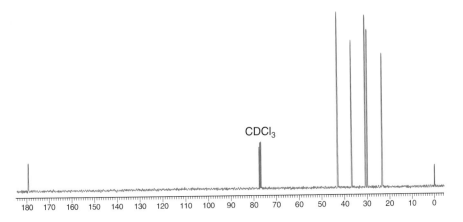

CDCl$_3$

Figure 17.2-4
^{13}C NMR spectrum of caprolactam (CDCl$_3$)

SOURCE: Reprinted with permission of Aldrich Chemical.

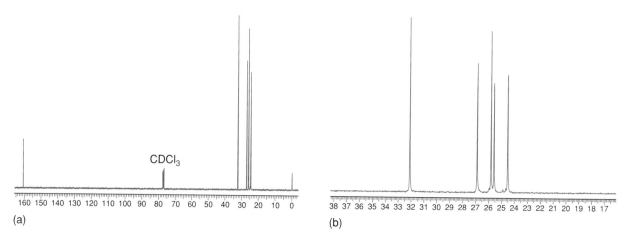

CDCl₃

(a)

(b)

Figure 17.2-5   (a) ^{13}C NMR spectrum of cyclohexanone oxime (CDCl$_3$); (b) ^{13}C NMR spectrum of cyclohexanone oxime (expanded region) (CDCl$_3$)

SOURCE: Reprinted with permission of Aldrich Chemical.

Results and Conclusions for Parts A and B

1. Determine the percent yield for the two-step sequence.
2. Why is pH control important in this reaction?
3. The last step in the mechanism of the Beckmann rearrangement is a tautomerization. Write a detailed mechanism for the tautomerization reaction to form caprolactam.

Cleanup & Disposal

The aqueous solutions and filtrates should be neutralized and poured down the drain with water. Hexanes and petroleum ether should be placed in a container labeled "nonhalogenated organic solvent waste."

Critical Thinking Questions *(The harder ones are marked with a ❖.)*

1. In a Beckmann rearrangement, the group anti (trans) to the oxime hydroxyl group migrates to form the amide product. Give the product expected in each case from the Beckmann rearrangement of the syn and the anti oximes derived from acetophenone.

❖ 2. Hydrolysis of caprolactam by refluxing with 6 M HCl gives a water-soluble product of formula $C_6H_{14}NO_2Cl$. Neutralization of this material leads to a compound of formula $C_6H_{13}NO_2$. Draw structures for these compounds and write equations showing their formation.

3. Give the product resulting from the Beckmann rearrangement of the oxime derived from 9-fluorenone.

4. The formation of seven-membered rings is generally not easy, yet a seven-membered ring forms readily in this experiment. Explain.

5. What is the role of sodium carbonate in the oxime preparation?

❖ 6. *Library project:* Find a reference to the original article by E. Beckmann on acid-catalyzed rearrangement of an oxime to form an amide.

❖ 7. There are six signals in the ^{13}C NMR spectrum of cyclohexanone oxime in Figure 17.2-5. Explain.

Experiment 17.3: Pinacol Rearrangement and Photochemical Synthesis of Benzopinacol

Molecular rearrangements, such as the acid-catalyzed pinacol rearrangement, are very common in organic chemistry. In this experiment, you will:

♦ perform a molecular rearrangement of pinacol or a pinacol derivative.

♦ prepare a pinacol derivative by bimolecular photochemical reduction.

Techniques

Technique F	Filtration and recrystallization
Technique G	Reflux
Technique K	Thin-layer chromatography
Technique M	Infrared spectroscopy
Technique N	Nuclear magnetic resonance spectroscopy

Background and Mechanism

Pinacol is the common name for 2,3-dimethylbutane-2,3-diol. The most well-known reaction of this compound is the acid-catalyzed pinacol rearrangement.

pinacol
(2,3-dimethylbutane-2,3-diol)

pinacolone
(3,3-dimethyl-2-butanone)

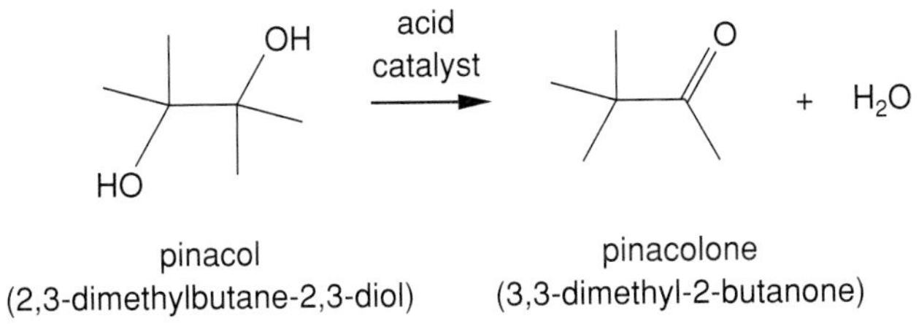

The first step is protonation of one of the hydroxyl groups by a protic acid, followed by loss of water to form a carbocation. With pinacol, there is a 1,2-shift of a methyl group, yielding a rearranged carbocation that is stabilized by an attached hydroxyl group. Loss of a proton gives pinacolone.

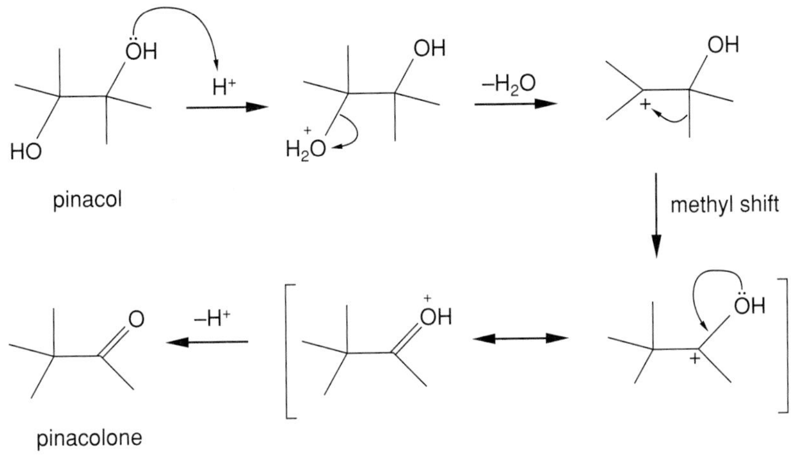

Aryl-substituted diols also undergo the pinacol rearrangement. An example is shown below for the pinacol rearrangement of benzopinacol (1,1,2,2-tetraphenylethane-1,2-diol).

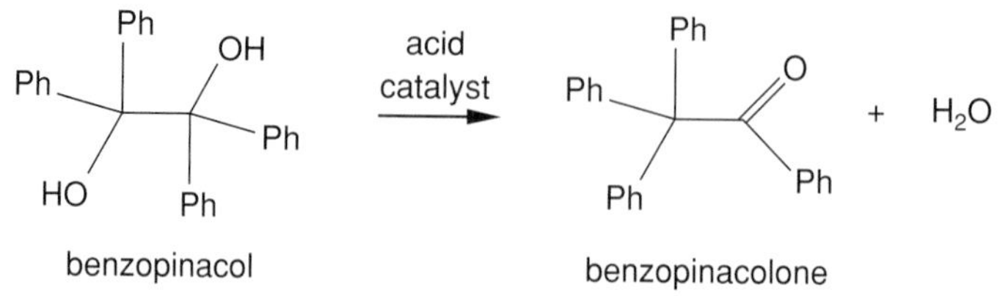

$$\underset{\text{benzopinacol}}{\text{Ph, Ph OH, Ph, HO, Ph}} \xrightarrow[\text{catalyst}]{\text{acid}} \underset{\text{benzopinacolone}}{\text{Ph, Ph, Ph, O, Ph}} + H_2O$$

Benzopinacol must first be synthesized photochemically from benzophenone. This synthesis is an example of a bimolecular photochemical reduction.

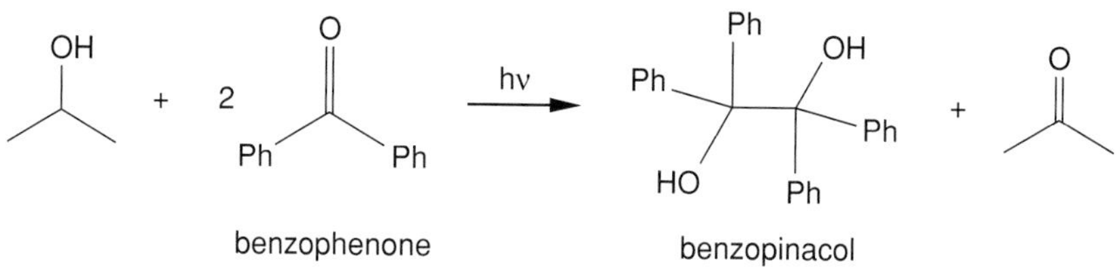

$$\text{OH} + 2 \underset{\text{benzophenone}}{\overset{O}{\underset{Ph \quad Ph}{\parallel}}} \xrightarrow{h\nu} \underset{\text{benzopinacol}}{\text{Ph, Ph OH, Ph, HO, Ph}} + \overset{O}{\parallel}$$

Prelab Assignment

1. Design a flow scheme for the workup procedure of the assigned pinacol rearrangement.
2. Explain how to prepare 2.0 mL of 3 M H_2SO_4 starting with concentrated H_2SO_4 (18 M) and water.
3. (Part B) Suggest alternate reaction conditions for this experiment if the weather is inclement.

Experimental Procedure

Safety First!

Always wear eye protection in the laboratory.

1. Wear eye protection at all times in the laboratory.
2. Wear gloves when handling sulfuric acid. In the event of skin contact with sulfuric acid, wash with copious amounts of cold water. Immediately notify the instructor. Carefully neutralize spills of sulfuric acid with sodium carbonate.
3. Work in the hood or in a well-ventilated area for this experiment.

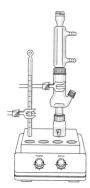

Part A: Microscale Synthesis of Pinacolone

To a dry, 5-mL conical vial containing a spin vane, add 750 mg of pinacol. Using a calibrated pipet, add 2.0 mL of 3 M sulfuric acid. Fit the vial with a Hickman still and a water condenser. Heat and stir the solution on a sand bath or heat block and collect distillate (boiling point 90–105°C) until about 1 mL has been collected. Use a Pasteur pipet to transfer distillate from the Hickman still to a clean conical vial. The distillate will consist of two layers: the top organic layer and an aqueous layer. Draw off the bottom layer with a Pasteur pipet and set aside for disposal. Add 1 mL of water to the vial. Cap the vial and shake to thoroughly mix the contents. Remove the cap and draw off the bottom layer. Repeat the washing with another 1-mL portion of water. After removing the water layer, transfer the organic layer to a clean vial containing a small amount of anhydrous sodium sulfate. Swirl the contents briefly and remove the liquid using a filter pipet. Place the liquid in a tared vial and reweigh the vial.

Characterization for Part A

Boiling Point: Measure the micro boiling point of the product. Correct for barometric pressure. The reported boiling point of pinacolone is 106°C.

Refractive Index: Place 3 or more drops of the product on a clean, dry stage of the refractometer and measure the RI. The reported value is 1.3960^{20}.

Infrared Spectroscopy: Obtain an IR spectrum of the liquid product. There should be strong carbonyl absorption near 1715 cm^{-1} and there should be little or no absorption at 3500–3300 cm^{-1}.

NMR Spectroscopy: Obtain ^{1}H and ^{13}C NMR spectra in CDCl$_3$ and interpret the spectra. NMR spectra are shown in Figures 17.3-1 and 17.3-2.

Results and Conclusions for Part A

1. Determine the percent yield of product.
2. Is there any evidence for the presence of starting material or side product? Explain.
3. Summarize spectroscopic evidence that supports the structure assigned to the product.

Part B: Miniscale Photoreduction of Benzophenone

Dissolve 2.0 g of benzophenone in 15 mL of warm 2-propanol in a 50-mL Erlenmeyer flask. Add 2 drops of glacial acetic acid and cork the flask. Place the flask in direct sunlight for 1 week. During this time crystals of benzopinacol should form on the sides and bottom of the flask. Uncover the flask and filter the solid using a Büchner funnel. Wash the crystals with 3 mL of cold 2-propanol and air dry the crystals.

Characterization for Part B

Melting Point: Determine the melting point of the product. The reported melting point of benzopinacol is 185–186°C.

Infrared Spectroscopy: Obtain an IR spectrum of the product using either KBr or Nujol and interpret the spectrum. Note the key bands characteristic of the structure of

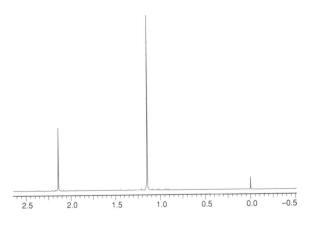

Figure 17.3-1 ^{1}H NMR spectrum of pinacolone (CDCl$_3$)

SOURCE: Reprinted with permission of Aldrich Chemical.

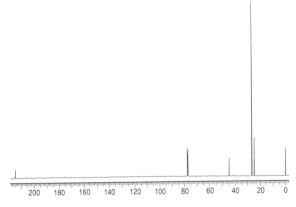

Figure 17.3-2 ^{13}C NMR spectrum of pinacolone (CDCl$_3$)

SOURCE: Reprinted with permission of Aldrich Chemical.

the product, including the strong absorption at 3500–3200 cm^{-1} as well as the absence of a carbonyl stretch near 1700 cm^{-1}.

NMR Spectroscopy: Obtain the ^{1}H NMR spectrum of the product as a solution in CDCl$_3$. Interpret the spectrum noting the chemical shift positions and the integration values. The spectrum should show two signals.

Results and Conclusions for Part B

1. Determine the percent yield of product.
2. Is there any evidence for the presence of starting material or side product? Explain.

Part C: Miniscale Synthesis of Benzopinacolone

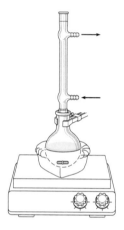

Place 1.5 g of dry benzopinacol in a 50-mL round-bottom flask fitted with a reflux condenser and a stir bar. Add 15 mL of glacial acetic acid and a few small iodine crystals. Heat the mixture at reflux for 15 minutes. Cool the solution and add 10 mL of ethanol. Cool the mixture on ice and collect the product by suction filtration using a Büchner funnel. Wash the crystals with a few drops of ice-cold ethanol. Air dry the crystals. Weigh the product.

Characterization for Part C

Melting Point: Determine the melting point of the product. The reported melting point of benzopinacolone is 182–184°C. Do a mixed melting point determination using a 1:1 mixture of starting material and product.

Infrared Spectroscopy: Obtain an IR spectrum of the product using either KBr or Nujol and interpret the spectrum. Note the key bands characteristic of the structure of the product, including the strong absorption near 1700 cm^{-1}.

NMR Spectroscopy: Obtain the ^{1}H NMR spectrum of the product as a solution in CDCl$_3$. Interpret the spectrum noting the chemical shift positions and the integration values. Note the large singlet at $\delta 7.2$ and a more complex aromatic signal for the phenyl protons of the benzoyl group.

Cleanup & Disposal

Pour the acetic acid filtrate down the drain with running water. Pour aqueous extracts down the drain. Place 2-propanol filtrate in a container labeled "nonhalogenated organic waste" in the hood.

Results and Conclusions for Part C

1. Determine the percent yield of product.
2. Is there any evidence for the presence of starting material or side product? Explain.
3. Explain why it is necessary to do a mixed melting point determination.

Critical Thinking Questions *(The harder ones are marked with a ❖.)*

1. Predict the ^{13}C NMR spectrum of pinacol.
2. Write the mechanism of the acid-catalyzed pinacol rearrangement of 1,1,2-triphenylethane-1,2-diol. Assume phenyl migration from C-1 to C-2.
❖ 3. Suggest a mechanism for the bimolecular reduction of benzophenone to give benzopinacolone.

4. Suggest possible side products that could be obtained during the reaction of pinacol with aqueous sulfuric acid.

5. Distillation of pinacolone gives a distillate consisting of two layers. What type of distillation is this? Why is the boiling point of the distillate below 100°C instead of the reported boiling point of pinacolone (bp 106–107°C)?

6. Consult the lecture text and write examples and equations for two other molecular rearrangement reactions. Categorize each according to the type of mechanism.

7. Are vicinal alkoxy alcohols expected to give acid-catalyzed pinacol rearrangement? If so, show a possible mechanism.

8. Is there a type of carbocation that is more stable than a tertiary carbocation? Explain.

❖ 9. Diazotization of vicinal amino alcohols is followed by a pinacol rearrangement. Explain using a structural example.

10. *Library project:* Use secondary literature sources from the library to research migratory behavior (migratory aptitude) of different substituents in the pinacol rearrangement.

Chapter 18

Enols, Enolates, and Enones

One important property of aldehydes and ketones is acidity. The α-hydrogens of aldehydes and ketones are relatively acidic, with pK_as in the range of 20–22. Reactions with strong bases such as amide ion can convert ketones completely to anions. Bases such as hydroxide or ethoxide are sufficiently strong to convert a collection of ketone molecules partially to their anions. This is the case in Experiment 18.1, which is an example of a mixed aldol condensation. α-Halogenation can also be accomplished in the presence of base, leading to a reaction known as the iodoform reaction. Replacement of all three α-hydrogens of the methyl group of a methyl ketone followed by reaction with strong base results in loss of a carbon from the carbon chain forming a carboxylate ion as in Experiment 19.2.

The α,β-unsaturated ketones, known as enones, can undergo reaction at the β-carbon or at the carbonyl carbon, depending upon the type of reaction and the nature of the reactant and nucleophile used. Experiment 18.2 is an example of the exploration of this principle. There can be selective reaction with either the carbonyl group or the alkene double bond.

Experiment 18.1: Preparation of α,β-Unsaturated Ketones Via Mixed Aldol Condensation

In this experiment, you will investigate the aldol condensation and its application in the synthesis of β-hydroxyaldehydes and ketones and α,β-unsaturated aldehydes and ketones. You will:

- perform a mixed aldol condensation.

- synthesize an α,β-unsaturated ketone.

- identify an unknown aldehyde and aldol condensation product.

Techniques

Technique F	Recrystallization and vacuum filtration
Technique M	Infrared spectroscopy
Technique N	Nuclear magnetic resonance spectroscopy

Background and Mechanism

The aldol condensation is the acid- or base-catalyzed self-condensation of an aldehyde to give a β-hydroxyaldehyde. In this reaction, a carbon-carbon bond is formed between the α-carbon of one molecule and the carbonyl carbon of the other. Dehydration of the aldol products to give α,β-unsaturated carbonyl compounds is easily accomplished upon heating.

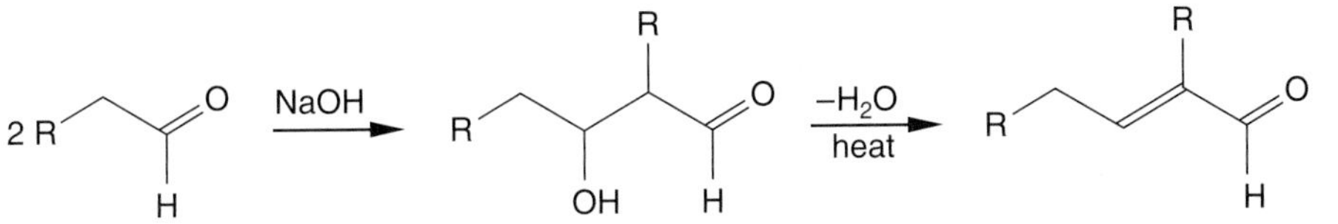

R = alkyl β-hydroxyaldehyde α,β-unsaturated aldehyde

The first step in the mechanism is deprotonation of an α-hydrogen by a strong base to form a resonance-stabilized enolate ion.

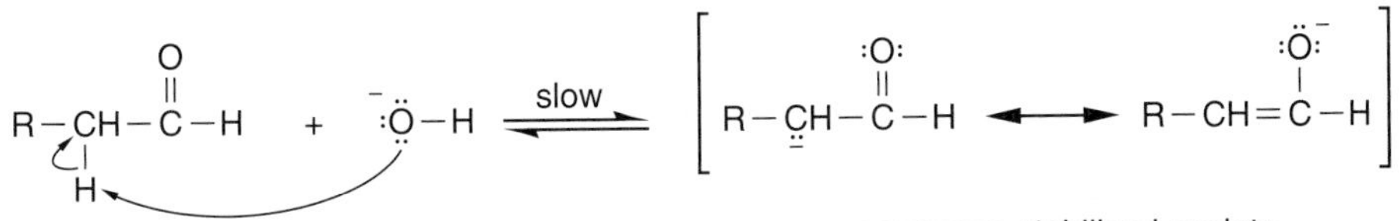

resonance-stabilized enolate

Although the aldehyde is only partially converted to the enolate ion, the enolate is very reactive and will add to the carbonyl of an aldehyde:

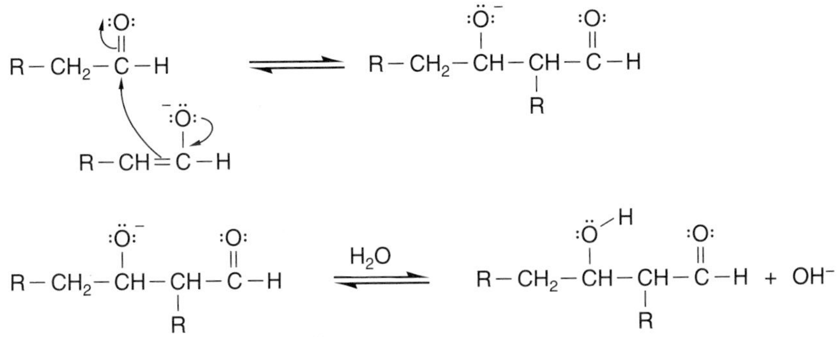

The initial product of the aldol condensation is a β-hydroxyaldehyde, which can undergo base-catalyzed elimination upon heating to yield an α,β-unsaturated aldehyde. Dehydration will occur readily if the new double bond is conjugated with an aromatic ring.

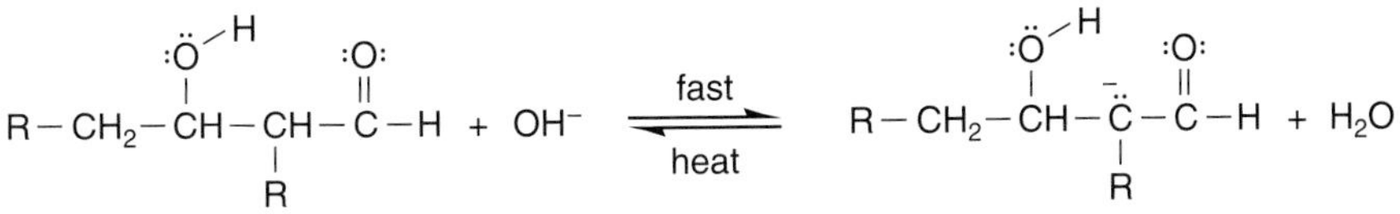

β-hydroxyaldehyde

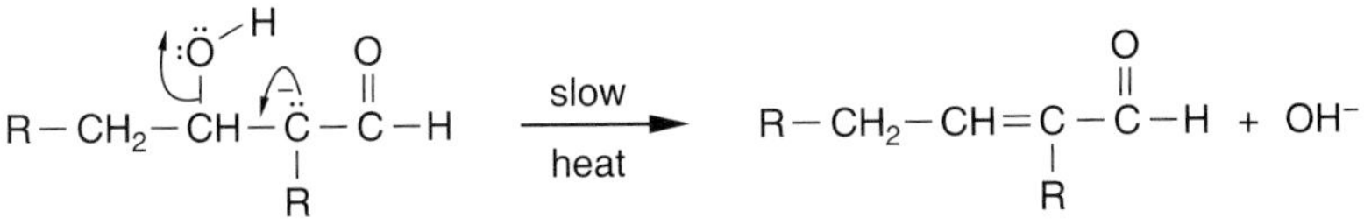

α,β-unsaturated aldehyde

A mixed aldol condensation is a reaction between two different aldehydes or ketones. At least one of the compounds should be an aldehyde. This condensation is generally practical only if one of the components has no acidic α-hydrogens. Otherwise, up to four products can be formed, lowering yields. In this experiment, a mixed aldol condensation between two moles of cinnamaldehyde (3-phenylpropenal) with acetone in the presence of base produces 1,9-diphenyl-1,3,6,8-nonatetraen-5-one (dicinnamalacetone), which is bright yellow.

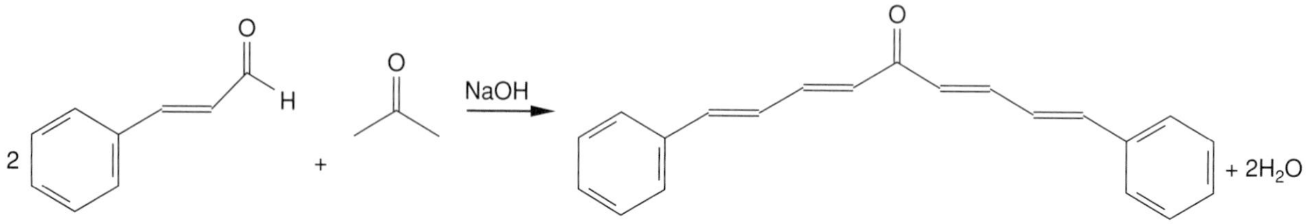

cinnamaldehyde   acetone   dicinnamalacetone

This reaction is frequently used as a "clock reaction" in demonstrations, since the reaction is easily monitored by the sudden and dramatic appearance of the bright yellow product. In the first condensation step, acetone reacts with one mol of cinnamaldehyde to produce 6-phenyl-3,5-hexadien-2-one. A second mol of cinnamaldehyde condenses with the 6-phenyl-3,5-hexadien-2-one to yield dicinnamalacetone.

In Part B of this experiment, you will prepare dibenzalacetone (1,5-diphenyl-1,4-pentadien-3-one) via a mixed aldol condensation between two equivalents of benzaldehyde and acetone.

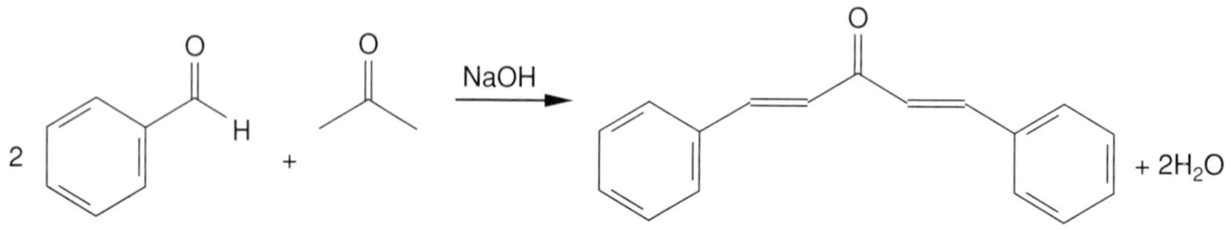

benzaldehyde   acetone   dibenzalacetone

In Part C, you will condense a ketone and an unknown aldehyde, purify the product, and identify the aldehyde from the melting point of the mixed aldol product. The possible aromatic aldehydes are benzaldehyde, *p*-anisaldehyde, *p*-chlorobenzaldehyde, *p*-ethylbenzaldehyde, *p*-tolualdehyde, and 2,4-dimethoxybenzaldehyde.

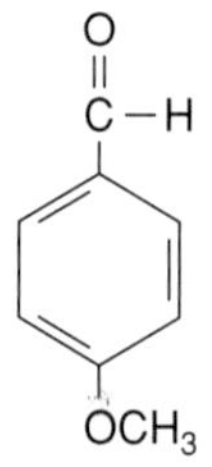

p-anisaldehyde 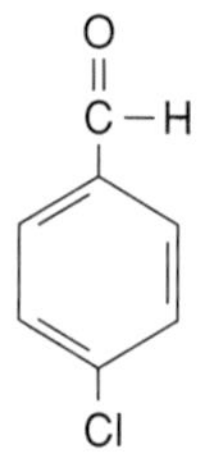   *p*-chlorobenzaldehyde 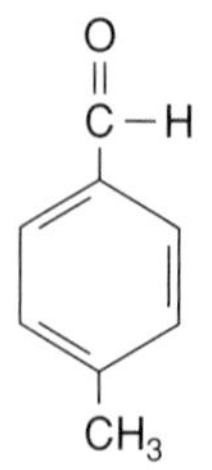   *p*-tolualdehyde 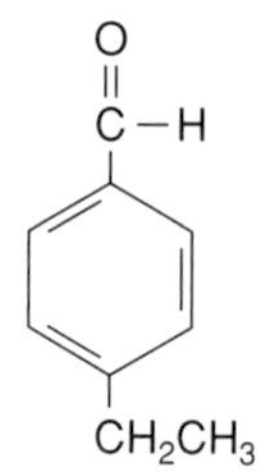   *p*-ethylbenzaldehyde 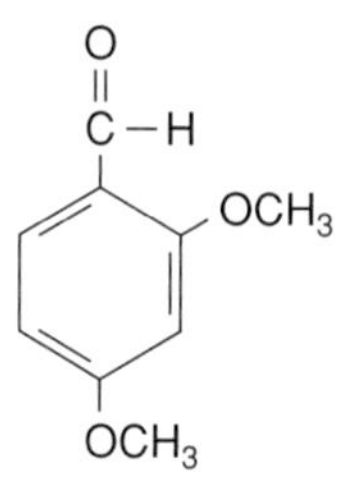   2,4-dimethoxybenzaldehyde

The ketones are acetone, cyclopentanone, and cyclohexanone. The structure of the product will be determined from the melting point. The melting points of the aldol products are given in Table 18.1-1.

Table 18.1-1   Melting Points of the Mixed Aldol Products

Aldehyde	Ketone		
	Acetone (°C)	Cyclopentanone (°C)	Cyclohexanone (°C)
Benzaldehyde	112	192	120
p-Anisaldehyde	131	212	159
p-Chlorobenzaldehye	195	225	148
p-Ethylbenzaldehyde	125	150	127
p-Tolualdehyde	177	238	173
2,4-Dimethoxybenzaldehyde	139	187	177

Prelab Assignment

1. Write a detailed mechanism for the formation of dicinnamalacetone or dibenzalacetone.
2. Calculate the quantities (mmol) of each reagent and determine which is the limiting reagent.
3. Several of the possible aldol products have similar melting points. Explain how IR or NMR might be useful in differentiating between these structures.
4. Design a flow scheme to show how the product is isolated from the starting materials and impurities during the workup of the reaction.

Experimental Procedure

1. Wear eye protection at all times in the laboratory.
2. Sodium hydroxide is corrosive and should be handled with gloves.
3. The organic liquids used in this experiment are volatile. Avoid breathing their vapors.
4. The aldol products are potential skin irritants and contact should be rigorously avoided. If contact is made, flush the skin with copious amounts of cool water. Notify the instructor.

Part A: Miniscale Synthesis of 1,9-Diphenyl-1,3,6,8-nonatetraen-5-one (Dicinnamalacetone)

Place 0.4 mL of cinnamaldehyde and 2.5 mL of 95% ethanol in a 10-mL Erlenmeyer flask. Add 1.5 mL 2 M KOH. Swirl to mix. To the pale yellow solution, add 0.11 mL of acetone. Swirl vigorously to mix, then let the flask stand undisturbed. After several minutes, the product will precipitate from solution, forming bright yellow crystals. Cool in an ice bath for an additional 15 minutes to ensure complete crystallization. Filter the crystals using a Hirsch or a Büchner funnel. Wash with several small portions of ice-cold 95% ethanol. The product may be recrystallized from 2-propanol or 95% ethanol. Weigh the dried crystals and determine the yield.

Characterization for Part A

Melting Point: Determine the melting point of the product. The reported melting point of dicinnamalacetone is 144–145°C.

Infrared Spectroscopy: Obtain an IR spectrum of the product using either the KBr or the Nujol mull technique and interpret the spectrum, noting the position of the carbonyl absorption. The IR spectrum of dicinnamalacetone as a Nujol mull is shown in Figure 18.1-1.

NMR Spectroscopy: Dissolve a sample of the dry product in $CDCl_3$. Obtain the ^1H NMR spectrum. Due to extensive conjugation, the aromatic and alkenyl protons all come into resonance around $\delta7$. Expand this region and interpret the spectrum. The ^1H NMR spectrum of dicinnamalacetone is shown in Figure 18.1-2.

Results and Conclusions for Part A

1. Calculate the percent yield of dicinnamalacetone.
2. Evaluate the stereochemistry of the product from the ^1H NMR and/or IR. The product is reported to have trans stereochemistry. From Figure 18.1-2(b), calculate the coupling constant. Explain why the trans product is expected to form preferentially.
3. Write a detailed mechanism for the formation of dicinnamalacetone.

Part B: Microscale Synthesis of Dibenzalacetone (1,5-Diphenyl-1,4-pentadien-3-one)

Place 204 µL of benzaldehyde and 73 µL of acetone in a 5-mL conical vial equipped with a spin vane. Add 0.5 mL of 5 M sodium hydroxide and 0.5 mL of 95% ethanol to the stirred

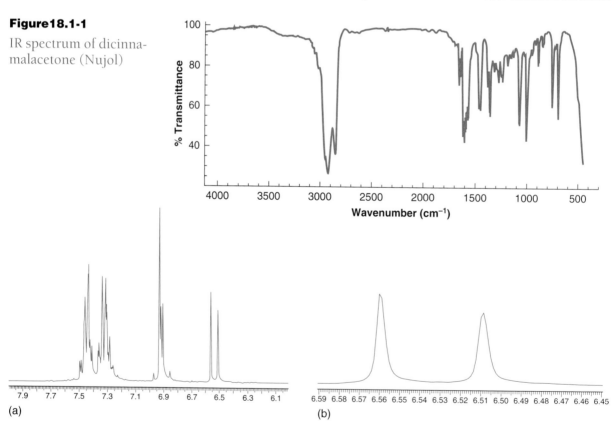

Figure 18.1-1

IR spectrum of dicinnamalacetone (Nujol)

Figure 18.1-2 ^1H NMR spectra of (a) dicinnamalacetone ($CDCl_3$); (b) dicinnamalacetone (expanded region) ($CDCl_3$)

SOURCE: Reprinted with permission of Aldrich Chemical.

mixture. Cap the vial. A yellow precipitate will form. Stir the reaction mixture at room temperature for 30 minutes. Isolate the product by vacuum filtration using a Hirsch funnel. Wash the product three times with 1-mL portions of cold water. Recrystallize from ethanol, 95% ethanol, or ethyl acetate. Weigh the dry product.

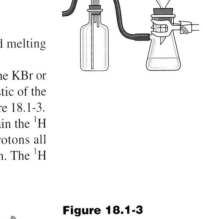

Characterization for Part B

Melting Point: Determine the melting point of dibenzalacetone. The reported melting point is 110–111°C.

Infrared Spectroscopy: Obtain an IR spectrum of the product using either the KBr or Nujol mull technique. Interpret the spectrum, noting the key bands characteristic of the functional groups present. The IR spectrum of dibenzalacetone is shown in Figure 18.1-3.

NMR Spectroscopy: Dissolve a sample of the dry product in CDCl$_3$. Obtain the ^{1}H NMR spectrum. Due to extensive conjugation, the aromatic and alkenyl protons all come into resonance around δ7. Expand this region and interpret the spectrum. The ^{1}H NMR spectrum of dibenzalacetone is shown in Figure 18.1-4.

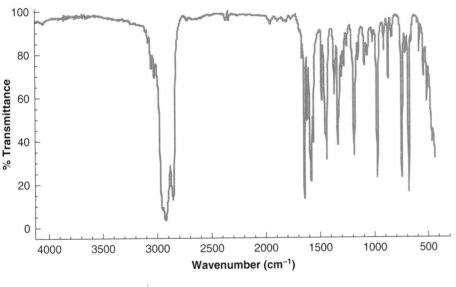

Figure 18.1-3

IR spectrum of dibenzalacetone (Nujol)

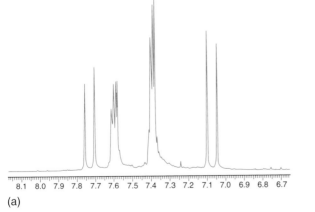

(a)

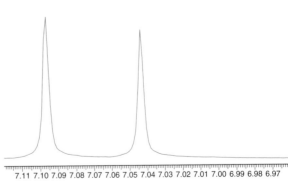

(b)

Figure 18.1-4 ^{1}H NMR spectra of (a) dibenzalacetone (CDCl$_3$); (b) dibenzalacetone (expanded region) (CDCl$_3$)

Results and Conclusions for Part B

1. Calculate the percent yield of dibenzalacetone.
2. Evaluate the stereochemistry of the product from the ^{1}H NMR and/or IR. The product is reported to have trans stereochemistry. From Figure 18.1-4(b), calculate the coupling constant. Explain why the trans product is expected to form preferentially.
3. Write a detailed mechanism for the formation of dibenzalacetone

Part C: Identification of a Mixed Aldol Product from the Miniscale Condensation of an Unknown Aromatic Aldehyde with a Ketone

Each pair of students will get an unknown aldehyde from the instructor. Each student should choose a different ketone to react with the aldehyde. Place 0.5 mL or 300 mg of the aldehyde and 2.5 mL of 95% ethanol in a 10-mL Erlenmeyer flask. Add 1.75 mL of 2 M KOH. Swirl to mix. Add 125 μL of the ketone. Swirl vigorously until precipitation is complete. If no precipitate forms after 10 minutes, heat the solution gently on a sand bath in the hood until precipitation occurs. Cool in an ice bath for 15 minutes to ensure complete crystallization. Filter the crystals using a Hirsch or a Büchner funnel. Wash with several small portions of ice-cold 95% ethanol. Rinse the crystals with a small portion of 10% acetic acid to neutralize excess base, then rinse again with cold 95% ethanol. Air dry briefly.

To find a recrystallization solvent, test the solubility of the aldol product in toluene and in 2-propanol. Use the better solvent to recrystallize. Recrystallize at least twice. Weigh the dry product.

Characterization for Part C

Melting Point: Determine the melting point of the recrystallized product. The reported melting point of each of the mixed aldol products is given in Table 18.1-1.

Results and Conclusions for Part C

1. Determine the identity of the unknown aldehyde and mixed aldol product.
2. Write a mechanism for the formation of the observed product.
3. Calculate the percent yield of the product.

C l e a n u p   &   D i s p o s a l

Neutralize the aqueous filtrate with dilute hydrochloric acid and wash down the drain with water. Place recrystallization solvents in a container labeled "nonhalogenated organic solvent waste."

Critical Thinking Questions *(The harder ones are marked with a ❖.)*

1. Dicinnamalacetone has four carbon-carbon double bonds. How many different geometric isomers are possible?
2. Is self-condensation of cinnamaldehyde a possibility in this experiment? Explain.
3. A possible side product of this reaction is the self-condensation of acetone. Draw the structure of the side product. Explain how the experimental procedure is designed to minimize the formation of this side product.

4. The reaction of an aldehyde and acetone can also be catalyzed by acid. Write a detailed mechanism for the acid-catalyzed aldol condensation of acetone and cinnamaldehyde.

5. Cinnamaldehyde and acetone are colorless, yet dicinnamalacetone is bright yellow. Explain.

6. Draw the structure of the expected product of the mixed aldol condensation of benzaldehyde and acetophenone.

❖ 7. The double mixed aldol condensation of 1,3-diphenylacetone with benzil in the presence of base gives an interesting product called 2,3,4,5-tetraphenylcyclopentadienone (tetracyclone). Write a mechanism for this reaction and draw a structure for the product. This dark purple compound is an excellent Diels-Alder diene. Write the structure of the product formed from the Diels-Alder reaction of this material with methyl propynoate.

8. There are a number of base-catalyzed condensations related to the aldol condensation. Write an example for each of these "name" reactions.
   a.   the Claisen ester condensation.
   b.   the Perkin condensation.
   c.   the Dieckmann condensation (intramolecular Claisen condensation).

9. Explain why the product of condensation of one mol of cinnamaldehyde with one mol of acetone is not isolated from the synthesis of dicinnamalacetone.

❖ 10. Literature reports of mixed aldol condensations between two different ketones are rare. Cite possible reasons.

Experiment 18.2: Reduction of Conjugated Ketones with Sodium Borohydride

Because conjugated enones have two potential electrophilic reaction sites, they can react with nucleophiles to give different products, depending upon the reaction conditions. In this experiment, the reaction of sodium borohydride with a conjugated ketone will be explored. You will:

◆   investigate the selectivity of hydride for 1,2-addition versus 1,4-addition.

◆   deduce the structure of the product(s) by spectroscopy.

◆   determine the percent composition of each of the products.

◆   investigate structure-reactivity relationships of the enone.

Techniques

Technique I	Extraction and drying
Technique J	Gas chromatography
Technique M	Infrared spectroscopy
Technique N	Nuclear magnetic resonance spectroscopy

Background

Sodium borohydride ($NaBH_4$) is a relatively mild reducing agent that will selectively reduce aldehydes and ketones, but not esters, unconjugated alkenes, carboxylic acids, or amides. The mechanism of the reaction of $NaBH_4$ with conjugated enones involves transfer of hydride from borohydride ion to either the carbonyl carbon or the γ carbon. Reduction at the carbonyl carbon and protonation of the resulting alkoxide by the solvent, ethanol, will yield the 1,2-product (the unsaturated alcohol). Reduction at the γ carbon will result in formation of a 1,4-product (the saturated carbonyl compound). When excess borohydride is used as in this experiment, the 1,4-product may be further reduced to the saturated alcohol.

To determine how the hydride ion reacts with conjugated systems, either 2-cyclohexenone or *trans*-4-phenyl-3-buten-2-one will be reduced and the product composition carefully examined using IR spectroscopy and ^{1}H NMR spectroscopy. Because the reaction is conducted with excess borohydride, the formation of the 1,4-product will be deduced from the presence of the saturated alcohol. The structures of the two starting materials are shown here:

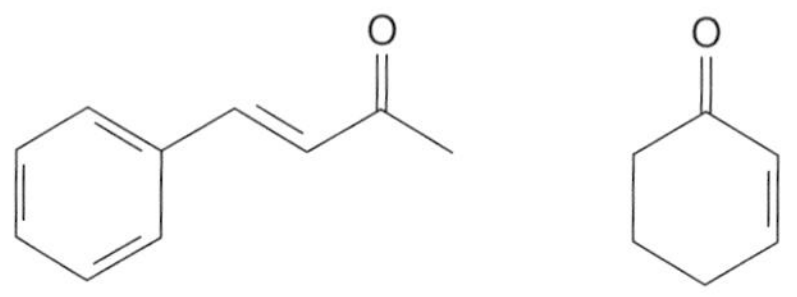

trans-4-phenyl-3-buten-2-one 2-cyclohexenone

Prelab Assignment

1. Draw the predicted 1,2-product, the 1,4-product, and the fully reduced product of the assigned conjugated ketone.
2. Refer to the assigned procedure and verify that an excess of $NaBH_4$ is to be used for these reactions. Explain using calculations.
3. Suggest mechanisms of reaction for 1,2-addition and for conjugate addition (1,4-addition) of $NaBH_4$ to a conjugated ketone such as 2-cyclohexenone.
4. For each of the predicted products of the assigned ketone, tabulate significant infrared absorption bands that would help identify the structure and predict the chemical shifts of each type of proton in the ^{1}H NMR spectra.
5. Write a flow sheme for the workup procedure of the assigned enone

Experimental Procedure

Safety First!

Always wear eye protection in the laboratory.

1. Wear eye protection at all times in the laboratory.
2. Sodium borohydride should be used under the hood. Wear gloves.

General Directions

Students will work in pairs, with each partner reducing a different conjugated ketone. After performing the reaction and isolating the products, students should work together to analyze and characterize the products.

Part A: Microscale Reduction of 2-Cyclohexenone

Weigh 96 mg (0.10 mL) of 2-cyclohexenone and add it to a 5-mL conical vial or 10-mL Erlenmeyer flask. Add 2 mL of 95% ethanol. Swirl to mix. Add 75 mg of sodium borohydride to the solution. Let stand 10 minutes with occasional stirring or swirling.

Add 2 mL of water. Heat the solution briefly to boiling. The solution should be clear and colorless.

Transfer the clear, warm solution to a centrifuge tube using a filter pipet and allow to cool to room temperature. Add 4 mL of methylene chloride. Swirl and shake to mix the layers. Transfer the lower (organic) layer to an Erlenmeyer flask. Add anhydrous Na_2SO_4 until no more clumping of the solid is observed. After letting stand for 10 minutes, transfer the solution using a Pasteur filter pipet to a clean, tared Erlenmeyer flask containing a boiling chip. Evaporate the solvent in the hood leaving an oily residue. Weigh the product. Record the IR spectrum and the NMR spectrum and deduce the structure of the product(s). If more than one product is formed, use NMR spectroscopy or GC to determine the product composition.

Part B: Microscale Reduction of trans-4-Phenyl-3-buten-2-one

Weigh out 147 mg of *trans*-4-phenyl-3-buten-2-one and add to a 5-mL conical vial or 10-mL Erlenmeyer flask. Add 2 mL of 95% ethanol. Swirl to mix. Add 75 mg of sodium borohydride to the solution. Let stand 10 minutes with occasional stirring or swirling. (If desired, run a TLC in 50% hexanes/ethyl acetate to determine if the starting material has been completely reacted.) Add 2 mL of water. Heat the solution briefly to boiling. The solution should be clear and colorless.

Transfer the clear, warm solution to a centrifuge tube using a filter pipet and allow to cool to room temperature. Add 4 mL of methylene chloride. Swirl and shake to mix the layers. Transfer the lower (organic) layer to an Erlenmeyer flask. Add anhydrous Na_2SO_4 until no more clumping of the solid is observed. After letting stand for 10 minutes, transfer the solution using a Pasteur filter pipet to a clean, tared Erlenmeyer flask containing a boiling chip. Evaporate the solvent in the hood leaving an oily residue. Weigh the product. Record the IR spectrum and the NMR spectrum and deduce the structure of the product(s). If more than one product is formed, use NMR spectroscopy to determine the product composition.

Characterization for Parts A and B

Infrared Spectroscopy: Record the IR spectrum of the oily product and interpret, noting key bands characteristic of functional groups present.

NMR Spectroscopy: Obtain the ^1H spectrum of the product in $CDCl_3$. Interpret the spectrum in terms of the predicted product(s).

Alternate Analysis of the 2-Cyclohexenone Reaction with Gas Chromatography: Use a GC instrument equipped with a polar capillary column to analyze the product(s) using a temperature-programmed protocol starting at a column temperature of 40–50°C. The final temperature should be 175–200°C and should be held for 2–3 minutes. Total analysis time should be 12–15 minutes. Determine the relative percentages of the product(s). Inject standards of 2-cyclohexenone and solutions of each of the possible products (the 1,4-product, the 1,2-product, and the fully reduced product).

Results and Conclusions

1. Examine the IR spectrum. Does it appear that the product is a mixture or is it primarily a single compound? Indicate which absorption bands were important in your decision. Explain.
2. Examine the ^1H NMR spectrum for evidence of the 1,2-product, the 1,4-product, and the fully reduced product. Correlate specific signals in the NMR spectrum to structural features of these compounds. From the NMR spectrum, does it appear that the product is a mixture or is it primarily a single compound? Explain.

3. From the IR and NMR spectra, indicate the structure of each product formed in the reaction.
4. If possible, determine the percent composition of any mixtures by comparing integrations of specific signals of the NMR spectrum.
5. Calculate the percent yield of the reaction.
6. Propose a mechanism for the formation of each of the observed products.
7. Compare the results for the two conjugated enones. Explain any differences in reactivity of the two enones used in this experiment and propose a hypothesis to explain the differences.

Cleanup & Disposal

Solutions of sodium borohydride should be neutralized with dilute HCl. Use caution since hydrogen gas may be evolved. Once neutralized, the aqueous solutions may be washed down the drain with water.

Critical Thinking Questions

1. When doing a reduction with sodium borohydride in aqueous ethanol, it is important to use a significant excess of the sodium borohydride. Why? (Hint: Consider what reaction might occur between water and sodium borohydride.)
2. Suggest factors that might cause a conjugated system to react via 1,4-addition instead of the normal 1,2-addition.
3. The reduction of enones by $NaBH_4$ can be monitored in several different ways. Suggest a few methods and their relative merits and drawbacks.
4. Draw the structure of the reduction products of the reaction of the assigned enone with:
   a. $NaBD_4$ in CH_3CH_2OH and H_2O
   b. $NaBH_4$ in CH_3CH_2OD and D_2O
   c. $NaBH_4$ in CD_3CD_2OH and H_2O
5. Draw a mechanism for 1,2-addition and for 1,4-addition (conjugate addition) of the reaction of excess $NaBH_4$ with 3-buten-2-one.
6. Based upon the results from this experiment, speculate on the nature of the product(s) expected for the reaction of *trans*-cinnamaldehyde with excess $NaBH_4$.
7. *Molecular modeling assignment:* Use a molecular modeling program to diagram the electron density surface for the assigned enone. Note where the blue regions (areas of electron deficiency) are located. Do the results agree with the observed product(s) you obtained in this experiment?

Experiment 18.3: Identification of Products of Catalytic Transfer Hydrogenation of an Enone

Conversion of alkenes to alkanes is accomplished using an organic hydrogen transfer agent in the presence of a catalyst. The reaction can be selective, allowing addition of hydrogen to take place in the absence of a hydrogen atmosphere. In this experiment you will:

♦ perform a catalytic transfer hydrogenation on an enone.

♦ determine the structure(s) and amount(s) of the product(s).

♦ deduce the mechanism of the reaction.

Techniques

Technique F	Filtration
Technique G	Distillation and reflux
Technique J	Gas-liquid chromatography
Technique N	Nuclear magnetic resonance spectroscopy
Technique P	Mass spectrometry

Background

Hydrogenation of alkenes using molecular hydrogen appears to be one of the simplest organic reactions. However, molecular hydrogen is a flammable gas and using it in large quantities could prove hazardous. In this experiment, catalytic transfer hydrogenation is used as an alternative to molecular hydrogen. In catalytic transfer hydrogenation, hydrogen is transferred from a donor substance in the presence of 10% Pd on carbon catalyst. Hydrogen gas is formed only in small amounts (if at all) so that special equipment to contain hydrogen gas is not required for the experiment. Hydrogenation of alkenes is selective. Alkenes can be reduced in the presence of a number of other functional groups, such as aromatic rings, ketones, and carboxylic acid esters.

In a typical example of the reaction, an aliphatic alkene such as 1-decene is submitted to transfer hydrogenation conditions. The alkene is heated under reflux with an excess of the transfer agent, cyclohexene, together with the catalyst, 10% Pd on carbon.

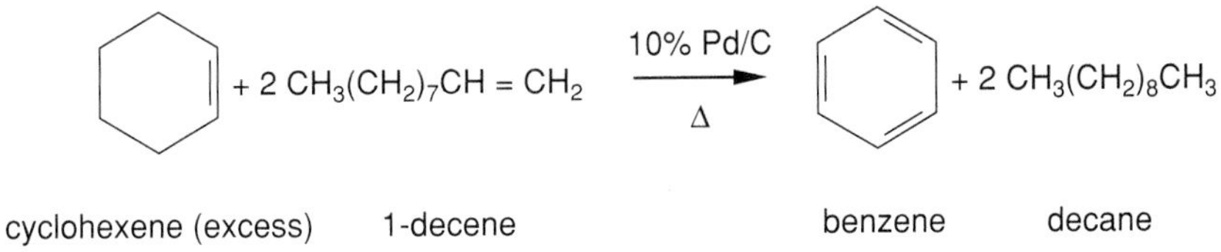

| cyclohexene (excess) | 1-decene | | benzene | decane |

During transfer hydrogenation, the equivalent of a hydrogen molecule is transferred from each cyclohexene donor molecule via the catalyst to the acceptor molecule, 1-decene. The equivalent of a hydrogen molecule is placed on the 1-decene receptor, converting it to decane. Cyclohexene is initally converted to 1,3-cyclohexadiene, a compound that readily donates a second molecule of hydrogen to form benzene.

In this example, two separate molecules are involved in the reaction, with cyclohexene acting as the hydrogen donor and 1-decene acting as the hydrogen acceptor. However, the same molecule can act as both the hydrogen donor and the hydrogen acceptor. In this experiment, you will investigate the disproportionation of 2-cyclohexenone. This is a transfer hydrogenation reaction in which the substrate acts as both a donor and an acceptor. This leads to two products, which you will identify by spectroscopic methods and physical and chemical properties. Characterization is possible using GC-MS and/or NMR spectroscopy (Schoffstall and Coleman, unpublished results). Based upon the product composition, you will propose a reasonable mechanism to account for the products.

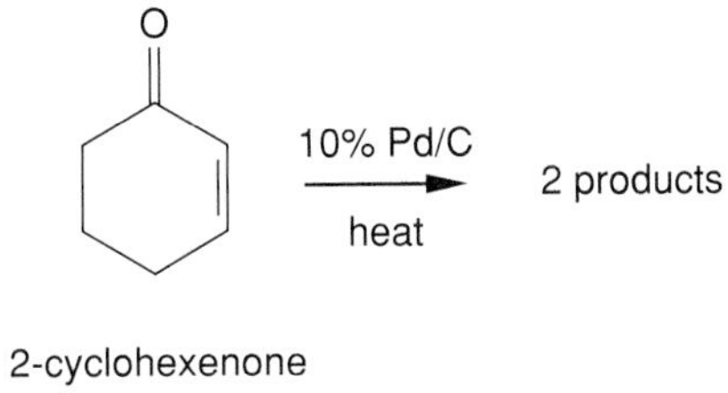

2-cyclohexenone

Prelab Assignment

1. Based upon the products of the reaction of cyclohexene and 1-decene, predict the possible products formed in this experiment.
2. Predict the relative amounts of the products that should be formed.
3. Does the experimental preparative procedure make sense in terms of the predicted products? Explain.
4. Write a flow scheme for the workup of the reaction.

Experimental Procedure

1. Wear safety goggles at all times in the laboratory.
2. Wear gloves when handling reagents and solvents in this experiment.
3. Notify the instructor if any chemicals get on your skin.
4. Avoid breathing solvent vapors. Evaporate solvents using a fume hood.

Part A: Microscale Reaction of 2-Cyclohexenone

Into a 5-mL conical vial containing a spin vane, transfer 0.50 mL of 2-cyclohexenone and 5–10 mg of 10% Pd/C catalyst. Attach a water-jacketed condenser and reflux the mixture for 45 minutes with stirring. Cool to room temperature and transfer the mixture to a second vial by filter pipet by passing the mixture through a second pipet containing a small wad of cotton and about 1 cm of Celite. Weigh the product mixture.

Part B: Miniscale Reaction of 2-Cyclohexenone

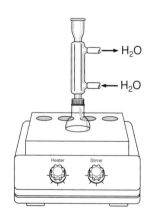

Measure out 2.0 mL of 2-cyclohexenone and transfer to a 10-mL round-bottom flask containing a spin bar. Add 30 mg of 10% Pd/C catalyst, fit the flask with a water-jacketed condenser, and reflux the mixture with stirring for 45 minutes. Cool the mixture and transfer to a centrifuge tube by filtering through a short column (2–3 cm) of Celite in a Pasteur pipet. Add 3.0 mL of 3M NaOH to the tube. Cap the tube and shake with venting. Draw off the bottom aqueous layer and transfer to a 25-mL Erlenmeyer flask. Add 2 mL of water to the organic layer in the centrifuge tube. Cap the vial and shake with venting. Combine the bottom aqueous layer with the aqueous solution in the Erlenmeyer flask. Add a small amount of anhydrous sodium sulfate to the organic liquid in the centrifuge tube and let stand for 5 minutes. If clumping of the drying agent occurs, add a small amount of additional drying agent. Transfer the liquid by filter pipet to a dry, tared vial and weigh. Label as Product A. The quantity of organic liquid should be about 0.75 mL. Clean the centrifuge tube.

Transfer the aqueous solution to the centrifuge tube. Carefully add 3.5 mL of 3 M HCl to the solution in the centrifuge tube, followed by 2.0 mL of methylene chloride. Cap and shake with venting. Transfer the bottom, organic layer to a 5-mL conical vial. Add 1.0 mL of methylene chloride to the centrifuge tube. Cap and shake with venting. Transfer the bottom organic layer and combine with the first methylene chloride extract in the conical vial. Add a small amount of anhydrous sodium sulfate to the conical vial and let stand for 5 minutes. Draw off the organic solution with a filter pipet and transfer to a dry, tared vial containing a boiling chip. Evaporate the solvent (35–40°C) on a heat block or sand bath in the hood. Apply a gentle stream of nitrogen or air to the inside of the vial if desired. Reweigh the cooled vial containing the solid product. Label as Product B.

Characterize Product A and Product B by GC-MS and/or ^{1}H FT-NMR spectroscopy. Any remaining starting material will appear as a contaminant in the spectra of Product A. Determine the percent yield of each product.

Characterization for Parts A and B

Melting Point: Determine the melting point of the solid product. Following NMR analysis, verify if the melting point is in agreement with the proposed structure of the solid product.

Infrared Spectroscopy: Obtain an IR spectrum of the products isolated during the preparative procedure. Determine which functional class can be assigned to each product. Verify if the IR spectrum is in agreement with the proposed structure.

NMR Spectroscopy: Analyze a small portion of the crude reaction mixture by ^{1}H NMR spectroscopy after removal of the Pd/C catalyst. Analyze the spectrum and identify the products formed during the reaction. For Part B, dissolve a sample of each of the separated products in $CDCl_3$. Filter each solution into a clean, dry NMR tube. Obtain the NMR spectra and interpret them, noting the chemical shift positions and the integration values.

Optional Analysis with Gas Chromatography-Mass Spectrometry (GC-MS): Analyze a small portion of the crude reaction mixture using GC-MS after removal of the Pd/C catalyst. Analyze the spectrum and identify the products formed during the reaction.

Results and Conclusions for Parts A and B

1. Draw the structures of the two products obtained in the experiment and explain how NMR and IR spectroscopy supported the structural assignments.
2. For Part B, determine the percent yield of products.
3. Propose a mechanism to account for the formation of the observed products. Using the mechanism, hypothesize the driving force for this reaction.
4. Was there evidence to suggest that starting material was present in the products? Explain.

Cleanup & Disposal

Place the filter pipet containing residual catalyst in a container in the hood labeled "filter pipets containing Pd catalyst."

Critical Thinking Questions

1. Predict the disproportionation products for catalytic transfer hydrogenation of 1,2-dimethylcyclohexene.
2. Suggest the structure of another compound that might be expected to give disproportionation via catalytic transfer hydrogenation.
3. In the reaction of 1-decene with cyclohexene, 1-decene acted as the hydrogen acceptor and cyclohexene was the hydrogen donor. Propose a reason the reaction proceeds this way rather than 1-decene acting as the hydrogen donor to form decadiene and cyclohexane.
4. Suggest a reason that oils are intentionally partially hydrogenated commercially rather than being totally hydrogenated.

5. Predict the product of catalytic transfer hydrogenation of methyl oleate with deuterated cyclohexene (C_6D_{10}).

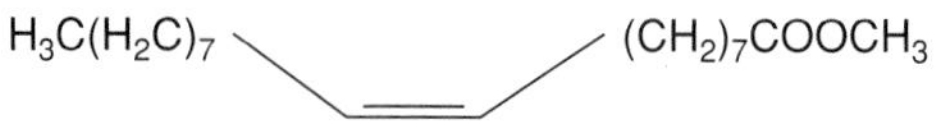

$$H_3C(H_2C)_7 \qquad\qquad (CH_2)_7COOCH_3$$

methyl oleate

6. Is the mechanism of heterogeneous catalytic hydrogenation the same as the mechanism of catalytic transfer hydrogenation? Explain.

7. Does this experiment effectively teach hydrogenation? Does it teach another concept in addition to hydrogenation? Explain.

Reference

Schoffstall, A., and Coleman, S. "Discovery Experiments Using Catalytic Transfer Hydrogenation," unpublished results.

Chapter 19

Carboxylic Acids

Carboxylic acids comprise the most commonly encountered class of organic acids. The pK_a values of carboxylic acids are in the 3–5 range, placing them in the weak acid category. Carboxylic acids are relatively strong when compared to the very weakly acidic character of a number of other classes of organic compounds. Phenols are somewhat less acidic than carboxylic acids. Others such as alcohols, ketones, and terminal alkynes are much less acidic. Alkyl and aryl acids, on the other hand, are strong sulfonic acids.

The relative acidities of carboxylic acids is an interesting topic focused on the effects of electron-donating and electron-withdrawing effects of nearby substituents on the acidity. This topic is examined in an analytical application involving the titration of carboxylic acids to determine pK_a as an option in Experiment 19.1. Later, in Chapter 28, carboxylic acids can also be characterized by titration to determine the molecular weight of an unknown acid.

The preparation of organic acids from Grignard reagents is a particularly useful method for adding a carbon atom and arriving at a carboxylic acid as the product. This is the topic of Experiment 19.1. Another method of synthesis of carboxylic acids is the oxidation of alcohols or alkylbenzenes with strong oxidants such as potassium permanganate, used in oxidation of toluene and xylenes in Experiment 11.1. An oxidative reaction resulting in loss of a carbon from the chain gives carboxylic acids via the iodoform reaction in Experiment 19.2.

Experiment 19.1: Synthesis and Identification of an Unknown Carboxylic Acid

Carboxylic acids can be prepared by reacting Grignard reagents with CO_2. In this experiment, you will:

- synthesize a carboxylic acid via the Grignard reaction of a substituted bromobenzene.

- identify the carboxylic acid from its melting point, from its acid dissociation constant (K_a), or its spectroscopic properties.

- determine the relationship between electron-donating properties and acidity of substituted benzoic acids.

Techniques

Technique F	Recrystallization and filtration
Technique G	Reflux
Technique I	Extraction
Technique M	Infrared spectroscopy
Technique N	Nuclear magnetic resonance spectroscopy

Background

Many naturally occurring carboxylic acids have biological and physiological activity, playing important roles as amino acids, fatty acids, metabolites, antibiotics, and vitamins. The structures of some biologically important carboxylic acids are shown here.

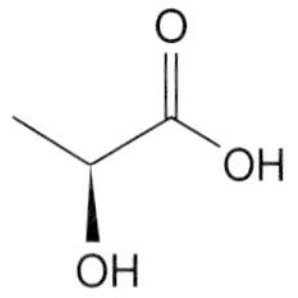

(S)-lactic acid 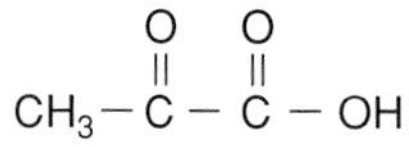   pyruvic acid 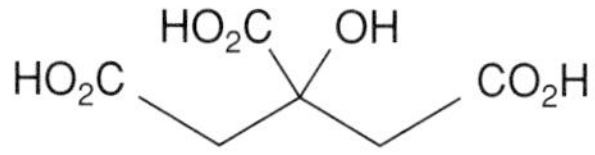   citric acid

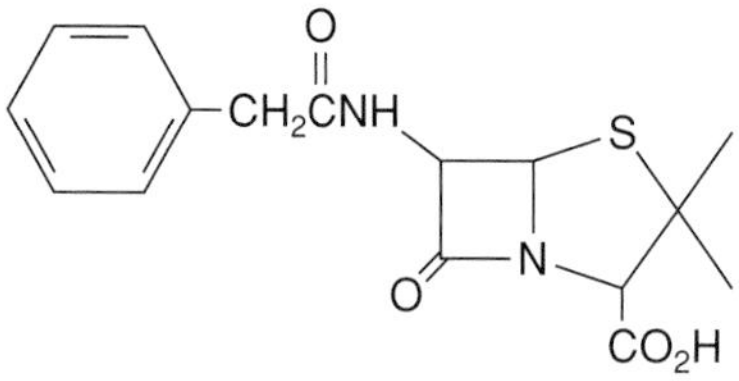

penicillin G 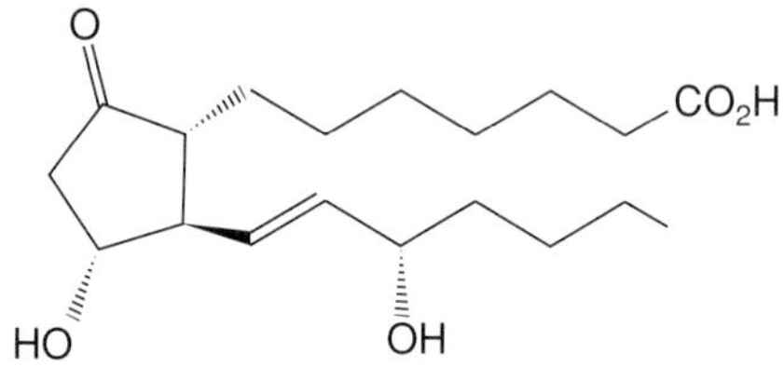   prostaglandin E₁

Carboxylic acids can be synthesized by oxidation of primary alcohols with Jones reagent or other aqueous oxidizing agent, by hydrolysis of the corresponding esters, amides, acyl halides, or nitriles, or by reaction of a Grignard reagent with carbon dioxide, followed by acidic hydrolysis. This latter reaction, which is the focus of this experiment, produces carboxylic acids containing one additional carbon. This is illustrated below for the reaction of phenylmagnesium bromide with carbon dioxide.

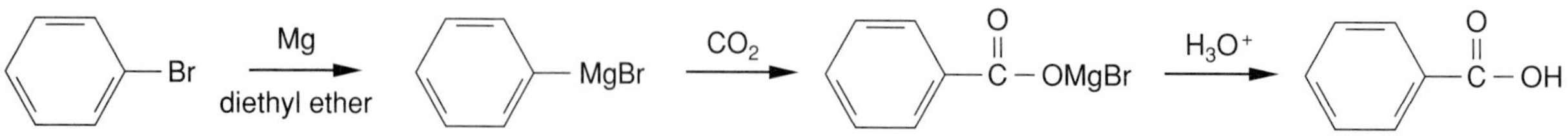

In this experiment, you will be given an unknown aryl bromide: *o*-, *m*-, or *p*-bromoanisole or *o*-, *m*-, or *p*-bromotoluene. You will prepare the Grignard reagent from the unknown bromobenzene and magnesium, react the Grignard reagent with dry ice (solid carbon dioxide), and then isolate and purify the product by acidic hydrolysis, followed by recrystallization. From the melting point and spectroscopic analysis, you will identify the unknown compound. The physical data for the carboxylic acid products are shown in Table 19.1-1.

Table 19.1-1 Physical Properties of Carboxylic Acids Prepared from Aryl Halides

Carboxylic acid	Properties of carboxylic acid	
	Melting point (°C)	pK_a
o-Anisic acid	98	8.2×10^{-5}
m-Anisic acid	106	8.2×10^{-5}
p-Anisic acid	182	3.3×10^{-5}
o-Toluic acid	103	1.3×10^{-4}
m-Toluic acid	108	5.4×10^{-5}
p-Toluic acid	180	4.2×10^{-5}

Carboxylic acids are weak acids that partially ionize in water to generate carboxylate anions and hydronium ions:

$$RCO_2H + H_2O \rightleftharpoons RCO_2^- + H_3O^+$$

The ionization (or acid dissociation) constant K_a for this reaction is a measure of the acid strength of the carboxylic acid. The stronger the acid, the farther the equilibrium lies to the right and the larger the K_a.

$$K_{eq}[H_2O] = K_a = \frac{[RCO_2^-][H_3O^+]}{[RCO_2H]}$$

The acidity of a carboxylic acid is dependent on a number of different factors including electronegativity of substituents, proximity of substituents to the carboxyl group, and number of substituents on the ring or carbon chain of the carboxylic acid. In this experiment, you will determine the pK_a of the acid. By comparing the value with pK_a values for other substituted aromatic carboxylic acids, you will deduce how substituents affect acidity. Alternatively, you may be given an unknown carboxylic acid to titrate.

Prelab Assignment

1. Write a detailed mechanism for the reaction of phenylmagnesium bromide with carbon dioxide, showing clearly the direction of electron flow.
2. Write a detailed flow scheme for the workup procedure of this reaction.
3. One of the by-products of the reaction is a coupling product formed from the reaction of the Grignard reagent with the aryl halide. Write the product of this reaction using bromobenzene as the aryl halide and explain where in the procedure the product would be separated from this by-product.
4. Identify regions of the IR spectrum that could be used to differentiate between *ortho*, *meta*, and *para*-substituted benzoic acids.
5. Explain how the NMR spectrum could be useful in distinguishing between the toluic acids and the anisic acids.
6. The K_a can be determined by titrating a solution of the carboxylic acid with standardized base and plotting the pH of the solution as a function of volume of added base. At exactly half the equivalence point volume, the pH of the solution is equal to the pK_a of the carboxylic acid. Explain this relationship mathematically and show how the K_a can be derived from the pK_a.

Experimental Procedure

1. Always wear eye protection in the laboratory.

2. Diethyl ether and petroleum ether are extremely volatile and flammable. Make certain that there are no flames anywhere in the lab when these solvents are being used. Use in a well-ventilated area and avoid breathing the vapors. Petroleum ether may be harmful if inhaled or absorbed through the skin. Avoid prolonged contact or inhalation.

3. Substituted aryl halides are toxic. Avoid contact with the liquids and do not breathe their vapors.

4. Dry ice causes severe burns. Wear insulated protective gloves when handling pieces of dry ice.

Miniscale Grignard Synthesis of Substituted Benzoic Acids

All glassware must be dried in an oven at 110°C for at least 1 hour prior to use. Do not put plastic parts in the oven. Dry a 10-mL round-bottom flask, a 5-mL conical vial, a water-jacketed condenser, a drying tube, and a Claisen adapter. Allow the glassware to cool to room temperature. Assemble glassware as pictured.

Use dry glassware.

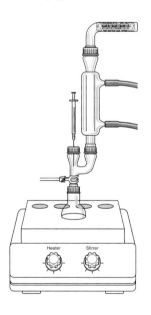

Weigh 190 mg of magnesium turnings and place in a dry 10-mL round-bottom flask containing a spin bar. Add a tiny crystal of iodine to the flask. In a dry 5-mL conical vial, dissolve 900 µL of the aryl halide in 4.0 mL of anhydrous diethyl ether from a freshly opened can. Cap the vial. Draw 0.2 mL of this solution into a dry syringe and add it dropwise through the septum into the flask containing the magnesium and iodine. Cloudiness and bubbles at the surface of the metal are indicators that the reaction has begun. This may take 5 minutes. If no reaction is observed at the end of 5 minutes, warm the flask with the palm of your hand or try adding another crystal of iodine. If the reaction still hasn't started, consult the instructor.

Once the reaction has started, add the remainder of the bromobenzene/diethyl ether solution dropwise over 5–10 minutes. Rinse the 5-mL conical vial with 1.2 mL of anhydrous diethyl ether and add the rinse to the reaction flask. Reflux the reaction mixture gently, with stirring, in a sand bath or heat block for 15 minutes after the addition of the aryl halide is complete. The final Grignard solution should be cloudy and most of the magnesium should be gone. Cool to room temperature and add more anhydrous diethyl ether as necessary to maintain the original volume. The Grignard reagent must be used immediately for the next step.

Transfer the solution of the Grignard reagent as quickly as possible over about 2 g of crushed dry ice in a 25-mL beaker. The dry ice sublimes (goes directly from the solid to the gas phase), so weigh it immediately prior to use. Rinse the flask with 2 mL of diethyl ether and add it to the beaker. Cover the beaker with a watch glass until the excess dry ice has sublimed.

Cool the beaker in an ice bath. Then hydrolyze the carboxylate salt by slowly adding 1.6 mL of 6 M HCl to the viscous solid while continually stirring. Work under the hood. Any excess magnesium metal will vigorously react. Magnesium pieces that do not react can be removed from the beaker with tongs. Check the pH of the aqueous phase to make sure that the solution is acidic. It is important that all solid be dissolved, so add more diethyl ether and/or more 6 M HCl as necessary in order to dissolve all the solid. Transfer the two-phase system to a centrifuge tube and shake it until all of the solid dissolves. Vent frequently to avoid pressure buildup. Remove the aqueous layer and set aside for later disposal.

The ether layer contains the benzoic acid and any coupling by-product such as biphenyl. Wash the organic layer with 1.0 mL of 10% sodium hydroxide to convert the ether-soluble benzoic acid to the water-soluble sodium benzoate. Remove the aqueous layer to a clean beaker. Add another 1.0-mL portion of 10% sodium hydroxide to the ether layer remaining in the centrifuge tube. Combine the aqueous layer with the previous aqueous layer in the clean beaker.

The aqueous layer will contain some dissolved diethyl ether. Heat the aqueous solution on a hot plate for a few minutes to evaporate the ether. Let the solution cool to room temperature. While stirring, add 6 M HCl dropwise to precipitate the benzoic acid. Check the pH to make sure that it is strongly acidic. Cool the mixture in an ice bath, then suction filter. Wash the crystals with several small portions of cold water.

The crude solid can be recrystallized from water. Let the crystals air dry thoroughly before taking a melting point. If instructed to do so, you will titrate a solution of the acid and determine the K_a. Otherwise, you should characterize the carboxylic acid and identify it from its melting point and spectroscopic analysis.

Optional Titration of Substituted Carboxylic Acid: Rinse and fill a burette with standardized 0.100 M NaOH. Adjust the level of NaOH until the meniscus is at 0.00 mL. (This makes it easier to work up the data.) Calibrate the pH meter to pH = 4.00 and pH = 7.00 with the standardized buffer solutions. The instructor will demonstrate this procedure. Weigh about 50 mg of the synthesized aromatic acid or other acid assigned by the instructor. Place the acid in a 100-mL beaker. Dissolve the acid in 15 mL of methanol. When the acid is dissolved, add 25 mL of distilled water. Don't worry if the solution looks cloudy; the acid will dissolve as the titration proceeds. Put a stir bar in the beaker. Immerse the electrode of the pH meter into the solution so that the tip of the electrode is covered by the solution. **Make certain that the tip of the electrode is covered by solution and does not come in contact with the spin bar.** Gently start the mixture stirring.

Record the initial pH in a table. Add small aliquots (3–4 drops) of the NaOH solution. After each addition, record the pH and the total volume of NaOH added. When close to the equivalence point (how will you know?), add the NaOH in 1–2 drop increments until the equivalence point is reached. After the equivalence point, add the NaOH in 3–4 drop increments until the pH of the solution is fairly constant. Discard the titrated acid sample down the drain. If time permits, repeat the titration with a second sample of the acid.

Characterization

Melting Point: Determine the melting point of the dry solid.

Infrared Spectroscopy: Run an IR spectrum of the product in either dry KBr or Nujol. Interpret the spectrum. The spectrum should show broad OH stretching around 3300–2400 cm^{-1} and carbonyl stretching around 1690 cm^{-1}. Several absorption bands may be helpful in differentiating between compounds with similar melting points: *m*-anisic acid has a sharp absorption band at 1044 cm^{-1}; *m*-toluic acid lacks this band, but shows absorption at 1167 cm^{-1}, while *o*-toluic acid lacks both of these absorption bands.

NMR Spectroscopy: Run an NMR spectrum of the product from $\delta 0$–12 and interpret the spectrum. The spectrum should show strong absorption around $\delta 7$ for the aromatic region and broad absorption around $\delta 11$–12 for the carboxylic acid proton.

Results and Conclusions

1. Determine the identity of the aromatic acid, using the melting point and spectro-scopic analysis.
2. Calculate the yield of the carboxylic acid.
3. Indicate how the IR spectrum was helpful in supporting the structure of the aro-matic acid.
4. Indicate how the NMR spectrum was useful in identifying the unknown.
5. (Optional) Make a plot of pH versus total volume of NaOH added. From the graph, find the equivalence point and the pH at one-half the equivalence point. Calculate the K_a of the acid. Calculate the percent error between the experimental K_a and the literature value. Suggest reasons for any discrepancies.

Cleanup & Disposal

Neutralize any excess NaOH with 10% HCl and wash the solution down the drain with water. Solutions of acids containing a halogen should be placed in a container labeled "halogenated organic waste." Other solutions of acid should be placed in a container labeled "nonhalogenated organic waste."

Critical Thinking Questions

1. Indicate which carboxylic acid in each group below is the strongest acid and explain why:
    a.   2-fluoroacetic acid, 2-bromoacetic acid, 3-fluoropropanoic acid.
    b.   benzoic acid, *p*-methylbenzoic acid (*p*-toluic acid), *p*-fluorobenzoic acid.
    c.   *p*-nitrobenzoic acid, 3,5-dinitrobenzoic acid, benzoic acid.

2. The acidity of *o*-hydroxybenzoic acid is 100 times greater than *p*-hydroxybenzoic acid. Explain.

3. Suppose that the pH meter consistently reads 0.5 pH units too high. How will this error affect:
    a.   the equivalence point?
    b.   the pK_a of the acid?
    c.   the pK_b of the base?

4. Rank the following anions in terms of decreasing base strength (strongest base = 1). Explain.

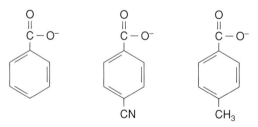

5. Referring to the procedure for the synthesis of substituted benzoic acids, write balanced chemical reactions to explain why:
    a.   HCl is added to the product of the reaction of the Grignard reagent with carbon dioxide.
    b.   The ethereal layer is extracted with aqueous sodium hydroxide.
    c.   HCl is added to the aqueous layer.

6. Benzene is a possible side-product when reacting phenylmagnesium bromide. Explain how benzene could be produced. Using this mechanism, draw the structure of the side-product that could be produced when using 2-methoxyphenylmagnesium bromide.

7. Classify the methoxy group, the methyl group, and the nitro group as electron-donating or electron-withdrawing. Using the literature K_a values found in Table 19.1-1 and the fact that the K_a value for *p*-nitrobenzoic acid is 3.2×10^{-4}, rank *p*-nitrobenzoic acid, benzoic acid, *p*-methoxybenzoic acid, and *p*-toluic acid in terms of decreasing acidity. Which substituent(s) increase the acidity relative to that of benzoic acid? Which substituent(s) decrease the acidity relative to benzoic acid? Explain how electron-withdrawing properties and electron-donating properties of the functional group affect the acidity of *para*-substituted benzoic acids relative to benzoic acid.

8. A student wished to prepare a Grignard reagent from *p*-bromophenol. The student carefully dried all the glassware and added the right reagents in the right amounts. However, after working up the reaction mixture, the student discovered the "product" was actually unreacted starting material. Explain why this experiment was unsuccessful.

9. Each of the following conversions can be accomplished using one or more of the methods used to synthesize a carboxylic acid (oxidation of a primary alcohol, hydrolysis of a nitrile, or Grignard reaction). Devise a synthesis of the indicated product and explain which method or methods could accomplish the desired transformation. Note that the conversions may require more than one step.

   a. 1-butanol to pentanoic acid

   b. 1-butanol to butanoic acid

   c. bromobenzene to benzoic acid

   d. *p*-hydroxybenzyl bromide to *p*-hydroxyacetic acid

   e. *p*-bromobenzyl bromide to *p*-bromoacetic acid

10. *Molecular modeling assignment:* Construct models of the benzoate anion and the *p*-nitrobenzoate anion. Then construct electron density maps for each anion. What conclusions can be drawn by comparing the electron density maps of the two compounds? Based upon these results, what conclusions can be drawn about the strengths of the conjugate acids?

Experiment 19.2: Synthesis of *trans*-Cinnamic Acid Via the Haloform Reaction

The haloform reaction is used to synthesize a carboxylic acid from a methyl ketone. In this experiment, the oxidizing agent is commercial bleach with hypochlorite as the active agent. You will:

♦ carry out a haloform reaction using hypochlorite as oxidizing agent.

♦ oxidize a methyl ketone to a carboxylic acid.

Techniques

Technique F	Vacuum filtration
Technique I	Extraction
Technique M	Infrared spectroscopy
Technique N	Nuclear magnetic resonance spectroscopy

Background

Carboxylic acids may be synthesized from methyl ketones by commercial bleach (Clorox). Bleach contains 5% sodium hypochlorite (NaOCl), which acts as an oxidizing agent. In solution, hypochlorite is in equilibrium with chlorine and sodium hydroxide:

$$NaOCl + H_2O + NaCl \;\rightleftharpoons\; 2\,NaOH + Cl_2$$

The oxidation reaction of a methyl ketone is shown here.

$$R-\overset{\overset{\displaystyle O}{\|}}{C}-CH_3 + NaOCl(Cl_2, NaOH) \longrightarrow CHCl_3 + R-\overset{\overset{\displaystyle O}{\|}}{C}-O^-Na^+ \xrightarrow{H^+} R-\overset{\overset{\displaystyle O}{\|}}{C}-OH$$

The reaction mechanism involves removal of a relatively acidic α-hydrogen from the methyl group by hydroxide to afford a resonance-stabilized enolate, which subsequently reacts with chlorine to give an α-chloroketone.

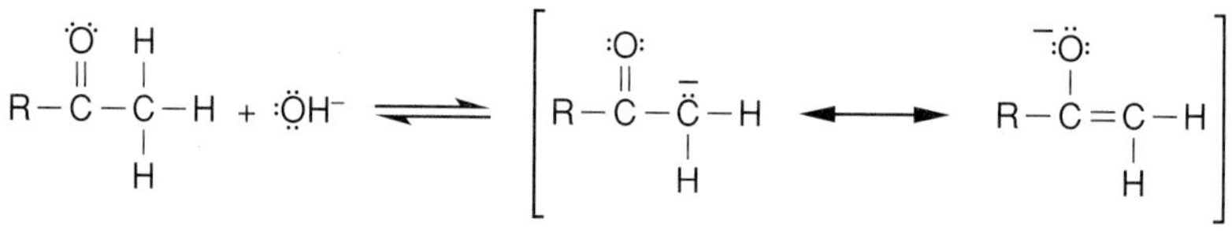

resonance-stabilized enolate

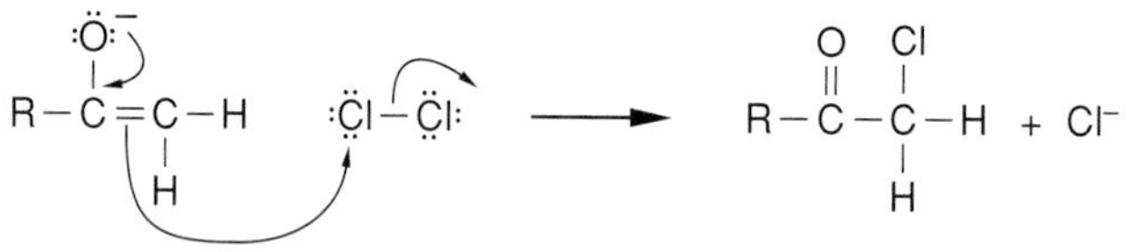

The process is repeated twice to give a trichloromethyl ketone. In the next stage, base adds to the ketone at the carbonyl carbon to yield a tetrahedral intermediate. The carbonyl π-bond is reestablished generating a carboxylic acid. The trichloromethyl carbanion is expelled.

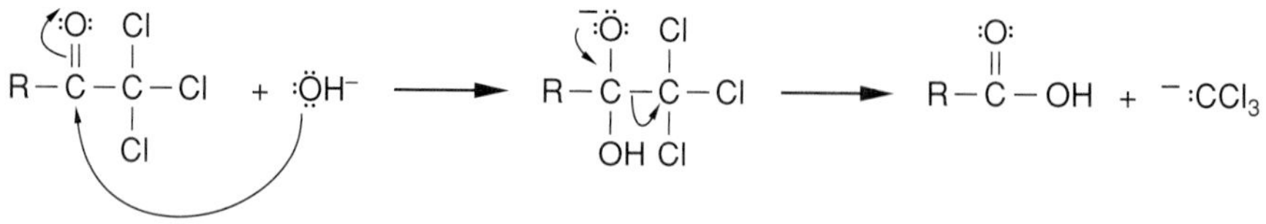

The anion abstracts an acidic hydrogen to form chloroform and the carboxylic salt. Acidification leads to the carboxylic acid.

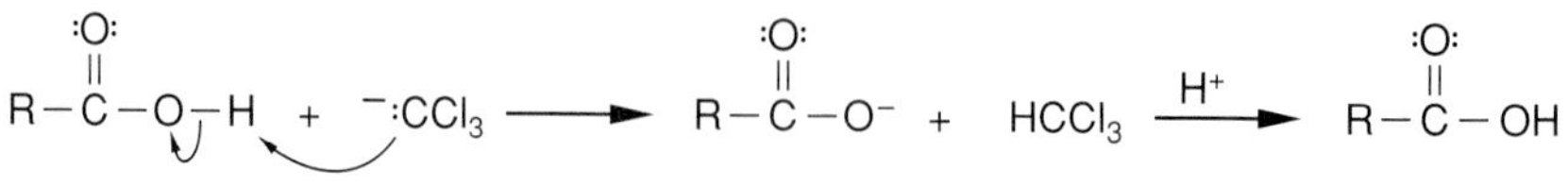

In this reaction, the haloform reaction will be used to synthesize *trans*-cinnamic acid from *trans*-4-phenyl-3-buten-2-one.

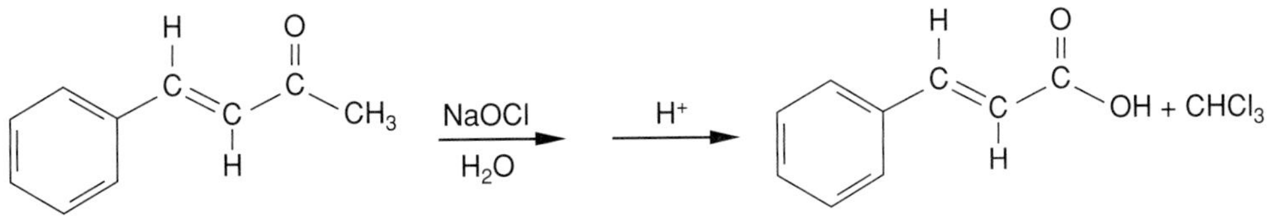

trans-4-phenyl-3-buten-2-one   trans-cinnamic acid

Prelab Assignment

1. Write a detailed reaction mechanism of the haloform reaction of *trans*-4-phenyl-3-buten-2-one.
2. Write a flow scheme for the reaction and workup procedure for this experiment.

Experimental Procedure

1. Wear eye protection at all times in the laboratory.
2. Concentrated NaOH is corrosive and toxic. Wear gloves when handling this reagent. Immediately notify the instructor about spills on the bench top.
3. *trans*-Cinnamic acid is an irritant. Wear gloves and avoid skin contact.
4. Avoid contact with bleach.

Microscale Synthesis of trans-Cinnamic Acid

Place 150 mg of *trans*-4-phenyl-3-buten-2-one (benzalacetone) in a 5-mL conical vial equipped with a spin vane. Add 1 mL of THF and stir to dissolve. Add 3.0 mL of 5% sodium hypochlorite (fresh Clorox or similar bleach) and 2 drops of 50% NaOH. Fit a water-jacketed condenser on the vial. Turn on the water to the condenser, vigorously stir, and gently heat the mixture for 30 minutes using a warm sand bath or heat block set at the lowest setting. Do not let the solution get too hot. Cool the flask to room temperature and transfer the contents to a centrifuge tube. Add 25 mg of sodium sulfite to remove any unreacted oxidizing agent and swirl briefly. Add 4 mL of diethyl ether to the tube and 1.5–2.0 mL of water. Cap the tube and shake. If necessary, add more water to help break up the emulsion. Transfer the bottom aqueous layer to a clean 10-mL Erlenmeyer flask. Properly dispose of the diethyl ether layer, which contains the chloroform by-product. Add 3 M HCl dropwise to the aqueous solution to pH 3 to precipitate the product. Collect the solid by vacuum filtration using a Hirsch funnel. Wash with cold water. Recrystallize from ethanol/water. Weigh the product.

Caution

Characterization

Melting Point: Determine the melting point of the product. The reported melting point of *trans*-cinnamic acid is 133–134°C.

   Infrared Spectroscopy: Obtain an IR spectrum of the product using either the KBr or the Nujol mull technique and interpret the spectrum noting the key bands characteristic of the functional groups present. The product should show broad absorption around

3400 to 3000 cm^{-1} and a carbonyl stretch around 1685 cm^{-1}. The IR spectrum of *trans*-cinnamic acid is shown in Figure 19.2-1.

NMR Spectroscopy: Dissolve the product in CDCl$_3$ and record the ^{1}H NMR spectrum from 0–13 ppm. The product should show a broad singlet around δ12 for the carboxylic acid proton, two doublets for the alkenyl protons at δ6.4 and δ7.8, and a complex multiplet for the aromatic protons around δ7.5. The ^{1}H NMR and ^{13}C NMR for *trans*-cinnamic acid are shown in Figures 19.2-2 and 19.2-3.

Results and Conclusions

1. Determine the percent yield of the product.
2. Measure the coupling constant between the alkenyl protons in the ^{1}H NMR and determine the stereochemistry of the double bond.

Figure 19.2-1

IR spectrum of *trans*-cinnamic acid

SOURCE: ©BIO-RAD Laboratories, Sadtler Division.

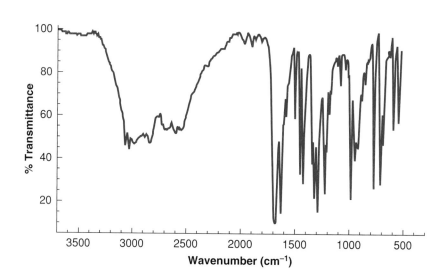

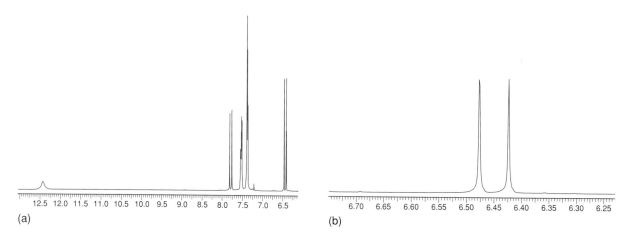

(a)   (b)

Figure 19.2-2 ^{1}H NMR spectra of (a) *trans*-cinnamic acid (CDCl$_3$); (b) *trans*-cinnamic acid (expanded region) (CDCl$_3$)

SOURCE: Reprinted with permission of Aldrich Chemical.

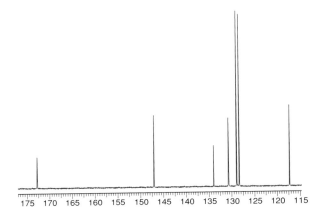

Figure 19.2-3

^{13}C NMR spectrum of *trans*-cinnamic acid (CDCl$_3$)

SOURCE: Reprinted with permission of Aldrich Chemical.

Cleanup & Disposal

The organic extract (diethyl ether and chloroform) should be placed in a container labeled "halogenated organic solvent waste." The aqueous filtrate should be neutralized with 10% sodium bicarbonate and then can be washed down the drain with water.

Critical Thinking Questions *(The harder one is marked with a ❖.)*

1. What would be the product of the haloform reaction on 2-pentanone using bromine and sodium hydroxide?

❖ 2. Dibenzoylmethane is not a methyl ketone but it gives a positive haloform test. Two molecules of benzoic acid are ultimately formed as well. Explain this result mechanistically.

3. Acetaldehyde and ethanol also give a positive haloform test. Explain.

4. Normally, carbon-carbon bonds are not easily broken. Explain why the trichloromethyl carbon-carbon bond in the tetrahedral intermediate is so easily broken.

5. Explain why the haloform reaction only works for methyl ketones and not for compounds such as 3-pentanone.

6. Predict the dibromination product of acetone and bromine in acidic solution and in basic solution. Propose mechanisms to explain why different products are obtained.

Chapter 20

Carboxylic Acid Esters

E sters are fragrant derivatives of carboxylic acids that are usually made by Fischer esterification or by reacting acid chlorides with alcohols. Esters are very common in nature, being present in many flowers and fruits. Some have familiar odors such as pineapple, banana, apple, and others.

The preparation of esters and their fruity properties are the topics of Experiment 20.1. The design of a combinatorial method of analysis takes advantage of differing odors to identify a particular ester to synthesize among a group of possible candidates. Like ketones, esters have relatively acidic α-carbons. This property is useful if there are two carboxyl groups attached to an α-carbon as in the synthesis starting with β-carbonyl esters in Experiment 21.1.

In addition to direct preparation from another carboxylic acid or carboxylic acid derivative, esters can be synthesized from ketones by a reaction known as the Baeyer-Villiger rearrangement in Experiment 20.2. The outcome of this reaction may be visualized by taking an oxygen from a peroxide reagent and inserting it between a carbonyl carbon of a ketone and one of its α-carbons. The reaction is useful because ketones are generally difficult to oxidize.

Experiment 20.1: Combinatorial Chemistry and the Synthesis of Fruity Esters

Carboxylic acid esters are frequently used as flavorants and perfumes. In this experiment, you will use combinatorial chemistry to synthesize an ester that produces an assigned scent. You will:

♦ mix combinations of alcohols and carboxylic acids to produce an assigned scent.

♦ devise a procedure for the synthesis of the ester.

♦ analyze the structure of the ester using spectroscopic techniques.

Techniques

Technique G	Reflux
Technique M	Infrared spectroscopy
Technique N	Nuclear magnetic resonance spectroscopy

Background

Esters are found in many natural products, contributing to the aromas of bananas, oranges, pineapples, and other fruits. The structure of the ester determines its scent. By reacting different alcohols and carboxylic acids you can produce esters of varying scents. In this experiment, you will be assigned a specific scent to prepare and you will experiment with various alcohols and carboxylic acids in order to produce the appropriate ester. You will combine five alcohols and three carboxylic acids in order to find the right combination to produce the assigned scent.

If you were to try one combination of alcohol and carboxylic acid, synthesize the product, test the scent, and then continue to try other combinations, many different experiments might have to be performed. Combinatorial chemistry is a technique used in the pharmaceutical industry to reduce the time required to identify a potential drug. Using combinatorial techniques, many small experiments are run all at one time, so the researcher can quickly screen the combinations of reagents that produce the desired physical or biochemical properties. In the pharmaceutical world, the property of interest might be enzyme activity or toxicity. In this experiment, the property to be selected will be odor. You will run test tube reactions to find the combination of alcohol and ester that produces the assigned scent. You will then design a larger scale synthesis to produce 100 mg of the given ester.

There are many different synthetic methods for synthesizing esters, but two of the most common are acid-catalyzed Fischer esterification of a carboxylic acid and alcohol and condensation of an acid chloride with an alcohol or alkoxide. These are described briefly below.

The Fischer esterification is an equilibrium reaction in which an acid and an alcohol combine to produce the ester and water. The reaction, which is acid catalyzed, is illustrated below for the formation of ethyl acetate from acetic acid and ethanol:

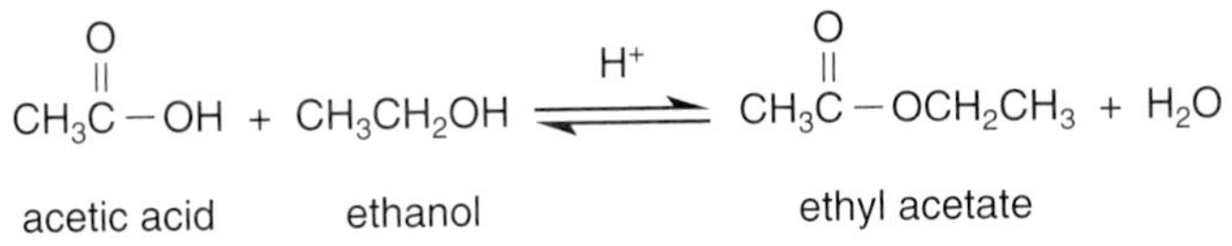

$$
\underset{\text{acetic acid}}{CH_3\overset{\displaystyle O}{\overset{\|}{C}}-OH} + \underset{\text{ethanol}}{CH_3CH_2OH} \underset{\xrightarrow{\hspace{1cm}}}{\overset{H^+}{\rightleftharpoons}} \underset{\text{ethyl acetate}}{CH_3\overset{\displaystyle O}{\overset{\|}{C}}-OCH_2CH_3} + H_2O
$$

To drive the equilibrium toward completion, either the carboxylic acid or the alcohol is used in excess. Alternately, if the ester has a significantly different boiling point than the alcohol or acid, the ester can be separated from the acid and alcohol by distillation.

A second synthetic route to esters is through an acid chloride. This method can be used for any carboxylic acid, but it is especially good for malodorous acids, such as butanoic acid. Workup and purification are simplified, since only stoichiometric amounts of the alcohol can be used. This method cannot be used to make esters of formic acid, since formyl chloride is unstable. Acid chlorides are very reactive, so moisture must be rigorously excluded from the reaction to avoid hydrolysis of the acid chloride. A tertiary amine base, such as pyridine or N,N-diethylaniline, is added to neutralize the HCl formed in the reaction. The preparation of ethyl acetate from acetyl chloride and ethanol is illustrated below:

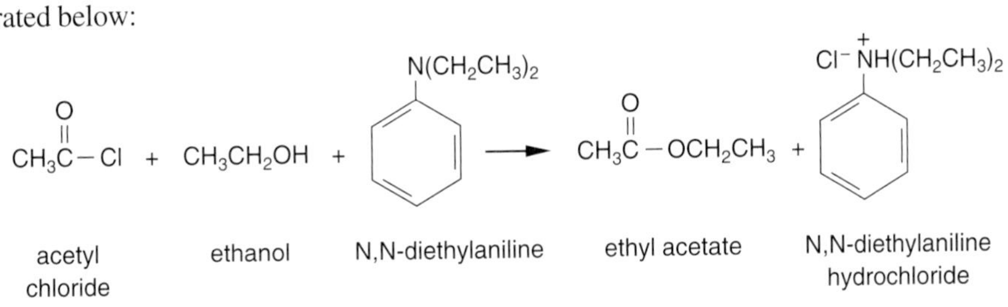

$$
\underset{\substack{\text{acetyl}\\\text{chloride}}}{CH_3\overset{\displaystyle O}{\overset{\|}{C}}-Cl} + \underset{\text{ethanol}}{CH_3CH_2OH} + \underset{\text{N,N-diethylaniline}}{\boxed{N(CH_2CH_3)_2}} \longrightarrow \underset{\text{ethyl acetate}}{CH_3\overset{\displaystyle O}{\overset{\|}{C}}-OCH_2CH_3} + \underset{\substack{\text{N,N-diethylaniline}\\\text{hydrochloride}}}{\boxed{Cl^-\overset{+}{N}H(CH_2CH_3)_2}}
$$

The instructor will assign each student or group of students a specific scent to be identified. Although it is generally poor (and unsafe!) laboratory technique to smell chemicals, the reagents for this experiment have been chosen because they are nontoxic and relatively nonhazardous. You will use the methanol, ethanol, 1-pentanol, 3-methyl-1-butanol, and 1-octanol as the alcohols and butanoic acid, propanoic acid, and acetic acid as the carboxylic acids.

Prelab Assignment

1. Write a detailed mechanism for a) the Fischer esterification of acetic acid with ethanol in the presence of sulfuric acid and b) the reaction of acetyl chloride with ethanol.
2. In this experiment, five different alcohols and three different carboxylic acids will be used. How many different esters could be synthesized from these starting materials?
3. Prepare a table to record the results of the scents for all combinations of alcohols and carboxylic acids.
4. Explain which reagent is used in excess and provide a reasonable explanation.
5. Would using more sulfuric acid cause the reaction to occur faster? Use the mechanism to explain your answer.

Experimental Procedure

Safety First!

Always wear eye protection in the laboratory.

1. Wear eye protection at all times in the laboratory.
2. Sulfuric acid is very caustic. Use gloves when handling this reagent.
3. Acid chlorides are corrosive. On contact with water, they produce HCl. Work under the hood or in a well-ventilated area and wear gloves.
4. Sulfuric acid is a highly toxic oxidizing agent. It will cause severe burns. Wear gloves and goggles when handling this reagent.
5. Many carboxylic acids have noxious odors. Work under the hood or in a well-ventilated area.

Part A: Combinatorial Selection

Before starting the experiment, each student or group of students will be assigned a specific scent to prepare. Although students may work individually, this experiment is best performed in groups of three, with each student testing a carboxylic acid with all five of the alcohols. Obtain a test tube rack and the appropriate number of 10×13 test tubes. Label each test tube at the top with the names or codes for the alcohol-carboxylic acid combination.

Set a water bath to 80–95°C and let it stand for 15–20 minutes to come to thermal equilibrium. (Alternately, an 800-mL beaker and a 250-mL beaker can be used. The smaller beaker should be inverted inside the larger beaker and the larger beaker filled with water. The individual test tubes will fit around the 250-mL beaker without tipping over.)

In the appropriate test tube, mix 10 drops (0.5 mL) of the carboxylic acid and 20 drops (1 mL) of the alcohol. Carefully add 5 drops of concentrated sulfuric acid to the test tube. Swirl the contents carefully to mix and place a boiling chip in each test tube. Carefully place the test tubes in the water bath and heat for 10 minutes. The test tubes may be heated all at one time or you may elect to heat one at a time. However the experiment is conducted, you should carefully observe the reaction mixtures to make sure that the liquids do not boil away. If this occurs, redo that specific combination and heat at a lower temperature.

After the heating period is over, remove the test tubes from the hot water bath and allow the liquids to cool to room temperature. Remove each test tube and check the odor by wafting your hand over the top of the test tube. Alternately, pour a little of the liquid in the test tube onto a watch glass. Record the observed scents for all of the assigned combinations in the table.

Results and Conclusions for Part A

1. Write a balanced reaction for the combination of alcohol and carboxylic acid that produced the assigned scent.
2. Write the structures of the esters formed by the other combinations of alcohols and carboxylic acids test. Indicate the observed scent for each of the esters.
3. For the combination of reagents that produced the assigned scent, look up the physical properties of the starting carboxylic acid, alcohol, and ester and determine which of the two methods (Fischer esterfication procedure or acid chloride esterification procedure) would be most appropriate. Write a paragraph describing why the selected procedure was chosen.
4. Design an experimental plan to prepare 100 mg of the assigned ester using either the Fischer esterfication procedure or the acid chloride procedure described in Part B. The plan should take into account the physical properties (including boiling points and solubilities) of the starting materials and products in the reaction and workup.
5. Decide which of the characterization methods will provide the most useful information about the structure of the ester synthesized. Write a short summary justifying the characterization methods selected.
6. Write a flow scheme to show how the product will be isolated during the workup procedure.
7. Explain how the product could be characterized using infrared spectroscopy, ¹H NMR spectroscopy, and/or ¹³C NMR spectroscopy. List the specific absorption bands in the IR spectrum and the expected chemical shifts and splitting patterns in the NMR spectra that would aid identification of the product.

Part B: Microscale Synthesis of an Ester ✓

Based upon the physical properties of the starting alcohol, carboxylic acid, and ester, select one of the methods below and devise a microscale synthesis of the assigned ester. Have the instructor approve the procedure before starting the experiment. Carry out the synthesis, then isolate, purify, and characterize the product.

General Procedure 1: Microscale Fischer Esterification. Place 2.5 mmol of the carboxylic acid, 1.25 mmol of the alcohol, and 1 drop of concentrated sulfuric acid in a 5-mL conical vial fitted with a water-jacketed reflux condenser and spin vane. Start water flowing through the reflux condenser and heat the solution to reflux using a heat block or sand bath. Reflux for 30 minutes. At the end of the reflux period, cool the solution in an ice bath. Add 1 mL of cold water and 1 mL of diethyl ether. Cap the vial and shake to mix the layers, venting frequently. Allow the layers to separate. Remove the aqueous layer to an Erlenmeyer flask for disposal. Add 0.5 mL of 5% sodium bicarbonate to the vial and swirl gently. Do not cap the vial until gas evolution has ceased. Cap the vial, and shake to mix the layers, venting often. Remove the lower aqueous layer to the flask. Repeat the extraction with a second 0.5-mL portion of 5% sodium bicarbonate. Cap the

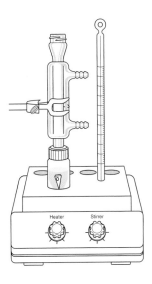

vial and shake to mix the layers, venting often. Remove the lower aqueous layer to the flask. Check the pH of the organic layer remaining in the vial. If the solution is still acidic, continue extraction with sodium bicarbonate until the solution is neutral to pH paper. Dry the organic layer over anhydrous sodium sulfate. Use a filter pipet to transfer the dried solution to a tared Erlenmeyer flask or vial. Evaporate the diethyl ether on a warm sand bath. Cool to room temperature. Weigh the flask and determine the yield of crude ester.

The crude product may contain both ester and any unreacted alcohol that was not extracted with the water. If the ester and alcohol are relatively low boiling (<150°C), the purity of the ester can be determined by gas chromatography. Alternatively, the reaction mixture can be analyzed by thin-layer chromatography using anisaldehyde or vanillin spray to visualize the compounds. If the product mixture contains a significant amount of the alcohol, the mixture may be distilled using a Hickman still or the ester may be separated from the more polar alcohol by chromatography on a silica gel column using hexane as an eluent. If there is not a significant amount of alcohol remaining, the crude mixture may be analyzed without further purification.

Use dry glassware.

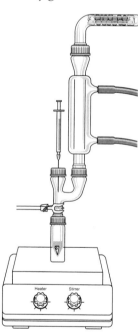

General Procedure 2: Microscale Esterification Using an Acid Chloride.   All of the equipment must be dried in an oven to remove traces of water prior to starting the experiment. Remove the glassware from the oven and let it cool to room temperature. Mix 1.0 mmol of the alcohol, 1.0 mmol of N,N-diethylaniline, and 0.2 mL of dry anhydrous diethyl ether in a dry 3-mL conical vial, fitted with a spin vane, Claisen adapter, and septum. Cool the solution in an ice bath. Using a syringe, slowly add 1.0 mmol of the appropriate acid chloride while swirling the vial. After addition is complete, fit the vial with a water-jacketed reflux condenser and reflux for 30 minutes using a warm sand bath or heat block. Cool the solution in an ice bath. Solid N,N-diethylaniline hydrochloride will precipitate. Use a filter pipet to transfer the ethereal solution to a clean vial. Add 0.2 mL of 5% HCl and swirl. Cap the vial and shake to mix, venting frequently. Transfer the lower aqueous layer to flask for disposal. Wash the organic layer with 0.2 mL of water and then with 0.2 mL of brine, removing the aqueous layers each time. Dry the ethereal solution over anhydrous sodium sulfate. Use a filter pipet to transfer the dried solution to a clean, tared vial. Remove the ether by evaporation on a warm sand bath to leave an oily residue. Determine the mass of the crude product. The crude product may be purified by column chromatography using hexane as the eluent, or it may be analyzed without further purification.

Characterization for Part B

Thin-layer Chromatography: Dissolve a small amount of the product mixture in methylene chloride and spot the solution on a silica gel TLC plate. Also spot solutions of the alcohol and carboxylic acid starting materials. Develop the plate in hexanes. Visualize the spots by spraying the dried plate with anisaldehyde spray. (Anisaldehyde forms colored complexes with oxygen-containing compounds.)

Gas Chromatography: Inject 1 μL of the crude reaction mixture on a nonpolar column (silicone or Apiezon column). Inject standards of the alcohol and the pure ester, if available. Do not inject the acid chloride or the carboxylic acid onto the column.

Infrared Spectroscopy: Record the IR spectrum of the product, noting characteristic absorbances of carboxylic esters, such as the carbonyl and C−O stretch. Analyze the spectrum for the presence of any starting materials, such as the alcohol and carboxylic acid.

NMR Spectroscopy: Record the ^{1}H NMR and/or ^{13}C NMR spectra of the product. Integrate the signals and correlate the structure of the ester with the spectrum.

Results and Conclusions for Part B

1. Calculate the percent yield of the ester.
2. Justify the structure of the ester, based on IR spectrum, NMR spectra, and TLC and/or GC analysis.
3. How effective was the chosen procedure in preparing the ester? Were there any obvious problems with the method chosen? Explain.
4. Write a detailed mechanism for the selected esterification method.

Cleanup & Disposal

The contents of the test tubes should be placed in a waste container. The aqueous filtrate may be washed down the drain with water. Chromatography solvents should be placed in a container labeled "nonhalogenated organic solvent waste."

Critical Thinking Questions

1. The ester responsible for the aroma of fresh raspberries is 2-methylpropyl formate. Propose a synthesis of this ester.
2. The aroma of rum is due to ethyl formate. Propose a synthesis of this ester.
3. A lactone is a cyclic ester. Propose a structure of the lactone derived from:
   a. 4-hydroxybutanoic acid
   b. 5-hydroxy-3,5-dimethylhexanoic acid
4. Acid chlorides can be prepared from carboxylic acid and thionyl chloride ($SOCl_2$). In the reaction, SO_2 and HCl are produced. Propose a mechanism for the formation of propanoyl chloride from propanoic acid and thionyl chloride.
5. Give the product of the reaction of butanedioic acid with two moles of 1-butanol.
6. In the Fischer esterification procedure, the ratio of alcohol to carboxylic acid was 1:2, while in the procedure using the acid chloride, the ratio of alcohol to carboxylic acid was 1:1. Explain why the procedures used differing amounts of the alcohol.
7. In the reaction of an acid chloride with an alcohol, a stoichiometric amount of N,N-diethylaniline was added to the reaction mixture. Write a reaction to show the purpose of the N,N-diethylaniline.
8. N,N-Diethylaniline is a tertiary amine. Why is a tertiary amine base used in the esterification reaction with an acid chloride rather than a primary or secondary amine?
9. *Library project:* Look up the structures of the compounds responsible for the odor of caraway and spearmint. Explain how these structures differ. Explain why (biochemically) these compounds have different odors.
10. *Molecular modeling assignment:* Consider the reaction of formic acid and methanol to produce methyl formate and water. Construct models of the starting materials and products of the assigned reaction, minimize their energies, then determine the energy differences between products and reactants. Assuming that the entropy change of the reaction is very small, determine whether or not this reaction is spontaneous. Use the results to explain the role of sulfuric acid in esterification reactions.
11. A possible mechanism for the Fischer esterification involves an S_N2-like attack of the hydroxyl oxygen of the carboxylic acid on the protonated alcohol, followed by deprotonation to yield the ester, as shown here. Propose an experiment you could perform using isotopically labeled substrates to determine whether this mechanism

is feasible. Explain specifically what reagents you would use and exactly what you would expect to observe in the reaction.

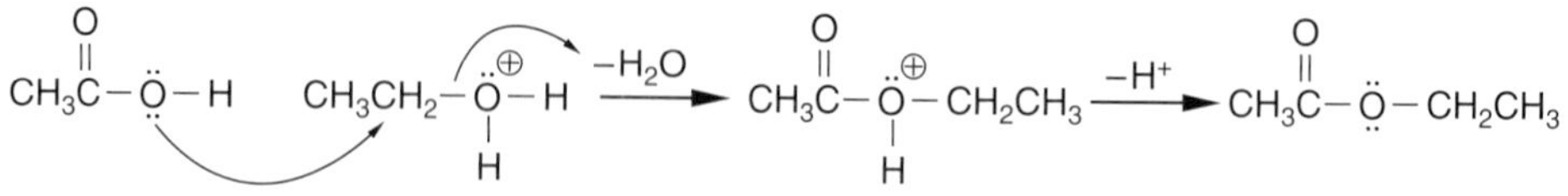

$$CH_3\overset{\overset{\displaystyle O}{\|}}{C}-\overset{\displaystyle \ddot{O}}{}-H \quad CH_3CH_2-\overset{\overset{\displaystyle \oplus}{\cdot\cdot}}{\underset{\displaystyle H}{O}}-H \xrightarrow{-H_2O} CH_3\overset{\overset{\displaystyle O}{\|}}{C}-\overset{\overset{\displaystyle \oplus}{\cdot\cdot}}{\underset{\displaystyle H}{O}}-CH_2CH_3 \xrightarrow{-H^+} CH_3\overset{\overset{\displaystyle O}{\|}}{C}-\overset{\displaystyle \ddot{O}}{}-CH_2CH_3$$

Experiment 20.2: Synthesis of Esters by Baeyer-Villiger Oxidation

Carbon-carbon single bonds resist oxidation unless they are part of a functional group such as a carbonyl group of a ketone. Ketones also resist oxidation by many oxidants including permanganate and dichromate. However, ketones can be oxidized by peracids and peroxides to give esters as shown in this experiment. You will:

◆ oxidize a ketone to an ester via a rearrangement reaction.

◆ determine the structure of the ester.

◆ justify the mechanism of reaction.

Techniques

Technique I	Extraction
Technique J	Gas chromatography
Technique K	Thin-layer chromatography
Technique M	Infrared spectroscopy
Technique N	Nuclear magnetic resonance spectroscopy
Technique P	Mass spectrometry

Background

Carbon-carbon single bonds are typically difficult to break, but there are some notable exceptions. For example, only the benzylic carbon is retained when alkyl-substituted aromatics are oxidized (as in Experiment 11.1). Ketones undergo α,β-carbon-carbon bond cleavage when subjected to strong oxidants. An example is the reaction of cyclohexanone with hot nitric acid to form hexanedioic acid (adipic acid). The Baeyer-Villiger reaction is a reaction of ketones with somewhat less powerful oxidants. The original reagent used by Baeyer and Villiger in 1899 was Caro's acid, H_2SO_5. Nowadays, reagents such as peroxyacetic acid and *m*-chloroperbenzoic acid are commonly used. These reagents, while effective, often require several days of reaction time. An oxidant that also gives the Baeyer-Villiger reaction is known as Oxone, a mixture of $KHSO_4$, K_2SO_4, and the oxidant, K_2SO_5 (Kennedy and Stock, 1960). The reagent, when added to water, yields a buffered persulfate oxidation mixture. The reaction of Oxone with cyclohexanone is shown below.

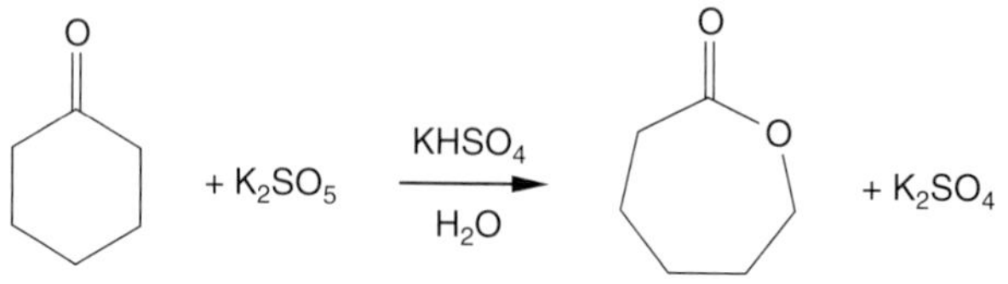

$$\text{cyclohexanone} + K_2SO_5 \xrightarrow[H_2O]{KHSO_4} \varepsilon\text{-caprolactone} + K_2SO_4$$

cyclohexanone   ε-caprolactone

In this experiment, you will receive an unknown ketone for reaction with Oxone. Following the reaction, separation and purification of the ester will enable identification of the ester will enable identification of the unknown ketone. The unknown will be chosen from cyclopentanone, cyclohexanone, cycloheptanone, 4-methylcyclohexanone, or another unknown ketone chosen by the instructor. Upon completion of the reaction and the workup procedure, you will identify the product by gas chromatography or gas chromatography/mass spectrometry.

The reaction of ketones with Oxone does not go to completion under the limited time frame given in the procedure. Rather, only some of the starting material will react. You will separate and analyze the mixture of starting ketone and ester. More rigorous reaction conditions than those used here will result in formation of a side product.

The reaction mechanism is an example of a molecular rearrangement reaction (Smith, 1963). The first step is postulated to be an attack on the carbonyl group by the peroxy reagent. The rearrangement consists of a migration of one of the carbon-carbon bonds to an oxygen with cleavage of the oxygen-oxygen single bond.

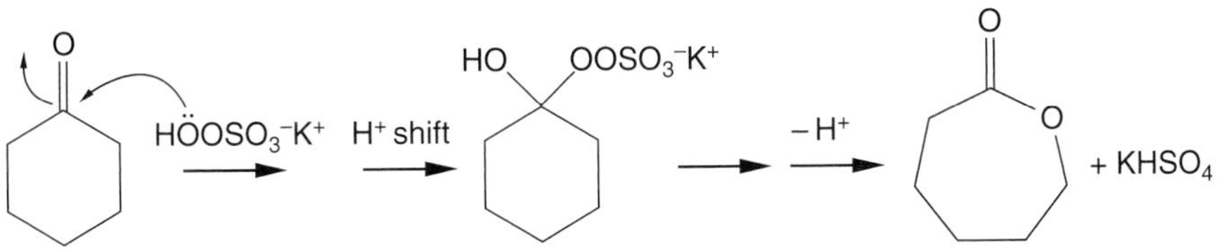

When there is a difference in types of carbon that can migrate to oxygen, the general order is tertiary > secondary ~ phenyl > primary > methyl. The more electron-rich carbons are better able to migrate to oxygen. Methyl ketones generally lead to acetate esters because the methyl group is least apt to migrate to oxygen during the rearrangement step.

Prelab Assignment

1. Write a balanced equation for the Baeyer-Villiger oxidation of cyclohexanone using Oxone. (Hint: Not all of the oxygens of K_2SO_5 have a -2 oxidation state.)
2. Propose a mechanism for the Baeyer-Villiger oxidation of cyclopentanone. Is the reaction acid catalyzed or is it base catalyzed?
3. What is the expected outcome of the reaction of 3-methylcyclohexanone with Oxone? How might the outcome cause a problem in analysis of the product?
4. Write a flow scheme for the workup in this experiment.

Experimental Procedure

1. Wear eye protection at all times in the laboratory.
2. Oxone and the ketones should be used under the hood. Wear gloves.

Part A: Microscale Oxidation of an Unknown Cyclic Ketone

Dissolve 300 mg of Oxone in 1.0 mL of water in a 5-mL conical vial equipped with a spin vane. Cool the solution to 0–5°C in an ice bath. Add 100 mg of an unknown ketone to the solution and stir for 30 minutes. Warm to room temperature and cautiously (foaming may occur) add 10% aqueous sodium carbonate dropwise with stirring to make the solution pH 7–8. Extract the solution with two 2-mL portions of methylene chloride. Combine the methylene chloride extracts in a small Erlenmeyer flask. Add a spatula tip of anhydrous sodium sulfate and swirl. If clumping occurs, add a little more of the drying agent. Let stand for 5 minutes. Transfer the clear solution into a tared 5-mL conical vial containing a boiling chip. Evaporate the solvent using a warm sand bath or heat block. (Do not overheat!) Cool the vial to room temperature and weigh the vial. An oil should remain in the vial.

Characterization for Part A

Infrared Spectroscopy: Record the IR spectrum and interpret, noting key bands characteristic of functional groups present. If more than one carbonyl stretching absorption is noted, this is an indication that the reaction has not gone to completion. Compare the spectrum with the IR spectrum of the starting unknown ketone.

Gas Chromatography: Obtain a sample of the mixture of possible ketones. This will be the standard. Inject a sample of the standard mixture. Using the same GC conditions inject a sample of the product. There should be some remaining unreacted ketone, allowing you to match the unknown ketone peak with one of the peaks of the standard. The lactone product will be eluted from the column later because it has a significantly higher boiling point than the starting ketone. It will also be possible to estimate the relative percentages of ketone and lactone in the mixture.

Gas Chromatography/Mass Spectrometry: If available, do a GC-MS analysis of the product. From the molecular ion peaks and fragmentation patterns it is possible to identify the starting material and product and to obtain a percentage of each component in the product mixture.

Results and Conclusions for Part A

1. Determine the structure of the product formed in this reaction.
2. Calculate the percent yield for the reaction based upon the weight of the product and the GC results.

Part B: Miniscale Oxidation of Cyclohexanone

Dissolve 3.0 g of Oxone in 8.0 mL of water in a 50-mL Erlenmeyer flask equipped with a spin bar. Cool the solution to 0–5°C in an ice bath. Add 1.0 g of cyclohexanone to the stirred solution. Continue stirring for 30 minutes. Warm to room temperature and cautiously (watch for foaming) add 10% aqueous Na_2CO_3 in small portions with stirring to make the solution pH 7–8. Transfer the solution to a small separatory funnel. Rinse with 2–3 mL of water and add the rinse to the funnel. Extract the solution with two 5-mL portions of methylene chloride. Combine the methylene chloride extracts in a small Erlenmeyer flask. Add a spatula or two of anhydrous sodium sulfate and swirl. If clumping occurs, add a little more of the drying agent. Let stand for 10 minutes. Gravity filter into a small, tared Erlenmeyer flask containing a boiling chip. Evaporate the solvent in the hood using a warm sand bath or heat block. (Do not overheat!) Cool the flask to room temperature and weigh the vial. An oil should be left in the vial.

Transfer the oil to a 3-mL conical vial containing a boiling chip. Place a Hickman still on the vial and add a thermometer. Distill the liquid boiling at 155°C. Remove all of the distillate from the Hickman still. Distill the higher boiling product fraction at about 210°C. Transfer the distillate to a tared vial. Weigh the product.

Characterization for Part B

Infrared Spectroscopy: Record and interpret the IR spectrum, noting key bands characteristic of functional groups present. If more than one carbonyl stretching absorption is noted, this is an indication that the product still contains some of the starting ketone. Compare the spectrum with the IR spectrum of cyclohexanone.

Gas Chromatography: Inject a sample of cyclohexanone, followed by a sample of the product. The product may contain some unreacted cyclohexanone, which will appear at the same retention time as the cyclohexanone standard. The ε-caprolactone will be eluted from the column later than cyclohexanone because it has a significantly higher boiling point than cyclohexanone. It will also be possible to determine the percentages of cyclohexanone and ε-caprolactone in the product based upon the integrated chromatogram.

Gas Chromatography/Mass Spectrometry: If available, do a GC-MS analysis of the product. From the molecular ion peaks and fragmentation patterns it is possible to identify the cyclohexanone and ε-caprolactone.

NMR Spectroscopy: Dissolve 5–10 mg of sample in $CDCl_3$ and obtain an integrated ^1H NMR spectrum of the product. Interpret the spectrum.

Results and Conclusions for Part B

1. Calculate the purity of the distilled product.
2. Calculate the percent yield for the reaction based upon the weight of the product.

Cleanup & Disposal

The neutralized aqueous solutions from the extraction may be washed down the drain with water.

Critical Thinking Questions

1. What side reaction could occur during the reaction of ketones with Oxone in aqueous solution at elevated temperature? (Hint: Think about what might happen to the ester under the reaction conditions.)

2. Reaction of a ketone with a peroxy acid such as *m*-chloroperbenzoic acid is typically carried out in an organic solvent. Of what advantage would this be over oxidizing in an aqueous solution?

3. Based upon the proposed reaction mechanism, predict the outcome of the reaction of the following ketones with aqueous Oxone: benzophenone (diphenyl ketone); *p*-nitrophenyl phenyl ketone; 2,2-dimethyl-3-pentanone.

4. Draw the structures of the oxidation products of the following ketones with Oxone.

    a. 3-pentanone

    b. 2-methyl-3-pentanone

    c. acetophenone

5.   It has been found that retention of configuration occurs during the rearrangement. Based upon the reaction mechanism, draw the structure of the oxidation product expected for the reaction of the following ketones with Oxone.

a.

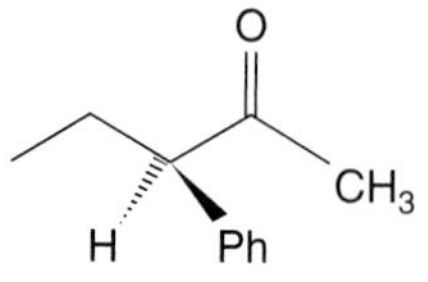

b.

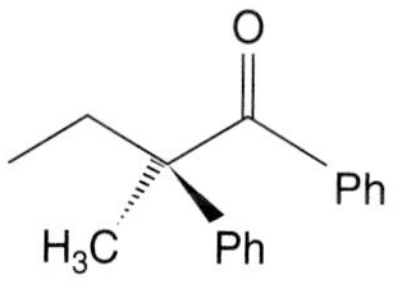

6.   What particular feature of the NMR spectrum of the product differentiates it from the spectrum of the starting ketone? How can this be used to advantage in estimating the percentage of product formed?

7.   It has been found that when ^{18}O-labeled ketones are used to study the mechanism of reaction, esters having ^{18}O completely retained in the carbonyl group are formed. Is this observation consistent with the proposed mechanism? Explain.

References

Hassall, C. H., The Baeyer-Villiger Oxidation of Aldehydes and Ketones in *Organic Reactions, Volume 9*, Chapter 3, Adams, R., ed. New York: John Wiley, 1957.

Kennedy, R. J., and Stock, A. M. The Oxidation of Organic Substances by Potassium Peroxymonosulfate. *J. Org. Chem.* 25 (1960): 1901.

Plesnicar, B., Oxidations with Peroxy Acids and Other Peroxides in *Oxidation in Organic Chemistry, Part C*, Chapter 3, Trahanovsky, W. S., ed. New York: Academic Press, 1978.

Smith, P. A. S., Rearrangements Involving Migration to an Electron-Deficient Nitrogen or Oxygen in *Molecular Rearrangements, Volume 1*, Chapter 8, de Mayo, P., ed. New York: Interscience Publishers, 1963.

Chapter 21

Dicarbonyl Compounds

This chapter and some experiments in Chapter 22 offer a glimpse at the versatility of dicarbonyl compounds in synthesis. Dicarbonyl compounds such as diethyl malonate and ethyl acetoacetate (acetoacetic ester) have acidic hydrogens, giving these β-dicarbonyl compounds acidities (pK_a = 10–12) that rival the acidity of phenols (pK_a = 10). One notable feature is that β-dicarbonyl compounds may be easily converted to anions using alkoxide bases.

In Experiment 21.1, you will investigate a condensation reaction of the α-carbon of β-diesters or β-ketoesters. This reaction involves condensation with an aromatic o-hydroxyaldehyde in a reaction known as the Knoevenagel condensation. Subsequent reaction leads to a class of organic heterocycles known as coumarins.

Dicarbonyl compounds are used in several experiments of Chapter 22 to prepare nitrogen heterocycles. Dicarbonyl compounds are very reactive with nucleophiles such as amines and amine derivatives and are therefore ideal starting materials.

Experiment 21.1 Base-Catalyzed Condensations of Dicarbonyl Compounds

Under base catalysis, 1,3-dicarbonyl compounds form stabilized anions that can undergo a variety of reactions, including alkylation and condensation. In this experiment, you will:

- condense an unknown 1,3-diester with an unknown substituted aromatic aldehyde.

- use a solvent-free system and microwave irradiation to synthesize a coumarin derivative.

- determine the identity of the coumarin based upon melting point and spectral characteristics.

- prepare a condensation product and use spectroscopy to determine its structure.

Techniques

Technique F	Recrystallization and filtration
Technique G	Reflux
Technique M	Infrared spectroscopy
Technique N	Nuclear magnetic resonance spectroscopy

Background

In the aldol reaction, an enolate of an aldehyde or ketone is condensed with another aldehyde or ketone to generate a β-hydroxy carbonyl compound, which can dehydrate to form an α,β-unsaturated ketone or aldehyde. Stabilized enolates, such as those formed from 1,3-dicarbonyl compounds, can also condense with aldehydes or ketones. When the dicarbonyl compound is a diester and the acceptor molecule is an aldehyde, the reaction is called the Knoevenagel reaction:

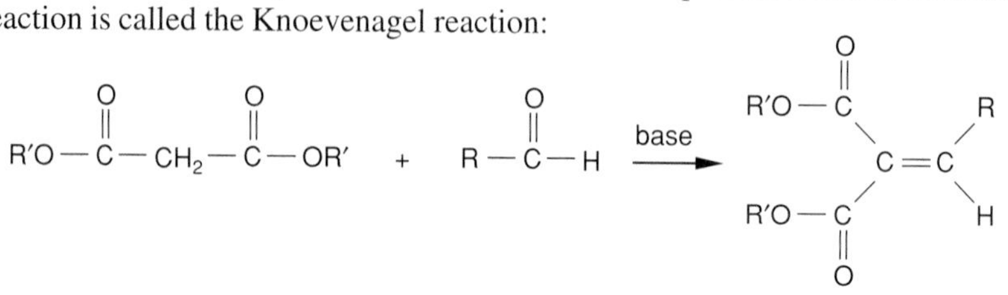

Dicarbonyl compounds will form an enolate with relatively weak bases like piperidine or pyridine because the α-protons are much more acidic than those of aldehydes and ketones. Higher yields are obtained when the aldehyde or ketone does not contain α-hydrogens. A typical Knoevenagel reaction involves heating the diester and aldehyde in ethanol at high temperatures for 2–3 hours, extracting the product from the solvent, then isolating the product by slow recrystallization. A relatively new innovation is the use of microwave-assisted organic synthesis. Using microwaves to heat the mixture not only shortens the reaction time considerably, but also alleviates the need for a solvent. The microwave-assisted solvent-free reactions offer significant advantages, including greatly shortened reaction time, decreased exposure to toxic chemicals, decreased thermal degradation, lower cost, and ease of working up the reaction mixture. Microwave irradiation also frequently increases the yield. For these reasons, microwave-assisted organic synthesis has been adopted into some college chemistry courses.

Microwave radiation corresponds to wavelengths between infrared radiation and radio frequencies (wavelengths between 1 centimeter and 1 meter). This corresponds to a frequency of about 10^9 cycles per second. Because the reactions are done in the absence of a solvent, the reactants absorb all of the microwave energy. Heating in a microwave oven occurs as polar molecules orient themselves to an electric field. The oscillation of the field causes molecules to continuously reorient themselves. This agitation and electrical field excitation creates high internal temperatures that increase the reaction rate.

In this experiment, you will perform a microwave-assisted synthesis of coumarin or a derivative of coumarin (Bogdal et al., 1996). Coumarins have diverse properties and functions. Coumarin (2H-1-benzopyran-2-one) is found in sweet clover, tonka beans, and lavender oil. Other coumarin derivatives are used as additives in food and cosmetics. One coumarin derivative, called Warfarin, is a strong anticoagulant. In low concentrations it is useful as a blood thinner in adults with high blood pressure. At higher concentrations, Warfarin kills rodents such as mice and rats by causing severe internal bleeding. The structures of coumarin and Warfarin are shown here:

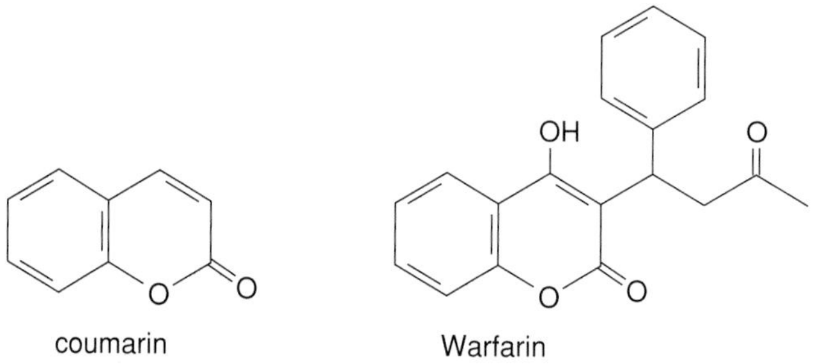

coumarin   Warfarin

In this experiment, you will obtain an unknown dicarbonyl compound and an unknown hydroxybenzaldehyde. After reaction, isolation, and purification of the product, you will identify the coumarin synthesized by melting point determination and IR and ^1H NMR spectroscopy. The generalized reaction is shown below. An alternate synthesis is provided in Part B for laboratories not equipped with a laboratory-grade microwave oven.

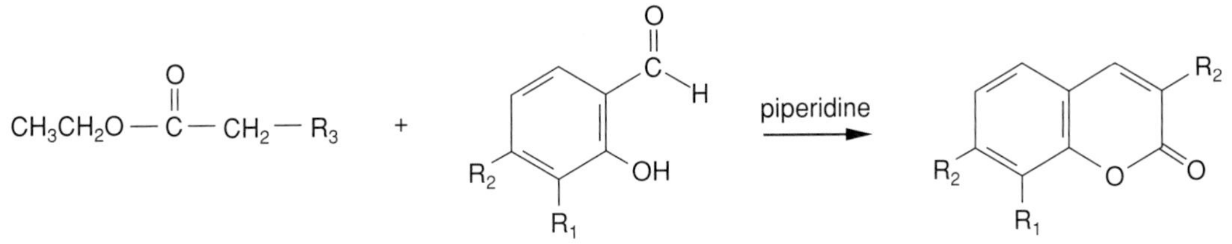

Prelab Assignment

1. Write a detailed mechanism of the condensation and subsequent cyclization of diethyl malonate ($R_3 = CO_2CH_2CH_3$) with salicylaldehyde ($R_1 = R_2 = H$).
2. Draw resonance forms to explain the acidity of 1,3-dicarbonyl compounds like diethyl malonate.
3. Look up the structure of piperidine and write a reaction to show its role in the synthesis.

Experimental Procedure

Safety First!

Always wear eye protection in the laboratory.

1. Wear eye protection at all times in the laboratory.
2. Explosions can occur if the microwave oven is not functioning properly. Use proper precautions. Be careful when removing vials from the oven as they may be extremely hot.
3. Make certain that the reaction vessels used do not have any cracks or chips or they may shatter during the reaction.

Table 21.1-1 Melting Points of Coumarin Derivatives

	Structure of dicarbonyl compound		
R_1	R_2	R_3	Melting point (°C)
H	H	$CO_2CH_2CH_3$	91–92
OCH_3	H	$CO_2CH_2CH_3$	89–91
$N(CH_2CH_3)_2$	H	$CO_2CH_2CH_3$	80–82
H	H	$COCH_3$	120–122
$N(CH_2CH_3)_2$	H	$COCH_3$	152–153
H	H	CN	182–184
$N(CH_2CH_3)_2$	H	CN	225–226

Part A: Microscale Preparation of Coumarin or a Coumarin Derivative

These reactions should be carried out in a laboratory-grade microwave oven, such as Synthwave Prolabo, Anton Paar Multiwave, CEM Monomode Reactor, or other brand. All reaction vessels used in the experiment should be free of chips and star cracks. The reaction vessels should have a much larger volume than the volume of the reagents used. For this reaction, a 100-mL beaker can be used as the reaction vessel.

Combine 100 mg of the unknown salicylaldehyde derivative, 180 mg of the unknown malonic ester, and 10 mg of piperidine in the 100-mL beaker. Set a watch glass on top to condense any vapors formed. Irradiate the reaction for 6–10 minutes at 10% power. The beaker and contents will be very hot. Let the reaction flask cool to room temperature before handling. Recrystallize the crude product in 95% ethanol. Filter the crystals using a Hirsch funnel. Air dry the crystals and weigh.

Part B: Miniscale Preparation of Coumarin or a Coumarin Derivative

Place 1.0 mL of the assigned hydroxybenzaldehyde and 1.5 mL of the dicarbonyl compound into a 25-mL round-bottom flask fitted with a stir bar and heating mantle or sand bath. Add 4.0 mL of ethanol, 1 mL piperidine, and 0.2 mL glacial acetic acid. Fit the flask with a water-jacketed condenser and a drying tube. Turn the water on to the condenser and reflux with stirring for 2 hours. Remove the heating mantle and let the solution slowly cool to room temperature. Put the solution in an ice bath until crystals begin to form. Stir for 5 minutes while the solution is in the ice bath. Filter the crystals using a Büchner or Hirsch funnel through two pieces of filter paper. Rinse with small aliquots of cold ethanol. Recrystallize from 95% ethanol. Air dry the crystals.

Characterization for Parts A and B

Melting Point: Determine the melting point of the product and compare it to the literature values in Table 21.1-1.

IR Spectroscopy: Obtain an IR spectrum using the Nujol mull or KBr technique.

NMR Spectroscopy: Dissolve about 50 mg of the sample in CDCl$_3$. Record the ^{1}H NMR spectrum. Integrate the spectrum. Interpret the spectrum.

Results and Conclusions for Parts A and B

1. Determine the identity of the synthesized coumarin.
2. Explain how the IR and/or ^{1}H NMR spectra aided in the structural identification.
3. Write a detailed mechanism for the formation of the coumarin derivative synthesized in this reaction.
4. Explain why piperidine (rather than hydroxide) is used in this reaction.

Cleanup & Disposal

Place all organic solutions in a container labeled "nonhalogenated organic solvent waste."

Critical Thinking Questions

1. When 3-carboethoxycoumarin is refluxed with aqueous sodium hydroxide, a new product is formed. Draw the structure of this product.

2. Which of the resonance structures of the enolate from the dicarbonyl compound would be expected to be in the major resonance contributor? Explain.

3. *Library project:* Look up the Perkin, Doebner, and Dieckmann "name" reactions and illustrate each reaction.

4. Explain why glass is a good material for use in microwave oven experiments.

5. When cooking with a home microwave oven, manufacturers often recommend keeping a glass of water in the microwave oven when it is not in use in case anyone accidentally turns on the microwave. Explain why this is recommended.

6. Propose a synthesis of Warfarin.

7. *Library project:* Find another example in the literature of microwave-assisted synthesis. Write a short paragraph summarizing its advantages over the traditional organic chemistry laboratory procedure.

8. Give the product of the following reaction:

$$
\text{(1-hydroxy-2-naphthaldehyde)} \quad + \quad CH_3CH_2O-\overset{\displaystyle O}{\overset{\|}{C}}-CH_2-\text{(}p\text{-}NO_2\text{ phenyl)} \xrightarrow{\text{piperidine}}
$$

Reference

Bogdal, D., Pielichowski, J., and Boron, A. *Synlett,* 37 (1996): 873–874.

Experiment 21.2: Reactions of Diketones: *Synthesis of 2,5-Dimethyl-1-phenylpyrrole and Preparation of an Unknown Pyrrole*

In this experiment, you will learn about the synthesis of pyrroles, an important class of nitrogen-containing heterocycles. You will:

◆   synthesize and characterize a pyrrole derivative.

◆   carry out a condensation of an amine with a ketone leading to ring closure.

Techniques

Technique F	Vacuum filtration
Technique G	Reflux
Technique M	Infrared spectroscopy
Technique N	Nuclear magnetic resonance spectroscopy

Background

Heterocycles are cyclic compounds containing at least one atom other than carbon in the ring. The most common heterocycles contain oxygen, nitrogen, or sulfur. Many biologically active molecules contain heterocycles. An example is chlorophyll, which is shown here. Chlorophyll contains four pyrrole rings.

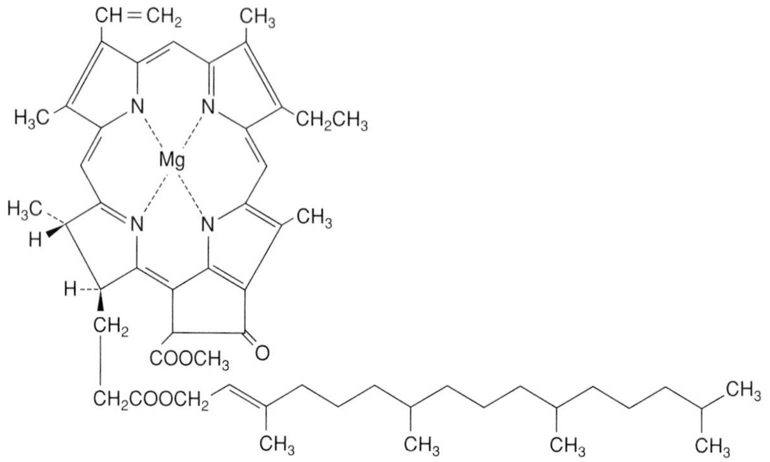

chlorophyll

Pyrrole is a five-membered ring in which nitrogen is the heteroatom. Pyrrole is aromatic, possessing six π electrons in a planar ring structure.

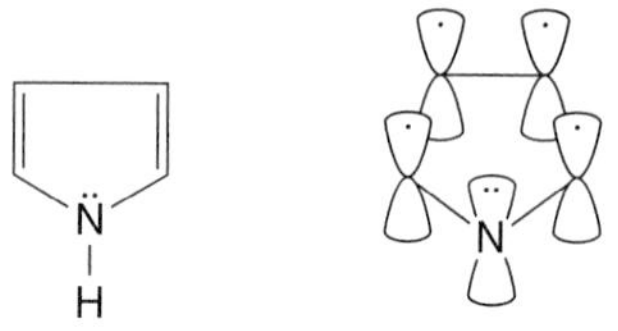

pyrrole

p orbitals of pyrrole
(containing 6 π electrons)

In this experiment, trisubstituted pyrroles will be synthesized. The initial reaction is an acid-catalyzed condensation of an amine with one of the carbonyl groups of a dialdehyde or a diketone to form an imine. The imine nitrogen of this initial product attacks the second carbonyl to form a five-membered ring. The driving force for this reaction is the formation of the aromatic ring. The reaction occurs with a variety of 1,4-diketones and primary amines. In Part A of this experiment, the 1,4-diketone is 2,5-hexanedione and the primary amine is aniline.

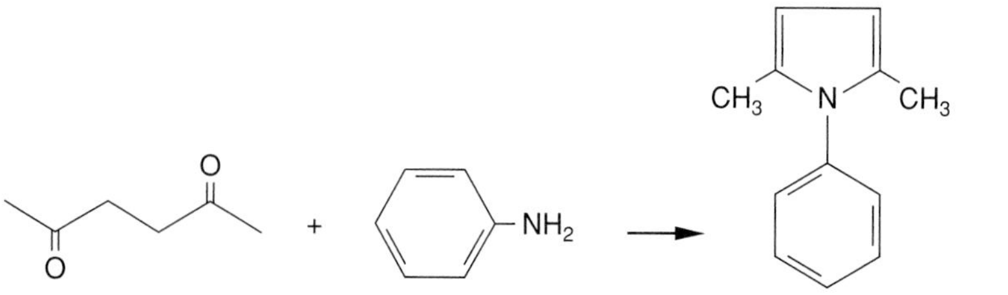

2,5-hexanedione   aniline   2,5-dimethyl-1-phenylpyrrole

In Part B of this experiment, you will prepare a pyrrole from 2,5-hexanedione and an unknown aromatic amine (*m*-bromoaniline, *p*-ethylaniline, *o*-toluidine, or *o*-anisidine). From the melting point of the synthesized pyrrole, you will determine the structure of the aromatic amine.

Prelab Assignment

1. Calculate the number of mmoles of aniline and 2,5-hexanedione to be used in this experiment (Part A).
2. The first step in the mechanism of the synthesis of pyrroles is the formation of an imine by condensation of the amine with one of the carbonyl groups of the diketone. Write a detailed mechanism for the reaction of aniline with 2,5-hexanedione to form the imine.
3. (Part B) Draw the structures of the pyrroles formed from the reaction of 2,5-hexanedione with *m*-bromoaniline, *p*-ethylaniline, *o*-toluidine, and *o*-anisidine.

Experimental Procedure

1. Wear eye protection at all times in the laboratory.
2. Wear gloves when handling all of the reagents for this experiment.
3. Aniline and aromatic amines are highly toxic and cancer suspect agents. They can be absorbed through the skin. Work under the hood and avoid breathing the vapors.
4. HCl is corrosive and toxic and can cause burns if it comes in contact with your skin. Wash any affected areas with copious amounts of water and notify the instructor immediately.

Part A: Microscale Synthesis of 2,5-Dimethyl-1-phenylpyrrole

Into a 3-mL conical vial equipped with a spin vane or boiling chip and a reflux condenser, place 0.4 mL of methanol, 122 μL of freshly distilled aniline, 158 μL of 2,5-hexanedione (acetonylacetone), and 1 drop of concentrated hydrochloric acid. Reflux the mixture for 20 minutes. Add the warm mixture via Pasteur pipet to a small ice-cooled Erlenmeyer flask or beaker containing 3.5 mL of 0.5 M HCl. Collect the crystals by suction filtration using a Hirsch funnel. Wash with a small amount of cold water. Weigh the crude product and then recrystallize from 10% aqueous methanol (about 0.75–1.5 mL). Cool the mixture in ice and collect the product. Dry and weigh the crystals.

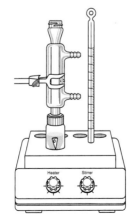

Characterization for Part A

Melting Point: Determine the melting point of the product. The reported melting point of 2,5-dimethyl-1-phenylpyrrole is 50–51°C.

Infrared Spectroscopy: Obtain an IR spectrum of the product using either the KBr or the Nujol mull technique and interpret the spectrum noting the key bands characteristic of the structural features, especially at 1600 cm^{-1}, 1499 cm^{-1}, and 1402 cm^{-1}.

NMR Spectroscopy: Obtain a ^{1}H NMR spectrum of the product in CDCl$_3$. Interpret the spectrum noting the chemical shifts of the signals and the integrated areas. The product should show two singlets near δ2 and δ7.2 and a multiplet at δ7.4.

Results and Conclusions for Part A

1. Determine the percent yield of the product.
2. Explain how this reaction could be followed using chromatography and spectroscopy.

Part B: Microscale Preparation of an Unknown Pyrrole

Obtain a sample of an aromatic amine from the instructor. If the amine is a liquid, pipette 130 μL of freshly distilled amine into a 3-mL conical vial. If the amine is a solid, weigh 160 mg and transfer to the 3-mL conical vial. Equip the 3-mL conical vial with a spin vane or boiling chip and a reflux condenser. Add 0.4 mL of methanol, 158 μL of 2,5-hexanedione (acetonylacetone), and 1 drop of concentrated hydrochloric acid. If the amine does not dissolve, add another 1–1.5 mL of methanol. Reflux the mixture for 20 minutes. Add the warm mixture via Pasteur pipet to a small ice-cooled Erlenmeyer flask or beaker containing 3.5 mL of 0.5 M HCl. Collect the crystals by suction filtration. Wash with a small amount of cold water. Weigh the crude product and then recrystallize from hot 10% aqueous methanol (about 1.0–3.0 mL). Do not add more than 3 mL of aqueous methanol. If the solid does not go completely into solution after 3.0 mL of aqueous methanol has been added, filter off the remaining solid and continue with the recrystallization. Cool the solution in an ice bath until crystallization is complete. Collect the product by vacuum filtration. Dry and weigh the crystals.

Results and Conclusions for Part B

1. Determine the identity of the product by comparing the melting point of the product with the melting points of the pyrroles below:

Pyrrole	Melting point (°C)
2,5-dimethyl-1-(*p*-ethylphenyl)pyrrole	51–53
2,5-dimethyl-1-(*o*-tolylphenyl)pyrrole	168–170
2,5-dimethyl-1-(*o*-methoxyphenyl)pyrrole	60–62
2,5-dimethyl-1-(*m*-bromophenyl)pyrrole	84–86

2. Draw the structure of the aromatic amine used in this experiment.
3. Determine the percent yield of the product.

Cleanup & Disposal

All residues and filtrates should be placed in a container labeled "nonhalogenated organic solvent waste" because of the possible presence of aniline.

Critical Thinking Questions

1. Acid is required as a catalyst in this reaction, but the reaction fails if the reaction mixture is too acidic. Explain.

2. What would be the expected product in this reaction if ammonia were used in place of aniline?

3. Explain why the ring-forming reaction fails if a secondary amine is used in place of a primary amine in this experiment.

4. Give the structures and names of the oxygen- and sulfur-containing analogs of pyrrole.

5. The phosphorus analogs of pyrroles are called phospholes. Suggest a starting material to prepare a phosphole.

6. Using the synthetic approach described in this experiment, give the starting materials required for the preparation of 1-methyl-2,5-diethylpyrrole.

7. Give the product expected from the reaction of *cis*-1,2-diacetylcyclohexane with aniline.

8. Why does the synthesis of a pyrrole fail if 1,4-cyclohexanedione is used as the diketone?

9. Suggest a reason for the unsuitability of 2,4-dinitroaniline as a replacement for aniline in this experiment.

10. *Library project:* A published procedure for the preparation of 2,5-dimethyl-1-phenylpyrrole is described in an article by Enno Wolthuis in the *Journal of Chemical Education* in 1979. Locate this reference and from it find the original research journal reference, also by Wolthuis, upon which it is based. Turn in a photocopy of the abstract from the original article or other information as directed by your instructor.

Chapter 22

Amines

A mines are very common in nature and many of them are physiologically active. Amines can be isolated from plants by extraction with aqueous HCl. The extracts can be neutralized with base and extracted from the aqueous solution with an organic solvent to furnish the free amine. Naturally occurring amines that can be isolated in this way are known as alkaloids, common in flowers, leafy shrubs, and trees.

A versatile reaction of amines is the diazotization of aniline or ring-substituted aniline derivatives. The resulting diazonium salts can undergo reaction with any one of several reagents to afford aromatic compounds, some of which cannot be prepared by electrophilic aromatic substitution. The multistep synthesis of 1-bromo-3-chloro-5-iodobenzene in Experiment 29.1 utilizes diazotization as a key step. An important use of diazonium salts is in the preparation of aromatic azo compounds in a reaction known as coupling. Coupling of diazonium salts to prepare azo dyes is illustrated in Experiment 22.1.

This chapter on amines is a sequel to the chapter on dicarbonyl compounds (Chapter 21). Ketones and diketones react with ammonia, amines, and amine derivatives such as hydrazine, phenylhydrazine, and hydroxylamine to form imines, hydrazones, and oximes, respectively. When dicarbonyl compounds react, the possibility of ring formation exists and this becomes an attractive way to prepare a number of classes of heterocyclic compounds, including pyrazoles, pyrimidines, and imidazoles. β-Diketones are very reactive with nucleophilic amines and amine derivatives giving five- and six-member heterocycles. Experiments 22.2–22.4 illustrate the versatility of the reaction of amines and amine derivatives with dicarbonyl compounds to prepare heterocyclic compounds.

Experiment 22.1: Relating Color to Structure: *Synthesis of Azo Dyes*

Azo dyes are important in industry and in the world economy. In this experiment you will synthesize an azo dye using a coupling reaction of a diazonium salt compound with an aromatic amine or phenol. You will:

♦ prepare a diazonium salt from an assigned aromatic amine.

♦ couple the diazonium salt with an assigned phenol or aromatic amine.

♦ use the prepared dye to color a fabric.

◆ evaluate the indicator properties of the synthesized dye under acidic and basic conditions.

◆ use microbiological techniques to evaluate the antibacterial properties of the synthesized dye with gram-positive and gram-negative bacteria (optional).

◆ evaluate the optical properties of the synthesized dyes.

Techniques

Technique F	Vacuum filtration
Technique O	UV-visible spectroscopy

Background

Using dyes to improve the appearance of plain fabrics is not new. According to history books, the first recorded use of dyes was in China over 4600 years ago, where artisans used plant extracts as dyes. In the early 1500s, the French, Dutch, and Germans developed a dye industry based upon the cultivation of plants. However, the use of synthetic dyes is relatively new, arising as a serendipitous discovery by an 18-year-old research assistant named William Henry Perkins. In a misguided attempt to synthesize quinine by the oxidation of *N*-allyltoluidine, Perkins produced a red-brown solid, which was definitely not quinine. He repeated the experiment using aniline, a simpler aromatic amine that was structurally less complex than *N*-allyltoluidine. The aniline sample he used was isolated from tar, and actually contained a mixture of aniline, *o*-toluidine, and *p*-toluidine.

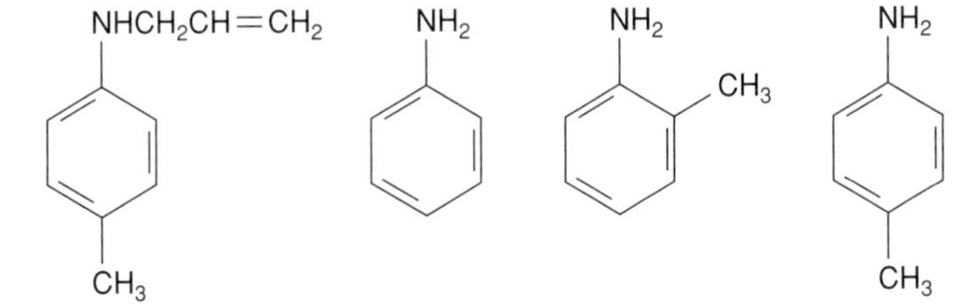

N-allyltoluidine   aniline   *o*-toluidine   *p*-toluidine

This time he isolated black crystals. However, when he dissolved the black crystals in ethanol, the solution formed a beautiful purple color, which is known as Aniline Purple. This was the first recorded preparation of a synthetic dye. This fortuitous discovery made Perkins a rich and famous man.

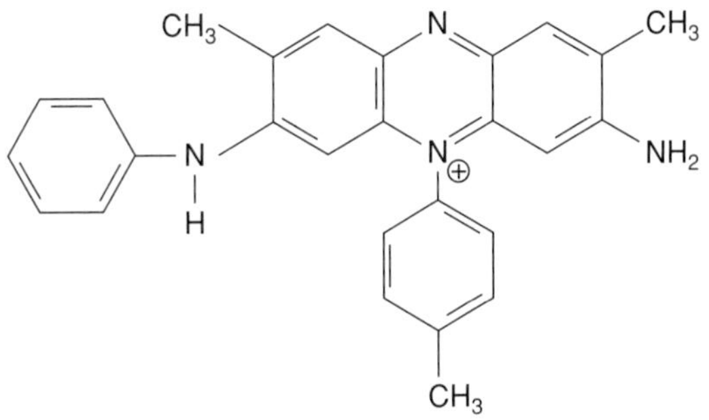

Aniline Purple

Other chemists were also interested in using aniline as a precursor to dyes. A brewer named Peter Griess discovered that aniline could be converted to a diazonium salt, which (if it didn't explode) could be coupled to other aromatic compounds to form azo dyes in a variety of colors. Azo dyes have the general structure shown here, where G is OH, NR_2, or other electron-donating group and Z can be a variety of different atoms or functional groups:

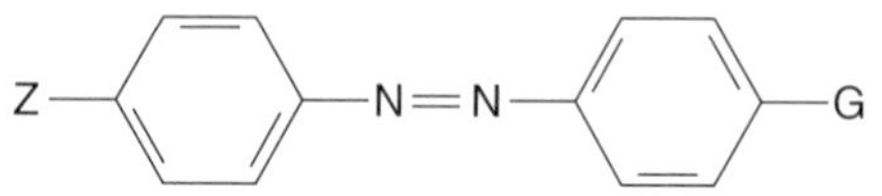

an azo dye

Virtually all azo dyes are synthesized through the process of diazotization and coupling. Diazotization involves treating aromatic amines with nitrous acid to yield diazonium salts. These salts are very versatile intermediates that can be used to prepare a wide variety of aromatic substitution products in which the amino group is replaced by some other substituent, such as F, Cl, Br, I, OH, H, and CN. However, when the diazonium salt is reacted with electron-rich aromatics, the result is a highly colored coupled product, called an azo dye. Azo dyes, containing an azo (N=N) group, are important as dyes for clothing and foods. They are also used as pigments in paints, printing inks, and printing processes. An example of the coupling reaction is shown here:

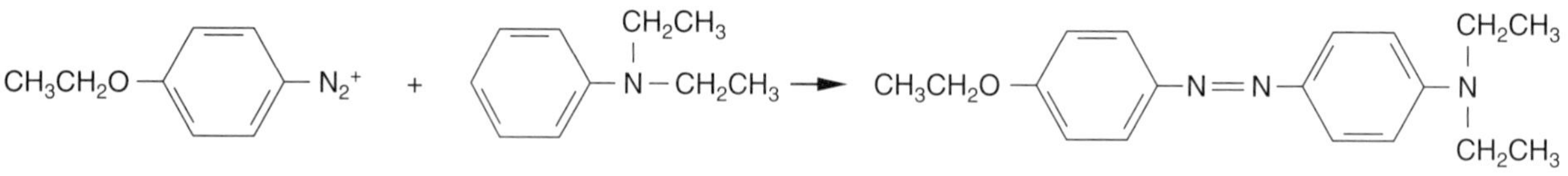

a diazonium ion   coupling component   an azo dye

In the diazotization reaction, nitrous (HNO_2) acid is generated *in situ* from sodium nitrite and a mineral acid, usually hydrochloric acid according to the reaction:

$$HCl + NaNO_2 \rightarrow HNO_2 + NaCl$$

Diazotization must be carried out at 0°C to minimize the reaction with water to produce a phenol. This reaction is significant at room temperature. Diazotization must be carried out as rapidly as possible and with good stirring and cooling. The pH is also critical. The diazonium salts produced are unstable and should not be isolated. Although diazonium salts are safe when dissolved in a solution, when dry, the diazonium salts are quite explosive! For that reason, diazonium salts are always freshly prepared and must be used immediately in the coupling reaction.

Diazonium salts should never be allowed to dry out!

The mechanism of the reaction involves protonation of the nitrous acid generated *in situ*. Nitrous acid then loses water to form an electrophilic nitrosonium ion (^+NO):

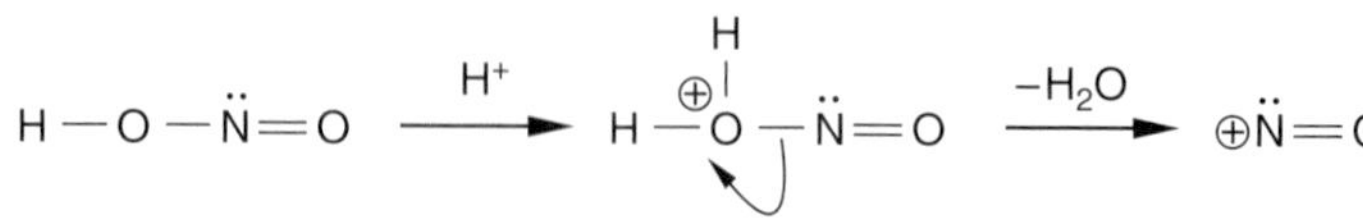

The nitrosonium ion promptly reacts with the lone pair of the nitrogen leading to an N-nitrosoanilinium ion. Proton transfer and tautomerization affords a diazohydroxide.

This intermediate is protonated and loses water to afford a diazonium salt. The sequence of steps for the diazotization of aniline is shown here:

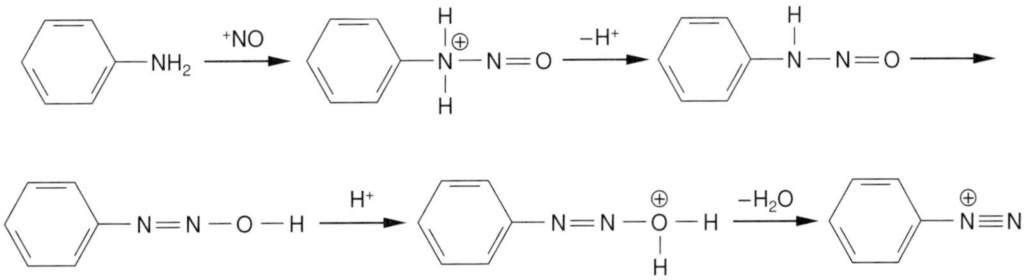

adiazohydroxide   benzenediazonium ion

The diazonium ion is a weak electrophile that will react with electron-rich aromatics such as phenols and anilines to give electrophilic aromatic substitution reactions. As with other electron-rich substituents, ortho and para substitution products predominate. In this experiment, you will be assigned an aromatic amine to diazotize and an electron-rich aromatic compound to couple from the list below:

Diazo Component	Coupling Agent
Aniline	
Sulfanilic Acid	N,N-Dimethylaniline
p-Nitroaniline	N-Methylaniline
m-Nitroaniline	2-Naphthol
p-Anisidine	1-Naphthol
m-Anisidine	Resorcinol
p-Toluidine	Phenol
m-Toluidine	m-Phenylenediamine
Sulfanilamide	

After you have synthesized the azo dye, you will analyze its color and dye a fabric using a direct dyeing process. In the direct dyeing method, the dye is directly applied to the fabric. Direct dyes work best on polar fabrics, such as wool or silk, although other materials like cotton can also be dyed in this manner. In direct dyeing, the acidic or basic groups on the dye interact with the fabric to form a salt. Other types of dyes may require a mordant, such as a metal hydroxide or a metal complex of tannic acid, to act as a binding agent for the dye. The mordant is coated on the fabric, and then the dye forms an insoluble complex called a lake. A third method of applying dye is called a vat method. In this process, a water-soluble form of the dye is applied to the fabric and the dye then undergoes a chemical reaction to produce an insoluble form of the dye. An example of this is indigo. The dye is applied in its colorless reduced form. Upon drying and subsequent oxidation by air, the dye turns blue.

You will also characterize the dye using UV-visible spectroscopy. The λ_{max} for each dye will be determined and the value correlated with the dye color.

Dyes can serve as pH indicators if they give a color change apparent to the eye over a narrow pH range (1–2 pH units). You will test your dye as a potential pH indicator. You will also test your synthesized dye as an inhibitor of bacterial growth (optional).

Prelab Assignment

1. Draw the product of the assigned combination of diazonium component and coupling component.
2. Calculate the masses of 10 mmol of the assigned diazonium component and 10 mmol of the coupling component.
3. Calculate the theoretical yield (in grams) of the azo dye.
4. Write a detailed mechanism for the synthesis of the assigned diazonium salt.
5. Write a detailed mechanism for the assigned coupling reaction.
6. Prepare a table to record the results of the colors for all assigned combinations of diazonium component and coupling component.

Experimental Procedure

Safety First!

Always wear eye protection in the laboratory.

1. Wear eye protection at all times in the laboratory.

2. Aromatic amines are harmful if ingested or inhaled. They can also be absorbed through the skin. Aromatic amines may cause allergic reactions and may be carcinogenic. Always wear gloves when handling these compounds. Work under the hood or in a well-ventilated area.

3. Phenols are generally toxic irritants. Some phenols are corrosive and cause severe damage to the skin. Other phenols are carcinogenic. Always wear gloves and work under the hood. Do not breathe the vapors or allow the compound to contact the skin.

4. Hydrochloric acid and sulfuric acid are corrosive. Neutralize any spills immediately with sodium bicarbonate. Should the acid come in contact with your skin, wash the affected area immediately with copious amounts of cool water.

5. Diazonium salts are explosive when dry. Never let the diazonium salt dry out. Use the prepared diazonium solution immediately in the next step.

Part A: Miniscale Diazotization of an Aromatic Amine

Wear gloves and work under the hood while performing this procedure. Dissolve 10 mmol of the assigned diazo component in 8 mL of 3 M HCl. Heat the solution gently. If necessary, add up to 10 mL of water in order to dissolve the solid. After most of the solid has dissolved, cool the solution to 5°C in an ice bath with stirring. Some solid may precipitate, but the reaction should still work well if stirred. While stirring, slowly add 10 mL of freshly prepared 1 M sodium nitrite solution. Adjust the rate of addition so that the temperature of the solution remains below 10°C. Test the solution with starch-iodide paper to make sure that enough sodium nitrite has been added. If the test paper does not immediately turn blue-violet, add 1 M sodium nitrite solution dropwise until the paper turns blue-violet. The diazonium salt must not be allowed to dry but should be used immediately. Keep the solution in the ice bath and proceed immediately to the next step, B.1 or B.2, depending upon the nature of the assigned coupling component. If the assigned coupling component is a phenol, go to Part B.1. If the coupling component is an amine, go to Part B.2.

Part B.1: Miniscale Coupling with a Phenol

Follow this procedure if the assigned coupling component is a phenol. Dissolve or suspend 10 mmol of the assigned phenol in 20 mL of 1 M NaOH in a 125-mL Erlenmeyer flask. Cool the solution in an ice bath. While stirring, slowly add the diazonium salt solution to the solution of the coupling component. Let the mixture stand in the ice bath for at least 15 minutes until crystallization is complete. If little or no colored solid appears, it may be necessary to adjust the pH of the solution with dilute HCl or NaOH in order to induce precipitation. Collect the azo dye by vacuum filtration, washing with cold water. Dry the azo dye overnight and weigh it. Save the crystals for subsequent direct dyeing (Part C), spectral analysis with UV-visible spectroscopy (Part D), and pH indicator testing (Part E). The antibacterial properties of the dye may also be tested (Part F).

Part B.2: Miniscale Coupling with an Amine

This procedure should be followed if the assigned coupling component is an aromatic amine. Dissolve or suspend 10 mmol of the amine in 10 mL of 1 M HCl. Cool the solution in an ice bath. With stirring, slowly add the diazonium salt solution to the solution of the coupling component. After addition is complete, let the mixture stand in the ice bath for 15 minutes. Neutralize the acidic solution with 3 M sodium carbonate until it is neutral to litmus. Let the solution stand in an ice bath until crystallization is complete. Collect the azo dye by vacuum filtration, washing with cold water. Dry the azo dye overnight and weigh it. Save the crystals for subsequent direct dyeing (Part C), spectral analysis with UV-visible spectroscopy (Part D), and pH indicator testing (Part E). The antibacterial properties of the dye may also be tested (Part F).

Part C: Direct Dyeing with the Azo Dye

Suspend 0.5 g of the dye in 100 mL of hot water. Acidify the solution with a few drops of concentrated sulfuric acid. Obtain two pieces of cloth to dye. Immerse the two pieces of cloth in the mixture for at least 10 minutes. Remove the dyed cloth, rinse off excess dye with water, and let air-dry. If the cloth does not appear to absorb the dye, it may be necessary to adjust the pH of the dyeing mixture with dilute HCl or NaOH. To test for fade resistance, wash one of the dyed cloths 5 times with a laundry detergent. Compare the colors of the cloths. If possible, obtain other types of cloth (wool, silk, nylon, or other fabrics) and test the dye solution.

Part D: Recording the UV-Visible Spectrum of the Prepared Dyes

Dissolve 3–5 mg of the dye in 10 mL of 95% ethanol. Record the UV-visible spectrum of the dye from 800–350 nm, diluting with more ethanol if necessary. Record the λ_{max} of the dye.

Part E: Determining the pH Indicator Range of the Prepared Dyes

Prepare a dye solution by dissolving 10 mg of the dye in 100 mL of 95% ethanol. Dissolve 2 drops of the dye solution in 20 mL of deionized water. Swirl to mix thoroughly. Divide the solution into three test tubes. After recording the color of the solution and the pH using pH paper, set one test tube aside. To the second test tube, add 0.1 M HCl dropwise, swirling well after each addition, until a color change occurs. Record the color and the pH. To the third test tube, add 0.1 M NaOH dropwise, swirling well after each addition, until a color change occurs. Then record the resulting color and the pH.

Determine whether or not the synthesized dye would be a good indicator. If so, indicate the range of indicator activity.

Dye	Initial Solution		0.1 M HCl		0.1 M NaOH	
	Color	pH	Color	pH	Color	pH

Part F: Determining Antibacterial Properties of Dye (optional)

Obtain 2 Mueller-Hinton agar plates, a broth culture each of *Staphylococcus aureus* and *Escherichia coli,* 2 sterile swabs, a pair of forceps, and a petri dish containing sterile blank disks. Swab each plate with the corresponding organism. Be sure to cover every square mm of surface area. Dissolve or suspend 30 mg of the dye in 10 mL of deionized water. Dip a sterile blank disk into the dissolved dye solution with forceps and place the disks equidistant from each other on the plate. Be sure to label the back of the plate with the dye number. Repeat for the second organism. Incubate the plates for 24–48 hours. Measure the zone of inhibition (zone of clearing) around each disk. The larger the zone of inhibition, the more effective the dye is at controlling bacterial growth. Dispose of all plates and broths in biohazard waste.

Results and Conclusions for Parts A through F

1. Calculate the percent yield of the azo dye.
2. Make a table of the dyes produced from each assigned combination and record the colors of the solid dyes.
3. Evaluate the effectiveness of the synthesized dye in the direct dyeing of the cloth. Was the dye colorfast? How did the assigned dye compare to other azo dyes? If other fabrics were also tested, evaluate the dye's performance with the various fabrics used.
4. Indicate whether or not the dye produced in this experiment functions as an indicator. Explain. If applicable, give the range of indicator activity.
5. Write structures of the dye under acidic conditions and under basic conditions.
6. Compile a table of results from other students, recording the colors of the synthesized dyes and the calculated λ_{max} values. Note and explain any structural differences between different colored compounds.
7. Measure the zone of inhibition for each plate in Part F. Compare the antibacterial effects of the dye with *Staphylococcus aureus* (a gram-positive organism) and *Escherichia coli* (a gram-negative organism). Is the dye selective for gram-positive or gram-negative bacteria? Compare the results of other synthesized azo dyes.

Cleanup & Disposal

Put all solutions containing aromatic amine or azo dye in a waste container under the hood. Dispose of all plates and broths in biohazard waste.

Critical Thinking Questions

1. Explain why the coupling of the diazonium salt with a phenol or an aromatic amine occurs at the para position. Justify your answer using resonance forms.
2. Why was concentrated sulfuric acid added to the dye solution in Part C?

3. When preparing the diazonium salt, the solution is tested with potassium iodide-starch paper. A positive test is the immediate formation of a blue color. What is the KI-starch paper testing for? Explain.

4. Look up the structure of the following dyes and propose a synthesis using a diazonium coupling procedure:

   a. Para Red   c.   Aniline Yellow

   b. Methyl Orange   d.   Congo Red

5. Explain why the addition of the HCl or NaOH improves the ability of azo dyes to dye fabrics.

6. Explain why silk and wool are more easily dyed with azo dyes than other fabrics like nylon or cotton.

7. Propose a mechanism for the formation of Aniline Purple, first synthesized by Perkins.

8. *Library project:* Some dyes cannot be applied directly, but must use a mordant to adhere to fabrics. Describe how this process differs from direct dyeing.

9. *Library project:* Indigo dye is an example of vat dyeing, in which the dye is applied in a colorless form and subsequent chemical reaction produces the color. Look up the structure of indigo and explain how vat dyeing produces the characteristic bluish color of indigo.

10. *Library project:* Investigate which dyes are used today to dye fabrics.

11. *Library project:* Some azo dyes have been used in food coloring. Which ones have been used? Are any dyes still used? What health problems were encountered as a result of using azo dyes in food coloring?

12. *Molecular modeling assignment:* Construct a model of the diazonium salt formed from the reaction of aniline with nitrous acid. Minimize the energy. Determine the electron density surface of the diazonium salt. Based upon the results, determine where a nucleophile is likely to attack.

13. *Molecular modeling assignment:* Construct a model of the azo dye synthesized in the experiment. Calculate energies for the highest occupied molecular orbital (HOMO) and the lowest unoccupied molecular orbital (LUMO). Use the result to explain why dyes absorb in the visible region of the spectrum.

Experiment 22.2: Synthesis of Pyrazole and Pyrimidine Derivatives

Heterocycles constitute a very important class of organic compounds. Many drugs and biochemical metabolites are heterocycles. In this experiment, you will:

- condense an unknown diketone with hydrazine or guanidine.

- prepare a five-member or six-member ring heterocycle and characterize the compound.

- identify the product (pyrazole or pyrimidine derivative).

- determine the structure of the unknown diketone.

- propose a mechanism of formation of the product.

Techniques

Technique F	Recrystallization and filtration
Technique M	Infrared spectroscopy
Technique N	Nuclear magnetic resonance spectroscopy

Background

Five-member ring and six-member ring heterocycles are widespread in nature and several of them are crucial to life. Of five-member ring heterocycles, pyrroles and imidazoles are very important (Behr et al., 1967). Pyrroles are components of hemoglobin and chlorophyll. Imidazoles are important as intermediates in biosynthesis of purines and as both acid and base catalysts. Pyrazoles are important synthetic intermediates and several derivatives have been marketed as drugs.

The synthesis of pyrazole derivatives will be investigated in Parts A and B of this experiment. You will receive an unknown diketone for reaction with the condensing agent, hydrazine. The possible unknown ketones are 2,4-pentanedione and 3-methyl-2,4-pentanedione. The product will be a substituted pyrazole listed in Table 22.1-1.

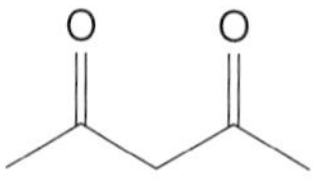

2,4-pentanedione 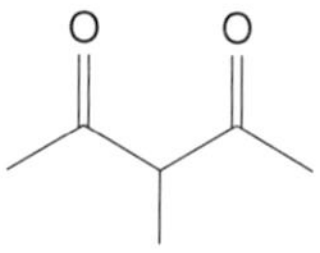   3-methyl-2,4-pentanedione

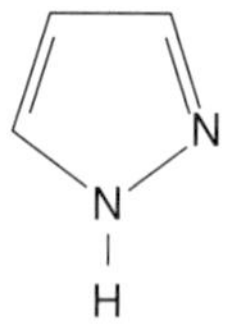

1H-pyrazole

Table 22.2-1 Melting Points of Pyrazoles

Heterocyclic product	Melting point (°C)
3,5-Dimethyl-1H-pyrazole	106–107
3,4,5-Trimethyl-1H-pyrazole	138–139

By considering the mechanism of reaction and the possible components leading to the heterocycle, it will be possible to determine the structure of the starting diketone. A typical reaction is shown here.

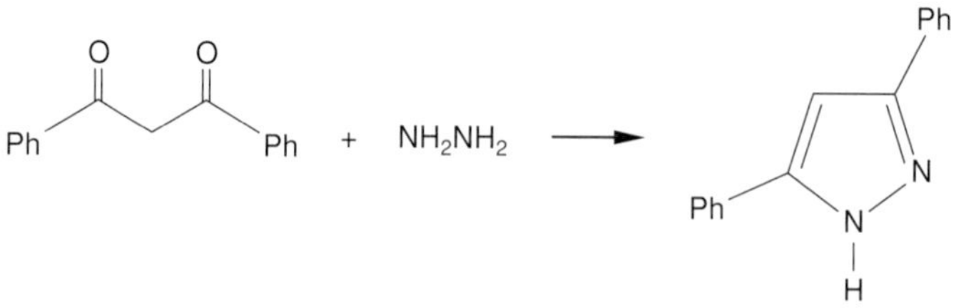

1,3-diphenyl-1,3-propanedione   3,5-diphenyl-1H-pyrazole

Additional starting diketones may be assigned by the instructor. Upon completion of the reaction and the workup procedure, identification of the product will be determined with the aid of the melting point with verification by IR and ^{1}H NMR spectroscopy.

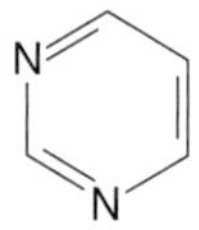

Pyrimidine derivatives will be prepared in Parts C and D of this experiment. Pyrimidines play an important role in nature. They are crucial components in the genetic material of life. Three pyrimidines are DNA and/or RNA bases. Uracil is found in RNA and thymine is found in DNA. Cytosine is present in both types of nucleic acids.

pyrimidine

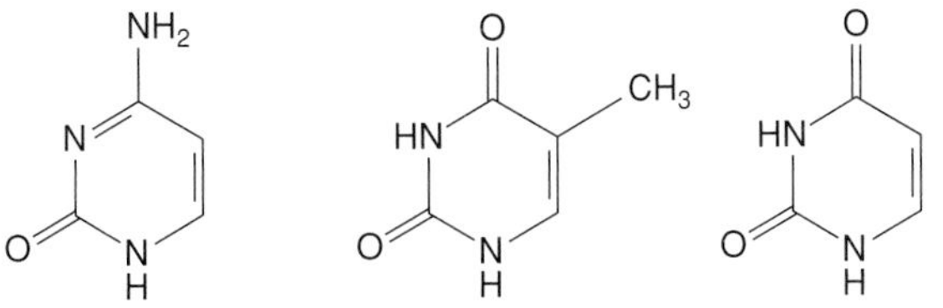

cytosine   thymine   uracil

These compounds can be synthesized in the laboratory using a "3 + 3" condensation (Brown, 1962; 1970). Two nitrogens from a carbonic acid derivative such as urea can be condensed to form cyclic amides with bifunctional reagents. One such synthesis of uracil is shown here.

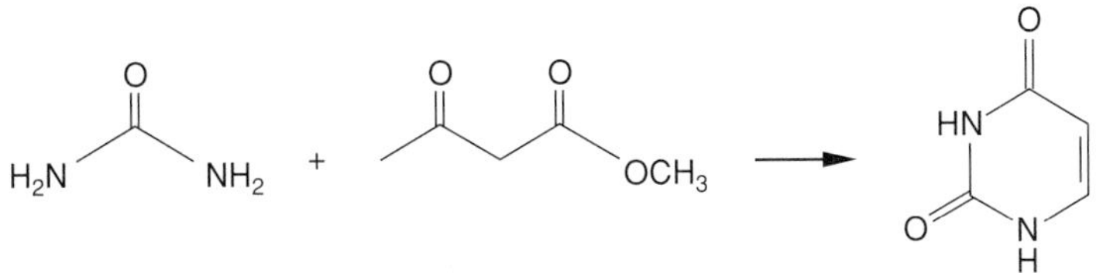

urea   methyl 3-oxobutanoate   uracil
            (methyl acetoacetate)

Uracil is an aromatic compound that consists of two tautomers. The thermodynamically more stable tautomer is the form containing the amide linkages. This is opposite to the situation with phenols, where the enol form is more stable.

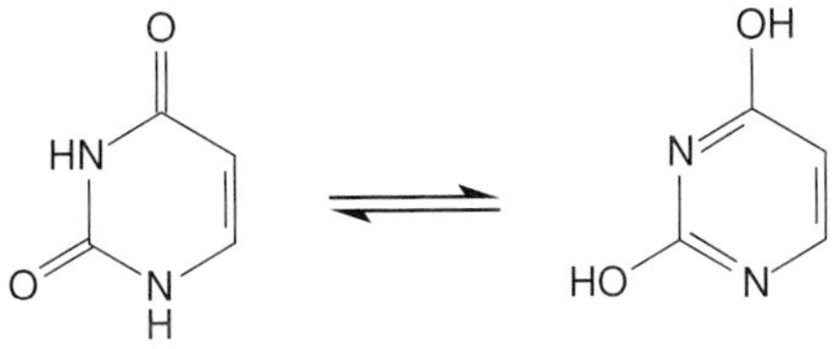

more stable tautomer   less stable tautomer

Synthesis of pyrimidines is important in the pharmaceutical industry. For example, some drug formulations are based on similarities with metabolites. One such compound is the cytosine derivative, cidofovir, an antiviral drug introduced in 1996. Another drug based on cytosine is an antineoplastic drug introduced to the market in 1998, known as capecitabine.

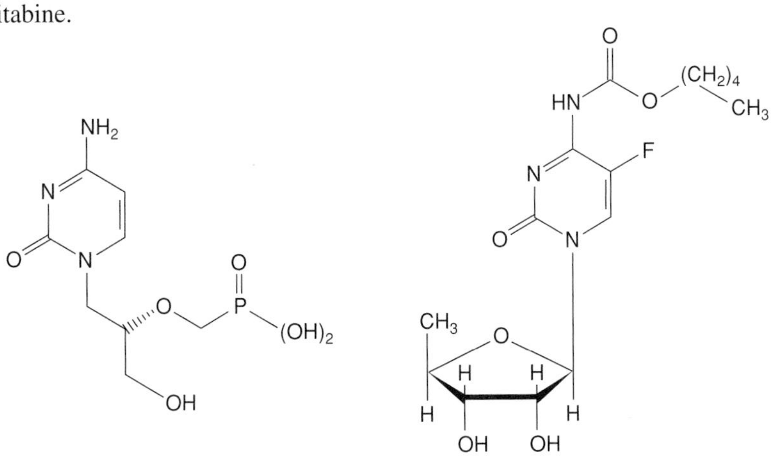

cidofovir   capecitabine

Development of new drugs like those above is an extremely expensive and time-consuming process. The structures are fairly complex and the cost of synthesis is considerable. The average time to get a drug to market is seven or more years. Drugs like cidofovir must be discovered first and then have their syntheses scaled up using the fewest and most economical steps. Drugs are very expensive because of the costs involved in producing them and having them approved for use on humans.

Synthesis of some simple pyrimidines can be done quickly and efficiently, particularly those for which reactive β-diketones can be employed as starting materials. An example of a synthesis of 2-amino-4-methyl-6-phenylpyrimidine is shown here.

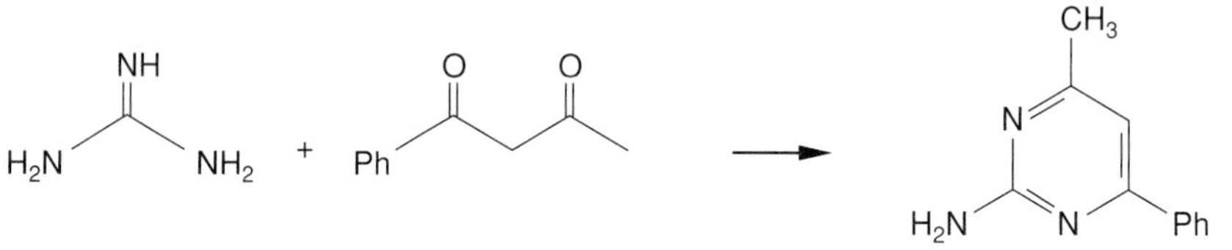

guanidine   1-phenyl-1,3-butanedione   2-amino-4-methyl-6-phenylpyrimidine
                        (1-benzoylacetone)

The mechanism of condensation consists of two nucleophilic addition-elimination reactions, both resulting in formation of imines after splitting out water. A third double bond is furnished from the imine of the guanidine condensing agent, rendering the compound aromatic.

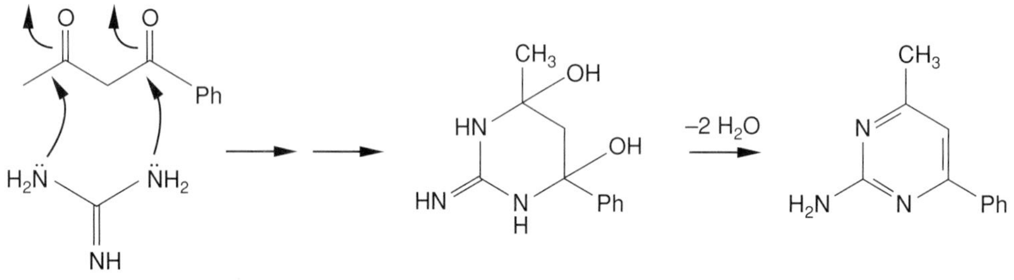

guanidine   2-amino-4-methyl-5-phenylpyrimidine

In this experiment, you will synthesize a pyrimidine by condensing guanidine and an unknown β-diketone. Each of the pyrimidines has a characteristic melting point that should permit identification. Spectroscopy can be used to see if the spectra are consistent with the proposed structure. The possible unknown starting materials are 2,4-pentanedione and 3-methyl-2,4-pentanedione. The instructor may assign additional unknowns.

The pyrimidines formed in these reactions are listed with reported melting points in Table 22.2-2.

Table 22.2-2   Melting Points of Pyrimidines

Pyrimidine	Melting point (°C)
2-Amino-4,6-dimethyl-	197
2-Amino-4,5,6-trimethyl-	206–207

Prelab Assignment

1. Draw the predicted organic product for each of the possible combinations of reactants.
2. Write a mechanism for the formation of one of the products drawn in the first question.
3. Tabulate significant infrared absorption bands that would help identify the structure of the possible product for the assigned reaction. Predict the chemical shifts of each type of proton in the ^{1}H NMR spectrum of the product for the assigned reaction.

Experimental Procedure

1. Wear eye protection at all times in the laboratory.
2. If any of the solutions contact the skin, wash immediately with cool tap water. Hydrazine solution is toxic and is a cancer suspect agent. Guanidine carbonate is an irritant. Use these reagents only in a well-ventilated hood. Wear gloves and work in the hood.

Part A: Microscale Synthesis of a Five-Member Ring Heterocycle from Hydrazine and an Unknown Diketone

Measure out 0.5 mL of hydrazine hydrate and add to a 5-mL conical vial containing a spin vane. Cool the vial in an ice bath. Add 0.30 mL of the unknown diketone. Add 2 mL of water. Swirl to mix. Start the stirrer and warm the solution or mixture in a heat block to about 50°C for 10 minutes. Cool to room temperature and cool further in an ice bath to complete precipitation. Filter the mixture and collect the solid on a Hirsch funnel. Wash with two small portions of ice water. Allow to air dry on the filter. Recrystallize the partially dry solid from a minimum amount of methanol. The hot solution should be clear and colorless. (If not, filter the hot solution and reheat to boiling.) Cool to room temperature and then cool further in an ice bath. Collect the solid on a Hirsch funnel and air dry. Weigh the dried product.

Part B: Miniscale Synthesis of a Five-Member Ring Heterocycle from Hydrazine and an Unknown Diketone

Measure out 2.0 mL of hydrazine hydrate and add to a 25-mL Erlenmeyer flask containing a spin bar. Add 5 mL of water. Cool the solution in an ice bath. Carefully add 1.0 mL of the unknown diketone. Swirl to mix. Remove the flask from the ice bath. Start the stirrer and warm in a water bath or hot plate to about 50°C for 10 minutes. Cool to room temperature and cool further in an ice bath to complete precipitation. Filter the mixture and collect the solid on a Büchner funnel. Wash with two 3-mL portions of ice water. Allow to air dry on the filter. Recrystallize the partially dry solid from a minimum amount of methanol. The hot solution should be clear and colorless. (If not, filter the hot solution and reheat to boiling.) Cool to room temperature and then in an ice bath. Collect the solid on a Büchner funnel and air dry. Weigh the dried product.

Characterization for Parts A and B

Melting Point: Obtain the melting point of the dry, recrystallized solid. Compare the value with the reported values in the table and identify the product.

Infrared Spectroscopy: Prepare a Nujol mull or a KBr pellet of the solid and record the IR spectrum. Interpret the spectrum.

NMR Spectroscopy: Obtain the ^{1}H NMR spectrum of the product in $CDCl_3$. Interpret the spectrum.

Results and Conclusions for Parts A and B

1. From the melting point, identify the pyrazole and the unknown diketone.
2. Examine the IR spectrum and indicate if the product gives data consistent with the structure of the expected product.
3. Examine the ^{1}H NMR spectrum and assign all signals based upon the structure of the expected product.
4. Calculate the percent yield of the reaction.
5. Propose a reaction mechanism for formation of the product.

Part C: Microscale Synthesis of a Substituted Pyrimidine

Weigh out 180 mg of guanidine carbonate and add it to a 5-mL conical vial containing a spin vane. Add 100 mg of powdered potassium acetate and 0.25 mL of diketone. Add 1 mL of water. Swirl to mix. Start the stirrer and warm in a heat block to about 50°C for 10 minutes. Cool to room temperature and cool further in an ice bath to complete precipitation. Filter the mixture and collect the solid on a Hirsch funnel. Wash with two small portions of ice water. Allow to partially air dry on the filter. Recrystallize the partially dry solid from a minimum amount of methanol. The hot solution should be clear and colorless. (If not, remove the solids using a filter pipet and reheat the solution to boiling.) Cool to room temperature and then cool further in an ice bath. Collect the solid on a Hirsch funnel and air dry. Weigh the dried product.

Part D: Miniscale Synthesis of a Substituted Pyrimidine

Weigh out 540 mg of guanidine carbonate and add it to a 10-mL Erlenmeyer flask containing a spin bar. Add 300 mg of powdered potassium acetate and 0.75 mL of diketone. Add 2 mL of water. Swirl to mix. Start the stirrer and warm on a hot plate or hot water bath to about 50°C for 10 minutes. Cool to room temperature and cool further in an ice bath to complete precipitation. Filter the mixture and collect the solid on a Büchner funnel. Wash with two small portions of ice water. Allow to partially air dry on the filter. Recrystallize the partially dry solid from a minimum amount of methanol. The hot solution should be clear and colorless. (If not, filter the hot solution and reheat to boiling.) Cool to room temperature and then in an ice bath. Collect the solid on a Büchner funnel and air dry. Weigh the dried product.

Characterization for Parts C and D

Melting Point: Obtain the melting point of the dry, recrystallized solid. Compare the value with the reported values in the table and identify the product.

Infrared Spectroscopy: Prepare a Nujol mull or a KBr pellet of the solid and record the IR spectrum. Interpret the spectrum.

NMR Spectroscopy: Obtain the ^{1}H NMR spectrum of the product in $CDCl_3$. Interpret the spectrum.

Results and Conclusions for Parts C and D

1. From the melting point identify the pyrimidine and the unknown diketone.
2. Examine the IR spectrum and indicate if the product from the melting point gives data consistent with the prediction. Indicate which absorption bands were important in your decision. Do the IR data support the structural assignment? Explain.

3. Examine the ^{1}H NMR spectrum and assign all signals in light of the product from the melting point. Do the NMR spectral data support the conclusion based on the melting point? Correlate specific signals in the NMR to structural features of both compounds if a clear decision was not possible from the melting point. Are the NMR data decisive?

4. Calculate the percent yield of the reaction.

5. Propose a mechanism for the observed product.

Cleanup & Disposal

Neutralized aqueous solutions may be washed down the drain with water.

Critical Thinking Questions

1. Suggest a pair of starting materials to synthesize pyrazole. Propose a mechanism.

2. Suggest a pair of starting materials to synthesize cytosine. Propose a mechanism.

3. What pyrazole(s) would be expected as the product(s) starting with hydrazine and ethyl acetoacetate? With methylhydrazine and benzoylacetone? Why might it be necessary to consider the possibility of obtaining more than a single product for these reactions?

4. Give the product of reaction of 2,4-pentanedione and acetamidine ($CH_3C(=NH)NH_2$).

5. What pyrimidine would be expected as the product starting with acetamidine and diethyl malonate? With urea and 2,4-pentanedione?

6. Propose a synthesis of 4,5-dimethyl-1H-pyrazole.

7. Propose a synthesis of 2-phenyl-4,5,6-trimethylpyrimidine.

8. Is it likely that both ring connections are made simultaneously in this experiment as was observed for the Diels-Alder synthesis in Chapter 9, or is it more likely that each connection is made separately? Explain.

9. Consider the two tautomeric forms of urea. Give a reason to explain the greater stability observed for the keto tautomer over the enolic form. Is this expected to be true for pyrimidines such as cytosine? Explain.

10. Suggest a spectroscopic method that could be used to show that the keto form of uracil predominates at room temperature. Why does the keto form predominate over the enol form?

11. *Molecular modeling assignment:* Calculate the electron density for 2,4-pentanedione and benzoylacetone. What do the results suggest concerning the reactivity of these compounds?

12. What driving force is likely responsible for formation of the products in this experiment?

13. Using another type of dicarbonyl compound not used in this experiment, suggest a synthesis of a heterocyclic compound that contains a five-member ring.

References

Behr, L. C., Fusco, R., and Jarboe, C. H. *Pyrazoles, Pyrazolines, Pyrazolidines, Indazoles and Condensed Rings,* Wiley, R. H., ed. New York: Wiley-Interscience, 1967.

Brown, D. J. *The Pyrimidines.* New York: Interscience, 1962.

Brown, D. J. *The Pyrimidines, Supplement I.* New York: Interscience, 1970.

Experiment 22.3: Synthesis of Heterocyclic Compounds

Heterocycles constitute a very important class of organic compounds. Most drug and many biochemical metabolites are heterocycles. In this experiment, the student will:

+ prepare a porphyrin derivative from pyrrole.

+ prepare a fused, bicyclic heterocycle containing an imidazole ring and characterize the compound.

+ determine the mechanism of formation of the heterocyclic compound.

Techniques

Technique F	Filtration and recrystallization
Technique G	Reflux
Technique K	Thin-layer chromatography
Technique M	Infrared spectroscopy
Technique N	Nuclear magnetic resonance spectroscopy
Technique O	Visible spectroscopy

Background

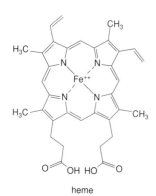

heme

Five-member ring heterocycles are found widely in nature. One example is heme, which contains four pyrrole rings per heme unit (Chance et al., 1966; Kim et al., 1978).

Heme derivatives are metal complexes of the cyclic structures comprising the porphyrins framework in biologically important compounds such as chlorophyll and hemoglobin. One of the subunits of hemoglobin is called heme, shown here. Heme contains four disubstituted pyrrole rings.

In Part A of this experiment, a purplish-blue porphyrin derivative known as *meso*-tetraphenylporphyrin will be synthesized by condensing pyrrole with benzaldehyde according to the remarkably straight-forward scheme below (Adler et al., 1967).

$$4 \text{ pyrrole} + 4 \text{ PhCHO} \longrightarrow \textit{meso-}\text{tetraphenylporphyrin} + 4 \text{ H}_2\text{O}$$

pyrrole   *meso*-tetraphenylporphyrin

The product, *meso*-tetraphenylporphyrin, will be characterized using visible spectroscopy. You will compare your experimental λ_{max} values to those from the literature. (Mullins et al., 1965).

In Part B of this experiment, you will synthesize a purplish-blue porphyrin derivative by condensing pyrrole with an unknown aromatic aldehyde. You will identify the product by TLC and IR spectroscopy. The possible aromatic aldehydes are benzaldehyde, anisaldehyde (*p*-methoxybenzaldehyde), *p*-methylbenzaldehyde, and *p*-ethylbenzaldehyde.

Other heterocycles also play a role in life processes, including 4,5-dimethylbenzimidazole, which exists in the form of a nucleotide component of vitamin B_{12}. The structures of 4,5-dimethylbenzimidazole and vitamin B_{12} are shown here.

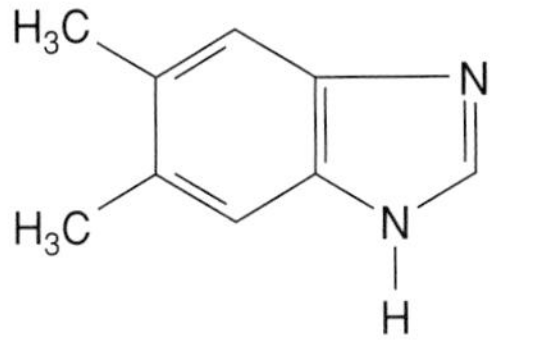

4,5-dimethylbenzimidazole

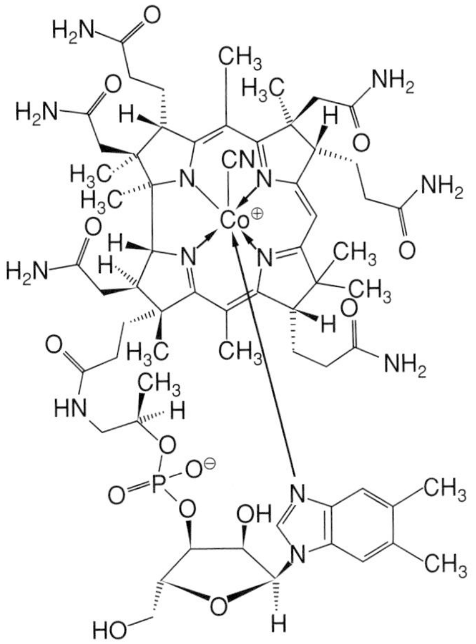

Vitamin B_{12} (Cyanocobalamin)

The 4,5-dimethylbenzimidazole ring is linked to a sugar phosphate through one nitrogen of the five-member ring and to a cobalt (III) ion at the other nitrogen. Vitamin B_{12} is required by animals and must be ingested. Animals having vitamin B_{12} deficiencies may exhibit nervous disorders and uncontrolled movements. Chickens with this anemia may produce eggs that don't hatch. Most vitamin B_{12} is produced synthetically by pharmaceutical companies. In Part B of this experiment, you will synthesize benzimidazole, a component of vitamin B_{12}.

Preparation of benzimidazoles is usually done by starting with a disubstituted benzene (Grimmett, 1997). In Part C of this experiment, synthesis of benzimidazole will be accomplished by heating *o*-phenylenediamine with formic acid. The ^{1}H NMR spectrum of benzimidazole is shown in Figure 22.3-1.

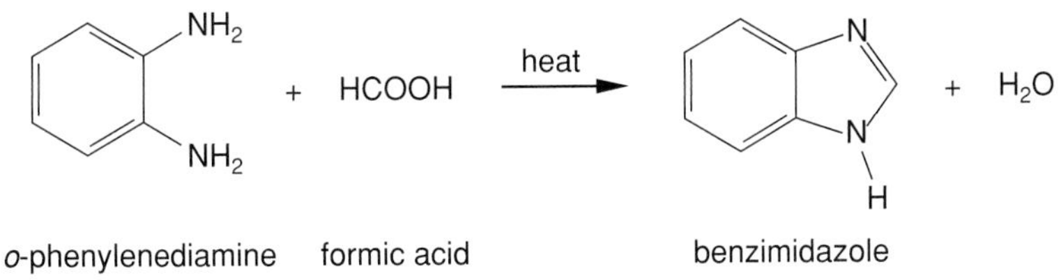

o-phenylenediamine   formic acid   benzimidazole

Prelab Assignment

1. Refer to the experimental procedure and identify the reagent that is used in excess. Explain using calculations.
2. Calculate the theoretical yield of the product.

3. Tabulate significant infrared absorption bands that would help identify the structure of the product or of one of the possible unknown products. Predict the chemical shifts of each type of proton in the ^{1}H NMR spectrum of the product.
4. Tabulate significant infrared absorption bands that would help identify the structure of the product for each combination in Part B. Predict the chemical shifts of each type of proton in the ^{1}H NMR spectrum of the product.

Experimental Procedure

1. Wear eye protection at all times in the laboratory.
2. If any of the solutions contact the skin, wash immediately with cool tap water. Pyrrole is an irritant. Propanoic acid is toxic and corrosive. *o*-Phenylenediamine is a toxic, cancer suspect agent. Use these reagents only in a well-ventilated hood. Wear gloves and work in the hood.

Part A: Miniscale Reaction of Benzaldehyde and Pyrrole

Measure out 8.0 mL of propanoic acid and add to a 25-mL round-bottom flask containing a spin bar. Attach a water-jacketed condenser and set up for reflux. Begin to warm the flask using a heat block or sand bath. Measure out 0.20 mL of benzaldehyde and add it to the heating liquid. Then measure out 0.13 mL of distilled pyrrole. Add it to the solution and reflux for 30 minutes. Cool the mixture and disassemble the apparatus. Further cool the mixture in an ice bath. Suction filter the cooled mixture using a Büchner funnel. Wash the colored solid with cold methanol (4 mL) and then with three 2-mL portions of cold water. (The odor of propanoic acid should be absent. If not, wash with two more portions of cold water.) Collect the solid on a Büchner funnel and air dry. Weigh the dried product.

Characterization for Part A

Melting Point: Do not obtain the melting point of the dry solid. The melting point is too high to determine.

Infrared Spectroscopy: Prepare a Nujol mull or a KBr pellet of the solid and record the IR spectrum. Interpret the spectrum, noting key bands characteristic of functional groups.

Figure 22.3-1

^{1}H NMR spectrum of benzimidazole

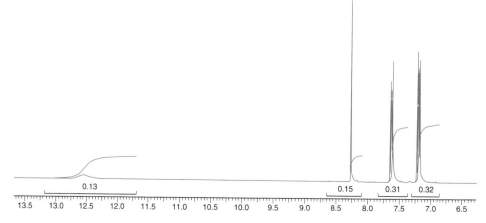

| 0.13 | | 0.15 | 0.31 | 0.32 |
| 13.5 13.0 12.5 12.0 11.5 11.0 10.5 10.0 9.5 9.0 8.5 8.0 7.5 7.0 6.5 |

Visible Spectroscopy: Dissolve 2–3 mg of the product in 5 mL of toluene. Warm the solution if necessary. Place the solution in one of a pair of matched cuvettes and fill the other cuvette with toluene. Measure the visible spectrum by scanning from 800–400 nm using a spectrophotometer. Dilute the solution if necessary. Reported λ_{max} values are at 418, 481, 515, 548, 592, and 648 nm.

Part B: Microscale Reaction of Pyrrole and an Unknown Aromatic Aldehyde

Pyrrole air-oxidizes easily and must be purified by distillation or passing through a silica gel column before using. Add 8.0 mL of propanoic acid to a 25-mL round-bottom flask containing a spin bar. Attach a water-jacketed condenser and set up for reflux. Begin to warm the flask using a heat block or sand bath. Add 0.20 mL of the unknown aromatic aldehyde to the heating liquid. Then add 0.13 mL of purified pyrrole to the solution. Reflux with stirring for 30 min. Cool the mixture and disassemble the apparatus. Further cool in an ice bath. Suction filter the cooled mixture on a Büchner funnel. Wash the colored solid with cold methanol (4 mL) and then with three 2-mL portions cold water. (The odor of propanoic acid should be absent. If not, wash with two more portions of cold water.) Collect the solid on a Büchner funnel and air dry. Weigh the dried product.

Characterization for Part B

Thin-layer Chromatography: Dissolve a few crystals of the product in methylene chloride. Spot at the origin. Also spot standard solutions of the four possible heme products. Develop in 50:50 methylene chloride:hexanes.

IR Spectroscopy: Prepare a Nujol mull of the solid and record the IR spectrum. Interpret, noting key bands characteristic of functional groups indicated in accordance with the predicted product or products.

Results and Conclusions

1. Calculate R_f values for the unknown and the four standards.
2. Examine the IR spectrum and indicate if the product gives data consistent with the structure of the expected product.
3. Examine the ^1H NMR spectrum and assign all signals based upon the structure of the expected product. Are there or should there be any signals below $\delta 6.0$?
4. Determine the identity of the unknown heme and write a reaction for its formation.
5. Calculate the percent yield of the reaction.

Part C: Microscale Reaction of o-Phenylenediamine and Formic Acid

Weigh out 3 mmol of *o*-phenylenediamine and add to a 5-mL conical vial containing a spin vane. Add 1 mL of 96% formic acid. Swirl to mix. Attach a water-jacketed condenser. Start the stirrer and reflux in a heat block for about 1 hr. Cool to room temp. and transfer the dark solution to a 25-mL Erlenmeyer flask in an ice bath. Wash the vial with 2 mL of water and add the washings to the Erlenmeyer flask. Add 1.5 mL of conc. ammonium hydroxide to the solution to precipitate the product. Filter the cold mixture and collect the solid on a Hirsch funnel. Wash with 1-2 mL of ice water. Allow to air dry briefly on the filter. Recrystallize the partially dry solid from a minimum amount of hot water. If the solution is dark, add a small amount of decolorizing charcoal to the warm (not boiling) solution. Gravity filter the hot solution. Allow to cool to room temp. and then cool further in an ice bath. Collect the solid on a Hirsch funnel and air dry. Weigh the dried product.

Characterization for Part C

Melting Point: Obtain the melting point of the dry, recrystallized solid. Compare the reading with the reported value of 172-174°C.

    IR Spectroscopy: Prepare a Nujol mull or a KBr pellet of the solid and record the IR spectrum. Interpret the spectrum, noting key bands characteristic of the predicted product.

    NMR Spectroscopy: Obtain the ^{1}H NMR spectrum of the product in $CDCl_3$. Interpret the spectrum.

Results and Conclusions for Part C

1. Examine the IR spectrum and indicate if the product gives data consistent with the structure of the expected product.
2. Calculate the percent yield of the reaction.
3. Propose a reaction mechanism for formation of the product.

Cleanup & Disposal

Neutralized aqueous solutions may be washed down the drain with water.

Critical Thinking Questions

1. Write out reactions that might be used to synthesize: 4,5-dimethylbenzimidazole, N-methylbenzimidazole, 2-methylbenzimidazole.

2. Is it likely that ring connections are made simultaneously for *meso*-tetraphenyl-porphyrin and benzimidazole ring formation in this experiment as was observed for the Diels-Alder synthesis in Chapter 9 or is it more likely that each connection is made separately? Explain.

3. Synthesis of benzimidazole in this experiment illustrates a method known as a "4 + 1" ring formation, indicating the number of atoms furnished to form the ring by each of the two starting materials. Suggest starting materials that might be used to prepare: pyrrole using a "4 + 1" synthesis; imidazole using a "3 + 2" synthesis.

4. *Molecular modeling assignment:* Use a molecular modeling program to calculate the electron density for formic acid; for *o*-phenylenediamine. What do the results suggest concerning the structures and potential reactivities of these compounds?

5. What driving force is likely responsible for formation of the compounds synthesized in this experiment? Explain.

References

Adler, A. D., Longo, F. R., Finarelli, J. D., Goldmacher, J., Assour, J., Korsakoff, L. *J. Org. Chem.* 32, (1967): 476.

Chance, B., Estabrook, R. W., and Yonetani, T. *Hemes and Hemoproteins.* New York: Academic Press, 1966.

Grimmett, M. R. *Imidazole and Benzimidazole Synthesis.* London: Academic Press, 1997.

Mullins, J., Adler, A. D., Hochstrasser, R. *J. Chem. Phys.* 43, (1965): 2548.

Experiment 22.4: Synthesis of Heterocycles and Kinetics by Spectroscopic Analysis

In this experiment, you will focus on reaction mechanisms of free radicals and reaction kinetics. You will:

♦ synthesize an aromatic heterocycle starting from acyclic precursors.

♦ dimerize an aromatic heterocycle and study its dissociation to give free radicals.

♦ use a spectrophotometer to study the decay rate of a long-lived, colored free radical.

Techniques

Technique F	Filtration and recrystallization
Technique G	Reflux
Technique M	Infrared spectroscopy
Technique N	Nuclear magnetic resonance spectroscopy
Technique O	Visible spectrophotometry

Background

This experiment consists of three parts: synthesis of 2,4,5-triphenylimidazole (lophine), or a substituted lophine; oxidation of lophine or lophine derivative to form a heterocyclic free radical and a dimer; and a kinetics study of the recombination reaction of the radical to form the dimer (Pickering, 1980).

Lophine and its derivatives are made from benzil, ammonium acetate, and benzaldehyde or a substituted benzaldehyde as shown here.

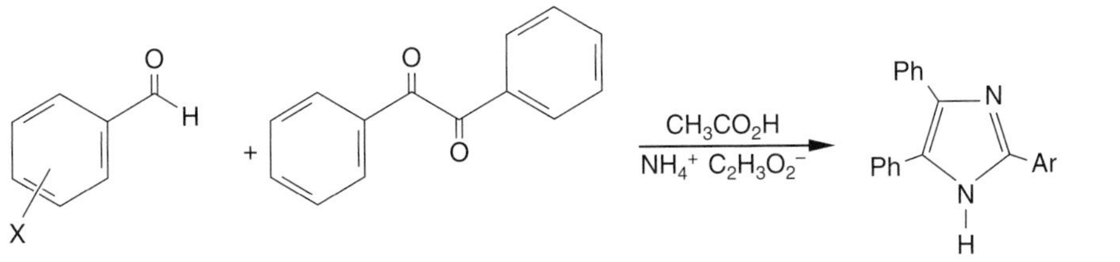

benzaldehyde or substituted benzaldehyde

benzil

CH_3CO_2H / $NH_4^+ \; C_2H_3O_2^-$

lophine or lophine derivative
Ar = phenyl or substituted phenyl

Treatment of lophine with N-iodosuccinimide (NIS) results in transfer of hydrogen from lophine to form succinimide and a dimer of lophine or lophine derivative (Schoffstall and Law, unpublished results).

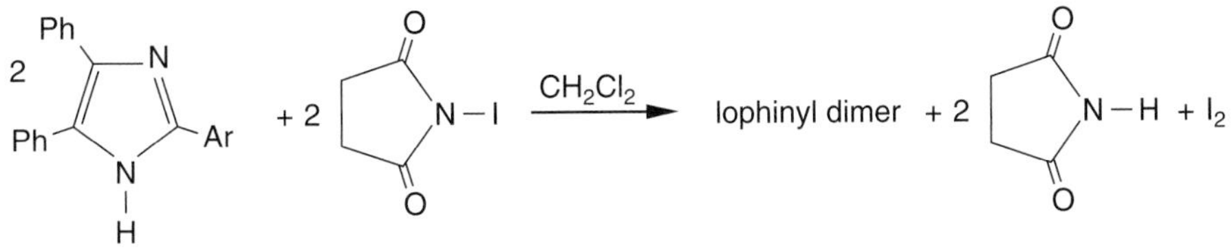

lophine or lophine derivative

N-iodosuccinimide (NIS)

CH_2Cl_2

lophinyl dimer + 2

succinimide

$+ I_2$

Exposure of the dimer to sunlight generates colored lophinyl radicals. The radicals show different characteristics, depending on the type of substitution. Substituted lophinyl dimers produce radicals of various colors, ranging from very light blue/green to very dark purple. The radicals gradually reform the colorless dimer. The rate of the radical recombination may be strongly influenced by substituents. The rate of this reaction can then be monitored by spectrophotometry. The reaction scheme is shown here.

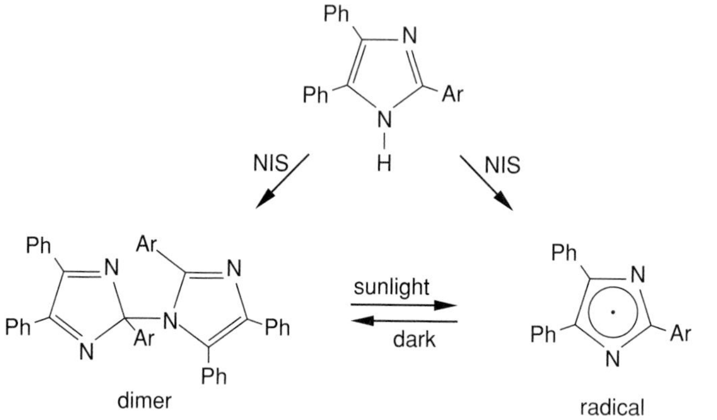

Ar = Ph, *p*-CH_3OPh, 2,4-$(CH_3O)_2Ph$, *p*-ClPh, *p*-FPh, *p*-BrPh, *o*-ClPh

Physical data for all lophines, dimers, and radicals formed are given in Table 22.4-1. Any of the benzaldehydes given in Table 22.4-1 may be chosen as reactants for the preparation of a substituted lophine (a 2-aryl-4,5-diphenylimidazole). Reaction kinetics will be investigated at temperatures below 40°C. Radicals that give slow recombination kinetics are studied between 25° and 40°C. Those radicals that recombine quickly (fast kinetics) are studied between 5° and 25°C.

Prelab Assignment

1. The instructor will assign a substituted benzaldehyde for investigation. Calculate the mass and mol required of all starting materials. Determine which reagent is limiting.
2. Write the reaction mechanism for the preparation of the assigned lophine. Hint: Think of the ammonium ion as furnishing a proton to a carbonyl group, followed by attack of ammonia on the carbon of the protonated carbonyl. This can occur for each carbonyl of benzil. Reaction with benzaldehyde or substituted benzaldehyde gives ring closure.
3. Write the balanced equation for reaction of sodium thiosulfate with molecular iodine.
4. What should be graphed to determine if a reaction is first order? What is graphed to determine if a reaction is second order?
5. What should be graphed to determine the energy of activation of a reaction?

Experimental Procedure

Safety First!

Always wear eye protection in the laboratory.

1. Wear eye protection at all times in the laboratory.
2. Wear gloves when handling solutions containing imidazole derivatives. Biological effects due to 4-substituted imidazoles are probably small, but skin contact should be avoided. Also avoid contact with methylene chloride, toluene, and solutions of N-iodosuccinimide.
3. Toluene is a toxic liquid. Work in a hood or in a well-ventilated area for this experiment.

Table 22.4-1   Physical Properties of Lophines, Dimers, and Lophinyl Radicals

Aryl groups	Lophine derivative M.P./°C	Dimer M.P./°C	λ_{max} Dimer (nm)[a]	λ_{max} radical (nm)[a]	Rate: rel. to lophine
Phenyl	273–275	202	300	560 purple	—
p-Methoxyphenyl	231–232	127–128	290	602 deep blue	slow
p-Chlorophenyl	264	207–209	306	568 wine red	similar
p-Bromophenyl	254–258	210–211	284	560 light blue	fast
2,4-Dimethoxyphenyl	164–165	124–128	332	620 blue/green	fast
o-Chlorophenyl	192–194	173–180	265	540 purple	—
p-Fluorophenyl	206–210	355–370	262	554 purple/red	similar

[a] In toluene.

(Cescon et al., 1971; White and Sonnenberg, 1966)

Part A: Miniscale Preparation of Lophines (2-Aryl-4,5-diphenylimidazoles)

Dissolve 0.5 g of benzil, 2.4 mmol of a substituted benzaldehyde, and 2 g of ammonium acetate in 20 mL of acetic acid in a 50-mL round-bottom flask equipped with a reflux condenser and stir bar. Reflux the mixture for 1 hour with stirring. Cool the reaction mixture to room temperature, allowing solids to settle to the bottom. Decant the solution into a 250-mL beaker. Place any solids remaining in the round-bottom flask, mainly unreacted benzil, in the appropriately labeled waste container. Add 40 mL of water slowly with stirring to the 250-mL beaker. The mixture should be cloudy. Allow the mixture to stand for 10 minutes in an ice bath. Collect the solid by suction filtration using a Büchner funnel. If only a small amount of solid has been collected on the filter, add concentrated ammonium hydroxide to the filtrate to induce further crystallization and collect solids by suction filtration. Combine the solid products. Recrystallize the combined solids from ethanol/water. Two or more recrystallizations are needed to obtain pure crystals and to get good results when used for Part B. Recrystallization may be carried out easily using hot ethanol-water. Not all of the crude substituted lophine may dissolve in ethanol. Filter off any solid that does not dissolve. Do not use more than 10 mL of ethanol. Add about 10–15 mL of water to reach the cloud point. Let the mixture cool to room temperature and then place in an ice bath. Vacuum filter the solid. Weigh the dried solid and record the melting point. Typical yields are 50–90%.

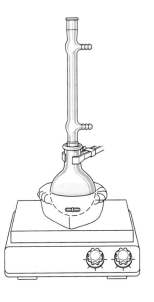

Part B: Miniscale Synthesis of Substituted Lophinyl (2-Aryl-4,5-diphenylimidazole) Dimers

Dissolve 1.0 mmol of a substituted lophine and 113 mg of NIS in 20 mL of methylene chloride in a 125-mL Erlenmeyer flask equipped with a magnetic stir bar. The characteristic color of the radical will appear, followed in less than 2 minutes by the color of iodine. Add 20 mL of 5% aqueous sodium thiosulfate. If the iodine color persists, add more thiosulfate solution until the color of the radical returns. (Note that the 2-(*p*-halophenyl) radicals have colors that are similar to the color of iodine.) After 15 minutes, pour the reaction mixture into a 125-mL separatory funnel. Stopper the funnel and shake carefully with venting. Remove the stopper and allow the layers to separate. The aqueous layer should be colorless. Draw off the bottom layer into an Erlenmeyer flask and dry over anhydrous sodium sulfate. Decant the dried organic solution into a dry Erlenmeyer flask. The color of the solution is due to the 2-lophinyl radical. This solution may be used directly for the kinetics study (Part C below) or the dimer can be isolated. To isolate the dimer, evaporate the methylene chloride solution in the dark. Recrystallize from ethanol.

Characterization for Parts A and B

Melting Point: Determine the melting points of lophine or the lophine derivative and the dimer. The reported melting point of lophine is 273–275°C. Refer to Table 22.4-1 for a complete list of melting points. Some of the melting ranges are broad.

    Infrared Spectroscopy: Prepare a Nujol mull or a KBr pellet of the lophine derivative and the corresponding dimer for IR analysis. Interpret the spectra in terms of the peaks expected. The IR spectrum of lophine is shown in Figure 22.4-1.

    NMR Spectroscopy: Dissolve a sample of a lophine derivative in $CDCl_3$ or in deuterated DMSO. Dissolve a sample of dimer in $CDCl_3$. Filter the solutions into clean, dry NMR tubes using filter pipets. Record the integrated NMR spectra and analyze the spectra. The ^1H NMR spectrum of lophine is shown in Figure 22.4-2.

Results and Conclusions for Parts A and B

1. Calculate the percent yield of the product.
2. Discuss similarities and differences between the IR and NMR spectra of the lophine derivative and the corresponding dimer.

Part C: Kinetics of the Recombination Reaction

General Directions

The same spectrophotometer should be used throughout the experiment. Set the Spectronic 20 instrument at the wavelength of the radical from Table 22.4-1 that should be used to give the maximum absorbance of the radical in toluene. Monitor the absorbance

Figure 22.4-1

IR spectrum of lophine (2,4,5-triphenylimidazole) (Nujol)

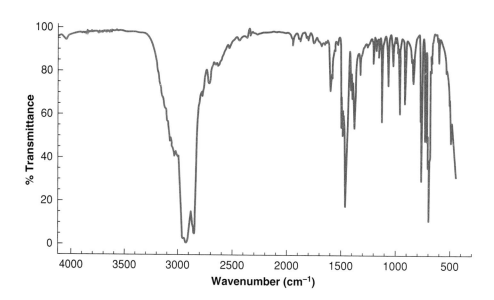

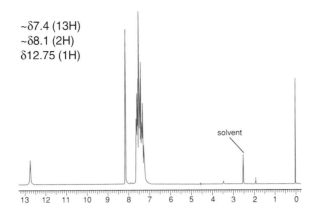

~δ7.4 (13H)
~δ8.1 (2H)
δ12.75 (1H)

solvent

13 12 11 10 9 8 7 6 5 4 3 2 1 0

Figure 22.4-2

^{1}H NMR spectrum of lophine (2,4,5-triphenylimidazole) (deuterated DMSO)

SOURCE: Reprinted with permission of Aldrich Chemical.

as a function of time. If the color disappears too rapidly, try a lower temperature to determine the reaction rate constant. Compile the data to determine the order of the reaction. Use data sets from different temperatures to determine the energy of activation of the recombination reaction.

Calibrating the Spectronic 20

Warm up the Spectronic 20 instrument while samples are being prepared. Dial in the desired wavelenth using the wavelength knob. To an 8-mL Spectronic 20 cuvette, add 6 mL of toluene and place the tube in the sample holder. Refer to Technique O for step-by-step directions. Fill a cuvette with a solution of a lophine radical and place the cuvette in the sample compartment. The absorbance should be between 0.5–0.8. If the reading is greater than 0.8, allow the solution to stand in the dark until the reading is within the desired range. If the reading is less than 0.5, add more of the lophine dimer to the solution and swirl to dissolve. Alternatively, expose the solution briefly to a high-intensity light source to bring out a more intense color.

Sample Preparation

Use the solution directly from Part B or dissolve about 20 mg of dimer in 20 mL of toluene. Filter the suspension through coarse filter paper into a Spectronic 20 cuvette. Irradiate the solution briefly using a sunlamp to develop the color of the radical. There is no need to be quantitative up to this point. Prepare three tubes containing about 6 mL each of solutions to be analyzed. One tube should be maintained at room temperature, one at about 15°C, and one at about 30°C. Use covered styrofoam cups as water baths. Measure the absorbance for each tube until the absorbance is half of its original value. Record absorbance and time every 2 to 3 minutes for the tube at 15°C, every minute for the tube at room temperature, and every 30 seconds for the tube at 30°C.

Reactions at lower temperature take longer (about 45 minutes), but give better results. The free radical may be regenerated by exposure to sunlight. The 2-(*p*-methoxyphenyl) substituted imidazolyl radical is more stable than the others and the kinetics should be studied between 20° and 30°C. The 2-(2,4-dimethoxyphenyl) and 2-(*p*-bromophenyl) substituted imidazolyl radicals are less stable than the others and the kinetics should be studied between 5° and 25°C.

Results and Conclusions for Part C

1. Determine the order of the reaction by plotting absorbance versus time, ln absorbance versus time, or 1/absorbance versus time. Determine the specific rate constant from the kinetic plot that gives a straight line.
2. Prepare a graph using data from the three different temperatures. Determine the energy of activation from the slope of the line. Typical E_a values range from 5 to 15 kcal/mol. Compare your results of kinetics studies with those of other students.
3. Is the reaction order consistent with that which might be expected for a radical recombination reaction? Explain.

Cleanup & Disposal

Place recovered benzil in a container labeled "recovered organic solids."
Aqueous acetic acid solutions may be poured down the drain. Place all methylene chloride solutions in a bottle labeled "halogenated solvent waste" in the hood. Place all toluene solutions in a bottle labeled "nonhalogenated solvent waste."

Critical Thinking Questions *(The harder ones are marked with a ❖.)*

1. Predict the structures of possible side products in the synthesis of lophine or substituted lophine. What is the driving force behind the formation of the desired lophine product?
2. Suppose that 4,4′-dichlorobenzil were substituted for benzil in the lophine synthesis in Part A. Draw the structure of the expected product of the reaction.
3. Why do melting points of many of the dimers in Table 22.4-1 have a broad range?
4. What qualitative analytical method might be used to monitor the synthesis of lophine or derivative of lophine along with any side products as the reaction progresses? Explain.
5. Explain which atom is oxidized and which is reduced during dimer formation.
❖ 6. The kinetic order of reaction may change during the latter stages of radical recombination. Suggest a possible reason.
❖ 7. Describe how the kinetics of radical recombination can be followed using the technique of electron spin resonance spectroscopy. Give a reason that this method was not used for this experiment.

References

Cescon, L. A., Coraor, G. R., Dessauer, R., Deutsch, A. S., Jackson, H. L., MacLachlan, A., Marcali, K., Potrafke, E. M., Read, R. E., Silversmith, E. F., and Urban, E. J. "Some Reactions of Triarylimidazolyl Free Radicals." *Journal of Organic Chemistry* 36 (1971): 2267.

Pickering, M. L. "Radical Recombination Kinetics." *J. Chem. Educ.* 37 (1980): 833.

Schoffstall, A. M., and Law, J. K., unpublished results.

White, D. M., and Sonnenberg, J. "Oxidation of Triarylimidazoles. Structures of the Photochromic and Piezochromic Dimers of Triarylimidazyl Radicals." *J. Amer. Chem. Soc.* 88 (1966): 3825.

Chapter 23

Aryl Halides

A ryl halides that contain one or more electron-withdrawing substituents may undergo a type of aromatic substitution reaction known as nucleophilic aromatic substitution. This reaction is different from an S_N1 or S_N2 reaction because a negatively charged intermediate is formed. The presence of two or more electron-withdrawing groups ortho and/or para to the carbon bearing the halo group causes the ring to be electron deficient and further facilitates attack by nucleophiles as detailed in Experiment 23.1. An example of nucleophilic aromatic substitution is shown below for the reaction of 1-chloro-2,4-dinitrobenzene with hydrazine to prepare 2,4-dinitrophenylhydrazine.

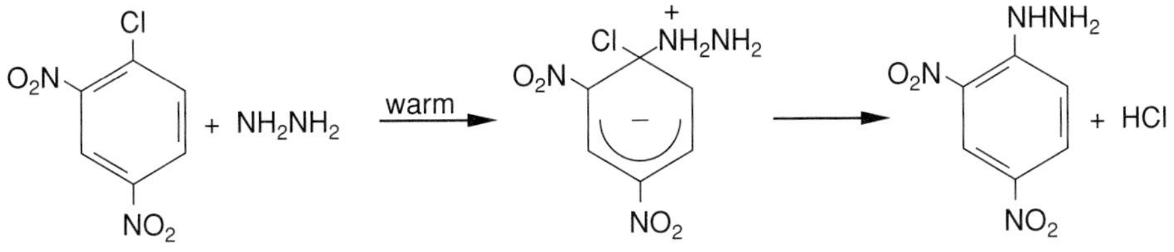

2,4-dinitrophenylhydrazine

Experiment 23.1: Nucleophilic Aromatic Substitution

In this experiment, you will study nucleophilic substitution of aromatic compounds and the types of substituents necessary for a successful reaction. You will:

♦   investigate the reaction of an aryl halide containing one or more electron-withdrawing subtituents.

♦   carry out a nucleophilic aromatic substitution reaction.

Techniques

Technique F	Filtration and recrystallization
Technique G	Reflux
Technique M	Infrared spectroscopy
Technique N	Nuclear magnetic resonance spectroscopy

Background and Mechanism

Nucleophilic aromatic substitution reactions occur by reaction of a nucleophile with an aromatic halide. The reactions are facilitated by the presence of one or more strongly electron-withdrawing substituents, such as nitro or cyano, placed ortho/para to the leaving group. Common nucleophiles are hydroxide and alkoxide ions, ammonia, and amines. Ethoxide is used as the nucleophile in Part A of this experiment to prepare 2,4-dinitrophenetole.

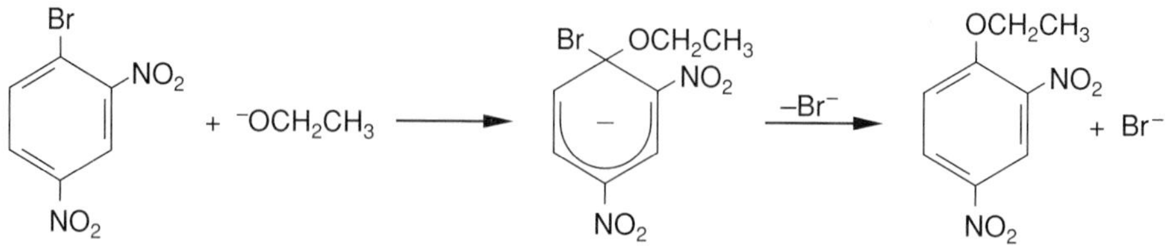

1-bromo-2,4-dinitrobenzene   intermediate   2,4-dinitrophenetole

Nucleophilic aromatic substitution occurs by an addition-elimination mechanism. Addition of a nucleophile gives an intermediate that has a negative charge distributed in the aromatic ring and on the electron-withdrawing substituents. The intermediate has a number of resonance forms. The activating role played by the substituents is pivotal. This is demonstrated by the fact that chlorobenzene gives no reaction with ethoxide at temperatures below 100°C.

Among the activated halobenzenes, fluorobenzenes work best because the intermediate is stabilized by the powerfully withdrawing fluoro substituent. Because the first step of the addition-elimination mechanism is the slow step, the strong electron-withdrawing effect of a fluoro substituent facilitates formation of the intermediate. Loss of halide from the intermediate in the second step is fast. These steps are summarized here for the reaction *p*-fluoronitrobenzene with sodium ethoxide (Part B).

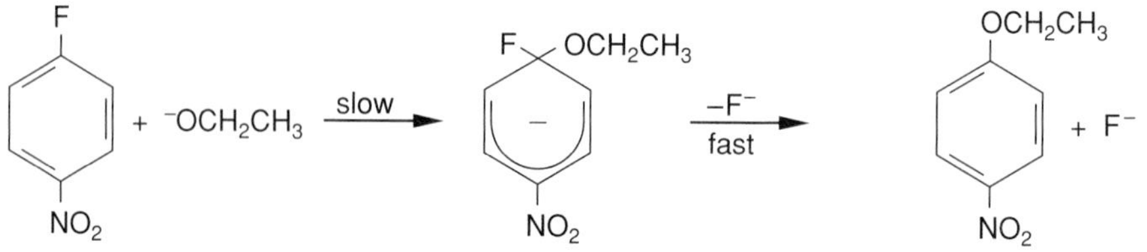

p-fluoronitrobenzene   *p*-nitrophenetole

Prelab Assignment

1. Calculate the amount (mmol) of ethoxide used in the procedure.
2. Write all important resonance forms for the intermediate formed in this reaction.
3. Explain why electron-withdrawing substituents must be ortho/para to the leaving group in order for the nucleophilic aromatic substitution reaction to work well.

Experimental Procedure

Safety First!

Always wear eye protection in the laboratory.

1. Wear eye protection at all times in the laboratory.
2. Wear gloves while handling reagents for this experiment.
3. Work in the hood or in a well-ventilated area for this experiment.
4. 1-Bromo-2,4-dinitrobenzene is a strong skin irritant. Keep reagents under the hood.

Part A: Microscale Reaction of Sodium Ethoxide with 1-Bromo-2,4-dinitrobenzene

Add 125 mg of 1-bromo-2,4-dinitrobenzene to 2 mL of absolute ethanol in a 5-mL con-ical vial fitted with a water-jacketed condenser and a spin vane. Add 100 mg of sodium ethoxide to the solution. Reflux the resulting solution with stirring for 30 minutes. Allow the solution to cool for a few minutes and then transfer it via Pasteur pipet to a beaker containing about 5 mL of cold water. Swirl the mixture briefly and collect the precipitate by suction with a Hirsch funnel. Wash the product with 1 mL of cold water and allow it to remain on the filter with suction for about 10 minutes. Recrystallize using an ethanol-water solvent pair. Weigh the dry product.

Characterization for Part A

Melting Point: Obtain the melting point of the product. The reported melting point of 2,4-dinitrophenetole is 86–87°C.

Infrared Spectroscopy: Obtain an IR spectrum using either KBr or Nujol and inter-pret the spectrum noting the key bands characteristic of the structure of the product. Char-acteristic absorptions for the aromatic ring and for the nitro groups should also be noted.

NMR Spectroscopy: Obtain the ^{1}H and ^{13}C NMR spectra of the product in deuter-ated DMSO. Interpret the spectrum noting the chemical shift positions.

Part B: Microscale Reaction of Sodium Ethoxide with p-Fluoronitrobenzene

Add 95 μL of *p*-fluoronitrobenzene via calibrated pipet to 2 mL of absolute ethanol in a 5-mL conical vial fitted with a water-jacketed condenser and a spin vane. Add 100 mg of sodium ethoxide to the solution and heat the resulting solution at reflux with stirring for 30 minutes. Allow the solution to cool for a few minutes and then transfer it via Pas-teur pipet to a beaker containing about 5 mL of cold water. Swirl the mixture briefly and collect the precipitate using a suction flask with a Hirsch funnel. Wash the solid product with 1 mL of cold water and allow the solid to remain on the filter with suction for about 10 minutes. Recrystallize using an ethanol-water solvent pair. Weigh the dry product.

Characterization for Part B

Melting Point: Determine the melting point of the *p*-nitrophenetole. The reported melt-ing point is 60°C.

Infrared Spectroscopy: Obtain an IR spectrum using either KBr or Nujol and inter-pret the spectrum noting the key bands characteristic of the structure of the product. There should be C—O absorption near 1200–1100 cm^{-1}. Characteristic absorptions for the *p*-disubstituted aromatic ring and for the nitro groups should also be noted.

NMR Spectroscopy: Obtain the ^{1}H and ^{13}C NMR spectra of the product in deuter-ated DMSO solvent and interpret noting the chemical shift positions. See Figures 23.1-1–23.1-3.

Results and Conclusions for Parts A and B

1. Determine the percent yield of the product.
2. Can the product be clearly differentiated from the starting material by IR spec-troscopy? Explain.
3. Use the spectral data to justify the structures of the product.
4. Using the chemical shift values (given in Hz) of the expanded spectrum in Figure 23.1-2, calculate the coupling constant for the CH$_2$ and CH$_3$ protons.

Figure 23.1-1

^{1}H NMR spectrum of
p-nitrophenetole
(deuterated DMSO)

SOURCE: Reprinted with
permission of Aldrich Chemical.

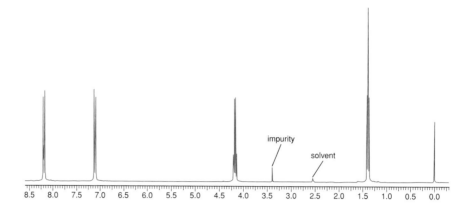

Figure 23.1-2

Expanded ^{1}H NMR
spectrum of
p-nitrophenetole
(deuterated DMSO)

SOURCE: Reprinted with
permission of Aldrich Chemical.

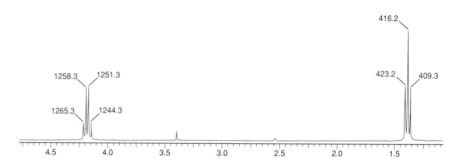

Figure 23.1-3

^{13}C NMR spectrum of
p-nitrophenetole
(deuterated DMSO)

SOURCE: Reprinted with
permission of Aldrich Chemical.

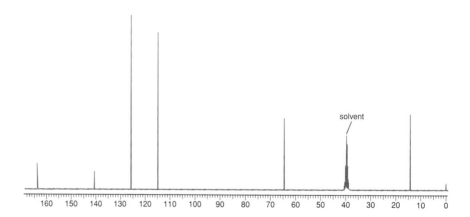

Cleanup & Disposal

Pour the neutralized aqueous filtrate down the drain.

Critical Thinking Questions

1. Draw the structures for products from reaction of 1-bromo-2,4-dinitrobenzene with
    each of the following nucleophiles: ammonia, sodium methoxide, and aniline.

2. Rank 1-bromo-4-nitrobenzene, 1-bromo-2,4-dinitrobenzene, and 1-bromo-2,4,6-
    trinitrobenzene in order of reactivity toward nucleophiles. Explain.

3. Explain why 4-methoxybromobenzene fails to react with sodium ethoxide under the reaction conditions.

4. Suggest three substituents other than nitro that should facilitate nucleophilic aromatic substitution.

5. 2,4-Dinitrofluorobenzene is used in N-terminal residue analysis to react with the free amino group of a peptide. Following the reaction, the peptide is hydrolyzed to give one amino acid (the N-terminal amino acid of the peptide) that contains a 2,4-dinitrophenyl group and various other amino acids. Write an equation for the reaction of the peptide shown below with 2,4-dinitrofluorobenzene. This reaction is the basis for the Sanger method of peptide structure analysis, which was instrumental in Sanger being awarded a Nobel Prize. (Sanger received Nobel Prizes in 1958 and 1980.)

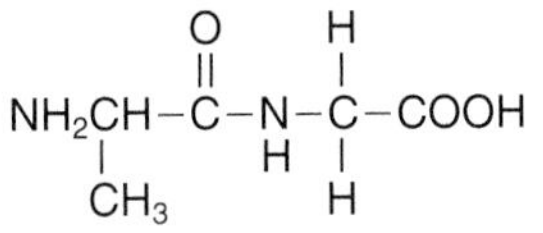

$$\underset{\underset{\displaystyle CH_3}{|}}{NH_2CH}-\underset{\underset{\displaystyle}{\overset{\displaystyle O}{\|}}}{C}-\underset{\underset{\displaystyle H}{|}}{N}-\underset{\underset{\displaystyle H}{|}}{\overset{\overset{\displaystyle H}{|}}{C}}-COOH$$

6. Look up the term "Meisenheimer complex" in an organic text. What is a Meisenheimer complex? How are these complexes related to this experiment?

7. Use reaction mechanisms to show why an aromatic nitro substituent enhances nucleophilic aromatic substitution but slows electrophilic aromatic substitution reactions, such as halogenation or sulfonation.

8. *Molecular modeling assignment:* Use molecular modeling to show why chlorobenzene is less reactive in nucleophilic aromatic substitution than 1-chloro-2,4-dinitrobenzene.

Chapter 24

Phenols

Alcohols and phenols both have hydroxyl groups and share some similar reactions such as acylation to form esters and alkylation to form ethers. In Experiment 24.1, phenol derivatives are acylated to synthesize vanillin acetate. However, several other reactions of vanillin demonstrate the resilience of the phenolic hydroxyl group in resisting change. The aldehydo group of vanillin can be both oxidized and reduced without affecting the phenolic hydroxyl group. The aldehydo group can also be derivatized to form an oxime.

Esterification of phenols is illustrated in Experiment 21.1 during ring closure to form coumarins. That reaction is an intramolecular reaction of a phenol with an ester giving transesterification, a process in which an ester is changed into a different ester.

Phenols are weak acids having pK$_a$ values typically of about 10. Use of this acidic property can be an advantage in separation procedures in the laboratory that require separation of a phenol from neutral organic compounds. For example, in Experiment 18.3, phenol can be separated from neutral organic compounds by extraction with aqueous base.

Phenols are highly activated toward electrophilic aromatic substitution reactions such as halogenation. This is illustrated in Experiment 12.1, where the rate of bromination of phenol relative to other aromatics was investigated. Phenols are readily brominated to form polybromo derivatives. This reaction is illustrated in Chapter 28 during derivatization of phenols for characterization of unknowns.

Experiment 24.1: Exploring Structure-function Relationships of Phenols: *Synthesis of Salicylic Acid, Aspirin, and Vanillin Derivatives*

The presence of one or more functional groups in a compound influences chemical and physical properties. Difunctional molecules offer two sites for reactivity. Modifying one or more of the functional groups may dramatically alter the reactivity and physical properties of the compound. In this experiment, you will explore reactions of vanillin, which is both a phenol and an aldehyde, and salicylic acid, which is both a phenol and a carboxylic acid. You will:

◆ convert vanillin to an ester, an alcohol, a carboxylic acid, and an oxime and evaluate the effect of the structural change on odor.

◆ hydrolyze an ester to a phenol.

◆   synthesize an analgesic by esterification of a phenol.

◆   use thin-layer chromatography to evaluate product composition.

◆   use chemical tests and/or spectroscopic techniques to identify the structure of the product.

Techniques

Technique F	Recrystallization and vacuum filtration
Technique G	Reflux
Technique K	Thin-layer chromatography
Technique M	Infrared spectroscopy
Technique N	Nuclear magnetic resonance spectroscopy

Background and Mechanism

Phenolic compounds are ubiquitous in nature, occurring in spices such as thyme (thymol), vanilla (vanillin), and cloves (eugenol). Phenols are responsible for the spiciness of peppers (capsaicin) and the itching of poison ivy (urushiol). Other naturally occurring phenols, such as methyl salicylate and salicylic acid, are analgesics and provide pain relief. The prevalence of biologically active phenolic compounds attests to the importance of the phenol group.

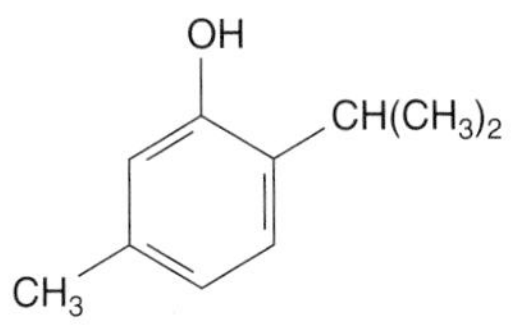

thymol 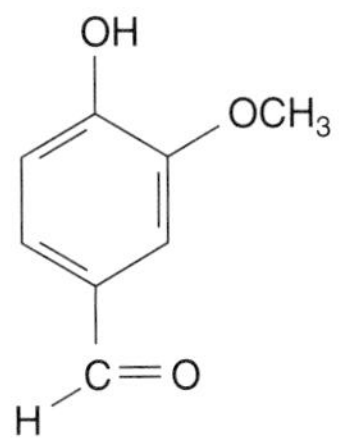   vanillin 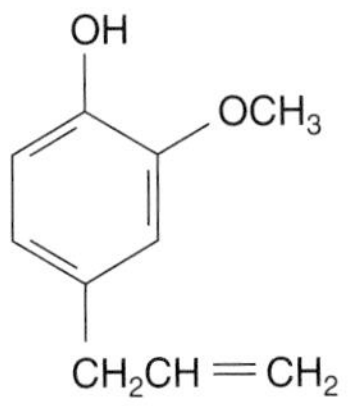   eugenol 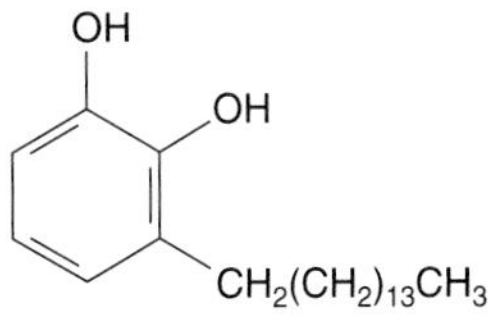   urushiol

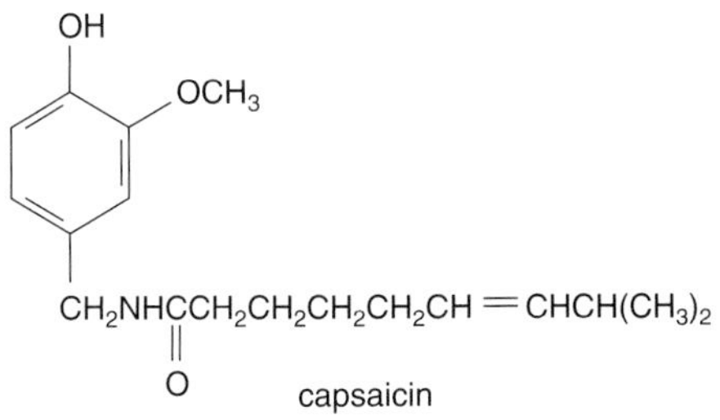

capsaicin 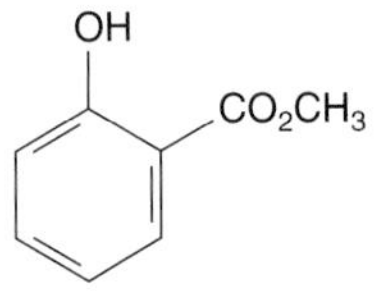   methyl salicylate 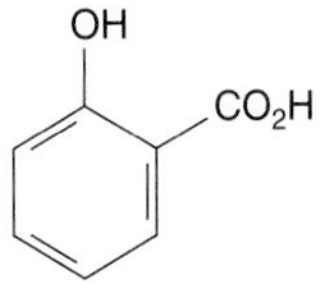   salicylic acid

Vanillin is a flavorant that is associated with the sweet smell of baked goods and desserts. Vanillin is extracted from vanilla pods that grow on a tropical orchid. Natural vanilla extract is made by cutting the beans into small pieces and soaking them in successive quantities of hot 65–70% alcohol. Many commercial vanilla extracts are now actually blends from natural and synthetic vanillin, which may contain vanillin derivatives.

In Part A of this experiment, you will modify the structure of vanillin through acetylation, reduction, oxidation, and derivatization. These structural changes may also alter physical properties. Of particular importance is odor. It is postulated that human beings are able to differentiate between ten thousand different odors. In this experiment, you will synthesize vanillin acetate, vanillyl alcohol, vanillic acid, and vanillin oxime and compare their odors to the odor of vanillin. You will determine how the functional group affects the physical property of odor. The four products of the vanillin reactions are given below.

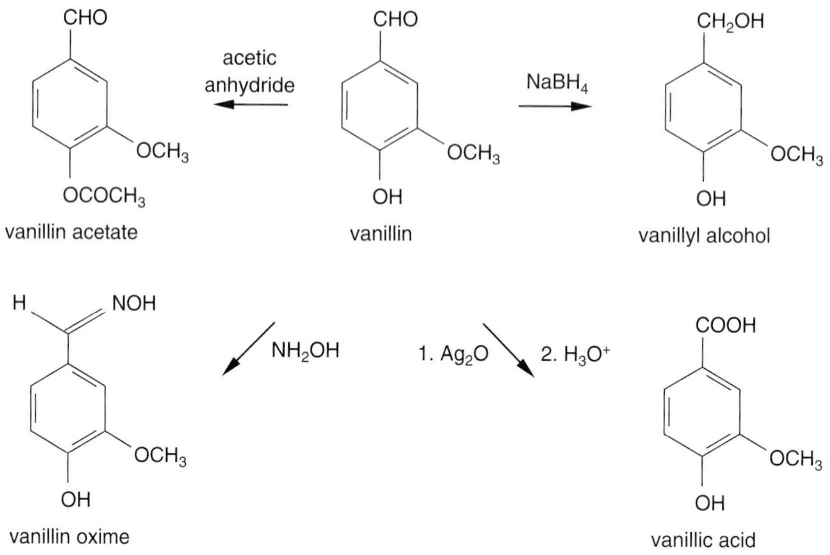

vanillin acetate   vanillin   vanillyl alcohol

vanillin oxime   vanillic acid

Other phenols, such as salicylic acid, play important roles as analgesics. In Part B of this experiment, methyl salicylate will be hydrolyzed to salicylic acid under basic conditions:

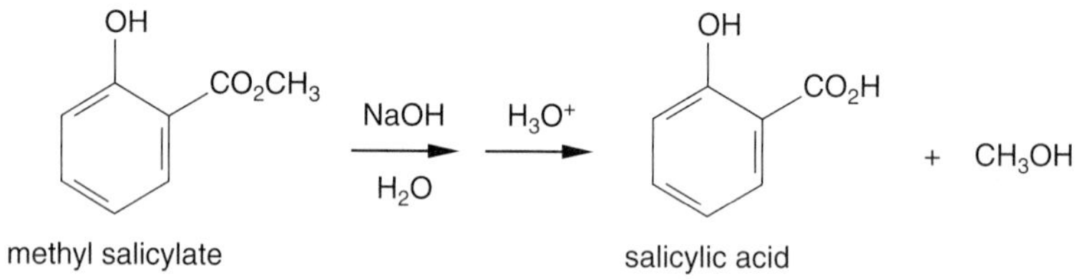

methyl salicylate   salicylic acid

Methyl salicylate, a major component in oil of wintergreen, is used in breath mints. Hydrolysis of this compound produces salicylic acid, an effective, but extremely harsh analgesic that is irritating to the mucous membranes, esophagus, and stomach due to its strongly acidic nature. Salicylic acid can be extracted from willow bark as a glycoside (a sugar bonded to the salicylic acid). This compound was used in ancient Greece and Rome to reduce fevers and alleviate pain. As an example of how modifying the structure of a phenol can dramatically change its properties, consider the reaction of salicylic acid with acetic anhydride to produce acetylsalicylic acid:

salicylic acid   acetic anhydride   acetylsalicylic acid   acetic acid
                                                  (aspirin)

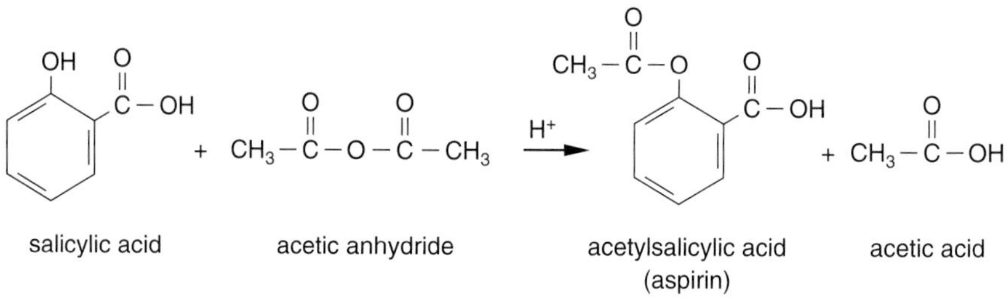

Felix Hoffman, a German chemist working at Bayer, had a father who suffered from arthritis. Hoffman set out to develop a drug that would be less irritating to the stomach. In 1897, Hoffman synthesized acetylsalicylic acid, which you commonly know as aspirin. Acetic anhydride preferentially reacts with the phenolic hydroxyl group rather than the carboxylate group.

Aspirin is now one of the largest selling analgesics, although other products, such as ibuprofen and acetaminophen, have been developed to alleviate side effects such as gastrointestinal problems and Reyes syndrome in children who have chicken pox or the flu. In Parts B and C of this experiment, you will isolate salicylic acid from methyl salicylate and synthesize aspirin.

Prelab Assignment

1. Calculate the quantity (mmol) of reagents to be used in the assigned experiment. Which reagent is the limiting reagent? Show by calculations.
2. Write a detailed mechanism for
    a. the hydrolysis of methyl salicylate in basic solution.
    b. the acetylation of salicylic acid in acidic solution.
    c. the acetylation of vanillin in basic solution.
3. Write a flow scheme for the workup procedure for the assigned reaction.
4. Predict the IR spectrum of the product in the assigned experiment. Indicate how IR could be useful in determining whether any unreacted starting material remains. Make a list of important absorption bands that would be useful in identifying the structure of the product.
5. Draw the predicted ^{1}H NMR spectrum for the product of the assigned procedure, noting relevant chemical shifts.
6. Predict the order of elution of vanillin and each of the products on a TLC plate developed in 50:50 ethyl acetate: hexane.
7. Write a flow scheme for the workup procedure for the assigned reaction.

Experimental Procedure

1. Wear eye protection at all times in the laboratory
2. Acetic anhydride is an irritant. Use under the hood and do not breathe the fumes.
3. Sulfuric acid causes severe burning. If any acid comes in contact with your skin, wash immediately with copious amounts of water. Notify your instructor.
4. Sodium hydroxide is caustic. Wear gloves when handling this reagent. If basic solutions come in contact with your skin, wash the affected area and notify your instructor.

S a f e t y   F i r s t !

Always wear eye protection in the laboratory.

Part A: Miniscale Reactions of Vanillin

The instructor will assign each student a specific reaction or will request students to do two or more of the vanillin experiments.

Reaction 1. Preparation of Vanillin Acetate. Dissolve 2.0 mmol of vanillin in 5 mL of 10% NaOH solution in a 50-mL Erlenmeyer flask. Heat the reaction mixture on a warm sand bath if necessary to dissolve the solid. Cool to room temperature. Add 6 g of crushed ice and 1.6 mL of acetic anhydride. Stopper the flask and swirl the reaction mixture intermittently for 15 minutes. Filter the white crystals using a Büchner or Hirsch funnel. Rinse the flask and crystals with three 1-mL portions of cold water. Recrystallize the crude product from an ethanol-water solvent pair. Weigh the dried crystals and obtain a melting point.

Reaction 2. Preparation of Vanillyl Alcohol.

Dissolve 2.5 mmol of vanillin in 2.5 mL of 1 M NaOH solution in a 25-mL Erlenmeyer flask. Swirl the flask to produce a homogeneous yellow solution. Swirl the flask in an ice-water bath for 1–2 minutes and cool the solution to approximately 10°C. While constantly swirling, add 1.95 mmol of $NaBH_4$ in three to four portions over a period of 3 minutes. Allow the solution to stand undisturbed for 30 minutes at room temperature. After that time, cool the flask in an ice-water bath and add 3 M HCl dropwise with swirling. Continue to add HCl dropwise until the pH of the solution is distinctly acidic to pH paper (pH ~3–4). Cool the solution and gently scratch the side of the flask with a glass rod to induce crystallization. Suction filter the crude product using a Büchner or Hirsch funnel and wash with three small portions of cold water. Let the crystals air dry for several minutes while continuing suction in order to eliminate excess water. Then recrystallize the crude product from ethyl acetate. Weigh the dried crystals and obtain a melting point.

Reaction 3. Preparation of Vanillic Acid.

The oxidizing agent should be prepared first. Dissolve 1.0 mmol of silver nitrate in 1.0 mL of distilled water in a 5-mL conical vial equipped with a spin vane. While stirring, add 0.5 mL of 2.5M NaOH dropwise from a calibrated pipet to precipitate the oxidizing agent, silver oxide. Stir for 5 minutes with a glass rod, then let the solid settle to the bottom. Use a Pasteur pipet to remove the aqueous layer. Wash the solid remaining with four 0.75-mL aliquots of distilled water. After each addition, stir the silver oxide and water with a glass rod, allow the solid to settle, then remove the aqueous layer with a pipet. On the last wash, carefully remove the last traces of water from the silver oxide in the conical vial.

Add 2.5 mL of 2.5 M NaOH to the silver oxide and stir. Heat the solution to 55–60°C on a hot water bath or hot plate and then add 1 mmol of vanillin in three or four small portions, stirring well after each addition. Stir while heating for 15 minutes. During this time, a silver mirror will form on the side of the vial. Silver particles will flake off the wall and fall to the bottom of the conical vial. Transfer the yellow reaction mixture to a 10-mL Erlenmeyer flask. Wash the metallic silver with four 0.75-mL portions of water. Add these washings to the reaction mixture. Filter the mixture if any silver flakes are in the reaction mixture.

Isolate the product by adding 6 M HCl dropwise until the reaction mixture is acidic to pH paper. Allow the reaction mixture to stand at room temperature for several minutes, then place the flask in an ice-water bath to complete the crystallization process. Scratch the wall of the flask to induce crystallization. Suction filter the white crystals using a Hirsch funnel. Rinse the flask and crystals with three 0.5-mL portions of cold water. Air dry for a few minutes. Recrystallize the crude product from water. Weigh the dried crystals and obtain a melting point.

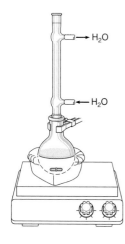

Reaction 4: Preparation of Vanillin Oxime.

In a 10-mL round-bottom flask fitted with a stir bar and a water-jacketed reflux condenser, dissolve 1.0 mmol vanillin, 150 mg hydroxylamine hydrochloride, 300 mg sodium acetate trihydrate in 4.5 mL of ethanol and 15 drops of water. Reflux with stirring for 30 minutes on a warm sand bath. While still hot, transfer the warm solution to an Erlenmeyer flask or beaker. Let cool to room temperature, then place in an ice-water bath until crystallization is complete. Suction filter using a Hirsch funnel. Wash the solid with several small portions of cold 50% aqueous ethanol. Recrystallize the crude solid using an ethanol-water solvent pair. Weigh the dried crystals and obtain a melting point.

Comparing the Aromas of Vanillin Derivatives.   Test the fragrances of each of the vanillin derivatives synthesized with respect to commercial vanillin. Dissolve 200 mg of each derivative in 10 mL of 25% ethanol. Open the bottle and waft the aroma of each derivative. After each analysis, waft the aroma of the commercial vanillin. Tabulate results. If time permits, ask others to evaluate the scents of the vanillin derivatives.

Characterization for Part A

Melting Point: Obtain the melting point of the products. According to the literature, the melting points are vanillin acetate (78°C), vanillyl alcohol (114°C), vanillic acid (210°C), and vanillin oxime (117°C).

Thin-layer Chromatography: Dissolve a few crystals of the product in a small amount of methylene chloride and apply a small spot to a silica gel plate. Also spot a solution of authentic vanillin and the pure product, if available. Develop the chromatogram with 1:1 ethyl acetate:hexanes.

Infrared Spectroscopy: Record the IR spectrum of the product as a Nujol mull or as a KBr pellet. Interpret the spectrum.

NMR Spectroscopy: Dissolve about 50 mg of the product in $CDCl_3$. Record the ^1H NMR and/or the ^{13}C NMR spectrum. Interpret the spectra.

Results and Conclusions for Part A

1. Calculate the percent yield of the product obtained in the reaction.
2. Use the IR spectrum and/or the ^1H NMR spectrum to justify the structure of the product.
3. Compare the odors of vanillin, vanillin acetate, vanillyl alcohol, vanillic acid, and vanillin oxime to the odor of pure vanilla. Rank the compounds with respect to the similarity of the odor to pure vanilla. Examine the structure of each compound and determine which functional groups are critical to the odor of vanilla.
4. Explain why vanillin was acetylated in base, rather than acid.
5. Write a detailed mechanism for the preparation of vanillin oxime.
6. Write a balanced reaction for the preparation of silver oxide from silver nitrate and sodium hydroxide.
7. In the reaction of vanillin with sodium borohydride, what gets oxidized and what gets reduced?
8. In the reaction of vanillin with silver oxide, what gets oxidized and what gets reduced?

Part B: Miniscale Hydrolysis of Methyl Salicylate

To a 25-mL round-bottom flask containing a spin bar, add 7.5 mL of 5 M NaOH and 1.0 mL of methyl salicylate. Swirl to mix. A white solid will form, which will eventually disappear as the reaction is heated. Fit the flask with a water-jacketed reflux condenser. Place a thermometer in the sand bath or heat block and heat to approximately 120°C. Turn the water on gently to the reflux condenser and lower the flask into the heated sand bath or heat block. Reflux the solution with stirring for 15 minutes. Then remove the flask from the heat source and cool to room temperature.

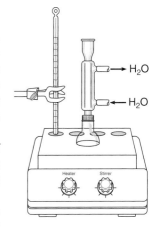

Transfer the solution to a 50-mL Erlenmeyer flask and carefully add 10–15 mL of 3 M sulfuric acid in small aliquots while swirling. A white precipitate will form. Check the pH to make sure that it is strongly acidic (pH ~2–3). Cool the mixture in ice and then suction filter. Wash the crystals with small amounts of cold water.

The crude product can be recrystallized from water. Filter the crystals using suction filtration. Wash the crystals with cold water. Let the crystals air dry before weighing and taking a melting point.

Characterization for Part B

Melting Point: Record the melting point of the product. The melting point of pure salicylic acid is 158°C.

Thin-layer Chromatography: Dissolve a few crystals of the product in a small amount of methylene chloride and apply a small spot to a silica gel plate. Also spot a solution of authentic methyl salicylate and salicylic acid. Develop the chromatogram with 1:1 ethyl acetate:hexanes.

Infrared Spectroscopy: Record the IR spectrum of the product as a Nujol mull or as a KBr pellet.

NMR Spectroscopy: Dissolve about 50 mg of the product in CDCl$_3$. Record the ^{1}H NMR spectrum.

Results and Conclusions for Part B

1. Based upon the TLC results, determine whether the product is pure.
2. Calculate the percent yield of salicylic acid.
3. From the IR spectrum and/or the ^{1}H NMR spectrum, determine the structure of the product. Justify your answer.
4. Write chemical equations and explain the following observations: A precipitate immediately forms upon mixing methyl salicylate and sodium hydroxide. This precipitate disappears as the solution is heated. When sulfuric acid is added to the cooled solution, a precipitate forms.

Part C: Miniscale Synthesis of Acetylsalicylic Acid

Use the air-dried product from Part B. Save enough for characterization of the product and use the remainder in this experiment. If necessary, scale down this procedure to accommodate the mass of salicylic acid obtained in Part B. Alternately, you may use purchased salicylic acid for this synthesis.

Place 1 g of salicylic acid, 2.5 mL of acetic anhydride, and 5 drops of concentrated sulfuric acid in a 50-mL Erlenmeyer flask containing a spin bar. Swirl to dissolve. Heat the reaction mixture with stirring on a warm sand bath or hot plate for 10 minutes. Cool to room temperature and then put in an ice bath to crystallize the aspirin. Add 20 mL of cold water to the flask and stir carefully. Vacuum filter the crystals using a Büchner funnel. Rinse the crystals with several small portions of ice-cold water. The crude aspirin can be recrystallized from ethyl acetate. Dissolve the crude product in a minimum amount of hot ethyl acetate and if necessary, do a hot gravity filtration to separate the hot solution from insoluble impurities. Allow the solution to cool to room temperature, then place in an ice bath to complete crystallization. Vacuum filter using a Büchner funnel. Weigh the dried crystals.

Characterization for Part C

Melting Point: Determine the melting point of the dry product. The reported melting point of acetylsalicylic acid is 138–140°C.

Infrared Spectroscopy: Record and interpret the IR spectrum of aspirin as a Nujol mull or as a KBr pellet. The IR spectrum of acetylsalicylic acid is shown in Figure 24.1-1.

NMR Spectroscopy: Dissolve about 50 mg of the product in deuterated DMSO. Record the ^{1}H NMR spectrum. The ^{1}H NMR spectrum of acetylsalicylic acid is shown in Figure 24.1-2.

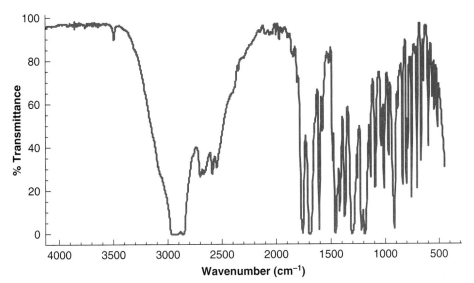

Figure 24.1-1

IR spectrum of acetyl-salicylic acid (Nujol)

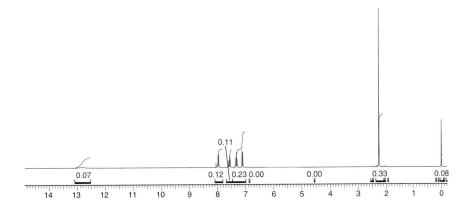

Figure 24.1-2

^{1}H NMR spectrum of acetylsalicylic acid (CDCl$_3$ and deuterated DMSO)

SOURCE: Reprinted with permission of Aldrich Chemical.

Results and Conclusions for Part C

1. Calculate the percent yield of aspirin.
2. Evaluate the purity of the product.
3. The preparation of aspirin is an example of a nucleophilic acyl substitution reaction of the nucleophilic phenol group on the protonated acetic anhydride electrophile. Write a detailed mechanism for this reaction.
4. Salicylic acid is both a phenol and a carboxylic acid. Explain why acetylation with acetic anhydride occurs at the phenolic hydroxyl group rather than at the carboxyl group.

Cleanup & Disposal

Neutralize all basic and acidic solutions and wash down the drain with water.

Critical Thinking Questions

1. Look up the Reimer-Tiemann reaction in an organic chemistry textbook and propose a mechanism for the formation of vanillin starting from 2-methoxyphenol.

2. Give the expected product of the acetylation of *p*-hydroxybenzyl alcohol with acetic anhydride.

3. Predict the acetylation products expected from each compound below using acetic anhydride and basic conditions:

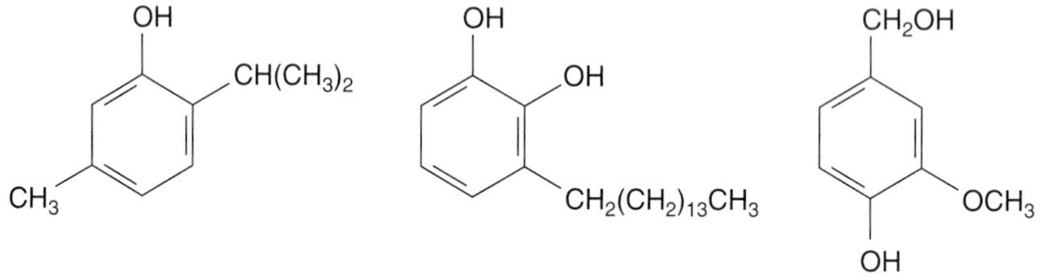

thymol   urushiol   vanillyl alcohol

4. If vanilla extract is exposed to air for a lengthy period, what reaction might occur? Why would this affect the flavor adversely? Explain.

5. What is the expected product if vanillyl alcohol were acetylated with acetyl chloride and pyridine?

6. Suppose that quality control at an aspirin manufacturing plant fell down on the job and left some aspirin tablets to be sold that contained salicylic acid. What might happen to persons who ingested these tablets as a result? Explain.

7. *Library project:* Explore the differences, if any, between vanillin and "artificial" vanillin. Are vanillin substitutes commonly used in baking? Are these substitutes acceptable?

8. *Molecular modeling assignment:* Construct models of vanillin and vanillyl alcohol. Minimize energies. Examine the most stable conformations. Explain how the conformations differ.

Chapter 25

Carbohydrates

Many students of organic chemistry are taking this course because it is required for a prehealth professional curriculum. Therefore, it is important to focus upon biomolecules in the organic laboratory, such as an experiment on testing for different types of sugars (Experiment 25.1).

Sugars have many hydroxy groups, making them quite water soluble and at the same time insoluble in common organic solvents. Esterification changes the solubility properties. An example is the acetylation of sugars in Experiment 25.2. In this experiment, a common sugar (glucose or sucrose) is completely acetylated in the presence of an excess of acetic anhydride. Since acetylation takes place at all hydroxyl groups, the acetylated products are soluble in organic solvents and insoluble in water. A notable feature of the acetylation reactions is that no bond breaking takes place directly at any of the stereogenic centers. The starting sugars are optically active, as are the acetylated products.

Qualitative organic analysis offers methods for identifying unknown organic compounds. The principles and methods used are discussed in Chapter 28. A qualitative scheme for the identification of sugars bears both similarities to and differences from the qualitative organic analysis scheme. All sugars are polyfunctional alcohols and can also be hemiacetals or acetals. Some are pentoses or hexoses. Others are disaccharides or polysaccharides. A scheme for distinguishing between these possibilities provides a method of identification of unknown sugars as presented in Experiment 25.1. For example, sugars may be reducing or nonreducing. Reducing sugars are more common and have cyclic hemiacetal or free aldehyde structures as shown here.

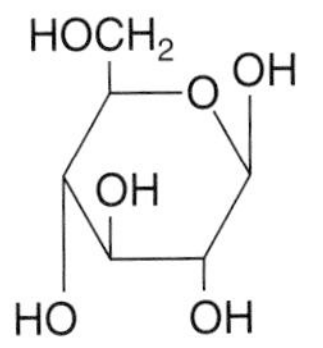

β-D-glucopyranose,
a reducing sugar
(a Haworth projection)

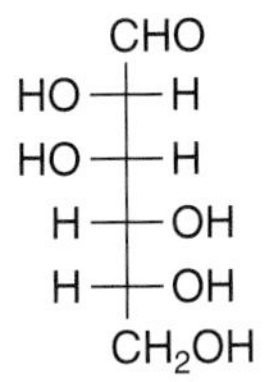

D-mannose,
a reducing sugar
(a Fischer projection)

Experiment 25.1: Classification of Sugars
and Identification of an Unknown Sugar

Sugars can be classified according to results from a series of chemical tests. In this experiment, you will:

- differentiate between pentose and hexose sugars.

- differentiate between ketohexose and aldohexose sugars.

- differentiate between monosaccharides, disaccharides, and polysaccharides.

- differentiate between reducing and nonreducing monosaccharides and disaccharides.

- identify an unknown sugar.

Background

A number of informative simple tests can be used to characterize sugars. The structure of a simple sugar depends upon the number of carbons present, the predominant ring form, and the stereochemistry of each stereogenic center. The structure of polysaccharides also depends upon the types of linkages connecting the sugar components and the number of sugar components in the polymer chain. (Refer to your organic text for a discussion of nomenclature of sugars.)

Classification of Sugars

A sugar that is an aldehyde is called an **aldose;** a sugar that is a ketone is called a **ketose.** A sugar with five carbons is called a **pentose;** one with six carbons is called a **hexose.** Structures of some simple sugars are shown here.

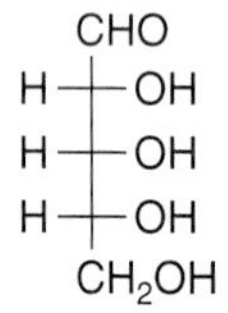

D-ribose (an aldopentose)	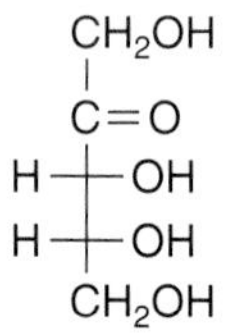 D-ribulose (a ketopentose)	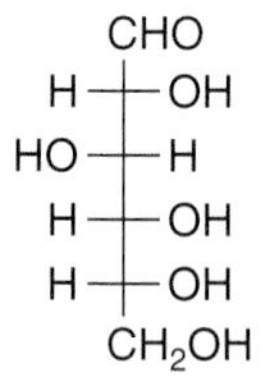 D-glucose (an aldohexose)	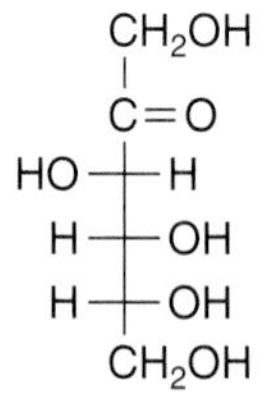 D-fructose (a ketohexose)
CHO H—OH H—OH H—OH CH₂OH	CH₂OH C=O H—OH H—OH CH₂OH	CHO H—OH HO—H H—OH H—OH CH₂OH	CH₂OH C=O HO—H H—OH H—OH CH₂OH

   In solution, pentose and hexose sugars can form a five-membered ring (called a furanose) or a six-membered ring (called a pyranose). The cyclization produces a hemiacetal with a new stereogenic center. This center is described as being either α or β, depending upon the conformation. The α- and β-forms can be interconverted by a process called mutarotation, which proceeds via the acyclic aldehyde or ketone.

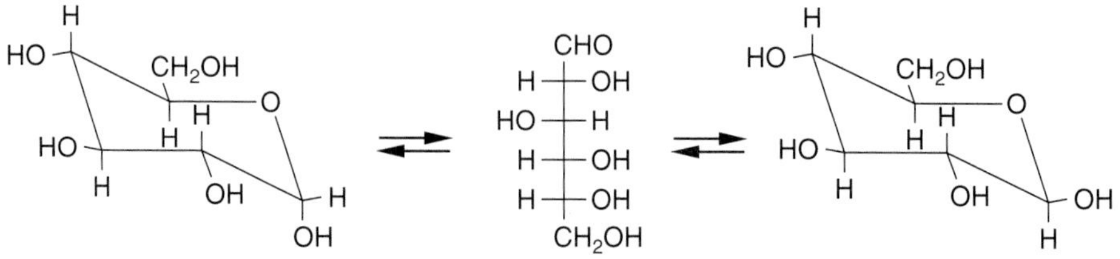

α-D-glucopyranose	D-glucose	β-D-glucopyranose

Sugars that form cyclic hemiacetals will react with oxidizing agents such as cupric ion. These sugars are called **reducing sugars** because they reduce cupric ion. The sugars shown here are both reducing sugars.

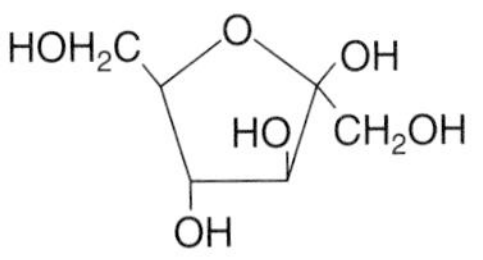

β-D-fructofuranose 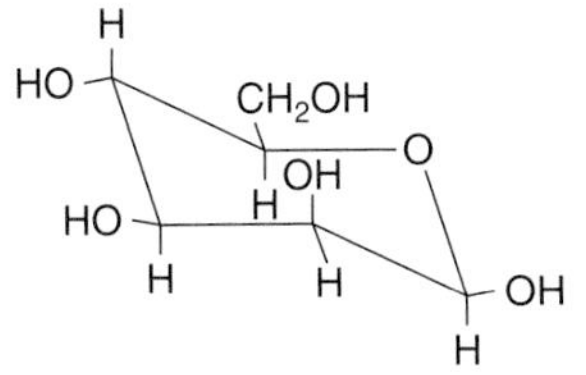   β-D-mannopyranose

Acetals do not undergo mutarotation, so they will not react with oxidizing agents. Sugars that are acetals are called **nonreducing sugars**. A nonreducing sugar and a nonreducing sugar derivative are shown here.

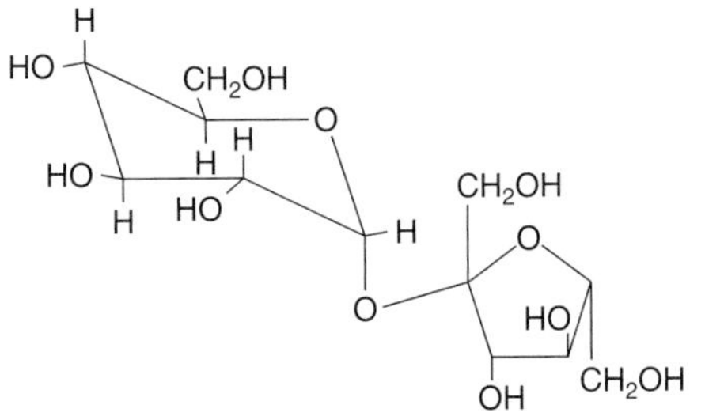

sucrose, α-D-glucopyranosyl-β-D-fructofuranoside
(or β-D-fructofuranosyl-α-D-glucopyranoside)

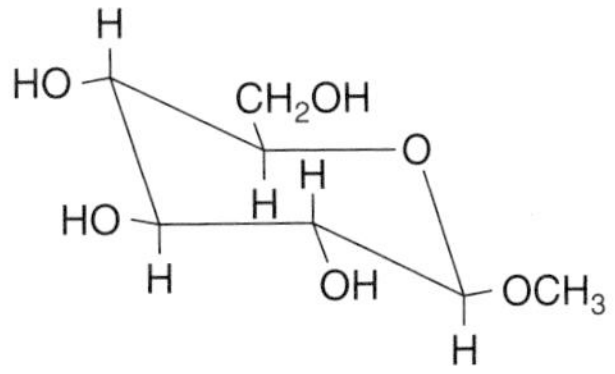

methyl β-D-glucopyranoside

Saccharides may contain more than one sugar component. Simple sugars are called **monosaccharides**. Glucose, galactose, fructose, and mannose are all monosaccharides. A **disaccharide** is formed by linking two monosaccharides. Sucrose is a disaccharide of glucose and fructose; lactose is a disaccharide of galactose and glucose. A **polysaccharide** contains many linked monosaccharides. Cellulose and amylose (a major component of starch) are both polysaccharides of glucose. The physical and chemical properties of these two polysaccharides are a consequence of the different configuration of the linkages between the glucose units. In cellulose, the glucose residues are held together by β-1,4′ linkages. The resulting structure is relatively linear.

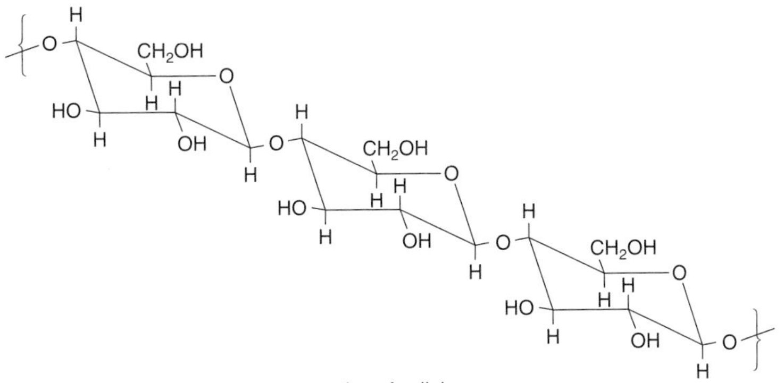

a portion of cellulose

In amylose, the glucose residues are held together by α-1,4′ linkages. Amylose is not linear, but forms an α-helix. Glycogen is an important energy-storage compound. In

addition to the α-1,4′ linkages, this polysaccharide is extensively cross-linked to glucose residues through 1,6′ linkages.

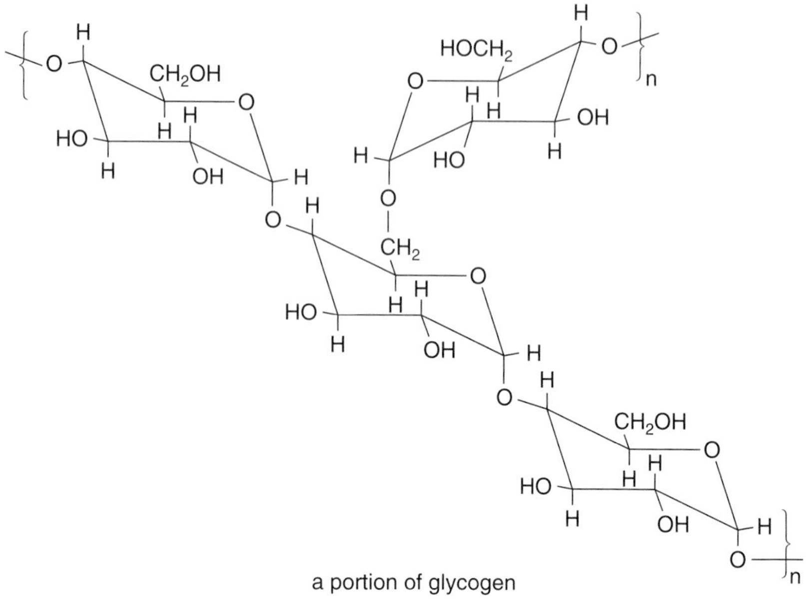

a portion of glycogen

Structural Determination

Chemical tests will be used to determine the structure of an unknown sugar. The tests to
be used are described here.

Bial Test

The Bial test is used to differentiate between pentose and hexose sugars. Under acidic
conditions, sugars containing five-carbons (aldopentoses and ketopentoses) are rapidly
dehydrated to give furfural.

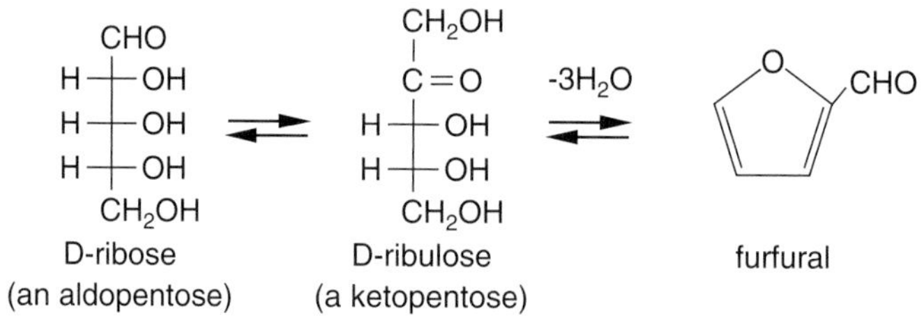

D-ribose
(an aldopentose)

D-ribulose
(a ketopentose)

furfural

Furfural reacts further with orcinol and ferric chloride to form a blue-green solution,
characteristic of pentose sugars.

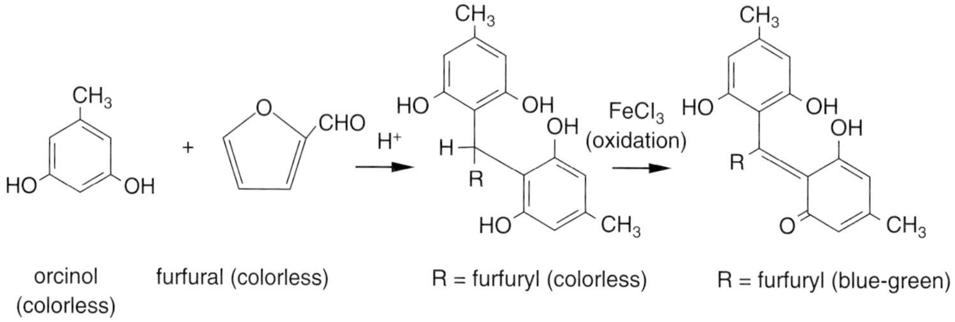

orcinol
(colorless)

furfural (colorless)

R = furfuryl (colorless)

R = furfuryl (blue-green)

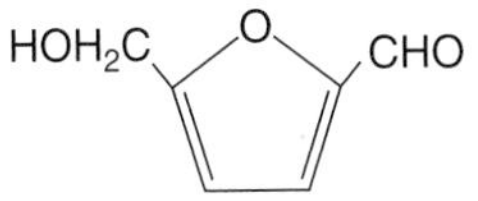

Six-carbon sugars are also dehydrated, but dehydration to 5-hydroxymethylfurfural and subsequent reaction with orcinol and ferric chloride occurs less rapidly. The rate of formation and the color of the solution can be used to distinguish pentoses from hexoses. Pentose sugars form blue-green solutions; hexose sugars form reddish brown complexes. Sugars containing five carbons react more rapidly than sugars with six carbons.

Disaccharides can also react with Bial reagent. Under the acidic reaction conditions, disaccharides are cleaved to give monosaccharides, which then react to form furfural and/or 5-hydroxymethylfurfural.

HOH₂C — O — CHO

5-hydroxymethylfurfural

Seliwanoff Test

The Seliwanoff test is used to differentiate between an aldehyde group and a ketone group in hexose sugars. The test is based on the differing rates of dehydration of aldohexoses and ketohexoses to form 5-hydroxymethylfurfural.

$$\text{D-glucose (an aldohexose)} \xrightarrow{H_3O^+} \text{D-fructose (a ketohexose)} \rightleftharpoons \alpha\text{-D-fructofuranose} \xrightarrow{-3H_2O} \text{5-hydroxymethylfurfural}$$

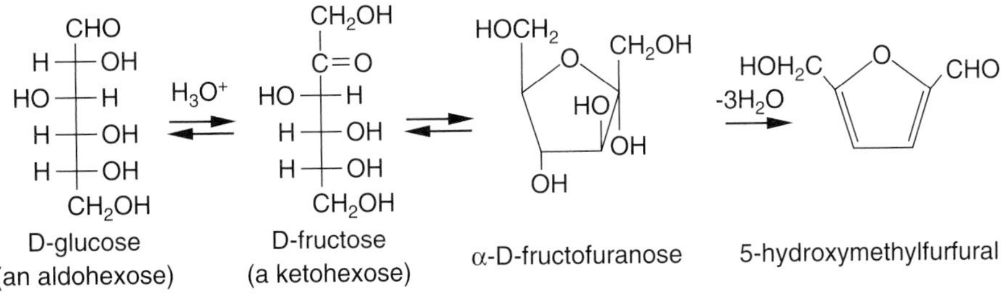

This product (5-hydroxymethylfurfural) then reacts with resorcinol to form a deep red solution.

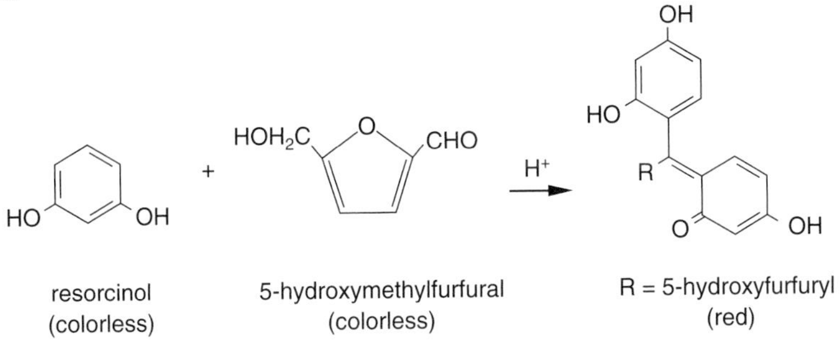

resorcinol (colorless) + 5-hydroxymethylfurfural (colorless) $\xrightarrow{H^+}$ R = 5-hydroxyfurfuryl (red)

The rate of reaction for a ketohexose is much faster than the rate of reaction for an aldohexose. Ketohexoses react within 60 seconds; aldohexoses react in 2–5 minutes. The rate of formation of the red product can be used to differentiate between aldohexoses and ketohexoses. This reaction is not useful for five-carbon sugars. Disaccharides, which are hydrolyzed to monosaccharides under the reaction conditions, may also give a reaction, but at a slower rate.

Benedict's Test

The Benedict's test is used to differentiate between reducing and nonreducing sugars. In this test, Cu^{2+} is reduced to Cu^+ and the reducing sugar is oxidized to a carboxylic acid. The appearance of copper (I) oxide, which is brick red, confirms the sugar to be a reducing sugar. Nonreducing sugars do not react.

Both ketoses and aldoses can be reducing sugars. Under the reaction conditions, ketose sugars undergo base-catalyzed rearrangement to aldose sugars, which subsequently react with the Benedict's reagent. This rearrangement for D-fructose is shown here.

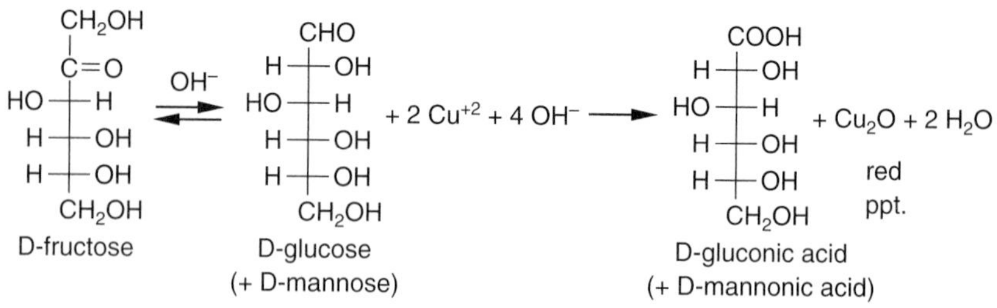

D-fructose ⇌ D-glucose (+ D-mannose) + 2 Cu^{+2} + 4 OH$^-$ ⟶ D-gluconic acid (+ D-mannonic acid) + Cu$_2$O + 2 H$_2$O, red ppt.

Barfoed's Test

The Barfoed's test also differentiates between reducing and nonreducing sugars, but because it is more sensitive than the Benedict's test, it can be used to differentiate between reducing monosaccharides and reducing disaccharides. The reduction of Cu^{2+} to copper (I) oxide occurs more rapidly for monosaccharides than for disaccharides. The rate of appearance of the red precipitate can distinguish reducing monosaccharides from reducing disaccharides.

D-fructose ⇌ D-glucose (+ D-mannose) + 2 Cu^{+2} + 2 H$_2$O ⟶ D-gluconic acid (+ D-mannonic acid) + Cu$_2$O + 4 H$^+$, red ppt.

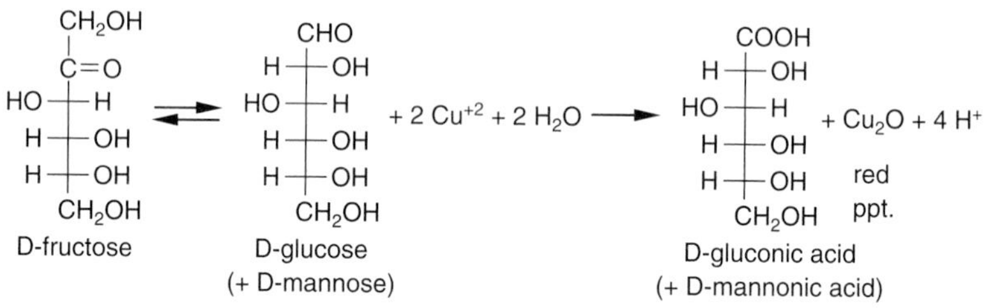

Osazone Formation

Osazones are formed by the reaction of a sugar with phenylhydrazine. The reaction is illustrated here for D-glucose.

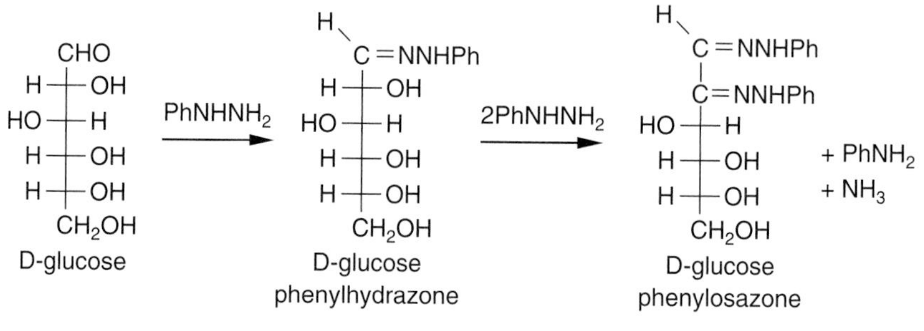

D-glucose →(PhNHNH$_2$) D-glucose phenylhydrazone →(2PhNHNH$_2$) D-glucose phenylosazone + PhNH$_2$ + NH$_3$

An osazone is a solid derivative of a sugar containing two phenylhydrazone groups. In theory, it should be possible to use the melting point of this derivative to identify the unknown sugar. In practice, this is not easily accomplished because osazone derivatives melt over a very wide range and the identical osazone is obtained for more than one sugar. D-glucose, D-mannose, and D-fructose each give the same osazone, so the melting point of the osazone could not distinguish between these sugars. However, careful observation of the rate at which the osazone forms and the appearance of the precipitate can differentiate between epimeric sugars. Osazones form at different rates for different sugars: fructose reacts very rapidly, while glucose takes longer to react. The appearance of the precipitate can also be different. The crystal structure ranges from coarse (for glucose) to very fine (for arabinose).

Iodine Test

The iodine test is used to differentiate between starch and glycogen. Starch contains amylose, a polymeric chain of glucose residues bonded together by α-1,4′ linkages. Amylose forms an α-helix. Molecular iodine can intercalate into the α-helical structure. The resulting complex is bluish black. The extensive branching of glycogen deters the formation of an α-helix. The addition of molecular iodine to a solution of glycogen does not give a bluish black color; instead, the color of the solution is reddish purple.

In this experiment you will perform these chemical tests on the unknown sugar and on a series of known sugars: xylose, arabinose, glucose, galactose, methyl α-D-glucopyranoside, fructose, lactose, sucrose, starch, and glycogen. Your instructor will provide a list of possible unknowns.

Prelab Assignment

1. Draw Fischer projections of xylose, arabinose, glucose, galactose, and fructose. Draw side-view projections (chair forms) or Haworth projections of methyl α-D-glucopyranoside, lactose, and sucrose. Draw a piece of starch (amylose) and glycogen using abbreviated structures.
2. Classify completely xylose, arabinose, glucose, galactose, fructose, methyl α-D-glucopyranoside, lactose, sucrose, starch, and glycogen as a monosaccharide, disaccharide, or polysaccharide. Classify each monosaccharide by number of carbons and whether it is a ketone or an aldehyde (i.e., aldopentose, ketohexose, etc.). Classify each monosaccharide and disaccharide as reducing or nonreducing.
3. Design a possible flow scheme of the chemical tests that could be used to identify the sugar. For each test, indicate what sugars would be expected to give a positive result with the reagent and which sugars would be expected to give a negative result.
4. Prepare a table in your lab notebook to record the results of each chemical test on the known sugars, the unknown sugar, and the control.

Experimental Procedure

1. Wear eye protection at all times in the laboratory.
2. Phenylhydrazine, used in the formation of the osazone, is a suspected carcinogen. Always wear gloves when working with this reagent and avoid breathing the vapors.

S a f e t y   F i r s t !

Always wear eye protection in the laboratory.

General Instructions for the Microscale Identification of an Unknown Sugar

Obtain two samples of an unknown sugar from the instructor. Record the number of this unknown. One of the vials will contain a 10% aqueous solution of the unknown sugar; the other will contain a 1% aqueous solution of this same unknown. The 10% solution will be used only for the formation of the osazone in Part A. Do this part of the experiment first, since it requires the most time. Use the 1% solution of the unknown for all other chemical tests. Make certain that the correct concentration is used for each test.

For each chemical test, obtain 12 clean 15 × 125 mm test tubes; label 10 with the name of a sugar: xylose, arabinose, glucose, galactose, methyl α-D-glucopyranoside, fructose, lactose, sucrose, starch, and glycogen. Be sure to shake up the starch solution prior to use, since starch is not very soluble. Use one test tube for the unknown sugar. Use the last test tube as a control. The control, which is distilled water, must be run along with each known in each chemical test. The labeled test tubes may be reused after each test, but make certain they are cleaned out thoroughly or misleading results will be obtained.

Record the result of each test. Indicate the color produced in the test (greenish-blue, red, etc.), a "+" or "–" to indicate reactivity or nonreactivity, the length of time for reaction, or the observed crystal structure (coarse, fine, etc.).

Part A: Osazone Formation

This test requires the use of 10% solutions of the known sugars and the 10% solution of the unknown sugar.

> Phenylhydrazine is a suspected carcinogen. Wear gloves and work under the hood!

To the appropriately labeled test tubes add 10 drops of 10% solutions of xylose, arabinose, glucose, galactose, methyl α-D-glucopyranoside, fructose, lactose, sucrose, starch, and glycogen. Place 10 drops of the 10% solution of the unknown sugar in a clean test tube. To the last test tube add 10 drops of distilled water. Add 2 mL of the phenylhydrazine reagent to every test tube. Mix thoroughly and place the test tubes in a boiling water bath all at the same time. Watch carefully for signs of cloudiness or pre-cipitation and record the time of appearance in the data table. After 30 minutes, remove the test tubes from the water bath and let cool to room temperature. Place in an ice bath. Note the time of formation and the appearance of each of the precipitates (coarse, very coarse, fine, very fine, etc.).

Part B: Chemical Tests

These tests require the use of 1% solutions of the known sugars and the 1% solution of the unknown sugar.

B.1: Bial Test

Place 1 mL of 1% solutions of xylose, arabinose, glucose, methyl α-D-glucopyranoside, galactose, fructose, lactose, sucrose, starch, and glycogen into the appropriately labeled test tubes. Add 1 mL of distilled water to the "control" test tube. Place 1 mL of the 1% solution of the unknown sugar in a clean test tube. Add 1 mL of Bial's reagent (a mix-ture of orcinol and ferric chloride) to each test tube. Add a boiling stone to each tube to prevent bumping. Heat in a boiling water bath all at the same time. Record any color changes. Repeat with the other samples. If it is difficult to determine the color of the solution, add 2.5 mL of water and 0.5 mL of 1-pentanol to the test tube. Shake the test tube vigorously, then let the layers separate. Observe the color in the 1-pentanol layer.

B.2: Seliwanoff Test

Add approximately 10 drops of 1% solutions of xylose, arabinose, glucose, methyl α-D-glucopyranoside, galactose, fructose, lactose, sucrose, starch, and glycogen to the appro-priately labeled test tube. To the "control" test tube, add 10 drops of distilled water. Obtain a clean test tube and measure out 10 drops of the 1% solution of the unknown. To each test tube add 2 mL of the Seliwanoff reagent (a solution of resorcinol in dilute HCl). Put the test tubes in the boiling water bath all at the same time. Leave for 1 minute. Remove the test tubes from the boiling water bath and record any color changes. Return all unreacted test tubes to the boiling water bath and watch carefully to determine how long it takes for each tube to change color. (It may be easier to do this with groups of two or three test tubes, rather than trying to monitor more.) Note the time at which the color changes and record this time in the table. If no color change has occurred after 5 minutes, consider the test to be negative.

B.3: Benedict's Test

Place 10 drops of 1% solutions of xylose, arabinose, glucose, methyl α-D-glucopyrano-side, galactose, fructose, lactose, sucrose, starch, and glycogen in the appropriately labeled test tubes. Place 10 drops of the 1% solution of unknown in a clean test tube. Use 10 drops of distilled water as the control. Add 2 mL of the Benedict's reagent (a basic solution of $CuSO_4$) to each test tube. Place all of the test tubes in a boiling water bath for 2–3 minutes. Remove the test tubes from the heat and carefully note the color of each precipitate (not the color of the solution!). Reducing sugars react to form red, brown, or yellow precipitates. If no precipitate forms (even if the solution is colored), the sugar is not a reducing sugar.

B.4: Barfoed's Test

Place 10 drops of 1% solutions of xylose, arabinose, glucose, methyl α-D-gluco-pyranoside, galactose, fructose, lactose, sucrose, starch, and glycogen in the appropriately labeled test tubes. Place 10 drops of the 1% solution of unknown in a clean test tube. Use 10 drops of distilled water as the control. Add 2 mL of the Barfoed's reagent which is an acidic solution of copper(II) acetate, to each test tube. Place the test tubes in a boiling water bath for 10 minutes, noting carefully the time of formation of any red precipitate.

B.5: Iodine Test

This test should be run on glucose, starch, glycogen, distilled water, and the unknown sugar. Place 1 mL of 1% solutions of each of these sugars in the appropriately labeled test tubes. Place 1 mL of distilled water in the "control" test tube. Place 1 mL of the 1% solution of unknown in a clean test tube. Add 1 drop of the iodine solution to each of the test tubes. Observe and record the color of the resulting solution.

Results and Conclusions

1. From the results of the chemical tests, determine the identity of the unknown sugar.
2. Write balanced equations for all positive reactions of the unknown.
3. Write a detailed, reasonable mechanism for the formation of the osazone derivative.
4. Explain why distilled water was used as a control in these chemical tests.

Cleanup & Disposal

Filter the osazone solution through filter paper into a container labeled "osazone waste." The filtrate may contain unreacted phenylhydrazine. Wear gloves. Transfer all of the solutions from the Benedict's and Barfoed's tests into a container labeled "aqueous copper waste."

Critical Thinking Questions *(The harder one is marked with a ❖.)*

1. Write a reasonable mechanism for the acid-catalyzed mutarotation of D-glucose.
2. Why do D-glucose, D-mannose, and D-fructose give identical osazones?
3. Write a reasonable mechanism for the acid-catalyzed conversion of D-fructose to D-glucose.
4. Give structures of two other sugars that will give the same osazone as galactose.

5. Sucrose generally does not form a red precipitate with Benedict's reagent. However, if a few drops of acid are added to the solution, the sucrose will react. Write a mechanism to explain this observation.

6. Suppose that the unknown sugar is mannose. Describe the behavior of this sugar in each of the following chemical tests:

   a. Benedict's test.
   b. Bial test.
   c. Seliwanoff test.

7. Draw the structures of α-D-glucopyranose and β-D-glucopyranose. Will an equimolar solution of these anomers exhibit optical activity? Explain.

8. Will L-glucose give the same results in all of these tests as D-glucose? Explain.

9. Look up the structure of cellulose. How does it differ from the structure of amylose?

❖ 10. Write a reasonable mechanism for the reaction of 5-hydroxymethylfurfural with resorcinol.

11. Why do disaccharides react with Cu^{2+} more slowly in the Barfoed's test than in the Benedict's tests?

12. Suppose that D-allose is added to the list of unknowns. Explain the predicted behavior of D-allose in each of the tests. Would it be possible to distinguish D-allose from each of the other unknowns? Explain. If not possible, suggest an additional test that could be used.

13. Why does the Seliwanoff test fail for five-carbon sugars?

14. Why don't polysaccharides react in the Barfoed's test or the Benedict's test?

Experiment 25.2: Esterification of Sugars: *Preparation of Sucrose Octaacetate and α- and β-D-Glucopyranose Pentaacetate*

Two of the most common sugars—D-glucose and D-sucrose—are used as starting materials for the two parts of this experiment. You will:

◆ esterify a reducing sugar, D-glucose.

◆ esterify a nonreducing sugar, D-sucrose.

◆ measure the optical rotation.

Techniques

Technique F	Recrystallization
Technique G	Reflux
Technique L	Polarimetry
Technique M	Infrared spectroscopy
Technique N	Nuclear magnetic resonance spectroscopy

Background

In addition to their familiar roles as sweeteners, sugars are often converted to sugar derivatives that may be used for other purposes. For example, esterification of D-sucrose with acetic anhydride affords D-sucrose octaacetate, which is used as an additive to animal feed. The feed is inedible for humans because of the bitter taste of D-sucrose octaacetate (Mann et al., 1992).

D-sucrose   + CH$_3$CO$_2$COCH$_3$ → D-sucrose octaacetate   + CH$_3$CO$_2$H   acetate = –OAc
                 acetic anhydride   acetic acid

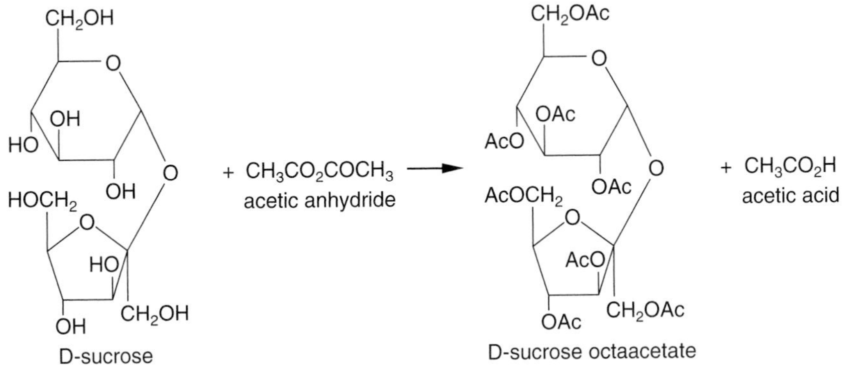

D-Sucrose is a nonreducing sugar. This means that the sugar is an acetal rather than a hemiacetal. The linkage between the two sugar rings of sucrose connects the anomeric C-1 carbon of the glucopyranose ring with the anomeric C-2 of the fructofuranose ring, leaving no free OH groups attached to either anomeric carbon. In Part A of this experiment, you will prepare D-sucrose octancetate. D-Glucose is a reducing sugar. Reducing sugars are acetylated at the anomeric carbon as well as at all other carbons bearing hydroxyl groups. Acetylation at the anomeric carbon may result in either the α or β form of D-glucopyranose pentaacetate. In Part B conditions are chosen to produce β-D-glucopyranose pentaacetate.

D-glucose; mixture   + CH$_3$CO$_2$COCH$_3$ → β-D-glucopyranose   + CH$_3$CO$_2$H
of anomers   acetic anhydride   pentaacetate   acetic acid

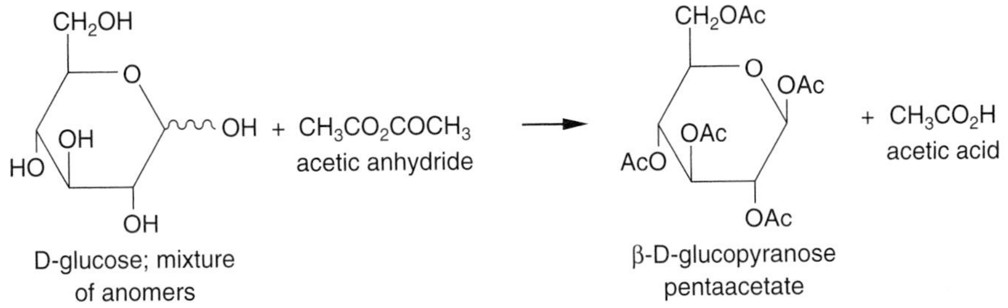

Because the anomeric acetate is more labile (reactive) than the other acetate groups, acetylated sugars are often useful in preparing other sugar derivatives. A reaction of β-D-glucopyranose pentaacetate is illustrated in Part C of this experiment, where isomerization of the β-anomer to the more stable α-anomer is observed. The reaction is catalyzed by zinc chloride.

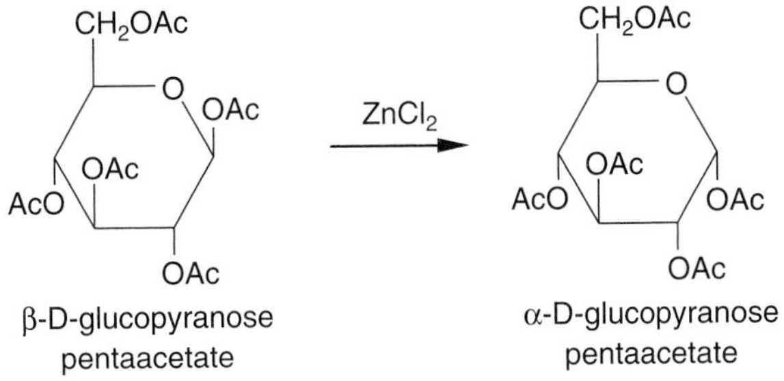

β-D-glucopyranose   → α-D-glucopyranose
pentaacetate   ZnCl$_2$ pentaacetate

Zinc chloride catalyzes loss of acetate ion from the anomeric carbon, leaving a resonance-stabilized carbocation. Attack by acetate from the bottom furnishes the α-anomer, which in this case happens to be thermodynamically more stable than the β-anomer.

It is also possible to prepare α-D-glucopyranose pentaacetate directly from D-glucose and acetic anhydride in the presence of zinc chloride. This is illustrated in Part D.

Prelab Assignment

1. Write the mechanism for the reaction of D-glucose or D-sucrose with acetic anhydride.
2. Write a flow scheme for the preparation and workup procedure of the assigned acylated sugar.

Experimental Procedure

1. Always wear eye protection in the laboratory.
2. Acetic anhydride is a noxious liquid and is a corrosive lachrymator. It is very reactive with water. Use under the hood. Avoid breathing the vapor. Treat skin exposure with copious amounts of cold water.
3. Pyridine is an irritating liquid. Use under the hood.
4. Zinc chloride is an acidic solid. Avoid skin contact.
5. Do not taste the products from this experiment.

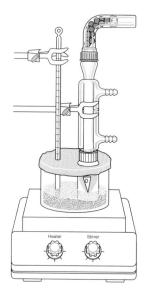

Part A: Microcale Preparation of D-Sucrose Octaacetate

To a dry, 5-mL conical vial containing a spin vane, add 207 mg of D-sucrose and 1 mL of pyridine. Next, add 2 mL of acetic anhydride. Fit the vial with a water-jacketed condenser and a calcium chloride drying tube. Turn on the water to the condenser. Heat at reflux with stirring for 30–45 minutes. Cool the flask until it is just warm to the touch. Transfer the solution into an Erlenmeyer flask that contains 5 g of ice and 5 mL of water. Use a stirring rod to stir the mixture until the product collects as a thick syrup on the bottom of the flask and most of the ice has melted. Decant the water. Add 10 mL of cold water to the Erlenmeyer flask and stir to wash the product. Decant the water and repeat the water washings with two additional 10-mL portions. Decant the last water wash leaving an oily product.

To crystallize the product, add hot 95% ethanol to the oil and heat on a sand bath or heat block until all of the oil has dissolved. Remove the hot solution from the bath and add warm water to the cloud point. If the oil starts to reform, add 2–3 drops of ethanol until the solution becomes clear and reheat the solution. Add warm water to the cloud point again. Repeat this process until the cloudy solution does not oil out. Allow the solution to cool slowly to room temperature. Cool further using an ice bath. Add a seed crystal, if available. The crystals of D-sucrose octaacetate are slow to form; it may be necessary to store the recrystallization solution in the lab drawer until the next lab period. Collect the crystals using suction filtration. Wash the crystals with cold 95% ethanol. Air dry the product for 10–15 minutes. Recrystallize from ethanol-water.

Characterization for Part A

Melting Point: Determine the melting point of the product. The reported melting point of pure D-sucrose octaacetate is 89°C. The melting point of the crude product is typically 50–60°C.

Infrared Spectroscopy: Record the IR spectrum of the product using the Nujol mull or KBr technique. Interpret the spectrum. The IR spectrum of D-sucrose octaacetate is shown in Figure 25.2-1.

NMR Spectroscopy: Dissolve a sample of the dried product in $CDCl_3$ and record the ^1H and ^{13}C NMR spectra. The ^1H and ^{13}C NMR spectra of D-sucrose octaacetate are shown in Figures 25.2-2 and 25.2-3, respectively.

Part B: Microscale Preparation of β-D-Glucopyranose Pentaacetate

To a dry, 5-mL conical vial containing a spin vane add 200 mg of D-glucose and 200 mg of anhydrous sodium acetate. Next, add 2 mL of acetic anhydride and fit the vial with a water-jacketed condenser and a calcium chloride drying tube. Turn on the condenser water. Heat at reflux with stirring until all solids have dissolved. This should take approximately 15 minutes. Stirring may be difficult at first. Cool the vial until it is just warm to the touch (about 1 minute). Use a Pasteur pipet to transfer the contents of the vial to a beaker containing 10 g of crushed ice and 10 mL of cold water. Stir the mixture with a glass rod and suction filter the solid product after all or nearly all of the ice has melted. Wash the crystals with three separate portions of cold water. Recrystallize from ethanol.

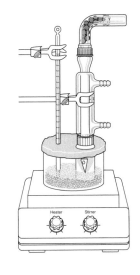

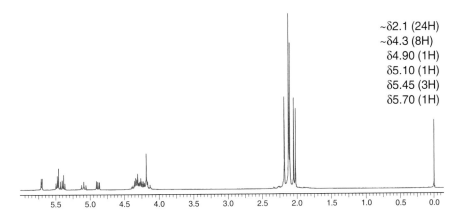

Figure 25.2-1

IR spectrum of D-sucrose octaacetate (Nujol)

Figure 25.2-2

^1H NMR spectrum of D-sucrose octaacetate ($CDCl_3$) ($C_{28}H_{38}O_{19}$)

SOURCE: Reprinted with permission of Aldrich Chemical.

~δ2.1 (24H)
~δ4.3 (8H)
δ4.90 (1H)
δ5.10 (1H)
δ5.45 (3H)
δ5.70 (1H)

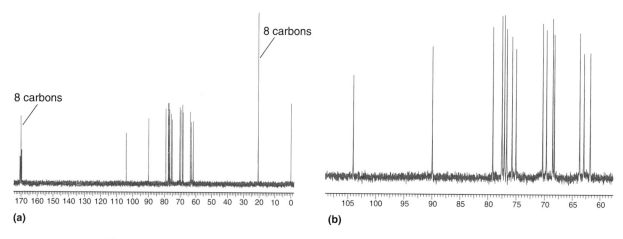

Figure 25.2-3 ^{13}C NMR spectrum of (a) D-sucrose octaacetate (CDCl$_3$) (C$_{28}$H$_{38}$O$_{19}$)
(b) D-sucrose octaacetate (expanded region)

SOURCE: Reprinted with permission of Aldrich Chemical.

Characterization for Part B

Melting Point: Determine the melting point of the product. The reported melting point of β-D-glucopyranose pentaacetate is 131–132°C. The melting point of the crude product is typically 110–120°C.

Infrared Spectroscopy: Record the IR spectrum of the product using the Nujol mull or KBr technique and interpret the spectrum. The IR spectrum of β-D-glucopyranose pentaacetate is shown in Figure 25.2-4.

NMR Spectroscopy: Dissolve a sample of the dried product in CDCl$_3$ and record the ^{1}H and ^{13}C NMR spectra. The ^{1}H and ^{13}C NMR of β-D-glucopyranose pentaacetate are shown in Figures 25.2-5, 25.2-6, and 25.2-7 respectively.

Figure 25.2-4

IR spectrum of
β-D-glucopyranose
pentaacetate (Nujol)

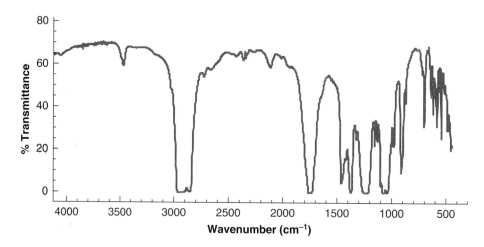

Part C: Microscale Conversion of β-D-Glucopyranose Pentaacetate to the α-Anomer

To a dry, 3-mL conical vial fitted with a spin vane, water-jacketed condenser, and drying tube, add 10 mg of anhydrous zinc chloride and 0.5 mL of acetic anhydride. Heat on a sand or water bath at 90–100°C with stirring until the solid has dissolved. Add about

δ2.05 (15H)
δ3.85 (1H)
δ4.10 (1H)
δ4.30 (1H)
δ5.20 (3H)
δ5.70 (1H)

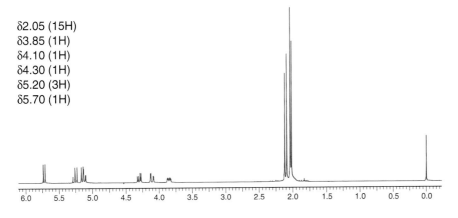

Figure 25.2-5

^{1}H NMR spectrum of β-D-glucopyranose pentaacetate (CDCl$_3$) C$_{16}$H$_{22}$O$_{11}$

SOURCE: Reprinted with permission of Aldrich Chemical.

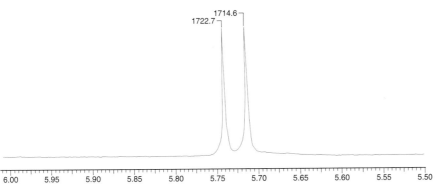

Figure 25.2-6

^{1}H NMR spectrum of β-D-glucopyranose pentaacetate (expanded) (CDCl$_3$) C$_{16}$H$_{22}$O$_{11}$

SOURCE: Reprinted with permission of Aldrich Chemical.

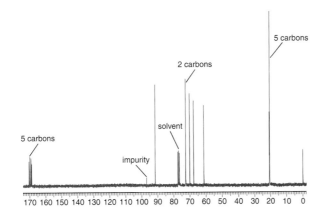

Figure 25.2-7

^{13}C NMR spectrum of β-D-glucopyranose pentaacetate (CDCl$_3$)

SOURCE: Reprinted with permission of Aldrich Chemical.

100 mg of β-D-glucopyranose pentaacetate and heat with stirring for an additional 10 minutes. Cool the reaction vessel until it is warm to the touch and transfer the solution with a pipet to a beaker containing 5 g of ice. Stir the mixture with a glass rod and suction filter the product after it has crystallized. Recrystallize the product using a minimum amount of ethanol.

Part D: Miniscale Preparation of α-D-Glucopyranose Pentaacetate and Measurement of Optical Rotation

To a dry, 50-ml, round-bottom flask, containing a stir bar, add 50 mg of anhydrous zinc chloride and 2.5 mL of acetic anhydride. Warm the mixture with stirring using a heating

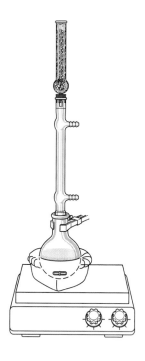

mantle to dissolve the zinc chloride. This will take approximately 5 minutes. Add 500 mg of α-D-glucose to the solution and fit the flask with a water-jacketed condenser and a drying tube. Turn on the water to the condenser and heat the stirred solution at reflux for 45 minutes. Cool the solution to room temperature and pour into 25 mL of ice water in a 100-mL beaker. Stir the mixture intermittently for 15–20 minutes using a stirring rod. The oil that forms initially should gradually solidify. Filter the solid product using a Büchner funnel. Wash the product several times with cold water. Draw air through the solid on the filter for 10–15 minutes. Recrystallize from ethanol. Record the mass of dry α-D-glucopyranose pentaacetate. Add 4–5 mL of methylene chloride to dissolve the product. Then add methylene chloride to adjust the volume of the solution to 10.0 mL. Measure the optical rotation using a polarimeter ($[\alpha]_D = +101.6°$).

Characterization for Parts C and D

Melting Point: Determine the melting point of the product. The reported melting point of α-D-glucopyranose pentaacetate is 110–111°C.

Infrared Spectroscopy: Record the IR spectrum as a Nujol mull or a KBr pellet. The IR spectrum of α-D-glucopyranose pentaacetate is shown in Figure 25.2-8.

NMR Spectroscopy: Dissolve a sample of the dried product in $CDCl_3$ and record the ^1H and ^{13}C NMR spectrum. The ^1H and ^{13}C NMR spectra of α-D-glucopyranose pentaacetate are shown in Figures 25.2-9, 25.2-10, and 25.2-11.

Results and Conclusions

1. Calculate the percent yield of the dried product.
2. Assign the IR absorptions at 2950 and 1745 cm^{-1}.
3. Calculate the specific rotation and determine the optical purity of the product.

Cleanup & Disposal

Aqueous filtrates and methanol and ethanol filtrates should be washed down the drain. Place methylene chloride solutions in a container in the hood labeled "halogenated organic solvent waste."

Figure 25.2-8

IR spectrum of α-D-glucopyranose pentaacetate (Nujol)

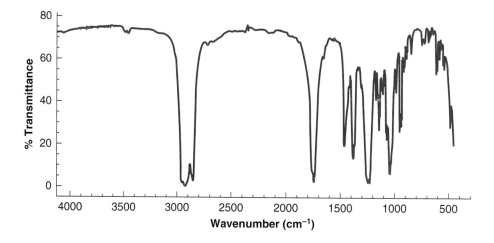

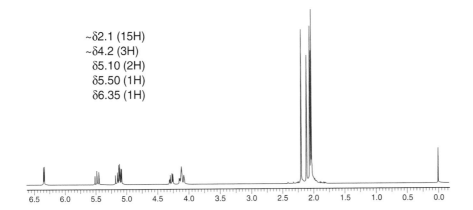

~δ2.1 (15H)
~δ4.2 (3H)
δ5.10 (2H)
δ5.50 (1H)
δ6.35 (1H)

Figure 25.2-9

¹H NMR spectrum of
α-D-glucopyranose
pentaacetate (CDCl₃)
C₁₆H₂₂O₁₁

SOURCE: Reprinted with
permission of Aldrich Chemical.

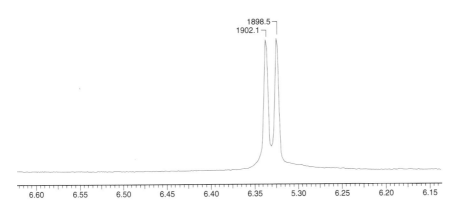

1898.5
1902.1

Figure 25.2-10

¹H NMR spectrum of
α-D-glucopyranose
pentaacetate
(expanded) (CDCl₃)
C₁₆H₂₂O₁₁

SOURCE: Reprinted with permission of Aldrich Chemical.

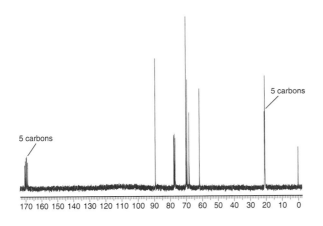

5 carbons

5 carbons

Figure 25.2-11

¹³C NMR spectrum of
α-D-glucopyranose
pentaacetate (CDCl₃)

SOURCE: Reprinted with permission of Aldrich Chemical.

Critical Thinking Questions *(The harder one is marked with a ❖.)*

1. Should the α- and β-anomers of glucopyranose pentaacetate have equal and opposite specific rotations? Explain.

2. Suggest an alternate reagent for acetylation of sugars.

3. One of the acetate groups of α- and β-glucopyranose pentaacetate can be removed
   readily. Explain.

4. What experimental conditions would remove all of the acetate groups to convert the product back to the original sugar? Explain.

5. The NMR spectrum of β-D-glucopyranose pentaacetate shows five unique methyl absorptions. Explain.

6. Explain the role of sodium acetate in the preparation of β-D-glucopyranose pentaacetate.

7. How does zinc chloride facilitate conversion of β-D-glucopyranose pentaacetate to the α-anomer? Propose a mechanism.

❖ 8. Explain why α-D-glucopyranose pentaacetate is more stable than the β-anomer.

9. Referring to Figure 25.2-3 (a) and (b), account for all 28 carbons in the structure of D-sucrose octaacetate. Note that three signals from δ76–78 are due to the solvent.

10. Explain the difference in coupling constants observed in Figures 25.2-6 and 25.2-10.

Reference

Mann, T. D., Mosher, J. D., and Woods, W. F. "Preparation of Sucrose Octaacetate— A Bitter-Tasting Compound." *J. Chem. Educ.* 69 (1992): 668.

Chapter 26

Lipids

L ipids are classified as a group of biomolecules that includes fats (triglycerides), waxes, steroidal vitamins and hormones, terpenes, prostaglandins, and other fatty, naturally occurring compounds. In the body, lipids serve as a source of energy and as protection.

In Experiment 26.1, a triglyceride contained in nutmeg is isolated by extraction, analyzed, and characterized. Hydrolysis of the triglyceride furnishes a soap, which upon acidification yields a fatty acid. Experiment 26.2 is an investigation of a fascinating property of certain cholesterol derivatives called the liquid crystalline state.

Experiment 26.1: Soap from a Spice: *Isolation, Identification, and Hydrolysis of a Triglyceride*

Natural products are often composed of complex mixtures of proteins, carbohydrates, and lipids. In this experiment, you will isolate a triglyceride from a natural product, nutmeg. In this experiment, you will:

♦ isolate and purify the triglyceride.

♦ determine the identity through melting point and spectroscopic analysis

♦ hydrolyze the triglyceride.

♦ saponify the triglyceride to prepare soap.

Techniques

Technique F	Recrystallization
Technique G	Reflux
Technique K	Thin-layer chromatography
Technique M	Infrared spectroscopy
Technique N	Nuclear magnetic resonance spectroscopy

Background

Contrary to its name, nutmeg is not a nut, but is the kernel of a fruit grown on the nutmeg tree primarily in Indonesia and Grenada. Originally used to flavor beer, the spice is used today to flavor vegetable dishes and desserts.

Like any natural product, nutmeg contains a number of components. Essential oils make up about 5–15% of the nutmeg seed. These oils include aromatic ethers (such as eugenol and safrole), terpenes (such as α-pinene), monoterpene alcohols (such as geraniol), sesqueterpenes (such as caryophyllene), terpinic esters (such as linalyl acetate), carboxylic acids (such as formic acid), and aromatic hydrocarbons (such as toluene). These essential oils are used as flavoring agents in food and pharmaceuticals and as additives in perfumes.

The other main component of the nutmeg seed is fixed oil (called the nutmeg butter). The major component of the nutmeg butter is a specific triglyceride, which makes up about 75% of the mass of the nutmeg butter. Triglycerides are types of lipids, an important class of natural products, providing protection and energy.

The structure of the triglyceride determines whether the lipid will be liquid (oil) or solid (fat). Triglycerides are composed of three long-chain fatty acids bonded to glycerol (1,2,3-propanetriol) in ester linkages. The three fatty acids may be identical or different. The basic structure of a triglyceride is shown here, where n can be the same number or different. When all three fatty acids are saturated, the resulting triglyceride is a solid (called a fat). When one or more of the fatty acid linkages is unsaturated, the triglyceride is a liquid (called an oil).

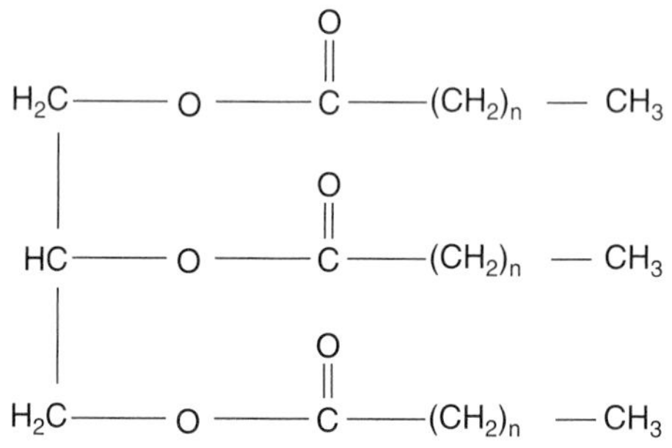

general structure of saturated fat

The purpose of Part A of this experiment is to isolate the triglyceride, to purify it, and then to identify it (De Mattos and Nicodem, 2002; Schelble, 2002). The triglyceride is composed of one or more of the following fatty acids: myristic acid (dodecanoic acid), stearic acid (octadecanoic acid), oleic acid (Z-9-octadecanoic acid), or linoleic acid (9Z,12Z-9,12-octadecadienoic acid).

In Part B of this experiment, you will hydrolyze the triglyceride isolated in Part A to yield glycerol and the individual fatty acids after saponification and acidic hydrolysis. The initial product of the reaction will be soap. A soap is sodium or potassium salt of a long-chain fatty acid, such as stearic acid. An example of soap is sodium stearate, shown below:

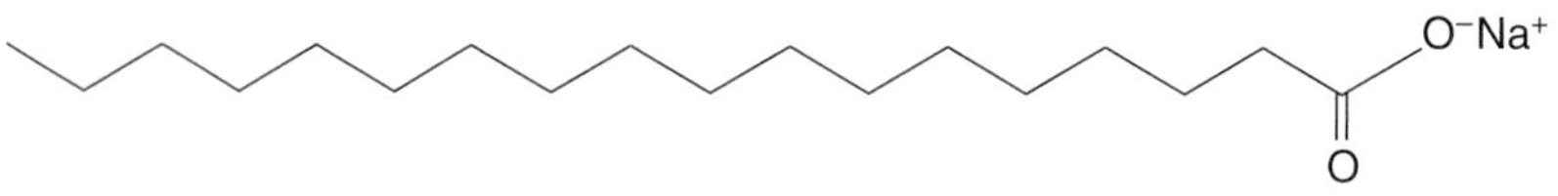

sodium stearate (sodium octadecanoate)

Soaps have long nonpolar hydrocarbon "tails" and polar carboxylate "heads." A soap thus has both lipophilic (fat-loving) and hydrophilic (water-loving) properties. In water, this dual nature is resolved by aggregation of 50–100 soap molecules into micelles. The carboxylate heads surround the exterior of the micelle and the hydrocarbon tails are packed into the interior. Grease dissolves in the nonpolar interior of the micelle, and is then washed away with the water. Unfortunately, soap is not a very effective cleaning agent in hard water. Hard water contains relatively high concentrations of Ca^{2+}, Mg^{2+}, and Fe^{3+} ions. Soap forms complexes with these divalent ions and precipitates, leaving a scummy residue.

The procedure used today for making soap is similar to the recipe used by American pioneer women. Soaps are made from saponification of vegetable oils, such as coconut oil, palm oil, olive oil, and cottonseed oil, and animal fats, such as lard and tallow. This process is illustrated below for the saponification of tristearin, a triglyceride containing three stearic acid fatty acids:

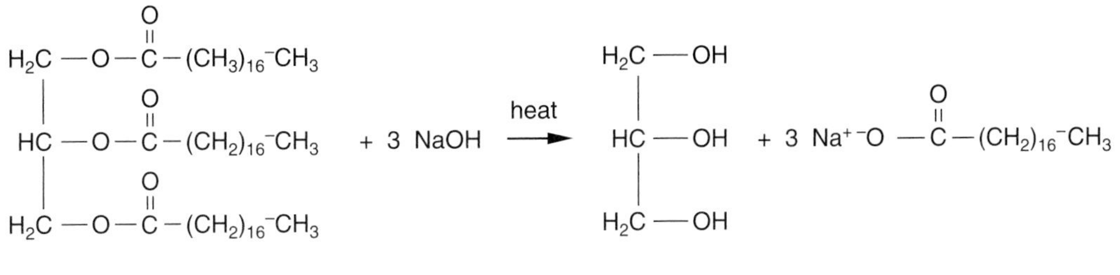

Dyes, perfumes, or other additives are mixed into the soap before it hardens. The properties of the soap depend upon the type of fat or oil used, including number of carbons in the chain and the degree of unsaturation in the fatty acid, and the metal ion: soaps made from KOH have softer consistency than soaps made from NaOH. The soap produced will be reacted with HCl to yield the fatty acid. If desired, the soap can be isolated and tested for cleaning properties.

Prelab Assignment

1. Draw the structures of myristic acid, stearic acid, oleic acid, and linoleic acid and classify each as being saturated, monounsaturated, or polyunsaturated.
2. Explain how IR and NMR spectroscopy could help differentiate between triglycerides composed of saturated or unsaturated fatty acids.
3. Write a detailed mechanism for the hydrolysis of tristearin with sodium hydroxide.
4. Vegetable oils have different properties from animal fats. Explain these properties of fats and oils in terms of intermolecular attractions.

Experimental Procedure

Safety First!

Always wear eye protection in the laboratory.

1. Wear eye protection at all times in the laboratory.
2. Diethyl ether is very flammable. Keep well away from flames or heat sources. Use under the hood.
3. Sodium ethoxide is caustic! Wear gloves when handling solid sodium ethoxide. If sodium hydroxide gets on your skin, wash the affected area immediately under cold running water. Notify your instructor.

Part A: Miniscale Isolation of a Triglyceride from Nutmeg

Weigh a 2.00-g piece of nutmeg and grind it into fine particles. Transfer the ground nutmeg to a 25-mL round-bottom flask containing a stir bar. Add 5 mL of diethyl ether. Fit the flask with a water-jacketed condenser. Turn the water on to the condenser and heat

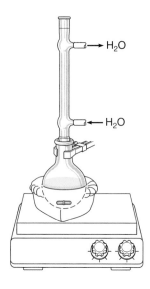

H₂O →

H₂O ←

Hood!

gently with a heating mantle or sand bath while constantly stirring. The mixture has a tendency to bump. This can be minimized by gently refluxing with stirring for 30 minutes. Let the flask cool to room temperature while the solids settle to the bottom of the flask. Trying to disturb the solid as little as possible, decant the diethyl ether solution through a fluted filter paper into a tared 50-mL Erlenmeyer flask that contains a boiling chip. To ensure quantitative transfer of the triglyceride, rinse the solids in the round-bottom flask with two 2-mL portions of diethyl ether and pour through the filter paper. Evaporate the diethyl ether on a warm sand bath or steam bath to leave a yellowish oily solid. Let cool to room temperature and weigh the crude product.

To recrystallize, dissolve the oil in acetone, using 1 mL of acetone per 50 mg of crude product. Warm the solution over a water bath. Pour the warm solution into a clean 25-mL Erlenmeyer flask and let it cool to room temperature. When crystals form, cool in an ice-water bath. Vacuum filter the crystals using a Büchner funnel. Wash the crystals with minimum amounts of ice-cold acetone and allow to air dry. Weigh the crystals.

Characterization for Part A

Melting Point: Determine the melting point of the product.

Thin-layer Chromatography: Dissolve a small amount of the triglyceride in methylene chloride and spot on a silica gel plate. Also spot a solution of the crude nutmeg. Develop the plate in hexanes.

Infrared Spectroscopy: Obtain an IR spectrum of the isolated product using either KBr or the Nujol mull technique and interpret the spectrum.

NMR Spectroscopy: Dissolve the product in CDCl₃. Record the ¹H NMR spectrum from 1–12 ppm and interpret the spectrum.

Bromine in Methylene Chloride Test: In a test tube, dissolve 20 mg of the solid in 5 drops of methylene chloride. Add dropwise a solution of freshly prepared 0.1 M solution of bromine dissolved in methylene chloride. Note the number of drops required for the bromine color to remain. A positive test is one in which more than 3–4 drops of bromine is required.

Results and Conclusions for Part A

1. Calculate the percent of the triglyceride in nutmeg.
2. From the IR spectrum, determine whether the triglyceride is saturated or unsaturated. Indicate important bands in the IR spectrum that supported this conclusion.
3. From the ¹H NMR spectrum, determine the type of fatty acids present in the triglyceride. Justify your conclusion.
4. Based upon the melting point and spectroscopic evidence, is it possible to rule out any of the possible fatty acids? Explain.
5. Evaluate the purity of the isolated triglyceride.

Part B: Microscale Hydrolysis of a Triglyceride

Dissolve 100 mg of the triglyceride in 2 mL of 95% ethanol in a 10-mL round-bottom flask with a stir bar. Cautiously add 34 mg of solid sodium ethoxide. Heat under reflux for 15 minutes. A white solid will form during heating. Let the mixture cool to room temperature, then carefully add 2.0 mL water and 4.0 mL of 35% NaCl to the solid. Suction filter the solid using a Büchner or a Hirsch funnel and wash with 3.0 mL of cold water. Separate the solid into two portions. Save half of the soap for testing in Part C, if desired. Otherwise, hydrolyze all of the solid.

Transfer the other half of the solid into a 50-mL beaker and add 2.0 mL of water. Cool the solution in an ice bath. Add 6 M HCl dropwise while stirring the solution with a stirring rod. Continue to add HCl until the pH is acidic. (Test the pH of the solution with litmus paper.) Suction filter the white precipitate using a Hirsch funnel. Wash the solid with a small amount of cold water. Let the solid air dry. If desired, the solid can be recrystallized from methanol.

Characterization for Part B

Melting Point: Determine the melting point of the fatty acid.

Thin-layer Chromatography: Dissolve a small amount of the fatty acid in methylene chloride and spot on a silica gel plate. Also spot solutions of the possible fatty acids. Develop the plate in hexane.

Results and Conclusions for Part B

1. Based upon the TLC and/or melting point, what conclusions can be drawn about the composition of the triglyceride? Was it composed of a single fatty acid or was there more than one type?
2. Based upon the melting point, determine the identity of the fatty acid(s) in the triglyceride.
3. Draw the structure of the triglyceride(s).
4. Write a flow scheme for the workup procedure for the hydrolysis of the triglyceride.

Part C: Determination of Properties of the Soap from Nutmeg

Prepare soap solutions by placing approximately 10 mg of the solid in 10 drops of water in each of four test tubes. Shake the test tubes to mix.

Sudsing Properties in Tap Water.   To the first test tube, add 10 drops of tap water. Cap the tube and shake. Observe the amount of suds formed.

Sudsing Properties in Hard Water (Ca^{2+}).   Into the second test tube, add 10 drops of a 1% solution of CaCl$_2$ (simulating hard water). Cap the tube and shake. Note the formation and the color of any precipitate formed.

Sudsing Properties in Hard Water (Fe^{3+}).   Into the third test tube, add 10 drops of a 1% solution of FeCl$_3$. Cap the tube and shake. Note the formation and the color of any precipitate formed.

Cleansing Ability.   To test cleaning ability, spread a thin layer of lard on a watch glass and a thin layer of vegetable oil on a second watch glass. Dip a cotton swab in the soap solution in the fourth test tube and observe how well the soap cleans the lard and oil.

Cleanup & Disposal

Wash any aqueous solution down the drain with water. Place nutmeg residues in the solid waste container. Place all ether solutions in a container labeled "nonhalogenated organic solvent waste."

Critical Thinking Questions

1. Explain why lipids are soluble in solvents like ether and not in water.
2. Explain why diethyl ether is a better choice for the extraction than ethanol.
3. Discuss why a plant produces a secondary natural product, such as the one isolated from nutmeg.

4. What is the purpose of adding saturated sodium chloride to the soap?

5. Glycerol is a by-product in the hydrolysis of the triglycercide. What happens to the glycerol in the workup procedure?

6. How would the properties of soaps change if short-chain fatty acids (less than six carbons) were substituted for the long-chain fatty acids?

7. Write the products of saponification of glycerol trioleate with sodium hydroxide.

8. *Library project*: Laundry balls is a new consumer product that claims to get clothes clean in an automatic washer without the use of laundry detergents or any other chemicals. The manufacturers claim that the laundry balls work by changing the molecular structure of water by infrared waves. Write a brief report, either supporting or refuting these claims.

9. *Library project:* For environmental reasons, phosphates may no longer be added to detergents. Explain the correlation between phosphates and eutrophication.

10. *Molecular modeling assignment:* Construct models of triolein and tristearin. Minimize energies. Examine the most stable conformations of each of these molecules. Explain why one triglyceride is an oil while the other is a fat.

References

De Mattos, M., and Nicodem, D. "Soap from Nutmeg: An Integrated Introductory Organic Chemistry Laboratory Experiment." *J. Chem. Ed.* 79 (2002): 94–95.

Schelble, S., Chemistry Department, University of Colorado at Denver, private communication (2002).

Experiment 26.2: Preparation of Esters of Cholesterol and Determination of Liquid Crystal Behavior

Most students are aware of the common properties of the three states of matter (gas, liquid, solid). However, some liquids exhibit interesting properties at temperatures near their melting points. In this experiment, you will:

• esterify cholesterol to prepare an ester

• examine the liquid crystal properties of the ester.

Techniques

Technique F	Recrystallization and filtration
Technique K	Thin-layer chromatography
Technique M	Infrared spectroscopy

Background

Liquid crystals have been known since 1850 when W. Heintz melted a triglyceride of stearic acid and observed the crystals changing from a solid state to a cloudy liquid (the liquid crystal state) to a clear liquid at different temperatures. Cholesterol myristate is another example of a liquid crystal.

cholesterol myristate

At temperatures below 71°C, cholesterol myristate is a crystalline solid. When the solid is heated up to 71°C, it becomes an opaque liquid and when heated to 86°C, it becomes a clear liquid. The state between 71°C and 86°C is an intermediate state, a state between being solid and being liquid. At this temperature range, the compound is a liquid crystal, with properties similar to the solid and the liquid phases, but also different.

In the solid phase, most compounds have a well-defined crystal structure, with molecules aligned in a crystal lattice in a regular repeating unit, held together by intermolecular forces. When solids are heated, the molecules vibrate more rapidly and eventually overcome the intermolecular forces. The molecules move in many directions and become randomly dispersed in the liquid phase (an anisotropic liquid). For cholesteric liquid crystals, however, the movement of molecules in the transformation of solid to liquid does not occur randomly. Because intermolecular attractions in some directions are stronger than in others, the resulting solution becomes relatively ordered, with alternating layers of randomly dispersed molecules. The differences in these states are illustrated in Figure 26.2-1.

These drawings illustrate the differences in arrangement of molecules in the cholesteric liquid crystal state as compared to the solid and liquid states. In cholesteric liquid crystals, the layers are composed of randomly situated rod-shaped molecules in organized and repeating layers. This organization, called twisted nematic orientation, accounts for the fluid properties observed for cholesteric liquid crystals, since heating can cause layers to slide over each other. This organization also accounts for the thermochromic property observed for some liquid crystals.

The actual orientation of liquid crystals affects their properties and subsequent applications. For example, cholesteric liquid crystals (called twisted nematics because of their organization) are used extensively in temperature-sensing devices and in LCD (liquid crystal displays) in calculators, clocks, watches, battery testing strips, and computer screens. In the twisted nematic orientation, the long axes of the rodlike crystals

Figure 26.2-1

Representations of solid, liquid, and liquid crystalline states of matter

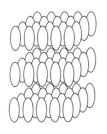

solid state
(highly ordered)

liquid state
(random, no order)

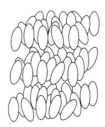

liquid crystalline state
(ordered)

Figure 26.2-2

Twisted nematic orien-
tation of cholesteric
liquid crystals

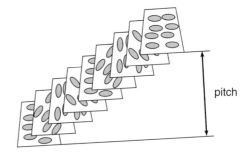

pitch

are oriented along the plane. Each layer is rotated from the adjacent layers by a small angle, with the resulting layers looking like a spiral staircase or helix. At some distance the rod layers will again be oriented in the identical direction (will be parallel), as can be seen in Figure 26.2-2. The pitch is the spacing or distance between a specific layer and its parallel layer. The pitch is defined as the distance it takes to rotate one full turn in the helix. The pitch affects the properties of the liquid crystal.

When light interacts with a twisted nematic liquid crystal, some of the light is reflected. Light with a wavelength that matches the distance between parallel layers will be reflected, causing the compound to appear colored. When the spacing between parallel layers matches the wavelength of blue light, the compound will reflect blue, while different colors will be reflected at other spacing differences.

When twisted nematic liquid crystals are heated, the spacing between layers slightly increases and the angles between layers rotate slightly. These changes decrease the spacing between parallel layers and alter the wavelength of light reflected. The color of the liquid crystal is thus a visible indication of the temperature of the liquid crystal.

Twisted nematic liquid crystals are also widely used in LCD displays. Passing polarized light through a twisted nematic liquid crystal causes the plane of polarized light to rotate by 90 degrees and to pass through to a rear filter, where the light is reflected back by a mirror. These areas appear bright. When an electric field is applied, the liquid crystal molecules reorient themselves to align with the field. Since the light is no longer rotated 90 degrees and cannot pass through the rear filter, the light is not reflected, making these areas appear dark.

Although the twisted nematic orientation is the most important economically, liquid crystals can adopt other orientations, such as smectic orientations. Smectic orientations are more highly ordered than twisted nematic orientations, and derive their name from the Greek word for "soap." The scummy substance formed when a bar of soap sits on a plate is a type of smectic liquid crystal. Smectic liquid crystals are finding uses as ferroelectric liquid crystals in display technology.

The importance of liquid crystals is not limited to industrial and technical applications. There are numerous biological compounds and organic compounds that are liquid crystals. For example, slugs secrete a mucus that is liquid crystalline; the viscosity of the secretions changes in response to ambient conditions. Cell membranes also exhibit liquid crystalline behavior, with proteins aligned accordingly. Many carbon compounds exhibit liquid crystal behavior.

As is true for many physical properties, the liquid crystal properties of a compound depend upon its structure. Many liquid crystals, like cholesterol, are organic molecules with a long nonpolar chain at one end and a polar group at the other end. These molecules form rodlike shapes that can orient themselves along an axis. Other liquid crystals, particularly those that exhibit smectic orientation, contain aromatic rings.

In this experiment, you will prepare an ester of cholesterol and examine the liquid crystal properties of the resulting product. After synthesizing and purifying the cholesteryl esters, you will explore the liquid crystal properties by heating the solid on a watch glass. By compiling class data, you will determine the effect of structure on liquid crystal properties. The acid chlorides to be used in this experiment are benzoyl chloride, *o*-, *m*-, and *p*-nitrobenzoyl chloride, *o*-, *m*-, and *p*-methoxybenzoyl chloride, 2,4-dinitrobenzoyl chloride, decanoyl chloride, dodecanoyl chloride, tetradecanoyl chloride, octadecanoyl chloride, or another assigned acid chloride.

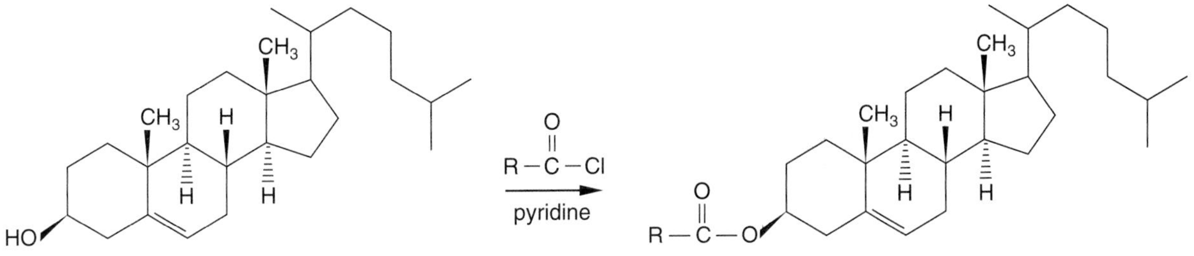

Prelab Assignment

1. Write a mechanism for the preparation of the assigned ester.
2. Calculate the volume of acid chloride to be used in this experiment.
3. Look up the physical properties of the assigned ester.
4. Write a flow scheme for the workup procedure.

Experimental Procedure

1. Wear eye protection at all times in the laboratory.
2. Acid chlorides are lachrymators. Use under the hood. Wear gloves at all times.
3. Pyridine has a strong unpleasant odor and is toxic. Keep under the hood and wear gloves at all times when using this reagent.

S a f e t y   F i r s t !

Always wear eye protection in the laboratory.

Miniscale Synthesis of Esters of Cholesterol

The instructor will assign a specific acid chloride in this experiment. The possible reagents to use are benzoyl chloride, *o*-, *m*-, and *p*-nitrobenzoyl chloride, *o*-, *m*-, and *p*-methoxybenzoyl chloride, 2,4-dinitrobenzoyl chloride, decanoyl chloride, dodecanoyl chloride, tetradecanoyl chloride, octadecanoyl chloride, or other assigned acid chloride.

All glassware should be dry before starting this experiment. Do all work under the hood. Weigh 500 mg of cholesterol (1.25 mmol) and place in a 50-mL Erlenmeyer flask. Add 1.5 mL of dry pyridine. Stopper the flask and swirl gently to dissolve the cholesterol. Add 1.7 mmol of the acid halide and swirl to mix. Heat the mixture on a steam bath or a hot plate for 10 minutes.

Cool the flask in an ice-water bath and add 7 mL of methanol while swirling. Suction filter the ester using a Büchner funnel. Wash the flask with a small volume of ice-cold methanol. Pour this methanol over the crystals to wash off adsorbed impurities.

Hood!

Recrystallize the crude ester from ethyl acetate. Heat the crude ester in ethyl acetate until all the solid has dissolved. If necessary, perform a hot gravity filtration to remove solid that will not dissolve in the hot ethanol. Let the clear solution cool to room temperature. When crystals have formed, cool the mixture in an ice-water bath. Vacuum filter the crystals using a Büchner funnel. Wash the crystals with minimum amounts of ice-cold ethyl acetate. Dry the crystals by continuing to draw a vacuum. After the crystals are dry, weigh the product.

To test the liquid crystal properties, spread 100 mg of the ester on one end of a glass microscope slide. Place the microscope slide on a hot plate set on high. (Be careful not to touch the hot plate.) Watch the plate closely to observe signs of transition, where the solid melts to a cloudy liquid and then when the cloudy liquid becomes clear. As soon as the liquid becomes clear, use tongs to move the slide to a bright light and observe the slide carefully as the ester cools. Record your observations. This heating and cooling process can be repeated multiple times.

To determine the temperatures at which transitions occur, place 200 mg of the solid in a small test tube and set the test tube in an aluminum heat block. Place a thermometer down into the liquid crystal. Begin heating the solid slowly and observe carefully. Record the temperature at which phase changes occur.

Characterization

Thin-layer Chromatography: Dissolve a small amount of the ester in methylene chloride and spot on a silica gel plate. Also spot solutions of cholesterol. Develop the plate in hexanes.

Infrared Spectroscopy: Obtain an IR spectrum of the product isolated using either KBr or the Nujol mull technique and interpret the spectra.

Results and Conclusions

1. Calculate the percent yield of the ester.
2. Identify absorption bands in the IR spectrum that support the structure of the product.
3. Estimate the purity of the product based upon thin-layer chromatography.
4. Tabulate the temperatures at which liquid crystal behavior is observed. Compare the behavior and temperatures with those of other prepared esters. Are there differences between the cholesteryl benzoates and the cholesteryl esters having long carbon chains? Are there differences between having electron-donating groups and electron-withdrawing groups on the benzoate esters? Summarize the effect of structure on liquid crystal behavior.

Cleanup & Disposal

Place organic solutions in a container labeled "nonhalogenated organic solvent waste." Place all NMR samples in the halogenated organic solvent waste.

Critical Thinking Questions

1. Do spherical molecules exhibit liquid crystalline behavior? Explain.
2. Explain two reasons why pyridine was used in this experiment.
3. Could either a primary or a secondary amine be used in this experiment in place of pyridine? Why or why not?

4. Another method of preparing esters is with Fischer esterification in which an alcohol and carboxylic acid are heated in the presence of an acid catalyst. Would this method have worked to synthesize the esters formed in this reaction? Justify your answer.

5. *Library project:* Find a reference about chiral liquid crystals and write a short paragraph explaining the theory.

6. *Library project:* Explain how cholesterol is esterified biologically. What effect (if any) does esterification of cholesterol have on biological function?

7. *Library project:* Look up representative structures for smectic liquid crystals. What common feature do many of these compounds have? Explain how this feature supports the liquid crystalline behavior.

Reference

Chemical of the Week: Liquid Crystals. See on Internet at http://scifun.chem.wisc.edu/chemweek/liqxtal/liqxtal.html (Accessed 11-9-02)

Chapter 27

Amino Acids and Derivatives

A mino acids contain both amino and carboxyl groups. They exist as high-melting solids because they are zwitterions (ammonium and carboxylate ions in the same molecule). This dipolar nature causes high-melting behavior and low solubility in organic solvents. In Experiment 27.1, p-aminobenzoic acid (PABA) is isolated by crystallization from aqueous solution adjusted to a particular pH range. PABA is physiologically active as a structural component of folic acid.

Amino acid derivatives can often be prepared in much the same way that amines or carboxylic acids can be derivatized. Amines can be acetylated with acetic anhydride or by acid chlorides. Carboxyl groups can be esterified without protection of the amino group by Fischer esterification, as described for the synthesis of alkyl esters of PABA in Experiment 27.1.

Experiment 27.1: Conversion of an Amino Acid to a Sunscreen: *Multistep Preparation of Benzocaine or a Benzocaine Analog*

Benzocaine, the ethyl ester of p-aminobenzoic acid, is a topical anesthetic and sunburn preventative. In this experiment, you will synthesize analogs of benzocaine and evaluate their effectiveness as sunscreens. You will:

♦ prepare benzocaine or an analog using a multistep synthesis.

♦ acetylate an aromatic amine.

♦ oxidize an aromatic alkyl side chain to a carboxyl group.

♦ hydrolyze an amide to an amine.

♦ convert a carboxylic acid to an ester using a Fischer esterification.

♦ analyze the effectiveness of the sunscreen preparation in filtering out ultraviolet radiation.

Techniques

Technique F	Recrystallization and vacuum filtration
Technique G	Reflux
Technique M	Infrared spectroscopy
Technique N	Nuclear magnetic resonance spectroscopy
Technique O	Ultraviolet spectroscopy

Background

Benzocaine is a topical anesthetic that also acts as a sunburn preventative. It is prepared by esterification of *p*-aminobenzoic acid, a vitamin B-complex factor necessary for nucleic acid synthesis in some bacteria.

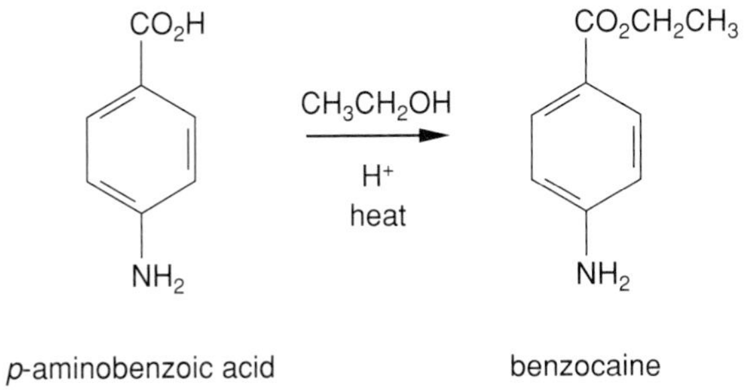

p-aminobenzoic acid   benzocaine

The structure of *p*-aminobenzoic acid (PABA) differs from the 20 naturally occurring α-amino acids in that the amino group of PABA is attached to the δ carbon. PABA is also classified as an amino acid because it has both an amino group and a carboxyl group in the same molecule. As an amino acid, PABA is physiologically active. It is incorporated by enzymes into folic acid along with pteridine and glutamic acid.

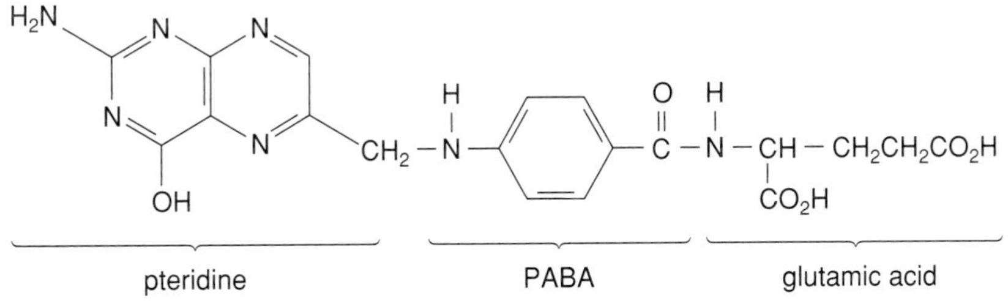

pteridine   PABA   glutamic acid

folic acid

Humans can ingest folic acid from leafy greens and vegetables, but bacteria must manufacture folic acid. Some drugs, like sulfanilamide (Experiment 29.2), have structures similar to that of PABA and act as competitive inhibitors.

Esterification of *p*-aminobenzoic acid changes its pharmacological properties. This illustrates the point that a relatively small change in structure can dramatically alter the physiological and biological properties of a molecule. Pharmaceutical companies frequently develop a new drug by making small alterations in the structure of a known drug, such as an antibiotic, and screening for biological activity.

In this experiment, you will synthesize *p*-aminobenzoic acid in a three-step sequence starting with *p*-toluidine. In the first step, *p*-toluidine is protected against oxidation by acetylation with acetic anhydride. In the second step, the methyl group is oxidized by potassium permanganate to give *p*-acetamidobenzoic acid. Hydrolysis of the acetamido group regenerates the amine as *p*-aminobenzoic acid (PABA). The sequence is shown here.

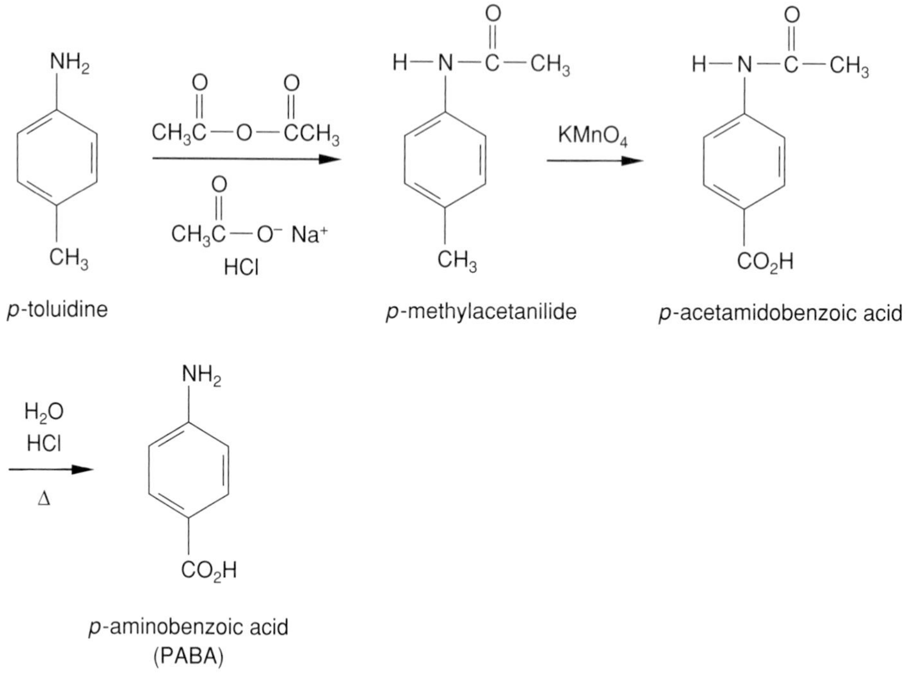

p-toluidine   *p*-methylacetanilide   *p*-acetamidobenzoic acid

p-aminobenzoic acid
(PABA)

In the last step, *p*-aminobenzoic acid and an assigned primary alcohol are heated with sulfuric acid to produce benzocaine or a benzocaine analog.

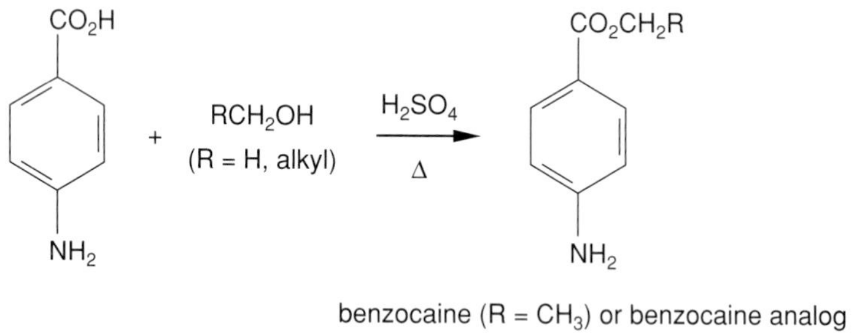

benzocaine (R = CH$_3$) or benzocaine analog

The effectiveness of each analog as a sunscreen will be assessed by measuring how well the compound absorbs ultraviolet radiation from the sun. Exposure to this radiation can cause sunburn. Prolonged or severe exposure can cause skin cancer and genetic mutations. Sunscreens function by absorbing or reflecting UV radiation of the appropriate wavelengths at the surface of the skin.

Prelab Assignment

1. Calculate the theoretical mass of benzocaine (the ethyl ester of *p*-aminobenzoic acid) that could be obtained starting with 1.5 mmol of *p*-toluidine.
2. Suppose that a 75% yield is obtained in each step of the reaction. What mass of benzocaine would be obtained?
3. Write a flow scheme for each reaction and workup procedure in the synthesis.

4. Look up the physical properties of the assigned *p*-aminobenzoic acid ester in a chemical reference book, such as *Dictionary of Organic Compounds, Beilstein,* or *Chemical Abstracts.*

5. Using your lecture text as a reference, write a detailed mechanism for each of the following reactions:
   a. acetylation of *p*-toluidine with acetic anhydride (Part A).
   b. acidic hydrolysis of *p*-acetamidobenzoic acid to *p*-aminobenzoic acid (Part C).
   c. Fischer esterification of *p*-aminobenzoic acid with ethanol to produce benzocaine (Part D).

6. Explain the role of each of the four nutrient agar plates when evaluating the effectiveness of the synthesized sunscreen. If the synthesized sunscreen proves to be effective, what would you expect to observe in Plates 1, 2, and 3 after incubation? If the synthesized sunscreen were more effective than the commercial sunscreen, how would Plates 3 and 4 differ? Explain.

Experimental Procedure

1. Wear eye protection at all times in the laboratory.

2. HCl and H_2SO_4 are corrosive and toxic and can cause burns. Wear gloves when handling these reagents. Wash with copious amounts of cool water if any acid is spilled on your skin. Neutralize acid spills on bench top with sodium bicarbonate and notify the instructor.

3. Potassium permanganate is a toxic oxidizer.

4. Methylene chloride is a toxic irritant and a suspected carcinogen. Wear gloves when handling this reagent. Avoid breathing the vapors; work in a hood.

5. *p*-Toluidine is a toxic, cancer suspect agent. Work under the hood and wear gloves when handling this reagent.

Part A: Miniscale Synthesis of p-Methylacetanilide

Prepare a solution of sodium acetate by mixing 2.15 g of sodium acetate trihydrate ($CH_3CO_2Na \cdot 3H_2O$) and 5–6 mL of water in a 10-mL Erlenmeyer flask. Swirl vigorously to dissolve. Set the sodium acetate solution aside for later use.

In a 125-mL Erlenmeyer flask fitted with a stir bar, add 1.61 g of *p*-toluidine to 40 mL of water. With stirring, add 1.3 mL of concentrated HCl. Stir for 2 minutes. With stirring, add 2.1 mL of acetic anhydride and immediately add the sodium acetate solution. Stir vigorously to mix the reagents. Cool the solution in an ice bath and continue to stir vigorously while the product crystallizes. Isolate the product by vacuum filtration, washing the crystals with several small portions of ice-cold water. Let the crystals air dry or place in a warm drying oven. Weigh the product. Save at least 25 mg of the product for characterization and spectral analysis. The remainder may be used without purification in the next step.

Characterization for Part A

Melting Point: Determine the melting point of the dry product. The reported melting point of *p*-methylacetanilide is 149–151°C.

Figure 27.1-1

¹H NMR spectrum of *p*-methylacetanilide in deuterated DMSO

Source: Reprinted with permission of Aldrich Chemical.

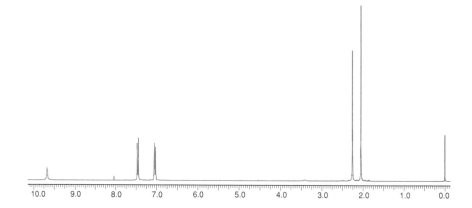

Infrared Spectroscopy: Prepare a Nujol mull or KBr pellet of the solid and record the IR spectrum. Interpret the spectrum.

NMR Spectroscopy: Dissolve the product in deuterated DMSO. Record the ¹H NMR spectrum from 0–10 ppm. Interpret the NMR spectrum. The ¹H NMR spectrum of *p*-methylacetanilide in deuterated DMSO is shown in Figure 27.1-1.

Results and Conclusions for Part A

1. Calculate the percent yield of *p*-methylacetanilide.
2. Assign all peaks in the NMR spectrum and calculate the coupling constants between the aromatic protons.
3. Identify the characteristic bands in the IR spectrum that support the structure of the product.
4. The next step in the procedure calls for the use of 1.1 g of *p*-methylacetanilide. If less than this amount is available, recalculate the amounts of all reagents and solvents that should be used in the procedure.

Part B: Miniscale Synthesis of p-Acetamidobenzoic Acid

Prepare a solution of potassium permanganate by dissolving 2.9 g of KMnO₄ in 15 mL of boiling water. Swirl to dissolve. Add additional boiling water, as necessary, to dissolve the potassium permanganate. Set aside for later use.

To a 250-mL Erlenmeyer flask fitted with a stir bar add 1.1 g of *p*-methylacetanilide, 2.9 g of magnesium sulfate heptahydrate (MgSO₄ · 7 H₂O) and 70 mL of water. Heat with stirring to 85°C on a sand bath or hot plate. While vigorously stirring the solution of *p*-methylacetanilide, slowly add the hot solution of potassium permanganate via Pasteur pipet. The addition should take approximately 30 minutes. It is important to add the permanganate solution slowly and uniformly to avoid local buildup of the oxidant. After all of the oxidant has been added, stir the solution vigorously for 5 minutes. Add 3–4 mL of ethanol to react with excess oxidant. Then, boil the solution until the color is dissipated. Filter off the brown manganese dioxide by pouring the hot solution through Celite on a fluted filter paper into a clean Erlenmeyer flask. Wash the brown precipitate with hot water to dissolve adsorbed product. If the filtrate is still colored, add 1–2 mL of ethanol boil, and refilter.

Cool the filtrate in an ice bath and acidify with 20% sulfuric acid until the pH is 3–4. Collect the product using vacuum filtration, rinsing the crystals with small amounts of ice-cold water. Dry the crystals as much as possible by continuing suction. The product does not need to be completely dry for the next step. Save at least 25 mg of the product and let it dry thoroughly for characterization and spectral analysis.

Characterization for Part B

Melting Point: Determine the melting point of the dry product. The reported melting point of *p*-acetamidobenzoic acid is 250–252°C.

Infrared Spectroscopy: Prepare a Nujol mull of the solid and record the IR spectrum. The product should show two strong C=O stretching absorptions at 1700–1660 cm^{-1} and broad OH stretching around 3300 cm^{-1}.

NMR Spectroscopy: Dissolve the product in deuterated DMSO. Record the ^{1}H NMR spectrum from 0–13 ppm. The product should show one singlet for the CH$_3$ protons at δ2.1, aromatic protons around δ7.8, a singlet for the NH proton, and a singlet for the carboxylic acid proton. The ^{1}H NMR spectrum of *p*-acetamidobenzoic is shown in Figure 27.1-2.

Results and Conclusions for Part B

1. In this reaction, what compound is oxidized and what compound is reduced? Write a balanced equation for the conversion of *p*-methylacetanilide and potassium permanganate to *p*-acetamidobenzoic acid and manganese dioxide.
2. In the workup procedure, ethanol is added to react with excess potassium permanganate. What are the products of this reaction? How are these products eliminated in the reaction workup?
3. Interpret the IR and/or NMR spectra and justify the structure of the product.
4. In the next step in the reaction, the acetamido group will be hydrolyzed to regenerate the amine. This reaction requires 5 mL of 6 M HCl per 1 g of *p*-acetamidobenzoic acid. Based on the mass of *p*-acetamidobenzoic acid obtained (after subtracting the amount saved for characterization), calculate the volume of 6 M HCl required.

Part C: Miniscale Synthesis of p-Aminobenzoic Acid

Weigh the *p*-acetamidobenzoic acid from Part B into a 25-mL round-bottom flask containing a stir bar. For every 1 g of *p*-acetamidobenzoic acid, add 5 mL of 6 M HCl to the flask. Attach a water-jacketed condenser to the round-bottom flask, start the stirrer and reflux gently, with stirring, for 30 minutes. A yellow solid will begin to form. After the heating time is over, let the reaction mixture cool to room temperature. Add 2.5 mL of cold water. Transfer the solution to a 50-mL Erlenmeyer flask. Then add 15 M aqueous ammonia solution dropwise, until the pH is just slightly basic (pH = 7–8). Do not go past pH 8. Add 0.5 mL of glacial acetic acid. Stir vigorously and cool the solution in an ice bath. It may be necessary to induce crystallization by adding a seed crystal. Suction filter the product. Air dry and weigh the crystals. Save 25 mg for characterization.

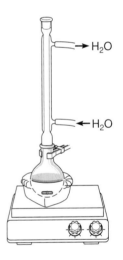

Figure 27.1-2

The ^{1}H NMR spectrum of *p*-acetamidobenzoic in deuterated DMSO

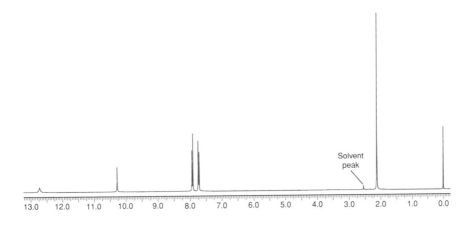

Solvent peak

13.0 12.0 11.0 10.0 9.0 8.0 7.0 6.0 5.0 4.0 3.0 2.0 1.0 0.0

Characterization for Part C

Melting Point: Determine the melting point of the dry product. The reported melting point of *p*-aminobenzoic acid is 188°C.

Infrared Spectroscopy: Prepare a Nujol mull or KBr pellet of the solid and record the IR spectrum.

NMR Spectroscopy: Dissolve the product in $CDCl_3$. Add a few drops of deuterated DMSO if necessary. Record and interpret the ^1H NMR spectrum from $\delta 0$–12. The ^1H NMR spectrum of *p*-aminobenzoic acid is shown in Figure 27.1-3.

UV Spectroscopy: Prepare a 0.001% solution of *p*-aminobenzoic acid in water and record the UV spectrum. Pure *p*-aminobenzoic acid has a wavelength maximum (λ_{max}) of 288 nm ($\varepsilon = 1070$).

Optional Analysis

p-Aminobenzoic acid is biologically active. Some bacteria incorporate it in the biosynthesis of folic acid, one of the B-complex vitamins. The biological activity of *p*-aminobenzoic acid can be tested by streaking a nutrient agar petri dish with folic-acid deficient bacteria. Prepare an aqueous solution of *p*-aminobenzoic acid by dissolving 1–2 mg in 0.5 mL of water. With sterilized forceps, dip a small, round piece of filter paper (about 1–2 mm diameter) in the solution, shake off excess water, and place the disk on the plate. Replace the plate lid and incubate at 37°C for 24–48 hours. Without opening the petri dish, examine the plate and record observations. Do not remove the lid from the petri dish. When finished, return the unopened petri dish to the instructor for disposal.

Results and Conclusions for Part C

1. Calculate the percent yield of *p*-aminobenzoic acid.
2. From the structure of *p*-aminobenzoic acid, the ^1H NMR spectrum might be expected to show a signal due to the carboxylic acid at $\delta 12$ and a signal due to the NH_2 protons. Explain why these signals are not observed.
3. The solubility of *p*-aminobenzoic acid changes depending upon the pH. This compound is more soluble at low pH or at high pH than it is at pH = 7. Explain this observation. Hint: Draw the structure that predominates at pH = 1, pH = 7, and pH = 12.

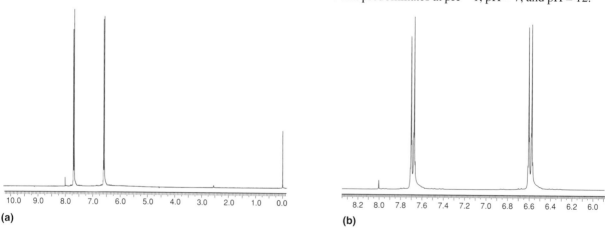

(a)   (b)

Figure 27.1-3   (a) The ^1H NMR spectrum of *p*-aminobenzoic acid ($CDCl_3$ and deuterated DMSO).
(b) The ^1H NMR spectrum of *p*-aminobenzoic acid ($CDCl_3$ and deuterated DMSO)
(expanded region)

4. The pK_a values for *p*-aminobenzoic acid are 2.36 and 4.92. Write the two acid dissociation steps for *p*-aminobenzoic acid and calculate the isoelectric point.
5. The next step in the procedure calls for the use of 330 mg of *p*-aminobenzoic acid. If less than this amount is available, recalculate the amounts of all reagents that should be used in the procedure.

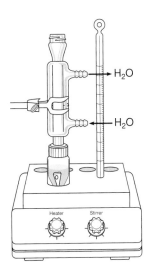

Part D: Microscale Esterification of p-Aminobenzoic Acid

The instructor will assign each student or pair of students a specific primary alcohol to be used to esterify *p*-aminobenzoic acid.

To a 5-mL conical vial containing a spin vane and water-jacketed condenser, add 330 mg of *p*-aminobenzoic acid and 2.5 mL of ethanol or other assigned alcohol. While stirring, add 0.25 mL of concentrated H_2SO_4 dropwise down through the condenser. The precipitate that forms upon the addition of sulfuric acid should dissolve when the solution is heated. Reflux, with stirring, for 1 hour. Cool to room temperature, then neutralize cautiously with dropwise addition of 10% Na_2CO_3 until the pH is around 8. (Gas evolution will be vigorous.) Transfer the solution into a centrifuge tube and extract with two 3-mL portions of methylene chloride. Wash the combined methylene chloride layers with two 8-mL portions of water. Dry the methylene chloride solution over anhydrous sodium sulfate. Gravity filter into a clean Erlenmeyer flask containing a boiling chip. Evaporate the solvent on a steam bath under the hood to leave a whitish residue. Recrystallize using a methanol-water solvent pair. Suction filter the product, and let it air dry. Weigh the white crystals.

Characterization for Part D

Melting Point: Compare the melting point of the product obtained with the expected melting point. Benzocaine (the ethyl ester of *p*-aminobenzoic acid) melts at 91–92°C. The methyl ester, the propyl ester, and the butyl ester melt at 114–115°C, 73–74°C, and 57–58°C, respectively.

   Infrared Spectroscopy: Prepare a Nujol mull or KBr pellet of the solid and record the IR spectrum. Interpret the spectrum. The IR spectrum of benzocaine is shown in Figure 27.1-4. Other benzocaine analogs will have similar IR spectra.

   NMR Spectroscopy: Dissolve the product in $CDCl_3$. Record the ^1H NMR spectrum.

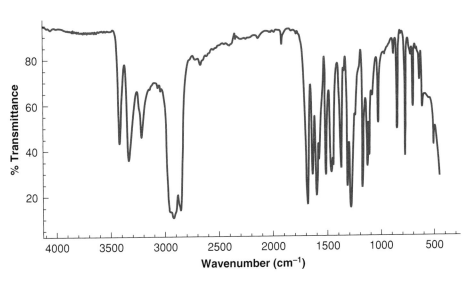

Figure 27.1-4

IR spectrum of benzocaine (Nujol)

Results and Conclusions for Part D

1. Calculate the yield of *p*-aminobenzoic acid ester.
2. Calculate the overall yield of *p*-aminobenzoic acid ester from the four-step sequence. What step was limiting? Explain.
3. Write a balanced equation to explain the purpose of adding sodium carbonate to the reaction mixture.

Part E: Determining the Effectiveness of the p-Aminobenzoic Acid Ester as a Sunscreen

There are two different methods of evaluating the effectiveness of the synthesized sunscreen in absorbing ultraviolet radiation. The first procedure uses a bacterium, *Serratia marcescens,* which is red when grown at room temperature. Obtain four commercial nutrient agar plates, four sterile swabs, a culture broth of *Serratia marcescens,* a cardboard cutout, clear plastic wrap, mineral oil, and a commercial sunscreen. Wear gloves and goggles at all times.

Label the nutrient agar plates as 1, 2, 3, and 4. Swab each plate thoroughly with the broth culture, making sure to cover the entire plate with the bacteria. Prepare each plate as directed below.

Plate 1 is the control. Place the lid on the plate and set aside. It will be incubated with no exposure to radiation.

For Plate 2, place the cardboard cutout on top of the opened plate. Place the plate in the ultraviolet chamber, about 12 inches from the source. Expose the plate to 260 nm ultraviolet radiation for 3 minutes. Remove the cutout and replace the lid. Set aside for incubation.

For Plate 3, prepare a Nujol mull of the product to make it easier to spread the sunscreen evenly across the plate. To do this, take a small amount of the product (the tip of a spatula) and grind to a paste with a mortar and pestle. Add a drop or two of Nujol (mineral oil) and mix thoroughly. The consistency should be that of a thick paste. Place a small piece of clear plastic wrap over the cutout and smear a thick layer of the Nujol mull on top of the plastic wrap. Place the plastic wrap and cutout on top of the plate. Place the plate in the ultraviolet chamber, about 12 inches from the source. Expose the plate to 260 nm ultraviolet radiation for 3 minutes. Remove the plastic and cutout and replace the lid. Set aside for incubation.

For Plate 4, smear a sample of commercial sunscreen on the cutout. Place the plate in the ultraviolet chamber, about 12 inches from the source. Expose the plate to 260 nm ultraviolet radiation for 3 minutes. Remove the plastic and cutout and replace the lid. Set aside for incubation.

Invert the plates and incubate at room temperature for 24 hours. At the end of this time, examine the plates for growth. Based upon the areas of growth, determine the effectiveness of the synthesized sunscreen as compared to the commercial brand.

A second method of evaluating the sunscreen's effectiveness at absorbing radiation is by measuring its spectral properties. Fill a quartz cuvette with ethanol and scan the UV spectrum from 240 to 400 nm. Dissolve 10 mg of the benzocaine derivative in 100.0 mL of ethanol. Rinse out the cuvette with the solution. Then fill the cuvette and scan the UV spectrum from 240 to 400 nm. If the absorbance is too strong, prepare a 1:10 dilution and rerun the spectrum. Similarly prepare a solution of 10 mg of *p*-aminobenzoic acid in 100.0 mL of ethanol, diluting if necessary, and record the spectrum from 240 to 400 nm. Record the results in your lab notebook and on the board to share the data with other students.

Results and Conclusions for Part E

1. Describe the growth on each of the nutrient agar plates. Which plate(s) had the most growth? The least? What do these observations reveal about the effectiveness of the synthesized sunscreen?
2. How did the synthesized sunscreen compare to the commercial sunscreen?
3. Compare the effectiveness of benzocaine and the various other benzocaine analogs. Which synthetic sunscreens were the most effective in absorbing radiation?
4. Determine the maximum wavelength of absorption for the *p*-aminobenzoic acid ester and *p*-aminobenzoic acid.
5. Evaluate the effectiveness of the *p*-aminobenzoic acid ester in absorbing ultraviolet radiation and compare it to *p*-aminobenzoic acid.

Cleanup & Disposal

Aqueous filtrates should be carefully neutralized before being washed down the drain with running water. Ethanol solutions should be placed in a container labeled "nonhalogenated organic solvent waste." Dispose of all nutrient and agar broths in the biohazard waste container.

Critical Thinking Questions

1. A student proposes to prepare benzocaine (the ethyl ester of *p*-aminobenzoic acid) by the following sequence starting with *p*-nitrotoluene: oxidation to *p*-nitrobenzoic acid, esterification with ethanol and acid, and finally reduction of the nitro group. What problems might the student encounter?
2. Why is Fischer esterification not an effective method of preparing esters of tertiary alcohols?
3. The pK_a of benzoic acid is 4.19. Explain why *p*-aminobenzoic acid is a weaker acid than benzoic acid. Explain why *p*-nitrobenzoic acid is a stronger acid than benzoic acid.
4. Which anion is more basic: *p*-nitrobenzoate or *p*-aminobenzoate? Explain.
5. Amino acids, such as *p*-aminobenzoic acid, exist as zwitterions. Write the structure for the zwitterion.
6. Explain why amino acids exhibit the least water solubility at their isoelectric point.
7. *p*-Aminobenzoic acid is used by many bacteria as a precursor to folic acid. Sulfanilamide has a similar structure to *p*-aminobenzoic acid and is a competitive inhibitor. Once sulfanilamide binds to the active site of the bacterial enzyme, the enzyme can no longer catalyze the synthesis of folic acid and the bacteria stops growing. Look up the structure of sulfanilamide and propose a reasonable synthesis of this molecule starting with aniline.
8. Although their use is decreasing in favor of naturally occurring antibiotics such as penicillin and streptomycin, sulfa drugs have been used as antibacterial agents. What is a sulfa drug? What functional group is common to all sulfa drugs?
9. *Library project:* Explain how sulfa drugs were discovered.

10. What purpose does the sodium acetate serve in the acetylation of *p*-toluidine in Part A?

11. *p*-Toluidine is acetylated in the first reaction of the sequence. During the third step, the compound is de-acetylated. Explain why this sequence of steps is necessary.

12. *Molecular modeling assignment:* Construct a model of *p*-aminobenzoic acid as a neutral molecule and as a zwitterion (with an ammonium cation and a carboxylate anion). Compare dipole moments for each molecule and examine the individual electron density maps. What are the relative charges? Determine which species has the lower energy in the gas phase. Then examine solvation effects by surrounding each of the molecules with water molecules. Examine the electron density maps and atomic charges. Determine which species has the lower energy when solvated.

Chapter 28

Qualitative Organic Analysis II

Q ualitative organic analysis is a process whereby compounds can be identified based upon their physical and chemical properties and by conversion to solid derivatives that have characteristic melting points. Students may be issued one or more unknown substances for identification. In the section that follows, an exploratory approach to qualitative organic analysis is given in Experiment 28.1, followed by an introduction to analysis of various functional groups and experimental procedures in Experiment 28.2.

Experiment 28.1: Designing a Classification Scheme for Characterizing an Organic Compound

Important steps in qualitative organic analysis are observation of chemical reactions of an unknown compound and use of that information to determine what functional groups are present or absent in the unknown compound. In this experiment, you will:

- perform a series of chemical tests on known aldehydes, ketones, alcohols, phenols, carboxylic acids, and amines.

- carefully observe and record the results of the chemical tests.

- develop a flow scheme that can be used to differentiate an aldehyde, a ketone, an alcohol, a phenol, a carboxylic acid, or an amine.

Background

The first step in the process of successfully identifying an unknown organic compound is to determine what functional groups are present or absent in the unknown. One method is to use careful observation of results of chemical reactions. In this experiment, you will perform a series of chemical tests on known compounds that have different functional groups. You will carefully note changes during the reaction and record the results for both a positive test (functional group present) and a negative test (functional group not present). Then you will develop a flow scheme that can be used to distinguish between an aldehyde, a ketone, an alcohol, a phenol, a carboxylic acid, and an amine. You should also indicate how to differentiate between a primary, a secondary, and a tertiary alcohol and between a primary, a secondary, and a tertiary amine.

The compounds to be used in this experiment are 2-pentanone, 3-pentanone, benzaldehyde, 1-hexanol, 4-methyl-2-pentanol, 2-methyl-2-pentanol, 3,5-dimethylphenol, *p*-anisic acid, *m*-anisidine, N-methylaniline, and N,N-diethylaniline.

Prelab Assignment

1. Draw the structure of each of the compounds used in this experiment. Classify each compound as a ketone, aldehyde, primary alcohol, secondary alcohol, tertiary alcohol, phenol, primary amine, secondary amine, tertiary amine, or carboxylic acid.
2. Prepare tables in your lab notebook, similar to the ones described in each procedure, to record the results of the solubility and chemical tests.

Experimental Procedure

Safety First!

Always wear eye protection in the laboratory.

General Safety Note: Some of the chemicals used in qualitative organic analysis are irritants. Some are suspected carcinogens. Always assume that the chemical being used is toxic and take necessary precautions. Always **wear eye protection** and always **wear gloves.** Perform all chemical tests under the hood and avoid breathing the vapors. Individual safety precautions will be given by each procedure.

General Instructions

Solubility tests are useful for determining what functional groups are present in a compound. Aldehydes, amines, carboxylic acids, ketones, and certain other oxygen-containing compounds having five or fewer carbons are totally or partially soluble in water. Test the solubility of one of the ketones, benzaldehyde, one of the alcohols, 3,5-dimethylphenol, *p*-anisic acid, *m*-anisidine, N-methylaniline, and N,N-diethylaniline. Record the results of all the solubility tests in a table in your lab notebook. It is assumed that compounds with similar functional groups will have similar solubilities. If a compound dissolves in a given solvent, record the other two compounds in that class as having similar solubilities.

General Procedure. Do the solubility tests before doing the chemical tests. Test the solubility of the compound in water: If the compound is insoluble in water, then test its solubility in 10% NaOH, 10% $NaHCO_3$, and 10% HCl. Based on the results of the solubility tests, perform chemical tests only as necessary. Carefully observe the results of the tests and record detailed information in the data table (fluffy orange precipitate, blue solution, etc.). The tests may be done in any order, but each specific chemical test should be run on all of the compounds to be tested at the same time.

Part A: Microscale Solubility Tests

Safety First!

Always wear eye protection in the laboratory.

1. Wear eye protection at all times in the laboratory.
2. NaOH and HCl are corrosive and toxic. Wear gloves when handling these reagents and wipe up any spills immediately.

Solubility in Water.   Do this solubility test first. Place 20 mg of a solid or 2 drops of a liquid compound in a small test tube or vial. Add 10 drops distilled water. Mix well. If the compound dissolves, test its pH by dipping the end of a glass rod into the solution and touching a piece of pH paper. Record the pH. Note that any compound that is soluble in water will also be soluble in 10% NaOH, 10% NaHCO$_3$, and 10% HCl. If a compound is water soluble, it should not be tested with these reagents.

Solubility in 10% NaOH.   Run this test only on the compounds that are not soluble in water. Place 20 mg of a solid unknown or 2 drops of a liquid unknown in a small test tube or vial. Add 10 drops 10% NaOH. Mix well. Solubility is noted by the formation of a homogenous solution, color change, or evolution of heat or gas. If some of the compound dissolves, but not all, add another 10 drops of 10% NaOH.

Solubility in 10% NaHCO$_3$.   Run this test only on the compounds that are not soluble in water. Place 20 mg of a solid unknown or 2 drops of a liquid unknown in a small test tube or vial. Add 10 drops 10% NaHCO$_3$. Mix well. Solubility is noted by the formation of a homogenous solution, color change, or evolution of heat or gas. If some of the compound dissolves, but not all, add another 10 drops of 10% NaHCO$_3$.

Solubility in 10% HCl.   Run this test only on the compounds that are not soluble in water. Place 20 mg of a solid unknown or 2 drops of a liquid unknown in a small test tube or vial. Add 10 drops 10% HCl. Mix well. Solubility is noted by the formation of a homogenous solution, color change, or evolution of heat or gas. If some of the compound dissolves, but not all, add another 10 drops of 10% HCl.

Results and Conclusions for Part A

1. Prepare a table in your lab notebook to record the results of the solubility tests. The table should have five columns titled "Compound," "Solubility in water/pH," "Solubility in 10% NaOH," "Solubility in 10% NaHCO$_3$," and "Solubility in 10% HCl." In the Compound column, list all of the compounds used in this experiment (including the ketone, two alcohols, and the two amines that weren't tested). In the table, write "+" if the compound dissolves or "−" if the compound does not dissolve in the given reagent. If the compound is soluble in water, record the pH of the solution.

2. Are there any compounds that can be differentiated from the other compounds based upon solubility? If so, what functional groups are present in these compounds? Write a flow scheme to show how these compounds could be differentiated from the other compounds using solubility tests.

Part B: Microscale Chemical Tests

General Procedure.   The chemical tests may be done in any order. It is unnecessary and a waste of time to run chemical tests on a compound if you were able to distinguish it using the solubility tests. Do not do chemical tests on any compound that was soluble in 10% HCl, in 10% NaOH, or in 10% NaHCO$_3$ unless specifically told to do so in the instructions for a specific test. For other tests, just write NA (not applicable) in each table for these compounds. **Perform chemical tests only on compounds that were not differentiated using solubility tests.**

B.1: Chromic Acid Test

Safety First!

Always wear eye protection in the laboratory.

1. Chromium salts are classified as carcinogenic (cancer causing). Always wear gloves when working with chromium-containing reagents and work in the hood.
2. Sulfuric acid is a highly toxic oxidizing agent. Wear eye protection at all times. Wash with copious amounts of cold water if skin comes in contact with acid. Neutralize any spills on the bench with sodium bicarbonate.

In one of the wells of a spot plate or a small test tube, place 2 drops of a liquid compound or about 20 mg of a solid compound. Dissolve the compound in about 5 drops of reagent-grade acetone by stirring with a glass rod. Add 1 drop of the chromic acid solution. Observe carefully. A positive reaction occurs if a blue-green precipitate forms within 2–3 seconds or if the solution changes color. Note the occurrence of a precipitate, its color, and the time it took to form the precipitate.

You should also perform this test on acetone alone to make certain there is no impurity in the acetone (such as isopropyl alcohol) that will react with chromic acid. Place 5 drops of the acetone on a spot plate. Add 1 drop of the chromic acid solution and observe any color change. If a blue-green precipitate does occur, the test must be rerun with purified acetone.

Results and Conclusions for the Chromic Acid Test

Prepare a table in your lab notebook to record the results of the chromic acid test. The table should have three columns titled "Compound," "Precipitate/Color," and "Time." In the Compound column, list all of the compounds used in this test. In the table, write "+" if a precipitate formed, note the color of the precipitate, and record the time for precipitation. Write "–" if a precipitate did not form. Write NA if a compound was not tested.

B.2: 2,4-Dinitrophenylhydrazine (DNP) Test

Safety First!

Always wear eye protection in the laboratory.

1. Many derivatives of hydrazine are suspected carcinogens. Always wear gloves when handling solutions of 2,4-dinitrophenylhydrazine.
2. This reagent is made with sulfuric acid, which is a strong acid. Always wear goggles. If the reagent comes in contact with skin, wash the affected area with copious amounts of cool water. Bench spills can be neutralized with sodium bicarbonate.

Hood!

In a well of a spot plate or a small test tube, add 1 drop of a liquid compound or about 20 mg of a solid compound. Dissolve the solid compound in 5 drops of ethanol. To the solution add 7–8 drops of the 2,4-dinitrophenylhydrazine reagent. Stir with a glass rod. A positive test is the formation of a yellow, orange, or red precipitate.

Results and Conclusions for the DNP Test

Prepare a table in your lab notebook to record the results of the DNP test. This table should have two columns titled "Compound" and "Precipitate/Color." In the Compound column, list all of the compounds tested. In the table, write "+" if a precipitate formed and note the color of the precipitate. Write "–" if a precipitate did not form. Write NA if a compound was not tested.

B.3: Ferric Chloride Test

Do this test only on 2-methyl-2-pentanol, 3,5-dimethylphenol, and *m*-anisic acid. In a well of a spot plate, add 1 drop of a liquid compound or about 20 mg of a solid compound. Dissolve each compound in 2 drops of water (if the compound is water soluble) or in 2 drops of ethanol (if the compound is not water soluble). Stir with a glass rod. To the solution add 1–2 drops of 2.5% ferric chloride solution and stir. A positive test is a color change from yellow (the color of the ferric chloride) to blue, green, violet, or red. Observe the solution closely, because the color change may disappear rapidly.

Results and Conclusions for the Ferric Chloride Test

Prepare a table in your lab notebook to record the results of the ferric chloride test. The table should have two columns titled "Compound" and "Color Change." In the Compound column, list the three compounds tested. In the table, write "+" if a color change occurred and note the color. Write "–" if a color change did not occur.

B.4: Iodoform Test

1. Iodine may cause burns. Wear gloves.
2. Sodium hydroxide solutions are corrosive. If reagent spills on skin, wash off with copious amount of cold water.

Note: Run this test only on 2-pentanone and 3-pentanone. Label two small test tubes. In one test tube, add 2 drops of 2-pentanone. In the other test tube, add 2 drops of 3-pentanone. To each tube, add 1 mL of water. Swirl vigorously to dissolve. To each tube, add 1 mL of 10% aqueous sodium hydroxide and 1.5 mL of 0.5 M iodine-potassium iodide solution. Observe each test tube. A positive test is the disappearance of the brown color and the formation of a yellow precipitate (iodoform).

Results and Conclusions for the Iodoform Test

Prepare a table in your lab notebook to record the results of the iodoform test. The table should have two columns titled "Compound" and "Yellow Precipitate." In the Compound column, list the two ketones you tested. In the table, write "+" if a yellow precipitate formed. Write "–" if there was no precipitate.

B.5: Lucas Test

This test can only be used for compounds that are at least partially water soluble. Run this test only on 1-hexanol, 4-methyl-2-pentanol, and 2-methyl-2-pentanol. Label three small, dry test tubes or 3-mL conical vials. To one of the test tubes, add 2 drops of 1-hexanol. In a similar manner, add 2 drops of the other alcohols to the two other test tubes. Add 10 drops of the Lucas reagent to each of the test tubes. Swirl rapidly and observe closely for the formation of cloudiness. Note the appearance of the solution and the time it takes for the homogenous solution to turn cloudy. A positive test is the formation of cloudiness. Observe the reaction for at least 10 minutes before concluding that the test is negative. Record the results and the time in the data table.

Results and Conclusions for the Lucas Test

Prepare a table in your lab notebook to record the results of the Lucas test. The table should have three columns titled "Compound," "Cloudiness," and "Time Required." In the Compound column, list the three compounds tested. In the table, write "+" if the solution turned cloudy and note the time required. Write "−" if, even after waiting for 10 minutes, the solution still did not turn cloudy.

B.6: Tollens Test

S a f e t y F i r s t !

Always wear eye protection in the laboratory.

The Tollens reagent must be prepared immediately before use. Dispose of any excess reagent and the products of the reaction by cautiously adding dilute nitric acid to the silver mirror, then washing down the drain with lots of cold running water. **Under no circumstances should you store the test tube in your locker. On standing, silver fulminates are formed, which are explosive.**

Do this test only on compounds that haven't been identified by the solubility tests. Make sure that the test tubes to be used in this test are very clean, since any dirt or oil may prevent the silver mirror from being easily observable. The test tubes may be cleaned by rinsing with 10% NaOH. To a clean test tube, add 1 mL of 5% $AgNO_3$ and 1 drop of 10% NaOH. Add concentrated aqueous NH_3 dropwise, with shaking, until the precipitate formed just dissolves. This will take 2–4 drops of NH_3. Add 1 drop of a liquid compound or about 20 mg of a solid compound. Swirl the tube to mix, then let stand for 10 minutes. If no reaction occurs, heat the test tube in a beaker of warm water for a few minutes. Do not heat to dryness. A positive test is the formation of a silver mirror coating on the sides of the test tube. **Do not store the test tube in the drawer! Explosive silver fulminates are formed.**

Results and Conclusions for the Tollens Test

Prepare a table in your lab notebook to record the results of the Tollens test. The table should have two columns titled "Compound" and "Silver Mirror." In the Compound column, list the compounds you used in this chemical test. In the table, write "+" if a silver mirror formed in the test tube. Write "−" if a silver mirror did not form.

B.7: Hinsberg Test

S a f e t y F i r s t !

Always wear eye protection in the laboratory.

Benzenesulfonyl chloride is a lachrymator. Use under the hood and avoid breathing the vapor. Wear gloves when handling this reagent.

Do this test only on *m*-anisidine, N-methylaniline, and N,N-diethylaniline. Do the test on one compound at a time and follow directions carefully to avoid misleading results.

In a test tube, place 2–3 drops of a liquid compound or 20 mg of a solid compound, 2 mL of 2 M KOH, and 4 drops of benzenesulfonyl chloride. Stopper the test tube with a cork and shake vigorously for 3–4 minutes or until the odor of benzenesulfonyl chloride is gone. Test the pH of the solution. It should be strongly basic. If not, add KOH   Hood! dropwise until it is strongly basic. Let the test tube stand for a few minutes and note carefully the appearance of the mixture in the test tube.

If the solution in the test tube is homogenous, note this in the table. Then carefully add 1.5 M HCl dropwise to adjust the pH to 4 and observe whether or not a precipitate forms. Record this observation in the table. Discard the contents of the test tube into a container for waste from the Hinsberg test. Start the test over with a new compound.

If the solution in the test tube is not homogenous, but forms two layers, separate the layers into two clean test tubes. To the organic layer, carefully add 1.5 M HCl dropwise until the solution is acidic to litmus paper. (Add the acid slowly in order to avoid adding too much.) Note whether or not a solid precipitates from the acidic solution. Discard the contents of the test tube into a container for waste from the Hinsberg test. Start the test over with a new compound.

Repeat the Hinsberg test until all three amines have been tested. Carefully note the observations in the table.

Results and Conclusions for the Hinsberg Test

Prepare a table in your lab notebook to record the results of the Hinsberg test. The table should have three columns titled "Compound," "Appearance of Reaction Mixture," and "Precipitate in Acid Solution?." In the Compound column list the three compounds tested. In the Appearance of Reaction Mixture column, write homogenous or two-layered. In the Precipitate in Acid Solution column, write "+" if a solid formed after acidifying the solution. Write "–" if a solid did not form after acidifying the solution.

Results and Conclusions for Part B

1. Consolidate the results from each solubility test and each chemical test in a table of 11 rows and 12 columns. In the first column, write the names of all the compounds tested in Part A and Part B of this experiment. In the second column, write the type of functional group represented by each compound (aldehyde, primary alcohol, etc.). In each of the other columns, write the results of each solubility test and each chemical test. Write a "+" if the compound gave a positive result and a "–" if the compound gave a negative result.
2. Based on the results of the solubility tests and the chemical tests, devise a flow scheme that could be used to differentiate between:
   a. an aldehyde and a ketone.
   b. an alcohol and a phenol.
   c. a carboxylic acid and a phenol.
   d. an alcohol and an amine.
   e. a primary alcohol, a secondary alcohol, and a tertiary alcohol.
   f. a primary amine, a secondary amine, and a tertiary amine.
3. Explain how it is possible to distinguish between 2-pentanone, 3-pentanone, and benzaldehyde.

4. Based on the results of the solubility tests and the chemical tests, devise a flow scheme that could be used to differentiate between 2-pentanone, 3-pentanone, benzaldehyde, 1-hexanol, 4-methyl-2-pentanol, 2-methyl-2-pentanol, 3,5-dimethylphenol, *p*-anisic acid, *m*-anisidine, N-methylaniline, and N,N-diethylaniline.

5. Use the flow scheme to indicate how the following compounds could be identified: 1-hexanol, 2-methyl-2-pentanol, cyclohexanone, acetophenone, pentanal, 2,4-dibromophenol, cyclohexyl amine, and benzoic acid.

Part C: Microscale Classification of an Unknown Compound

Experimental Procedure

Have your laboratory instructor approve the flow chart for solubility determination and the flow chart for the chemical tests. Obtain 1–2 mL of a liquid unknown or 400 mg of a solid unknown from the laboratory instructor. Record the number of the unknown. Following the flow scheme prepared in Part B, perform solubility tests and chemical tests as necessary to identify the functional group(s) present in the unknown. Record the results of the solubility and chemical tests in your lab notebook. Indicate the functional group(s) of the unknown compound.

Cleanup & Disposal

Wash aqueous solutions down the drain with water. **The silver mirror in the test tube from the Tollens test cannot be stored in the drawer since explosive silver fulminates are formed. The silver mirror must be destroyed immediately.** Cautiously add dilute nitric acid to the silver mirror in the test tube, then wash it down the drain with copious amounts of cold running water.

Critical Thinking Questions

1. List the compound(s) that gave a positive reaction with CrO_3. What type of reaction did these compounds undergo? What type of reagent is CrO_3? Propose a possible product from the reaction of each of the compounds with CrO_3.
2. List the compound(s) that gave a positive test with the Lucas reagent. Was there a difference in the amount of time required for the reaction to occur? What can you conclude about the nature of this reaction? Write a detailed mechanism for each positive reaction.
3. Why did the solution turn cloudy in the Lucas test?
4. Write the structure of the compound(s) that gave a positive test with the iodoform reagent. Explain any differences in structure between compounds that gave a positive iodoform test and those that gave a negative iodoform test.
5. Look up the structure of iodoform. Using the compound(s) that gave a positive test, propose a mechanism.
6. List the compound(s) that gave a positive DNP test. What similarities exist between the structures of these compounds?
7. Write the structures of the products formed from the reaction of 2,4-dinitrophenylhydrazine with each of the compounds that gave a positive test.

8. Propose a mechanism for the reaction of 2-pentanone with 2,4-dinitrophenyl-hydrazine.

9. List the compound(s) that reacted with silver nitrate (Tollens test). In this reaction, silver ion is reduced to metallic silver. What is the organic product?

10. List the compound(s) that dissolved in NaOH, but not in water. Write a balanced reaction for each positive test and explain why the compound dissolved.

11. List the compound(s) that dissolved in $NaHCO_3$, but not in water. Write a balanced reaction for each positive test and explain why the compound dissolved.

12. Draw the structures of the benzenesulfonamides that formed between benzenesulfonyl chloride and *m*-anisidine, N-methylaniline, and N,N-diethylaniline. Draw the structures of the benzenesulfonamides at pH = 4 and at pH = 10. Using these structures, explain how the Hinsberg test differentiates between primary, secondary, and tertiary amines.

13. List the compound(s) that dissolved in HCl, but not in water. Write a balanced reaction for each positive test and explain why the compound dissolved.

14. Explain why silver fulminates should not be stored in the desk drawer.

Experiment 28.2: Experimental Methods of Qualitative Organic Analysis

Introduction to Qualitative Organic Analysis

Before the development of modern spectroscopic techniques such as nuclear magnetic resonance spectroscopy (NMR) and infrared spectroscopy (IR), the structure of an organic compound was determined by physical properties and chemical tests and by conversion of the compound to a solid derivative with a known melting point. Today IR spectroscopy makes it relatively easy to identify the functional groups present in a molecule and NMR spectroscopy establishes the carbon-hydrogen framework of the molecule. Yet qualitative organic analysis—the process of identification of a compound by chemical tests and derivative preparation—plays an important role in organic chemistry laboratory courses. There are four main benefits of performing qualitative organic analysis in the lab.

First, qualitative organic analysis fosters the development of a methodical approach to problem solving. Second, the results obtained correlate highly with skillful and careful laboratory technique. This will show up when preparing derivatives of an unknown. Third, performing wet chemical tests helps you learn the reactions introduced in lecture. Fourth, successfully completing qualitative organic analysis will make you appreciate the early pioneers in multistep natural product organic synthesis and make you grateful for the development of NMR and IR spectroscopic techniques.

In Experiment 10.1, simple alkyl halides, alkenes, dienes, and alkynes were analyzed. The unknowns in Experiment 28.2 are aldehydes or ketones (Part E.1), alcohols or phenols (Part E.2), carboxylic acids or amines (Part E.3) or amines, alcohols, carboxylic acids, phenols, aldehydes, or ketones (Part E.4). Other functional groups, such as halogens, nitro groups, double bonds, triple bonds, ethers, and cyano groups, may also be present in any of the unknowns. In Part E.5, spectroscopic techniques will be used to identify a general unknown.

Overall Approach to Identifying the Unknown

The approach to identifying the unknown is similar in all cases and involves an eight-step process:

Step 1: Purify the unknown (distill if it is a liquid; recrystallize if it is a solid).

Step 2: Determine the physical properties of the unknown (boiling point, melting point, refractive index, density, etc.).

Step 3: Examine the physical state of the unknown (color, crystal structure, etc.).

Step 4: Perform solubility tests.

Step 5: Use wet chemical tests to identify the functional group(s) present.

Step 6: Select an appropriate derivative.

Step 7: Prepare and purify the derivative.

Step 8: Identify the unknown.

Each of these steps is discussed below.

Step 1: Purify

If the unknown is a liquid, purify the unknown by distillation using a Hickman still. Discard the first few drops of distillate. Distill until the temperature of the distillate starts to decrease or until just a few drops of liquid remain in the vial. The boiling range should be no more than 3–4°C. If the range is greater than this, consult your instructor. Use a capillary tube to determine the micro boiling point of the purified liquid, remembering to make a correction for atmospheric pressure.

If the unknown is an impure solid, recrystallize from an appropriate solvent. Air dry the crystals. Set aside a sample of the crystals to dry and begin testing the remaining crystals for solubility. When the sample is dry, determine the melting point. The melting point range should be no greater than 2°C. If the range is greater than 2°C, the sample is either still wet or impure and must be recrystallized again.

Step 2: Determine the Physical Properties

Record the corrected boiling point or melting point of the unknown. Other physical properties, such as refractive index or density, may also be measured. Make any necessary corrections for temperature and record the data.

Step 3: Examine the Physical State

Examine the physical characteristics of the unknown. Does the compound have an odor? If so, is it fruity? Acrid? Skunky? Functional groups may confer characteristic odors. Is the compound colored? If solid, carefully examine the crystal form. Are the crystals light and fluffy? Feathery? Such information may prove helpful when trying to identify the compound.

Step 4: Perform Solubility Tests

The solubility of an organic compound provides information about the functional groups that are present in the molecule and/or the length of the carbon chain. The unknown can be tested for solubility in water, 10% HCl, 10% NaOH, 10% NaHCO$_3$, and concentrated H$_2$SO$_4$.

In general, compounds that are soluble or partially soluble in water tend to be short-chain organic compounds (compounds containing five or fewer carbons that have a functional group capable of hydrogen bonding, such as hydroxy, carboxyl, amino, or carbonyl) or higher molecular weight compounds that contain several of these polar functional groups. If a compound is water soluble, testing the aqueous solution with litmus paper will indicate whether the unknown is acidic (low-molecular-weight carboxylic acid or phenol that contains several electron-withdrawing groups), basic (low-molecular-weight amine), or neutral (low-molecular-weight alcohol, ketone, or aldehyde).

Primary, secondary, and tertiary amines will dissolve in dilute HCl due to the formation of a water-soluble ammonium salt.

$$RNH_2 + HCl \longrightarrow RNH_3^+ \, Cl^-$$

$$RR'NH + HCl \longrightarrow RR'NH_2^+ \, Cl^-$$

$$RR'R''N + HCl \longrightarrow RR'R''NH^+ \, Cl^-$$

Other organic compounds (unless they are water soluble) will not dissolve in dilute HCl. Hence, solubility in HCl is indicative that the compound is an amine.

Acidic organic compounds, such as carboxylic acids and phenols, will dissolve in dilute NaOH due to the formation of a water-soluble salt.

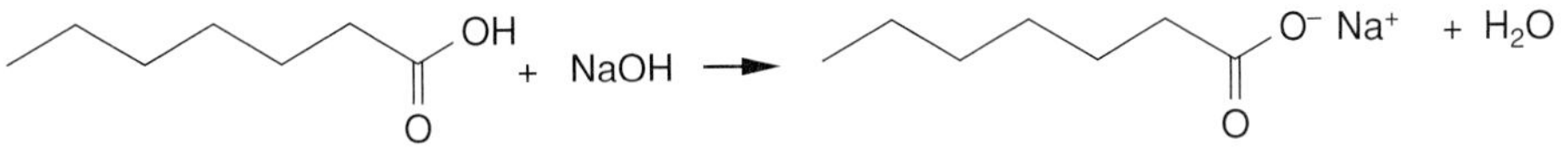

carboxylic acid   water-soluble carboxylate salt

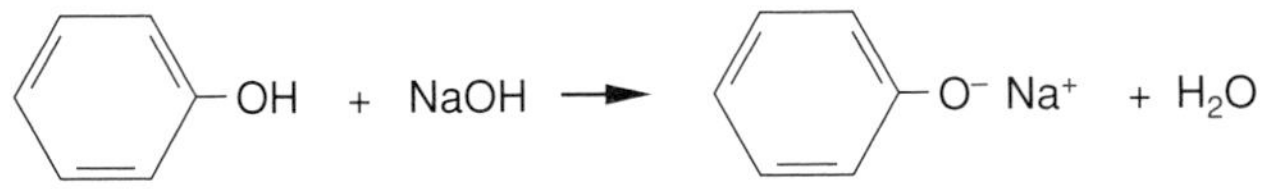

phenol   water-soluble phenoxide salt

If the organic compound is insoluble in water but soluble in dilute NaOH, the compound is either a carboxylic acid or a phenol. It is possible to differentiate between these two functional classes by testing the solubility of the unknown in dilute sodium bicarbonate. Being stronger acids than phenols, carboxylic acids will dissolve in the weaker base, sodium bicarbonate.

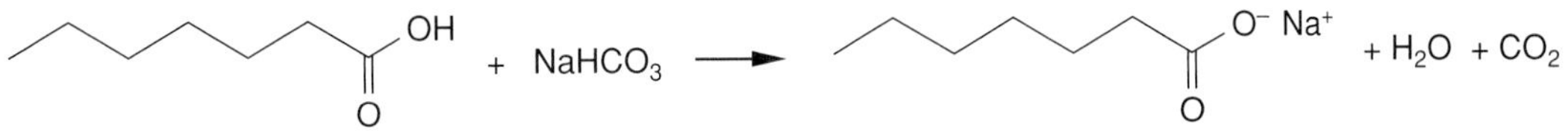

carboxylic acid   water-soluble carboxylate salt

Since most phenols are weaker acids than carboxylic acids, they do not dissolve in aqueous bicarbonate except when they contain electron-withdrawing groups such as cyano or nitro. If an unknown compound dissolves in dilute NaOH, but not $NaHCO_3$, it is probably a phenol. If the unknown dissolves in both NaOH and $NaHCO_3$, it is probably a carboxylic acid, but it could be a nitrophenol or similar compound.

Compounds that contain oxygen or nitrogen (such as amines, alcohols, carboxylic acids, phenols, ketones, and aldehydes) dissolve in concentrated sulfuric acid due to the formation of protonated oxonium or ammonium salts. This is illustrated here for 2-butanol and 2-butanone.

Amines, carboxylic acids, and phenols can be identified using the previously discussed solubility tests. Ketones, aldehydes, and alcohols can be differentiated from alkanes and alkyl halides by testing their solubilities in concentrated sulfuric acid. Note that other functional groups may appear to dissolve in concentrated sulfuric acid due to reaction with the reagent. Knowledge of functional group reactivity is essential.

Evidence of solubility in concentrated sulfuric acid is dissolving, evolution of gas, formation of a different precipitate, or color change. All of these physical changes are considered "soluble."

Notice that it is not possible through solubility tests to differentiate between alcohols, ketones, and aldehydes. Fortunately, there are many chemical tests that can easily do the task.

Step 5: Identify the Functional Groups Present

Using appropriate chemical tests and/or spectroscopic analysis, it is possible to determine the functional groups present in a compound. It is important to approach this task in a methodical and organized fashion. Performing random wet chemical tests will lead to confusion, cause misinterpretation of results, and waste your time. The spot tests may be used either to verify the solubility tests or to differentiate between alcohols, phenols, ketones, and aldehydes or other functional groups.

Spot tests can be done quickly. The tests are designed to show a change (color change or precipitation) when a specific functional group is present. A negative test is one that does not give the indicated change. **It is critical that every test be run simultaneously with a "known" compound for comparison.** A "known" is one that contains the functional group being tested. Being able to observe a positive test side by side with the unknown simplifies interpretation of the results. Sometimes colors will look slightly different or rates of reactions will vary from the known. This can be due to the presence of other functional groups or aromatic rings. If the results of a test are inconclusive, repeat the test with fresh samples of the known and unknown. Note carefully all changes that occur and write them down in your lab notebook.

The tests are classified based on the functional group they are used to identify: aldehydes and ketones (C-3), methyl ketones (C-6), aldehydes (C-11 and C-2), phenols (B-3 and C-4), alcohols (C-2 and C-7), amines (C-5 and C-8), carboxylic acids (B-3, B-4, and C-8), alkenes and alkynes (C-1 and C-9), and heteroatoms (C-10). The designation refers to the specific experimental procedure for conducting the chemical test.

The instructor may include spectroscopic analysis of the unknown. If so, run infrared spectra on the unknowns and carefully analyze the spectra for the presence or absence of functional groups. In your lab notebook, carefully tabulate important absorption bands and note their significance in identifying the functional group. With the instructor's permission, students may record and analyze the ^{1}H NMR spectrum, ^{13}C NMR spectrum, and/or mass spectrum and carefully note the relevant information obtained about the functional group(s) present or absent.

Identifying Aldehydes and Ketones

2,4-Dinitrophenylhydrazine Test—DNP (C-3)

The first step in the process of identifying the functional group is to differentiate between compounds containing hydroxyl groups (alcohols and phenols) and carbonyls (aldehydes and ketones). This can easily be accomplished by testing with 2,4-dinitrophenylhydrazine (DNP). Ketones and aldehydes react quickly to form a yellow, orange, or red 2,4-dinitrophenylhydrazone.

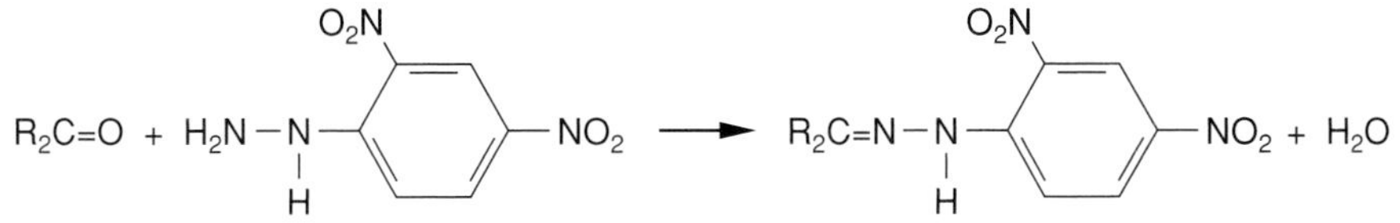

2,4-dinitrophenylhydrazine   a 2,4-dinitrophenylhydrazone

The color of the precipitate can give an indication of the degree of conjugation. Nonconjugated carbonyl compounds generally form yellow precipitates while conjugated carbonyl compounds form orange or red crystals. Because this reaction is so quick and easy, it is also used to prepare derivatives of aldehydes and ketones. Alcohols generally do not react with DNP unless they are allylic or benzylic and are oxidized to the carbonyl compound under the reaction conditions. Other carbonyl-containing compounds (such as carboxylic acid or acid derivatives) do not react with DNP under conditions of the test.

Identifying Aldehydes

Tollens Test (C-11)

Once it has been established that the unknown is an aldehyde or ketone, the Tollens test can be used to differentiate between these two functional groups. Aldehydes are oxidized by the Tollens reagent (a silver ammonia complex) to produce elemental silver (a silver mirror) on the glass wall of the test tube.

$$R-\overset{\overset{\displaystyle O}{\|}}{C}-H + 2\,Ag(NH_3)_2^+ + 2\,OH^- \longrightarrow \underset{\text{(silver mirror)}}{2\,Ag} + R-\overset{\overset{\displaystyle O}{\|}}{C}-O^- + NH_4^+ + NH_3 + H_2O$$

Ketones generally do not react with the Tollens reagent.

Chromic Acid Test (C-2)

A second method for distinguishing between aldehydes and ketones uses chromic acid oxidation. Aldehydes are oxidized to carboxylic acids and the color changes from orange to blue-green.

$$3\,R-\overset{\overset{\displaystyle O}{\|}}{\underset{\text{orange}}{C}}-H + 2\,CrO_3 + 3\,H_2SO_4 \longrightarrow 3\,R-\overset{\overset{\displaystyle O}{\|}}{\underset{\text{blue-green}}{C}}-OH + Cr_2(SO_4)_3 + 3\,H_2O$$

Ketones are not oxidized under the reaction conditions. The presence of a blue-green precipitate or emulsion is indicative of a positive test. Primary and secondary alcohols are also oxidized by chromic acid. In fact, the rate at which the alcohol reacts is the basis for determining whether the alcohol is primary or secondary.

Identifying Methyl Ketones

Iodoform Test (C-6)

The iodoform test identifies ketones that have a methyl group adjacent to the carbonyl. Hydrogens on carbons adjacent to a carbonyl (called α-hydrogens) are acidic. A methyl group has three α-hydrogens. In the presence of base, deprotonation and reaction of the enolate with iodine results in substitution of a halogen for a hydrogen. This process is repeated until all three hydrogens have been replaced. Subsequent reaction of the organic compound with base yields iodoform, which precipitates from the reaction mixture.

$$R-\overset{\overset{\displaystyle O}{\|}}{C}-CH_3 + 3\,I_2 + 4NaOH \longrightarrow R-\overset{\overset{\displaystyle O}{\|}}{C}-O^-\,Na^+ + CHI_3 + 3\,NaI + 3\,H_2O$$

a methyl ketone   iodoform
                                                       (yellow precipitate)

Any aldehyde or ketone with α-hydrogens will react in this manner. However, iodoform can only be formed from reaction of a methyl group attached to a carbonyl. Methyl carbinols (such as 2-propanol, 2-butanol, etc.) can also react to give iodoform, because they are oxidized to the corresponding methyl ketones under the reaction conditions. The only aldehyde that will give a positive test is acetaldehyde (CH_3CHO).

Infrared Spectroscopy

Aldehydes and ketones can be distinguished by infrared spectroscopy. Both types of functional groups show strong carbonyl absorption. For aldehydes, the $C{=}O$ stretch occurs around $1725\ cm^{-1}$ while for ketones, this band is around $1715\ cm^{-1}$. Conjugation lowers the absorption frequency for both aldehydes and ketones. In addition, aldehydes show two sharp but weak bands for the aldehydic $C-H$ stretch at $2850\ cm^{-1}$ and $2750\ cm^{-1}$, although the higher frequency band is often obscured. The appearance of these bands distinguishes aldehydes from ketones.

Nuclear Magnetic Resonance Spectroscopy

The ^1H NMR spectrum can also aid in identifying an aldehyde as the aldehydic proton has a chemical shift of δ9–10. Because protons adjacent to the carbonyl have chemical shifts near δ2, methyl ketones can be distinguished from many other ketones or aldehydes by the appearance of a singlet near δ2.1 that integrates to three protons. Although ^{13}C NMR is helpful in identifying whether the unknown contains a ketone or aldehyde group by the appearance of signal at δ180–220, ^{13}C NMR spectroscopy is not particularly useful in distinguishing between aldehydes and ketones.

Mass Spectrometry

In general, both aldehydes and ketones show a molecular ion peak. A common peak for aldehydes is an M-1 peak from loss of the aldehydic hydrogen. Both aldehydes and ketones often undergo α-cleavage to form resonance-stabilized carbocations. Aldehydes and ketones that contain γ-hydrogens undergo the McLafferty rearrangement to split out a neutral molecule. Methyl ketones often show a prominent peak at 43 m/z for $[CH_3C{=}O]^+$.

Identifying Phenols

Solubility in NaOH (B-3)

Phenols are soluble in dilute sodium hydroxide solutions; water-insoluble alcohols are not. This is probably the best method of differentiating between alcohols and phenols.

Ferric Chloride Test (C-4)

Phenols react with ferric chloride to form green, blue, purple, or red complex ions, the color determined by the structure of the phenol.

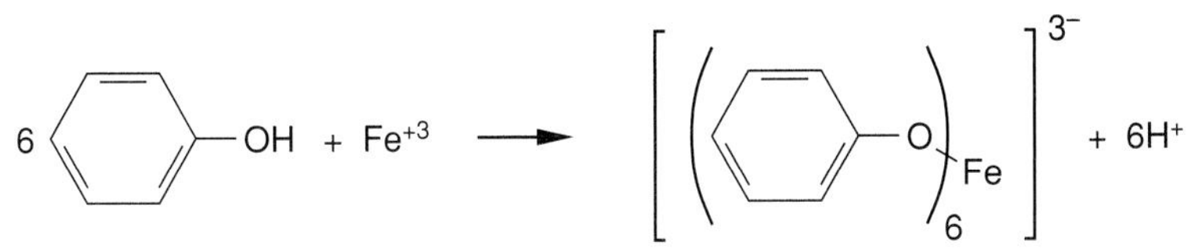

Alcohols do not react with ferric chloride, so this is a good method to differentiate between these functional groups. The color of the solution may be very faint. For this reason, it is good to run this test with a blank that contains all the reagents except the phenol. Because the color of the complex disappears rapidly, it is important to observe the color directly after mixing the reagents. Some phenols do not form a colored complex with $FeCl_3$.

Infrared Spectroscopy

Phenols are identified by IR spectroscopy through the presence of an OH band near 3300 cm^{-1}, $C-O$ stretching absorption around 1220–1200 cm^{-1}, and aromatic absorption near 3100 cm^{-1} (sp^2C-H stretching), 2000–1667 cm^{-1} (overtone bands), 1600–1475 cm^{-1} (C=C ring stretching), and 900–690 cm^{-1} (out-of-plane bending vibrations).

Nuclear Magnetic Resonance Spectroscopy

The ^1H NMR spectrum of a phenol shows signals around $\delta7$ for the aromatic protons and a small, usually broad singlet for the $O-H$ proton that varies from $\delta4$ to 7. The ^{13}C NMR shows signals for the aromatic carbons around $\delta110–165$.

Mass Spectrometry

Phenols generally show molecular ion peaks. The M-1 peak due to loss of the phenolic hydrogen is often the base peak. Other important peaks in the mass spectra of phenols are the M-28 peak (loss of CO) and M-29 peak (loss of HCO).

Identifying Alcohols

Chromic Acid Test (C-2)

Primary and secondary alcohols can be oxidized by chromic acid to aldehydes and ketones. Tertiary alcohols do not react.

$$3 \ R\!-\!\underset{\underset{\text{H}}{|}}{\overset{\overset{\text{OH}}{|}}{C}}\!-\!H + 4 \ CrO_3 + 6 \ H_2SO_4 \longrightarrow 3 \ R\!-\!\overset{\overset{O}{\|}}{C}\!-\!OH \ + \ 2 \ Cr_2(SO_4)_3 \ + \ 9 \ H_2O$$

a primary alcohol — orange — a carboxylic acid — blue-green

$$3 \ R\!-\!\underset{\underset{\text{R}'}{|}}{\overset{\overset{\text{OH}}{|}}{C}}\!-\!H + 2 \ CrO_3 + 3 \ H_2SO_4 \longrightarrow 3 \ R\!-\!\overset{\overset{O}{\|}}{C}\!-\!R' \ + \ Cr_2(SO_4)_3 \ + \ 6 \ H_2O$$

a secondary alcohol — orange — a ketone — blue-green

$$R\!-\!\underset{\underset{\text{R}'}{|}}{\overset{\overset{\text{OH}}{|}}{C}}\!-\!R'' + 2 \ CrO_3 + 3 \ H_2SO_4 \longrightarrow \text{No Reaction}$$

a tertiary alcohol — orange

Lucas Test (C-7)

This test can be very useful for determining whether an alcohol is primary, secondary, or tertiary. The Lucas reagent (concentrated HCl/ZnCl$_2$) converts alcohols into alkyl chlorides.

$$ROH + HCl \xrightarrow{\ ZnCl_2\ } RCl + H_2O$$

These are S_N1 conditions with ZnCl$_2$ functioning as a Lewis acid catalyst. As might be expected, tertiary alcohols react faster than secondary alcohols. Primary alcohols do not react under these conditions. A positive test is noted by the appearance of a water-insoluble layer (the formation of the alkyl chloride). It is important to run this test with known alcohols to facilitate comparison of rate of reaction. Tertiary alcohols react almost immediately to form a cloudy solution, which quickly separates into two layers. Secondary alcohols react more slowly. After 3–5 minutes, the solution will generally become cloudy. Primary alcohols do not react unless they are allylic or benzylic.

A serious drawback of this test is its limitation to alcohols that are water soluble or partially water soluble (alcohols of fewer than six carbons or alcohols with more than one polar functional group). Alcohols that are insoluble in water will also be insoluble in the Lucas reagent, making it impossible to see that a new insoluble layer is formed.

Infrared Spectroscopy

The distinguishing characteristic in the neat IR spectra of alcohols is the broad absorption due to O$-$H stretching around 3600–3200 cm^{-1}. The substitution pattern of the alcohol can be determined by the C$-$O stretching frequency: for primary alcohols, this band appears at 1050 cm^{-1}; for secondary alcohols, the band is at 1100 cm^{-1}, and for tertiary alcohols, the band is at 1150 cm^{-1}.

Nuclear Magnetic Resonance Spectroscopy

The hydroxylic proton shows a variable signal in the ^{1}H NMR spectrum. Depending upon concentration and acidity, the signal for the O$-$H proton varies from $\delta 2$–5.

Splitting may or may not be observed depending upon the purity and concentration of the sample. With 60–90 MHz instruments, the hydroxylic signal is generally a singlet, while with high-field FT-NMR spectrometers, for which very dilute concentrations are used, the hydroxyl proton is typically split by neighboring protons. The ^{13}C NMR spectrum of alcohols shows a signal for the alcohol carbon at $\delta50–75$.

Mass Spectrometry

Primary and secondary alcohols generally show only a weak molecular ion peak, while tertiary alcohols typically do not give a molecular ion peak. Two important fragmentation patterns are dehydration leading to loss of a neutral molecule of water (M-18) and α-cleavage, where the largest group is usually the one lost most frequently. This is illustrated below, where R_1 is the largest alkyl group.

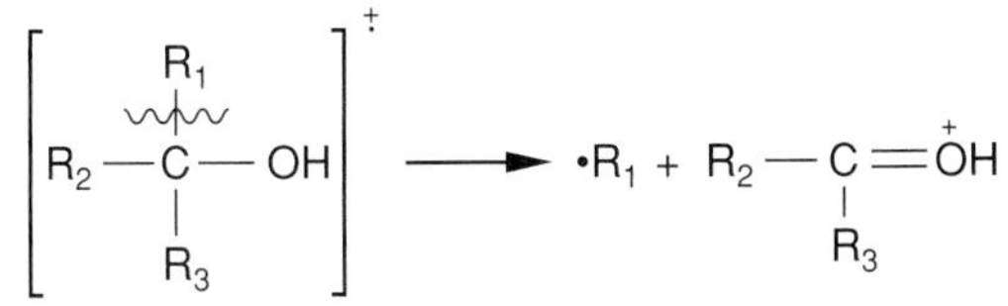

Alcohols that contain four or more carbons may also fragment to split out water and ethylene simultaneously. This results in an M-46 peak.

Identifying Amines

Solubility in HCl (B-2)

Water-soluble amines are basic to litmus paper; water-insoluble amines are soluble in dilute hydrochloric acid solutions. This is usually a good method for classsifying a compound as an amine.

Hinsberg Test (C-5)

The solubility of a compound in dilute HCl can identify the compound as being an amine (B-2). The Hinsberg test distinguishes between primary, secondary, and tertiary amines, which all react differently with benzenesulfonyl chloride. Primary amines react to form insoluble sulfonamides. In the presence of base, the sulfonamide is deprotonated to yield the water-soluble sodium sulfonamide salt. When treated with acid, the water-insoluble sulfonamide is reformed, yielding a second layer.

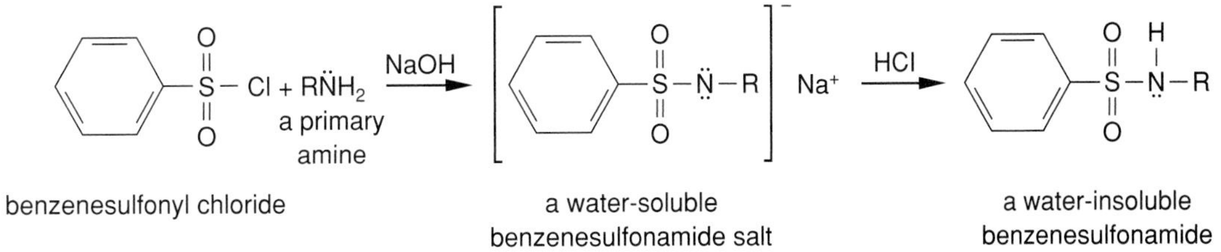

Secondary amines also react with benzenesulfonyl chloride to yield insoluble sulfonamides. Lacking an acidic proton, the sulfonamide cannot react with excess sodium hydroxide and the sulfonamide will not be soluble in the aqueous solution.

benzenesulfonyl
chloride

a water-insoluble
benzenesulfonamide

Tertiary amines do not react with benzenesulfonyl chloride.

Careful observations must be made with this test, which involves mixing the amine and the benzenesulfonyl chloride. If no reaction occurs, then the amine is tertiary. If a homogenous solution occurs after the careful addition of NaOH, the amine is primary. This can be verified by adding HCl. The formation of a precipitate indicates that the amine is primary. If a precipitate forms after the addition of NaOH, the amine is secondary. The tests should also be run on known primary, secondary, and tertiary amines for comparison.

Neutralization Equivalent (C-8)

The equivalent weight of the amine can easily be determined by titrating a measured amount of the amine with standardized HCl. Knowing the mass of the sample, the molarity of the HCl solution and the volume of HCl used, the equivalent weight can easily be determined:

$$\text{Equivalent Weight} = \frac{\text{mg amine}}{(\text{mL HCl})(\text{molarity HCl})}$$

The molecular weight of the sample will be a multiple of the equivalent weight. For example, if the unknown is a monobasic amine, the equivalent and molecular weights will be the same. If the unknown is a dibasic amine, the molecular weight will be twice the equivalent weight.

Infrared Spectroscopy

The number of N−H stretching absorptions can be used to determine if an amine is primary or secondary. Primary amines show two overlapping bands at 3550–3250 cm^{-1}, while secondary amines show only one band in this region. Tertiary amines do not have any protons bonded to nitrogen so do not absorb in this region. Primary amines show broad absorption at 1640–1560 cm^{-1} due to N−H bending. Amines show C−N stretching absorption at 1250–1000 cm^{-1}.

Nuclear Magnetic Resonance Spectroscopy

Primary and secondary amines show broad signals for the N−H protons in the ^{1}H NMR spectrum. The signal varies from $\delta 1$ to 4. Protons on carbons bonded to nitrogen (α-hydrogens) come into resonance from $\delta 2.5$ to 3. Splitting is not observed. The integration of the signal can differentiate primary from secondary amines. The ^{13}C NMR spectrum of amines may be useful in providing supporting evidence for a suggested structure.

Mass Spectrometry

The mass spectrum can be useful in determining whether the unknown is an amine or contains nitrogen. For compounds containing an odd number of nitrogen atoms, the molecular ion peak is odd numbered while the masses of the fragment ions are even. For compounds containing an even number of nitrogens and compounds without nitrogen, the mass of the molecular ion peak is even and fragment ions are odd numbered. In theory, it should be very easy to determine if the unknown contains nitrogen; however, aliphatic amines typically do not show a molecular ion peak. The base peak for amines is typically due to β-cleavage. When more than one alkyl group can be lost, typically the largest group (R_2 in the example shown here) is preferentially cleaved.

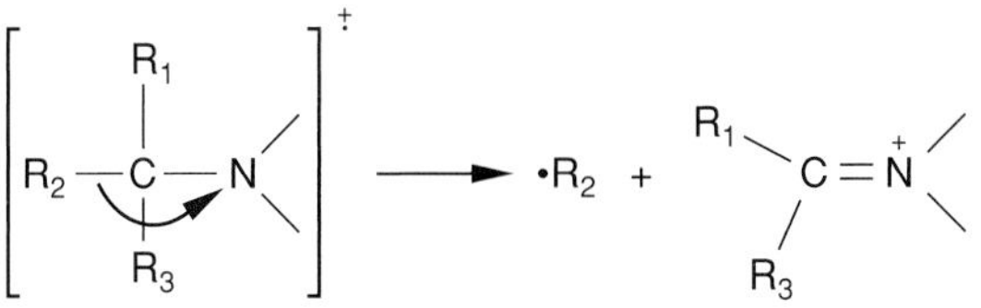

Identifying Carboxylic Acids

Neutralization Equivalent (C-8)

The best method for determining that the unknown is a carboxylic acid is through the solubility tests: carboxylic acids are soluble in both 10% NaOH and 10% $NaHCO_3$. The equivalent weight of the acid can easily be determined by titrating a measured amount of the acid with standardized NaOH. Knowing the mass of the sample, the molarity of the NaOH solution, and the volume of NaOH used, the equivalent weight can easily be determined.

$$\text{Equivalent Weight} = \frac{\text{mg acid}}{(\text{mL NaOH})(\text{molarity NaOH})}$$

The molecular weight of the sample will be a multiple of the equivalent weight. For example, if the unknown is a monocarboxylic acid, the equivalent and molecular weights will be the same. If the unknown is a dicarboxylic acid, the molecular weight will be twice the equivalent weight.

Infrared Spectroscopy

Carboxylic acids can be easily identified by infrared spectroscopy due to the presence of both the carbonyl and the hydroxyl group. Carbonyl (C=O) stretching absorption is broad and occurs near 1710 cm^{-1}. The O−H stretching frequency occurs at 3300–2400 cm^{-1} and is generally broad enough to obscure the C−H stretching. Carboxylic acids also have C−O stretching at 1320–1210 cm^{-1}.

Nuclear Magnetic Resonance Spectroscopy

Carboxylic acids show a broad singlet downfield ($\delta 10$–12) in the ^{1}H NMR spectrum. The signal integrates to one proton. The carboxyl carbon shows a signal at $\delta 165$–180 in the ^{13}C NMR spectrum.

Mass Spectrometry

The mass spectra of aliphatic carboxylic acids show weak molecular ion peaks while the molecular ion peak of aromatic carboxylic acids is quite intense. Short-chain carboxylic acids fragment via α-cleavage to lose OH (M-17) and COOH (M-45). Carboxylic acids also frequently show a peak for the $COOH^+$ ion at m/z 45. Longer chain carboxylic acids containing a γ-hydrogen undergo McLafferty rearrangements to split out neutral alkenes.

Identifying Alkenes and Alkynes

Potassium Permanganate (C-9)

In addition to the primary functional groups already mentioned, the unknown may also contain a halogen, a nitro group, a carbon-carbon double bond, or a carbon-carbon triple bond. The presence of carbon-carbon double and triple bonds can be detected easily by testing the sample with an aqueous solution of potassium permanganate. If a double or triple bond is present, the purple color of the permanganate will disappear and a brown precipitate (MnO_2) will form.

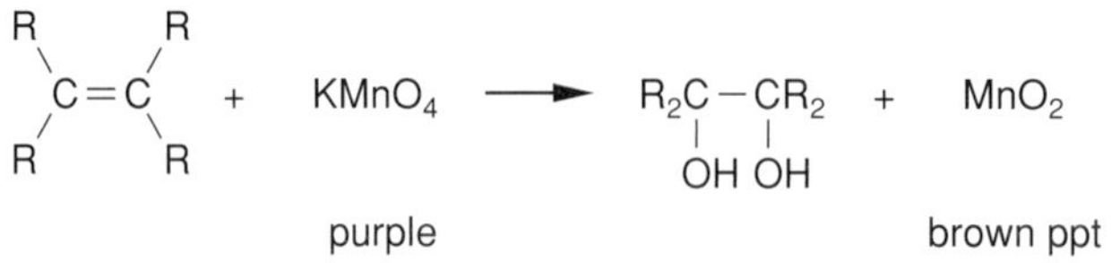

$$R_2C=CR_2 + KMnO_4 \longrightarrow \underset{\underset{\text{OH OH}}{|\quad|}}{R_2C-CR_2} + MnO_2$$

purple   brown ppt

$$RC{\equiv}CR' + KMnO_4 \longrightarrow RCO_2H + R'CO_2H + MnO_2$$

purple   brown ppt

Other easily oxidizable groups, such as aldehydes, will also react with potassium permanganate to give a brown precipitate.

Bromine in Methylene Chloride (C-1)

A second test for unsaturation is the reaction of bromine. The rapid disappearance of the red color of molecular bromine is taken as evidence for an alkene or alkyne. With bromine, alkenes form the colorless dibromoalkanes and alkynes form the colorless tetrabromoalkanes.

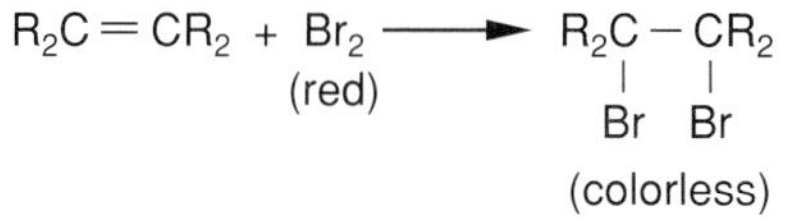

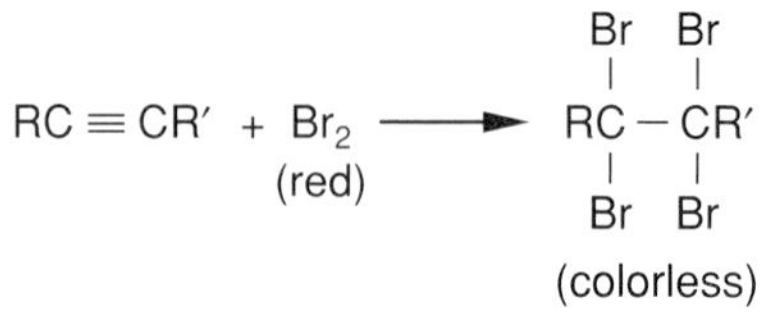

$$R_2C=CR_2 + Br_2 \longrightarrow \underset{\underset{\text{Br Br}}{|\quad|}}{R_2C-CR_2}$$

(red)   (colorless)

$$RC{\equiv}CR' + Br_2 \longrightarrow \overset{\overset{\text{Br Br}}{|\quad|}}{\underset{\underset{\text{Br Br}}{|\quad|}}{RC-CR'}}$$

(red)   (colorless)

Infrared Spectroscopy

Alkenes and alkynes can often be identified from their infrared spectra. Alkenes having a vinyl hydrogen show sharp absorption at $3100 \ cm^{-1}$ due to C=C–H stretching. For nonsymmetrical alkenes, absorption due to C=C stretching is found at $1650 \ cm^{-1}$. Symmetrical alkenes do not show absorption in this region. The out-of-plane bending region from $1000–650 \ cm^{-1}$ provides useful information about the substitution pattern of the alkene. This absorption occurs around $700 \ cm^{-1}$ for *cis*-alkenes and $890 \ cm^{-1}$ for *trans*-alkenes. Terminal alkynes show a sharp absorption band at $3300 \ cm^{-1}$ for sp C–H stretching and a small band at $2150–2100 \ cm^{-1}$ for C≡C stretching. This band is absent in symmetrical internal alkynes.

Nuclear Magnetic Resonance Spectroscopy

Vinyl protons can be distinguished in the ^1H NMR spectrum, coming into resonance around δ4.5–6.5. The signals are generally complex due to vicinal, geminal, and allylic coupling. Integration of the signals can indicate the substitution pattern of the alkene. Allylic protons come into resonance around δ1.6–2.6. The proton on a terminal alkyne comes into resonance near $\delta 2.5$. In the ^{13}C NMR, alkenyl carbons come into resonance from $\delta 100-150$ and alkynyl carbons show signals at $\delta 70-90$.

Mass Spectrometry

Most alkenes give distinct molecular ion peaks. Fragmentation occurs to form allyl carbocations:

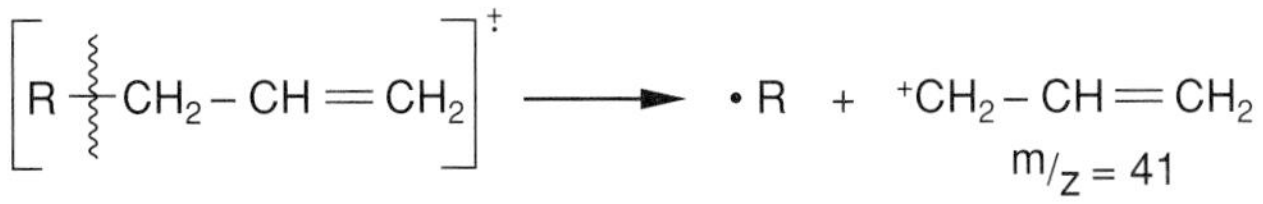

$$\left[R \overset{\text{\}}{-} CH_2-CH=CH_2 \right]^{\ddagger} \longrightarrow \cdot R \;\; + \;\; {}^{+}CH_2-CH=CH_2$$
$$m/_z = 41$$

Because the double bond migrates during fragmentation, mass spectrometry is not generally useful for identifying the position of the double bond or for determining stereochemistry of the double bond. Alkynes usually give strong molecular ion peaks. Terminal alkynes often show M-1 peaks for loss of the terminal hydrogen. Similar fragmentation occurs for alkynes as for alkenes.

Identifying Heteroatoms (Nitrogen, Sulfur, Chlorine, Bromine, and Iodine)

Sodium Fusion Test (C-10)

Some of the unknowns may contain atoms other than carbon, hydrogen, and oxygen. If there is reason to suspect that the unknown may contain a halogen, nitro group, or sulfur atom, the sodium fusion test may be done. In this test, the compound is decomposed by heating in the presence of sodium metal, which converts the heteroatoms to ions. Sulfur is converted to sulfide ion, nitrogen is converted to cyanide ion, and the halogens are converted to halide ions. The solution is then tested for the presence of these ions.

Test for Sulfur

A portion of the solution is carefully acidified. This converts sulfide to hydrogen sulfide, which then reacts with lead acetate to form a black precipitate of lead sulfide.

$$Na_2S + 2HCl \longrightarrow 2\,NaCl + H_2S$$

$$H_2S + Pb(O_2CCH_3)_2 \longrightarrow PbS + 2\,CH_3CO_2H$$
$$\text{(black)}$$

Under these conditions, the appearance of a black precipitate is evidence for the presence of sulfur in the compound.

Test for Nitrogen

A portion of the solution is carefully acidified. The addition of Fe^{2+} and Fe^{3+} to a solution containing nitrogen (in the form of cyanide) results in a deep blue precipitate of sodium ferric ferrocyanide (called Prussian blue).

$$6\,NaCN + Fe(NH_4)_2(SO_4)_2 \longrightarrow Na_4Fe(CN)_6 + Na_2SO_4 + (NH_4)_2SO_4$$

$$Na_4Fe(CN)_6 + Fe^{3+} \longrightarrow NaFe[Fe(CN)_6] + 3Na^{+}$$
$$\text{(deep blue)}$$

Under these conditions, the appearance of a deep blue precipitate is taken as evidence that the unknown contains nitrogen.

Test for Halogen

A portion of the solution is boiled to remove sulfide and cyanide, which interfere with the test for the halogens. Then the solution is treated with $AgNO_3$ to precipitate any halides present in the form of silver halide. The colors of the silver halides are slightly different.

$$NaCl + AgNO_3 \longrightarrow NaNO_3 + AgCl$$
$$\text{(white)}$$

$$NaBr + AgNO_3 \longrightarrow NaNO_3 + AgBr$$
$$\text{(off-white)}$$

$$NaI + AgNO_3 \longrightarrow NaNO_3 + AgI$$
$$\text{(yellow)}$$

The halide can be identified by filtering the precipitate and attempting to dissolve it in concentrated ammonium hydroxide. Silver chloride dissolves readily due to the formation of the complex ion.

$$AgCl(s) + 2\ NH_4OH(aq) \longrightarrow Ag(NH_3)_2Cl\ (aq) + 2\ H_2O$$

Silver bromide dissolves only with great difficulty even with stirring after addition of excess ammonium hydroxide. Silver iodide will not dissolve in concentrated ammonium hydroxide.

Since a great many of the unknowns do not contain sulfur, halogen, or a nitro group, performing a sodium fusion test may not be necessary. Do this test only after the functional group has been identified and the possible unknown contains one of these heteroatoms.

Infrared Spectroscopy

For halides, infrared spectroscopy is useful only for alkyl or aromatic chlorides. Iodides and bromides absorb only in the far infrared regions. Alkyl chlorides show strong $C-Cl$ stretching at 800–600 cm^{-1} while for aryl chlorides, this $C-Cl$ stretching occurs at 1096–1089 cm^{-1}. Alkyl fluorides show $C-F$ stretching at 1400–1000 cm^{-1}. However, alkyl fluorides tend to be relatively uncommon as unknowns. Compounds containing nitro groups show two strong absorption bands due to NO_2 stretching at 1550 and 1350 cm^{-1}. The IR spectra of sulfur-containing compounds vary greatly depending upon the specific functional group. The student is referred to the references at the end of Chapter 28 for additional information.

Nuclear Magnetic Resonance Spectroscopy

The presence of heteroatoms can generally be inferred in the ^1H NMR spectra from the chemical shifts of protons α to the heteroatom. These protons come into resonance around $\delta 3$–4. In the ^{13}C NMR, carbons directly bonded to the chlorine, bromine, or iodine give signals at $\delta 20$–50, while carbons directly bonded to nitrogen come into resonance around $\delta 30$–60.

Mass Spectrometry

Compounds containing chlorine or bromine can be identified by their isotopic peaks in their mass spectra. Monochlorinated compounds give an M+2 peak (the mass of molecular ion peak plus 2 more) that is one-third the intensity of the molecular ion peak. This peak is due to the presence of two isotopes of chlorine, ^{35}Cl and ^{37}Cl. Monobrominated compounds have an M+2 peak that is of approximately equal intensity to the molecular ion peak. Compounds that contain more than one chlorine or bromine atom give peaks at M+4, M+6, and so on, depending upon the numbers of halogens. Alkyl iodides do not show isotopic peaks. The most common fragmentation for halides is loss of the halide to form a carbocation and a halogen radical. This pattern is most important for alkyl iodides and bromides. Chlorinated and brominated compounds can also fragment to cleave $H-Cl$ or $H-Br$. Alkyl nitro compounds generally show only weak molecular ion peaks, while aromatic nitro compounds show strong molecular ion peaks. The molecular ion peaks occur at odd mass units, assuming that there is an odd number of nitrogen atoms in the molecule. Many nitro compounds show a peak at $m/z = 30$ (for NO^+) and $m/z = 46$ (for NO_2^+).

Step 6: Select an Appropriate Derivative

Once the functional group has been identified and the presence or absence of other heteroatoms has been confirmed, the next step is to narrow the list of possible unknowns and determine what derivative or derivatives will be the most helpful to pinpoint the identity of the unknown. The usual procedure is to prepare derivatives that are solids. This is because it is easier to purify a small amount of a solid and to obtain an accurate melting point than it is to purify a small amount of liquid and obtain an accurate boiling point.

When deciding upon a derivative, look over the table of unknowns for the given functional group, find the melting point of the solid unknown, and bracket the possible compounds. For solid unknowns, include compounds that melt 5–6° above the observed melting point. (Be aware that sometimes the reported literature values of physical properties differ by 1–2°C.) For liquid unknowns, bracket the range of the corrected boiling point of the unknown by 5° on either side of the observed boiling point (go 5° higher and 5° lower than the observed corrected boiling point).

There are three things to consider when selecting which derivative to prepare:

1. The derivative should melt between 50° and 250°C. Compounds that melt lower than 50°C tend to be oily semisolids that are hard to recrystallize. It is difficult to measure an accurate melting point for compounds that melt above 250°C.
2. The melting point of the derivative should be significantly different from the melting point of the unknown. It is then possible to conclude that a reaction between the unknown and the derivatizing reagent has actually occurred.
3. The melting point of the derivatives of the other possible unknowns should be significantly different from one another. For practical purposes, this means that the melting points of the derivatives should differ by at least 5°–10°C.

Several examples will illustrate the process of selecting an appropriate derivative to prepare.

Example 1: An unknown sample has an observed melting point of 76–78°C. Below is a list of possible compounds with the melting points of two derivatives, X and Y. Which derivative will be more useful?

Compound	Melting point (°C)	Melting point (°C): Derivative X	Melting point (°C): Derivative Y
A	78	95	38
B	79	106	110
C	80	140	108
D	82	132	112

Solution: Derivative X will give the most useful information, since the melting points are all significantly different. Derivative Y is a poor choice for two reasons: the derivative of compound A melts too low and the melting points for the derivatives of compounds B, C, and D are too close together. Even with the best laboratory technique, there is no way to differentiate between these compounds and it would be a waste of time to prepare derivative Y.

Example 2: An unknown sample has a recorded melting point of 165–167°C. There are three derivatives that can be made: X, Y, and Z. How do you proceed?

	Melting point (°C)	Melting point(°C): Derivative X	Melting point(°C): Derivative Y	Melting point(°C): Derivative Z
A	167	43	195	137
B	168	210	121	138
C	168	—	175	152
D	170	37	199	89
E	172	—	122	92

Solution: Derivative X can be eliminated quickly: there are "holes" in the list with missing values for some of the compounds; also two of the derivatives have melting points that are too low. There are also problems with both the Y and the Z derivatives. For derivative Y, the melting points are too close together for Compounds A and D (195° and 199°C) and Compounds B and E (121° and 122°C). For derivative Z, the melting points are also too close together for Compounds A and B (137° and 138°C) and Compounds D and E (89° and 92°C).

This will probably entail the preparation of two different derivatives. Prepare one derivative (derivative Y, for example) and take a melting point. If the derivative melts at 175°C, the work is done: Compound C is the unknown. If Derivative Y melts at 120–122° or 194–196°C, then a second derivative (in this case derivative Z) must be prepared in order to identify the unknown.

The goal should be to identify the unknown with the minimum amount of work. Do not prepare more derivatives than necessary; however, be aware that it will not always be possible to identify the unknown based on only one derivative. Also, if time allows, a second derivative can often furnish supporting information.

Step 7: Prepare and Purify the Derivative

Sometimes it will take several attempts to prepare a solid derivative. Be aware that the directions for preparing the derivative are general and geared for a sample that has an average molecular weight of about 100 g/mol. It may be necessary to adjust the amounts of reagents or the time required for reaction to occur.

The derivative must be purified until the melting point range is 2°C or less. Obtaining a sharp and accurate melting point may require several recrystallizations of the solid derivative. Be prepared to invest the time and effort to prepare a pure derivative; the benefit will be an accurate identification of the unknown.

Derivatives for Selected Functional Groups

The reactions of the various functional groups with the derivatizing agent are given here. Procedures will be given in the experimental procedure section.

Derivatives of Aldehydes and Ketones

Three of the most useful derivatives are the 2,4-dinitrophenylhydrazone, semicarbazone and oxime.

2,4-Dinitrophenylhydrazone (D-4). This is also the basis of the chemical test to differentiate ketones and aldehydes from alcohols and phenols.

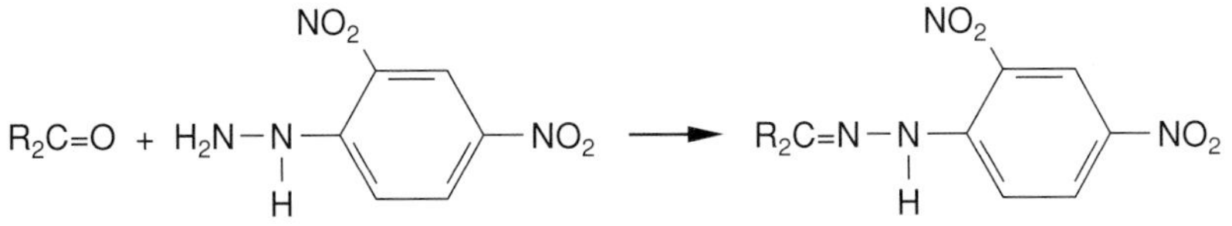

$$R_2C=O + H_2N-N \overset{H}{\underset{}{}} \text{—DNP} \longrightarrow R_2C=N-N \text{—a 2,4-dinitrophenylhydrazone}$$

If appropriate for a given unknown, it is best to choose to make a DNP derivative: the reaction is simple and quick and the product is relatively easy to purify. Hindered ketones take longer to react than unhindered aldehydes and ketones. Other hydrazones, such as *p*-nitrophenylhydrazone and phenylhydrazone, may also work well.

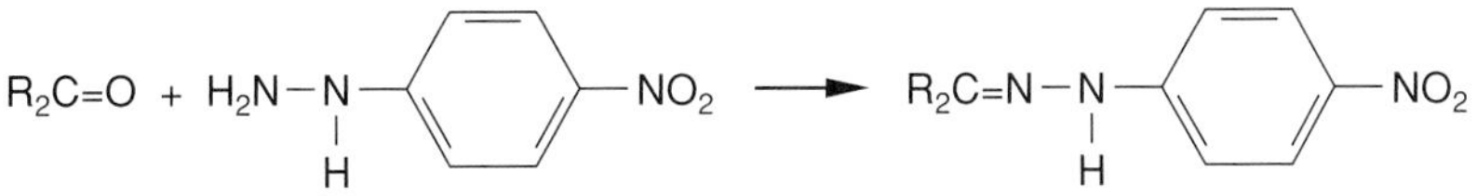

$$R_2C=O + H_2N-N \text{—} p\text{-nitrophenylhydrazine} \longrightarrow R_2C=N-N \text{—a } p\text{-nitrophenylhydrazone}$$

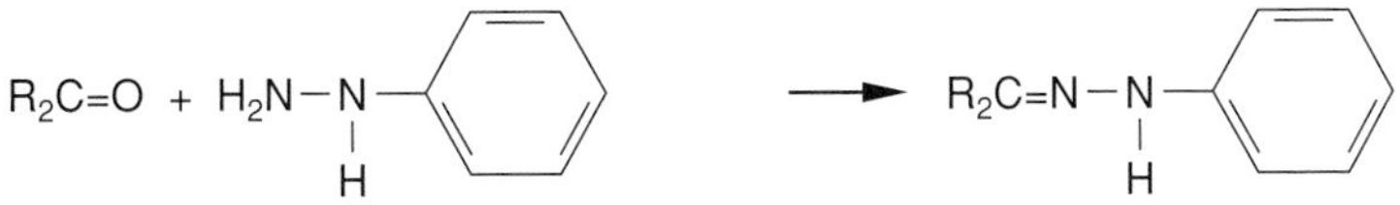

$$R_2C=O + H_2N-N \text{—phenylhydrazine} \longrightarrow R_2C=N-N \text{—a phenylhydrazone}$$

Semicarbazone (D-5). Semicarbazones are a little more difficult to isolate than DNP derivatives, but most aldehydes and ketones form crystalline solids when reacted with semicarbazide hydrochloride in acetic acid/sodium acetate buffer.

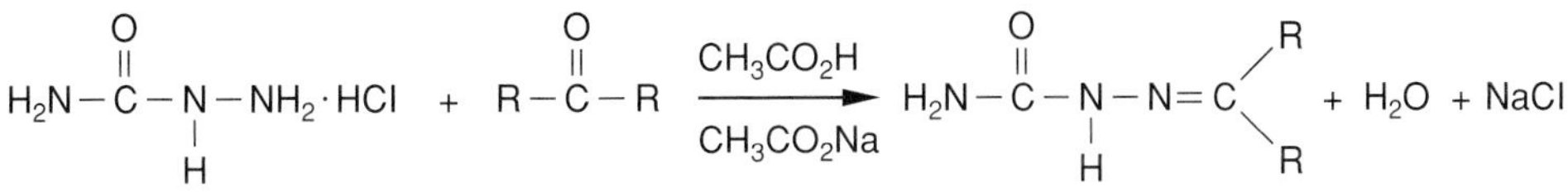

$$H_2N-\overset{O}{\overset{\|}{C}}-N-NH_2 \cdot HCl + R-\overset{O}{\overset{\|}{C}}-R \xrightarrow[CH_3CO_2Na]{CH_3CO_2H} H_2N-\overset{O}{\overset{\|}{C}}-N-N=C\overset{R}{\underset{R}{}} + H_2O + NaCl$$

semicarbazide hydrochloride / a semicarbazone

Sometimes oils are obtained in this reaction. Scratching the side of a glass flask while cooling the solution may cause the oil to solidify. It is crucial to remove all traces of acetic acid from the semicarbazone in order to prevent decomposition. Washing the precipitate with water will accomplish this adequately.

Oximes (D-6). These are generally useful derivatives for ketones and aldehydes. Oximes are prepared by the reaction of the aldehyde or ketone with hydroxylamine hydrochloride.

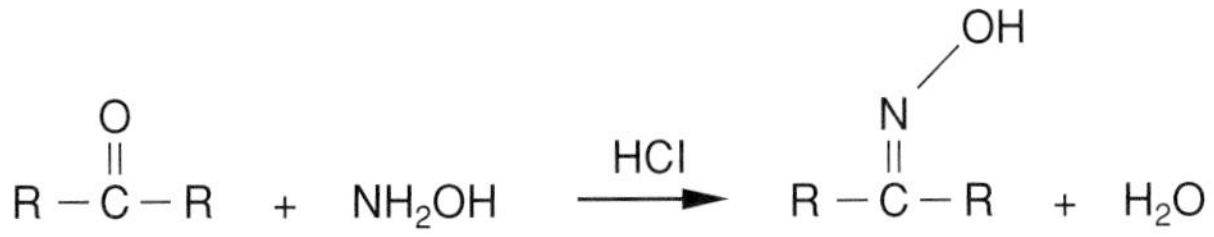

$$R-\overset{O}{\overset{\|}{C}}-R + NH_2OH \xrightarrow{HCl} R-\overset{N-OH}{\overset{\|}{C}}-R + H_2O$$

hydroxylamine / an oxime

Some oximes with low melting points tend to form oils. Before preparing an oxime derivative, make certain that the melting point will be sufficiently high to be useful.

Derivatives of Alcohols

The three most common derivatives to prepare for alcohols are phenylurethanes, α-naphthylurethanes, and the 3,5-dinitrobenzoates.

α-Naphthylurethane and Phenylurethane (D-2 and D-3). These are discussed together because the reactions are essentially the same.

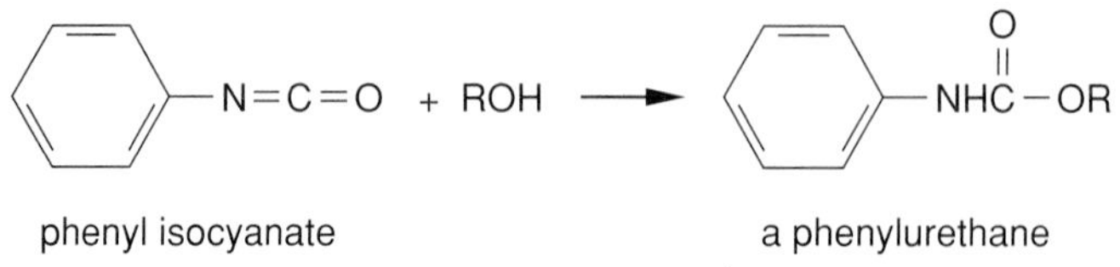

phenyl isocyanate   a phenylurethane

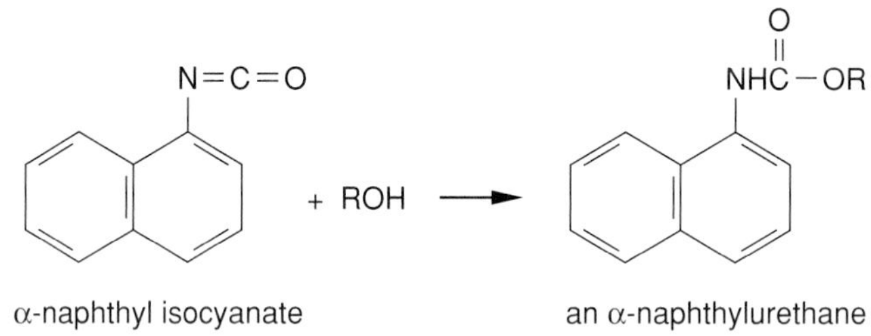

α-naphthyl isocyanate   an α-naphthylurethane

This procedure works best for primary and secondary alcohols. This derivative also works with phenols. It is important that all traces of water be removed from the alcohol or phenol unknown before reacting with the reagent. The presence of water hydrolyzes the isocyanate to an amine, which then reacts with more isocyanate to yield a disubstituted urea. This makes it difficult to purify the product.

3,5-Dinitrobenzoate (D-1). This procedure works well for alcohols and phenols. This is an example of an esterification reaction between an acid chloride and an alcohol.

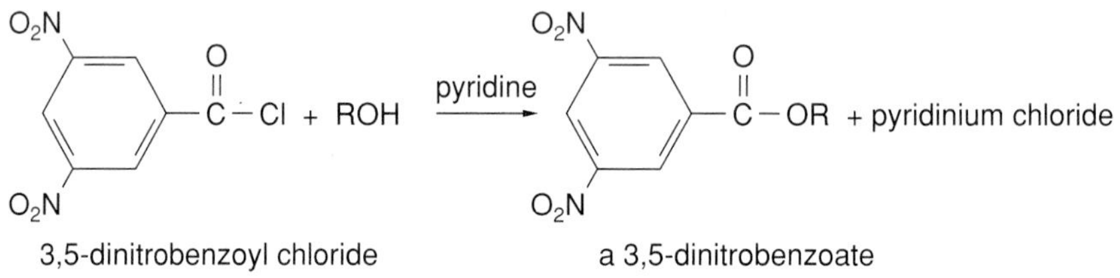

3,5-dinitrobenzoyl chloride   a 3,5-dinitrobenzoate

Pyridine is added to the reaction mixture to neutralize the HCl as it is formed. The reaction occurs within a few minutes with primary alcohols. For secondary alcohols and phenols, the reaction may take an hour. Tertiary alcohols are very slow to react.

Derivatives of Phenols

The three primary derivatives of phenols are the α-naphthylurethane, the brominated phenol, and the aryloxyacetic acid.

Aryloxyacetic Acid (D-20). This reaction works with most phenols, although sterically hindered phenols may be very slow to react. The phenol is deprotonated with base; the phenoxide then reacts with chloroacetic acid to yield an aryloxyacetic acid. Phenol is used to illustrate this reaction as shown on next page.

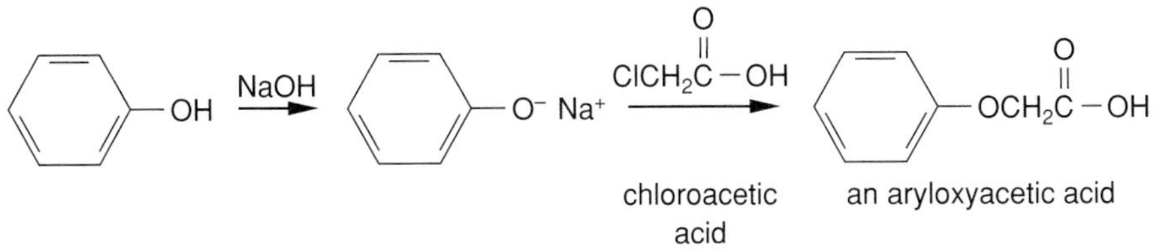

This procedure will not work with an alcohol; it is specific for phenols.

α-Naphthylurethane (D-18).   Phenols react with α-naphthyl isocyanate to yield crystalline α-naphthylurethanes. Phenol is used to illustrate this reaction as shown here.

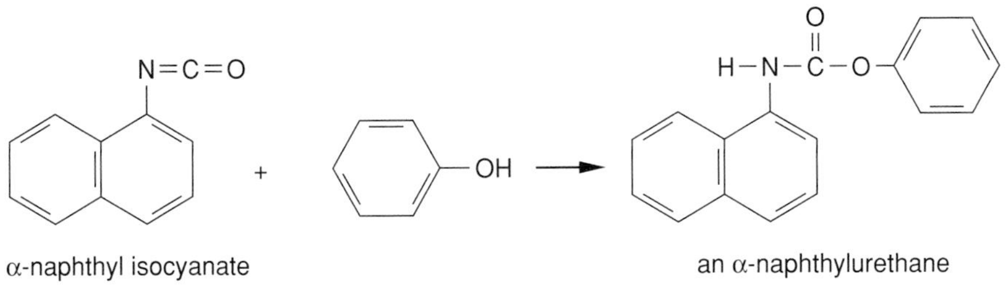

Phenols having electron-withdrawing groups tend to react slowly or not at all. Reaction of phenols with phenyl isocyanate is not very useful, since oils are obtained predominantly.

Brominated Phenols (D-19).   Phenols are very reactive toward electrophilic aromatic substitution with bromine. The phenol will be polybrominated: all unsubstituted ortho and para positions will be brominated. Phenol is used to illustrate this reaction as shown here.

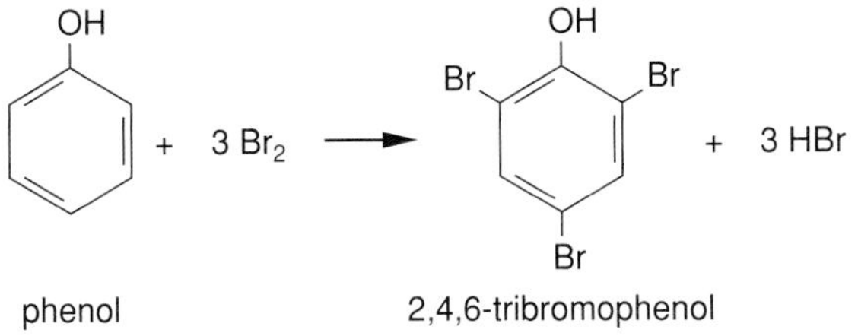

This reaction does not always work, especially if the ring contains groups that are easily oxidized; when it does, however, the derivative is usually quick to form and easy to recrystallize.

Derivatives of Amines

The two main derivatives of amines are amides (acetamides and benzamides) and picrates.

Benzamide (D-9).   Primary and secondary amines form solid amides when reacted with benzoyl chloride.

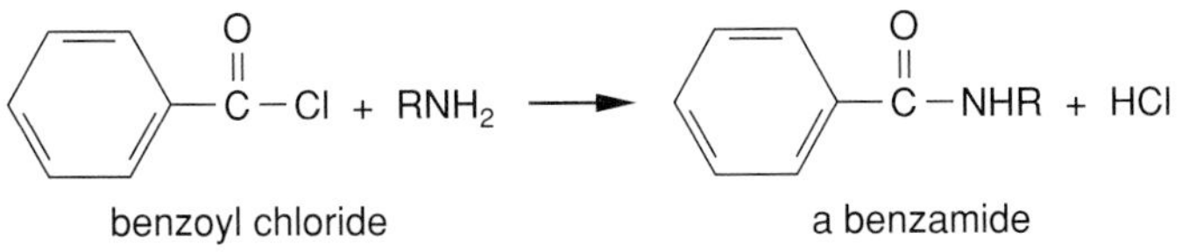

Addition of NaOH neutralizes the acid formed in the reaction. This reaction does not work with tertiary amines.

Acetamides (D-8).   This reaction works only for primary and secondary amines. The acetylating agent is acetic anhydride.

$$RNH_2 \quad + \quad CH_3\overset{\overset{O}{\|}}{C} - O - \overset{\overset{O}{\|}}{C}CH_3 \quad \xrightarrow{CH_3CO_2Na} \quad CH_3\overset{\overset{O}{\|}}{C} - NHR \quad + \quad CH_3\overset{\overset{O}{\|}}{C} - OH$$

an amine   acetic anhydride   an acetamide

Tertiary amines do not react with acetic anhydride to form acetamides.

Picrates (D-10).   This reaction is useful for all amines—primary, secondary, and tertiary. The reaction is an acid-base reaction between the basic amine and picric acid. The picrate salt is an easily isolable yellow crystalline solid.

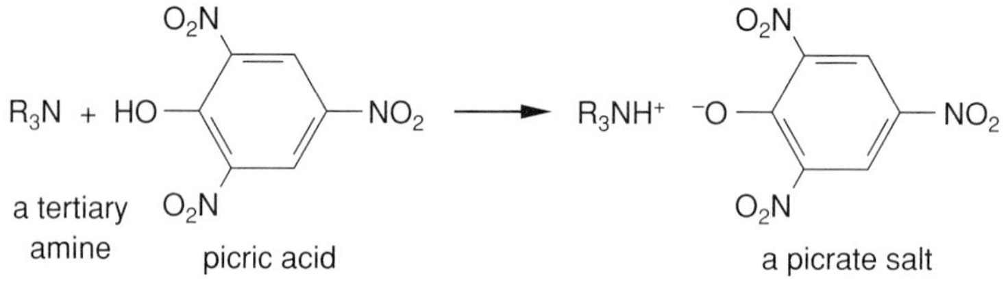

a tertiary amine   picric acid   a picrate salt

Caution!

Solid picric acid can be difficult to handle. It is generally unstable. Picric acid may explode when heated too rapidly. However, it remains one of the only ways to make derivatives of tertiary amines. Solutions of picric acid will be used in these experiments instead of solid picric acid.

Derivatives of Carboxylic Acids

The best derivatives of carboxylic acids are amides, which are easily prepared by converting the acid to the acid chloride and reacting with ammonia or an amine. The first step is the formation of the acid chloride.

$$R\overset{\overset{O}{\|}}{C} - OH \quad + \quad Cl - \overset{O}{S} - Cl \quad \longrightarrow \quad R\overset{\overset{O}{\|}}{C} - Cl \quad + \quad HCl \quad + \quad SO_2$$

a carboxylic acid   thionyl chloride   a carboxylic acid chloride

Amide (D-14).   Amides are formed from the reaction of ammonia with the acid chloride. They tend to be slightly more water soluble than the other amide derivatives.

$$R\overset{\overset{O}{\|}}{C} - Cl \quad + \quad 2\,NH_3 \quad \longrightarrow \quad R\overset{\overset{O}{\|}}{C} - NH_2 \quad + \quad NH_4Cl$$

a carboxylic acid chloride   a carboxamide

p-Toluidides and Anilides (D-16 and D-15).   These derivatives are discussed together since the reactions are the same. Amides are formed from the reaction of *p*-toluidine or aniline with the acid chloride.

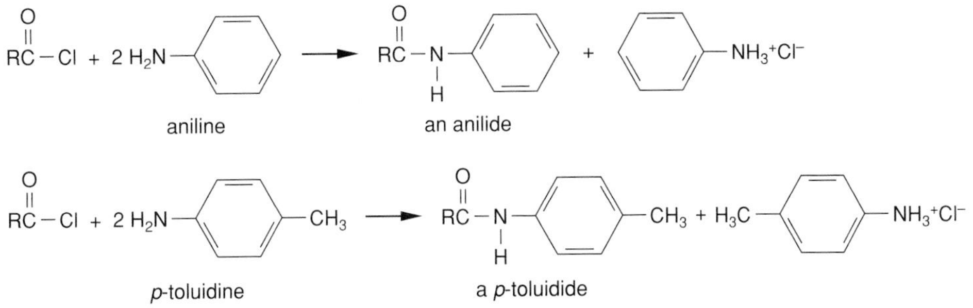

aniline   an anilide

p-toluidine   a *p*-toluidide

Anilides and *p*-toluidides are nicely crystalline derivatives that are easy to recrystallize.

p-Nitrobenzyl Esters and p-Bromophenacyl Esters (D-17). These are formed by the reaction of carboxylic acids with either *p*-nitrobenzyl chloride or *p*-bromophenacyl bromide.

p-nitrobenzyl bromide   a *p*-nitrobenzyl ester

p-bromophenacyl bromide   a *p*-bromophenacyl ester

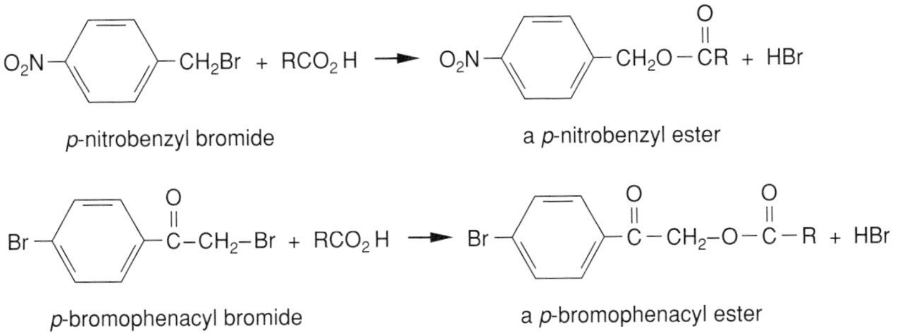

The pH must be carefully controlled, since the starting materials and/or product esters are easily hydrolyzed under basic conditions.

Step 8: Identify the Unknown

In addition to all of the relevant physical data for the unknown and all the derivatives prepared, your lab report should include a thorough explanation of how you identified the unknown.

A summary of the chemical tests and derivative preparation by functional group is shown in Table 28.2-1. The designation beside each entry gives a reference to where the procedure may be found later in this chapter.

Experimental Procedure

General Safety Note: Some of the chemicals used in qualitative organic analysis are irritants. Some are carcinogenic. Always assume that the chemicals used are toxic and take necessary precautions. This means always wear eye protection and always wear gloves. Perform all chemical tests under the hood and avoid breathing the vapors. Individual safety precautions will be given by each procedure.

Cleanup and Disposal: Because of the very small amounts of waste generated, the filtrates and test solutions may be washed down the drain unless instructed otherwise.

Table 28.2-1   Summary of Chemical Tests and Derivatives

Functional group	Chemical test and derivatives	Test	Specificity
Alcohol	Chromic acid	C-2	Will distinguish 1° and 2° from 3° alcohols
	Lucas test	C-7	Will distinguish 3° and 2° from 1° alcohols; only useful for water-soluble alcohols
	Derivatives of Alcohols		
	3,5-Dinitrobenzoate	D-1	Good for all alcohols
	α-Naphthylurethane	D-2	For 1° and 2° alcohols only
	Phenylurethane	D-3	For 1° and 2° alcohols only
Aldehyde	Chromic acid	C-2	Differentiates between aldehydes and ketones
	2,4-Dinitrophenylhydrazine	C-3	Selective for aldehydes and ketones
	Tollens test	C-11	Differentiates between aldehydes and ketones
	Derivatives of Aldehydes		
	2,4-Dinitrophenylhydrazone	D-4	
	Oxime	D-6	Derivatives may be oils
	p-Nitrophenylhydrazone	D-7	
	Phenylhydrazone	D-7	Derivatives may be oils
	Semicarbazone	D-5	
Amine	10% HCl	B-2	Most amines will dissolve in acid.
	Hinsberg test	C-5	Distinguishes 1°, 2°, and 3° amines
	Neutralization equivalent	C-8	Gives the equivalent weight of amine
	Derivatives of Amines		
	Acetamide	D-8	For 1° and 2° amines only
	Benzamide	D-9	For 1° and 2° amines only
	Benzenesulfonamide	D-12	For 1° and 2° amines
	Picrate	D-10	For 3° amines
	p-Toluenesulfonamide	D-11	For 1° and 2° amines
Carboxylic Acid	10% NaOH	B-3	Acids are soluble
	10% NaHCO$_3$	B-4	Acids are soluble
	Neutralization equivalent	C-8	Gives the equivalent weight of the acid
	Derivatives of Carboxylic Acids		
	Amide	D-14	Derivatives are often water soluble
	Anilide	D-15	
	p-Bromophenacyl ester	D-17	
	p-Nitrobenzyl ester	D-17	
	p-Toluidide	D-16	
Ketone	2,4-Dinitrophenylhydrazine	C-3	Distinguishes aldehydes and ketones from alcohols
	Iodoform	C-6	Specific for methyl ketones
	Derivatives of Ketones		
	2,4-Dinitrophenylhydrazone	D-4	
	p-Nitrophenylhydrazone	D-7	
	Oxime	D-6	Derivatives may be oils
	Phenylhydrazone	D-7	Derivatives may be oils
	Semicarbazone	D-5	
Phenol	10% NaHCO$_3$	B-4	Most phenols are soluble in NaOH, but not NaHCO$_3$
	Ferric chloride	C-4	Distinguishes between phenols and alcohols

Table 28.2-1 *(continued)*

Functional group	Chemical test and derivatives	Test	Specificity
Phenol (contd.)	*Derivatives of Phenols*		
	Aryloxyacetic acid	D-20	
	Brominated phenols	D-19	
	α-Naphthylurethane	D-18	Doesn't work well for phenols with electron-withdrawing groups
Other Functional Groups			
Alkenes and Alkynes	Bromine in methylene chloride	C-1	Red color disappears in presence of unsaturation
	Potassium permanganate	C-9	Purple color disappears and a brown precipitate forms in presence of unsaturation
Halogen	Sodium fusion test	C-10	For Cl, Br, or I only—forms white, off-white, or yellow precipitate with silver nitrate
Nitrogen	Sodium fusion test	C-10	Forms a blue precipitate
Sulfur	Sodium fusion test	C-10	Forms a black precipitate of lead sulfide

Part A: Purification of the Unknown

Obtain a sample of unknown from the laboratory instructor. Be sure to record the number of the unknown. Purify a liquid unknown by distillation with a Hickman still. Perform a micro boiling point determination, making a correction for atmospheric pressure. The refractive index may also be measured, making a correction for temperature. If the unknown is a solid, determine the melting point. If the melting point range is greater than 2°C, the solid must be recrystallized. Record all physical properties and observations.

Part B: Solubility Tests

Safety First!

Always wear eye protection in the laboratory.

1. Always wear eye protection and gloves.
2. Concentrated sulfuric acid is a highly toxic oxidizing agent. If acid comes in contact with skin, wash immediately with cool running water. If acid spills on the bench, notify the instructor immediately. The acid can be neutralized with sodium bicarbonate.
3. Even though dilute solutions are used, HCl and NaOH may harm the skin. Wipe up any spills immediately.

B-1: Solubility in Water

Put 20 mg of a solid unknown or 2 to 3 drops of a liquid unknown in a small test tube or vial. Add 10 drops of distilled water. Mix well. If the unknown dissolves, test its pH by dipping the end of a glass rod into the solution and touching a piece of litmus paper. If the pH is 3 or below, the unknown is an acid or a phenol with electron-withdrawing groups. If the pH is 9 or above, the unknown is an amine. If the pH is neutral (pH = 6–8), the unknown could be an alcohol, aldehyde, or ketone with fewer than six carbons. If the unknown is water soluble, it is not necessary to test its solubility in the other solutions. (Why not?)

B-2: Solubility in 10% HCl

Put 20 mg of a solid unknown or 2 to 3 drops of a liquid unknown in a small test tube or vial. Add 10 drops of 10% HCl. Mix well. Solubility is noted by the formation of a homogenous solution or color change or evolution of heat. This is evidence that the unknown is probably an amine.

B-3: Solubility in 10% NaOH

Put 20 mg of a solid unknown or 2 to 3 drops of a liquid unknown in a small test tube or vial. Add 10 drops of 10% NaOH. Mix well. Solubility is noted by the formation of a homogenous solution, color change, or evolution of heat or gas. Since both carboxylic acids and phenols are soluble in dilute NaOH, the unknown must be tested for solubility in 10% $NaHCO_3$.

B-4: Solubility in 10% NaHCO₃

Put 20 mg of a solid unknown or 2 to 3 drops of a liquid unknown in a small test tube or vial. Add 10 drops of 10% $NaHCO_3$. Mix well. Solubility is noted by the formation of a homogenous solution, color change, or evolution of heat or gas. Solubility in dilute sodium bicarbonate indicates that the unknown is either a carboxylic acid or a phenol that contains electron-withdrawing groups.

B-5: Solubility in Concentrated H₂SO₄

Caution!

Put 20 mg of a solid unknown or 2 to 3 drops of a liquid unknown in a small test tube or vial. Add 5 drops of concentrated sulfuric acid.

Solubility is indicated by a homogenous solution, color change, or evolution of heat or gas. Solubility indicates that the unknown contains oxygen, contains nitrogen, or is unsaturated.

Part C: Chemical Tests (Arranged Alphabetically)

C-1: Bromine in Methylene Chloride

Safety First!

Always wear eye protection in the laboratory.

1. Always wear eye protection and gloves, and work under a hood or in a well-ventilated area.
2. Bromine is a highly toxic oxidant. Contact, whether by inhalation or spilling, can cause severe burns. Be very careful when handling this reagent.
3. Methylene chloride is toxic and is a cancer-suspect agent. Do not breathe the vapors.

Fume hood

To a small test tube add 2 to 3 drops of a liquid unknown or about 20 mg of a solid unknown. Dissolve in 5 drops of methylene chloride. With a dropper bottle, add drop-wise a freshly prepared solution of bromine dissolved in methylene chloride. Shake well after each addition. A positive test will require the addition of 3–4 drops of the bromine solution before the red color of the bromine persists. A positive test is indication of unsaturation, such as a double or triple bond. However, easily oxidizable functionalities such as aldehydes will give a false positive test.

Suggested knowns to run for comparison are styrene and 2-pentyne.

C-2: Chromic Acid

1. Always wear eye protection and gloves, and work under a hood or in a well-ventilated area.
2. Chromium salts are classified as carcinogenic (cancer-causing). Be careful when working with chromium-containing reagents.
3. This reagent is prepared with sulfuric acid, a highly toxic oxidizing agent. Wash with copious amounts of cold water if acid comes into contact with your skin. If spills occur on bench, neutralize with sodium bicarbonate.

In one of the wells of a spot plate, put 2 to 3 drops of a liquid unknown or about 20 mg of a solid unknown. Dissolve the unknown in about 5 drops of reagent-grade acetone by stirring with a glass rod. Add 1 drop of the chromic acid solution. Observe carefully. A positive reaction occurs if a blue-green precipitate forms within 2 to 3 seconds. A positive reaction is indicative of a primary or secondary alcohol or an aldehyde. Tertiary alcohols do not react.

In addition to running this test on known compounds, this test should be performed on acetone alone to make certain there is no impurity in the acetone (such as isopropyl alcohol) that is oxidizable. Put 5 drops of the acetone on a spot plate. Add 1 drop of the chromic acid solution and observe any color change. If a blue-green precipitate does occur, the test must be rerun with purified acetone.

Suggested knowns to run for comparison are 1-propanol, 2-propanol, and pentanal.

C-3: 2,4-Dinitrophenylhydrazine (DNP) Test

1. Always wear eye protection and gloves, and work under a hood.
2. Many derivatives of hydrazine are suspected carcinogens. Be careful when handling solutions of 2,4-dinitrophenylhydrazine.
3. This reagent is made with sulfuric acid, which is a highly toxic oxidizer. If the reagent comes in contact with skin, wash the affected area with copious amounts of cool water. Bench spills can be neutralized with sodium bicarbonate.

In a well of a spot plate, add 2 to 3 drops of a liquid unknown or about 20 mg of a solid unknown. Dissolve the unknown (if solid) in 5 drops of ethanol. To the solution add 7–8 drops of the 2,4-dinitrophenylhydrazine reagent. Stir with a glass rod. A positive test is the formation of a yellow, orange, or red precipitate. Aldehydes and ketones will give a positive test. The color of the precipitate may yield information about the degree of conjugation in the molecule. The DNP products of aromatic aldehydes and ketones (or ones that are highly conjugated) tend to be orange to red; aliphatic and unconjugated aldehydes tend to give yellow DNP products.

Suggested knowns to run for comparison are benzaldehyde and 2-pentanone.

C-4: Ferric Chloride Test

Always wear eye protection and gloves, and work under a hood.

In a well of a spot plate, add 2 to 3 drops of a liquid unknown or about 20 mg of a solid unknown. Dissolve the unknown in 2 drops of water (if the unknown is water soluble) or in 2 drops of ethanol (if the unknown is not water soluble). Stir with a glass rod. To the solution add 1–2 drops of 2.5% ferric chloride solution and stir. A positive test is a color change from yellow (the color of the ferric chloride) to blue green, violet, or red. Observe the solution closely, because the color change may disappear rapidly. Phenols give a positive test with ferric chloride solution.

A suggested known to run for comparison is 3,5-dimethylphenol.

C-5: Hinsberg Test

1. Always wear eye protection and gloves, and work under a hood.
2. Many amines have objectionable odors. Do not breathe the fumes.
3. Benzenesulfonyl chloride is a lachrymator. Use under the hood.
4. Potassium hydroxide solutions are corrosive. If reagent spills on skin, wash off with copious amount of cold water.

Hood!

In a test tube, put 2 to 3 drops of a liquid compound or 20 mg of a solid compound, 2 mL of 2 M KOH, and 4 drops of benzenesulfonyl chloride. Stopper the test tube with a cork and shake vigorously for 3–4 minutes or until the odor of benzenesulfonyl chloride is gone. Test the pH of the solution. It should be strongly basic. If not, add KOH dropwise until it is strongly basic. Let the test tube stand for a few minutes and note whether the mixture in the test tube is homogenous or whether it has two layers.

If the solution is homogenous, add 1.5 M HCl dropwise to pH 4. Primary sulfonamides are insoluble in an acidic solution; precipitation from an acid solution is evidence that the unknown amine is primary.

If the solution is not homogenous, but forms two layers, separate the layers. To the organic layer, carefully add 1.5 M HCl dropwise until the solution is acidic to litmus paper. (Add the acid slowly in order to avoid adding too much.) Secondary sulfonamides are insoluble and will precipitate upon addition of acid, while tertiary amines remain soluble.

It is crucial to run the Hinsberg test on known amines for comparison. Suggested knowns to run for comparison are butylamine, *sec*-butylamine, and *tert*-butylamine.

C-6: Iodoform Test

1. Always wear eye protection and gloves, and work under a hood or in a well-ventilated area.

2. Iodine may cause burns. Wear gloves.

3. Potassium hydroxide solutions are corrosive. If the reagent spills on your skin, wash off with copious amount of cold water.

In a small test tube or 5-mL conical vial, add 2 to 3 drops of liquid unknown or 20 mg of a solid unknown. Add 1 mL of water. (If the unknown is not water soluble, add 5 drops of 1,2-dimethoxyethane to the water.) Swirl vigorously to dissolve. Add 1 mL of 10% aqueous sodium hydroxide and 1.5 mL of 0.5 M iodine-potassium iodide solution. Observe closely. A positive test is the disappearance of the brown color and the formation of a yellow precipitate (iodoform). If desired, the yellow precipitate can be filtered and air dried, and the melting point can be taken. Iodoform melts at 118–119°C. A positive test will be obtained for methyl ketones (ketones having the functionality CH_3CO) and for other compounds that can be oxidized to a methyl ketone (methyl carbinols).

Suggested knowns to run for comparison are 2-butanone, and acetaldehyde (for a positive tests), and 3-pentanone (to see what a negative iodoform test looks like).

C-7: Lucas Test

This test can be used only for alcohols that are at least partially water soluble. In a dry test tube or 3-mL conical vial, add 2 to 3 drops of a liquid unknown or 20 mg of a solid unknown. Add 10 drops of the Lucas reagent. Swirl rapidly, and observe closely for the formation of cloudiness or a separation of two layers, which is the formation of the insoluble alkyl halide. The length of time it takes for the homogenous solution to turn cloudy can be used to differentiate between primary, secondary, and tertiary alcohols. Tertiary alcohols should react almost immediately, secondary alcohols may require 5–10 minutes to react, and primary alcohols should not react.

Suggested knowns to run for comparison are butyl alcohol, *sec*-butyl alcohol, and *tert*-butyl alcohol.

C-8: Neutralization Equivalent

Molecular Weight Determination of a Carboxylic Acid. Rinse and fill a 5-mL buret or graduated pipet with standardized 0.100 M NaOH. Record the exact molarity and the initial volume of the NaOH solution.

Weigh accurately (to the nearest mg) about 100 mg of the unknown carboxylic acid. Dissolve in 20–30 mL of distilled water. If necessary, add 95% ethanol dropwise until all of the carboxylic acid dissolves. It may be necessary to warm the mixture to get it to dissolve completely. Swirl well to mix. Add 1 drop of phenolphthalein indicator. Titrate with NaOH to the faint pink end point. Record the volume of NaOH used. Repeat this titration at least two more times and average the results.

Molecular Weight Determination of an Amine. This procedure will work best for amines that have pK_b values around 4 to 6. Rinse and fill a 5-mL buret or graduated pipet with standardized 0.100 M HCl. Record the exact molarity and the initial volume of the HCl solution.

Weigh accurately (to the nearest mg) about 100 mg of the amine. Dissolve in 20–30 mL of distilled water. If necessary, add ethanol dropwise until all of the amine dissolves. Add 1–2 drops of methyl red indicator. While stirring, add small aliquots (2–3 drop increments) of HCl to the solution of amine. When close to the yellow end point, add HCl dropwise until the equivalence point is reached. Calculate the molecular weight of amine. Repeat the titration at least two more times and average the results.

C-9: Potassium Permanganate Test

1. Always wear eye protection in the laboratory.
2. Potassium permanganate solutions stain clothing and skin. Wear gloves.

To a small test tube, add 2 to 3 drops of a liquid unknown or about 20 mg of a solid unknown. Dissolve in 5 drops acetone. With a dropper bottle, add dropwise a 1% solution of $KMnO_4$ until the purple color persists. Shake well after each addition. A positive test is one in which more than 1 drop of the permanganate solution is required to sustain the purple color and a brown precipitate is formed. A positive test is indication of unsaturation, such as a double or triple bond.

Keep under the hood.
Suggested knowns to run for comparison are styrene and 2-pentyne.

C-10: Sodium Fusion Test

1. Always wear eye protection and gloves, and work under a hood or in a well-ventilated area.
2. Sodium metal reacts violently with water to liberate hydrogen gas! Be extremely careful when working with this metal. Metallic sodium should be stored under oil to prevent its reaction with moisture from the air (and air oxidation). Avoid skin contact.
3. Hydrogen sulfide and cyanide salts are very toxic. Be very careful.
4. Lead salts are toxic. Dispose of them in properly labeled jar only.

Caution!
Go to a separate lab area or a different lab where there are no organic vapors present. Obtain a small, disposable glass test tube that is made of soft glass (non-Pyrex tube). Obtain a pea-sized piece of sodium from the storage bottle and use a paper towel to wipe off the oil from the surface of the sodium. Always handle the sodium with forceps or tweezers. Drop the piece of sodium into the test tube. Heat the test tube with a Bunsen burner until the sodium melts. Carefully add 2 to 3 drops of a liquid unknown or 20 mg of a solid unknown to the test tube. Make certain that the unknown drops vertically into the sodium and is not caught on the side of the test tube. There may be a burst of flame as the unknown comes in contact with the sodium. While holding the test tube with a test tube holder, heat the test tube gently until there is no more reaction and then heat strongly for about 2 minutes until the tube is red hot. Cool the tube all the way to room temperature and then carefully add methanol dropwise to react with the remaining sodium metal. Reheat gently until all of the methanol is gone. Reheat the tube until red hot. Drop the test tube into a small beaker containing 20 mL of deionized water. If the tube has not already broken, break the tube using a glass rod. Heat to boiling and gravity filter into a small Erlenmeyer

flask giving a clear solution. The filtrate can now be tested for nitrogen, sulfur, and halogens. If nitrogen or sulfur is expected to be present, test first for these elements.

Test for Sulfur. Put 2–3 drops of the filtrate from the fusion solution into a spot plate well. Work under the hood. Add 5 drops acetic acid. Stir with a glass rod. Add 2–3 drops of 0.1 M lead acetate solution. A positive test for sulfur is the immediate formation of a brown-black precipitate of lead sulfide. Dispose of the precipitate in a specially labeled jar under the hood.

Test for Nitrogen. In a small test tube, place 10 drops of the filtrate from the sodium fusion test. Add 50 mg of powdered iron(II) sulfate. Swirl to dissolve. Heat to boiling. While still hot, cautiously add dilute sulfuric acid until the solution is acidic. The presence of a blue precipitate is evidence of the presence of a nitrogen atom. The solution may be filtered to make the presence of a precipitate easier to observe.

Test for Halogen. To a small test tube, add 10 drops of the filtrate from the sodium fusion test. Carefully acidify the solution with 3 M nitric acid. Work under the hood. If the tests for sulfur or nitrogen were positive, boil the solution gently **under the hood** to get rid of H_2S or HCN gas. Cool to room temperature. Add 2 drops of 0.1 M $AgNO_3$ to the solution. Observe closely for the formation of a voluminous precipitate. This is a positive test for a halogen other than fluoride. (A faint cloudiness is not a positive test and is probably due to the reaction of silver nitrate with chloride from the water.) Filter the precipitate and observe the color: silver chloride is white; silver bromide is off-white or cream-colored; silver iodide is yellow.

Hood!

The identity of the halogen can be further delineated by trying to dissolve the solid in dilute ammonium hydroxide. Remove the liquid from the precipitate by either filtration through a Hirsch funnel or centrifugation followed by decanting the solvent. To the solid, add 10 drops of dilute ammonium hydroxide. Stir with a glass rod. Silver chloride will readily dissolve due to the formation of a complex silver amine salt; silver bromide will partly dissolve with repeated stirring; silver iodide will not dissolve.

If the results of the test for halogen are inconclusive, a second procedure on the filtrate can be run. Add 10 drops of the acidified filtrate from the sodium fusion to a clean test tube and add 5 drops of 1% $KMnO_4$ solution. Stopper and shake the tube. Add about 15 mg of oxalic acid (or enough to discharge the purple color due to excess permanganate). Add 1 mL of methylene chloride to the tube. Stopper and shake the tube. Allow the layers to separate and observe the color of the bottom layer. A clear colorless layer indicates chloride ion, a brown layer indicates bromide, and a purple layer indicates iodide.

Suggested knowns to run for comparison are butyl chloride, butyl bromide, butyl iodide, butylamine, and thiophene.

C-11: Tollens Test

1. Always wear eye protection and gloves, and work under a hood or in a well-ventilated area.

2. The Tollens reagent must be prepared immediately before use. Dispose of any excess reagent and the products by cautiously adding 6 M nitric acid to the silver mirror, then washing it down the drain with lots of cold running water. Under no circumstances should the test tube be stored in the locker. **On standing, silver fulminates are formed, which are explosive.**

Caution!

Make sure that the test tubes to be used in this test are very clean, since any dirt or oil may prevent the silver mirror from being easily observable. The test tubes may be cleaned by rinsing with 10% NaOH. To a clean test tube, add 1.0 mL of 5% $AgNO_3$ and 1 drop of 10% NaOH. Add concentrated aqueous NH_3 dropwise, with shaking, until the precipitate formed just dissolves. This will take 2–4 drops of NH_3. Add 2 to 3 drops of a liquid unknown or about 20 mg of a solid unknown. Swirl the tube to mix, then let stand for 10 minutes. If no reaction occurs, heat the test tube in a beaker of warm water for a few minutes. Do not heat to dryness. A positive test is the formation of a silver mirror coating on the side of the test tube. **Do not store the test tube in the drawer! Explosive silver fulminates are formed.**

Suggested knowns to run for comparison are pentanal (positive) and 2-pentanone (negative).

Part D: Derivatives

Derivatives of Alcohols

D-1: 3,5-Dinitrobenzoate and p-Nitrobenzoate

S a f e t y   F i r s t !

Always wear eye protection in the laboratory.

Hood!

1. Always wear eye protection and gloves, and work under a hood or in a well-ventilated area.

2. 3,5-Dinitrobenzoyl chloride and *p*-nitrobenzoyl chloride release HCl upon contact with water. All glassware must be dry before starting this experiment. Be careful when handling these reagents.

3. Pyridine is a flammable, toxic liquid with an unpleasant odor. Do not breathe the vapors.

General Directions.   Make certain that all the equipment used is completely dry. If any water is present, the 3,5-dinitrobenzoyl chloride and *p*-nitrobenzoyl chloride will be hydrolyzed to the corresponding benzoic acid and no reaction will occur.

Preparation of 3,5-Dinitrobenzoate Derivative.   Add 100 mg (about 4–5 drops) of the unknown alcohol and 200 mg of 3,5-dinitrobenzoyl chloride to a small, dry conical vial equipped with spin vane and air condenser. If the alcohol or phenol is a solid, add approximately 1 mL of dry pyridine. Gently heat the stirred solution to boiling. Let the solution cool slightly. The ester will precipitate. Add 0.3 mL of distilled water to hydrolyze the remaining 3,5-dinitrobenzoyl chloride. Cool in an ice bath. Filter the product through a Hirsch funnel. Wash the precipitate with several small portions of 10% sodium carbonate to dissolve the impurity of 3,5-dinitrobenzoic acid, then wash the precipitate with a small portion of distilled water. Let the precipitate air dry. The solid product can be recrystallized from an ethanol-water solvent pair. Dissolve the product in about 0.5 mL of ethanol; add water dropwise until a faint cloudiness is evident. Cool in an ice bath. Filter using a Hirsch funnel. Alternately, the ester can be recrystallized using a Craig tube. Let the crystals air dry and take a melting point.

Preparation of *p*-Nitrobenzoate Derivative.   Add 100 mg (about 4–5 drops) of the unknown alcohol and 200 mg of *p*-nitrobenzoyl chloride to a small, dry conical vial equipped with spin vane and air condenser. If the alcohol or phenol is a solid, add approximately 1 mL of dry pyridine. Gently heat the stirred solution to boiling. Let the solution cool slightly. The ester will precipitate. Add 0.3 mL of distilled water to hydrolyze the remaining *p*-nitrobenzoyl chloride. Cool in an ice bath. Filter the product through a Hirsch funnel. Wash the precipitate with several small portions of 10% sodium carbonate to dissolve the impurity of *p*-nitrobenzoic acid. Then wash the precipitate with a small portion of distilled water. Let the precipitate air dry. The solid product can be recrystallized from an ethanol-water solvent pair. Dissolve the product in about 0.5 mL of ethanol; add water dropwise until a faint cloudiness is evident. Cool in an ice bath. Filter using a Hirsch funnel. Alternately, the ester can be recrystallized using a Craig tube. Let the crystals air dry and take a melting point.

D-2: α-Naphthylurethane

1. Always wear eye protection and gloves, and work under a hood or in a well-ventilated area.
2. α-Naphthyl isocyanate is a lachrymator. Do not breathe the vapors. After isolating the derivative, chill the filtrate (which may contain unreacted isocyanate) over ice and add 10% sodium bicarbonate while stirring.

All equipment used must be dry since water will cause hydrolysis of the isocyanate reagent. To a clean, dry conical vial containing a spin vane, add 100 mg of the unknown and 3 drops of a α-naphthyl isocyanate. Heat the solution to 60–70°C on a sand bath or heat block for 5 minutes. Let it cool to room temperature, then cool the solution in an ice bath to effect complete crystallization. Filter the urethane using a Hirsch funnel. Treat the filtrate as described in Safety First! Recrystallize from petroleum ether. Dissolve the product in warm petroleum ether and do a hot gravity filtration to remove any residual dinaphthylurea. Cool and collect crystals by vacuum filtration using a Hirsch funnel.

Caution: lachrymator!

D-3: Phenylurethane

1. Always wear eye protection and gloves, and work under a hood or in a well-ventilated area.
2. Phenyl isocyanate is a lachrymator. Do not breathe the vapors. After isolating the derivative, chill the filtrate (which may contain unreacted isocyanate) over ice and add 10% sodium bicarbonate while stirring.

All equipment used must be dry since water will cause hydrolysis of the isocyanate reagent. To a clean, dry conical vial containing a spin vane, add 100 mg of the unknown and 100 mg of phenyl isocyanate. Heat the solution to 60–70°C on a sand bath or heat block for 5 minutes. Let cool to room temperature, then cool the solution in an ice bath to effect complete crystallization. Filter the urethane using a Hirsch funnel. Treat the filtrate as described in Safety First! Recrystallize from petroleum ether. Dissolve the product in warm petroleum ether and do a hot gravity filtration to remove any residual dinaphthylurea. Cool and collect crystals by vacuum filtration using a Hirsch funnel.

Caution: lachrymator!

Derivatives of Aldehydes and Ketones

D-4: 2,4-Dinitrophenylhydrazone

Safety First!

Always wear eye protection in the laboratory.

1. Always wear eye protection and gloves, and work under a hood or in a well-ventilated area.
2. Many derivatives of hydrazine are suspected carcinogens. Always be careful when handling solutions of 2,4-dinitrophenylhydrazine.
3. This reagent is made with sulfuric acid, which is a highly toxic oxidizer. Always wear goggles and gloves. If the reagent comes in contact with your skin, wash the affected area with copious amounts of cool water. Bench spills can be neutralized with sodium bicarbonate.

To a clean vial or small Erlenmeyer flask, add 100 mg of the unknown. Dissolve the unknown in a minimum amount of 95% ethanol. Add 10 drops of 2,4-dinitrophenylhydrazine reagent. Swirl to mix. A precipitate should form immediately. If it does not, let the solution stand for 15 minutes, then put it in an ice bath. Filter the precipitate using a Hirsch funnel. Let air dry. Recrystallize using 95% ethanol.

D-5: Semicarbazone

Safety First!

Always wear eye protection in the laboratory.

1. Always wear eye protection in the laboratory.
2. Semicarbazide hydrochloride is toxic. Wear gloves when handling this reagent.

To a 10-mL Erlenmeyer flask, add 100 mg of the unknown, 100 mg of semicarbazide hydrochloride, 150 mg of sodium acetate, 1 mL of water, and 1 mL of 95% ethanol and swirl. Heat the solution on a sand bath for 5–10 minutes. Cool in an ice bath. Filter the precipitate using a Hirsch funnel. Wash with several small portions of cold water. Recrystallize from methanol. Alternately, recrystallize in a Craig tube using an ethanol-water solvent pair.

D-6: Oxime

Safety First!

Always wear eye protection in the laboratory.

1. Always wear eye protection in the laboratory.
2. Hydroxylamine hydrochloride is corrosive and toxic. Wear gloves when handling this reagent.

In a 3-mL conical vial fitted with a water-jacketed condenser and spin vane, mix 100 mg of the unknown, 100 mg of hydroxylamine hydrochloride, 200 mg of sodium acetate trihydrate, 1 mL of ethanol, and 10 drops of water. Reflux with stirring for 30 minutes on a warm sand bath. Let cool to room temperature. Put the vial in ice until crystallization is complete. Filter using a Hirsch funnel. Wash with several small portions of cold aqueous ethanol. Recrystallize using an ethanol-water solvent pair.

D-7: p-Nitrophenylhydrazone or Phenylhydrazone

1. Always wear eye protection and gloves, and work under a hood or in a well-ventilated area.
2. Phenylhydrazine and *p*-nitrophenylhydrazine are toxic and suspected carcinogens. Be careful when handling these reagents.

p-Nitrophenylhydrazone. Combine 100 mg of *p*-nitrophenylhydrazine, 100 mg of the unknown ketone or aldehyde, and 2–3 mL of ethanol in a 10-mL Erlenmeyer flask. Heat to boiling and add a drop of glacial acetic acid. Add up to 1 mL more of ethanol, if necessary, to obtain a clear solution, but do not add more. Cool. Collect the crystals in a Hirsch funnel. Recrystallize from ethanol. If no product precipitates on cooling, heat to boiling. Add hot water until the solution becomes cloudy. Cool and collect the crystals.

Phenylhydrazone. Combine 100 mg of phenylhydrazine, 100 mg of the unknown ketone or aldehyde, and 2–3 mL of ethanol in a 10-mL Erlenmeyer flask. Heat to boiling and add a drop of glacial acetic acid. Cool. Collect the crystals in a Hirsch funnel. Recrystallize from ethanol. If no product precipitates on cooling, heat to boiling. Add hot water until the solution becomes cloudy. Cool and collect the crystals.

Hood!

Derivatives of Amines

D-8: Acetamide

1. Always wear eye protection and gloves, and work under a hood or in a well-ventilated area.
2. HCl is a strong acid. If your skin comes in contact with the acid, wash with copious amounts of cold water. If acid spills on the bench, it should be neutralized immediately with sodium bicarbonate.
3. Amines are malodorous. Do not breathe the fumes.

Place 100 mg of the unknown amine in a conical vial or small Erlenmeyer flask. Add 2 mL of water and 3 drops of concentrated HCl. Swirl to mix. Add 150 mg of sodium acetate and swirl to mix. Add 4 drops of acetic anhydride all at once and swirl vigorously to mix. Let stand at room temperature for 5–10 minutes, then place in an ice bath to effect complete crystallization. Collect the white crystals by vacuum filtration using a Hirsch funnel. Wash the precipitate with several small portions of cold water. The amides may be recrystallized from 95% ethanol.

Hood!

D-9: Benzamide

1. Always wear eye protection and gloves, and work under a hood or in a well-ventilated area.
2. Benzoyl chloride is a lachrymator.
3. Amines are malodorous. Do not breathe the vapors.

Hood!

Place 100 mg of the unknown in a small vial. Add 1 mL of 10% NaOH and swirl vigorously to mix. Add 8 drops of benzoyl chloride, swirling after each drop. Let stand for 10 minutes. Adjust the pH of the solution to pH = 8 by cautious addition of 3 M HCl. Collect the precipitate by vacuum filtration using a Hirsch funnel. Wash the amide with several small portions of cold water. Air dry the crystals, then recrystallize using an ethanol-water solvent pair.

D-10: Picrate

1. Always wear eye protection and gloves, and work under a hood or in a well-ventilated area.
2. Picric acid is very unstable and may explode if heated too rapidly. Never store large quantities of picric acid. **Do not work with solid picric acid.**
3. Amines are malodorous. Do not breathe the vapors.

Caution! Picric acid can be explosive if it is heated too rapidly.

Place 30 mg of the unknown and 1 mL of 95% ethanol in a small, conical vial fitted with a spin vane and a water-jacketed condenser. Stir vigorously to mix. Add 1 mL of a saturated solution of picric acid in 95% ethanol. Carefully heat to reflux with stirring for 1 minute, then let cool slowly to room temperature. Collect the yellow crystals by vacuum filtration using a Hirsch funnel. Wash with a small portion of cold ethanol. Air dry the crystals. **Do not recrystallize picrate derivatives.** Exercise care when taking the melting point of a picrate, since picrates can be explosive. Avoid friction when preparing the sample for the capillary tube.

D-11: p-Toluenesulfonamide

1. Always wear eye protection and gloves, and work under a hood or in a well-ventilated area.
2. *p*-Toluenesulfonyl chloride is a lachrymator.

Add 2 mL of 10% NaOH, 100 mg of the amine, and 8–10 drops of *p*-toluenesulfonyl chloride to a 5-mL conical vial. Cap the vial and shake over a 10-minute period. Collect crystals in a Hirsch funnel. Wash with 3 M HCl and then with water. Recrystallize using an ethanol-water solvent pair.

D-12: Benzenesulfonamide

1. Always wear eye protection and gloves, and work under a hood or in a well-ventilated area.
2. Benzenesulfonyl chloride is a lachrymator.

Add 2 mL of 10% NaOH, 100 mg of the amine, and 8–10 drops of benzenesulfonyl chloride to a 5-mL conical vial. Cap the vial and shake over a 10-minute period. Collect crystals in a Hirsch funnel. Wash with 3 M HCl and then with water. Recrystallize using an ethanol-water solvent pair.

Derivatives of Carboxylic Acids

D-13: Preparation of Acid Chloride

1. Always wear eye protection and gloves, and work under a hood or in a well-ventilated area.
2. Thionyl chloride is a severe irritant. Do not breathe the fumes.
3. In the experiment, sulfur dioxide and hydrogen chloride are evolved. Do not breathe the vapors.

All amide derivatives of carboxylic acids are prepared from the acid chloride. In a 5-mL conical vial fitted with a water-jacketed condenser and a spin vane, add 100 mg of the unknown and 0.5 mL thionyl chloride. Stir to mix. Heat on a sand bath to reflux with stirring for 15–30 minutes. Cool to room temperature and then chill in an ice bath. The acid chloride is not isolated, but should be used directly with a solution of amine to form the amide derivatives in D-14, D-15, or D-16.

D-14: Amide

To the cooled solution of acid chloride (prepared in D-13 above), add 28 drops of ice-cold concentrated ammonium hydroxide dropwise. Swirl to mix thoroughly. Let stand until crystallization is complete. Vacuum filter the white crystals using a Hirsch funnel. Recrystallize the crude product from water or use an ethanol-water solvent pair.

D-15: Anilide

To a 5-mL conical vial equipped with a spin vane, add 7 drops of aniline and 0.5 mL of diethyl ether. Stir to mix. Cool in an ice bath. Add dropwise by pipet, with stirring, a cooled solution of the acid chloride prepared in D-13 above. Remove from the ice bath and let the solution stir for 10 minutes. Cautiously add 1 mL of water dropwise. Cap the vial and shake, venting if necessary. Remove the bottom aqueous layer. Wash the diethyl ether layer consecutively with 1-mL portions of 5% HCl, 5% NaOH, and water, removing the lower aqueous layer after each washing. Dry the organic layer with anhydrous sodium sulfate. Transfer the liquid by filter pipet to a clean, dry conical vial or Erlenmeyer flask. Evaporate the solvent on a warm sand bath or heat block. Recrystallize the crude product with an ethanol-water solvent pair.

Caution: Splattering may occur if water is added too fast.

D-16: p-Toluidide

Caution: Splattering may occur if water is added too fast.

To a 5-mL conical vial equipped with a spin vane, add 80 mg of *p*-toluidine and 0.5 mL of diethyl ether. Stir to mix. Cool in an ice bath. Add dropwise by pipet, with stirring, a cooled solution of the acid chloride prepared in D-13 above. Remove from the ice bath and let the solution stir for 10 minutes. Cautiously add 1 mL of water dropwise. Cap the vial and shake, venting if necessary. Remove the bottom aqueous layer. Wash the diethyl ether layer consecutively with 1-mL portions of 5% HCl, 5% NaOH, and water, removing the lower aqueous layer after each washing. Dry the organic layer with anhydrous sodium sulfate. Transfer the liquid by filter pipet to a clean, dry conical vial or Erlenmeyer flask. Evaporate the solvent on a warm sand bath. Recrystallize the crude product with an ethanol-water solvent pair.

D-17: p-Nitrobenzyl Ester and p-Bromophenacyl Ester

Safety First!

Always wear eye protection in the laboratory.

1. Always wear eye protection in the laboratory.
2. *p*-Nitrobenzyl chloride and *p*-bromophenacyl bromide are both lachrymators.
3. Do not breathe the fumes. Do all lab work under the hood in a well-ventilated room.

p-**Nitrobenzyl Ester.**   Dissolve 100 mg of the acid in 2 mL of 95% ethanol in a 5-mL conical vial fitted with a spin vane and reflux condenser. Then add two drops of 5% HCl. Add 200 mg of *p*-nitrobenzyl chloride. Stir to mix. Heat under reflux, with stirring for 30 minutes. Cool. Collect the crystals. Wash the crystals with 5% NaHCO$_3$ and water. Recrystallize from ethanol or an ethanol-water solvent pair.

p-**Bromophenacyl Ester.**   Dissolve 100 mg of the acid in 2 mL of 95% ethanol in a 5-mL conical vial fitted with a spin vane and reflux condenser. Add 2 drops of 5% HCl. Add 200 mg of *p*-bromophenacyl bromide. Stir to mix. Heat under reflux with stirring for 30 minutes. Cool. Collect the crystals. Wash the crystals with 5% NaHCO$_3$ and water. Recrystallize from ethanol or an ethanol-water solvent pair.

Derivatives of Ketones

These are the same as derivatives of aldehydes. See Sections D-4 through D-6.

Derivatives of Phenols

D-18: α-Naphthylurethane

Safety First!

Always wear eye protection in the laboratory.

1. Always wear eye protection and gloves, and work under a hood or in a well-ventilated area.
2. α-Naphthyl isocyanate is a lachrymator. Do not breathe the vapors.
3. Pyridine has a very strong unpleasant odor. Pyridine has been shown to cause sterility in males. Do not breathe the vapors.

4. Phenyl isocyanate is a lachrymator. Do not breathe the vapors. Chill the filtrate (which may contain unreacted isocyanate) over ice and add 10% sodium bicarbonate while stirring.

All equipment used must be dry since water will cause hydrolysis of the isocyanate reagent. To a dry 3-mL conical vial containing a spin vane, add 100 mg of the unknown phenol, 3 drops of α-naphthyl isocyanate, and 1 drop of reagent-grade pyridine. Heat the solution with stirring to 60–70°C on a sand bath or heat block for 5 minutes. Let cool to room temperature, then cool the solution in an ice bath to effect complete crystallization. Filter the urethane using a Hirsch funnel. Treat the filtrate as described above. Recrystallize from petroleum ether. Dissolve the product in warm petroleum ether and perform a hot gravity filtration to remove any residual dinaphthylurea. Cool and collect the crystals by vacuum filtration using a Hirsch funnel.

Hood!

D-19: Brominated Phenols

Safety First!

Always wear eye protection in the laboratory.

1. Always wear eye protection and gloves, and work under a hood or in a well-ventilated area.
2. Bromine is very toxic. Contact with skin, whether by inhalation or spilling, can cause severe burns. Always wear gloves when handling this reagent. This reaction must be done under the hood with good ventilation.

In a 10-mL Erlenmeyer flask dissolve 50 mg of the unknown phenol in 10 drops of methanol and 10 drops of water. Swirl to dissolve. Add 15 drops of the brominating solution. (This is a mixture of bromine in acetic acid that has already been prepared by the laboratory instructor.) Swirl well. Continue adding the brominating solution dropwise until the reddish color of the bromine persists. Add 2 mL of water, stopper the flask and swirl vigorously. Add 10% $NaHSO_3$ dropwise until the reddish color is discharged. Vacuum filter the precipitate using a Hirsch funnel. Wash the product with cold water. Recrystallize using an ethanol-water solvent pair.

D-20: Aryloxyacetic Acid

Safety First!

Always wear eye protection in the laboratory.

1. Always wear eye protection and gloves, and work under a hood or in a well-ventilated area.
2. Chloroacetic acid is corrosive and toxic. Wear gloves and goggles at all times. Work under the hood and do not breathe the vapors.
3. Sodium hydroxide is caustic. If the reagent comes in contact with skin, wash with copious amounts of cold water.

In a dry, 5-mL conical vial fitted with a spin vane and a water-jacketed condenser, combine 100 mg of the unknown, 100 mg of chloroacetic acid, and 12 drops of 8 M sodium hydroxide. Heat the stirred mixture on a hot sand bath or heat block for 30 minutes. Let the solution cool to room temperature, then cautiously add 12 drops of distilled water. Add 1 drop of Congo red indicator to the solution, then add 6 M HCl dropwise until the solution turns blue (pH = 3). Extract with 4 mL of diethyl ether. Remove the lower aqueous layer. To the diethyl ether solution, add 2 mL of 0.5 M Na_2CO_3. Shake and let the layers separate. The product is now in the aqueous layer as the carboxylate salt. Remove this lower aqueous layer to a clean Erlenmeyer flask and add 6 M HCl dropwise. The product will precipitate. Filter using a Hirsch funnel. Recrystallize the crude product from water.

Part E: Qualitative Organic Analysis of Unknowns

Prelab Assignment

1. Devise a flow scheme to determine the functional group(s) present in the unknown. (Alternately, use the flow scheme developed in Experiment 28.1.) Outline the relevant solubility tests (if any) and the chemical test that will be carried. Note that other heteroatoms, such as nitrogen, sulfur, or halogen, may be present and the compound may contain one or more double or triple bonds.
2. Reference each chemical test in the flow scheme with the procedure number and page number in Experiment 28.2 where that procedure is found. Directions for the solubility tests are given in Solubility Tests B1–B5. For chemical test procedures for aldehydes and ketones, refer to Chemical Tests C-2, C-3, C-6, and C-11. For alcohols and phenols, refer to Chemical Tests C-2, C-4, and C-7. For amines and carboxylic acids, refer to Chemical Tests C-5 and C-8.

Experimental Procedure

General Directions.   Have the laboratory instructor approve your flow scheme before starting the experiment. Obtain 1–2 mL of a liquid unknown or 200 mg of a solid unknown from the laboratory instructor. Record the number of the unknown. Purify the unknown by distillation or recrystallization. Record all relevant physical properties, making corrections for atmospheric pressure or temperature as necessary. Perform all necessary chemical tests, recording the results in your lab notebook. Tables of functional goups are given in Appendix A. Prepare as many solid derivatives as necessary to identify the unknown. Determine which derivative or derivatives will be the most useful to identify the unknown. Prepare the derivative, recrystallizing as many times as necessary to obtain a pure product with a sharp melting point. Be sure to record all of the results. If directed to do so by the instructor, record IR, ^{1}H NMR, ^{13}C NMR spectra, and/or the mass spectrum of the unknown.

Part E.1: Qualitative Analysis of Aldehydes and Ketones

Obtain an unknown compound, which will be either an aldehyde or ketone. Chemical test procedures for aldehydes and ketones are given in C-2, C-3, C-6, and C-11. The relevant tables are given in Appendix A: Table 3 for aldehydes and Table 4 for ketones. Procedures for preparing the derivatives of aldehydes and ketones are given in D-4 and D-7.

Part E.2: Qualitative Analysis of Alcohols and Phenols

Obtain an unknown compound, which will be either a primary, secondary, or tertiary alcohol or a phenol. Procedures for chemical tests for alcohols and phenols are given in Experiment 28.2 C-2, C-4, and C-7. Tables of possible unknown alcohols and phenols are given in Table 1 and Table 2, respectively, in Appendix A. The references for preparing the derivatives are given in D-1 through D-3, and D-18, D-19, and D-20. Refer to these sections for the detailed instructions of the experimental procedures.

Part E.3: Qualitative Analysis of Amines and Carboxylic Acids

Obtain an unknown compound, which will be either a primary, secondary, or tertiary amine or a carboxylic acid. Procedures for chemical tests for carboxylic acids and amines are given in C-5 and C-8. Perform a neutralization equivalent test on the carboxylic acid or amine by titrating a sample of the unknown. Assume that the carboxylic acid is monoprotic and that the amine is monobasic. A detailed procedure can be found in C-8. A table of possible unknown carboxylic acids is given in Table 5; tables of possible unknown primary, secondary, and tertiary amines are given in Table 6 and Table 7 in Appendix A. The references for preparing the derivatives are given in D-8 through D-11 and D-14 through D-16. Refer to these sections for the detailed instructions of the experimental procedures.

Part E.4: Qualitative Analysis of a General Unknown

Obtain an unknown compound, which will be an alcohol, aldehyde, amine, carboxylic acid, ketone, or phenol. The alcohol and amine can be primary, secondary, or tertiary. Procedures for chemical tests are given in C-1 through C-11. Prepare as many solid derivatives as necessary to identify the unknown. The tables of possible unknowns are given in Appendix A, Tables 1–7. References for preparing the derivatives are given in D-1 through D-16. Refer to these sections for the detailed instructions of the experimental procedures.

Part E.5: Qualitative Analysis and Spectroscopic Analysis of a General Unknown

Obtain an unknown compound, which will be an alcohol, aldehyde, amine, carboxylic acid, ketone, or phenol, or other functional group as directed by the instructor. Procedures for chemical tests are given in C-1 through C-11. Record the IR spectrum, the ^{1}H NMR spectrum, the ^{13}C NMR spectrum, and/or the mass spectrum of the unknown compound. Verify the result of the chemical tests in deciding what functional groups are present or absent in the unknown. The unknown may be any compound in Tables 1–7 (Appendix A) or any compound in the CRC Handbook of Organic Compounds. Prepare as many solid derivatives as necessary to identify the unknown. References for preparing the derivatives are given in D-1 through D-16. Refer to these sections for the detailed instructions of the experimental procedures.

Results and Conclusions for Part E

1. Tabulate the physical properties (corrected for atmospheric pressure and temperature) of the unknown.
2. Tabulate the results of any solubility tests performed and explain how the results were important in determining what type of functional group was present.

3. Tabulate the results of all chemical tests performed and explain how the results were important in determining what type of functional group was present or absent in the unknown.
4. Based on the physical properties of the unknown, the results of the solubility tests, the results of the chemical tests, the melting point of the derivative, or the spectroscopic analysis, determine the identity of the unknown.
5. Write balanced chemical reactions for all positive chemical tests and all derivative preparations.
6. If an IR spectrum was recorded, summarized the results of the spectroscopic analysis, describing exactly how the IR spectrum was useful in verifying the structure of the unknown. List the specific frequency (in cm^{-1}) and intensity in the IR spectrum that correlate to the structure of the known.
7. If a ^1H NMR spectrum was recorded, write the structure of the compound and correlate the structure to the ^1H NMR spectrum. Give the chemical shift, multiplicity, and integration for each proton in the molecule.
8. If a ^{13}C NMR spectrum was recorded, correlate each carbon in the unknown to a specific signal. Give the chemical shift for every carbon in the molecule.
9. If the mass spectrum was recorded, interpret the base peak and the molecular ion peak and isotopic peaks (if present).
10. Summarize the results of the spectroscopic analysis (if performed). Alternately, describe exactly how the IR spectrum and the NMR spectrum might have been useful in determining the identity of the unknown.
11. Write a summary paragraph that clearly explains how the identity of the unknown was determined.

Cleanup & Disposal

Wash the solutions from the solubility tests and the chemical tests down the drain with water. Place solutions from derivitization reactions in labeled containers.

Critical Thinking Questions

1. A student claims that a given unknown is a carboxylic acid because the sample disolves in water, 10% NaOH, and 10% $NaNCO_3$. Critique this conclusion.
2. Explain how solubility tests could be used to differentiate between:
   a. a phenol and a carboxylic acid.
   b. a phenol that contains electron-withdrawing groups and a phenol that contains electron-donating groups.
   c. a carboxylic acid and an amine.
   d. 1-bromooctane and dibutyl ether.
   e. 2-pentyne and pentane.
   f. 2-nitropentane and 2-aminopentane.
3. What chemical tests could be used to differentiate between:
   a. cyclohexanone, cyclohexanol, cyclohexylamine, and cyclohexanecarboxylic acid.
   b. 4-methyl-1-pentanol, 4-methyl-2-pentanol, and 2-methyl-2-pentanol.
4. In the Tollens test, what substance is being oxidized and what substance is being reduced?

5. An unknown compound (A) forms a yellow DNP derivative when reacted with 2,4-dinitrophenylhydrazine. The unknown forms a silver mirror with the Tollens reagent and gives a positive iodoform test. From this information alone, the unknown can be identified. What is the structure of the unknown?

6. An unknown compound (D) is soluble only in cold, concentrated sulfuric acid. The compound reacts with chromic acid to yield a blue-green precipitate. The 2,4-dinitrophenylhydrazone is bright red. Reaction with sodium hydroxide, ammonium hydroxide, and silver nitrate produces a silver mirror on the side of the test tube. What functional group(s) is(are) present?

7. Suggest one or more spectral methods to distinguish between:
   a. 2,4-dichlorobenzoic acid and *p*-chlorobenzoic acid.
   b. (Z)- and (E)-3-chloropropenoic acid.

c. 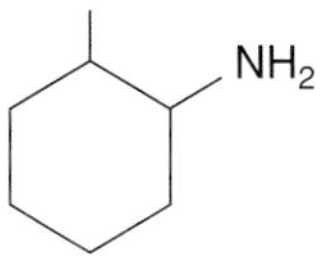 and

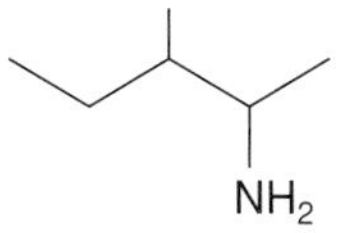

d. 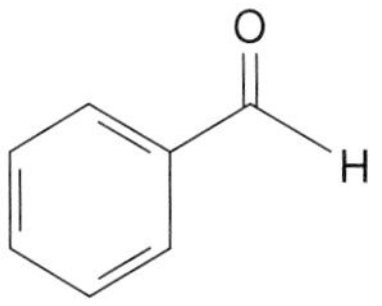 and

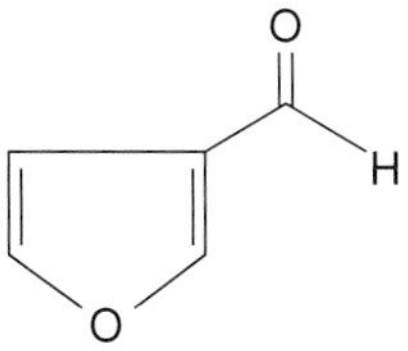

References

Furniss, B. S., Hannaford, A. J., Smith, P. W. G., Tatchell, A. R. *Vogel's Textbook of Practical Organic Chemistry*, 5th Edition. Essex: Longman Scientific & Technical, 1989.

Rappoport, Z. *CRC Handbook of Tables for Organic Compound Identification.* Boca Raton, FL: CRC Press, 1981.

Shriner, R. L., Hermann, C. K. F., Morrill, T. C., Curtin, D. Y., Fuson, R. C. *The Systematic Identification of Organic Compounds.* New York: John Wiley & Sons, Inc., 1997.

Chapter 29

Projects

One of the characteristics of organic synthesis is that many synthetic targets often require more than one step starting with a readily available starting material. Many syntheses require a number of steps and isolation of several intermediate products along the way. Experiments 29.1 and 29.2 are designed to afford you an opportunity to experience synthesis of a target in multiple steps using procedures and reaction times that permit dividing the laboratory work into reasonable periods using procedures that do not require special equipment or reaction conditions.

1-Bromo-3-chloro-5-iodobenzene is synthesized in Experiment 29.1 in six steps. Protective modification of the amino group of aniline as an acetamido group gives acetanilide. Acetanilide can be monobrominated and then monochlorinated. Removal of the protecting group followed by iodination and deamination gives a unique product that contains a bromo group, a chloro group, and an iodo group in the same molecule and all meta to one another.

Synthesis of one of several possible sulfanilamide derivatives is described in Experiment 29.2. The syntheses require four steps. A bacterial testing procedure for the sulfanilamide makes up an optional section at the end of the experiment.

Experiment 29.3 is a spectral project designed to distinguish between possible structures that are very similar. When traditional methods are not conclusive, special NMR methods of analysis can often offer definitive information to make structural assignments. These methods include 2D-NMR techniques that can be done using FT-NMR instrumentation.

Library searching is a key element in the preparation of most any compound. It is important to know if the target molecule has been synthesized previously by other workers. If it has been made, what were the conditions of the synthesis and what reagents were employed? Often, a number of different reagents and conditions can be found. After choosing what appears to be the most viable and convenient method to make the compound on the scale that is required, the researcher can try out the various steps required as reported to see what yield of product can be achieved. Often, the researcher will want to pick different procedures from various reported works because the desired target may not have been made previously or it may have been done using different reagents requiring more steps than what might be necessary using newer methodologies.

Much time can be saved by being familiar with past work on a topic or a synthetic target. The literature is built upon past work and is a valuable resource. Synthetic procedures done decades ago are still valid and are often cited by recent authors. For this reason, searching the older literature is often a requirement. Being able to translate from German, French or another language is frequently required. Experiment 29.4 is a library exercise on use of library databases to locate syntheses of compounds and to research topics in chemistry. Both hard-copy and electronic databases may be searched, but electronic database searching is rapidly becoming the method of choice.

Experiment 29.1: Multistep Synthesis of 1-Bromo-3-chloro-5-iodobenzene from Aniline

In a multistep synthesis, the product of one step of a reaction becomes the starting material for another reaction. In this experiment, you will:

- acetylate aniline to moderate its reactivity.
- convert acetanilide into 4-bromoacetanilide by electrophilic aromatic bromination.
- prepare 4-bromo-2-chloroacetanilide by electrophilic aromatic chlorination.
- hydrolyze an amide to an amine.
- prepare 4-bromo-2-chloro-6-iodoaniline by electrophilic aromatic iodination.
- diazotize an aromatic amine.
- analyze products by IR and NMR spectroscopy and mass spectrometry.

Techniques

Technique F	Recrystallization and vacuum filtration
Technique G	Reflux and steam distillation
Technique M	Infrared spectroscopy
Technique N	Nuclear magnetic resonance spectroscopy
Technique P	Mass spectrometry

Background

Although practicing good laboratory technique is always important, it is critical in a multistep synthesis. This experiment is actually a combination of six reactions, and in each reaction you will use the product of the previous step. The yield and the purity of the product of an earlier step will have crucial implications for subsequent reactions. For this reason, the amount of starting material used in the first step is large, so that a significant amount of product can be obtained at the end of the six-step sequence. The amounts of reagents required for each step are indicated in the experimental procedures. However, the actual amounts used will depend upon the amount of product isolated in the previous step. You will have to calculate the quantities of reagents based upon your yield in the previous step and scale the procedure accordingly.

In this experiment, aniline will be converted through a series of reactions into 1-bromo-3-chloro-5-iodobenzene, a compound that has three different *o,p*-directors meta to one another (Ault, 1976).

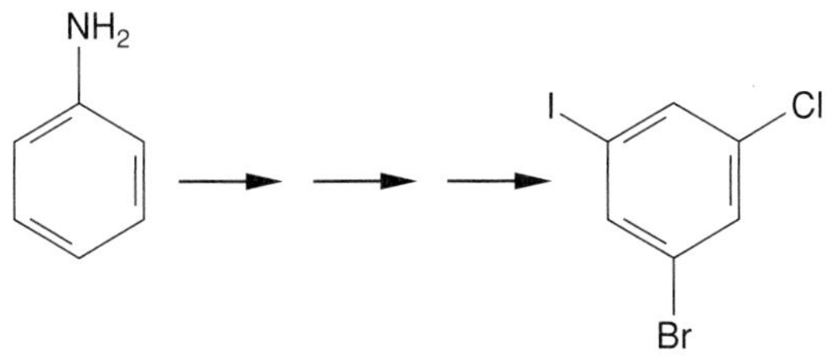

aniline   1-bromo-3-chloro-5-iodobenzene

The overall reaction scheme is shown here.

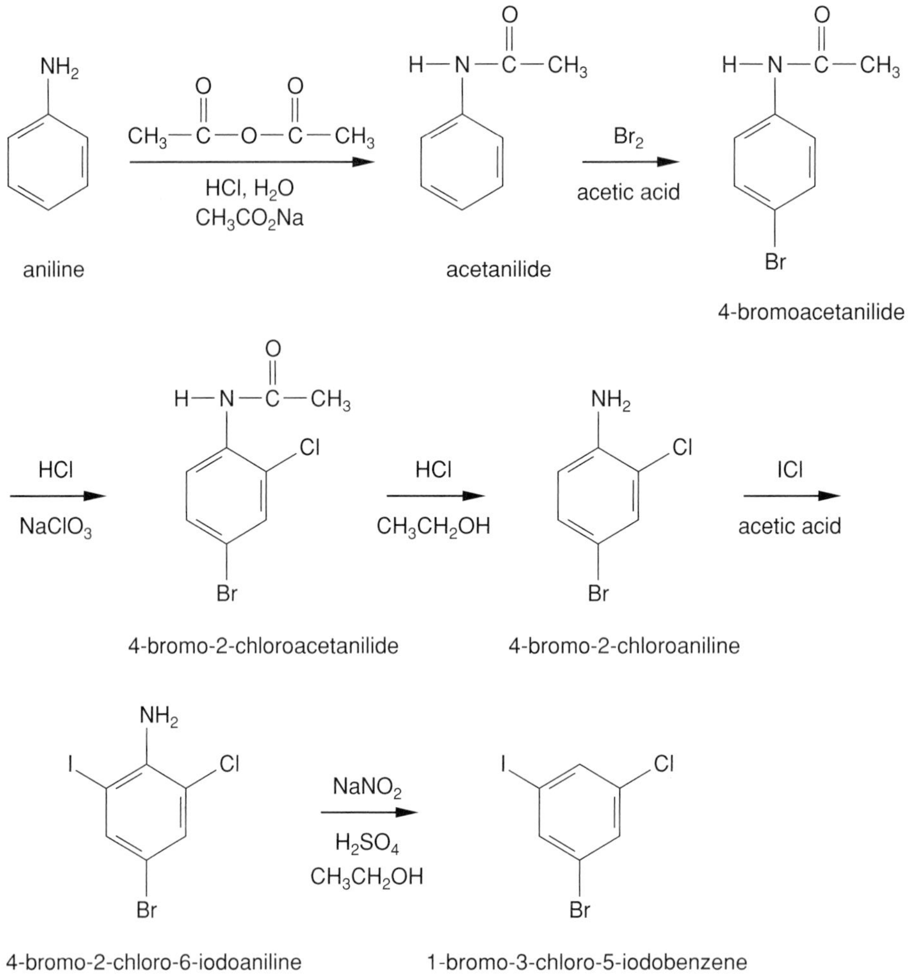

aniline → acetanilide → 4-bromoacetanilide

4-bromo-2-chloroacetanilide   4-bromo-2-chloroaniline

4-bromo-2-chloro-6-iodoaniline   1-bromo-3-chloro-5-iodobenzene

Step 1: Synthesis of Acetanilide.   In this first step of the multistep synthesis, aniline (a highly activated aromatic compound) is converted to acetanilide (a moderately activated aromatic compound).

aniline   acetanilide

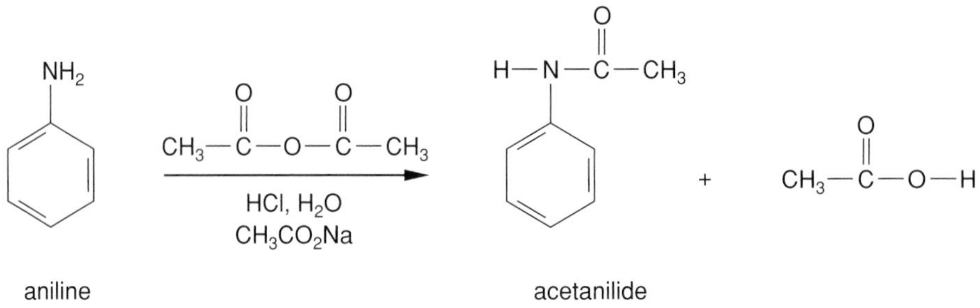

Although it may seem like a waste of time to put on the acetyl group, only to remove it in a later reaction, this is a critical step: aniline is so highly activated toward electrophilic aromatic substitution that it would undergo polybromination to produce 2,4,6-tribromoaniline. Thus the acetyl group acts to moderate the reactivity. The mechanism of the acetylation reaction involves protonation of acetic anhydride, nucleophilic addition of aniline, and elimination of acetic acid. The first step is protonation of the acetyl group in acetic anhydride to produce a resonance-stabilized intermediate.

Step 2: Electrophilic Aromatic Bromination of Acetanilide.   In the next step of the reaction, acetanilide is converted to 4-bromoacetanilide using molecular bromine in acetic acid.

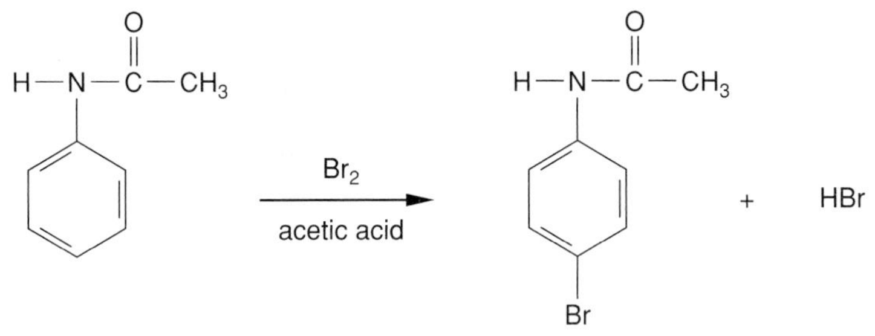

acetanilide   4-bromoacetanilide

Because acetanilide is activated toward electrophilic substitution, a Lewis acid catalyst is not required. The acetamido group is an electron-donating group and a moderate activator; therefore bromination is expected to occur at the ortho and para positions. Due to the size of the acetamido group, the para product is expected to predominate.

Step 3: Electrophilic Aromatic Chlorination of 4-Bromoacetanilide.   In this step of the synthesis, 4-bromoacetanilide reacts with chlorine to produce 4-bromo-2-chloroacetanilide. The ring is still activated, so again no Lewis acid catalyst is required. Because chlorine gas is toxic, the chlorine for this reaction will be generated *in situ* by an oxidation-reduction reaction of HCl and $NaClO_3$:

$$NaClO_3 + 6\ HCl \longrightarrow 3\ Cl_2 + 3\ H_2O + NaCl$$

The chlorine gas then reacts directly with 4-bromoacetanilide.

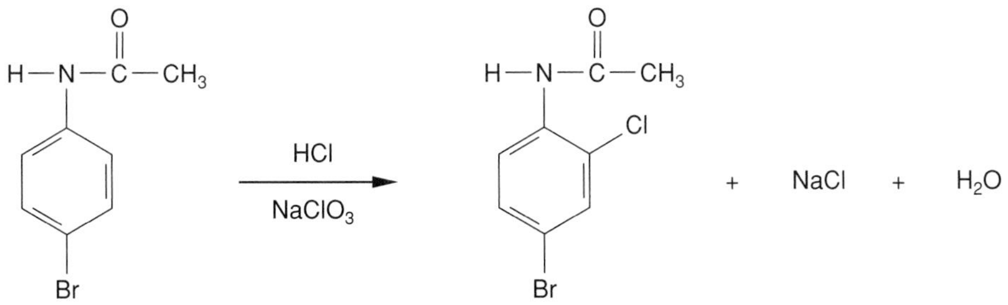

4-bromoacetanilide   4-bromo-2-chloroacetanilide

Step 4: Hydrolyis of Acetamido Group.   It was important in the first step of the synthesis to acetylate aniline to prevent polybromination. Now the aromatic ring is less activated, having two electron-withdrawing halogens on the ring. Also, iodine is not as reactive as bromine or chlorine. To activate the ring to further halogenation, it is necessary to remove the acetamido protecting group by acid hydrolysis. This is accomplished by refluxing with HCl.

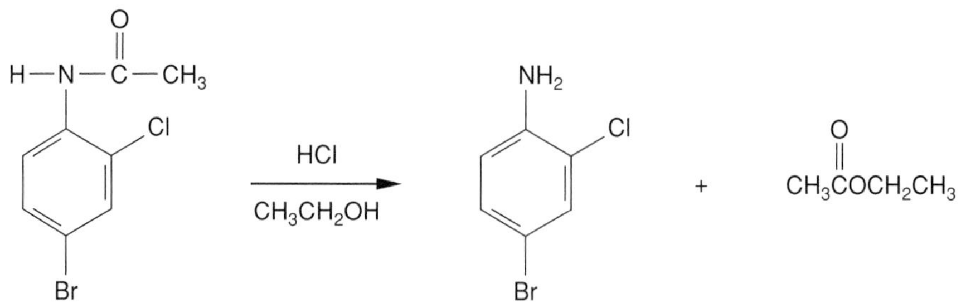

4-bromo-2-chloroacetanilide   4-bromo-2-chloroaniline

The mechanism involves protonation of the carbonyl oxygen by HCl, nucleophilic attack by ethanol, and subsequent loss of ethyl acetate to regenerate the amino group.

Step 5: Electrophilic Aromatic Iodination. In this step, the aromatic ring is iodinated with iodine monochloride. Iodine monochloride (ICl) is a polarized molecule, with the less electronegative iodine ion having a partial positive charge.

$$\overset{\delta+ \ \ \delta-}{I - Cl}$$

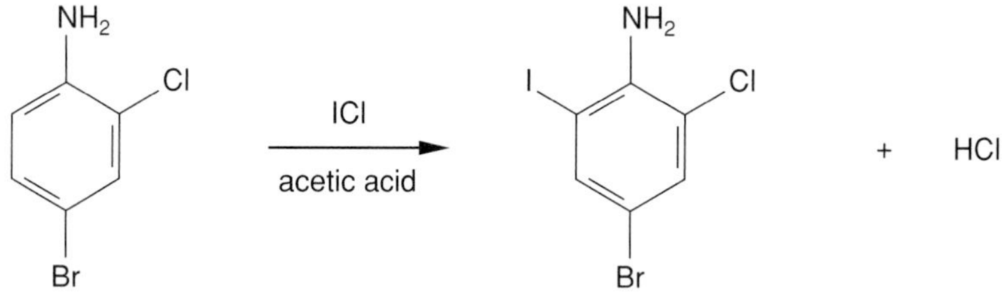

4-bromo-2-chloroaniline   4-bromo-2-chloro-6-iodoaniline

Step 6: Diazotization. In this last step of the reaction, the amino group is diazotized with $NaNO_2$ and sulfuric acid, and the resulting diazonium salt is reduced by a hydride from ethanol. The diazotization reaction is an important method of converting aromatic amines to a wide variety of functional groups. In this reaction, diazotization will be used to replace the amino group with a hydrogen. The crude product can be isolated from the acidic reaction mixture by steam distillation, extraction, and removal of the solvent.

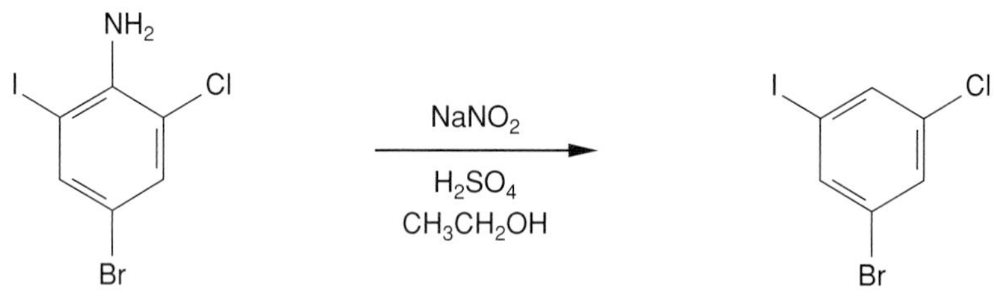

4-bromo-2-chloro-6-iodoaniline   1-bromo-3-chloro-5-iodobenzene

The reaction mechanism involves protonation of nitrous acid and subsequent loss of water to form a nitrosonium ion. The electrophilic nitrosonium ion reacts with an aromatic amine to form an N-nitrosoanilinium ion. Proton transfer and tautomerization, followed by protonation and loss of water, gives an aryldiazonium salt. The diazonium group can be replaced with a hydrogen with a reducing agent such as H_3PO_2 or ethanol. In this experiment, ethanol acts as the reducing agent.

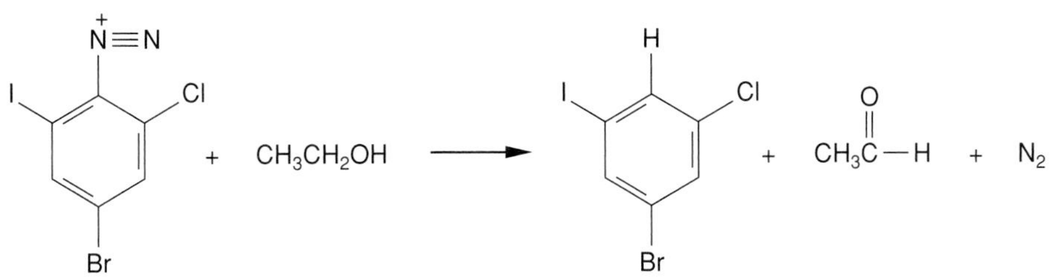

The mechanism involves donation of a hydride from ethanol to reduce the diazonium salt as shown here.

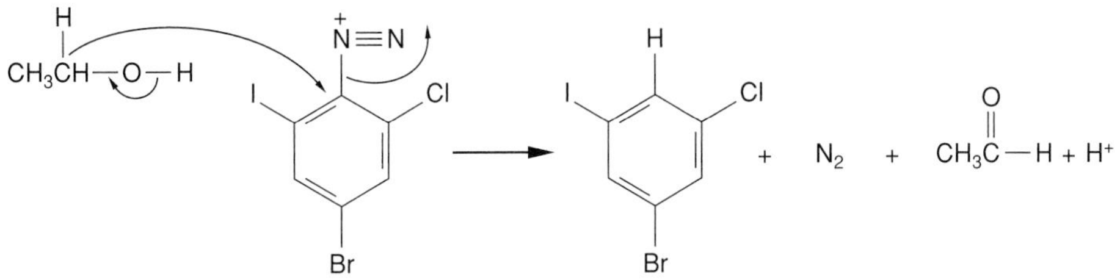

After the first reaction (Step 1), each subsequent reaction uses the product of the previous step. The amounts of all reagents used, including solvents, must be scaled according to the amount of product obtained in the previous step (after saving at least 25 mg of the product for characterization and spectral analysis in each step).

Prelab Assignment

1. Calculate the theoretical mass of 1-bromo-3-chloro-5-iodobenzene starting from 5 g of aniline.
2. Suppose that a 75% yield is obtained in each step of the reaction. What mass of 1-bromo-3-chloro-5-iodobenzene would be obtained?
3. Write a flow scheme for each reaction and workup procedure in the synthesis.
4. Write detailed reaction mechanisms for the following reactions:
   a. reaction of aniline with acetic anhydride.
   b. reaction of acetanilide with bromine.
   c. reaction of 4-bromoacetanilide with chlorine.
   d. hydrolysis of 4-bromo-2-chloroacetanilide with HCl and ethanol.
   e. reaction of 4-bromo-2-chloroaniline with ICl.
   f. preparation of the diazonium salt.

Experimental Procedure

Part A: Synthesis of Acetanilide from Aniline

Safety First!

Always wear eye protection in the laboratory.

1. Wear eye protection at all times in the laboratory.
2. HCl is corrosive and toxic. Use gloves when handling this reagent.
3. Aniline is a highly toxic irritant. Acetic anhydride is a corrosive lachrymator. Avoid skin contact and inhalation of the vapors.

Prepare a solution of sodium acetate by dissolving 3.5 g of anhydrous sodium acetate in 30 mL of water. Set aside for later use.

In a 250-mL Erlenmeyer flask containing a stir bar, dissolve 5.0 g of aniline in 135 mL of water and 4.5 mL of concentrated HCl. With stirring, add 6.6 g of acetic anhydride. Then immediately add the sodium acetate solution and mix thoroughly. Remove the stir bar and cool the mixture in an ice bath for 10 minutes. Collect the white precipitate by vacuum filtration, washing the crystals with several small portions of ice-cold water. Allow the product to air dry or put it in a warm drying oven. Weigh the dried product.

Characterization for Part A

Melting Point: Determine the melting point of the dry product. The reported melting point of acetanilide is 114°C.

Infrared Spectroscopy: Prepare a Nujol mull or KBr pellet of the solid and record the IR spectrum. Interpret the spectrum. The IR spectrum of acetanilide is shown in Figure 29.1-1.

NMR Spectroscopy: Dissolve the product in $CDCl_3$ or deuterated DMSO. Record and interpret the ^1H NMR spectrum. The ^1H NMR spectrum of acetanilide is shown in Figure 29.1-2.

Results and Conclusions for Part A

1. Calculate the percent yield of acetanilide.
2. The next step in the procedure calls for the use of 4.0 g of acetanilide. If less than this amount is available, recalculate the amounts of all reagents that should be used in the next procedure.
3. Acetic anhydride is the acylating agent in this reaction, yet the presence of sodium acetate is critical to product formation. Explain why sodium acetate is used in this procedure and explain what problems might be encountered if it were inadvertently omitted from the reaction.
4. Assign all peaks in the NMR spectrum and calculate the coupling constants between the aromatic protons.
5. Assign the important bands in the IR spectrum that verify the product identity.

Figure 29.1-1   IR spectrum of acetanilide (Nujol)

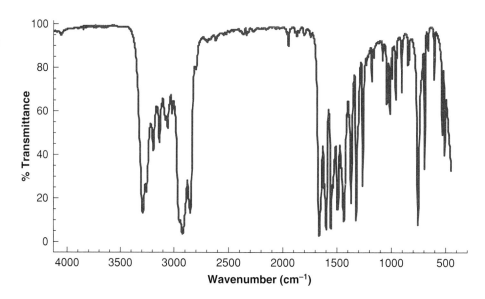

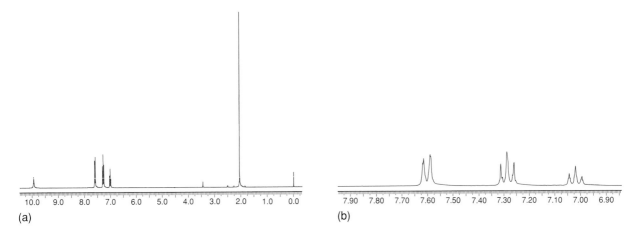

Figure 29.1-2 (a) ^{1}H NMR spectrum of acetanilide (deuterated DMSO). (b) ^{1}H NMR spectrum of acetanilide (deuterated DMSO) (expanded region)

Source: Reprinted with permission of Aldrich Chemical.

Part B: Miniscale Synthesis of 4-Bromoacetanilide from Acetanilide

Safety First!

Always wear eye protection in the laboratory.

1. Wear eye protection at all times in the laboratory.
2. Bromine is a highly toxic oxidizer. It is very corrosive to skin and lungs. Always wear gloves when handling this reagent. Work under the hood. Do not breathe bromine vapors. If inhaled, immediately breathe fresh air. If spilled on your skin, immediately wash off under cold running water and seek medical attention.

Caution!

Hood!

Prepare the brominating solution under the hood. In a 10-mL Erlenmeyer flask fitted with a cork, add 3 mL of acetic acid. The bromine will be added from the buret directly into the flask. Put the flask under the buret. Lower the buret so that the tip of the buret is well inside the lip of the Erlenmeyer flask. Slowly open the stopcock and add bromine until 1.6 mL of bromine have been added. Cork the flask until ready to use.

Dissolve 4.0 g of acetanilide in 15 mL of glacial acetic acid in a 250-mL Erlenmeyer flask fitted with a stir bar. Work under the hood. Start the solution stirring and slowly add the bromine solution using a Pasteur pipet. Rinse the flask with 1 mL of acetic acid and add this rinse to the solution of acetanilide. Stir the reaction mixture for 5 minutes, then slowly add 120 mL of water. Using a pipet, add enough 30% sodium bisulfite solution to discharge the reddish color. Collect the product by suction filtration, washing with several small portions of ice-cold water. The product can be recrystallized from ethanol, if desired, or it can be used directly in the next step. Allow the crystals to dry as much as possible by continuing suction or by pressing out water with a rubber sheet. Weigh the crystals and take a melting point on a dry sample of 4-bromoacetanilide.

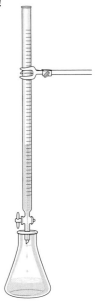

Characterization

Melting Point: Determine the melting point of the dry product. The reported melting point of 4-bromoacetanilide is 168–169°C.

Infrared Spectroscopy: Prepare a Nujol mull or KBr pellet of the solid and record the IR spectrum. Interpret the spectrum.

NMR Spectroscopy: Dissolve the product in CDCl$_3$; if necessary, add 1–2 drops of deuterated DMSO to dissolve. Record and interpret the ^{1}H NMR spectrum.

Results and Conclusions for Part B

1. Calculate the percent yield of 4-bromoacetanilide.
2. The next step in the procedure calls for the use of 3.6 g of 4-bromoacetanilide. If less than this amount is available, recalculate the amounts of all reagents that should be used in the next procedure.
3. A possible side product in this reaction is 2-bromoacetanilide. Where in the reaction workup would this product be separated?
4. Why is glacial acetic acid used as the solvent in this reaction?
5. In the workup procedure, sodium bisulfite (NaHSO$_3$) reacts with molecular bromine to produce Na$_2$SO$_4$ and NaBr. Write the balanced equation.

Part C: Synthesis of 4-Bromo-2-chloroacetanilide from 4-Bromoacetanilide

1. Wear eye protection at all times in the laboratory.
2. Chlorine gas is toxic. Wear gloves. Work under the hood and do not breathe the vapors.
3. HCl is a strong acid. Wear gloves when handling it. If HCl comes in contact with your skin, rinse with cold running water.

Hood!

In a 125-mL Erlenmeyer flask fitted with a stir bar, add 3.6 g of 4-bromoacetanilide to 9 mL of glacial acetic acid and 8 mL of concentrated HCl. Heat the mixture on a steam bath, swirling occasionally, until all of the solid dissolves. Cool the solution to around 0°C in an ice bath. While the solution is cooling, prepare the chlorinating solution: dissolve 0.94 g of NaClO$_3$ in 2.5 mL of water in a 10-mL Erlenmeyer flask fitted with a cork. Add the sodium chlorate solution slowly to the cold stirred solution of 4-bromoacetanilide. Some chlorine gas may be evolved, so work under the hood. As the sodium chlorate solution is added, a yellow precipitate will form. When the addition is complete, stir the solution at room temperature for 1 hour under the hood. Remove the stir bar and collect the product by suction filtration, washing with small quantities of ice-cold water. Dry the crystals by continuing suction. Record the weight. The crude product may be recrystallized from methanol or it may be used directly in the next step.

Characterization for Part C

Melting Point: Determine the melting point of the dry product. The melting point of 4-bromo-2-chloroacetanilide is 154–156°C.

Infrared Spectroscopy: Prepare a Nujol mull or KBr pellet of the solid and record the IR spectrum. Interpret the spectrum. The IR spectrum of 4-bromo-2-chloroacetanilide is shown in Figure 29.1-3.

NMR Spectroscopy: Dissolve the product in CDCl$_3$. Record and interpret the ^{1}H NMR spectrum from 0–10 ppm.

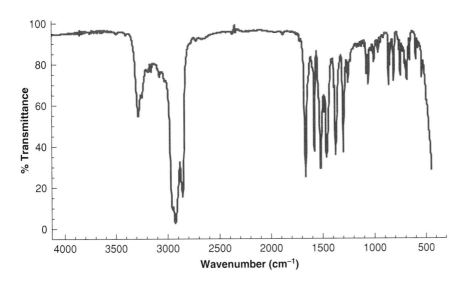

Figure 29.1-3

IR spectrum of 4-bromo-2-chloroacetanilide (Nujol)

Results and Conclusions for Part C

1. Calculate the yield of 4-bromo-2-chloroacetanilide.
2. Chlorine gas is generated *in situ* from HCl and $NaClO_3$ in an oxidation-reduction reaction. What compound is oxidized and what compound is reduced? Write balanced half-reactions for the oxidation and the reduction reactions.
3. The next step in the procedure calls for the use of 4.1 g of 4-bromo-2-chloroacetanilide. If less than this amount is available, recalculate the amounts of all reagents that should be used in the next procedure.
4. Assign the important bands in the IR spectrum that verify the product identity.

Part D: Synthesis of 4-Bromo-2-chloroaniline from 4-Bromo-2-chloroacetanilide

Safety First!

Always wear eye protection in the laboratory.

1. Wear eye protection at all times in the laboratory.
2. HCl is a strong acid. Wear gloves when handling it. If HCl comes in contact with your skin, wash off with lots of cold running water.
3. The 50% NaOH solution is very caustic. Wear gloves at all times. Clean up any spills immediately.

Into a 50-mL round-bottom flask fitted with a reflux condenser and a magnetic stirrer, add 4.1 g of 4-bromo-2-chloroacetanilide, 7 mL of 95% ethanol, and 4.3 mL of concentrated HCl. Heat the solution, with stirring, to a gentle boil, using a heating mantle as the heat source. Reflux for at least 1 hour. The solid will dissolve into a clear solution as it is heated, and then a white precipitate will form. Heat 26 mL of water on a hot plate. Pour the hot water into the reaction mixture and swirl the flask to redissolve the solid. Pour the reaction mixture onto 50 g of ice. While stirring, add 4 mL of 50% sodium hydroxide solution. Use pH paper to check the pH of the solution. If the pH is still acidic, add a little more NaOH until the stirred solution is basic to litmus paper. Collect the crude product by suction filtration, washing well with cold water. The product may be recrystallized from hexanes. Dry in air. Do not use a drying oven. Record the weight.

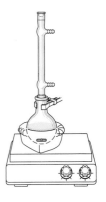

Characterization for Part D

Melting Point: Determine the melting point of the dry product. The reported melting point of 4-bromo-2-chloroaniline is 70–71°C.

Infrared Spectroscopy: Prepare a Nujol mull or KBr pellet of the solid and record the IR spectrum. The IR spectrum of 4-bromo-2-chloroaniline is shown in Figure 29.1-4.

NMR Spectroscopy: Dissolve the product in CDCl₃. Record and interpret the ¹H NMR spectrum. The ¹H NMR spectrum of 4-bromo-2-chloroaniline is shown in Figure 29.1-5.

Results and Conclusions for Part D

1. Calculate the yield of 4-bromo-2-chloroaniline.
2. Write chemical reactions to explain:
   a. why the reaction mixture of 4-bromo-2-chloroacetanilide and HCl forms a clear solution and then a precipitate forms.
   b. why the precipitate dissolves when hot water is added to the reaction mixture.
   c. why the product precipitates upon the addition of 50% NaOH to the reaction mixture.
3. The next step in the procedure calls for the use of 2.5 g of 4-bromo-2-chloroaniline. If less than this amount is available, recalculate the amounts of all reagents that should be used in the next procedure.

Figure 29.1-4

IR spectrum of
4-bromo-2-
chloroaniline (Nujol)

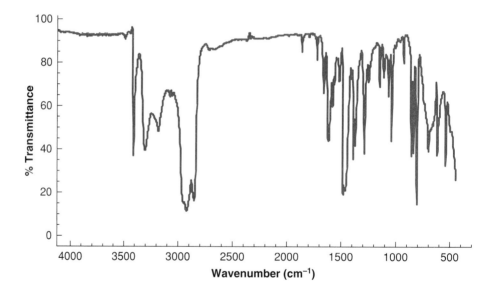

Figure 29.1-5

¹H NMR spectrum of
4-bromo-2-
chloroaniline (CDCl₃)

Source: Reprinted with
permission of Aldrich Chemical.

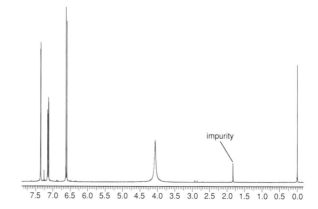

4. Calculate the mmol of 4-bromo-2-chloroaniline to be used in the next procedure. Calculate the mmol and mass of ICl needed to provide a molar ratio of 1.25 to 1 of ICl to 4-bromo-2-chloroaniline. Because of the hazards associated with its use, ICl will be dispensed from a buret as a 1.5 M of ICl in glacial acetic acid. Calculate the volume of solution needed to provide the calculated mass of ICl.

5. Assign important bands in the IR spectrum that verify the product identity.

Part E: Miniscale Synthesis of 4-Bromo-2-chloro-6-iodoaniline from 4-Bromo-2-chloroaniline

Safety First!

1. Wear eye protection at all times in the laboratory.

2. Iodine monochloride is a noxious reagent, much like molecular bromine. Wear gloves when handling it and work under the hood. If ICl comes in contact with your skin, wash with cold running water, notify your instructor immediately, and seek medical attention.

Always wear eye protection in the laboratory.

Hood!

In a 125-mL Erlenmeyer flask, dissolve 2.5 g (0.012 mol) of 4-bromo-2-chloro-aniline in 40 mL of glacial acetic acid. Add 10 mL of water and swirl to mix. Set aside.

Technical-grade ICl will be dispensed as a 1.5 M solution of ICl in glacial acetic acid. Calculate the mass of ICl/glacial acetic acid solution required to provide 0.015 mol of ICl. Tare a dry, 50-mL Erlenmeyer flask on the balance and add the calculated amount of ICl/glacial acetic acid solution from the buret. Cork the flask until ready to use. With a Pasteur pipet, add the ICl/glacial acetic acid solution to the solution of 4-bromo-2-chloroaniline over a period of 5 minutes. Rinse the flask with 1–2 mL of glacial acetic acid and add the rinse to the reaction mixture. Heat on a steam bath for 20 minutes. While hot, add saturated sodium bisulfite solution until the color is bright yellow. (You should not need to use more than 10 mL of sodium bisulfite solution.) Then add an amount of water so that the amount of sodium bisulfite solution plus that amount of water equal 12.5 mL total volume. (This should take approximately 2.5 mL of water.) Remove from the heat and allow the solution to cool slowly to room temperature. Then cool the solution in an ice bath. As the solution cools, the product will crystallize as long, colorless needles. Collect the product by suction filtration. Wash the crystals with 33% acetic acid and then with water. The crude product can be recrystallized by dissolving 1 g in 20 mL of glacial acetic acid. Heat the solution on a steam bath or hot plate and slowly add 5 mL of water to the hot solution. Allow the solution to cool slowly to room temperature and chill further in an ice bath before suction filtration. Air dry the crystals and weigh them.

Characterization for Part E

Melting Point: Determine the melting point of the dry product. The reported melting point of 4-bromo-2-chloro-6-iodoaniline is 97–98°C.

Infrared Spectroscopy: Prepare a Nujol mull or KBr pellet of the solid and record the IR spectrum. Interpret the spectrum. The IR spectrum of 4-bromo-2-chloro-6-iodoaniline is shown in Figure 29.1-6.

NMR Spectroscopy: Dissolve the product in CDCl$_3$. Record and interpret the ^{1}H NMR spectrum.

Figure 29.1-6

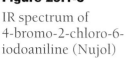

IR spectrum of
4-bromo-2-chloro-6-
iodoaniline (Nujol)

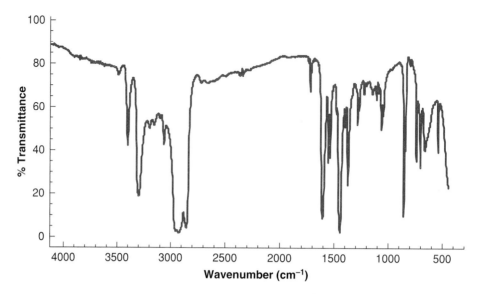

Results and Conclusions for Part E

1. Calculate the yield of 4-bromo-2-chloro-6-iodoaniline.
2. The next step in the procedure calls for the use of 2.0 g of 4-bromo-2-chloro-6-iodoaniline. If less than this amount is available, recalculate the amounts of all reagents that should be used in the next procedure.
3. Assign the important bands in the IR spectrum that verify the product identity.
4. Why is ICl used in the electrophilic aromatic iodination reaction rather than I_2?
5. Write a reaction to show why $NaHSO_3$ is used in the workup procedure.

Part F: Synthesis of 1-Bromo-3-chloro-5-iodobenzene from 4-Bromo-2-chloro-6-iodoaniline

S a f e t y   F i r s t !

Always wear eye protection in the laboratory.

1. Wear eye protection at all times in the laboratory.
2. Concentrated sulfuric acid is a highly toxic oxidizing agent that can cause severe burns. Always wear gloves when handling this reagent.
3. Diethyl ether is flammable. Keep it well away from flames.

    In a 250-mL round-bottom flask containing a stir bar, add 2.0 g of 4-bromo-2-chloro-6-iodoaniline and 10 mL of absolute ethanol. Stirring, add dropwise 4.0 mL of concentrated sulfuric acid. After the addition is complete, attach a water-jacketed condenser. Measure out 0.70 g of powdered sodium nitrite. Stirring, add the sodium nitrite down the reflux condenser in four portions. After the addition is complete, fit the round-bottom flask with a heating mantle and heat gently for 10 minutes. Heat 50 mL of water on a hot plate. Pour the hot water down the reflux condenser and mark the level of the water on the flask with a wax pencil. Remove the reflux condenser and fit the round-bottom flask with a stillhead, a condenser with rubber hoses attached, and a receiving flask. Steam distill the product until no more water-insoluble product is distilled (approximately 100 mL of distillate). Add additional hot water as necessary to replenish the original amount. The product has a tendency to solidify in the condenser, so it may be necessary to stop the water flow in the condenser several times during the distillation to allow the warm distillate to melt the product into the receiving flask.

Pour the distillate into a separatory funnel. Rinse the receiving flask with 25 mL of diethyl ether and add to the separatory funnel. Shake vigorously, venting often. After the layers have separated, drain off the lower aqueous layer into a clean flask. Drain the diethyl ether layer into a clean Erlenmeyer flask. Return the aqueous layer to the separatory funnel and extract with a second 25-mL portion of diethyl ether. Combine the diethyl ether layers. Dry over anhydrous magnesium sulfate. Gravity filter into a tared Erlenmeyer flask containing a boiling stone. Evaporate the diethyl ether on a steam bath to leave an off-white residue of 1-bromo-3-chloro-5-iodobenzene. Weigh the crude product.

Recrystallize from methanol, using 6 mL of methanol per gram of product. The product will crystallize in long, colorless needles. Suction filter the crystals using a Büchner funnel and air dry.

Characterization for Part F

Melting Point: Determine the melting point of the dry product. The reported melting point of 1-bromo-3-chloro-5-iodobenzene is 85–86°C.

Gas Chromatography/Mass Spectrometry: Obtain GC/MS data for the crude product (before recrystallization) and verify the identity of the compound. Identify impurities where possible.

Infrared Spectroscopy: Prepare a Nujol mull or KBr pellet of the solid and record the IR spectrum. Interpret the spectrum. The IR spectrum of 1-bromo-3-chloro-5-iodobenzene is shown in Figure 29.1-7.

NMR Spectroscopy: Dissolve the product in $CDCl_3$. Record and interpret the ^1H NMR spectrum.

Results and Conclusions for Part F

1. Calculate the yield of 1-bromo-3-chloro-5-iodobenzene.
2. Calculate the overall yield of the reaction, starting from aniline.
3. The reaction of the aryldiazonium salt with ethanol is an oxidation-reduction reaction. What compound is oxidized and what compound is reduced?
4. Use the spectral data to verify the structure of the product.

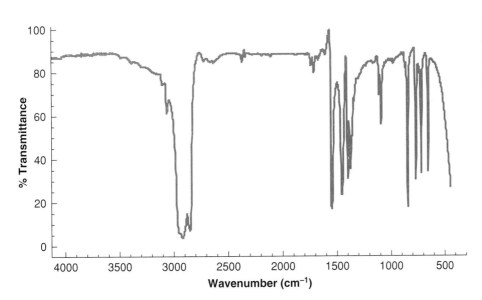

Figure 29.1-7

IR spectrum of 1-bromo-3-chloro-5-iodobenzene (Nujol)

Cleanup & Disposal

Neutralize all aqueous filtrates and wash down the drain with running water. Place nonhalogenated organic solutions in a container labeled "nonhalogenated organic solvent waste."

Critical Thinking Questions

1. In this multistep synthesis, the starting reagent was aniline. Propose a synthesis of aniline starting with benzene, showing all of the necessary reagents and showing the product of each step of the reaction.

2. Predict the product of the reaction of acetanilide with the following sequence of reagents: (1) ICl; (2) ethanol, HCl, and heat; (3) Br_2 and acetic acid; (4) $NaNO_2$, H_2SO_4, and ethanol.

3. Why does ICl react in electrophilic aromatic substitution reactions to give an iodinated product rather than a chlorinated product?

4. Would infrared spectroscopy, nuclear magnetic resonance spectroscopy, or mass spectrometry be the most useful in differentiating between the products of a specific reaction and the side products or impurities? Explain specifically what would be observed for each pair of compounds below.

    a. 4-bromoacetanilide and 2-bromoacetanilide
    b. 4-bromoacetanilide and 2,4-dibromoacetanilide
    c. 4-bromo-2-chloroacetanilide and 4-bromo-2,6-dichloroacetanilide
    d. 4-bromo-2-chloroacetanilide and 4-bromo-2-chloroaniline

5. Draw an orbital picture of a phenyl cation. How does a phenyl cation compare in stability to a benzyl cation? Explain.

6. *Molecular modeling assignment:* Determine the energies of the phenyl cation and benzyl cation. Explain any differences.

7. Investigate an industrial or a practical application of aryl halides.

Reference

Ault, A. *Techniques and Experiments for Organic Chemistry,* 2nd ed. Boston: Holbrook Press, 1976.

Experiment 29.2: Multistep Synthesis of Sulfanilamide Derivatives as Growth Inhibitors

In this experiment you will synthesize various sulfanilamide derivatives and test their activity as growth inhibitors of microorganisms. You will:

♦   synthesize a sulfa drug.

♦   evaluate the effect of the compound on bacterial growth.

Techniques

Technique F	Recrystallization and filtration
Technique G	Reflux
Technique K	Thin-layer chromatography

Technique M   Infrared spectroscopy
Technique N   Nuclear magnetic resonance spectroscopy
Technique O   UV-visible spectroscopy

Background

Desperation can lead to innovation. When Dr. Gerhard Domagk's daughter became deathly ill with a streptococcal infection that was nonresponsive to other treatments, the doctor administered an oral dose of Prontosil, a red dye manufactured in Germany. The researcher had been investigating properties of dyes as antibacterial agents. Prontosil had been found to inhibit the growth of *Streptococcus*. This desperate measure not only saved his daughter's life, but also resulted in the researcher being awarded the Nobel Prize for medicine in 1939 for use of the drug as an antibacterial agent. The active ingredient in Prontosil was not the drug itself, but was a metabolite of the drug, called sulfanilamide.

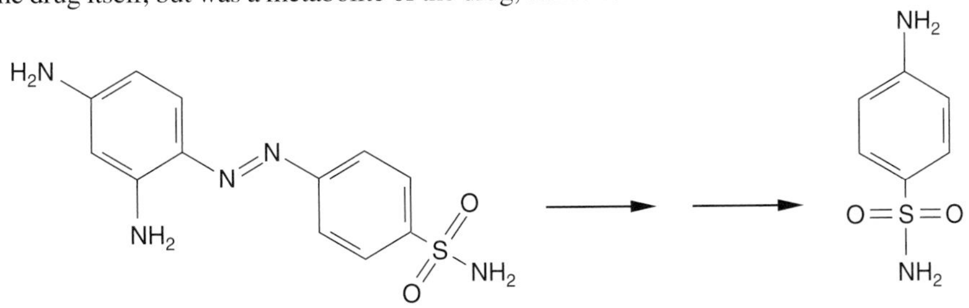

Prontosil   sulfanilamide

Sulfanilamide was the precursor to the sulfonamide family of drugs as further researchers prepared derivatives of sulfanilamide in an effort to improve its properties. These sulfa drugs revolutionized medicine and saved countless lives through treatment of bacterial diseases such as pneumonia and meningitis. The structures of selected sulfa drugs are shown here. Notice that the structures of these drugs are very similar, differing only in the sulfanomide group bonded to the sulfonyl group.

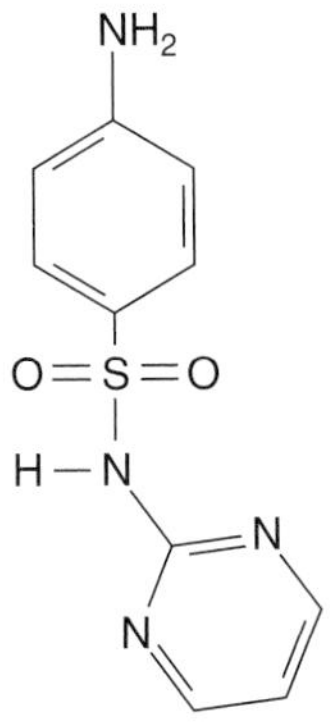

sulfadiazine 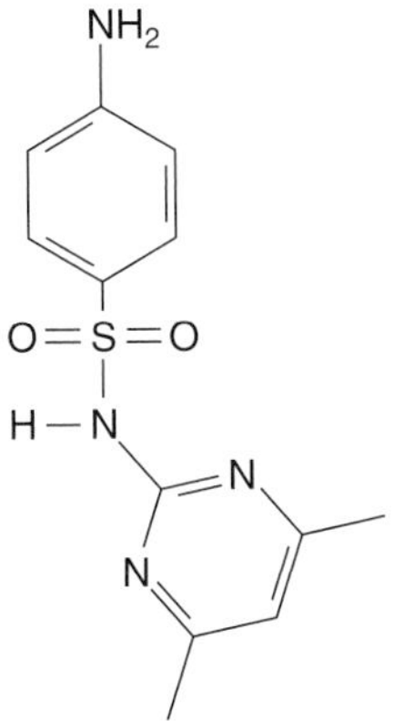   sulfamethazine 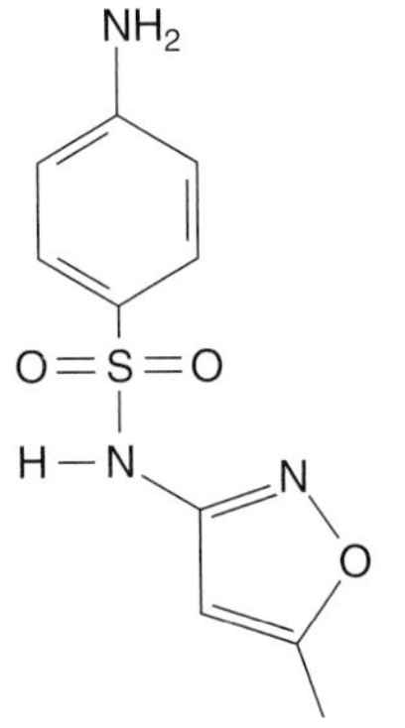   sulfamethoxazole 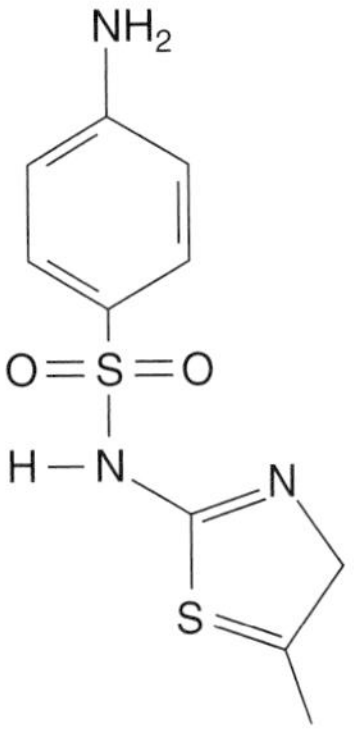   sulfathiazole

Sulfa drugs are not bacteriocides, but are bacterial growth inhibitors. Bacterial cells produce folic acid, an essential component for cell growth. The structures of the sulfa drugs are similar to the structure of *p*-aminobenzoic acid. The sulfa drug is a competitive inhibitor, binding to the active site on the enzyme that synthesizes folic acid and preventing *p*-aminobenzoic acid from being incorporated into folic acid. Since humans get their folic acid from their diets rather than biosynthetically, the sulfa drugs interfere with bacterial growth only.

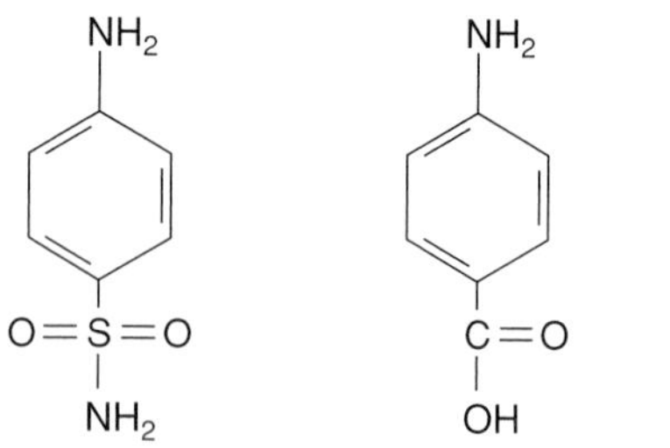

sulfanilamide   *p*-aminobenzoic acid

Although sulfa drugs were used extensively in World War II to combat infections, they do have side effects, such as kidney damage and possible allergic reactions. Because of these problems, sulfa drugs have largely been replaced by modern antibiotics, such as penicillin.

In this experiment, you will follow in Dr. Domagk's footsteps and prepare derivatives of sulfanilamide to develop an effective growth inhibitor. By varying the structure of the sulfanilamide and observing its effect, you will analyze how changes in structure can affect physiological properties and make recommendations for future drug development. This synthesis of a sulfa drug derivative is an example of a multistep synthesis, where the product of one step serves as the reactant for the subsequent step. Yields for a multistep synthesis may be low, so it is important to use good lab technique to optimize the yield for each step of the reaction. This synthesis involves four steps. Assuming that you are able to obtain a 60% yield on each step, the overall yield of the final product would be $0.60 \times 0.60 \times 0.60 \times 0.60$ or 13%. Because of the reduction in yield for subsequent steps, multistep syntheses generally start with larger amounts of starting materials than are typically used in the small-scale organic laboratory. It may be necessary to scale down a reaction if the amount of product obtained in a previous step is less than what is called for. Remember that all reagents should be scaled accordingly.

Synthesis of the sulfanilamide derivatives will be identical for all students until the third step, at which time students will vary the amine. Starting from aniline, you will acylate the amino group to form acetanilide, sulfonate the aromatic ring to form *p*-acetamidobenzenesulfonyl chloride, react an amine to form a benzenesulfonamide, then hydrolyze the acetamido group to form the sulfanilamide derivative. These four steps are discussed below. Alternatively, you may be instructed to use commercial acetanilide as the starting material for the synthesis, thereby reducing the synthesis to three steps.

Part A: Preparation of Acetanilide

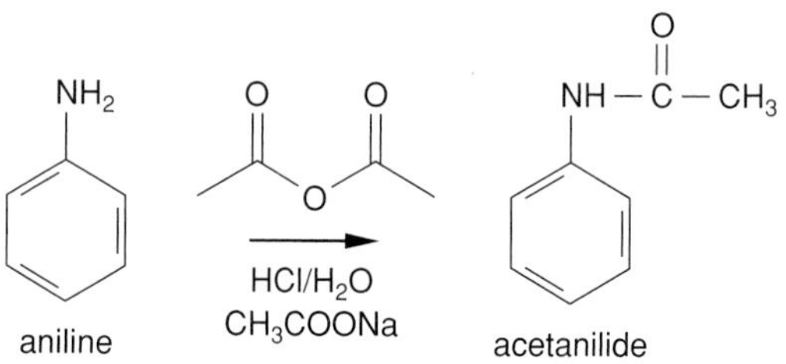

aniline   acetanilide

Acetylation of aniline serves a dual purpose in this experiment. First, acetylation serves to mediate the reactivity of the basic amino group, so that in the next step, the amine will not be protonated and become a *meta*-director. Second, the acetyl group prevents the amino group from reacting with the benzenesulfuryl chloride formed in the next step. The mechanism of this reaction involves protonation of the carbonyl group of the anhydride, addition of the nucleophilic amino group, and subsequent loss of the acetate group.

Part B: Preparation of p-Acetamidobenzenesulfonyl Chloride

p-acetamidobenzenesulfonyl chloride

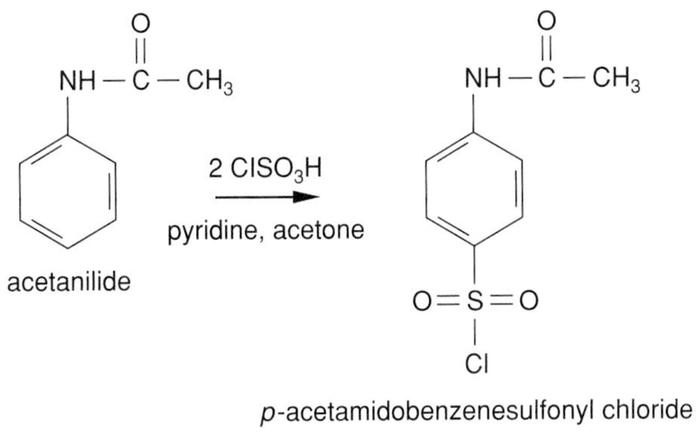

This electrophilic substitution reaction requires two equivalents of chlorosulfonic acid. The first equivalent furnishes the electrophile SO_3 *in situ* to yield *p*-acetamidosulfonic acid, while the second equivalent converts the sulfonic acid to the sulfonyl chloride.

$$HOSO_2Cl \rightleftharpoons SO_3 + HCl$$

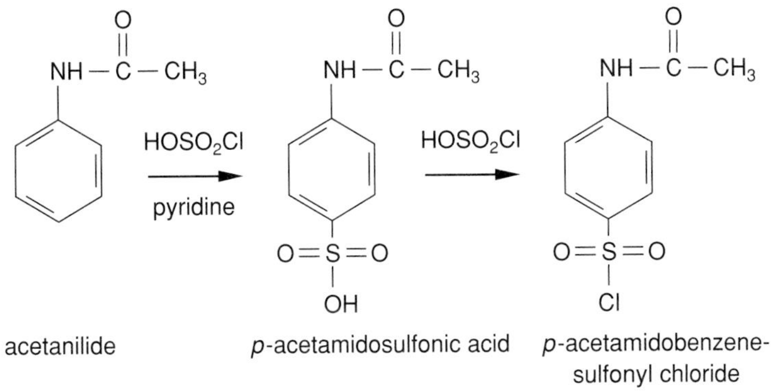

acetanilide   *p*-acetamidosulfonic acid   *p*-acetamidobenzene-
                                                  sulfonyl chloride

The acetamido group is activating and ortho/para-directing. However, because of steric hindrance, the *para*-isomer is formed preferentially.

The product is isolated by pouring the reaction mixture into ice water and collecting the solid product. The water also hydrolyzes any remaining chlorosulfonic acid to hydrochloric acid and sulfuric acid. Sulfonyl chlorides are not quite as moisture sensitive as carboxylic acid chlorides; however, they will eventually hydrolyze to sulfonic acids if not used quickly. For that reason, this product must be used immediately in the next step.

Part C: Reaction of p-Acetamidobenzenesulfonyl Chloride with an Amine

Students will be assigned a specific amine to use in this experiment. Reaction conditions may vary slightly depending upon the choice of amine and its relative expense and reactivity.

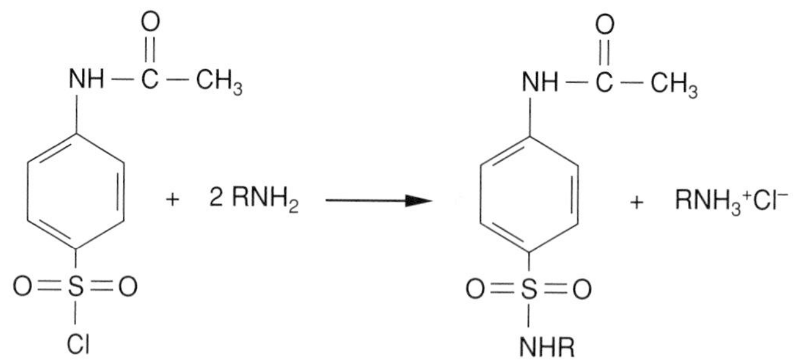

The reaction involves nucleophilic addition of ammonia or the amine to the sulfonyl chloride, followed by loss of HCl to form the sulfonamide. The possible amines are 2-aminopyridine, 3-aminopyridine, and 2-aminothiazole, whose structures are shown here.

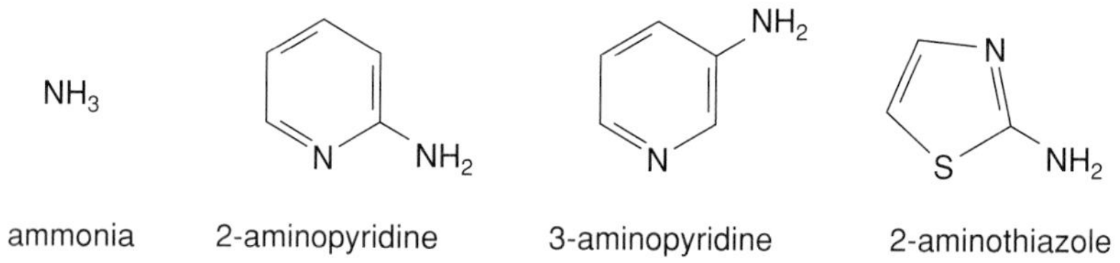

| ammonia | 2-aminopyridine | 3-aminopyridine | 2-aminothiazole |

Pyridine is a weak base. Its role in the reaction is to neutralize the HCl formed in the reaction. It is also non-nucleophilic and will not react with sulfuryl chloride.

Part D: Hydrolysis of the Acetamido Group

After formation of the sulfonamide group, the acetyl protecting group will be hydrolyzed to yield the free amino group. Amides can be hydrolyzed in either acidic or basic conditions. Here, basic conditions will be used for the sulfonamide derivatives, while acidic conditions will be used for sulfanilamide.

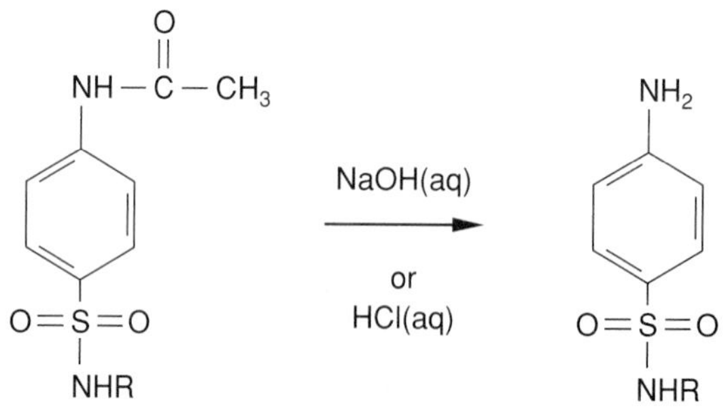

Part E: Bacterial Testing

Once you have synthesized and isolated sulfanilamide or the sulfanilamide derivative, you will test its effectiveness as a growth inhibitor. You will prepare antibiotic disks, place the disks in cultures of growing bacteria, and measure the zone of inhibition around each disk. The results will help you evaluate the effectiveness of the synthesized antibiotic in inhibiting bacterial growth.

Prelab Assignment

1. Calculate the mmoles used for each of the reactants in Part A and determine the limiting reagent. What is the theoretical yield of acetanilide?
2. Draw the structure of the antibiotic produced from the assigned amine.
3. Write a detailed mechanism for acetylation of aniline to form acetanilide.

4. In Part A, the acetylation occurs in buffered solutions. What are the components of the buffer solution? Explain how this buffer system works.

5. Explain why controlling the pH is critical in the acetylation reaction. Explain why acid is necessary to the reaction, but why too much acid is detrimental.

6. For each product of each step of the reaction, tabulate infrared absorption bands that might be helpful in identifying and verifying the structure.

7. For each reaction in the synthesis, write a flow scheme for the workup procedure.

8. The reaction in Part B calls for the use of 2.20 g acetanilide and 6.5 mL of chlorosulfonic acid. This reaction is later to be poured into 30 mL of ice water. Suppose that only 1.6 g of acetanilide is available (after saving out 25 mg for characterization). Calculate how much of the other reagents should be used.

9. Prepare a table to record the results of the antibacterial properties of the control, synthesized sulfa drug, commercial sulfanilamide, and commercial tetracycline on two different bacteria, *E. coli* and *S. aureus*.

Experimental Procedure

1. Wear eye protection at all times.

2. Acetic acid anhydride is a lachrymator. Wear gloves and work under the hood when handling this reagent.

3. Chlorosulfonic acid is very corrosive. Wear double-layered gloves when working with this noxious reagent. Chlorosulfonic acid can cause severe burns. If this reagent spills on you, wash immediately with cold running water and notify the instructor.

4. Chlorosulfonic acid reacts violently with water, even with moisture in the air, to form HCl and SO_3. Work under the hood at all times. To clean glassware that contains traces of chlorosulfonic acid, put the glassware under the hood, add ice chips, and let the ice melt. Wear gloves and rinse the glassware with copious amounts of water.

5. Concentrated ammonium hydroxide has a pungent odor. Care should be used in handling it. 2-Aminopyridine and 3-aminopyridine are toxic irritants. Use in a hood.

Part A: Miniscale Synthesis of Acetanilide

Prepare a solution of sodium acetate by dissolving 1.75 g of anhydrous sodium acetate in 15 mL of water. Set aside for later use. Dissolve 2.5 g of aniline in 65 mL water and 2.25 mL concentrated HCl in a 125-mL Erlenmeyer flask containing a stir bar. Start the solution stirring and then add 3.3 g acetic anhydride. While stirring, immediately add the sodium acetate solution and mix thoroughly. Cool the mixture in an ice bath for 10 minutes. Vacuum filter the white crystals using a Büchner funnel. Wash the crystals with several small portions of ice-cold water. Air dry the crystals. Weigh the dry product and take a melting point. Save 25 mg of the product for characterization and use the rest of the product for Part B.

Part B: Miniscale Synthesis of p-Acetamidobenzenesulfonyl Chloride

General Instructions

If the amount of product obtained in the previous step is less than the amount specified for this reaction, scale all of the reagents, including solvents, accordingly. Alternately, the instructor may have students use commercial acetanilide and start the synthesis from this point.

During this procedure, wear goggles at all times and wear double layers of acid-resistant gloves. Chlorosulfonic acid is a very strong acid that reacts violently with water. Exercise extreme caution when working with this reagent and when cleaning glassware containing residual amounts of this acid. Before starting this experiment, prepare a gas trap to collect the HCl formed in this reaction. The gas trap will consist of a one-hole rubber stopper that fits on top of the Erlenmeyer flask, glass and rubber tubing, a funnel, and a 400-mL beaker containing 100 mL of 1 M NaOH (see Figure 29.2-1). The funnel should be about **2 cm above** the surface of the liquid.

Place 2.20 g of acetanilide in a dry 125-mL Erlenmeyer flask. Place the flask on a hot plate under the hood and heat until the acetanilide is melted. Allow the liquid to solidify and then chill in an ice bath. Add 6.5 mL of chlorosulfonic acid via graduated cylinder and immediately connect the gas trap to the flask, as HCl will be evolved. Let the flask stand at room temperature until the solid has dissolved. This may require 30–45 minutes. In the meantime, take any glassware that contained chlorosulfonic acid to the hood and cautiously add ice chips to decompose the residual chlorosulfonic acid. **(Caution! Chlorosulfonic acid is a very strong acid that reacts violently with water. Exercise caution whenever this reagent comes in contact with water.)**

Hood!

After the solid has dissolved, gently warm the flask on a hot plate for 10 minutes. Then allow the flask to cool to room temperature. Slowly and carefully pour the solution into 30 mL of ice water in a 150-mL beaker while stirring vigorously. **(Caution! Any excess chlorosulfonic acid will react violently with water. This is a strongly exothermic reaction. Work under the hood.)** The product will precipitate as a milky-white solid. Suction filter the solid using a Büchner funnel and wash with two 3-mL portions of cold water. Press the solid with a spatula while under vacuum in order to remove as much water as possible. Use the product immediately in the next reaction.

Part C: Miniscale Synthesis of Sulfonamides

General Instructions

If the previous experiment was scaled by any factor, use that same factor to determine the amounts of reagents and solvents to use. Note that it may be necessary to alter the size of the glassware and equipment used. The procedures for this part of the experiment will differ depending upon the assigned amine. All solvents used in this experiment should be dry. Pyridine is best dried over KOH pellets for 24 hours. Acetone should be dried over anhydrous sodium sulfate and then gravity filtered.

Procedure for 2-Aminopyridine, 3-Aminopyridine, and 2-Aminothiazole.   Do all work under the hood. Dissolve the *p*-acetamidobenzenesulfonyl chloride from the previous

Figure 29.2-1

Setup for Experiment 29.2, Part B

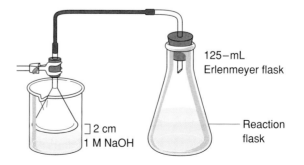

125–mL
Erlenmeyer flask

2 cm
1 M NaOH

Reaction
flask

step in 10-mL of dry acetone in a 25-mL Erlenmeyer flask. In a separate 25-mL Erlenmeyer flask, dissolve 10 mmole of the assigned amine in 2.5 mL of dry acetone, followed by 1.25 mL of dry pyridine.

While stirring, add the solution of *p*-acetamidobenzenesulfonyl chloride in acetone to the solution of the amine. Swirl to mix thoroughly and watch the reaction carefully. If a precipitate forms immediately, use a glass rod to break up lumps. Heat gently (up to 40°C) for 15 minutes with swirling on a hot plate. After the reaction is over, stopper the flask and place in a refrigerator for 24 hours or more in order to crystallize. For the sulfathiazole derivative, add 5–10 mL of water if the product has oiled out. Stir the mixture using a glass stirring rod until the oil has solidified.

After crystallization is complete, cool the reaction mixture in an ice-water bath. Suction filter the crystals using a Büchner funnel, washing the crystals with small amounts of ice-cold water. In order to improve the yield, a second crop of crystals should be isolated. Pour the filtrate from the suction filtration into a beaker and heat on a hot plate to reduce the volume by half. Cool the solution in an ice bath and filter the crystals. Combine the two crops of crystals. Weigh the combined air-dried product and determine the melting point. Save 25 mg of the product for characterization. Use the remainder in Part D. The sulfanilamide derivative is best hydrolyzed in base.

Procedure for Ammonia. Do all work under the hood. In a 125-mL Erlenmeyer flask containing a stir bar, add the *p*-acetamidobenzenesulfonyl chloride prepared in the last step. Add 13 mL of concentrated ammonia (15 M) to the flask. Heat gently on a hot plate while stirring for 15 minutes. Watch carefully and do not let the solution boil. By the end of the heating period, the initial solid should have dissolved. If not, heat a little longer, taking care not to boil the solution. When the solid has all dissolved, let the solution cool to room temperature and then cool further in an ice bath. Suction filter the new solid that forms. Wash the crystals with 10 mL of cold water. Let the crystals air dry overnight. Weigh the product and determine the melting point. Save 25 mg for characterization and use the remainder in Part D. Sulfanilamide is best hydrolyzed in acid.

Part D: Miniscale Synthesis of Sulfanilamides

General Instructions

If the previous experiment was scaled by any factor, use that same factor to determine the amounts of reagents and solvents to use. Note that it may be necessary to alter the size of the glassware and equipment used.

Basic Hydrolysis (for sulfonamides formed from 2-aminopyridine, 3-aminopyridine, and 2-aminothiazole). Mix the product from the previous reaction with 7.5 mL of 3 M NaOH in a 25-mL round-bottom flask fitted with a water-jacketed condenser and a stir bar. Reflux the mixture with stirring for 30 minutes. Let the solution cool to room temperature and pour the contents into a 50-mL beaker. Add 6 M HCl dropwise, until the solution is just acidic to litmus paper. Do not overacidify. Cool the mixture in an ice bath. Suction filter the crystals using a Büchner funnel, washing with a small amount of cold water. Recrystallize from water or an acetone-water solvent pair. If colored, add activated charcoal and do a hot gravity filtration. Cool the mixture in an ice bath and suction filter. When dry, take the melting point. It is important that the sulfonamide be very pure in order to determine its effectiveness in inhibiting bacterial growth. If necessary, recrystallize a second time. Save 25 mg for characterization.

Acidic Hydrolysis (for sulfonamide formed from ammonia). In a 25-mL round-bottom flask fitted with a water-jacketed condenser and a stir bar, add the dry *p*-acetamidobenzenesulfonamide and 5 mL of 6 M HCl. Stir to mix thoroughly. Heat the mixture at reflux for 60 minutes. After the solution has refluxed for an hour, cool the solution to room temperature. Pour the solution carefully into a 150-mL beaker. Rinse the round-bottom flask with 5 mL of water and add to the contents of the beaker. Chill the solution in an ice bath. While stirring, add approximately 0.1 g of solid $NaHCO_3$. A vigorous reaction will occur. After the reaction has subsided, add another 0.1 g-portion of $NaHCO_3$ and stir until the reaction subsides. Continue adding $NaHCO_3$ until the pH is basic to litmus (pH = 8). This will require approximately 3 g of $NaHCO_3$. During this time, the reaction mixture will have the consistency of a frothy paste. Vacuum filter the solid. Wash with two 5-mL portions of cold water. Recrystallize from water.

Characterization for Parts A–D

Melting Point: Determine the melting point for the product of each of the reactions. The respective literature melting points are sulfanilamide (165°C), 2-sulfanilamidopyridine (190°C), 3-sulfanilamidopyridine (258°C), and sulfathiazole (202°C).

Infrared Spectroscopy: Obtain an IR spectrum of each product isolated using either the KBr or the Nujol mull technique. Interpret the spectra.

NMR Spectroscopy: Obtain a ^{1}H NMR spectrum of each of the products isolated and interpret the spectra noting chemical shift positions and the integration.

UV/Vis Spectroscopy: Obtain UV and visible spectra in ethanol for each product. Record the absorption maxima and the molar absorptivities.

Thin-layer Chromatography: Dissolve a few crystals of each product in methylene chloride and apply a small spot to a silica gel plate. Develop the plate with methylene chloride. Measure R_f values and interpret the results.

Results and Conclusions for Parts A–D

1. Determine the percent yield for the product of each reaction.
2. Calculate the overall yield of the final product.
3. From the melting point or results of TLC, determine the purity of each product.
4. Use spectroscopic analysis to verify the structures of the products formed in each reaction.
5. Write detailed mechanisms for each reaction.
6. Acidity and basicity play a crucial role in all of the reactions in this synthesis. For each reaction, write chemical reactions to explain specifically why the pH is important.
7. Using a chemical supply catalog, calculate the cost per gram to synthesize the assigned sulfa drug (excluding labor costs).

Part E: Bacterial Testing of Antibiotic Susceptibility

In this part of the experiment, you will use two large commercial Mueller-Hinton plates, forceps, a broth culture of *E. coli* and a broth culture of *Staphylococcus aureus,* two sterile swabs, a petri dish containing sterile paper disks, and an incubator set to 37°C. On the bottom of each plate, use a marking pen to divide the plate into four quadrants. Label each quadrant as "control," "commercial sulfanilamide," "synthesized sulfanilamide," and "reference tetracycline," respectively. Using sterile technique, swab the nutrient agar plates with the bacteria. Be sure to cover all the surface area of the agar. Label one plate as *E. coli* and the other as *S. aureus.*

Control.   Sterilize the forceps by running the tips through a flame. Dip a sterile paper disk in acetone. Transfer the disk to a piece of filter paper to absorb excess solution, then let the solvent evaporate. Transfer the paper disk to the middle of the quadrant labeled "control" and tap down. Repeat for the other bacterial plate.

Tetracycline. Sterilize the forceps. Place a tetracycline disk in the center of the appropriate quadrant and tap down. Repeat for the other bacterial plate.

Commercial Sulfanilamide. Prepare a solution of the commercial sulfanilamide by dissolving 50 mg of the drug in 5 mL of acetone. Sterilize the forceps and dip a sterile paper disk into the solution. Transfer the disk to a piece of filter paper to absorb excess solution and let the solvent evaporate. Transfer the paper disk to the middle of the appropriate plate. Tap down. Repeat for the other bacterial plate.

Synthesized Sulfanilamide. Prepare a solution of the sulfanilamide by dissolving 50 mg of the sulfanilamide in 5 mL of acetone. Sterilize the forceps and dip a sterile paper disk into the solution. Transfer the disk to a piece of filter paper to absorb excess solution, and then let the excess solvent evaporate. Transfer the paper disk to the middle of the appropriate quadrant. Tap down. Repeat for the other bacterial plate.

Incubate the plates at 37°C for 24 hours. Do not open the plates. Measure the diameter (in millimeters) of the zone of inhibition (zone of clearing) for each disk on each bacterial plate. Record the results for each sulfanilamide on the board. Divide the zones by the zone of the reference antibiotic, tetracycline. The larger the ratio, the more effective the sulfa drug is at inhibiting growth of bacteria.

Cleanup & Disposal

Dispose of all plates and broths in biohazard waste. For Part B, all glassware containing residual chlorosulfonic acid should be cleaned very carefully. Destroy the excess chlorosulfonic acid by dropping ice chips into the glassware. HCl gas will be evolved, so work under the hood and wear two layers of gloves. Neutralize basic or acidic solutions and pour carefully down the drain with lots of running water.

Results and Conclusions for Part E

1. What was the purpose of the disk labeled as "control"? How did it aid in analyzing the effectiveness of the sulfanilamides?
2. Evaluate the effectiveness of the synthesized sulfanilamide in inhibiting the growth of each type of bacteria tested. How did the synthesized sulfanilamide compare to the commercial sulfanilamide? To the tetracycline? Rank the order of effectiveness of the antibacterial agent on each type of bacteria.
3. Compare the effectiveness of the various synthesized sulfanilamides in inhibiting bacterial growth. What types of amino groups appear to be the most effective? Are there other amines that might be similarly effective?
4. Were the synthesized sulfanilamides equally effective against the two different types of bacteria?

Critical Thinking Questions

1. For Part D, explain why the acetyl group was hydrolyzed in base rather than in acid.
2. 2-Aminopyridine and 3-aminopyridine both contain two nitrogen atoms. Explain why the amino group reacts rather than the pyridine nitrogen.

3. Propose syntheses for each of the following sulfa drugs: sulfadiazine, sulfadimidine, sulfamethoxazole

4. Write resonance structures for 2-aminothiazole and explain what other by-products might be formed in the synthesis of sulfathiazole.

5. Explain why the acetyl group can be selectively removed in the presence of the sulfonamide group.

6. Based upon its structure, suggest a reason that Prontosil is colored.

7. *Library project:* To what general class of dyes does Prontosil belong? How might this drug be synthesized?

Reference

Molecule of the Month: February 1998, Sulfanilamide, Tebbutt, P., Cherwell Scientific Publishing. See on Internet at http://www.chm.bris.ac.uk/motm/sulfanilamide/sulfanih.html.

Experiment 29.3: Structural Determination of Isomers Using Decoupling and Special NMR Techniques

In this experiment you will investigate NMR spectroscopic techniques that can be used when interpretation of ^{1}H and ^{13}C NMR spectra alone fails to provide definitive structural identification of an unknown compound. You will:

♦ determine the structure of an unknown compound using special NMR techniques.

Technique

Technique N   Nuclear magnetic resonance spectroscopy

Background and Mechanism

Imagine that an analysis is being done to identify a compound from a plant for its potential use as a natural pesticide. The compound will first have to be separated from other compounds and purified. Then what? Physical constants, such as refractive index and melting or boiling points, can be measured. Wet qualitative tests can be run to identify functional groups. Obtaining IR, NMR, and MS spectra can often provide information sufficient to solve the structure. Occasionally, a definitive answer proves to be elusive. This is the case when trying to distinguish between possible isomers with very similar spectroscopic properties. This is important for biologically active compounds because stereoisomers can often exhibit very different physiological activities.

Compounds extracted from plants may have very complex structures. However, the situation can be modeled by analyzing some relatively simple compounds. Suppose that an unknown organic compound is found by MS analysis to have m/z 112. The compound shows strong absorption in the IR at 1735 cm^{-1}, characteristic of an ester or a cyclic ketone. The ^{1}H NMR spectrum is complex, but shows a doublet near δ1.0, suggestive of a methyl group with one adjacent, nonequivalent hydrogen. Assignment of the other protons is more difficult. The ^{13}C NMR spectrum shows six signals. The NMR spectra are shown in Figures 29.3-1 and 29.3-2.

The ^{1}H NMR signal at δ1.1 is a doublet and integrates as three hydrogens. No signals are observed in the δ3–4 region, ruling out H−C−O absorption expected of an

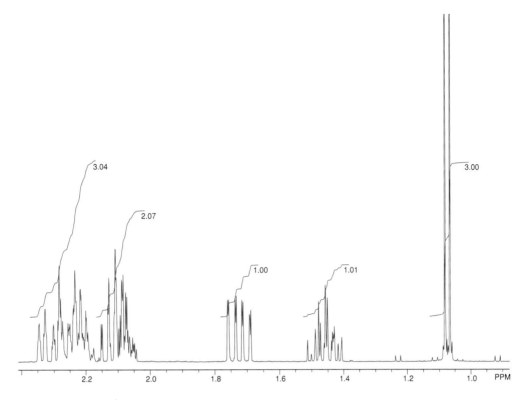

Figure 29.3-1 400 MHz ^{1}H NMR spectrum of an unknown ketone

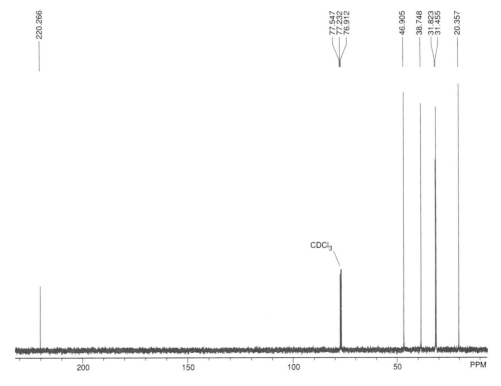

Figure 29.3-2 ^{13}C NMR spectrum of an unknown ketone

ester. The analyst concludes that the compound is a methylcyclopentanone. There are only two possibilities: 2-methylcyclopentanone and 3-methylcyclopentanone. Which is it? The physical constants of the two isomers are quite similar, as shown in Table 29.3-1.

Table 29.3-1 Physical Constants for Methylcyclopentanones

Compound	Refractive index (n_D)	Density (g/mL)	Boiling point (°C)
2-Methylcyclopentanone	1.4364	0.914	139.5
3-Methylcyclopentanone	1.4329	0.913	144.0

^{13}C DEPT NMR Spectra

Because the spectra do not furnish distinguishing features and because the physical properties of the isomers are very similar, alternate methods of analysis must be used. A number of NMR techniques may be helpful. One special NMR technique that gives useful information about the number of protons attached to each carbon is DEPT, distortionless enhancement by polarization transfer (Lambert et al., 1998). The ^{13}C DEPT NMR spectrum of the unknown compound is shown in Figure 29.3-3. The upper three plots are subspectra of the usual ^{13}C NMR spectrum shown at the bottom of the figure.

The lower frame of the DEPT spectrum shows carbon signals of all carbons bonded to at least one hydrogen atom. The frame labeled CH carbons indicates that the carbon at δ31.8 has only one attached hydrogen atom. The frame labeled CH$_2$ carbons indicates three methylene carbons. The frame labeled CH$_3$ carbons indicates only one methyl carbon at δ20.4.

For purposes of comparison, the ^{1}H and ^{13}C NMR spectra of 2-methylcyclopentanone are shown in Figures 29.3-4 and 29.3-5. There are some differences in the ^{1}H NMR

Figure 29.3-3

DEPT spectrum of an unknown ketone

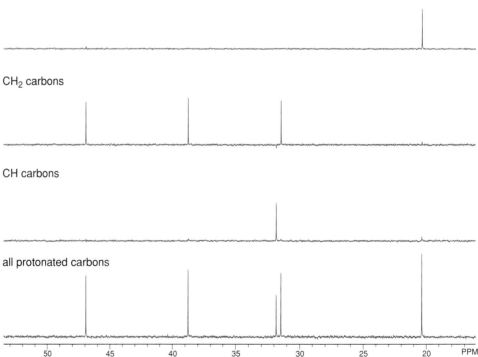

CH$_3$ carbons

CH$_2$ carbons

CH carbons

all protonated carbons

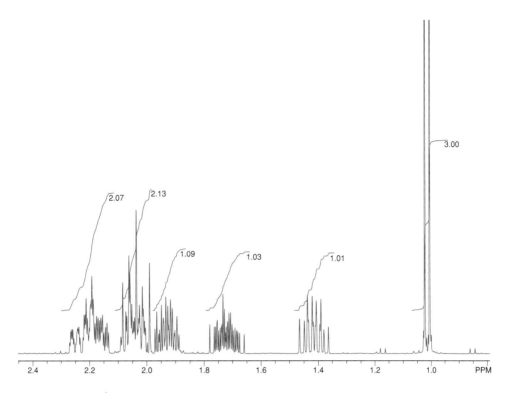

Figure 29.3-4   400 MHz ¹H NMR spectrum of 2-methylcyclopentanone

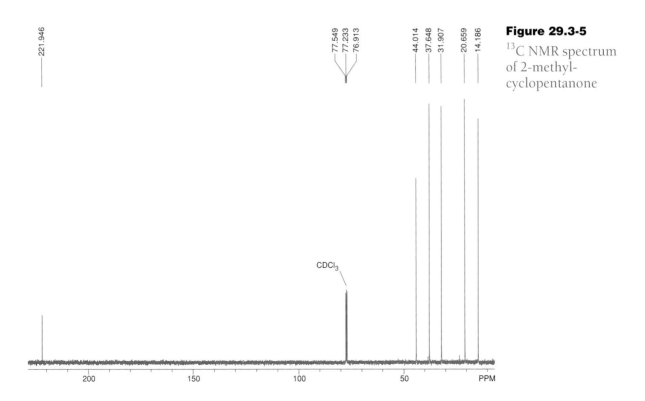

Figure 29.3-5

¹³C NMR spectrum of 2-methyl-cyclopentanone

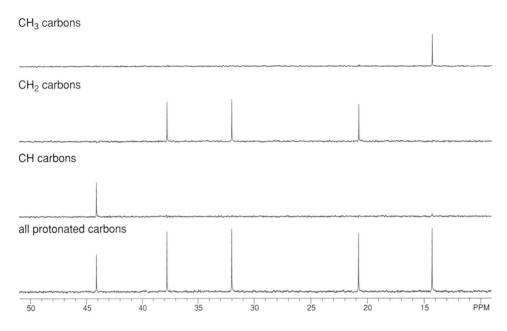

Figure 29.3-6 ^{13}DEPT spectrum of 3-methylcyclopentanone

spectra of the methylcyclopentanones shown in Figures 29.3-1 and 29.3-4. Careful analysis and expansion of the region of the spectrum between δ2.0 and δ3.0 might be helpful. However, the evidence provided by the DEPT spectra in Figures 29.3-3 and 29.3-6 is much more conclusive.

The ^{13}C DEPT NMR spectrum of 2-methylcyclopentanone given in Figure 29.3-6 provides definitive differences from the DEPT spectrum of 3-methylcyclopentanone (Figure 29.3-3). Notably, the signal at δ44.0 corresponds to a type of carbon having one attached proton. This carbon (C-2) appears downfield because it is attached directly to the carbonyl carbon. The CH carbon frames of the DEPT spectra are pivotal in allowing the definitive assignment of structures in Figures 29.3-1 to 29.3-3 as being due to 3-methylcyclopentanone and in Figures 29.3-4 to 29.3-6 as being due to 2 methylcyclopentanone.

NMR Prediction Software

For the case of the unknown methylcyclopentanone, the DEPT spectra were all that we needed in addition to the ^{1}H and ^{13}C NMR spectra to distinguish between the two possible methylcyclopentanone structures. The ^{13}C NMR spectra can be analyzed further to assign each signal to a particular carbon atom. Using ACD/CNMR chemical shift prediction software from Advanced Chemistry Development(ACD) Labs or similar software, it is possible to estimate carbon chemical shifts for organic compounds. For example, the methylcyclopentanones give the estimated chemical shifts (δ values) shown below using the ACD/CNMR software.

These data agree well with the experimental values for the chemical shifts of the signals in Figures 29.3-2 and 29.3-5. Without these correlations, it would be difficult to assign with certainty the signals for C-2 and C-5 in 3-methylcyclopentanone or for C-3 and C-4 in 2-methylcyclopentane. The numerical calculations are based upon "contributions" from various connectivities of each carbon in a compound and stereochemical contributions. Similarly, the assignment of proton shifts (δ values) in the ^{1}H NMR spectra may be estimated based upon empirical calculations using ACD/HNMR chemical shift prediction software. The chemical shifts are shown here.

2.19, 2.32
1.38, 1.78
2.13
1.03
1.71, 2.16

2.03, 2.22
1.38, 1.78
1.69, 1.89
2.03
1.07

The ^{1}H NMR spectra shown in Figures 29.3-1 and 29.3-4 are complex because of the nonequivalence of geminal, diastereotopic protons of the methylene groups of the methylcyclopentanes. In practice, it is best to ascribe signals in a ^{1}H NMR spectrum to particular protons using homonuclear decoupling and 2-D NMR techniques. These techniques are discussed in the next sections.

^{1}H–^{1}H Homonuclear Decoupling

Neighboring nonequivalent protons may often be identified in simple compounds by examining coupling patterns (doublets, triplets, etc.) and coupling constants (Technique N). For example, the neighboring C-2 and C-3 protons of *cis*- and *trans*-3-chloroacrylic acids show vicinal coupling. The coupling constants vary due to cis or trans relationships of coupled protons. As shown in Figures 29.3-7 and 29.3-8, the coupled protons in each compound have identical or nearly identical coupling constants, $J_{cis} = 8.2$ Hz and $J_{trans} = 13.6$ Hz.

Identities of signals of nonequivalent protons on adjacent carbons may be determined using decoupling techniques. Homonuclear decoupling techniques may be used to determine coupling relationships between nonequivalent protons. To perform homonuclear decoupling, an intense radio frequency is applied at the resonance frequency of a

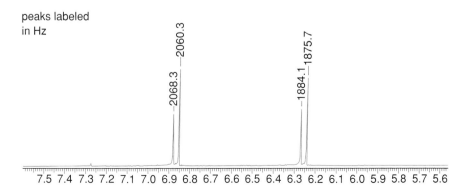

peaks labeled
in Hz

2068.3
2060.3
1884.1
1875.7

7.5 7.4 7.3 7.2 7.1 7.0 6.9 6.8 6.7 6.6 6.5 6.4 6.3 6.2 6.1 6.0 5.9 5.8 5.7 5.6

Figure 29.3-7

^{1}H NMR spectrum of *cis*-3-chloroacrylic acid (expanded in alkene region)

$J_{cis} = 2068.3 - 2060.3 = 8.0$ Hz
or
$1884.1 - 1875.7 = 8.4$ Hz
(Avg = 8.2 Hz)

Figure 29.3-8

^{1}H NMR spectrum of *trans*-3-chloroacrylic acid (expanded in alkene region)

J_{trans} = 2263.6 − 2250.0 = 13.6 Hz

or

1888.2 − 1874.6 = 13.6 Hz

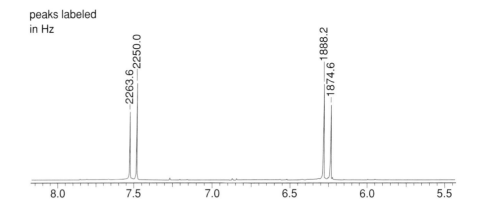

peaks labeled in Hz

particular signal. This decouples the proton or protons of the signal from adjacent protons that normally give coupling. An example of analysis using decoupling is shown in the ^{1}H NMR spectrum of 2-bromobutane. The resonance signals in 2-bromobutane are clearly separated for purposes of analyzing the effects of decoupling as shown in Figure 29.3-9.

Irradiation at δ4.08 (the frequency of the C-2 proton) gives a new NMR spectrum having a triplet at δ1.0, a singlet at δ1.69, a quartet at δ1.82, and little or no absorption at δ4.08. The signal that was at δ4.08 is saturated (equal population of ground and excited state populations), the signal at δ1.82 is coupled only to the adjacent methyl group, and the signal at δ1.69 is no longer coupled to any adjacent protons. The triplet at δ1.0 remains unaffected because the C-4 methyl group is not adjacent to C-2. Irradiation resulting in decoupling clearly provides the proximities of signals of coupled protons as shown in Figure 29.3-10.

Irradiation of the methyl signal at δ1.0 causes the C-2 protons to appear as a triplet in another decoupled spectrum shown in Figure 29.3-11. The C-2 protons are coupled only to the adjacent C-3 protons. Decoupling experiments can be used to help establish positions of neighboring protons or groups of protons.

Figure 29.3-9

^{1}H NMR spectrum of 2-bromobutane

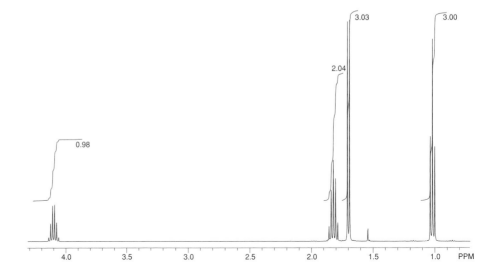

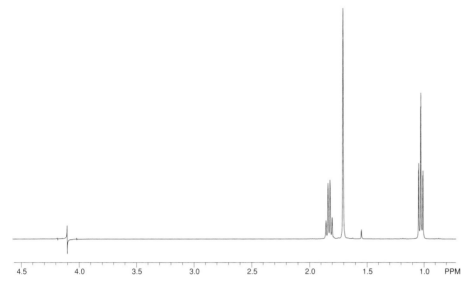

Figure 29.3-10
400 MHz ^{1}H NMR spectrum of 2-bromobutane (decoupled)

Irradiation at δ4.08

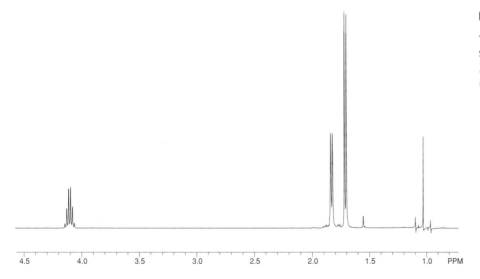

Figure 29.3-11
400 MHz ^{1}H NMR spectrum of 2-bromobutane (decoupled)

Irradiation at 51.0

COSY Spectra

Returning to the assignment of proton resonance signals for the methylcyclopent-anones, correlation NMR spectroscopy is useful for this purpose. An NMR experiment known as an ^{1}H-^{1}H COSY (correlation spectroscopy) experiment gives a 2-D plot of proton spectra yielding information like that obtained from the decoupling experiments above, but produced on a single diagram (Breitmaier, 1998; Silverstein and Webster, 1998). The COSY experiment gives a 2-D plot of the proton spectrum of the compound plotted on both the x and y axes. Signals on the y axis are "correlated" with those on the x axis by reading across to a "contour" and then down. Coupled protons appear as "cross peaks" off the diagonal. Contours on the diagonal are disregarded because these are for identical protons. That is, reading across to a contour on the diagonal will correlate with the same signal in the spectrum on the x axis. This is illustrated for 2-methylcyclopentanone in Figure 29.3-12.

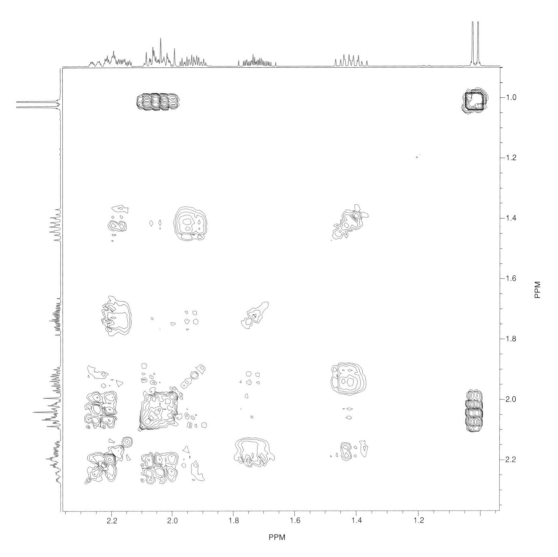

Figure 29.3-12 ^{1}H-^{1}H COSY spectrum of 2-methylcyclopentanone

The contour below the methyl protons at δ1.05 in the COSY spectrum indicates that the signal is split into a doublet by a proton at δ2.0–2.1. Using a similar analysis, the proton at δ1.4 is coupled with a proton at δ1.90–2.0 and with a proton at δ2.2. Other correlations can be made if desired.

HETCOR (heteronuclear chemical shift correlation) is another useful 2-D NMR technique. In HETCOR (also known as ^{1}H-^{13}C COSY), a proton spectrum is plotted against the spectrum of another nucleus, such as ^{31}P or ^{13}C as in the example here, and a 2-D plot is obtained by plotting the ^{1}H NMR on one axis and the ^{13}C NMR on the other axis. Contours mark intersections of perpendicular lines drawn from a signal in the proton spectrum and from a signal in the carbon spectrum corresponding to the carbon that is attached to that proton. HETCOR, along with DEPT, is useful in assigning specific, related signals in the ^{1}H NMR spectrum. The HETCOR spectrum for 2-methylcyclopentanone is shown in Figure 29.3-13.

The HETCOR spectrum of 2-methylcyclopentanone shows that the proton attached to carbon C-2 is correlated at a chemical shift of δ2.0–2.1. One of the protons

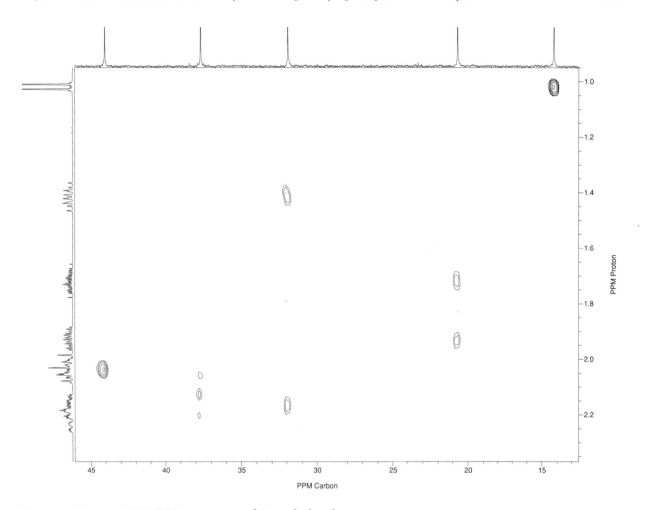

Figure 29.2-13   HETCOR spectrum of 2-methylcyclopentanone

of C-5 also has a chemical shift of δ2.0–5.1. The other C-2 proton has a chemical shift of δ2.2–2.3. The methyl protons have a chemical shift of δ1.10. The spectrum shows that each of the C-3 and C-4 protons have different chemical shifts.

The complex coupling patterns of the protons correlated with the carbon at δ20.5 suggest that this carbon is C-4 and that the remaining methylene carbon (C-3) gives the signal at δ32.0. To verify this assignment, it is possible to predict the δ values of the carbons in the ^{13}C NMR spectrum with the aid of chemical shift predictions software discussed earlier in this section.

In Part A of this experiment, you will measure homonuclear coupling constants for an assigned alkene. In Parts B and C of the experiment, you will characterize an unknown compound using ^{1}H NMR, ^{13}C NMR, COSY and HETCOR spectra.

Prelab Assignment

1. (Part A) Predict the appearance of the ^{1}H NMR spectrum of 2-bromobutane irradiated at δ1.70; at δ1.82.
2. (Part B) Analyze the correlations in the COSY spectrum of 2-methylcyclopentanone.
3. (Part C) Analyze the HETCOR spectrum of 2-methylcyclohexanone.

Experimental Procedure

1.   Use gloves when dissolving samples in CDCl₃, a cancer-suspect agent.
2.   Wear eye protection at all times in the laboratory.

Part A: Homonuclear Decoupling of a Known Alkene and Determination of Coupling Constants

Dissolve 5–10 mg of the known compound in $CDCl_3$. Record the integrated ^1H NMR spectrum. Measure coupling constants where possible (where overlapping signals do not interfere with the measurement). Using the same sample and sample tube, use scale expansion to magnify the region where coupling constants are to be measured. If possible, select an obvious signal that is fairly isolated from other signals. Set up the decoupling experiment applying power over the narrow frequency range of the chosen signal. Record the spectrum. If possible, record another spectrum using irradiation of the signal where decoupling had an effect on the first spectrum. Irradiation of this signal should give simplification of the signal that was irradiated when obtaining the first spectrum. (For example, in the case of signals for an ethoxy group, irradiation at the frequency of the methyl will cause the methylene quartet to collapse to a singlet. Irradiation at the methylene frequency will cause the methyl triplet to collapse to a singlet.)

Results and Conclusions for Part A

1.   Assign signals in the NMR spectrum to specific protons in the compound.
2.   Calculate the coupling constants (in Hz) between the alkenyl protons.

Part B: Characterization of an Unknown Compound Using Homonuclear Decoupling in ^1H NMR

Select an obvious signal that is fairly isolated from other signals. Set up the decoupling experiment applying power over the narrow frequency range of the chosen signal. Record the spectrum. If possible, record another spectrum using irradiation of the signal where decoupling had an effect on the first spectrum. Irradiation of this signal should give simplification of the signal that was irradiated when obtaining the first spectrum. (For example, in the case of signals for an ethoxy group, irradiation at the frequency of the methyl will cause the methylene quartet to collapse into a singlet. Irradiation at the methylene frequency will cause the methyl triplet to collapse into a singlet.)

Results and Conclusions for Part B

1.   Based on the spectral information, identify the structure of the unknown compound.
2.   Correlate specific carbons and specific protons with signals in the ^1H and ^{13}C NMR spectra.
3.   Measure the coupling constants (in Hz) between as many protons in the unknown compound as possible.
4.   Use an NMR software chemical shift prediction program to estimate the chemical shifts in the ^1H and ^{13}C NMR spectra of the compound. Compare the estimates with the actual chemical shifts.

Part C: Characterization of an Unknown Compound Using ^1H NMR, ^{13}C NMR, ^1H-^1H COSY and HETCOR (^1H-^{13}C COSY)

Dissolve 5–10 mg of solid or liquid unknown sample in $CDCl_3$ and obtain an integrated ^1H NMR spectrum. Next, obtain the 2-D COSY spectrum. Prepare a new sample by dissolving 20–25 mg of the unknown sample in $CDCl_3$. Record the noise-decoupled ^{13}C NMR spectrum. Set up the ^{13}C DEPT experiment. If only one signal is observed far downfield in the ^{13}C NMR spectrum, omit the signal when specifying the scan width. Record the DEPT spectrum. Next, set up the experiment and record the HETCOR spectrum.

Results and Conclusions for Part C

1. Based on the spectral information, identify the structure of the unknown compound.
2. Correlate specific carbons and specific protons with signals in the ^1H and ^{13}C NMR spectra.
3. Measure the coupling constants (in Hz) between as many protons in the unknown compound as possible.

Cleanup & Disposal

Place $CDCl_3$ solutions containing analyzed compounds in a bottle labeled "halogenated waste." Clean NMR tubes by rinsing the tubes and plastic tops with acetone and inverting them in a beaker at room temperature.

Critical Thinking Questions

1. For COSY spectra, what is the significance of the contours that appear on the diagonal?
2. Suggest a type of carbon that could give two different contours in a HETCOR spectrum. Explain or propose a partial structure.
3. Look up the Karplus equation in one of the references at the end of the experiment. Use the equation to explain the difference in coupling constants between cis and trans protons in chloroethene.
4. Use chemical shift prediction software to calculate the predicted δ values of the ^{13}C chemical shifts for the C-2 and C-5 carbons of 3-methylcyclopentanone. Compare these values with those in Figure 29.3-2.
5. Verify the peak assignments for all protons in 2-methylcyclopentanone.
6. Are the two frequencies used to plot HETCOR spectra the same or different? Explain.
7. For 3-methylcyclopentanone, explain the observed ^{13}C NMR chemical shifts.
8. Explain why the ^1H NMR spectrum of 3-methylcyclopentanone is so complex.

References

Breitmaier, E. *Structure Elucidation by NMR in Organic Chemistry.* Chichester, England: John Wiley and Sons Ltd., 1993.

Lambert, J. B., Shurvell, H. F., Lightner, D. A., and Cooks, R. G. *Organic Structural Spectroscopy,* Upper Saddle River, NJ: Prentice-Hall, 1998.

Silverstein, R. M., and Webster, F. X. *Spectrometric Identification of Organic Compounds,* 6th ed. New York: John Wiley and Sons, Inc., 1998.

Experiment 29.4 The Library-Laboratory Connection

Rapid developments in the coverage of chemical topics in electronic databases and web-based resources are changing the ways that chemists search for information about compounds, reactions, and applications. However, most journals and books are still located in the library in printed form. In this experiment, you will learn to:

- access the library: book section, reference book section, periodicals section, abstracts section, and computer-based-information-access section.

- use the printed version of *Chemical Abstracts.*

- use the electronic *Chemical Abstracts* database and other databases.

- locate recent and older citations from chemical periodicals.

- locate and derive information from advanced organic chemistry books and review articles.

- search the web to access chemical information.

- use computer databases available through the library to locate topical articles from recent periodicals.

Background

Library holdings include original research as reported in the primary literature, summaries of important scientific contributions in review articles and monographs (known as secondary sources), and articles in the primary chemical literature and certain secondary sources. Chemical knowledge is stored in books in the book collection and reference books in the reference section. It is important to know where these collections are located in the library. A useful guide to the chemical literature is given in March's *Advanced Organic Chemistry,* Appendix A (5th ed., M. Smith and J. March, Wiley, 2001). A growing reliance upon accessing electronic databases and electronic journals requires library users to be familiar with the electronic access options of the library.

When researching a specific chemical topic, relevant information is often readily accessible from chemical databases. There are two types: computer databases and printed databases. Two important computer databases are Chemical Abstracts Service and Beilstein. These databases lead the inquirer to the appropriate periodicals in the primary journals and review articles. The two most important printed databases are *Chemical Abstracts* and the *Beilstein Handbook of Organic Chemistry.*

It is important to understand how to access both computer and printed databases. This exercise emphasizes both approaches.

Part A: Chemical Abstracts (CA) and Beilstein

Chemical Abstracts

This collection is the singularly most important reference for chemists. The Chemical Abstracts Service (CAS) abstracts every article even remotely connected to chemistry. There is an increasing tendency for more and more online searching of CA. STN and SciFinder are the access routes for searching CA online. The hard copy of CA still has great value because not all information is available electronically.

CA was first published in 1907. (For references prior to 1907, the best source is Beilstein.) CAS publishes more than a half million new abstracts each year. Two volumes of

CA have been published per year since 1962. In addition to yearly indices, collective (semidecennial) indices have been published since 1956. CA has a number of types of indices, and these have changed by name and format over the years. The older Subject Index lists subjects and specific chemical compounds. In 1972 the Chemical Substances Index was added as a new type of index. Others are an Author Index, a Formula/Ring Index, and a Patent Index. An Index Guide, published periodically, gives terms that are used to index subjects in the Subject Index. The Index Guide should be consulted for hard copy searches each time prior to researching a topic in the Subject Index. The Formula/Ring Index lists formulas as C_xH_y, with elements other than C and H following H alphabetically. The Formula Index is used if there is difficulty naming the compound and the structure is known. CA names are often different than IUPAC names.

To find a compound using the Chemical Substances Index you should be familiar with the rules of organic nomenclature. Look under the parent name of the compound and then locate the appropriate substituents alphabetically. For example, 2-bromonaphthalene is listed under **naphthalene, 2-bromo.** When the correct citation is found, check the CA references following. For example, 5-bromo-2-hydroxybenzaldehyde is indexed in the Chemical Substances Index under **Benzaldehyde, 5-bromo-2-hydroxy [1761-61-1].** The number in brackets is the registry number for this compound. A CAS Registry Number is assigned in CA for every substance. Over 20 million organic and inorganic compounds have been assigned CAS Registry Numbers. Registry Numbers are particularly useful for online searching.

An example of a citation for 5-bromo-2-hydroxybenzaldehyde is a CA reference, which appears as follows: **128: 308424d.** The number **128** is the volume number; **308424d** is the abstract number. Looking up abstract 308424d in CA volume 128 gives reference to a paper that contains information about the solid phase synthesis of isoxazolidines using 5-bromo-2-hydroxybenzaldehyde.

In very old volumes of CA, the citations refer to page numbers, and small superscript numbers refer to location on the page. The citation **24: 5345[6], 1930,** is the year 1930, volume 24, page 5345. The raised 6 indicates that the abstract begins about six-tenths or two-thirds down the page. If **P** appears in the CA citation, this is a patent abstract. If an **R** appears, it is an abstract of a review article.

After locating the volume and abstract number of an article, go to the abstract in CA. Abstracts are listed in numerical order by volume. Read the abstract, note the reference to the primary review literature or patent, and then go back to consult the original article. Not all journals are available in all libraries, and some of the articles may appear in languages other than English.

A few ground rules apply on how to proceed. When given a common name of a chemical or a chemical topic, first check the Index Guide published by *Chemical Abstracts.* In order to get the correct name or correct synonym, always check the Index Guide prior to using the Subject Index. If the structural formula is available, locate the compound using the Formula Index, then switch to the Chemical Substances Index (since 1972) or Subject Index (before 1972).

For locating journals, proceed from an abbreviation of the title to the actual title. The titles of all journals indexed by CA are given in the *Chemical Abstracts Service Source Index* (CASSI). Some of the more common journal titles may be inferred—*J. Am. Chem. Soc.* is the abbreviation for the *Journal of the American Chemical Society. J. Org. Chem.* is the abbreviation for the *Journal of Organic Chemistry.* Others may range from being a bit difficult to being totally impossible to decipher without resorting to CASSI.

Most journals consist of primary literature, that is, reports of work done for the first time. There are a few journals that are reserved for publication of review articles, while others contain both reports of new work and review articles. Reviews are summaries of work on important topics. Review articles are considered to be secondary literature. Review journals include *Accounts of Chemical Research, Chemical Reviews,* and *Chemical Society Reviews.* Other secondary literature includes books and monographs such as *Organic Reactions.*

Searching CA

Online searching using the CA database can be done using SciFinder, SciFinder Scholar, or STN. Using computer databases can facilitate searching for specific compounds and their preparation. For example, entering the CAS Registry Number 1761-61-1 and chemical synthesis will allow access to several citations. Computer accessible references for chemical abstracts go back to 1967. Another databse called CAOLD covers in part the period from 1957–1966.

CASREACT is another useful CAS database. Entering a starting material and product or their CAS Registry Numbers will reveal methods of conversion, if known, and pertinent references. Online database searching is gaining in importance as electronic databases expand and searching techniques are updated.

It is important to understand that experimental procedures and physical properties of new compounds are reported in the primary literature when the compounds are first synthesized. If such a report occurred in 1920, subsequent authors who use the procedure generally refer to the original procedure. Therefore, many procedures for preparation of simple compounds appear in the older literature. Unless new procedures were recently discovered or if revised physical properties are given for a compound, it may be necessary to go to the early literature to find the original procedures and physical properties.

Because modifications and improvements for known procedures have been reported for many reactions over the years, it is recommended that CA searches should start with the most recent indices. When searching electronically, the database is accessible back to 1967. If necessary, search the printed version of CA starting with the semidecennial indices from 1966. References to compounds used recently will give citations of early literature listing physical properties of compounds. Working back sequentially is done most efficiently using the collective (semidecennial) indices. Prior to the ninth collective index, these were decennial indices. Searches done in this manner will likely turn up modern variations of the reaction as well as references to the original procedure.

CA gives references for a compound that may have to do with its application or spectroscopic properties that have nothing to do with its synthesis. CA does not list all compounds reported in an article in the abstract. However, CA indexes each compound prepared.

Beilstein

The other major database is the *Beilstein Handbook of Organic Chemistry* (Beilstein). The database is particularly useful for finding preparations and physical constants of organic compounds. Often, it is not necessary to refer to the primary literature.

The *Beilstein Handbook of Organic Chemistry* covers literature published since 1779. The printed database is published in series. Each series covers a particular time period. All volumes are in German up to the Fifth Supplement. Volumes since the Fifth Supplement are in English. Compounds are arranged according to their structures. Information provided about each compound includes preparation and purification, physical and chemical properties, and other relevant information, such as natural occurrence and isolation from natural products.

An important feature of Beilstein is that it goes back further in time than CA. For directions on using Beilstein, see *How to Use Beilstein,* Beilstein Institute, Frankfurt, FRG

1978. See also *Beilstein Dictionary,* also published by the Beilstein Institute, and *A Programmed Guide to Beilstein's Handbuch* by Olaf Runquist (Burgess Publishing Company, Minneapolis, 1966). A guide to using Beilstein is *The Beilstein System: Strategies for Effective Searching* (S. R. Heller, American Chemical Society, Washington, D.C. [1997]).

Beilstein is organized according to the main work (*hauptwork*) and five supplements (*erganzwerke*). The main work covers literature through 1909. The First Supplement covers 1910–1919; the Second Supplement, 1920–1929; the Third Supplement, 1930–1949; the Fourth Supplement, 1950–1959; and the Fifth Supplement, 1960–1979.

Searching Beilstein

Beilstein is organized in four major divisions starting from simple compounds and working toward more complex compounds. These are acyclic (noncyclic) compounds, isocyclic (cyclic) compounds, heterocyclic compounds, and natural products. Each of these divisions is subdivided. For example, the first two divisions are each subdivided into twenty-eight functional classes (functional groups). Hydroxy compounds (alcohols) are the second subdivision. Carboxylic acids are the fourth subdivision.

Substances are indexed in the formula index (*formelregister*) and subject index (*sachregister*) for each volume (*band*). For many common organic compounds, it is possible to locate one or more preparations for a substance, its physical properties, main chemical reactions, derivatives, and references. This information might be found in the main work or in one of the supplements, depending upon the year when the compound was first reported. After this first report, additional data may be cited in later supplements.

Like CA, Beilstein is available as a computer database, asccessible as CrossFire Belstein. CrossFire Beilstein searches reactions and properties of millions of organic compounds reported since 1980. CrossFire Beilstein is a product of MDL Information Systems, Inc. Searching can be done by drawing in the desired structure or a section of a structure, by submitting the name of the compound, or by submitting the name of the compound or its CAS Registry Number.

Some Representative Searches

Example I: Find a procedure for the oxymercuration-demercuration of a simple alkene such as 1-methylcyclohexene.

Locate one or more articles in a periodical or periodicals where an experimental procedure is given for the reaction; locate the article where the method was first used and obtain information about reaction conditions, yields, and side products (if any).

The first time a method is developed to perform an organic reaction, it is likely to be the subject of an entire research publication, or at least the major part of one. An example is an important method for hydration of alkenes: oxymercuration-demercuration. Oxymercuration-demercuration is the method recommended for Markovnikov addition of water to alkenes. Here are three approaches to obtaining necessary information. Electronic database searching is fastest.

Approach 1:   Search CAS online using STN or SciFinder. Searching for oxymercuration and 1-methylcyclohexene gives pertinent references very quickly. Searching CASREACT and CrossFire Beilstein can give many references to oxymercuration that will lead indirectly to earlier articles. The original reference is by H. C. Brown and P. Geoghagen, Jr. *J. Am. Chem. Soc. 89* (1967):1522.

Approach 2:   Use the printed version of *Chemical Abstracts* (CA) for this assignment. This approach will be more time-consuming than Approach 1. The subject is common enough to be found in monographs and review articles. However, if the choice is made to go to CA, the first step is to check the most recent CA Index Guide for **oxymercuration.** Go to the most recent collective general subject index and locate review articles on oxymercuration. Recall that review articles have an **R** in the CA reference. A review article will list several periodical citations dealing with the topic. Searching the printed version of CA is generally done only when online searching does not provide the desired information.

Approach 3:   An indirect method is to locate advanced texts that list references for important organic reactions. Any topic that is found in the introductory text will be covered in greater detail in advanced reference books. Find references for "oxymercuration-demercuration." Citations for review articles are particularly helpful because they may give a wealth of information about a particular topic.

Some of the advanced organic reference works are March's *Advanced Organic Chemistry* (5th ed., M. Smith and J. March, Wiley, 2001); *Advanced Organic Chemistry* (4th ed., Parts A & B, F. Carey and R. Sundberg, Plenum Press, 2001); and *Comprehensive Organic Transformations,* (2nd ed., R. Larock, VCH Publishers, New York, 1999). *Organic Syntheses,* collective volumes I–IX (Wiley), also available in an electronic searchable format (http://www.orgsyn.org), give procedures for the preparation of many organic compounds.

Example II: Find a procedure for the preparation of 3-ethyl-2,4-pentanediol.

Approach 1:   Search CAS STN online to locate the CAS Registry Number. Doing so turns up a registry number of 66225-33-0, but no citations. This does not mean that no citations can be found. Searching by the chemical name 3-ethyl-2,4-pentanediol revealed several references unrelated to preparation or synthesis of the compound. Interesting, however, one preparative procedure was given from Compt. rend. 224 (1947): 1234 and another from Ber. 42 (1909): 2500. These citations obviously are from years before the advertised capabilities of the online database. The compound is pretty simple. Checking for it in the Aldrich Chemical catalog gave no listing. If a listing had been found a reference to Beilstein might have been listed. If a preparation had not been found in the online search, a choice would have been to go to the 1966 collective index of CA using the Chemical Substances Index under **2,4-pentanediol, 3-ethyl** and work backward sequentially. Locate the 3-ethyl compound and look for an article that describes the preparation. There may be more than one type of preparation reported, but the assignment is to locate one of these. Remember that CAS online only covers from 1967 to the present and that preparations of simple compounds may have been done decades earlier.

Approach 2:   Consult Beilstein. The compound appears to be simple, and it may be listed in the main work or in one of the supplements. Start by consulting the formula index or the subject index. For such a simple compound, using Beilstein can often be the best approach.

Example III: Find an article on ozonolysis of 8-hydroxyquinoline and determine the nature of the reaction product(s).

Approach 1:   Use the CAS online database through STN or SciFinder. Enter the search terms **8-hydroxyquinoline** and **ozonolysis.** Obtain all references since 1967.

The database holds information from 1967 to the present. Obtain abstracts of citations dealing with the search.

Approach 2: Use the printed version of *Chemical Abstracts* to look up **8-hydroxy-quinoline** in the Index Guide to see if this name and **ozonolysis** are legitimate search words. (This was not necessary for the online search.)

CA has used the term **ozonization,** but not **ozonolysis,** since CA was first published. Consulting the 2002 Index Guide, it should be apparent that searching the reactions of ozone in CA could be complicated. For example, there are separate categories for **ozone, reactions; ozonization catalysts; heat of ozonization;** and **kinetics of ozonization.** There is a separate category under **water treatment.**

When looking for a type of reaction of a particular compound, such as **ozonolysis of 8-hydroxyquinoline,** it is better to search under the name of the compound because there is just one name used by CA. Go to the most recent collective Subject Index or to the 1966 collective index if you did an online search first that did not give any pertinent citations. Look for **quinoline, 8-hydroxy** and if necessary, work backward sequentially. Look under **quinoline, 8-hydroxy,** and locate **ozonization.** Take down the CA references. Go to the periodical and get the necessary information about products of the reaction. Note that the CA summary has a brief paragraph about each article cited. Sometimes there is sufficient information in that paragraph, making it unnecessary to consult the original reference.

Assignment for Part A

1. Carry out the example searches above.
2. Locate a literature preparation for one of the compounds below (or one assigned by your instructor). Turn in photocopies of the first page of the article and the page containing the experimental procedure. The compounds are:

3-azido-1,2-propanediol
Duocarmycin 5A, total synthesis
hexachloronaphthalene
2-ethylthiazolidine
2-aziridinecarbonitrile
1,4,5,8-tetrahydroxy-9,10-anthracenedione
1-ethyl-2-thiourea

Part B: Searching Science Library Databases

Science Direct, Wiley Interscience Journals, KluwerOnline, Dekker, and Academic Search Premier are among the databases that contain information on terms used in titles or keywords taken from journal articles from the recent scientific literature. To locate information about a particular topic, searching a database, such as one of these, may furnish some up-to-date references and even full-text articles.

There are no strict rules concerning searches on these electronic databases. However, you should set your search terms to be specific enough to obtain a reasonable number of references. For example, a search to determine effective drug treatment for Tourette's syndrome, a fairly common genetic disorder that particularly affects children, gave 171 records using **Tourette** as a keyword search term in one database. The same search in another database produced 736 citations. To narrow down the listings, one approach is to scan several of the most recent citations (listed first) and then to do a revised search, linking keywords or phrases of interest. For example, the search on Tourette was combined with a search on therapeutics to yield 43 listings. This is a more manageable number and the citations are more relevant.

The complete journal reference is given for each citation. Sometimes the full article may be available online. Older references often have to be accesed by consulting the printed journal.

Exercise to Part B

1. Use a science library database to find current information about each of the following topics. Write a brief report on each, citing references for:
    a.   drugs used in the treatment of Tourette's syndrome.
    b.   the genetics of Tourette's syndrome.
    c.   the treatment of Addison's disease.

Part C: Searching on the Web

Web searching for scientific information is possible from several uniform resource locators (URLs). For example, accessing information about the preparation of organic compounds is possible from http://www.orgsyn.org. This is the same information that is contained in Organic Syntheses, Collective Volumes I–IX (Wiley). The web version is supported by Wiley, DataTrace, and CambridgeSoft. Another useful resource, the Aldrich Catalog, can be found at http://www.sigmaaldrich.com/Brands/Aldrich.html.

Accessing reliable scientific information using a search engine such as Yahoo or Google is often a hit-or-miss proposition. From Yahoo, a very nice site is available on the synthesis of Duocarmycin SA. However, your time is usually better spent by logging on to STN, SciFinder, or CrossFire Beilstein to locate specific, reliable scientific information.

Exercise to Part C

1. Do a web search on Tourette's syndrome using Yahoo or Google. Compile a brief report on the nature of Tourette's syndrome. Compare the information retrieved from the web search with information obtained using the science library databases.
2. Do a web search on the process of decaffeinating coffee. Compile a brief report on the various methods used to decaffeinate coffee.
3. Do a web search to locate a procedure utilizing N-methyl C-phenyl nitrone. A procedure can also be found using CA online via STN or SciFinder. Compare and contrast these methods of searching.

Critical Thinking Questions

1. For each of the topics below, decide what method of searching (*Chemical Abstracts,* online searching of CAS, searching a science library database, or web searching) will be the most effective in obtaining the requested information. Locate an original periodical article describing the topic and turn in a photocopy of the first page of the article.

    a.   preparation of sodium alkyl sulfonates
    b.   physiological effects of procaine
    c.   total synthesis of (+)– compactin

2. Locate a literature preparation for one of the substances below or one assigned by your instructor. Turn in photocopies of the article and the page containing the experimental procedure. The substances are:

    a.   saccharin
    b.   cyclamate
    c.   Ibuprofen
    d.   Prozac

3. Locate a *Chemical Abstracts* citation on how the antibiotic bacitracin can aid in the absorption of insulin.

4. Record the *Chemical Abstracts* citation by Stefan O. Mueller in 1996 on genotoxicity.

5. Locate the *Chemical Abstracts* citation **127:204941h.** To what product does the article refer?

6. Locate the U.S. patent for enteric-coated nitric oxide tablets and record the *Chemical Abstracts* citation.

7. Record *Chemical Abstracts* citations for each of the following topics:

   a. regioselective oligomerization of thiophenes with ferric chloride
   b. synthesis and properties of combinatorial libraries of phosphoramidates
   c. synthesis of head-tail hydroxylated nylons from glucose
   d. synthesis of substituted indoles from aromatic amines
   e. preparation of acrylonitrile via regioselective conversion of terminal alkynes
   f. preparation of bicyclic compounds via Diels-Alder reaction of vinyl boranes
   g. regioselective bromination of alkylnaphthalenes with N-bromosuccinimide in acrylonitrile.

8. The drugs in this question each contain strained carbocyclic rings.

   a. locate the CAS Registry Number for the antiobesity drug sibutramine (trade name, Meridia). Give the complete citation for a preparation of the drug or for its hydrochloride salt.
   b. Locate the CAS Registry Number and chemical name for the drug called efavirenz. Give the citation for the patent on the preparation of crystalline forms of the drug assigned to the DuPont Pharmaceutical Company.

Experiment 29.5: Stereochemistry, Molecular Modeling, and Conformational Analysis

The purpose of this experiment is to practice drawing and rotating three-dimensional structures using a molecular model kit or a molecular modeling program. You will:

♦ represent optically active compounds as 3-D structures (wedge drawings), Newman projections, side-view (andiron) or sawhorse structures, and Fischer projections.

♦ identify stereogenic carbons and assign configuration (R or S).

♦ identify relationships between stereoisomers (meso, enantiomers, diastereomers).

♦ evaluate stabilities of conformational isomers and make predictions about the position of equilibrium.

Being able to envision molecules in three dimensions is a critical skill in organic chemistry, where the difference in how atoms are bonded can drastically alter the chemical and physical properties of the molecule. Consider the two molecules shown here, which differ only in the placement of the hydrogen and isopropyl substituent at C-4. One isomer ((S)-carvone) smells like caraway or licorice, while the other isomer ((R)-carvone) smells like spearmint. The small difference in how atoms are bonded has a large effect on the physical properties as the compounds bind to specific and different receptors on olfactory neurons.

A more tragic example of the importance of stereochemistry is thalidomide, a sedative developed in Germany in 1957. Thalidomide (N-phthalimidoglutarimide) has one stereogenic center (one carbon bonded to four different groups). One stereoisomer

(S)-carvone   (R)-carvone
caraway   spearmint

((S)-thalidomide) is an effective sedative with antinausea properties. The other stereoisomer ((R)-thalidomide), is a tetragenic agent. This isomer caused severe birth defects in 12,000 babies born to mothers who took the drug during pregnancy to ease morning sickness. This isomer blocks fetal growth, causing pharcomelia, or the development of flipperlike arms. The structures of thalidomide are shown below:

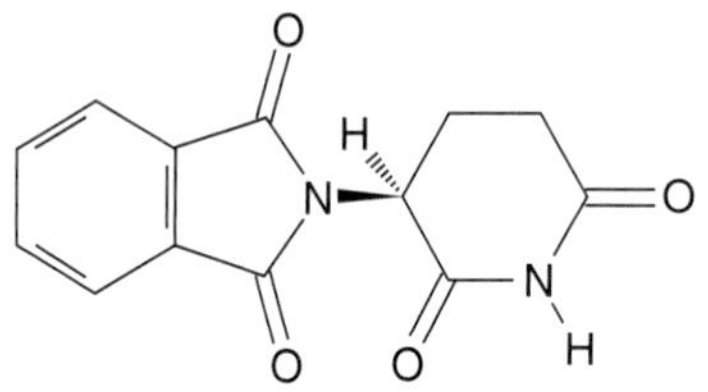

(S)-thalidomide
sedative

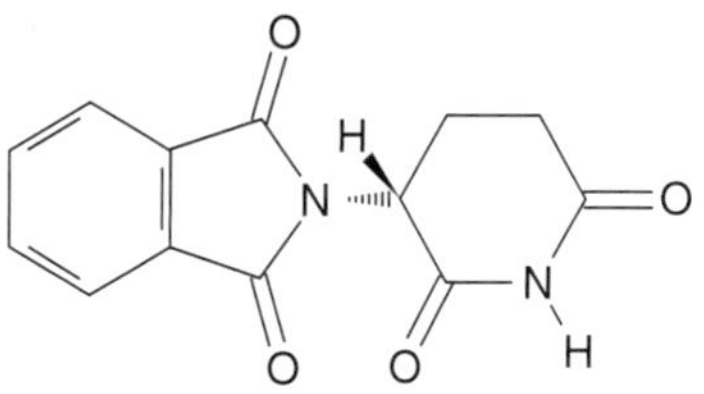

(R)-thalidomide
tetragen

Because stereochemistry is such a critical component of the organic chemistry curriculum, various methods of representing three-dimensional molecules have been developed to help students envision spatial relationships. The most common methods are side-view structures, Newman projections, and Fischer projections. These are discussed below.

Another method of analyzing three-dimensional structures is with molecular-modeling computer software, such as Spartan, CAChe, Chem3D, or similar modeling programs. Constructing computer models helps students visualize molecular structures. The programs can calculate energies, ring strain, and other properties to evaluate preferred conformations.

For complete descriptions of optical activity, stereoisomerism, conformational isomers, ring strain, and other concepts relating to stereochemistry and conformational analysis, refer to your organic chemistry textbook.

3-D Structures

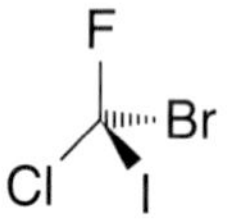

These drawings use three types of lines to simulate three-dimensional tetrahedral molecules: straight lines indicate that the bond lies in the plane of the paper; solid wedged lines indicate that the bond is coming out above the paper; dashed or dotted lines indicate that the bond is going away below the plane of the paper. In the example shown here, the fluorine and chlorine atoms lie in the plane of the paper, the iodine atom is coming out and the bromine atom is going back.

Side-view Structures

Just as the name implies, a side-view structure shows the carbon skeleton as a straight line with the substituents branching off from it. The staggered conformation of *meso*-1,2-dibromo-1,2-dichloro-1,2-difluoroethane is shown here. (The plane of symmetry in this compound can be most easily seen by looking at an eclipsed conformation.)

is equivalent to

staggered conformation

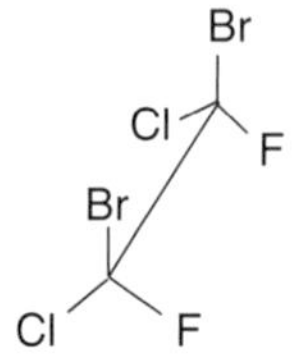

eclipsed conformation

Newman Projections

A Newman projection focuses on the interactions between substituents on adjacent carbon atoms. For the Newman projection of *meso*-1,2-dibromo-1,2-dichloro-1,2-difluoroethane, look down the C_1–C_2 bond axis. The junction of the three bonds in the center of the circle represents the C_1 atom and the circle the C_2 atom. This compound has a plane of symmetry, which is most easily seen in the eclipsed conformation (shown slightly off eclipsed for easier visualization).

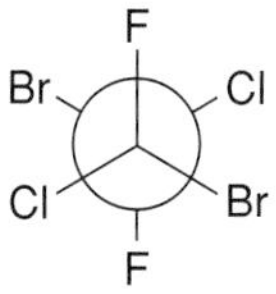

F
Br   Cl
Cl   Br
F
staggered

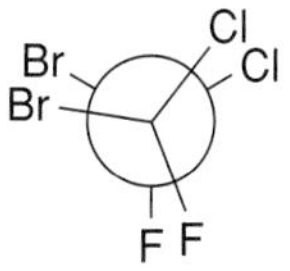

Cl
Br   Cl
Br
F F
eclipsed

Fischer Projections

In a Fischer projection, a cross represents a stereogenic center:

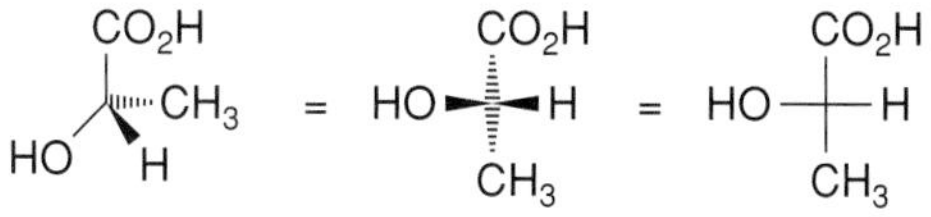

$$CO_2H \quad \quad CO_2H \quad \quad CO_2H$$

HO⟍   ⟋CH₃   =   HO ◄─ H   =   HO ── H

HO   H   CH₃   CH₃

The horizontal lines of the Fischer projection indicate that the bonds project toward the viewer and the vertical line indicates that the bonds project away from the viewer. There are certain rules that must be obeyed when using a Fischer projection: (1) a Fischer projection must be kept in the plane of the paper (i.e., it cannot be turned over); and (2) a Fischer projection cannot be rotated by 90°, but can only be rotated by 180°.

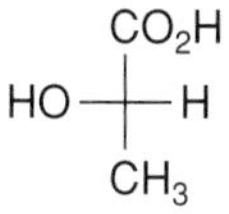

CO_2H
HO ── H   is NOT equivalent to
CH_3

OH
CH₃ ── CO_2H
H

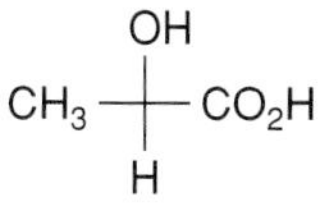

Rotation by 90° does not produce an equivalent structure.

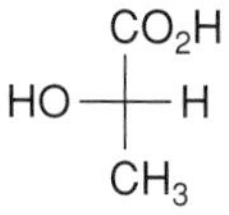

CO_2H
HO ── H   is NOT equivalent to
CH_3

CO_2H
H ── OH
CH_3

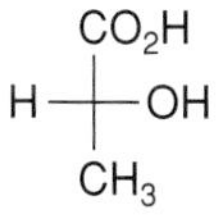

Flipping the Fischer projection over does not produce an equivalent structure.

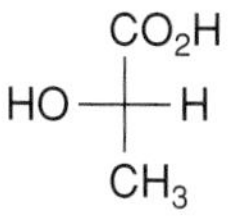

CO_2H
HO ── H   IS equivalent to
CH_3

CH_3
H ── OH
CO_2H

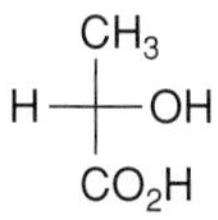

Rotating the Fischer projection by 180° does produce an equivalent structure.

If a structure has more than one stereogenic center, each stereogenic center is indicated with a cross. Below are the Fischer projections for (2S,3S)-2,3-butanediol and (2S,4S)-2,4-pentanediol, respectively:

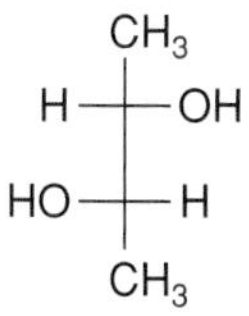

CH_3
H ── OH
HO ── H
CH_3

(2S,3S)-2,3-butanediol

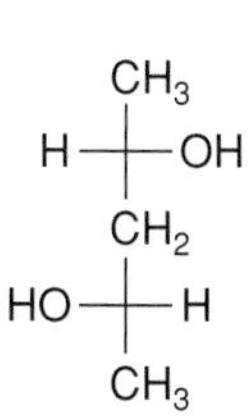

CH_3
H ── OH
CH_2
HO ── H
CH_3

(2S,4S)-2,4-pentanediol

In Part A of this experiment, you will use a model kit to assemble molecular models of compounds having one or more chiral carbons (also called stereogenic center or stereogenic carbon). You will rotate the models in the specified conformations and draw them in the lab notebook. In Part B, you will construct models on the computer, using a chemical drawing program such as ChemDraw, ChemWindow, ISIS, or other software program. After rotating the compounds into the specified conformations, you will paste the drawings into an electronic lab report. The drawings can also be imported into a molecular modeling program (such as Chem3D) for further analysis.

Molecular Modeling

In Part C of this experiment, you will use a computerized molecular modeling program (such as Chem3D) to analyze strain energy of an alkane and a cycloalkane. Strain energy is dependent upon bond stretching, angle bending, torsional strain, and van der Waals strain:

$$E_{strain} = \Sigma E_{bond\ stretching} + E_{angle\ bending} + E_{torsional} + E_{van\ der\ Waals}$$

Bond stretching strain refers to the increase in potential energy when sp^3 carbon-sp^3 carbon and sp^3 carbon-hydrogen bonds are distorted from their ideal bond distances of 153 pm and 111 pm, respectively. Angle bending refers to the strain resulting from varying bond angles from ideal, such as sp^3 carbons deviating from the tetrahedral bond angle of $109.5°$. Torsional strain results from eclipsing interactions. Van der Waals strain results from steric interactions between nonbonded atoms, such as gauche or 1,3-diaxial interactions in cyclohexanes. In this exercise, you will use a computer modeling program to investigate the relative energies of rotational isomers of butane and disubstituted cyclohexane. Based on the energy differences between the Newman projections, you will plot a potential energy diagram for the rotation about the C_2-C_3 bond in butane, calculate the energy differences between eclipsed and staggered conformations, and determine the lowest energy conformation. Similarly, you will determine the energy differences between the chair and boat rotational isomers for cyclohexane and methylcyclohexane, construct potential energy diagrams, and determine the lowest energy conformations.

In Part D of this experiment, you will use a molecular modeling program such as Spartan, ChemDraw, CAChe, Argus, or similar modeling program to explore properties of stereoisomers and conformational isomers. You will calculate relative energies of the various conformers and determine conformer distributions.

Prelab Assignment

1. Review the sections in the lecture textbook that deal with optically active compounds. Be able to define the following terms and give examples: enantiomer, racemic mixture, optical activity, stereogenicity, stereogenic center, meso, diastereomer, conformational isomers, and geometric isomers.
2. Draw Newman projections for *gauche*-butane and *anti*-butane. Indicate which conformation is more stable.
3. Draw methylcyclohexane in the two chair conformations. Indicate which conformation is more stable.

Part A: Molecular Models Using a Model Kit

Check out a molecular model kit from the stockroom or bring your own. Construct models for each of the following and draw the structures in your lab notebook.

1. Construct models of (R)-2-chlorobutane and its enantiomer, (S)-2-chlorobutane, or other structures as assigned by the instructor. For each drawing, label each stereogenic center as R or S.
   a. Draw a 3-D (wedge) structure for each enantiomer.
   b. Draw a Newman projection for each enantiomer.
   c. Draw a Fischer projection for each enantiomer.
2. Construct models of 2-bromo-3-chlorobutane or other assigned compound. Construct as many stereoisomers as possible. (How many stereoisomers are possible?) For each drawing, label each stereogenic center as R or S.
   a. Draw a 3-D (wedge) structure for each of the stereoisomers.
   b. Draw side-view projections for each pair of enantiometers.
   c. Draw Newman projections for each pair of enantiomers.
   d. Draw Fischer projections for each stereoisomer.
   e. Give the relationship (enantiomer or diastereomer) for each pair of drawings.
3. Construct models of 2,3-dibromobutane or other assigned compound. Construct as many stereoisomers as possible. (How many stereoisomers are possible?) Label each stereogenic center.
   a.. Draw a 3-D (wedge) structure for each of the stereoisomers.
   b. Draw side-view projections for each pair of enantiomers.
   c. Draw Newman projections for each pair of enantiomers.
   d. Draw Fischer projections for each stereoisomer.
   e. Give the relationship (enantiomer or diastereomer) for each pair of drawings. One of the stereoisomers has a plane of symmetry. Label it as meso and explain why it is unique.

Part B: Computer Modeling

Open a web browser and a chemical structure drawing program like ChemDraw, ChemWindow, or ISIS. Type the following URL in the address line:

http://www.mhhe.com/schoffstall

Follow the directions on the web page. When finished, submit the document to your instructor.

Part C: Conformational Analysis of Rotational Isomers

Open a web browser, a chemical structure drawing program (like ChemDraw, ChemWindows, or ISIS), and a molecular modeling program (like Chem3D). Type the following URL in the address line:

http://www.mhhe.com/schoffstall

Follow the directions on the web page. When finished, submit the document to your instructor.

Part D: Using Molecular Modeling to Analyze Rotational Conformations of Stereoisomers

Amino acids are important biological molecules that often contain more than one stereogenic center. Enzymes that metabolize amino acids are very selective, reacting with only one specific stereoisomer. In this exercise, you will analyze the geometries of four stereoisomers of the essential amino acid threonine. Under physiological conditions,

threonine

threonine (2-amino-3-hydroxybutanoic acid) is ionized, so that the amino group is protonated to form the ammonium group and the carboxylic acid group is deprotonated to form the carboxylate group.

Open a molecular modeling program and construct three-dimensional models of all four stereoisomers of threonine. (It may be easier to build one molecule, do the complete exercise, and then change the molecule accordingly.) Orient the R,R-stereoisomer so that the hydroxyl group (OH) and the carboxylate group (CO_2^-) are gauche to one another. Minimize the energy and record. Measure the $C-N$ bond length. Examine dipole moments of each of the bonds in the molecule and record. Repeat with the S,S, the R,S, and the S,R stereoisomers. Compare the energies of the stereoisomers. Do enantiomers have the same energy or different energies? Do diastereomers have the same energy or different energies? Analyze the similarities and differences in the $C-N$ bond distances of enantiomers and diastereomers. Finally, examine the dipole moments of the enantiomers and diastereomers and account for similarities and differences.

Critical Thinking Questions

1. Convert a Fischer projection of (S)-2-bromopentane into a side-view projection.
2. Convert a Newman projection of (R)-2-butanol into a side-view projection.
3. Convert a side-view projection of (R)-2-butanol into a Fischer projection.
4. Determine whether each of the following cyclic structures has one or more planes of symmetry. Label each stereogenic center as R or S.

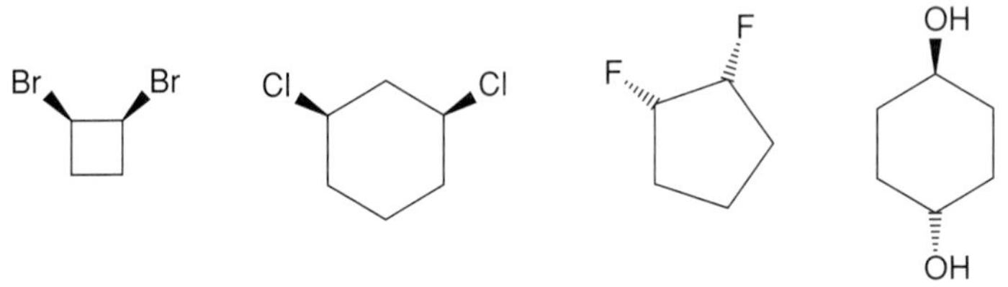

5. Cholesterol has the structure shown below. Label each stereogenic center as R or S.

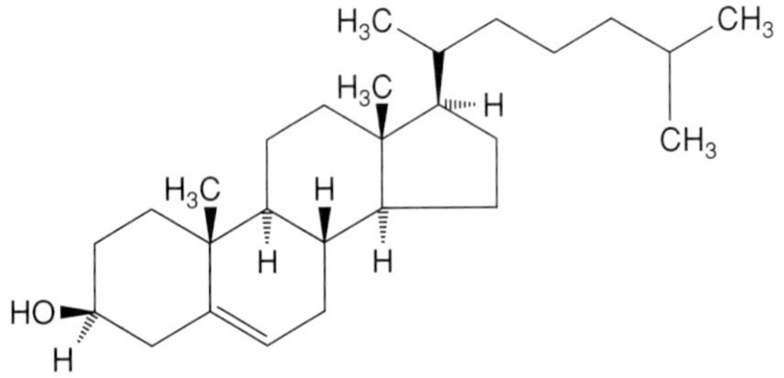

cholesterol

6. *Molecular modeling assignment:* Construct models of cyclohexanol, with the $O-H$ group in the axial position and the other with the OH group in the equatorial position. Minimize energies, then determine the energy difference between the two conformations of cyclohexanol. Which conformer is more stable?
7. (S,S)-Tartaric acid has a twofold axis of symmetry. Does this mean that it can't be optically active? Explain.

Acknowledgments

Dr. David Anderson, CU–Colorado Springs; Dr. Dabney Dixon, Georgia State University.

Appendix A

Tables of Derivatives for Qualitative Organic Analysis

Table A1 Alcohols

Name of compound	B.P °C	M.P °C	Phenyl-urethane	α-Naphthyl-urethane	p-Nitro-benzoate	3,5-Dinitro-benzoate
			Derivatives (M.P.) °C			
2-Propanol	82	—	75	106	110	123
2-Methyl-2-propanol	82	25	136	101	—	142
2-Propen-1-ol	97	—	70	108	28	49
1-Propanol	97	—	57	80	35	74
2-Butanol	99	—	64	97	25	76
2-Methyl-2-butanol	102	—	42	72	85	116
2-Methyl-1-propanol	108	—	86	104	69	87
3-Pentanol	116	—	48	95	17	99
1-Butanol	117	—	61	71	—	64
2-Pentanol	119	—	—	74	17	62
3,3-Dimethyl-2-butanol	120	—	77	—	—	107
2,3-Dimethyl-2-butanol	120	—	65	101	—	111
3-Methyl-3-pentanol	123	—	43	83	—	96
2-Methyl-1-butanol	128	—	31	82	—	70
2-Chloroethanol	131	—	51	101	—	98
3-Methyl-1-butanol	132	—	56	68	21	61
4-Methyl-2-pentanol	132	—	143	88	—	65
2,2-Dimethyl-1-butanol	136	—	65	80	—	—
1-Pentanol	138	—	46	68	11	46
2,4-Dimethyl-3-pentanol	140	—	95	95	155	—
Cyclopentanol	140	—	132	118	—	115
1-Hexanol	158	—	42	59	—	58
Cyclohexanol	161	25	82	129	50	112
2-Furfuryl alcohol	172	—	45	129	76	81
1-Heptanol	176	—	60	62	10	46
1-Octanol	195	—	74	67	12	61
Benzyl alcohol	205	—	77	134	85	113
2-Phenylethyl alcohol	220	—	78	119	62	108
Cinnamyl alcohol	257	33	90	114	78	121
1-Menthol	—	42	111	119	61	153
1-Hexadecanol	—	50	73	82	58	66
1-Octadecanol	—	60	80	—	64	66
Benzhydrol	—	68	139	135	131	141
Benzoin	—	133	165	140	123	—
Cholesterol	—	148	168	176	185	—
Ergosterol	—	165	185	202	—	202

Table A2 Phenols

Name of compound	B.P. °C	M.P. °C	Phenyl-urethane	α-Naphthyl-urethane	p-Nitro-benzoate	3,5-Dinitro-benzoate	Aryloxy-acetic acid	Bromo-derivative
					Derivatives (M.P.) °C			
2-Chlorophenol	175	7	121	120	115	143	145	mono: 48–49 di: 76
2-Hydroxybenzaldehyde	197	2	133	—	128	—	132	—
Phenol	180	42	126	133	127	146	99	tri: 95
3-Methylphenol	203	12	125	128	90	165	—	tri: 84
4-Allyl-2-methoxyphenol	255	19	95	122	81	131	—	tetra:118
2-Methoxyphenol	205	32	136	118	93	141	—	trio: 116
4-Chlorophenol	—	43	149	166	171	186	156	mono: 34 di: 90
2-Nitrophenol	—	45	—	113	141	155	158	di: 117
4-Ethylphenol	219	47	120	128	81	133	—	—
4-Bromophenol	—	64	140	169	180	191	157	tri: 171
3,5-Dimethylphenol	—	68	148	—	109	196	111	tri: 166
2,5-Dimethylphenol	212	75	162	172	87	137	—	trio:178
α-Naphthol	—	94	177	152	143	217	193	di: 105
3-Nitrophenol	—	97	129	167	174	159	156	di: 91
Catechol	—	104	169	175	169	152	136	tetra: 192
Resorcinol	—	110	164	—	175; 182	201	175; 195	tri: 112
4-Nitrophenol	—	114	156	150	159	186	187	di: 142
2,4-Dinitrophenol	—	114	—	—	139	—	—	mono: 118
β-Naphthol	—	122	155	156	169	210	95	84
Pyrogallol	—	133	173	—	230	205	198	di: 158

Table A3 Aldehydes

Name of compound	B.P. °C	M.P. °C	Semicar-bazone	2,4-Dinitro-phenyl-hydrazone	p-Nitro-phenyl-hydrazone	Phenyl-hydrazone	Oxime
				Derivatives (M.P.) °C			
2-Methylpropanal	64	—	125	187	130	Oil	Oil
Butanal	74	—	95; 106	123	87	93	Oil
3-Methylbutanal	92	—	107	123	110	Oil	48
Pentanal	103	—	—	98; 106	—	—	52
trans-2-Butenal	104	—	199	190	184	56	119
Ethoxyethanal	106	—	—	117	114	—	—
4-Methylpentanal	121	—	127	99	—	—	—
Hexanal	131	—	106	104	—	—	51
3-Methyl-2-butenal	135	—	223	182	—	—	—
Benzaldehyde	179	—	222	237	190	158	35
2-Hydroxybenzaldehyde	197	—	231	248	227	142	57
Geranial	228d	—	164	108	—	—	143
Cinnamaldehyde	252d	—	215	255d	195	168	64

Table A3   *(continued)*

Name of compound	B.P. °C	M.P. °C	Derivatives (M.P.) °C				
			Semicar-bazone	2,4-Dinitro-phenyl-hydrazone	*p*-Nitro-phenyl-hydrazone	Phenyl-hydrazone	Oxime
1-Naphthaldehyde	—	34	221	—	224	80	98
Piperonal	—	37	230	265d	199	102	110; 146
2-Aminobenzaldehyde	—	40	—	—	220	221	135
2-Nitrobenzaldehyde	—	44	256	265	263	156	102; 154
4-Chlorobenzaldehyde	—	48	230	254	237	127	110
4-Bromobenzaldehyde	—	57	228	128; 257	207	113	157; 111
3-Nitrobenzaldehyde	—	58	246	293d	247	120	120
2-Naphthaldehyde	—	60	245	270	230	205; 217	156
Vanillin	—	80	230	271d	227	105	117
4-Nitrobenzaldehyde	—	106	221	320	249	159	133; 182
4-Hydroxybenzaldehyde	—	115	224	280d	266	177	72;112

Table A4   Ketones

Name of compound	B.P. °C	M.P. °C	Derivatives (M.P.) °C				
			Semicar-bazone	2,4-Dinitro-phenyl hydrazone	*p*-Nitro-phenyl-hydrazone	Phenyl-hydrazone	Oxime
2-Propanone	56	—	190	126	148	42	59
2-Butanone	80	—	146	116	128	Oil	Oil
3-Methyl-2-butanone	94	—	113	120	108	Oil	Oil
3-Pentanone	102	—	138	156	144	Oil	Oil
2-Pentanone	102	—	106; 112	143	117	Oil	Oil
Pinacolone	106	—	157	125	—	Oil	75
4-Methyl-2-pentanone	118	—	132	95	—	—	Oil
3-Hexanone	125	—	113	130	—	—	Oil
2-Hexanone	128	—	125	106; 110	88	Oil	49
Cyclopentanone	130	—	210	146	154	55	56
4-Methyl-2-hexanone	142	—	120; 128	—	—	—	—
4-Heptanone	144	—	132	75	—	—	Oil
2-Heptanone	151	—	123	89	—	207	—
Cyclohexanone	156	—	166	160	146	81	91
2-Methylcyclohexanone	165	—	191	135	132	—	43
2,6-Dimethyl-4-heptanone	168	—	122; 126	66; 92	—	—	—
4-Methylcyclohexanone	171	—	199	—	128	109	37
2-Octanone	173	—	122	58	92	—	—
Cycloheptanone	181	—	163	148	137	—	23
Acetophenone	205	—	198	238	184	105	60
4-Methyoxyacetophenone	—	38	197	220	195	142	86
Benzalacetone	—	41	187	227	166	157	115
Benzophenone	—	48	164	238	154	137	142

(continued)

Table A4   *(continued)*

Name of compound	B.P. °C	M.P. °C	Semicar-bazone	2,4-Dinitro-phenyl-hydrazone	p-Nitro-phenyl-hydrazone	Phenyl-hydrazone	Oxime
4-Bromoacetophenone	—	51	208	230	—	126	128
2-Acetylnaphthalene	—	53	234	262d	—	170; 176	145; 149
4-Methoxybenzophenone	—	62	—	180	198	90; 132	115; 140
4-Bromobenzophenone	—	82	350	230	—	126	110; 116
9-Fluorenone	—	83	—	283	269	151	195
Benzil	—	95	174; 243	189	192; 290	134; 235	137; 237
4-Hydroxyacetophenone	—	109	199	210	—	151	145
Benzoin	—	133	205	245	—	158	99; 151
4-Hydroxybenzophenone	—	134	194	242	—	144	81
d,l-Camphor	—	178	237; 247	164; 177	217	233	118

Table A5   Carboxylic Acids

Name of compound	B.P. °C	M.P. °C	p-Toluidide	Anilide	Amide	p-Nitro-benzyl ester	p-Bromo-phenacyl ester
Ethanoic acid	118	—	153	114	82	78	86
2-Methylpropanoic acid	154	—	108	105	128	—	76
Butanoic acid	162	—	75	96	115	35	63
3-Methylbutanoic acid	176	—	106	109	135	—	68
Pentanoic acid	186	—	74	63	106	—	75
2,2-Dimethylpropanoic acid	—	35	119	132	155	—	75
Dodecanoic acid	—	44	87	78	100	—	76
Trichloroethanoic acid	—	57	113	97	141	80	—
Hexadecanoic acid	—	62	98	90	106	42	86
Octadecanoic acid	—	70	102	95	109	—	92
Phenylethanoic acid	—	76s	135	117	156	65	89
Phenoxyacetic acid	—	98	—	99	101	—	148
2-Toluic acid	—	104	144	125	142	90	57
3-Toluic acid	—	111	118	126	94	86	108
Benzoic acid	—	122	158	160	130	89	119
trans-Cinnamic acid	—	133	168	109; 153	147	116	145
3-Nitrobenzoic acid	—	140	162	154	143	141	132
2-Chlorobenzoic acid	—	142	131	114	142; 202	106	106
Diphenylethanoic acid	—	148	172	180	167	—	—
2-Nitrobenzoic acid	—	146	—	155	176	112	107
Adipic acid	—	153	241	151; 240	125; 220	106	154
4-Anisic acid	—	184	186	169	167	132	152
3,5-Dinitrobenzoic acid	—	204	—	234	183	157	159

Table A6   Primary and Secondary Amines

			Derivatives (M.P.) °C			
Name of compound	B.P. °C	M.P. °C	Acetamide	Benzamide	Benzene-sulfonamide	*p*-Toluene-sulfonamide
Propylamine	49	—	—	84	36	52
Diethylamine	56	—	—	42	42	60
Butylamine	77	—	—	42	—	65
Piperidine	106	—	—	48	93	96
Dipropylamine	109	—	—	—	51	—
N-Hexylamine	130	—	—	40	96	—
Aniline	184	—	114	160	112	103
N-Methylaniline	196	—	102	63	79	94
2-Bromoaniline	—	32	99	116	—	—
N-Benzylaniline	—	37	58	107	119	149
4-Methylaniline	—	45	147	158	120	118
Diphenylamine	—	53	101	180	124	141
2,4-Dichloroaniline	—	63	145	117	128	126
4-Bromoaniline	—	66	168	204	134	—
2-Nitroaniline	—	71	92	98; 110	104	142
4-Chloroaniline	—	72	179	192	122	95; 119
4-Methyl-3-nitroaniline	—	78	148	172	160	164
2-Bromo-4-nitroaniline	—	105	129	160	—	—
Triphenylmethylamine	—	105	207	160	—	—
3-Nitroaniline	—	114	155; 76	155	136	138
4-Chloro-2-nitroaniline	—	116	104	—	—	110
4-Methyl-2-nitroaniline	—	117	99	148	102	146
2-Methyl-4-nitroaniline	—	130	202	—	158	174
2,5-Dimethyl-4-nitroaniline	—	144	168	—	162	185
4-Nitroaniline	—	147	215	199	139	191

Table A7   Tertiary Amines

			Derivatives (M.P.) °C	
Name of compound	B.P. °C	M.P. °C	Methyl *p*-Toluene-sulfonate	Picrate
Triethylamine	89	—	—	173
Pyridine	116	—	139	167
3-Methylpyridine	143	—	—	150
4-Methylpyridine	143	—	—	167
3-Chloropyridine	149	—	—	135
Tripropylamine	156	—	—	116
2,4,6-Trimethylpyridine	172	—	—	156
Quinoline	239	—	126	203
2,6-Dimethylquinoline	—	60	175	186
Tribenzylamine	—	91	—	190

Appendix B

Laboratory Skills and Calculations

I n this section, you will learn how to do laboratory calculations. Directions for balancing oxidation-reduction reactions can be found on the course website.

Determining Quantities of Reagents Used

Mass to Mol Conversion
To convert between mass and mol (or mmol), use the molecular weight.

$$\text{mol} = \frac{\text{mass (g)}}{\text{MW (g / mol)}} \qquad \text{mmol} = \frac{\text{mass (mg)}}{\text{MW (mg / mmol)}}$$

Mass from Density
If the compound is a liquid, use the density of the liquid to calculate the mass before determining mol:

$$\text{mass} = \text{density (g/mL)} \times \text{volume (mL)}$$

Mol from Molarity
For a solution, use the molarity to calculate the number of moles:

$$\text{mol} = \text{M (mol/L)} \times \text{V(L)}, \qquad \text{mmol} = \text{M (mmol/mL)} \times \text{V(mL)}$$

Sample Calculation 1:
Calculate the number of moles in 125 µL of bromobenzene.

Solution: Since bromobenzene is a liquid, look up the density and the molecular weight in a chemical reference book. The density is 1.491 g/mL and the molecular weight is 157.02 g/mol.

$$\text{mass (g)} = 125\ \mu L = \frac{1.491\ \text{g}}{\text{mL}} \times \frac{1000\ \text{ml}}{1L} \times \frac{1L}{1 \times 10^6\ \mu L} = 0.1864\text{g}$$

$$\text{mol} = \frac{0.1864\ \text{g}}{157.02\ \text{g / mol}} \quad \text{or } 0.1864\ \text{g} \times \frac{1\ \text{mol}}{157.02\ \text{g}} = 1.187 \times 10^{-3}\text{mol}$$

This volume of bromobenzene contains 1.19×10^{-3} mol or 1.19 mmol (to the correct number of significant figures). Notice that one extra digit was carried along until the end of the calculation. This helps avoid round-off error.

Sample Calculation 2:

A procedure calls for using 1.75 mmol of bromobenzene. Calculate the volume of bromo-benzene to use.

Solution:

$$\mu L = 1.75 \text{ mmol} \times \frac{1 \text{ mol}}{1000 \text{ mmol}} \times \frac{157.02 \text{ g}}{\text{mol}} \times \frac{1 \text{ mL}}{1.491 \text{ g}} \times \frac{1000 \text{ } \mu L}{1 \text{ mL}} = 184 \text{ } \mu L$$

The required volume is 184 μL of bromobenzene.

Sample Calculation 3:

What volume of concentrated (18 M) H_2SO_4 contains 25 mmol of H_2SO_4? Report the answer in μL.

Solution:

$$\mu L \text{ } H_2SO_4 = 25 \text{ mmol} \times \frac{1 \text{ mol}}{1000 \text{ mmol}} \times \frac{1 \text{ L}}{18 \text{ mol}} \times \frac{1 \times 10^6 \text{ } \mu L}{1 \text{ L}} = 1.4 \text{ } \mu L$$

Determining the Limiting Reagent

The limiting reagent is the one that is consumed during a reaction. It is the one that determines the theoretical amount of product that can be formed.

Sample Calculation 4:

Determine the limiting reactant and the moles of product if 200 mg of 3-pentanone and 75 mg of sodium borohydride are used. The balanced reaction is:

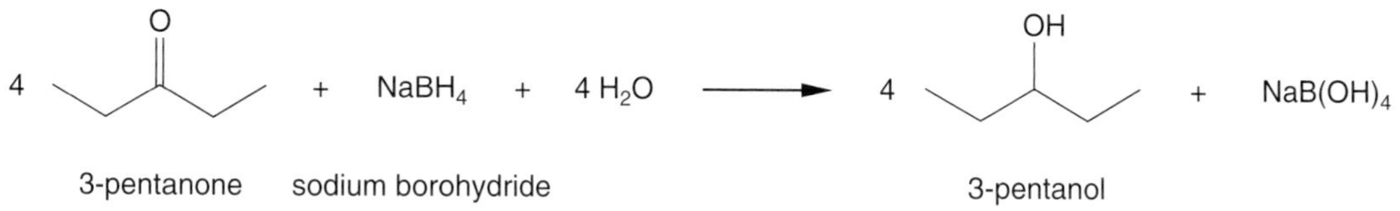

3-pentanone	sodium borohydride	3-pentanol

Solution:

Calculate the moles of 3-pentanol that could theoretically be obtained from each reactant:

$$200 \text{ mg 3-pentanone} \times \frac{1 \text{ g}}{1000 \text{ mg}} \times \frac{1 \text{ mol 3-pentanone}}{86.13 \text{ g}} \times \frac{4 \text{ mol 3-pentanol}}{4 \text{ mol 3-pentanone}} = 2.32 \times 10^{-3} \text{ mol of 3-pentanol}$$

$$75 \text{ mg NaBH}_4 \times \frac{1 \text{ g}}{1000 \text{ mg}} \times \frac{1 \text{ mol NaBH}_4}{37.83 \text{ g}} \times \frac{4 \text{ mol 3-pentanol}}{1 \text{ mol NaBH}_4} = 7.9 \times 10^{-3} \text{ mol of 3-pentanol}$$

Since 3-pentanone would produce the lesser amount, it is the limiting reactant: the theoretical yield of 3-pentanol is 2.32×10^{-3} mol or 2.32 mmol.

Calculating Percent Yield

Percent yield is defined as the mol of product actually obtained in an experiment divided by the number of mol theoretically possible from the balanced equation.

$$\% \text{ yield} = [\text{actual yield/theoretical yield}] \times 100$$

Solids should be thoroughly dry before weighing and calculating a yield. Typical yields for experiments in this lab book range from 25% to 95%, but individual results will vary based upon technique, purity of starting materials, and other experimental conditions. Since the yield can never be greater than 100%, if the observed yield is higher than this, the product may be wet or there may be a calculation error.

Sample Calculation 5:

When the borohydride reduction of 3-pentanone described in Calculation 4 was performed in the laboratory, a student obtained 187 mg of 3-pentanol. Calculate the percent yield.

Solution:

Convert the mass of 3-pentanol to mol and compare it to the theoretical yield. The molecular mass of 3-pentanol is 88.15 g/mol. Then 187 mg is 2.12 mmol or 2.12×10^{-3} mol. The theoretical yield was 2.32 mmol or 2.32×10^{-3}. The percent yield is 91.4%:

$$\% \text{ yield} = \frac{2.12 \times 10^{-3} \text{ mol}}{2.32 \times 10^{-3} \text{ mol}} \times 100 = 91.4\%$$

Calculating Percent Error

This calculation is used to determine the difference between a measured (experimental) value and the known or reported (literature) value.

$$\% \text{ error} = \left| \frac{\text{literature value} - \text{experimental value}}{\text{literature value}} \right| \times 100$$

Sample Calculation 6:

The corrected refractive index was measured in the lab as 1.4988. The literature value for this compound is 1.5186. Calculate the percent error.

Solution:

$$\% \text{ error} = \left| \frac{1.5186 - 1.4988}{1.5186} \right| \times 100 = 1.30\%$$

Extraction Calculations

The distribution coefficient (K_D) can be calculated from the solubilities of a solute in an organic solvent and in water:

$$K_D = \frac{\text{solubility in organic solvent (g / mL)}}{\text{solubility in water (g / mL)}}$$

Sample Calculation 7:

The solubility of an organic solute in hexane is 7.3 g/100 mL; the solubility of the solute in water is 1.2 g/50 mL. Calculate the K_D.

Solution:

$$K_D = \frac{7.3 \text{ g} / 100 \text{ mL of hexane}}{1.2 \text{ g} / 50 \text{ mL of water}} = 3.0$$

Knowing the K_D for a given solvent system, the amount of solute extracted into a given volume can be calculated. The amount of solute extracted into the organic solvent depends upon the distribution coefficient (and therefore the relative solubilities in the organic and aqueous layers). If "x" represents the mass of solute extracted into the organic solvent, then the amount extracted into water will be "initial mass – x." This equation then becomes:

$$K_D = \frac{x / \text{volume of organic solvent}}{(\text{initial mass } - x) / \text{volume of water}}$$

Sample Calculation 8:

200 mg of an organic solute is dissolved in 10 mL of water. How many mg of the organic solute will be extracted using 10 mL of hexane? The K_D for the system is 3.0.

Solution: Let x = mg of solute in hexane. Then 200 – x is the amount of solute remaining in the water layer. Solve for x:

$$3.0 = \frac{x / 10 \text{ mL of hexane}}{(200 - x) / 10 \text{ mL of water}} = 150 \text{ mg}$$

Doing the extraction with equal portions of hexane and water should result in a 75% recovery of the organic solute (150 mg/200 mg × 100 = 75%).

When doing an extraction, it is usually better to do multiple extractions with smaller volumes of solvent rather a single extraction with the same total volume. This is demonstrated in Sample Calculation 9.

Sample Calculation 9:

What mass of the original 200 mg of organic solute dissolved in 10 mL of water could be extracted doing two extractions, each with 5 mL of hexane:

Solution: For the first extraction, let x = mg of solute in hexane and 200 – x = mg of solute in water.

$$3.0 = \frac{x / 5 \text{ mL of hexane}}{(200 - x) / 10 \text{ mL of water}} = 120 \text{ mg}$$

After the first extraction, 120 mg of the organic solute has been taken up in the hexane and 80 mg remains in the water. For the second extraction, x can still represent the amount of solute in hexane, but the amount of solute that will remain in the water is 80–x (since that's all that remains in the water after the first extraction):

$$3.0 = \frac{x / 5 \text{ mL of hexane}}{(80 - x) / 10 \text{ mL of water}} = 48 \text{ mg}$$

The total amount of solute extracted into hexane from the two 5-mL extractions is $120 + 48 = 168$ mg. The amount extracted with one 10-mL extraction was only 150 mg, so doing multiple extractions increases the amount of product obtained in an extraction up to a point. The usual procedure is to do two to three extractions; the amount obtained in doing more is probably not worth the time it takes to do the extraction.

Appendix C

Designing a Flow Scheme

The purpose of a flow scheme is to help you understand why each step in the procedure is performed and how the product is purified. A flow scheme indicates where unwanted side products and unreacted starting materials are separated from the product. As you construct the flow scheme, ask yourself why each specific technique is being used. For example, if you are distilling a liquid product, think about the boiling points of the product and starting materials, and determine which fractions will contain the product. Extraction separates components based on preferential solubility in organic and aqueous layers. Compounds that are water soluble dissolve in the aqueous layer; water-insoluble organic compounds dissolve in the organic solvent. Solutes may have partial solubility in both water and the organic solvent; in this case the solute would be distributed between both phases. Gravity filtration, vacuum filtration, and steam distillation are other techniques used to help isolate the product from side products and by-products.

Part of designing a flow scheme is knowing the structure and physical properties of side products and by-products that can form in the reaction. Consider the conversion of cyclohexanol to cyclohexyl chloride.

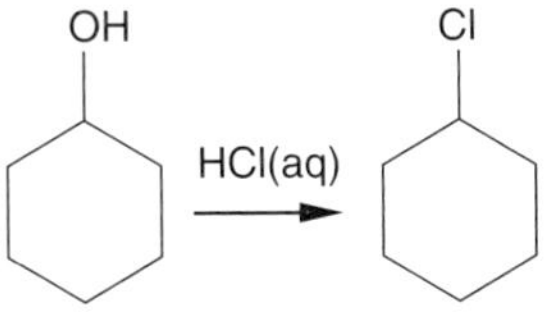

Cyclohexanol is a liquid with a boiling point of 161°C. It is insoluble in water. The product, cyclohexyl chloride, boils at 142°C and is also insoluble in water. A probable side product is cyclohexene, formed by elimination. Cyclohexene is a liquid that boils at 83°C. It is insoluble in water. Other side products are possible, but not as likely. Since reactions rarely go to completion, the flow scheme must indicate how excess starting materials are separated from the product.

The experimental procedure is given here. The numbered parts of the procedure refer to specific places in the flow scheme where separations occur.

Procedure

Place 1 mL of the cyclohexanol and 2.5 mL of concentrated HCl in a 5-mL conical vial. **(1)** Cap the vial and shake vigorously for 1 minute. Vent by loosening the cap. Repeat this process three times. The product will separate as an upper layer when the mixture is allowed to stand. Remove the lower aqueous layer **(2)** and discard. Add 1 mL of saturated sodium bicarbonate to the organic layer. Mix gently and note the evolution of gas. **(3)** Cap and shake gently for 1 minute. Allow the layers to separate and remove the aqueous layer **(4)**. After separation, dry the organic layer with anhydrous sodium sulfate.

Use a filter pipet to transfer the dried solution (**5**) to a conical vial fitted with a spin vane and a Hickman still. Distill. Discard distillate that boils below 100°C (**6**). Collect the fraction that boils between 138 and 142°C (**7**). Stop distilling when the temperature drops or the temperature goes higher than 142°C (**8**). Transfer the distillate to a clean, tared vial. Weigh the product and calculate the percent yield.

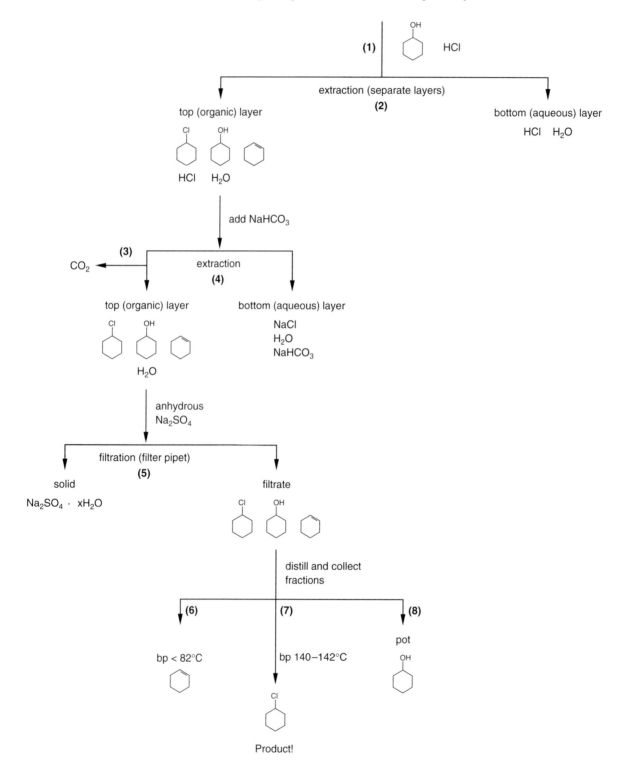

Appendix D

Material Safety Data Sheet

A Material Safety Data Sheet (MSDS) is kept in the laboratory for each chemical being used in that laboratory. The sheet is available prior to, during, and after use of each chemical. This permits checking on a chemical during planning stages of an experiment, during use, and during an emergency situation. The MSDS can be very useful in preparing for an experiment, so that adequate precautions can be taken. The MSDS will also be handy in case of an accident. The policy of requiring chemical suppliers to have an MSDS available for laboratory workers is in accordance with federal law under the "right to know laws" in the United States.

Each MSDS lists extensive data for each of the common organic chemicals used in the laboratory. MSDS sheets for less common organic chemicals usually contain less information because these chemicals may not have been evaluated thoroughly. The MSDS for acetic anhydride issued by Fisher Scientific Company is given as an example of the MSDS of a common organic chemical used in this book. The MSDS sections of most frequent interest to students are listed here for Section 3, Hazards Identification; Section 4, First Aid Measures; Section 5, Fire Fighting Measures; Section 8, Exposure Controls/Personal Protection; and Section 11, Toxicological Information. These sections are described in more detail here.

Section 3, Hazards Identification: This section lists possible harmful effects to the body.

Section 4, First Aid Measures: This section lists what to do if a person is exposed to the chemical by skin or eye contact, inhalation or ingestion by mouth.

Section 5, Fire Fighting Measures: This section gives measures to be followed in case the particular chemical is burning. Of note in this section are the National Fire Protection Association (NFPA) ratings for health, flammability, and reactivity of a chemical. These ratings are often given on labels in the shape of a diamond divided into three colored sections: blue for the health rating; red for the flammability rating, and yellow for the reactivity rating. Numbers from 0 to 4 in each colored section alert the user to the specific hazard involved for using that chemical. The higher the rating, the greater the hazard. The bottom diamond is white and lists any special properties of the chemical or special precautions to be taken. An example is shown in the Introduction on p. 5.

Section 8, Exposure Controls/Personal Protection: This section gives precautions for handling the chemical and hazardous properties of the chemical. Personal protective equipment for handling the chemical is listed in this section. This section will also contain information from the Occupational Safety and Health Administration (OSHA) for the permissible exposure level (PEL) specified by OSHA for common organic chemicals. The lower the PEL the more hazardous the vapor. For acetic anhydride, the PEL rating is 20 mg/m^3 TWA. This means that OSHA recommends no higher an exposure level than 20 mg of acetic anhydride per cubic meter of air averaged over a

24-hour period, known as a time-weighted average (TWA). The American Conference of Government Industrial Hygienists similarly recommends its maximum exposure level, called threshold limit value (TLV), as 21 mg/m^3 for acetic anhydride. PEL and TLV are the maximum levels of exposure to a particular chemical on a routine basis such as during a 24-hour period. Compared to other common chemicals, these PEL and TLV ratings are quite low and indicate that breathing the vapors of acetic anhydride is definitely to be avoided. Values for a common solvent such as acetone are 2400 and 1188 mg/m^3 for the PEL and TLV, respectively. The numbers are obviously not identical, but they are fifty to a hundred times higher than the values for acetic anhydride. These relatively higher values indicate that breathing acetone vapors is not regarded by these agencies to be as hazardous as breathing vapors of acetic anhydride. However, it is generally recommended that you avoid breathing all organic vapors. Your laboratory should be well ventilated and should be equipped with special hoods to limit exposure to chemicals, particularly ones with low PEL and TLV values.

Section 11, Toxicological Information: This section lists effects that may be observed following exposure or contact with the chemical and toxicity data for mice or rats. The LC_{50} is the median lethal atmospheric concentration or the concentration necessary to kill 50% of a population of animals tested. For acetic anhydride, the LC_{50} is given as 1000 ppm/4H, meaning that half of a rat population is expected to be killed when exposed to a concentration of 1000 ppm (parts per million) of acetic anhydride for a period of four hours.

The LD_{50} is the median lethal dose or the amount of the chemical necessary to kill 50% of the population of animals tested. The LD_{50} may be given as oral dosage or as injection beneath the skin. If acetic anhydride is administered orally to each rat in a population of rats in a concentration of 1780 mg per kg rat, half of the rat population is expected to be killed. There is no direct correlation between lethal dosage for laboratory animals and humans, but the LC and LD values are good indicators of the potential toxicity of a chemical to humans.

The MSDS issued for a chemical by different suppliers will not look the same. However, each of the sheets will contain the same 16 sections and provide most of the same information. The MSDS of acetic anhydride issued by Fisher Scientific Company in 1999 is shown on the following pages.

**** MATERIAL SAFETY DATA SHEET ****

Acetic anhydride
00130

**** SECTION 1 - CHEMICAL PRODUCT AND COMPANY IDENTIFICATION ****

MSDS Name: Acetic anhydride
Catalog Numbers:
AC4000 60000, AC400060040, S70072, S70078, S71903, A10, 1,
A10 100, A10 4, A10 500, A10-1, A10-100, A10-4, A10-500, A101, A10100, A104
A10500, A10\4, NC9556617, XXA10207LI

Synonyms:
Acetic oxide, acetyl oxide, ethanoic anhydride, acetic acid anhydride.
Company Identification: Fisher Scientific
    1 Reagent Lane
    Fairlawn, NJ 07410
For information, call: 201-796-7100
Emergency Number: 201-796-7100
For CHEMTREC assistance, call: 800-424-9300
For International CHEMTREC assistance, call: 703-527-3887

**** SECTION 2 - COMPOSITION, INFORMATION ON INGREDIENTS ****

CAS#	Chemical Name	%	EINECS#
108-24-7	Acetic anhydride	97.0	203-564-8

Hazard Symbols: F C
Risk Phrases: 10 34

**** SECTION 3 - HAZARDS IDENTIFICATION ****

EMERGENCY OVERVIEW

Appearance: Not available. Flash Point: 126 deg F.
Danger! Flammable liquid. Corrosive. Water-Reactive. May be harmful
if swallowed. Causes digestive and respiratory tract burns. May
cause severe eye and skin irritation with possible burns.
Target Organs: None.

Potential Health Effects
Eye:
    May cause irreversible eye injury. Eye damage may be delayed.
    Vapor or mist may cause irritation and severe burns. Contact with
    liquid is corrosive to the eyes and causes severe burns.
Skin:
    May cause severe irritation and possible burns. Prolonged skin
    contact may be painless with reddening of the skin followed by a
    white appearance of the skin. Skin burns may be delayed.
Ingestion:
    Causes digestive tract burns with immediate pain, swelling of the
    throat, convulsions and possible coma. May cause lung
    damage. Vapors may cause dizziness or suffocation.
Inhalation:
    Causes severe irritation of upper respiratory tract with coughing,
    burns, breathing difficulty, and possible coma. May cause lung
    damage.
Chronic:
    No information found.
    Prolonged skin contact may cause redness and
    subsequently a white appearance of the skin accompanied by wrinkling.
    Skin burns may be delayed.

**** SECTION 4 - FIRST AID MEASURES ****

Eyes:
    Immediately flush eyes with plenty of water for at least 15 minutes,
    occasionally lifting the upper and lower eyelids. Get medical aid
    immediately.
Skin:
    Get medical aid immediately. Immediately flush skin with plenty of
    soap and water for at least 15 minutes while removing contaminated
    clothing and shoes. Discard contaminated clothing in a manner which
    limits further exposure. Destroy contaminated shoes.
Ingestion:
    Do NOT induce vomiting. If victim is conscious and alert, give 2-4
    cupfuls of milk or water. Never give anything by mouth to an
    unconscious person. Get medical aid immediately.
Inhalation:
    Get medical aid immediately. Remove from exposure to fresh air
    immediately. If breathing is difficult, give oxygen. If breathing has
    ceased apply artificial respiration using oxygen and a suitable
    mechanical device such as a bag and a mask.
Notes to Physician:
    Treat symptomatically

**** SECTION 5 - FIRE FIGHTING MEASURES ****

General Information
    As in any fire, wear a self-contained breathing apparatus in
    pressure-demand, MSHA/NIOSH (approved or equivalent), and full
    protective gear. Vapors may form an explosive mixture with air.
    Vapors can travel to a source of ignition and flash back. Will burn
    if involved in a fire. Use water spray to keep fire-exposed
    containers cool. Wear appropriate protective clothing to prevent
    contact with skin and eyes. Wear a self-contained breathing apparatus
    (SCBA) to prevent contact with thermal decomposition products.
    Containers may explode in the heat of a fire. Water Reactive
    Flammable Liquid.
Extinguishing Media:
    For small fires, use dry chemical, carbon dioxide, water spray or
    alcohol-resistant foam. For large fires, use water spray, fog or
    alcohol-resistant foam. Use water spray to cool fire-exposed
    containers. Water may be ineffective. Do NOT use straight streams of
    water.
Auto Ignition Temperature: 630 deg F (332.22 deg C)
Flash Point: 126 deg F (52.22 deg C)
NFPA Rating: (estimated) Health: 3; Flammability: 2; Reactivity: 2
Explosion Limits, Lower: 2.9%
    Upper: 10.3%

**** SECTION 6 - ACCIDENTAL RELEASE MEASURES ****

General Information: Use proper personal protective equipment as indicated
    in Section 8.
Spills/Leaks:
    Use water spray to dilute spill to a non-flammable mixture. Avoid
    runoff into storm sewers and ditches which lead to waterways. Clean
    up spills immediately, observing precautions in the Protective
    Equipment section. Use water spray to disperse the gas/vapor. Remove
    all sources of ignition. Use a spark-proof tool. Spill may be
    neutralized with lime. Cover with material such as dry soda ash or
    calcium carbonate and place into closed container for disposal. A
    vapor suppressing foam may be used to reduce vapors.

**** SECTION 7 - HANDLING and STORAGE ****

Handling:
    Wash thoroughly after handling. Remove contaminated clothing and
    wash before reuse. Ground and bond containers when transferring
    material. Use spark-proof tools and explosion proof equipment. Do not
    get in eyes, on skin, or on clothing. Empty containers retain product
    residue, (liquid and/or vapor), and can be dangerous. Avoid contact
    with heat, sparks and flame. Do not ingest or inhale. Use with
    adequate ventilation. Do not allow contact with water. Do not
    pressurize, cut, weld, braze, solder, drill, grind, or expose empty
    containers to heat, sparks, or open flames.
Storage:
    Keep away from heat, sparks, and flame. Keep away from sources of
    ignition. Do not store in direct sunlight. Keep from contact with
    oxidizing materials. Store in a cool, dry, well-ventilated area away
    from incompatible substances. Keep away from water. Flammables-area.

**** SECTION 8 - EXPOSURE CONTROLS, PERSONAL PROTECTION ****

Engineering Controls:
    Use adequate general or local exhaust ventilation to keep airborne
    concentrations below the permissible exposure limits.

Exposure Limits

Chemical Name	ACGIH	NIOSH	OSHA - Final PELs
Acetic anhydride	5 ppm; 21 mg/m3	200 ppm IDLH	5 ppm TWA; 20 mg/m3 TWA

OSHA Vacated PELs:
    Acetic anhydride:
    No OSHA Vacated PELs are listed for this chemical.

Personal Protective Equipment
    Eyes:
        Wear appropriate protective eyeglasses or chemical
        safety goggles as described by OSHA's eye and face
        protection regulations in 29 CFR 1910.133 or European
        Standard EN166.
    Skin:
        Wear appropriate protective gloves to prevent skin
        exposure.
    Clothing:
        Wear appropriate protective clothing to prevent skin
        exposure.
    Respirators:
        Follow the OSHA respirator regulations found in 29CFR
        1910.134 or European Standard EN 149. Always use a

Source: Reproduced with permission of Fisher Scientific.

Hazard Class: 8(9.2)
UN Number: UN1715

""" SECTION 15 - REGULATORY INFORMATION """

UN FEDERAL
TSCA
  CAS# 108-24-7 is listed on the TSCA inventory.
Health & Safety Reporting List
  None of the chemicals are on the Health & Safety Reporting List.
Chemical Test Rules
  None of the chemicals in this product are under a Chemical Test Rule.
Section 12b
  None of the chemicals are listed under TSCA Section 12b.
TSCA Significant New Use Rule
  None of the chemicals in this material have a SNUR under TSCA.
SARA
  Section 302 (RQ)
    CAS# 108-24-7: final RQ = 5000 pounds (2270 kg)
  Section 302 (TPQ)
    None of the chemicals in this product have a TPQ.
  SARA Codes
    CAS # 108-24-7: acute, chronic, flammable, reactive.
  Section 313
    No chemicals are reportable under Section 313.
Clean Air Act:
  This material does not contain any hazardous air pollutants.
  This material does not contain any Class 1 Ozone depletors.
  This material does not contain any Class 2 Ozone depletors.
Clean Water Act:
  CAS# 108-24-7 is listed as a Hazardous Substance under the CWA.
  None of the chemicals in this product are listed as Priority
  Pollutants under the CWA.
  None of the chemicals in this product are listed as Toxic Pollutants
  under the CWA.
OHSA: None of the chemicals in this product are considered highly hazardous
  by OSHA.

STATE
Acetic anhydride can be found on the following state right to know
lists: California, New Jersey, Florida, Pennsylvania, Minnesota,
Massachusetts.
California No Significant Risk Level:
  None of the chemicals in this product are listed.
European/International Regulations
  European Labeling in Accordance with EC Directives
    Hazard Symbols: F C
    Risk Phrases:
      R 10 Flammable.
      R 34 Causes burns.
    Safety Phrases:
      S 16 Keep away from sources of ignition - No
        smoking.
      S 26 In case of contact with eyes, rinse immediately
        with plenty of water and seek medical advice.
      S 33 Take precautionary measures against static
        discharges.
      S 45 In case of accident or if you feel unwell, seek
        medical advice immediately (show the label where
        possible).
      S 9 Keep container in a well-ventilated place.
      S50A Do not mix with acids.
WGK (Water Danger/Protection)
  CAS# 108-24-7: 1
Canada
  CAS# 108-24-7 is listed on Canada's DSL/NDSL List.
  This product has a WHMIS classification of B3, D1A, D2B, E.
  CAS# 108-24-7 is not listed on Canada's Ingredient Disclosure List.
Exposure Limits

""" SECTION 16 - ADDITIONAL INFORMATION """

MSDS Creation Date: 2/23/1995 Revision #16 Date: 10/09/1998

The information above is believed to be accurate and represents the best
information currently available to us. However, we make no warranty of
merchantability of any other warranty, express or implied, with respect to
such information, and we assume no liability resulting from its use, users
should make their own investigations to determine the suitability of the
information for their particular purposes. In no way shall the company be
liable for any claims, losses, or damages of any third party or for lost
profits or any special, indirect, incidental, consequential or exemplary
damages, howsoever arising, even if the company has been advised of
the possibility of such damages.

NIOSH or European Standard EN 149 approved respirator
when necessary.

""" SECTION 9 - PHYSICAL AND CHEMICAL PROPERTIES """

Physical State: Liquid
Appearance: Not available.
Odor: strong odor - pungent odor - acetic odor
pH: Not available.
Vapor Pressure: 3.9 mm. Hg @68F
Vapor Density: 3.5 (air = 1)
Evaporation Rate: 0.46 (n-butyl acetate = 1)
Viscosity: Not available.
Boiling Point: 137 deg C
Freezing/Melting Point: 0 deg C
Decomposition Temperature: Not available.
Solubility: Not available.
Specific Gravity/Density: 1.0820g/cm3
Molecular Formula: C4H6O3
Molecular Weight: 102.09

""" SECTION 10 - STABILITY AND REACTIVITY """

Chemical Stability:
  Stable.
Conditions to Avoid:
  Incompatible materials, ignition sources, contact with water, excess
  heat.
Incompatibilities with Other Materials:
  Strong oxidizing agents.
Hazardous Decomposition Products:
  Carbon monoxide, carbon dioxide.
Hazardous Polymerization: Has not been reported.

""" SECTION 11 - TOXICOLOGICAL INFORMATION """

RTECS#
  CAS# 108-24-7: AD1925000
LD50/LC50:
  CAS# 108-24-7: Inhalation, rat: LC50 = 1000 ppm/4h; Oral, rat: LD50 =
  1780 mg/kg; Skin, rabbit: LD50 = 4 gm/kg.
Carcinogenicity:
  Acetic anhydride -
  Not listed by ACGIH, IARC, NIOSH, NTP, or OSHA.
Epidemiology:
  No data available.
Teratogenicity:
  No data available.
Reproductive Effects:
  No data available.
Neurotoxicity:
  No data available.
Mutagenicity:
  No data available.
Other:
  No data available.

""" SECTION 12 - ECOLOGICAL INFORMATION """

Ecotoxicity:
  Not available.
Environmental Fate:
  Not available.
Physical/Chemical:
  Not available.
Other:
  Not available.

""" SECTION 13 - DISPOSAL CONSIDERATIONS """

Dispose of in a manner consistent with federal, state, and local regulations.
RCRA P-Series: None listed.
RCRA U-Series: None listed.

""" SECTION 14 - TRANSPORT INFORMATION """

US DOT
  SHIPPING NAME: ACETIC ANHYDRIDE
  Hazard Class: 8
  UN Number: UN1715
  Packing Group: II
IMO
  Not regulated as a hazardous material.
IATA
  Not regulated as a hazardous material.
RID/ADR
  Not regulated as a hazardous material.
Canadian TDG
  Shipping Name: ACETIC ANHYDRIDE

Source: Reproduced with permission of Fisher Scientific.

Appendix E

Tables of Common Organic Solvents and Inorganic Solutions

Table E1 Common Organic Solvents

Solvent	Boiling point (°C)	Density[20] (g/mL)	Miscibility with H_2O
Acetic acid	118	1.049	yes
Acetone	56	0.791	yes
1-Butanol	118	0.810	yes
Cyclohexane	81	0.778	no
Diethyl ether	35	0.713	no
Ethanol (absolute)	78	0.789	yes
95% Ethanol	78	0.816	yes
Ethyl acetate	77	0.902	slight
Hexane	69	0.660	no
Hexanes	68–70	0.672	no
Ligroin	60–80	0.656	no
Methanol	65	0.792	yes
Methylene chloride	40	1.325	no
Pentane	36	0.626	no
Petroleum ether	35–60	0.640	no
1-Propanol	97	0.805	yes
2-Propanol	82	0.785	yes
Tetrahydrofuran (THF)	66	0.889	yes
Toluene	111	0.866	no

Table E2 Common Concentrated Acids and Bases

Solution	Molarity (mol/L)	Density[20] (g/mL)	Percent by weight
Inorganic acids:			
Hydrochloric acid, concentrated	12	1.18	37
Nitric acid, concentrated	16	1.42	71
Phosphoric acid, concentrated	14.7	1.70	85
Sulfuric acid, concentrated	18	1.84	96.5
Inorganic base:			
Ammonia, concentrated	15.3	0.90	28.4

Index

ic acid,

ne, 290–97